Plants and People in the African Past

Anna Maria Mercuri · A. Catherine D'Andrea
Rita Fornaciari · Alexa Höhn
Editors

Plants and People in the African Past

Progress in African Archaeobotany

IWAA

Springer

Editors
Anna Maria Mercuri
Laboratorio di Palinologia e Paleobotanica, Dipartimento di Scienze della Vita
Università di Modena e Reggio Emilia
Modena
Italy

Rita Fornaciari
Laboratorio di Palinologia e Paleobotanica, Dipartimento di Scienze della Vita
Università di Modena e Reggio Emilia
Modena
Italy

A. Catherine D'Andrea
Department of Archaeology
Simon Fraser University
Burnaby, BC
Canada

Alexa Höhn
Institut für Archäologische Wissenschaften
Goethe-Universität
Frankfurt
Germany

ISBN 978-3-030-07872-0 ISBN 978-3-319-89839-1 (eBook)
https://doi.org/10.1007/978-3-319-89839-1

Softcover re-print of the Hardcover 1st edition 2018

Printed on acid-free paper

This Springer imprint is published by the registered company Springer Nature Switzerland AG
The registered company address is: Gewerbestrasse 11, 6330 Cham, Switzerland

Contents

Plants and People in the African Past: Themes and Objectives of Archaeobotany

Anna Maria Mercuri[1(✉)], A. Catherine D'Andrea[2], Rita Fornaciari[3], and Alexa Höhn[4]

[1] Università di Modena e Reggio Emilia, Laboratorio di Palinologia e Paleobotanica, Dip. Scienze della Vita, Modena, Italy
annamaria.mercuri@unimore.it

[2] Department of Archaeology, Simon Fraser University, Burnaby, BC, Canada
adandrea@sfu.ca

[3] Università di Modena e Reggio Emilia, Laboratorio di Palinologia e Paleobotanica, Modena, Italy
rita.fornaciari@unimore.it

[4] Institut für Archäologische Wissenschafte, Goethe-Universität, Frankfurt, Germany
a.hoehn@em.uni-frankfurt.de

Africa with its impressive, deep history and ecosystem diversity continues to offer an ideal setting to expand our frontiers of understanding plants and people in the past. Early and unique interrelationships between humans and plants make Africa a critical reference point for interdisciplinary studies of cultural developments and environmental transformations. On-going research in many fields is evaluating the independent patterns of phytogeography and ethnography of Africa's plants and peoples. On the continent, climatic oscillations together with population growth and migration have sparked continual innovations and adaptation to changes over the entire duration of human existence. Palaeoethnobotanical and palaeoecological studies, focusing on vegetation development and natural resources exploitation, are investigating these fascinating issues through interdisciplinary and careful analyses of archaeological and botanical records.

The study of plant micro- and macroremains from archaeological sites, and other human-influenced contexts, is of key importance to investigate the ethnobotany and ecology of the past. The most common themes within this field of research are the knowledge of plants for food and medicine and their processing, the selection and manipulation of wild plants leading to cultivation and in some cases to domestication, foodways, and subsistence strategies of humans living in complex ecosystems, environmental change, the role of human agency in vegetation change, and plant migrations mediated by climate and human activity. Archaeobotany aims at understanding food, fuel and other plant uses as well as environment/habitat transformations, and situated at the nexus of human and the natural spheres, is one of the few reliable approaches to understanding the historical development of plant use and landscapes. The long-term reciprocal relationships between humans and the environment have become the motor of adaptation of living organisms and of the transformation of ecosystems and habitats, in Africa and beyond (Harlan and Stemler 1976; Fuller 2007, 2012; Fuller and Colledge 2008; Mercuri 2008a; Pearsall and Hastorf 2011; Zohary et al. 2012; Pelling 2013; Stevens et al. 2014).

A. M. Mercuri et al. (Eds.): *Plants and People in the African Past*,
https://doi.org/10.1007/978-3-319-89839-1_1

In Africa, ancient relationships between plants and people may differ from other regions of the world (Garcea 2004; Fuller and Hildebrand 2013; Larson et al. 2014), especially from those where plant domestication has been early (Zohary and Hopf 2000; Aiello 2011). In Africa, plant foods were gathered and foraged for millennia and perhaps they were cultivated in some areas (Wasylikowa and Dahlberg 1999; Mercuri et al. 2018), but unambiguous signs of domestication appear rather recently in the African Middle-Late Holocene archaeobotanical record (D'Andrea et al. 2001; Fuller 2007). On the pathway to food production, other choices were made, other turns taken: wild cereals supplied basic carbohydrate-protein resources, while cattle and small livestock were the main sources of domesticated food for much of the Holocene (Marshall and Hildebrand 2002). Moreover, pottery preceded food production: wild plants, including grasses and aquatic plants, were cooked in vessels as early as 10,000 years ago in the central Sahara (Dunne et al. 2016). In West Africa, tree fruits such as oil palm (*Elaeis guineensis* Jacq.) and incense tree (*Canarium schweinfurthii* Engl.) were significant resources which were likely managed in arboricultural systems as far back as the Early Holocene (D'Andrea et al. 2006; Oas et al. 2015).

Despite the later beginning of plant domestication, the conviction of some early researchers that Africa was a continent without history, implying that its environment had not been shaped to a great extent by ancient peoples, was prejudiced. In recent years, it has become increasingly evident that humans have had an important role in shaping African environments: tropical forests were by no means pristine, and savannas, while not entirely man-made had to a large extent been shaped by human activities. Archaeobotany and palaeoethnobotany are making impressive contributions to this body of research (e.g. van Gemerden et al. 2003; Mercuri 2008b; Höhn and Neumann 2012; Morin-Rivat et al. 2014).

Africa covers about one-fifth of the earth's land surface and encompasses several phytogeographical communities which developed under equatorial, tropical and Mediterranean climates (White 1983). Environmental and ecological conditions are extremely diverse not only passing from rainforest to desert and sea coast but also within similar ecosystems and latitudes. In addition, the continent hosts an immense diversity of peoples, traditions and histories. Together this necessitates more investigations of cultural-environmental changes to provide an extended chronological perspective.

The research presented in this volume includes studies comprising many of these diverse environments and cultures, presenting a clear demonstration of the variability and richness of data relating to African studies in the field of archaeobotany. The subject is crucial for sustainable development by increasing our knowledge of the links between natural resources and exploitation, opportunities and flexibility, changes and adaptation, as evident in long-term interactions of humans and the environment. If the study of one archaeological site can demonstrate local space-time events and the behaviour of a few people, the integration of data from many sites investigated with an interdisciplinary approach will enable regional and cross-cultural landscape reconstructions (Mercuri 2008a, 2014). In the pages of this book, archaeobotanists as botanists, archaeologists, anthropologists, ethnoarchaeologists and palaeoecologists compare and discuss the evidence from the African past with varied scientific languages, approaches, and research objectives. The targets, however, are the same and involve an interest in deepening the complex and fascinating history of this continent, and in the desire to learn from the 'nature and culture' of Africa the lessons on how *to exploit our lands without destroying them* (Plotkin 1994).

IWAA Workshops

The Workgroup for African Archaeobotany was founded by Professor Krystyna Wasylikowa (Kraków) in 1994 with the 1st IWAA meeting held in Mogilany, Poland in May 1994. Following this, International Workgroup for African Archaeobotany meetings have been held regularly every three years as key events where data and knowledge on African archaeobotany are exchanged on a wide range of topics. In following years, IWAA workshops were held in Leicester, England (1997); Frankfurt, Germany (2000); Groningen, Netherlands (2003); London, England (2006); Cairo, Egypt (2009); Vienna, Austria (2012) and most recently in Modena, Italy (2015) (Fig. 1).

Initially the focus of the workshop was on northern African countries and palaeoethnobotany, and included mainly macro-botanical remains. In later meetings the geographical and methodological boundaries enlarged. Today, presented research typically covers the entire African continent from South Africa to Morocco and from Mediterranean environments to the tropical forest zone. Topics have broadened considerably to include the relationships between humans and plants and ancient environments. In addition to macroremains, phytoliths, pollen, and non-pollen palynomorphs, analysed in multidisciplinary archaeo-environmental perspective, are now considered substantial components of archaeobotanical research. Ethnological and linguistic studies are also interdisciplinary elements of ancient plant use studies, as well as molecular and biochemical (for example, nucleic and fatty acids) analyses. The abstracts of conference papers presented in Modena provide a concise overview of the diversity and range of subjects and themes that are currently under investigation in African archaeobotany. They have been published as peer-reviewed long abstracts in Atti della Società dei Naturalisti e Matematici di Modena, Volume 146.

The workshop in Modena was dedicated to the memory of our colleague Prof. Ahmed Gamal-El-Din Fahmy, who organized and hosted the IWAA6 meeting at Helwan University in 2009. His untimely death in December 2013 has left us deeply saddened. In his honour the "Fahmy Memorial Speaker" award was created and awarded to the top student presentation in Modena. The winner was Ahmed's former student Dr. Mennat-Allah El-Dorry for her presentation on *Grapes, raisins and wine: archaeobotanical finds from the Monastery of St John the Little in Wadi El Natrun (Egypt)*. The second prize was given to Dr. Elshafaey A. E. Attia, and the third prize was given to Dr. Emuobosa A. Orijemie. We also commemorated the memory of Lydia

Fig. 1. The logo of the IWAA 8 congress (created and drawn by Serena Ferretti, in Reggio Emilia 2015)

Zapata Peña, another enthusiastic and extraordinary researcher and teacher, working on plant use within and beyond Africa. Indeed she also left us far too soon in January 2015.

Published Volumes from Previous IWAA Workshops

This volume is a compilation of peer-reviewed papers presented to the 8th International Workgroup for African Archaeobotany, held in Modena, Italy in June 2015. Such reference books have been published after each meeting of the International Workgroup for African Archaeobotany. With titles reflecting the main thematic lines of the presented studies, they provide an important outcome of each workshop and guidance for further studies and research in African archaeobotany.

Papers from the 1st meeting in Mogilani, Poland were published in volume 35(1) of *Acta Palaeobotanica* by Stuchlik and Wasylikowa (1995) without a synoptic title. The titles of the following two volumes, by Van der Veen (1999) and Neumann et al. (2003), pointed to plant exploitation, plant use for food and fuel, and trends toward agriculture as the 'green world' has been, and still is, at the base of resource procurement and economy in ancient and modern Africa (e.g., Cunningham 1997). The important subheading "Progress in African Archaeobotany" appeared for the first time with the book by Neumann et al. (2003). It mirrors the vivid and incessant advancement that this science has been able to achieve and contributes to the study of Africa's past.

"Fields of Change" by Cappers (2005) recalls that our cultural—agrarian—landscapes are subjected to adaptations and transformations under environmental—climate—as well as socio-economic changes, a remarkably contemporary theme. With the same awareness, the titles of the books by Fahmy et al. (2011) and Thanheiser (2016) underline the interest of archaeobotanists to open windows to new perspectives and employ diverse methods of deciphering human-plant relationships. The volume by Stevens et al. (2014) points to flora as an archaeological record and identifies it as a fundamental partner in the dyadic cultural-botanical relationship: the African flora, which has peculiar properties and a unique role in human and cultural history of the continent.

The lessons from previous books and IWAA conferences have been incorporated into the themes of this volume which illuminates human-plant interactions in the African past.

The Organization and Themes of This Book

The papers included in this volume represent the most current archaeobotanical research completed by international scholars from Africa, Europe and North America. The book presents a pan-African perspective, including study locales from Mediterranean Africa, the Horn, West, East, Central and Saharan Africa, as well as synthesis papers covering more than one region (Fig. 2). Contributions are data-rich and employ some of the most advanced methods available in archaeobotanical research. General themes touched on by papers in this volume include agricultural systems of sub-Saharan Africa, agricultural history, status of archaeobotanical studies, culinary practices, food and drink production/processing, plant domestication history, landscape reconstruction, anthropogenesis and Holocene vegetation history. Chapters focussing on studies of important species are presented including wild grasses, cotton, barley, sorghum, millet and rice. Much of the presented research focuses on macro-botanical remains, mainly seeds/fruits, but in addition, contributions on anthracology as well as micro-remains, including pollen, starch grains and phytoliths are also presented. Many chapters report on studies completed on key archaeological sites in Africa: Luxor, Helwan, Dakhleh Oasis and Hierakonpolis in Egypt; Takarkori in Libya; Tin Hanakaten Cave in Algeria; Kassala and Sai Island in Sudan; Mezber in Ethiopia; the Niger River basin in Benin; Pangwari and Motako in Nigeria; the Middle Senegal River

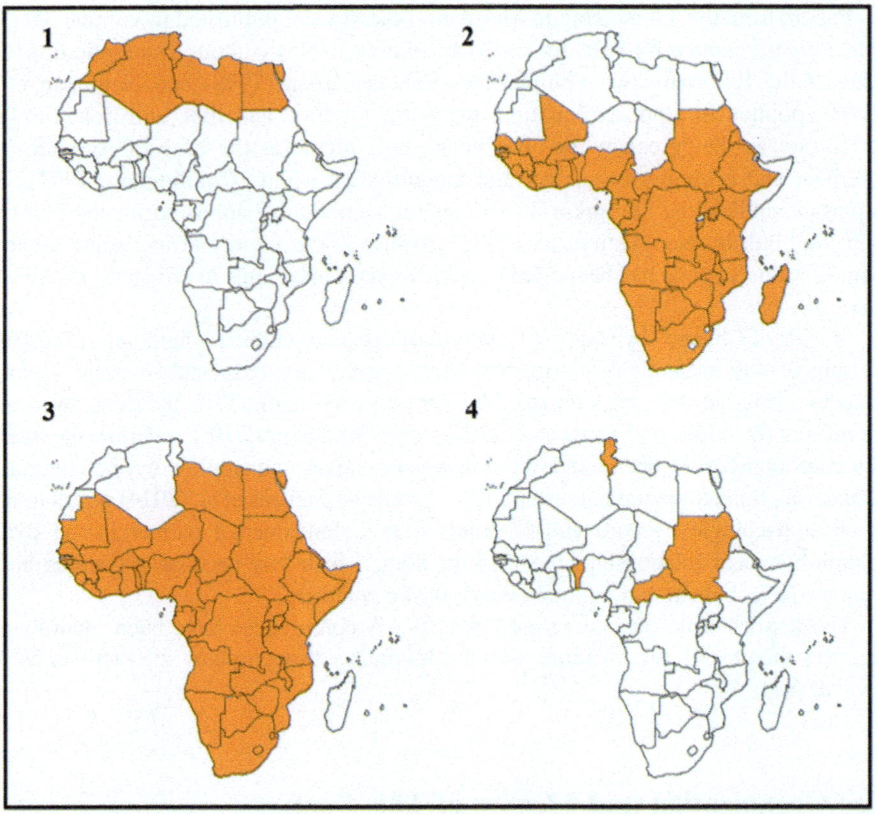

Fig. 2. Map of the country studied in the research papers included in this book, divided into the four themes represented at the Modena conference: (1) Mediterranean Africa; (2) Archaeology and Palaeoecology: Integrated Methods; (3) Plant Use, Agricultural History and Ethnoarchaeology: Foods and Fields; and (4) Climate and Agrarian-Cultural Landscapes.

Valley, and Mtwara Creek in Kenya. All papers present maps and geographical descriptions to make clear the phytogeographical context of the archaeobotanical study.

The volume is divided into four themes represented at the Modena conference: (1) Mediterranean Africa; (2) Archaeology and Palaeoecology: Integrated Methods; (3) Plant Use, Agricultural History and Ethnoarchaeology: Foods and Fields; and (4) Climate and Agrarian-Cultural Landscapes.

The first section, ***Mediterranean Africa,*** includes eight contributions on current research in Egypt, Libya, Tunisia, Algeria and Morocco, covering a temporal period ranging from the Palaeolithic to Late Antiquity (c. 2nd–8th centuries AD). Themes touched upon by these papers include ancient Egyptian Predynastic, New Kingdom and Medieval archaeobotany, and plant use by Holocene forager inhabitants of Mediterranean Africa.

The second section, ***Archaeology and Palaeoecology: Integrated Methods,*** presents five papers that emphasise the use of integrated methods in archaeology and palaeoecology. Contributions from West and East Africa and the Nile region deal with

Fig. 3. Photo of some participants to the IWAA8 in Modena (1. Rossella Rinaldi; 2. Alice Fietta; 3. Chris J. Stevens; 4. Senna Thornton-Barnett; 5. Sahbi Jaouadi; 6. Amanda Logan; 7. Welmoed A. Out; 8. Shannon Hardwick; 9. Sarah Walshaw; 10. Daphne Gallagher; 11. Elshafaey Abdelatif Elshafaey Attia; 12. Federica Riso; 13. Christopher A. Kiahtipes; 14. Roger Blench; 15. Kingsley Daraojimba; 16. Alexa Höhn; 17. Justin Bradfield; 18. Steven Brandt; 19. Ursula Thanheiser; 20. Phylippa Ryan; 21. Ryan M. Szymanski; 22. Claire Malleson; 23. Mats Widgren; 24. Lucia Veronica Collura; 25. Barbara Eichhorn; 26. Charlène Bouchaud; 27. Amy Styring; 28. Samira Amrani; 29. Laurent Bremond; 30. Alemseged Beldados; 31. René T. J. Cappers; 32. Louis Champion; 33. May Lesley Murungi; 34. Elisabeth Hildebrand; 35. Catherine A. D'Andrea; 36. Fabrizio Buldrini; 37. Rim Hamdy; 38. Adel M. Ahmed Ali; 39. Mennat-Allah El Dorry; 40. Christine Sievers; 41. Elisa Nervo; 42. Yolanda Carrión Marco; 43. Katharina Neumann; 44. Sonja Filatova; 45. Anna Maria Mercuri; 46. Marie-Pierre Ruas; 47. Rita Fornaciari; 48. Assunta Florenzano; 49. Paola Torri; 50. Maria Chiara Montecchi).

issues of taphonomy and the difficulties in differentiating anthropogenic and natural environmental changes in the prehistoric archaeological and palaeoecological records.

Section three, ***Plant Use, Agricultural History and Ethnoarchaeology*: *Foods and Fields***, presents six papers dealing with plant use, foodways and agricultural history, and ethnoarchaeological approaches to African archaeobotany with contributions from West, East Africa and the Horn. Examples are presented of plant use in agrarian and foraging societies, plant domestication and the evolution of ancient agricultural systems in Africa.

The final section, ***Climate and Agrarian-Cultural Landscapes***, presents four contributions on climate change and agricultural-mediated landscapes as documented in various study sites across northern and central Africa. Themes include human-environmental interactions, anthropogenic impacts and new evidence on the spread of early agricultural communities.

In more than twenty years of research activity (1994–2018), archaeobotanists have played a significant role in interdisciplinary cooperation that is needed to understand the detailed and complex history of Africa. This book reflects a portion of these cooperative efforts, and we hope it will encourage further studies of the relationships between plants and people by considering processes from a long-term perspective, and from local to regional scales in Africa.

Acknowledgements. The 8th IWAA was the meeting of about 70 scholars in Modena (Fig. 3—Photo of the participants—group under the big holm oak in Modena). We would like to express our gratitude to the referees whose efforts have contributed to the quality of papers presented in this volume: Aziz Ballouche, Marion K. Bamford, Ortrud Monika Barth, Koen Bostoen, Terry M. Brncic, Alan J. Clapham, Charles R. Clements, Savino di Lernia, Barbara Eichhorn, Mennat-Allah El Dorry, Rodolfo Fattovich, Assunta Florenzano, Daphne Gallagher, Renate Germer, Lara Carretero Gonzalez, Elisabeth Hildebrand, Christopher Hunt, Sarah Ivorra, Andrea Kay, Christopher A. Kiathipes, Chap Kusimba, Amanda Logan, Marco Madella, Evi Margaritis, Elena Marinova, Naomi Miller, Katharina Neumann, Ruth Pelling, Peter Rowley-Conwy, Rita Scheel-Ybert, Christine Sievers, Chris Stevens, Ursula Thanheiser, Mariano Ucchesu, Tania Valamoti, Sarah Walshaw, John-Peter Wild, Martin Williams, Michèle Wollstonecroft, Elsa Yvanez, Andrea Zerboni.

References

Aiello LC (ed) (2011) Origin of agriculture: New Data, New Ideas. Curr Anthrop 52 (S4): S161–S162

Cappers RTJ (ed) (2005) Fields of change: progress in African archaeobotany. Barkhuis & Groningen University Library, Groningen

Cunningham AB (1997) An Africa-wide overview of medicinal plant harvesting, conservation and health care. In: Global Initiative for Traditional Systems of Health & FAO (eds) Medicinal plants for forest conservation and health care. Non-Wood Forest Products 11, FAO, Rome

D'Andrea AC, Klee M, Casey J (2001) Archaeobotanical evidence for pearl millet (*Pennisetum glaucum*) in sub-Saharan West Africa. Antiquity 75(288):341–348

D'Andrea AC, Logan AL, Watson DJ (2006) Oil palm and prehistoric subsistence in tropical West Africa. J Afr Archaeol 4(2):63–71

Dunne J, Mercuri AM, Evershed RP et al (2016) Earliest direct evidence of plant processing in prehistoric Saharan pottery. Nat Plants 3(16194)

Fahmy AG, Kahlheber S, D'Andrea AC (eds) (2011) Windows on the African past. Current approaches to African archaeobotany. Africa Magna Verlag, Frankfurt

Fuller DQ (2007) Contrasting patterns in crop domestication and domestication rates: recent archaeobotanical insights from the Old World. Ann Bot 100(5):903–924

Fuller DQ (2012) New archaeobotanical information on plant domestication from macro-remains: tracking the evolution of domestication syndrome traits. In: Biodiversity in agriculture: domestication, evolution, and sustainability. Cambridge University Press, Cambridge, pp 110–135

Fuller DQ, Colledge S (2008) Recent lessons from Near Eastern archaeobotany: wild cereal use, pre-domestication cultivation and tracing multiple origins and dispersals. Praghara 18:105–134

Fuller DQ, Hildebrand E (2013) Domesticating Plants in Africa. In: Mitchell P, Lane P (eds) The Oxford handbook of African archaeology. Oxford University Press, Oxford, pp 507–525

Garcea EA (2004) An alternative way towards food production: the perspective from the Libyan Sahara. J World Prehist 18(2):107–154

Harlan JR, Stemler AB (1976) The races of Sorghum in Africa. In: Harlan JR, De Wet JMJ, Stemler ABL (eds) Origins of African plant domestication. Mouton Press, The Hague, Netherlands, pp 465–478

Höhn A, Neumann K (2012) Shifting cultivation and the development of a cultural landscape during the Iron Age (0–1500 AD) in the northern Sahel of Burkina Faso, West Africa: insights from archaeological charcoal. Quatern Int 249:72–83

Larson G, Piperno DR, Allaby RG et al (2014) Current perspectives and the future of domestication studies. P Natl Acad Sci 111(17):6139–6146

Marshall F, Hildebrand E (2002) Cattle before crops: the beginnings of food production in Africa. J World Prehist 16(2):99–143

Mercuri AM (2008a) Human influence, plant landscape evolution and climate inferences from the archaeobotanical records of the Wadi Teshuinat area (Libyan Sahara). J Arid Environ 72:1950–1967

Mercuri AM (2008b) Plant exploitation and ethnopalynological evidence from the Wadi Teshuinat area (Tadrart Acacus, Libyan Sahara). J Archaeol Sci 35(6):1619–1642

Mercuri AM (2014) Genesis and evolution of the cultural landscape in central Mediterranean: the 'where, when and how' through the palynological approach. Landsc Ecol 29(10):1799–1810

Mercuri AM, Fornaciari R, Gallinaro M et al (2018) Plant behaviour through human imprints and the cultivation of wild cereals in Holocene Sahara. Nat Plants 4:71–81

Morin-Rivat J, Fayolle A, Gillet J et al (2014) New evidence of human activities during the Holocene in the lowland forests of the Northern Congo Basin. Radiocarbon 56(1):209–220

Neumann K, Butler A, Kahlheber S (eds) (2003) Food, fuel and fields: progress in African archaeobotany. Heinrich Barth Institut, Köln

Oas S, D'Andrea AC, Watson DJ (2015) 10,000 year history of plant use at Bosumpra Cave, Ghana. Veg Hist Archaeobot 24(5):635–653

Pearsall DM, Hastorf CA (2011) Reconstructing past life-ways with plants II: human-environment and human interactions. In: Ethnobiology. Wiley-Blackwell, Hoboken, NJ, pp 173–187

Pelling R (2013) Patterns in the archaeobotany of Africa. In: Archaeology of African plant use, vol 61, p 205

Plotkin MJ (1994) Tales of a Shaman's apprentice: an ethnobotanist searches for new medicines in the rain forest. Penguin, New York, p 328

Stevens CJ, Nixon S, Murray MA et al (eds) (2014) Archaeology of African plant use, vol 61. Left Coast Press, Walnut Creek

Stuchlik L, Wasylikowa K (eds) (1995) Acta Palaeobotanica 35(1)

Thanheiser U (ed) (2016) News from the past: progress in African archaeobotany. Barkhuis, Groningen

Van der Veen M (ed) (1999) The exploitation of plant resources in ancient Africa. Kluwer/Plenum, New York

van Gemerden BS, Olff H, Parren MPE et al (2003) The pristine rain forest? Remnants of historical human impacts on current tree species composition and diversity. J Biogeogr 30:1381–1390

Wasylikowa K, Dahlberg J (1999) Sorghum in the economy of the Early Neolithic nomadic tribes at Nabta Playa, Southern Egypt. The exploitation of plant resources in ancient Africa. Springer, US, pp 11–31

White F (1983) The vegetation of Africa, a descriptive memoir to accompany the UNESCO/AETFAT/UNSO vegetation map of Africa (3 Plates, Northwestern Africa, Northeastern Africa, and Southern Africa, 1: 5,000,000)

Zohary D, Hopf M (2000) Domestication of plants in the Old World, 3rd edn. Oxford University Press, Oxford

Zohary D, Hopf M, Weiss E (2012) Domestication of plants in the Old World: The origin and spread of domesticated plants in Southwest Asia, Europe, and the Mediterranean Basin. Oxford University Press on Demand, Oxford

Mediterranean Africa

Archaeobotanical Study at the Early Dynastic Cemetery in Helwan (3100–2600 BC), Egypt: Plant Diversity at Early Dynastic Memphis

Adel Moustafa[1(✉)], Ahmed G. Fahmy[1], and Rim S. Hamdy[2]

[1] Department of Botany and Microbiology, University of Helwan, Helwan, Egypt
Adel.m.eg@gmail.com
[2] Department of Botany, University of Cairo, Giza, Egypt
rimhamdy@yahoo.com

Abstract. Recently, several studies have been published dealing with the analysis of plant remains from archaeological sites in the Memphite region. These investigations covered different archaeological periods. However, until now, no research has be completed on plant remains from the Early Dynastic period in this area. This paper will focus on analysis of macro-botanical remains from sealed pots discovered inside ten tombs at the Early Dynastic Cemetery in Helwan, 30 km south of Cairo (3100–2600 BC). The main goal is to study the economy and ecology relating to the site rather than the offering practices of ancient Egypt. In addition, this research will reconstruct agricultural practices, to shed light on plant diversity and to increase our understanding about non-elite lifeways that prevailed during Early Dynastic times in Memphis. Analysis of the plant remains retrieved from Helwan Cemetery has yielded a total of 25,743 fragments of charred plant macroremains. These remains were classified into two major groups: cultivated crops, including cereals, flax and legumes; and wild/weedy plants comprising wild edible fruits, field weeds, plants of moist habitats, plants of dry habitats and other indeterminate taxa. Cereal remains from the studied tombs consisted of chaff and grains. The study revealed that the arable economy of Memphis during this period was based on the cultivation of cereals primarily *Hordeum vulgare* L. subsp. *vulgare* (hulled barley) which was associated with *Triticum turgidum* L. ssp. *dicoccum* (Schrank) Thell. (emmer wheat).

Keywords: Archaeobotany · Cemetery · Early dynastic · Egypt
Helwan · Plant macroremains

Introduction

In Egypt, archaeobotanical studies have greatly contributed to our knowledge of plant diversity, economy, agricultural production and the environment that prevailed in the past with the help of substantial finds of plants within Pharaonic tombs and also by representations of plants, livestock and agriculture-related activities in ancient paintings and reliefs depicted in tombs and temples. In addition, many archaeobotanical

A. M. Mercuri et al. (Eds.): *Plants and People in the African Past*,
https://doi.org/10.1007/978-3-319-89839-1_2

investigations from settlement sites have increased our understanding of human diet and nutrition in the Egyptian past.

Research into Egyptian archaeobotany has a long history. Kunth (1826) identified fruit remains found by the French archaeologist J. Passalacqua in a XIXth dynasty tomb (1570–1070 BC) at Thebes. Unger (1866) drew our attention to the significance of mud bricks as a useful source of archaeobotanical information. He studied plant macro remains preserved in mud bricks from the Pyramids of Dahshur and from the City of Ancient Rameses (Fahmy 2004). Schweinfurth (1882, 1883, 1884, 1887) included brief morphological descriptions of more than 100 plant species represented in archaeological sediments by different types of plant fragments, including grains, seeds, floral heads and leaf fragments.

During the 20th century, considerable advances were made in Egyptian archaeobotanical research by Keimer (1924), Täckholm (1939, 1940, 1951, 1961) and Greiss (1949, 1957). The works of Täckholm et al. (1941), Täckholm and Drar (1950, 1954) provided ample evidence of cultivated and wild plant remains retrieved from different sites across the country. In addition, Murray (2000) provided an extensive summary of earlier archaeobotanical works. Recent studies have increased our knowledge of past relationships between humans and plants in ancient Egypt through different historical periods including Maadi (Kroll 1989; van Zeist and Roller 1993; van Zeist et al. 2003), Hieraknopolis (Fahmy 1995, 2003, 2004; El Hadidi et al. 1996; Fahmy et al. 2011), Adaïma (Newton 2004, 2007) and Giza (Murray and Gendy 2015; Malleson 2016). Nevertheless, at present the only publications dealing with the analysis of plant remains include data from the Neolithic period 'El-Omari' (Barakat 1990), the Predynastic period 'Maadi' (van Zeist et al. 2003) and the Old Kingdom period sites of Abusir and Giza (Murray 2004, 2005, 2007a, b; Gerisch 2007a, b; Pokorný et al. 2009; Pokorná and Pokorný 2010; Beneš 2011a, b; Pokorná 2011; Pokorná and Beneš 2014) in the Memphite region. Accordingly, our analysis and publication of plant remains from the Helwan cemetery as representative of the Early Dynastic period will help to fill a considerable gap in the agricultural and subsistence history of the Memphite region (Fig. 1 shows archaeological sites mentioned in the text).

Archaeological excavations at the cemetery of Helwan (Protodynastic—early Old Kingdom, c. 3200–2700 BC) revealed the presence of sealed pots inside tombs containing ash and soil which included botanical materials (seeds, fruits and charcoal) as well as a variety of plant remains from other contexts (wood associated with burials, coffin wood and organics in stone vessels). Most plant materials were preserved by charring and others, such as wood, were desiccated. Also, a considerable quantity of animal bones have been retrieved from the cemetery. In this regard, the Helwan cemetery is a noteworthy archaeological site because it has not only produced seeds and grains of both cultivated and wild plants, but also considerable quantities of charred and desiccated materials (charcoal, desiccated wood and other organic materials). This paper will focus on the analysis of plant macroremains (seeds, fruits and other vegetative parts) retrieved from sealed pots from 10 tombs excavated in Operation 4 to understand the economic and ecological characteristics of the site in the past. The aim of the present study is to elucidate the agricultural economy that flourished in

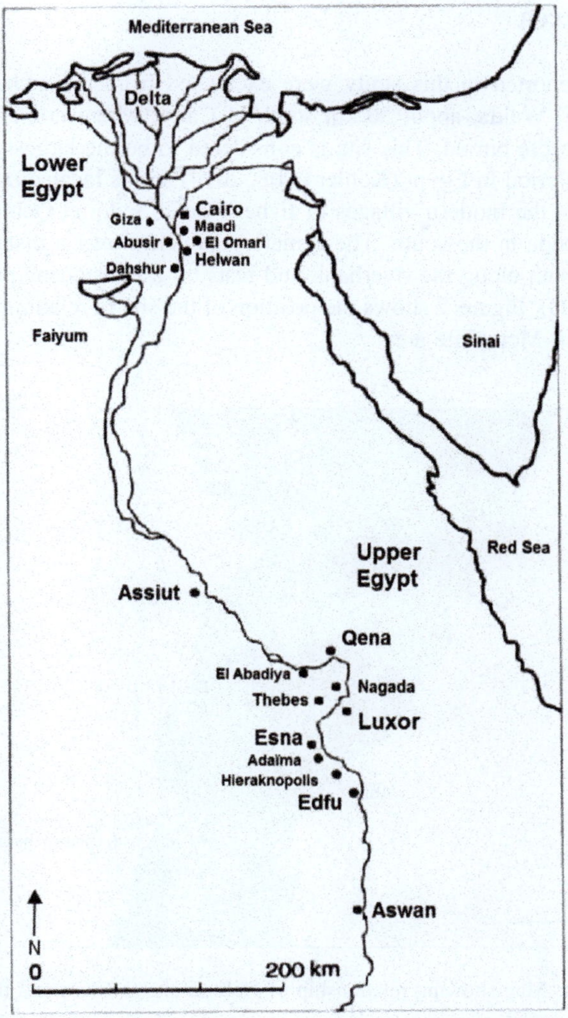

Fig. 1. Map showing archaeological sites mentioned in the text.

the Memphis area during Early Dynastic times and to shed light on past agricultural systems and the main sources of plant foods. It also will examine the weed flora to inform on Early Dynastic agricultural practices, including irrigation and harvesting technique. A detailed study of all botanical materials from Operation 4 at Helwan cemetery including charcoal and wood as well as details of the archaeological contexts will be published at a future date, with archaeobotanical raw data in a separate bioarchaeological volume.

The Study Area

The materials reported in this study were excavated from the archaeological site of Helwan/Ezbet El Walda, about 30 km south of Cairo, where non-elite inhabitants of early Memphis were buried. This site is considered to be the largest necropolis of the Early Dynastic Period in Egypt (Köhler 2008, 2014) and is located on the east bank of the Nile between the modern villages of Ezbet Kamel Sedqi el-Qebleyah in the north and Ezbet el-Walda in the south. The cemetery stretches over a distance of c. 1.5 km from north to south along the riverbank and rests on a Pleistocene palaeofan of Wadi Hof (Köhler 2008). Figure 2 shows the position of the site in relation to Cairo, the Nile and other sites in Memphite area.

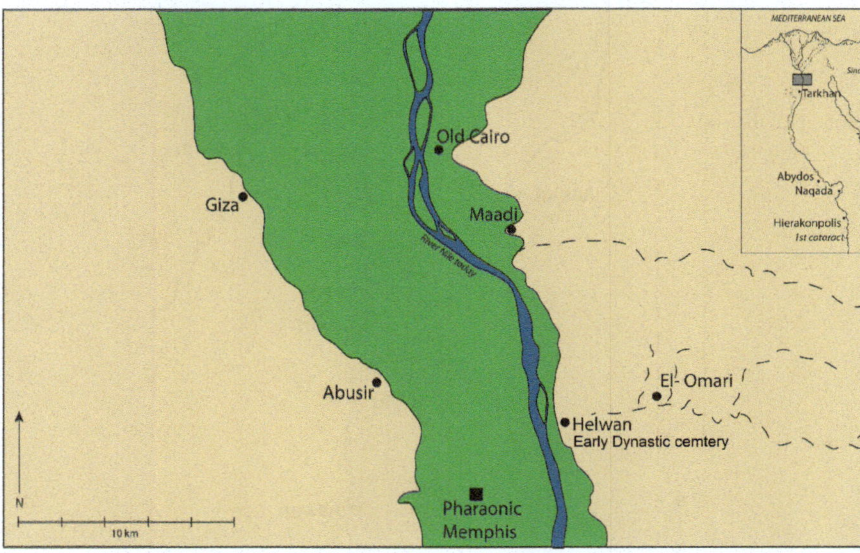

Fig. 2. Map showing relationship of Helwan site to Cairo and the Nile.

Six graves of the cemetery were excavated by Larsen in the 1920s followed by more than 12 years of excavation by Zaki Saad between 1942 and 1954 where more than 10,000 graves were uncovered. This was followed by two more excavation seasons organized by the Egyptian Antiquities Organization in 1966 and 1975 (Köhler 2008). Köhler began fieldwork at the cemetery in 1997 with support of Macquarie University in Sydney. Currently the work has continued as a joint project, led by Köhler for Macquarie University, Sydney and the University of Vienna (Köhler 2008). The excavations have now ceased and the project has entered a phase of analysis and publication.

The period of occupation probably began during Naqada IIIA (c. 3300 BC), continued well beyond Naqada IIIC and IIID (3100–2700 BC), and into the Old Kingdom (after 2700 BC) (Köhler 2015). There is evidence to suggest that certain parts of the site were also occupied throughout much of the Pharaonic period and into the Late Roman and Medieval ages (Köhler 2008). The 218 excavated tombs since 1997 primarily date from Late Naqada IIIC to early Old Kingdom (1st–4th Dynasties) times (Köhler 2015). A series of well-preserved samples of plant macroremains were obtained from ten tombs in the Early Dynastic Cemetery at Helwan/Ezbet El Walda (Operation 4). Table 1 presents the number of samples taken from each tomb, its date, and the age and sex of the occupant. The data reported here should be considered preliminary and pending final analysis. Figure 3 shows the grid plan of the excavated area and illustrates the distribution of the tombs under study.

Table 1 Occupant data and date of the Operation 4 tombs included in this study. Dating is according to Köhler (2014 & in preparation).

Tomb	Number of samples	Cultural dating	Relative chronology	BC dating	Age	Sex
48	1	Late Dynasty 1/early Dynasty 2	Group IIIC3-D1	3100–2900	Adult	Female
68	3	Mid Dynasty 2/Late Dynasty 2	Group IIID2-3	2700	Adolescent	Unknown
71	1	Late Dynasty 2/early Dynasty 3	Group IIID4	2700–2600	Mature Adult	Female
76	4	Dynasty 1-2/Mid Dynasty 2	Group IIID1-2	2700	Adult	Unknown
77	1	Dynasty 1-2/Mid Dynasty 2	Group IIID1-2	2700	Adult	Male?
83	11	Dynasty 1-2/Mid Dynasty 2	Group IIID1-2	2700	Adult	Male
88	4	Mid Dynasty 2	Group IIID2	2700	Mature Adult	Female
91	2	Dynasty 1-2	Group IIID1	3100–2900	Adult	Male?
100	2	Mid Dynasty 2	Group IIID2 (-3?)	2700	Adult	Unknown
104	1	Dynasty 2	–	3100–2900	Adult	Male

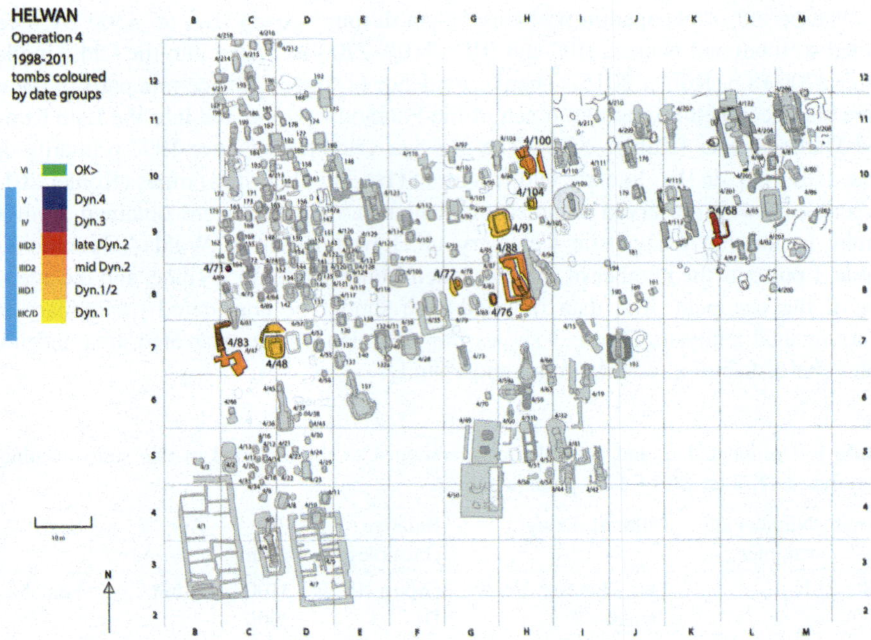

Fig. 3. Grid plan showing the distribution of 10 tombs from Operation 4 mentioned in the text (with kind permission from Prof. Dr. E. Christiana Köhler).

Materials and Methods

Thirty sediment samples were collected during the excavation seasons between 2003 and 2006 from sealed pots inside tombs. The volume of these samples range between 0.3 and 7.1 of ash as well as ash mixed with clay.

Water flotation processing was used for the separation of plant macroremains from the archaeological sediments found inside the pots. The volume of each sample was measured and samples were processed using bucket flotation. Floating macro-botanical remains were skimmed from the surface using a tea strainer (0.5 mm) and placed onto fine cheesecloth. This process was repeated until no more floating charred remains were observed. The cheesecloth was carefully folded over and set aside to dry while the heavy fraction was processed. After drying, each sample was stored in a small plastic box. The volume of plant macroremains was recorded in ml. Only charred materials were retrieved from these thirty samples and they appeared to be in a homogenous state of preservation.

All charred remains of cultivated and wild plants were sorted into categories using a binocular microscope (magnification: 10x–20x) and counted. Only nine samples were subsampled because of their large size, where 50% by volume of each sample was analyzed.

For Scanning Electron Microscope (SEM) studies, seeds and grains were coated with gold and examined by a JEOL JXA-840 SEM. It operated at an accelerating voltage of 15 kV at the Electron Microscope Unit, National Research Center, Dokki, Egypt.

Identification of cereal remains was based on Jacomet (2006) and Zohary et al. (2012). Remains of wild plants were identified using morphological descriptions and illustrations in the Flora of Egypt (Täckholm 1974; Boulos 1999, 2000, 2002, 2005), in addition to Willcox et al. (2002), Cappers et al. (2006) and online data bases and websites including those at the United States Department of Agriculture (USDA 2016). Specimens were identified according to their gross morphological features and compared to a modern reference collection of seeds and fruits held at the Helwan University Herbarium. The nomenclature of the wild taxa is according to Boulos (2009), while that of cultivated plants follows Zohary et al. (2012).

Results

Analysis of Helwan samples has yielded a total of 25,743 fragments of charred plant macroremains including grains, seeds, achenes, nutlets, capsules, pods and vegetative parts of cultivated and wild plants. Table 2 and Fig. 4 summarize the plant macro remains which were classified into two major groups: cultivated crops including 10,303 remains of cereals (40.02%), 6 flax (0.02%) and 2530 pulses (9.83%); and wild plants including 2 remains of wild edible fruits (0.01%), 11,480 field weeds (44.59%), 644 plants of moist habitats (2.50%), 10 plants of dry habitats (0.04%) and 768 indeterminate records (2.98%) (Fig. 4). Cereals and field weeds together dominate plant remains with a combined percentage of 84.61%, which are attributed to 40 species, 47 genera and 18 families.

Table 2 Plant macroremains separated from the Operation 4 tombs in Helwan Early Dynastic cemetery. (A) Number of plant macroremains, (F) Number of samples in which a taxon occurs.

Types/Families	Taxa	Plant element	A	F
1. Crops				
1.1 Cereals				
Poaceae	*Hordeum vulgare* L. subsp. *distichum*	Grain	37	9
		Rachis internode	62	16
	Hordeum vulgare L. subsp.*vulgare* (hulled)	Grain	706	29
		Rachis internode	6299	29
	Hordeum vulgare L. subsp. *vulgare* (naked)	Grain	9	6
	Hordeum sp.	Grain	144	16
		Spikelet	3	2
	Triticum turgidum L. ssp. *dicoccum* (Schrank) Thell.	Grain	75	20
		Glume base	697	23
		Fork	75	11
		Rachis internode	92	14
		Spikelet	3	3
	cf. *Triticum durum* Desf./*aestivum* L.	Rachis internode	2	2

(*continued*)

Table 2 (*continued*)

Types/Families	Taxa	Plant element	A	F
	Triticum sp.	Grain	19	12
	Cereal indeterminate	Grain	9	2
		Glume base	4	1
		Culm	1233	24
		Grain fragment	825	27
		Internodes unknown	9	2
		Cereal total	**10303**	
1.2 Flax				
Linaceae	*Linum usitatissimum* L.	Seed	3	2
		Capsule	3	3
		Flax total	**6**	
1.3 Pulses				
Fabaceae	*Lens culinaris* Medik.	Seed	5	3
	Trifolium alexandrinum L.	Seed	2460	26
		Fruit	65	7
		Pulse total	**2530**	
2. Wild plants				
2.1 Wild edible fruit				
Rhamnaceae	*Ziziphus spina-christi* (L.) Desf.	Fruit	2	2
		Wild plant total	**2**	
2.2 Field weeds				
Asteraceae	*Anthemis retusa* Delile, Descr.	Fruit	7	2
	Senecio aegyptius L.	Capitula	34	4
	Centaurea sp.	Capitula	1	1
	Pulicaria sp.	Fruit	2	2
Brassicaceae	cf. *Camelina* type	Seed	1	1
	Lepidium sativum L.	Fruit	4	2
Chenopodiaceae	*Chenopodium murale* L.	Seed	8	5
Fabaceae	*Lathyrus hirsutus* L.	Seed	12	6
	Lathyrus aphaca L.	Seed	1	1
	Lens sp.	Seed	4	3
	Lupinus type	Seed	4	3
	Medicago polymorpha L.	Seed	151	18
		Fruit	32	7
	Medicago lupulina L.	Seed	1245	25
	Melilotus type	Seed	25	7
	Pisum sp.	Seed	60	12
	Scorpiurus muricatus L.	Fruit	1	1
	Trifolium type 1	Seed	54	6

(*continued*)

Table 2 (*continued*)

Types/Families	Taxa	Plant element	A	F
	Trifolium type 2	Seed	24	5
	Trifolium type 3	Seed	7	3
	Vicia sativa L.	Seed	184	22
	Vicia ervilia (L.) Willd.	Seed	55	18
	Vicia type	Seed	7	5
	Vicia type 2	Seed	13	6
Malvaceae	*Malva* cf *parviflora* L.	Mericarp	1	1
	Malvaceae type1	Seed	11	8
	Malvaceae type2	Seed	13	5
Poaceae	*Agropyron* type	Grain	4	2
	Avena sp.	Grain	1	1
		Rachilla	7	6
	Bromus sp.	Grain	296	26
	Digitaria sanguinalis (L.) Scop.	Grain	2	2
	Lolium temulentum L.	Florets	1984	19
		Grain	870	10
		Internode	57	8
	Phalaris minor Retz.	Spikelet	45	10
		Grain	2981	30
	Phragmites type	Leaves	26	1
	Setaria type	Grain	2	2
	Grasses indeterminate	Grain	319	18
		Rhizome	47	13
		Fragments	871	23
		Culm fragment	75	6
		Internode	3	1
		Glume base	219	12
		Small branch	2	1
		Branches	721	21
Polygonaceae	*Persicaria senegalensis* (Meisn.)	Seed	2	2
	Polygonum equisetiform Sm.	Seed	12	8
	Polygonum sp.	Seed	4	3
	Polygonaceae type	Seed	6	4
	Rumex dentatus L.	Seed	692	25
		Fruiting bodies	183	13
	Rumex sp.	Seed	3	3
Portulacaceae	*Portulaca oleracea* L.	Seed	15	5
Solanaceae	*Solanum nigrum* L.	Seed	3	1
		Field weed total	**11480**	

(*continued*)

Table 2 (*continued*)

Types/Families	Taxa	Plant element	A	F
2.3 Plants of moist habitats				
Asteraceae	*Ceruana pratensis* Forssk.	Capitula	11	5
Brassicaceae	*Coronopus niloticus* (Delile) Spreng.	Fruit	1	1
Cyperaceae	*Carex divisa* Huds.	Fruit	1	1
		Seed	3	2
	Cyperus articulatus L.	Seed	3	3
	Cyperaceae type 1	Seed	225	23
	Cyperaceae type 2	Seed	3	1
	Eleocharis palustris (L.) Roem. & Schult.	Seed	63	11
	Fimbristylis bisumbellata (Forssk.)	Seed	1	1
	Scirpus type 1	Seed	29	1
	Scirpus type 2	Seed	4	1
Juncaceae	*Juncus rigidus* Desf.	Seed	1	1
Rosaceae	*Potentilla supina* L.	Seed	299	9
		Plants of moist places total	**644**	
2.4 Plants of dry habitats				
Asphodelaceae	*Asphodelus fistulosus* L.	Seed	1	1
Asteraceae	*Nauplius graveolens* (Forssk.)	Capitula	5	5
Boraginaceae	*Echium* sp.	Seed	3	3
Brassicaceae	*Zilla spinosa* (L.) Prantl	Fruit	1	1
		Plants of dry places total	**10**	
2.5 Indeterminate taxa				
Boraginaceae	Boraginaceae type	Seed	16	1
Brassicaceae	Cruciferae type	Seed	4	4
Euphorbiaceae	*Euphorbia* type	Seed	2	2
Fabaceae	Fabaceae 1 (*Vicia* type 1)	Seed	1	1
	Fabaceae 2	Seed	44	12
	Fabaceae 4 (*Vicia* type 3)	Seed	20	7
	Fabaceae 5 (*Vicia* type 4)	Seed	2	1
	Fabaceae unidentified	Seed	332	18
	Fabaceae fragments	Seed	149	7
	Fabaceae broken seed	Seed	194	19
Lamiaceae	Labiatae type 1	Fruit	2	2
	cf. Labiatae type 2	Seed	1	1
Poaceae	Poaceae type	Spikelet	1	1
		Indeterminate total	**768**	
		Total remains	25,743	

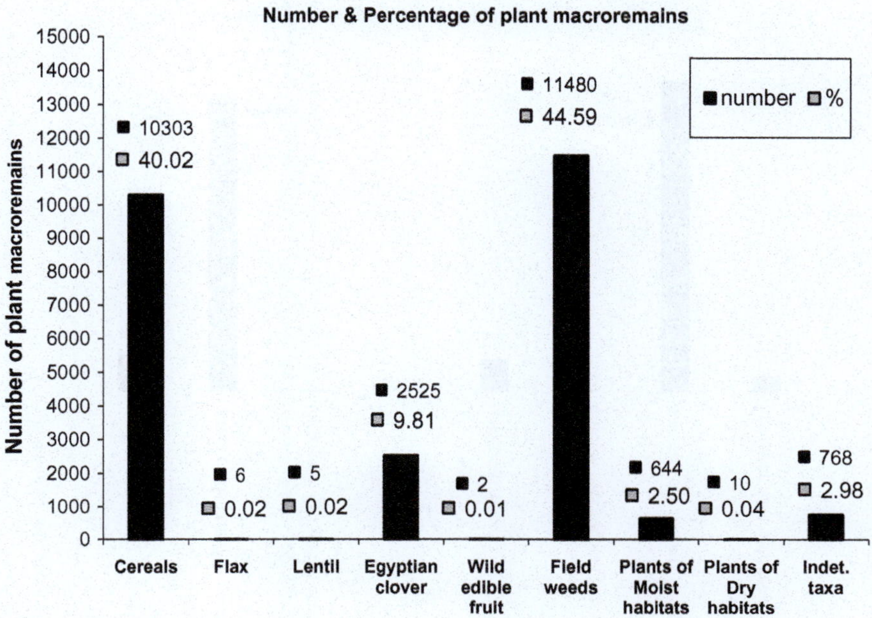

Fig. 4. Frequencies and percentages of plant macroremains from 10 tombs of Operation 4 at the Helwan Early Dynastic cemetery.

Crop Plants

Crops represent 49.87% of the total number of plant macroremains (Fig. 4), which are mostly of cereals, 40.02%, while 0.02% comprises flax and 9.83% are legumes.

Cereals

Cereal remains appeared in all studied samples and included grains, rachis fragments, glume bases, and spikelet forks. These elements consist of 90.3% chaff and 9.7% of grains. Remains of five cereals have been identified: *Hordeum vulgare* L. ssp. *vulgare* (hulled barley), *Hordeum vulgare* L. ssp. *distichum* (two-row barley), *Hordeum vulgare* L. ssp. *vulgare* (naked barley), *Triticum turgidum* ssp. *dicoccum* (Schrank) Thell. (emmer wheat), *Triticum aestivum/durum* (free threshing bread/macaroni wheat). The latter category was represented only by articulated rachis segments. As shown in Fig. 5, the remains of hulled barley dominate the macro-botanical assemblage with 70.7% of the total number of cereal grains and 67.7% of the total number of chaff elements. On the other hand, emmer, free threshing wheats, two-rowed and naked barley are represented by comparatively lower percentages.

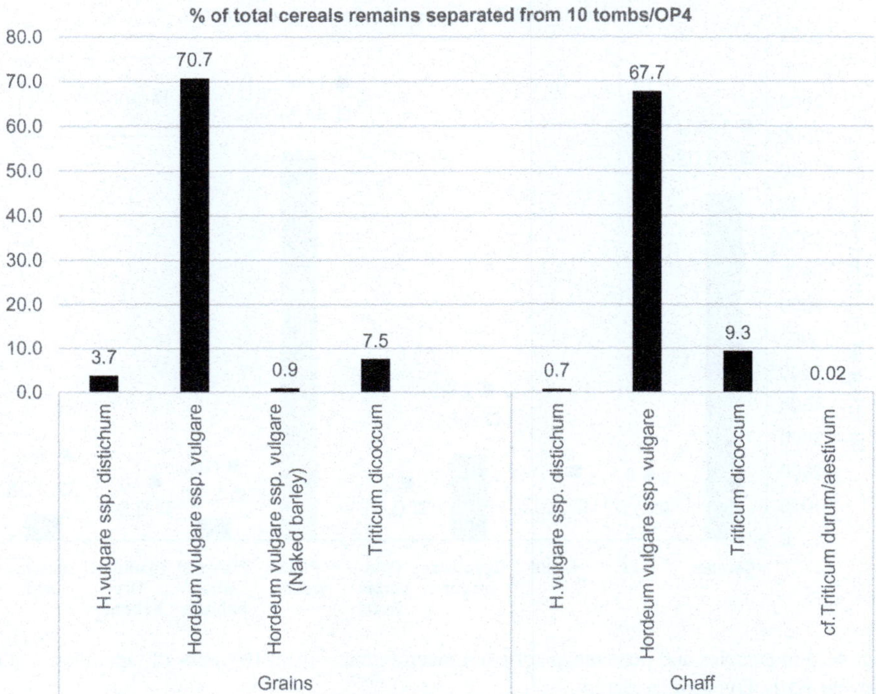

Fig. 5. Percentages of total cereal remains (grain and chaff) separated from 10 tombs of Operation 4 in the Helwan Early Dynastic cemetery.

Hulled barley remains vary in percentage among the 10 tombs, where chaff ranges 23.0–90.3% and grains range 0.2–23.2% of the total number of cereals remains retrieved from each tomb. Rachis internodes (Fig. 6) of barley were found in all studied tombs. The dominance of barley in all samples may suggest its significant role as a symbol of resurrection. Also, short and plump grains of naked barley which are characterised by the presence of transverse ridges on their dorsal side (Fig. 7) were recorded in low numbers, which may reflect the minor role of this crop in the Early Dynastic agricultural economy at Memphis. It also may represent a contaminant in fields of hulled barley. Van Zeist and Roller (1993) reported that all barley grains from the site of Maadi were of the hulled type. A total of 37 grains of *Hordeum distichum* were recorded from 9 samples with average dimensions of 3.1 × 1.5 × 1.05 mm (length x breadth x thickness). Rachis internodes of this variety (Fig. 8) were recorded in 16 samples. Although it can be difficult to distinguish between two and six-row barley (*Horedum vulgare* ssp. *hexastichum* L.), based on grain shape and the presence of associated rachis internodes, we conclude these grains are the two-rowed form (Table 2).

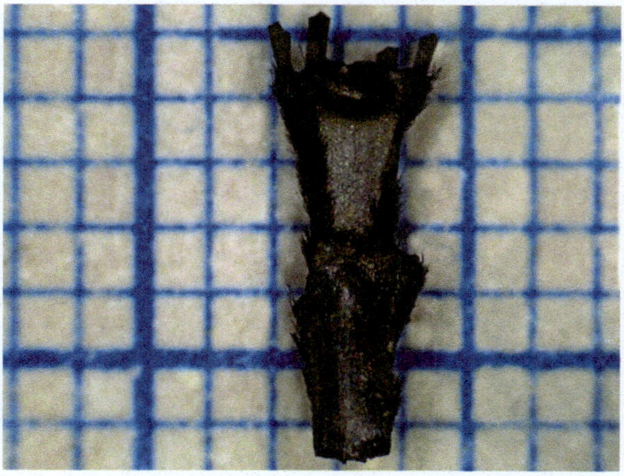

Fig. 6. Rachis internode of hulled barley. Each unit = 1 mm.

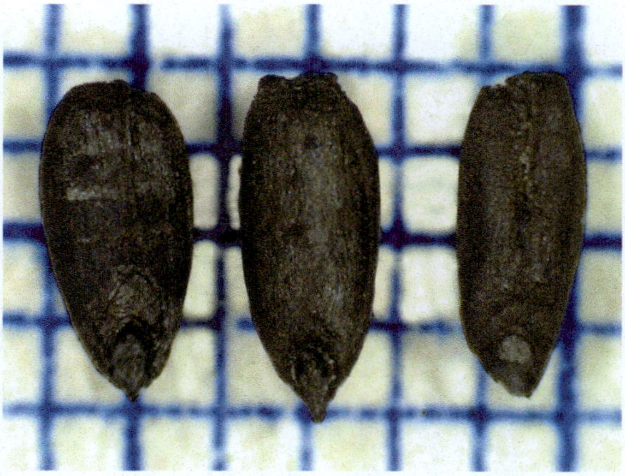

Fig. 7. Grains of naked barley. Each unit = 1 mm.

Emmer wheat remains were not abundant in the Helwan samples. They represent only 7.5% of the total number of cereals grains and 9.3% of the total number of chaff elements (Fig. 5). In addition, emmer grains were recorded in low numbers with percentages ranging 0.3–2.3% of the total number of cereals remains in each tomb while emmer chaff occurred in very small percentages ranging 0.2–26.1% of the total number of cereals remains retrieved from each tomb. Emmer chaff included glume bases, spikelet forks (Fig. 9), rachis segments and whole spikelets.

Rachis internodes of free threshing wheat were represented by 2 elements. This demonstrates that it was likely a contaminant of emmer/barley fields. As reported by van Zeist and Roller (1993), van Zeist et al. (2003) and Cappers et al. (2004), free-threshing wheat was not cultivated in Predynastic and Pharaonic Egypt.

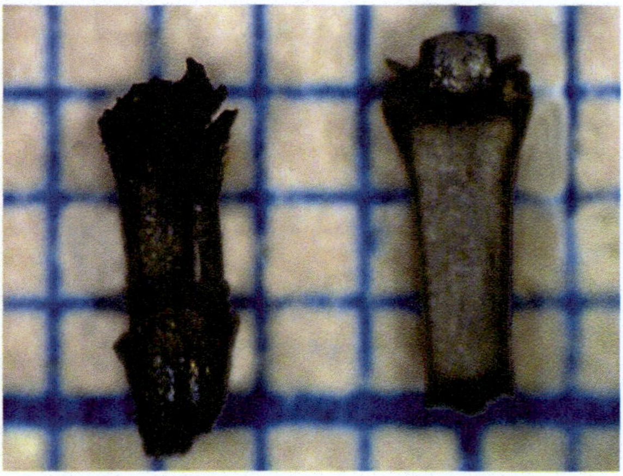

Fig. 8. Rachis internodes of *Hordeum distichum.* Each unit = 1 mm.

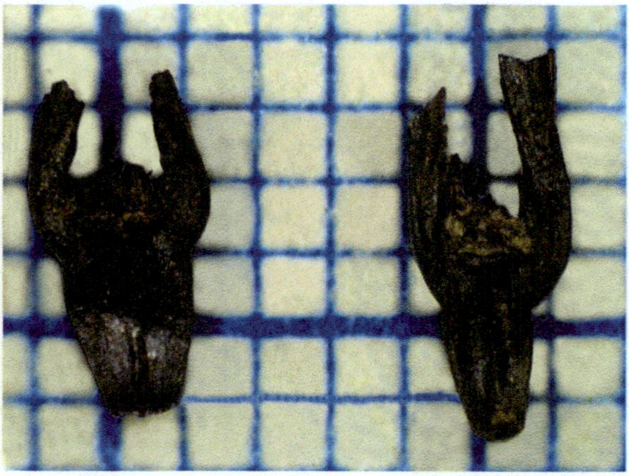

Fig. 9. Spikelet forks of emmer wheat. Each unit = 1 mm.

Flax

Very few seeds and capsules of *Linum usitatissimum* L. (Fig. 10) were recorded from the tombs, representing 0.02% of the total number of the plant macroremains. Flax has been cultivated in Egypt as both a fiber and oil crop.

Pulses

Pulses represent 9.83% of the total number of plant remains and include seeds of lentil which comprise only 0.02% of the total number of plant remains. In contrast, large numbers of *Trifolium alexandrinum* L. seeds and fruits were recorded, which represent 9.81% of the total number of plant remains (Fig. 4). Clover remains appeared in most examined samples (Table 2) and are in an excellent state of preservation (Fig. 11).

Fig. 10. SEM micrograph of a capsule of *Linum usitatissimum*.

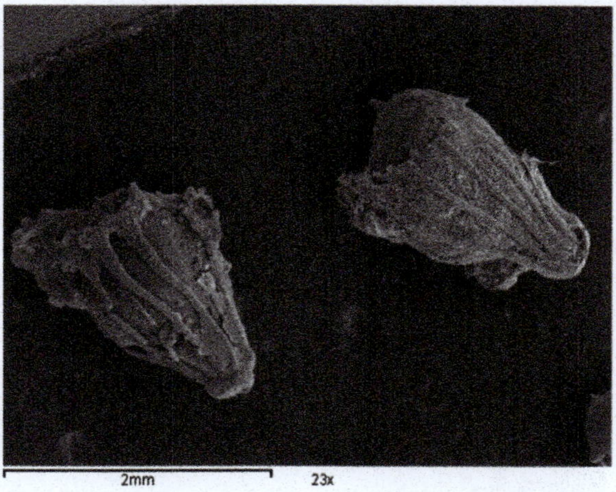

Fig. 11. SEM micrograph of *Trifolium alexandrinum*.

Wild/Weedy Plants

A total of 32 species of wild and weedy plants were retrieved from Early Dynastic flotation samples from the Helwan tombs. These include one species of wild fruit, 20 species of field weeds, eight species of moist habitats and three species occupying dry habitats.

Wild Fruits

The fruit of *Ziziphus spina-christi* (L.) Desf. (sidder) was recorded in the Early Dynastic Helwan cemetery. The fruit remains of sidder appear repeatedly in the

Egyptian archaeological contexts from Predynastic times onwards (Zohary and Hopf 1993; Fahmy 1995). These fruits were often recorded among offering dishes in tombs (Täckholm 1976; Wetterstrom 1984; Fahmy 1995). In addition, there is an interesting record of desiccated sidder fruits in the annex of the tomb of Tutankhamun, which was probably part of a royal meal and stored in a separate container (Vartavan 2002). Moreover, the sidder tree itself was considered sacred by ancient Egyptians (Täckholm 1976; Fahmy 2003).

Field Weeds

Twenty species of field weeds were recorded in the botanical assemblage from the Early Dynastic cemetery at Helwan with percentage value of 44.59%. Well-preserved specimens of *Scorpiurus muricatus* L. and *Lepidium sativum* L. are illustrated in Figs. 12 and 13, respectively. We noted that most of the weed species were represented by seeds, grains, fruit fragments or vegetative parts and consist of 10 winter weeds, two summer weeds and eight all year round weeds. As shown in Table 2, eight species dominate the weeds and were recorded in over 50% of the examined samples, namely, *Medicago polymorpha* L., *M. lupulina* L., *Vicia sativa* L., *V. ervilia* (L.) Willd., *Bromus* sp, *Lolium temulentum* L., *Phalaris minor* Retz., and *Rumex dentatus* L. In addition, among the 20 species are *Malva parviflora* L., *Medicago polymorpha*, *Solanum nigrum* L., *Portulaca oleracea* L. (Fig. 14), *Chenopodium murale* L., *Rumex dentatus*, *Senecio aegyptius* L., *Polygonum equisetiforme* Sm., which are still found as field weed assemblages associated with wheat and clover cultivation in modern Egypt (El Hadidi et al. 1999: Table 1). *Phalaris minor* and *V. sativa* are common species in modern wheat fields while *R. dentatus* and *M. polymorpha* are indicator species of clover (*T. alexandrinum*) (El Hadidi et al. 1999). The greater number of *P. minor*

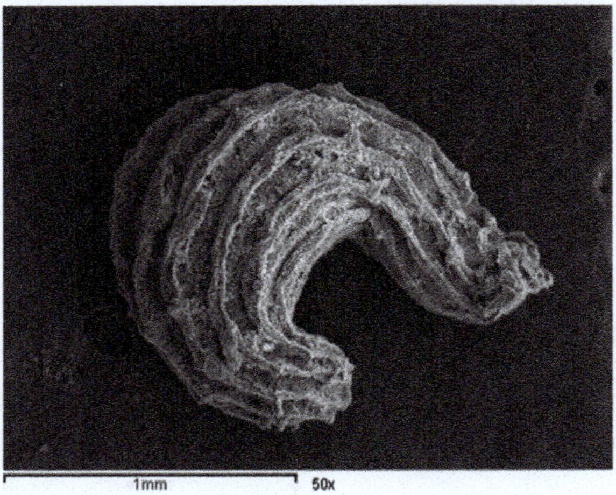

Fig. 12. SEM micrograph of *Scorpiurus muricatus*.

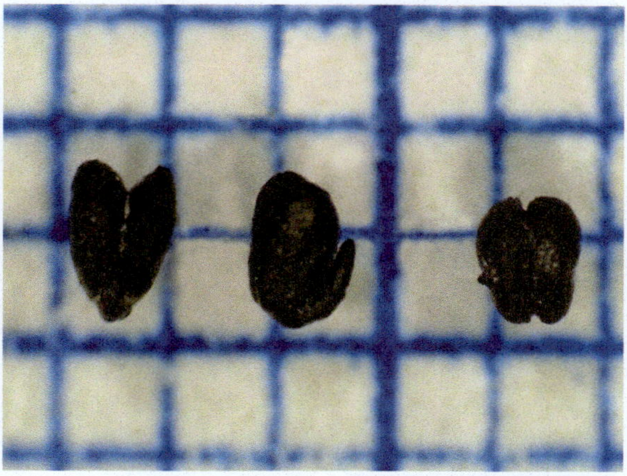

Fig. 13. Fruits of *Lepidium sativum*. Each unit = 1 mm.

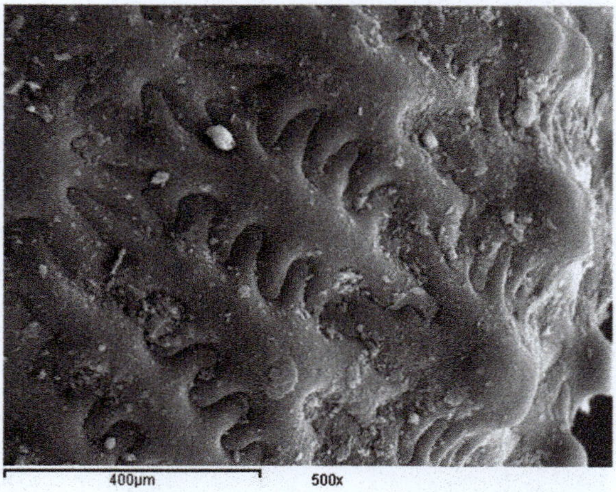

Fig. 14. SEM micrograph of the seed surface of *Portulaca oleracea* subsp. *nitida*.

(Fig. 15) and *L. temulentum* remains could suggest that both species were common field weeds in Early Dynastic wheat fields near the site. They may be considered as an indication that farming activities were practised during the winter months in contrast with the low number of summer weeds such as *P. oleracea* and *Digitaria sanguinalis* (L.) Scop., which sprout during late summer and continue during winter.

Fig. 15. SEM micrograph of the grain of *Phalaris minor* (lateral view).

Plants of Moist Habitats

Remains of eight species commonly found in moist habitats (marshes and muddy banks of water bodies) were recovered in the Helwan tomb samples with percentage 2.23% of the total number of the plant macroremains (Fig. 4). As examples, Figs. 16 and 17 are SEM micrographs of *Carex divisa* Huds. and *Eleocharis palustris* (L.) Roem. & Schult., respectively. Today, these species typically grow on wet soil near irrigation canals and water bodies. It is interesting to note that one of the earliest examples of pictorial evidence for irrigation canals in ancient Egypt is found on the Scorpion King mace-head showing the king ceremonially holding a hoe and digging a canal (Late Predynastic, c. 3100 BC), indicating that irrigation canals were present by at least the

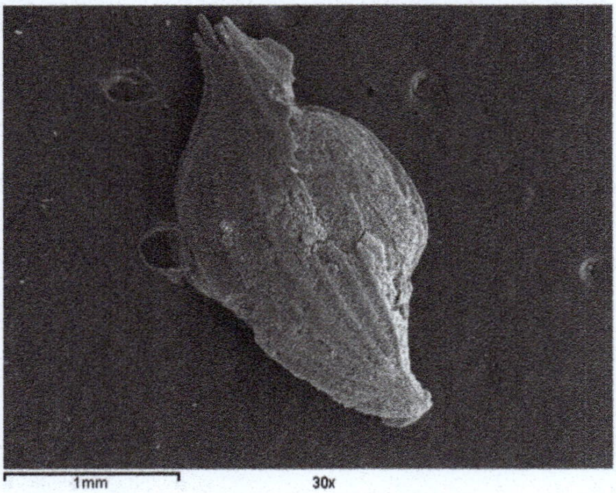

Fig. 16. SEM micrograph of a *Carex divisa* utricle.

Fig. 17. SEM micrograph of a nutlet of *Eleocharis palustris*.

Fig. 18. Capitula of *Ceruana pratensis*. Each unit = 1 mm.

Late Predynastic period. These moist conditions are a suitable habitat for the growth of pasture plants of moist habitat including *Fimbristylis bisumbellata* (Forssk.), *Ceruana pratensis* Forssk. (Fig. 18), and *Juncus rigidus* Desf. The latter two species are also important material in the manufacture of baskets and mats (El Hadidi et al. 1996).

Plants of Dry Habitats

Remains of 3 species which are adapted to dry conditions were identified from the 10 tomb samples. The following taxa have been attributed to this desert vegetation: *Zilla spinosa* (L.) Prantl, *Nauplius graveolens* (Forssk.) and *Asphodelus fistulosus* L. (Fig. 19).

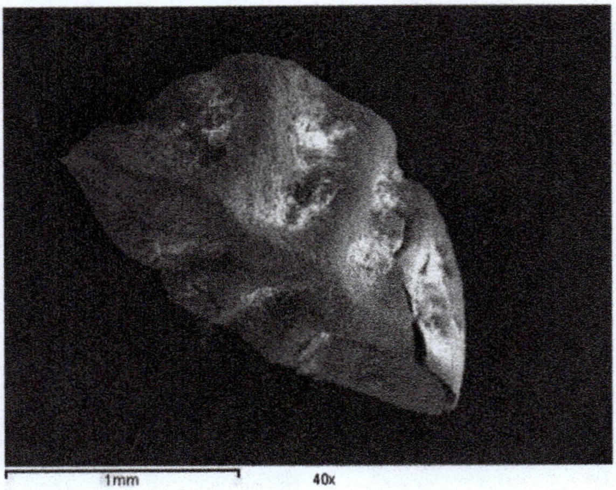

Fig. 19. SEM micrograph of a seed of *Asphodelus fistulosus*.

Indeterminate Taxa

The category indeterminate taxa includes plant macroremains which have been identified only to the family level. As such, it was not possible to deduce specific information about the significance of these specimens in the economic or ecological reconstructions. This category makes up only 2.86% of the total number of the plant macroremains (Fig. 4). It includes the plant macroremains of the following families: Boraginaceae, Brassicaceae, Fabaceae, Lamiaceae and Poaceae types.

Discussion

The overall analysis of the Helwan tomb samples suggests that this botanical assemblage is comprised of cereal chaff refuse with associated field weeds which were transported from the area where crop processing activities took place. This is similar to many rural settlements across the Near East and Europe where the processing remains of cereals and other crops have been recovered. But in the case of the Helwan tomb samples, it is difficult to determine if the cereal chaff refuse was transported to the settlement, directly to the cemetery or transported to the settlement then to the cemetery to serve as a fuel for fires made during funerary rites.

It is generally agreed that crop processing usually took place either at or near the point of deposition and their by-products were brought in for specific purposes such as fodder, bedding, fuel, or as building materials (van der Veen 2007). In addition, cereal crop by-products are used in rural Egypt today as kindling in hearths, kilns and furnaces in villages and nomadic settlements. Similar practices may have occurred in Early Dynastic Memphis, namely kindling the fire with chaff placed between branches of acacia and tamarisk wood possibly during burial ceremonies.

In addition, the analysis suggests that samples also represent accumulations of debris during food preparation. This debris was transferred from ovens used in cooking into pots which were later sealed by mud stoppers and placed beside the dead as a part of ritual ceremonies. However, it is not clear if these debris were put into jars at the settlement and then transferred to the cemetery and deposited into tombs or if they were placed into pots at the cemetery. The issue of funerary practices, deposition of these material and their significance will be the subject of future research on the Helwan tomb samples.

The results from this study demonstrate that the two major cereals retrieved from the Helwan botanical assemblage were hulled barley and emmer wheat. However, the remains of hulled barley, including grains and chaff elements, overwhelmingly dominate the botanical assemblage which suggests a significant role of barley in Memphis during Early Dynastic times (Late Naqada IIIC-early Old Kingdom). Similar results have been recorded from Adaïma, end Naqada II (Predynastic) to Naqada IIIB/C (Protodynastic, I and II dynasties) in Upper Egypt where Newton (2007) made a quantitative comparison between emmer wheat and barley remains and concluded that barley remains are 2.5 times more abundant than emmer.

Van Zeist and Roller (1993) have concluded that the arable economy of Predynastic Maadi (Naqada IIC/D) was largely based upon the cultivation of wheat, probably emmer, and hulled barley (two and six-rowed forms). Fahmy et al. (2011) reported that emmer wheat represented 56% of the total number of retrieved plant remains while barley represented only 6.1% of the botanical assemblage from Hierakonpolis (Naqada IIC/D). They concluded that the economy of Predynastic Hierakonpolis was based on the cultivation of cereals: emmer wheat and barley which were consumed by humans as well as animal populations. In addition, Wetterstrom (unpublished report) concluded that barley production was increasing during Predynastic times and the larger portion of the harvest was probably used as fodder and the rest for making beer. She recorded a displacement of emmer wheat by barley over time at Naqada KH3 and Naqada South Town (3800–3400 BC).

It seems that there is a marked reduction in the use of emmer wheat which was replaced by barley moving from Predynastic times to the early Old Kingdom. This trend may reflect the growing significance of barley, which may be attributed to beer making, the extensive use of barley as animal fodder and/or the success of barley cultivation under basin irrigation (Murray 2000). This observation was also made by Murray (2000) where textual and linguistic evidence suggested that barley was the predominant cereal during the Old and Middle Kingdom with emmer wheat rising to become the most important cereal during the New Kingdom to Ptolemaic period. This observation requires further research, including more archaeobotanical sampling at other archaeological sites dating to the later Pharaonic period, which should examine the relative abundance of barley and emmer wheat over time.

The role of flax and lentil in the economy of Memphis remain obscure. The samples from the Helwan tombs could have been contaminants from earlier processing episodes. Similar low quantities of flax and lentil were observed by Kroll (1989) and van Zeist and Roller (1993) at Maadi.

One interesting finding is the remarkably high numbers of *Trifolium alexandrinum* seeds and fruits, which may suggest that clover was a cultivated crop in Early Dynastic Memphis. Moreover, weeds of *R. dentatus* and *M. polymorpha* present in the same sample are indicator species for clover cultivation (El Hadidi et al. 1999). The large quantity of

Trifolium in the Helwan tomb samples may be the result of informal intercropping of legumes with cereals crops (Malleson 2016). Also the absence of dung remains coupled with an abundance of well preserved charcoal suggests that the botanical assemblage resulted from a field to fire pathway, with legumes and weeds derived from using cereal processing by-products directly as fuel and tinder. This assemblage did not arise from a fodder-dung-fuel pathway, and given the exceptional preservation, it is unlikely to have undergone mastication and digestion by livestock (Malleson 2016). Furthermore, Malleson (2016) suggested that *Trifolium* sp. and other weeds may have been an integral part of cereal fields, because farmers were likely aware of the benefits that they added to the crop. The findings of this study suggest that the remains of clover retrieved from the Early Dynastic cemetery at Helwan constitutes an interesting earlier example of what was suggested by Malleson (2016) at Old Kingdom site of Giza.

Sidder fruit was probably part of the diet on a regular basis: the fleshy mesocarp is slightly acidic, and may be consumed fresh, dried, as a drink or as a sweet cake. Their importance in the diet is also evidenced by their presence in tombs as food offerings and they could be interpreted as a bread-like food ingredient (Newton 2007). The presence of sidder fruit in the plant macroremains within ashy deposits at Helwan may be attributed to food preparation or bread making.

Another interesting finding is the exceedingly large number of weeds in the Helwan tomb assemblage. This could be attributed to the practice of informal intercropping by ancient farmers where they allowed clover and other weeds to grow amongst the cereals because they would increase the value of the harvest by improving the quality and quantity of valuable by-products (Malleson 2016). Moreover, this may indicate that Early Dynastic farmers did not remove weeds from their fields.

The large number of recorded winter weeds retrieved from the Helwan sample could be the result of the classic basin irrigation system which was applied in ancient Egypt. This practice involved the complete inundation of fields during the summer, and consequently the crop rotation was simple and confined to the winter season (Fahmy 1997). The 10 species of winter weeds are typically found in temperate regions of the Old world consisting of Mediterranean and Irano-Turanian elements. These field weeds were introduced to Egypt along with cereal crops from the Near East (El Hadidi 1993; Fahmy 1997) during the Predynastic period. The presence of summer weeds in this assemblage may simply be the result of their occurrence in near-by tropical ecosystems (Kosinová 1975).

The majority arable weeds retrieved from the Early Dynastic cemetery at Helwan grew to over 40 cm, however the presence of low growing weeds such as *Portulaca oleracea*, suggests that cereal crops and possible clover fields were harvested by cutting the cereals culms just above the soil. Thanheiser (1996) reported that the cereals were harvested by the same method in Pre and Early Dynastic periods by using sickles and the culms were cut just above the soil.

The occurrence of plants of moist habitats could reflect the presence of herbaceous species on the Nile floodplain near Memphis. Furthermore, the presence of such an assemblage of taxa demonstrates the existence of wet soils along water courses that would have penetrated cultivated fields especially during the winter. A similar observation was made in archaeobotanical analyses at Predynastic Maadi (Van Zeist and Roller 1993) and Hierakonpolis (El Hadidi et al. 1996). Among the recorded plants of moist habitats were and still today include taxa of the palaeotropical kingdom and comprise 6 species which are mainly Sudano–Zambezian (African elements). These are principally riverine species and are part of the natural vegetation of Egypt.

The plants of dry habitats derive from the desert vegetation near Helwan which is comprised of perennials and complementary assemblages of therophytes. Also, *Asphodelus fistulosus,* one of the co-dominant species of irrigated winter crop vegetation, typically invades uncultivated dry lands and areas under fallow. The presence of such an assemblage among cereal chaff remains could indicate that ruderal vegetation grew in the vicinity within cultivated fields near the site. Similar situations have been recorded at Predynastic Hierakonpolis (Fahmy 1995) and Adaïma (Newton 2007). The present vegetation around the site is sparse and dominated by a few stands of *Alhagi graecorum* Boiss. and scarce individuals of *Tamarix nilotica* Bunge and *Zygophyllum coccineum* L.

Conclusions

The assemblage of plant macroremains retrieved from the Early Dynastic cemetery at Helwan has shed some light on the agricultural history and relationships between humans and diverse plants that grew during this period. This was possible partly because of the excellent preservation plant macroremains in the tombs. The purpose of this study was to determine the plant diversity and agriculture practices that prevailed during Early Dynastic times in Memphis, in addition, to assess food consumed by inhabitants. The results indicate that the Early Dynastic inhabitants of Memphis cultivated cereals including hulled barley and emmer wheat as major sources of food. In addition they applied different agriculture practices such as intercropping legumes with cereal crops, basin irrigation, and specific harvesting techniques during the Early Dynastic period. The study also suggested the occurrence of different vegetation types and habitats of xerophytes and swampy stands in the general vicinity of the site resulting from a high water table.

This study represents a preliminary analysis of macro-botanical remains from 10 tombs in the Helwan Cemetery. The sample represents a fraction of the total botanical assemblage available from Helwan which includes more than 200 tombs already sampled for macroremains and awaiting analysis. Further study will certainly add to the emerging picture of the Early Dynastic agricultural economy of the Memphis region. Additional research will focus on the type, significance and importance of archaeological contexts as well as funerary practices to consider why these plant remains were kept in jars and placed in tombs, and the significance of these activities. Study of the entire botanical assemblage also will elucidate changes in the botanical assemblage during the long time span of the site occupation.

Acknowledgements. We wish to express our deepest gratitude and appreciation to Prof. Dr. Christiana Köhler, Institute for Egyptology, University of Vienna, for giving the opportunity to work with her excavation team at Helwan and for using this material. Many thanks are due to the Academy of Scientific Research and Technology (ASRT), Egypt for the grant (June 2007–June 2011) which has helped in the accomplishment of this study.

References

Barakat H (1990) Appendix IV–plant remains from El-Omari. In: Debono F, Mortensen B (eds) El-Omari: a neolithic settlement and other sites in the vicinity of wadi hof, vol 82. Archäologische Veröffentlichungen, von Zabern, Mainz am Rhein, Helwan pp 109–114

Beneš J (2011a) Analysis of wooden finds from burial shafts of AS 38. In: Vymazalová H et al (2011) Abusir XXII, The tomb of Kaiemtjenenet (AS 38) and the surroundings structures (AS 57-60). Charles University, Prague, pp 168–171

Beneš J (2011b) Xylotomic analysis of wooden objects from the embalmer's deposit of the shaft tomb of Menekhibnekau. In: Bareš L, Smoláriková K (2011) Abusir XXV. The shaft tomb of Menekhibnekau, vol 1, Archaeology. Charles University, Prague, pp 182–184

Boulos L (1999) Flora of Egypt I. Al Hadara Publishing, Cairo

Boulos L (2000) Flora of Egypt II. Al Hadara Publishing, Cairo

Boulos L (2002) Flora of Egypt III. Al Hadara Publishing, Cairo

Boulos L (2005) Flora of Egypt IV. Al Hadara Publishing, Cairo

Boulos L (2009) Flora of Egypt checklist. Al Hadara Publishing, Cairo

Cappers RTJ, Van Thuyne T, Sikking L (2004) Plant remains from Predynastic El Abadiya-2 (Nagada area, Upper Egypt). In: Hendrickx S, Friedman RF, Clalowicz KM, Chlodnicki M (eds) Egypt at its origins: studies in memory of Barbara Adams. Orientalia Lovaniensia Analecta 138. Peeters, Leuven, pp 275–293

Cappers RTJ, Bekker RM, Jans JEA (2006) Digital seed atlas of the Netherlands. Groningen Archaeological Studies 4. Barkhuis Publishing, Eelde, Netherlands

El Hadidi MN (1993) A historical flora of Egypt, a preliminary survey. In: Davies WV, Walker R (eds) Biological anthropology and the study of Ancient Egypt. British Museum Press, London, pp 144–155

El Hadidi MN, Fahmy AG, Willerding W (1996) The palaeoethnobotany of locality 11C, Hierakonpolis (3800–3500 cal. BC); Egypt. Taeckholmia 16:45–60

El Hadidi MN, Hasan LM, Moustafa MF (1999) The phytosociological relationship between the weed flora and cultivated crops in Qena governorate. Bull Fac Sci Assiut Univ 28(2):261–278

Fahmy AG (1995) A historical flora of Egypt, Preliminary Survey. PhD Dissertation, University of Cairo, Faculty of Science, Botany Department

Fahmy AG (1997) Evaluation of the weed flora of Egypt from the Predynastic to Graeco-roman times. Veg Hist Archaeobot 6(4):242–247

Fahmy AG (2003) Palaeoethnobotanical studies of Egyptian Predynastic cemeteries: new dimensions and contributions. In: Neumann K, Butler A, Kahlheber S (eds) Food, fuel and fields: progress in African archaeobotany. contributions of the 3rd international workshop on African archaeobotany, Frankfurt, 5–7 July 2000. Heinrich-Barth-Institut, Köln, pp 95–106

Fahmy AG (2004) Insights on the development of archaeobotanical and palaeoethnobotanical studies in Egypt. In: Hendrickx S, Friedman RF, Clalowicz KM, Chlodnicki M (eds) Egypt at its origins: studies in memory of Barbara Adams. Orientalia Lovaniensia Analectavol 138. Peeters, Leuven, pp 711–730

Fahmy AG, Friedman R, Fadl M (2011) Economy and ecology of Predynastic Hierakonpolis. In: Fahmy AG, Kahlheber S, D'Andrea C (eds) Windows on the African past: current approach to African archaeobotan, Helwan, 13–15 June 2009. Africa Magna Verlag, Frankfurt, pp 91–117

Gerisch R (2007a) Main street charcoal. In: Lehner M, Wetterstrom W (eds) Giza reports: the Giza Plateau mapping project, vol 1. Ancient Egypt research associates. Hanson, Mass, pp 167–169

Gerisch R (2007b) Gallery III.4 charcoal. In: Lehner M, Wetterstrom W (eds) Giza reports: the Giza Plateau mapping project, vol 1. Ancient Egypt Research Associates, Hanson, Mass, pp 259–262

Greiss EA (1949) Anatomical identification of plant material from Ancient Egypt. Bull Institut Égyptien 3:249–283

Greiss EA (1957) Anatomical identification of some Ancient Egyptian plant materials: Mémoires présentés à l'Institut de l'Égypte 55. Cairo

Jacomet S (2006) Identification of cereal remains from archaeological sites, 2nd edn. Basel University, IPAS

Keimer L (1924) Die gartenpflanzen im Alten Ägypten. Hamburg, Berlin

Köhler EC (2008) The Helwan cemetery. Archéo-Nil 18:113–130

Köhler EC (2014) Helwan III–Excavation in Operation 4, tombs1-50.VML, Rahden, Westf

Köhler EC (2015) Helwan–A necropolis of Ancient Memphis. http://egyptology.univie.ac.at/fileadmin/user_upload/i_egyptology/forschung/helwan/ProjHelwanen.pdf. Accessed 15 Dec 2017

Kosinová J (1975) Weed communities of winter crops in Egypt. Preslia 47:58–74

Kroll H (1989) Die pflanzenfunde von maadi. In: Rizkana I, Seeher J (eds) Maadi III: The non-lithic small finds and the structural remains of the Predynastic settlement. Ägyptologische Abhandlungen, vol 80. Mainz, pp 129–136

Kunth C (1826) Examen botanique des fruits et des plantes. In: Passalacqua J (ed) Catalogue raisonné et historique des antiquités découvertes en Égypte. Paris, pp 227–229

Malleson C (2016) Informal intercropping of legumes with cereals? A re-assessment of clover abundance in ancient Egyptian cereal processing by-product assemblages: archaeobotanical investigation at Khentkawes Town, Giza (2300–2100 BC). Veg Hist Archaeobot 25(5): 431–442

Murray M (2000) Cereal production and processing. In: Nicholson P, Shaw I (eds) Ancient Egyptian materials and technology. Cambridge University Press, Cambridge, pp 505–536

Murray M (2004) Provisions for the pyramid builders: new evidence from the ancient Egyptian site of Giza. Archaeol Intern 8:38–42

Murray M (2005) Feeding the town: new evidence from the complex of the Giza pyramid builders. General Anthropol 12(1–2):1–9

Murray M (2007a) Main street plant remains. In: Lehner M, Wetterstrom W (eds) Giza reports: the Giza Plateau mapping project, vol 1. Ancient Egypt research associates. Hanson, Mass, pp 163–165

Murray M (2007b) Gallery III.4 plant remains. In: Lehner M, Wetterstrom W (eds) Giza reports: the Giza Plateau mapping project, vol 1. Ancient Egypt Research Associates, Hanson, Mass, pp 257–258

Murray M, Gendy R (2015) A report on the Khentkawes Town—House E archaeobotanical remains. In: Sadarangani F, Witsell A (eds) Settlement and cemetery at Giza: reports from the AERA-ARCE field school. Ancient Egypt Research Associates, North Manchester, pp 237–251

Newton C (2004) Plant tempering of predynastic pisé at Adaïma in Upper Egypt: building material and taphonomy. Veg Hist Archaeobot 13(1):55–64

Newton C (2007) Growing, gathering and offering: predynastic plant economy at Adaïma (Upper Egypt). In: Cappers RTJ (ed) Fields of change: progress in African archaeobotany. Barkhuis, Groningen, The Netherlands, pp 139–155

Pokorná A (2011) Organic materials: plant remains. In: Vymazalová H et al. (2011) Abusir XXII: The tomb of Kaiemthenenet (AS 38) and the surrounding structures (AS 57-60). Charles University, Prague, pp 171–172

Pokorná A, Beneš J (2014). Plant macroremains from the old kingdom mud brick construction of the Werkaure tomb. Results of carpological and xylotomical analyses. In: Krejčí J, Arias Kytnarová K, Vymazalová H, Pokorná A, Beneš J (eds) Abusir XXIV. Mastaba of Werkaure, vol. I (Old Kingdom strata). Charles University, Prague, pp 283–293

Pokorná A, Pokorný P (2010) Organic materials: plant remains from the eastern group of mastabas. In: Bárta M (ed) Abusir XIX: tomb of Hetepi (AS 20), Tombs AS 33-35 and AS 50-53. Charles University, Prague, pp 361–370

Pokorný P, Kočár P, Sůvová Z, et al. (2009) Paleoecology of Abusir South according to plant and animal remains. In: Bárta M et al (eds) Abusir XIII: tomb complex of the Vizier Qar, His Sons Qar Junior and Senedjemib, and Iykai. Abusir South 2. Charles University, Prague, pp 29–41

Schweinfurth GA (1882) De la flore pharaonique. Bull Institut Égyptien 3:51–76

Schweinfurth GA (1883) The flora of Ancient Egypt. Nature 28:91–114

Schweinfurth GA (1884) Further discoveries in the flora of Ancient Egypt. Nature 29:312–315

Schweinfurth GA (1887) Die letzten botanischen entdeckungen in den gräbern Ägyptens. In: Engler A (ed) Botanisches jahrbücher systematic 8. Schweizerbart, Stuttgart, pp 1–15

Thanheiser U (1996) Local crop production versus import of cereals during the Predynastic Period in the Nile Delta. In: Kobusiewicz M, Kroeper K, Krzyżaniak L (eds) Interregional contacts in the later prehistory of Northeastern Africa. Archaeological Museum, Poznán, pp 291–302

Täckholm V (1939) A note on the plants found at Tanis. In: The Horticultural Review, vol 27. p 101

Täckholm V (1940) A mummy coffin in the Egyptian museum, stockholm, and its plant remains. Sven Bot Tidskr 34:141–161

Täckholm V (1951) Faraos blomster. Natur och Kultur, Stockholm

Täckholm V (1961) Botanical identification of the plants found at the monastery of Phoebammon in Thebes. Bull Société Archéologie Copte 3:1–38

Täckholm V (1974) Student's Flora of Egypt. Cairo University Herbarium, Cairo

Täckholm V (1976) Ancient Egypt, landscape, flora and agriculture. In: Rzóska J (ed) The Nile: biology of an ancient river. Monographiae Biologicae, vol 29. Springer Netherlands, Dordrecht, pp 51–68

Täckholm V, Drar M (1950) Flora of Egypt, vol 2. Bulletin of the Faculty of Science. Fouad I University 28, Cairo

Täckholm V, Drar M (1954) Flora of Egypt, vol 3. Bulletin of the Faculty of Science. Cairo University 30, Cairo

Täckholm G, Täckholm V, Drar M (1941) Flora of Egypt, vol 1. Bulletin of the Faculty of Science. Fouad I University 17, Cairo

Unger F (1866) Botanische streifzüge auf dem gebiete der kulturgeschichte. vii. Ein ziegel der dashurpyramide in Ägypten nach seinem inhalte an organischen einschlüssen. Sitzungsberichte der Mathematisch-Naturwissenschaftlichen klasse der kaiserlichen akademie der Wissenschaften, vol 54. Wien, pp 33–62

USDA, NRCS (2016) The PLANTS database. http://plants.usda.gov. Accessed 15 Dec 2017

Van Zeist W, Roller G (1993) Plant remains from Maadi, a Predynastic site in Lower Egypt. Veg Hist Archaeobot 2:1–14

Van Zeist W, Roller G, Fahmy AG (2003) An archaeobotanical study of Maadi, a Predynastic site in Lower Egypt. In: Van Zeist W (ed) Reports on archaeobotanical studies in the old world. Groningen, Netherlands, pp 167–207

Van der Veen M (2007) Formation processes of desiccated and carbonized plant remains. The identification of routine practice. J Archaeol Sci 34:968–990

Vartavan de C (2002) Hidden fields of Tutankhamun, from identification to interpretation of newly discovered plant material from the pharaoh's grave, 2nd edn. Triade Exploration, London

Wetterstrom W (1984) The plant remains. In: Wenke RJ (ed) Archaeological investigation at El-Hibeh 1980: preliminary report. American Research Center in Egypt reports, vol 9. Undena Publications, Malibu, California, pp 50–79

Willcox G, Fornite S, Herveux L (2002) Atlas of images of charred archaeobotanical finds. http://g.willcox.pagesperso-orange.fr/archaeobotanical%20images/index1.htm. Accessed 15 Dec 2017

Zohary D, Hopf M (1993) Domestication of plants in the Old World, 2nd edn. Clarendon Press, Oxford

Zohary D, Hopf M, Weiss E (2012) Domestication of plants in the Old World, 4th edn. Clarendon Press, Oxford, New York

Study of Plant Remains from the Embalming Cache KV63 at Luxor, Egypt

Rim Hamdy[1(✉)] and Ahmed G. Fahmy[2]

[1] Department of Botany, Cairo University, Giza, Egypt
rimhamdy@yahoo.com

[2] Department of Botany and Microbiology, Helwan University, Helwan, Egypt

Abstract. Plant remains from tomb KV63 dating to the XVIIIth dynasty of ancient Egypt (c. 16th–13th BC) have been analyzed using morphological characters. A total of 17 species were identified, and among these ten were used in the production of floral collars, including leaves of pomegranate (*Punica granatum*), olive (*Olea europaea*) and willow (*Salix subserrata*); flowers of low cornflower (*Centaurea depressa*), safflower (*Carthamus tinctorius*) and blue lotus (*Nymphaea caerulea*); fruits of ashwagandha (*Withania somnifera*), date palm (*Phoenix dactylifera*), and papyrus (*Cyperus papyrus*). Linen strings (*Linum usitatissimum*) were used for binding together the various components of the collars. Additional plant fragments identified included barley chaff (*Hordeum vulgare*), pine (*Pinus* sp.), juniper (*Juniperus* sp.) and edible fruits of persea (*Mimusops laurifolia*), doum palm (*Hyphaene thebaica*), Christ's thorn (*Ziziphus spina-christi*), grape (*Vitis vinifera*), olive (*Olea europaea*) and date palm (*Phoenix dactylifera*). The leaves of the latter two trees were observed to be present in all recovered collars. There is a great similarity between the species used in making the collar of the third coffin of Tutankhamun (buried in KV62) and those found in KV63 as well as KV54, which represents an embalming cache associated with KV62. A gilded collar (E4) and many safflower heads were recorded for the first time in KV63. The large number and variety of collars and associated plant remains found in the tomb supports the theory that KV63 was used as an embalming cache.

Keywords: Archaeobotany · Egypt · Embalming cache · Fruits
Flowers · Gilded collar · Leaves · Seeds · Safflower · Tutankhamun

Introduction

The tomb known as KV63 (XVIIIth dynasty, Kings' Valley) was accidentally discovered in 2005 by the University of Memphis Amenmesse Tomb Project, led by Dr. Otto Schaden, in the Valley of the Kings, Luxor (Fig. 1). The archaeological team found KV63 while excavating the remains of workmen's quarters at the entrance to KV10, which was constructed during the reign of Amenmeses of the XIXth dynasty (who was buried elsewhere). Both Theodore Davis and Howard Carter excavated in the area in the early 20th century without removing these particular structures. While exploring a layer of dark rock, excavators suddenly came across chips of white stone, which represented the last level excavated by Carter. Further exploration revealed a

A. M. Mercuri et al. (Eds.): *Plants and People in the African Past*,
https://doi.org/10.1007/978-3-319-89839-1_3

straight edge of cut stone, which formed the upper lip of a vertical shaft leading to the tomb KV63. The tomb had been concealed beneath workers' quarters that were built on the site a short time after KV63 was sealed.

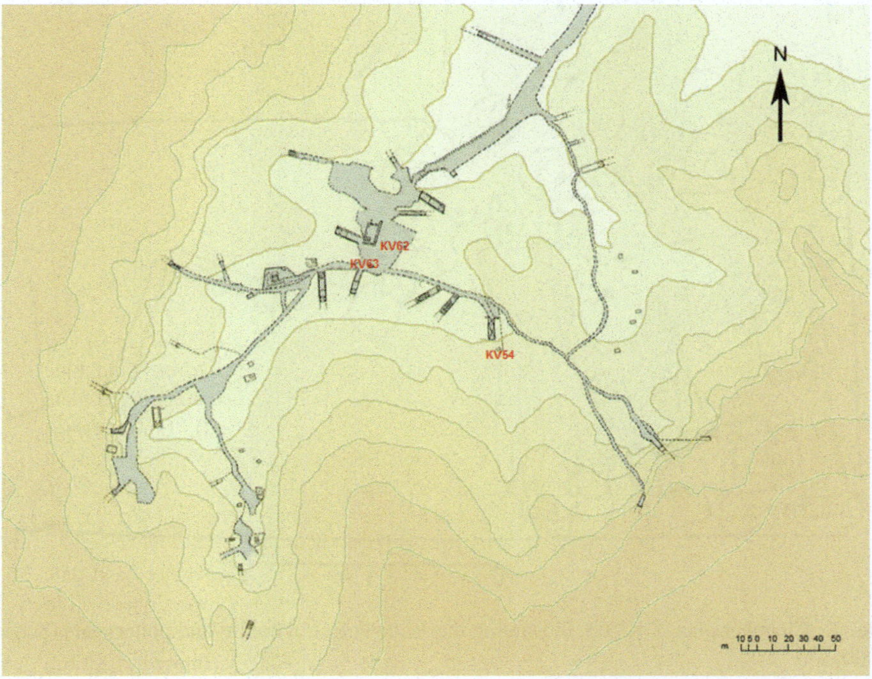

Fig. 1. Location map of KV63, Valley of the kings, Luxor, Egypt.

The discovery that the shaft led to a chamber was announced on 8 February 2006 by the Supreme Council of Antiquities. The shaft descends 5 m and is connected to a doorway that blocked the entrance to an undecorated L-shaped chamber, measuring 4 × 5 m. The chamber (Fig. 2) contained 28 large jars and seven wooden coffins (A, B, C, D, E, F and G). A golden anthropoid coffinette was discovered inside coffin G. No human remains were found inside the coffins, but floral collars and mummification debris were discovered in coffin E, the only sealed coffin which was found in the back corner of the chamber behind other coffins (Schaden 2010). It was initially believed that KV63 was a royal tomb, however, it is now considered to have been a chamber used for the storage of embalming materials.

Fig. 2. Chamber plan of KV63, illustrating the arrangement of coffins and pottery jars (Schaden 2009).

Floral Collars in Ancient Egypt

Floral collars in ancient Egypt are usually made of papyrus sheets sewn together to form a backing for strands of individual flowers, petals, fruits or leaves arranged one above the other (Winlock 1941). During the XIXth dynasty (13th–12th BC), 'Nedjemger,' the overseer of the garden of the Ramesseum at Thebes, is depicted in his tomb inspecting the manufacture of collars from fresh plant material (Fig. 3).

In ancient Egypt, floral collars were used in religious and celebratory events (Manniche 1989), placed around the necks of the dead. They were probably worn at the funerary banquet or draped around statuettes (Manniche 1989). The most famous intact floral collar was found upon the innermost coffin around the golden mask of Tutankhamun (Carter number 255a). The ceremonial or banquet dress worn by both sexes was not complete without a floral collar and until the Late Period anthropoid sarcophagi are regularly painted with collars (Kantor 1999). Plants were usually chosen on the basis of colour, fragrance, shape or size, and a particular craft was involved such as twisting, braiding or wrapping the plant materials (Barakat 1986). In addition to floral parts, faience or glass sequin beads were incorporated in the manufacture of collars and

Fig. 3. The picture shows Nedjemger, inspecting the manufacture of collars, Tomb TT138, Sheikh Abd el-Qurna on the west bank of the Nile, opposite to Luxor (Feucht 2006).

were found either in separate strands or alternately arranged with plants. It later became fashionable to wear collars of faience usually made in the form of flowers, leaves or fruits.

According to Kantor (1999), XVIIIth dynasty floral collars retrieved from archaeological sites in Egypt, especially those of Tutankhamun, have interested many botanists to identify botanical and other materials used in the design and production of collars, including Newberry (1927), Germer (1989), Manniche (1989), Hepper (1990) and Vartavan (1999). In addition, Winlock (1941, 2010) described collar-like floral garlands held by the New York Metropolitan Museum of Arts, which were found in pottery jars discovered by Davis (1907–1908) in a pit near the entrance of Tutankhamun's tomb. These are believed to have been accessories used during burial ceremonies of the young king, and were gathered afterwards and packed away in jars, which seems to have been the custom in Egyptian royal burials (Carter 1927).

This study aims to identify the botanical materials found in tomb KV63, especially those used in making floral collars and to integrate these records with existing studies. This will help to shed some light on the function of this tomb, located approximately 14 m from the southern edge of Tutankhamun's tomb (KV62).

Materials and Methods

Desiccated plant macroremains were clearly visible when the tomb was opened, therefore all visible samples were collected. However, natron samples beneath the collars have been identified while searching smaller seeds and fruits. Dry sieving was used to separate plant macroremains from the archaeological natron, using a 0.5 mm mesh sieve. Each sample was investigated using a binocular microscope with 10–50x magnification. Identifications were completed using the floras of Egypt, Africa, and the Middle East (Boulos 1999, 2000; Zohary and Feinbrun-Dothan 1966, 1972). Digital images have been taken of almost all macroremains in order to check and confirm identifications using local plant specimens and the reference collection kept at the archaeobotanical laboratory of Cairo University Herbarium (ABL). It should be mentioned that the collars retrieved from KV63 during the 2006 season were intact and in situ, preserved only by drying inside the coffin when they were discovered. However, Schweinfurth (1883) mentioned that some garlands were broken and reduced to powder due to careless handling when the coffins were first opened. The fragile collars were handled and transported from KV63 into KV10 for investigation, and were treated with Paraloid for consolidation which unfortunately caused damage to fragile plant structures.

Results and Discussion

The plant species identified in KV63

A total of seventeen species of seed plants were identified according to their gross morphology. These are listed below and arranged in alphabetic order. For each species the scientific name is followed by its family and English common name. Short notes concerning origin, habit and use are also provided.

1. *Carthamus tinctorius* L., Asteraceae–Safflower

Safflower is an annual herb with yellow-red flowers, probably indigenous to Persia and NW India. It was well known in ancient Egypt (Zohary and Hopf 1993), where it has been recorded since the XIIth dynasty as a dye on the tomb of the two brothers at Tell el Dab'a (Vartavan and Amorós 1997). Ancient Egyptians found its flower pleasing to the eye and included it in garlands during New Kingdom and Late Period (Hamdy 2015). A few seeds were found in Tutankhamun's tomb (Germer 1989; Hepper 1990; Zohary and Hopf 1993) compared with the large number of flowering heads recorded in KV63.

2. *Centaurea depressa* M. Bieb., Asteraceae–Low cornflower

An annual herb with bright-blue flowers, low cornflower was introduced to Egypt from Western Asia probably as a fields weed (Newberry 1889). Cornflower has been recorded since the Predynastic period at Hierakonpolis (El Hadidi 1982), while the flowering heads were recorded in Tutankhamun's tomb (Vartavan and Amorós 1997) and KV54 (Winlock 2010). Only a few flower heads were recorded in KV63.

3. *Cyperus papyrus* L., Cyperaceae–Papyrus

A perennial herb with triangular culms, papyrus is native to Egypt, and grew in abundance along the Nile River in ancient times. It was used as a twine and fibers were used as thread in sewing on or binding and the pith for manufacturing artificial flowers (Täckholm and Drar 1950). The feathery flowers born on long stems were an ideal base for tall composite bouquets used to decorate temples and tombs (Manniche 1989). First recorded at Naqada during Predynastic times, papyrus is considered the core material for all collars, and was abundant in KV63.

4. *Hordeum vulgare* L., Poaceae–Barley

Barley was one of the main cereals of Egypt since Predynastic times onward. It was used for making bread and beer and as fodder. Barley sprouts found during excavations were part of burial rites, being symbolic for the resurrection of Osiris (Manniche 1989). The presence of eigth bags full of chaff in KV63 is very interesting. It is believed to have been used as stuffing in body cavities during the embalming process (Schaden 2010).

5. *Hyphaene thebaica* L., Arecaceae–Doum palm

The doum is a fan palm with a dichotomously branched trunk and is native to Egypt. The fruit is ovoid, glossy and dark brown in colour. It has been recorded in Egypt since Predynastic times or earlier. Almost every part of the palm was used in antiquity. The hard trunks were used for roofing and supports, the leaves were used for basketry, ropes and similar fibrous items, and the fruit stone served as an ivory substitute to make buttons, rings and beads. The spongy pericarp is edible and is of medicinal value (Darby et al. 1977). A few fruits were recorded within the sealed jars in KV63.

6. *Juniperus* sp., Cupressaceae–Juniper

The only juniper in Egypt is *Juniperus phoenicea*, a shrub or small tree, today growing in restricted mountainous areas in Sinai, Jordan and is widespread in the Mediterranean (Boulos 1999). The fleshy megastrobili of juniper (commonly known as berries) were recovered inside Tutankhamun's tomb in a basket mixed with dates, raisins and coriander (Germer 1989; Hepper 1990). They also have been mixed with salt in which the bodies were laid for dehydration (Manniche 1989). The oil from seeds was used for anointing the dead body (Darby et al. 1977). Ten berries were recorded in KV63.

7. *Linum usitatissimum* L., Linaceae–Linseed/Flax

Linseed or flax is an annual herb, cultivated for its stems, which are used in the manufacture of linen, and for its seeds, which produce linseed oil (Manniche 1989). The earliest record of flax fibers dates to Predynastic times (Vartavan and Amorós 1997), and it was represented in KV63 by remnants of fibers and textiles.

8. *Mimusops laurifolia* (Forssk.) Friis, Sapotaceae–Persea

Persea is an evergreen tree which was introduced from highland Ethiopia and Yemen and cultivated in Egypt (Friis et al. 1986; Germer 1989; Hepper 1990). It was recorded at Saqqara in IIIrd dynasty contexts, where fruits were deposited as offerings (Vartavan and Amorós 1997). Since the 12th Dynasty onwards, the leaves were used in making garlands and collars, and its branches were used to make the large bouquets found in the tomb of Tutankhamun. Only a few leaves and seeds of persea were recorded in KV63.

9. *Nymphaea caerulea* Sav., Nymphaeaceae–Blue waterlily

Blue water lily is a perennial aquatic herb native to the Nile valley (Boulos 1999). Its petals were recorded first at Deir el-Bahari in the garland adorning Pharaoh Ahmose's mummy dating to the XVIIIth dynasty (Vartavan and Amorós 1997). To the ancient Egyptians, the blue waterlily symbolized the sun and the renewal of life (Manniche 1989). Petals were recorded among the collars in KV63.

10. *Olea europaea* L., Oleaceae–Olive

Olive is an evergreen shrub or tree characterized by silvery leaves and edible fruits. Its kernels were recorded in Kom Rab'ia during the Middle and New Kingdom. The presence of olive leaves in the KV63 floral collars indicates that they were cultivated locally. Olive trees grow in the Mediterranean region where the winters are cool and moist and summers are hot and dry. Leaves and fruits also have been recovered inside the tomb of Tutankhamun (Hepper 1990). A few desiccated stones were recorded in KV63.

11. *Phoenix dactylifera* L., Arecaceae–Date palm

The date palm is a feather palm, with an unbranched trunk. The fruit is variable in shape and colour. It is a multipurpose species, and today is known only in cultivation. Trees were used in ancient Egypt for thatching, basketry, sandals, nets and similar items (Germer 1989). It has been present in Egypt since Palaeolithic times at the site of Nabta Playa in the Nubian Desert (El Hadidi and El Fayoumi 1996). Fifteen stones were recovered in KV63.

12. *Pinus* sp., Pinaceae–Pine

Pine is an evergreen tree, and its wood, strobili (cone) fragments and seeds have been recorded in many archaeological sites in Egypt (Vartavan and Amorós 1997). Several species of pine are found in the Eastern Mediterranean region, where the climate is moister than in Egypt. Pine timber was used in making the doorway into the burial chamber of Tutankhamun's tomb, in addition to the wood of date palm and ebony (Hepper 1990). A few pine seeds were recorded in KV63.

13. *Punica granatum* L., Punicaceae–Pomegranate

The pomegranate is a shrub or small tree, with scarlet red flowers present in clusters or solitary. It has been recorded in XIIth dynasty contexts by Schweinfurth at Dra Abu el-Naga (Vartavan and Amorós 1997). Each part of the plant has medicinal value. Pomegranate flowers and leaves were set in many garlands, especially the leaves in the KV63 samples. It can be surmised that pomegranate was used in a large quantities during this period.

14. *Salix subserrata* L., Salicaceae–Willow

The willow is a deciduous tree native to Egypt, common along the banks of the Nile and in the Delta (Hepper 1990). Its leaves have been recorded in many preserved garlands, including those of Ahmose I, Amenhotep I and Nsi-Khonsou (Vartavan and Amorós 1997). It is frequently mentioned in ancient religious texts, and was represented in festivals held at Dendrah, Habu, Kom Ombo (Keimer 1931) and Ineni (Baum 1988) temples. Large amounts of willow leaves have been recorded in KV63.

15. *Vitis vinifera* L., Vitaceae–Grape vine

Grape is a deciduous vine climbing with tendrils. The leaves are lobed, and fruits appear as clusters of sweet globoid berries with pear shaped pips. In Pharaonic Egypt, the best vineyards were found in the Delta and in the oases. Grapes were eaten, used in making wine, formed a component of medical recipes (El Hadidi and El Fayoumi 1996), and used in embalming (Baumann 1960). Grape leaves were commonly found among offerings to the dead in Egyptian tombs (Hamdy 2015). Twenty four pips were recorded in KV63.

16. *Withania somnifera* L., Solanaceae–Ashwagandha

Ashwagandha is a perennial to semi-woody shrub. Its red fruits were commonly threaded on thin strips of date palm leaflets. Due to its showy appearance, it was included in many garlands and collars since the XVIIIth dynasty. Many berries were found in KV63.

17. *Ziziphus spina-christi* L, Rhamnaceae–Christ's thorn

Christ's thorn is an evergreen tree, and one of the most important and well known native plants of ancient Egypt. It was used as food, medicine and timber. Fruits were eaten raw or fermented with wine. Its fruits, seeds and wood were recorded from the tomb of Tutankhamun (Hepper 1990). A few stones were recorded in KV63.

The Plant Remains Arranged in KV63: Floral Collars and Other Plant Remains

The KV63 macro-botanical samples can be classified into two major groups: floral collars and other plant remains.

Floral Collars

An accumulation of semi-circular floral collars was found intact and in situ inside coffin E. One collar was placed near the knee (E1) and three in the chest area (E2, E3 and E4) (Fig. 4).

The examination of the fragmentary collars in KV63 reveals a similarity in most plant species used in constructing these collars where papyrus sheets are used as a foundation for either the leaves of olive (Fig. 5a), willow (Fig. 5b), or pomegranate (Fig. 5c), strung together on thin strips of papyrus culms or date palm leaflet strips. The leaves served as clasps for either the petals or sepals of blue lotus (Fig. 5b), or the flower heads of safflower, fruits of ashwagandha (Fig. 5d), cornflower, grass culms (Fig. 6) or gold foils (Fig. 7). Moreover, fibers of flax were used to tie the collar around the neck of the mummy.

Fig. 4. Coffin E after removing the lid with E1–E4 floral collars; E1: at the knee area and E2, E3 and E4: at the chest area and embalming refuses; large swabs E5 and E6 (photo taken by Sandro Vanini, personal communication).

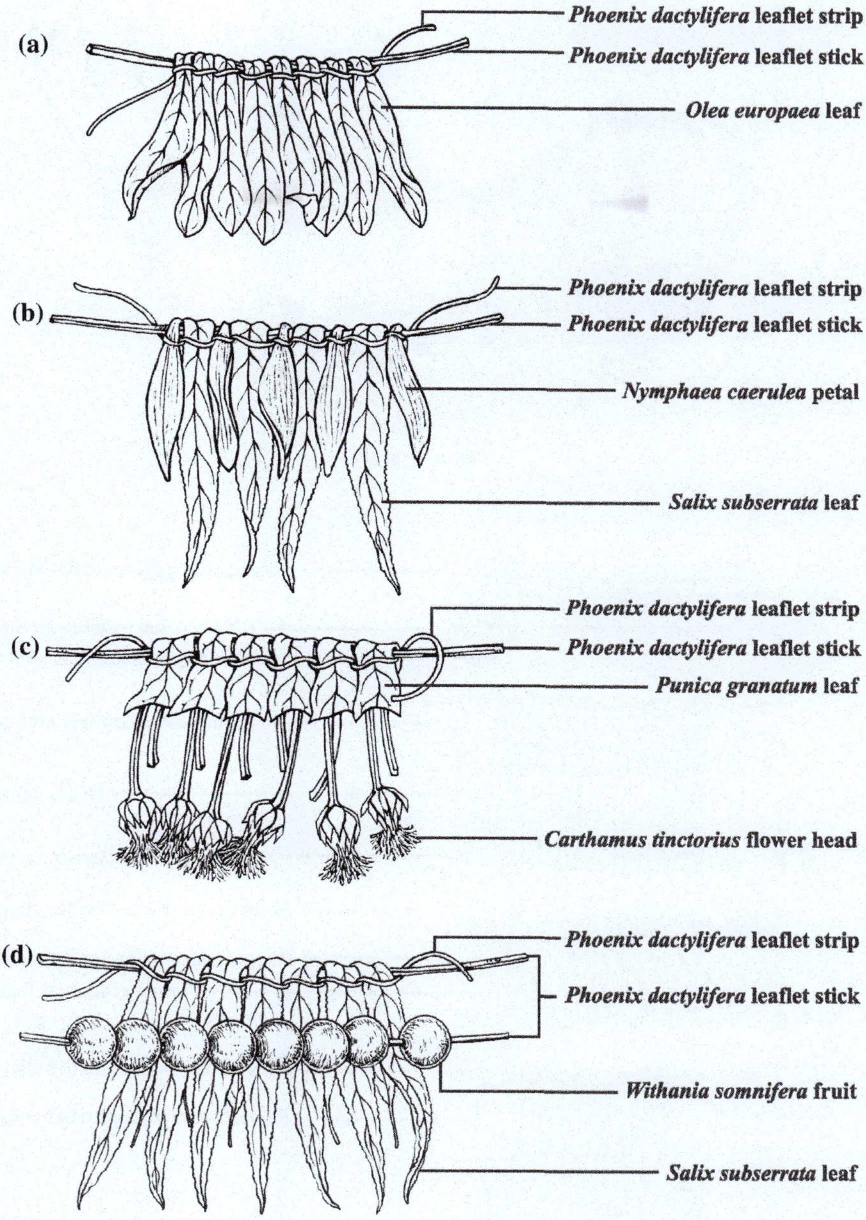

Fig. 5. Diagrammatic reconstruction of floral collar: **a**—*Olea europaea* leaves strung on thin strips of *Phoenix dactylifera* leaflet; **b**—Fragmentary part of collar (E1); **c**—Fragmentary part of collar (E2-a); **d**—Fragmentary part of collar (E2-b).

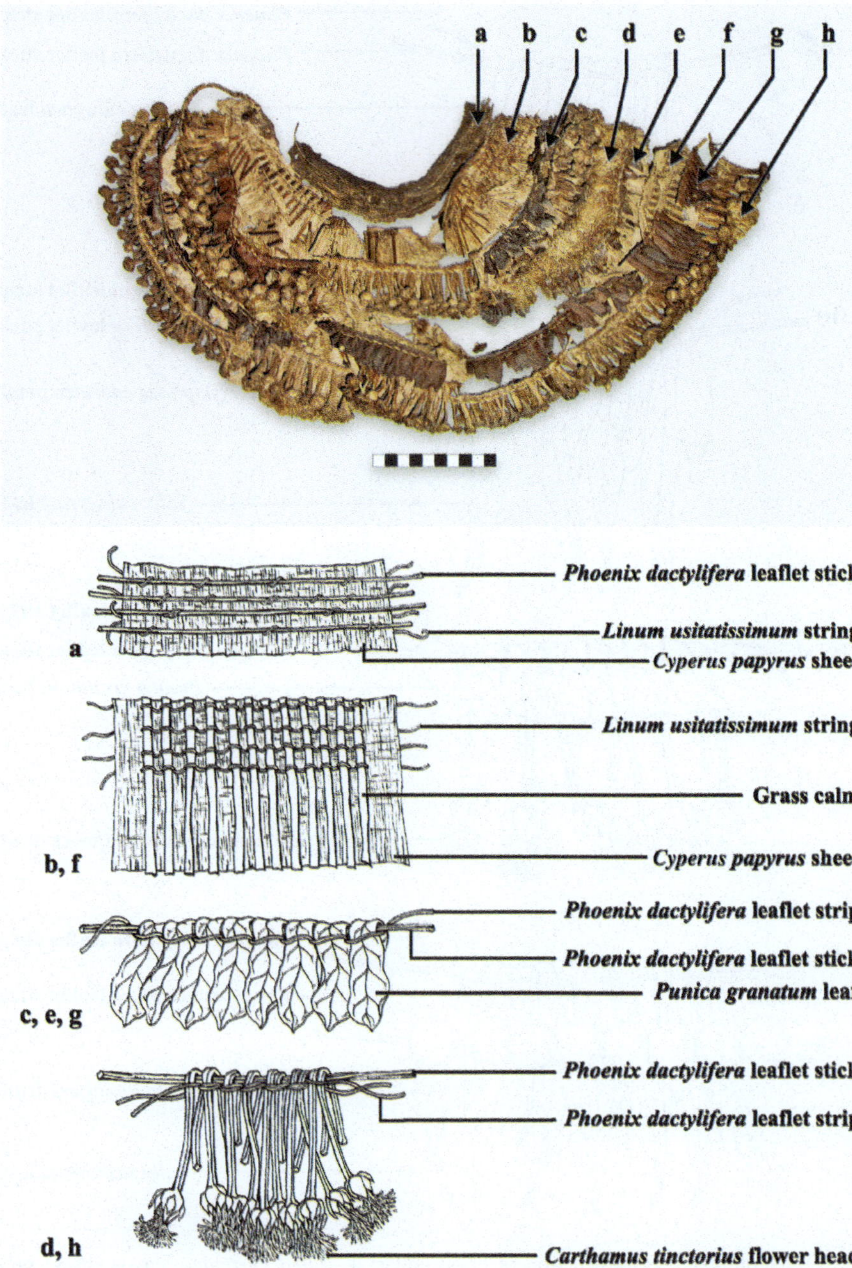

Fig. 6. Part of floral collar E3 and its diagrammatic reconstruction.

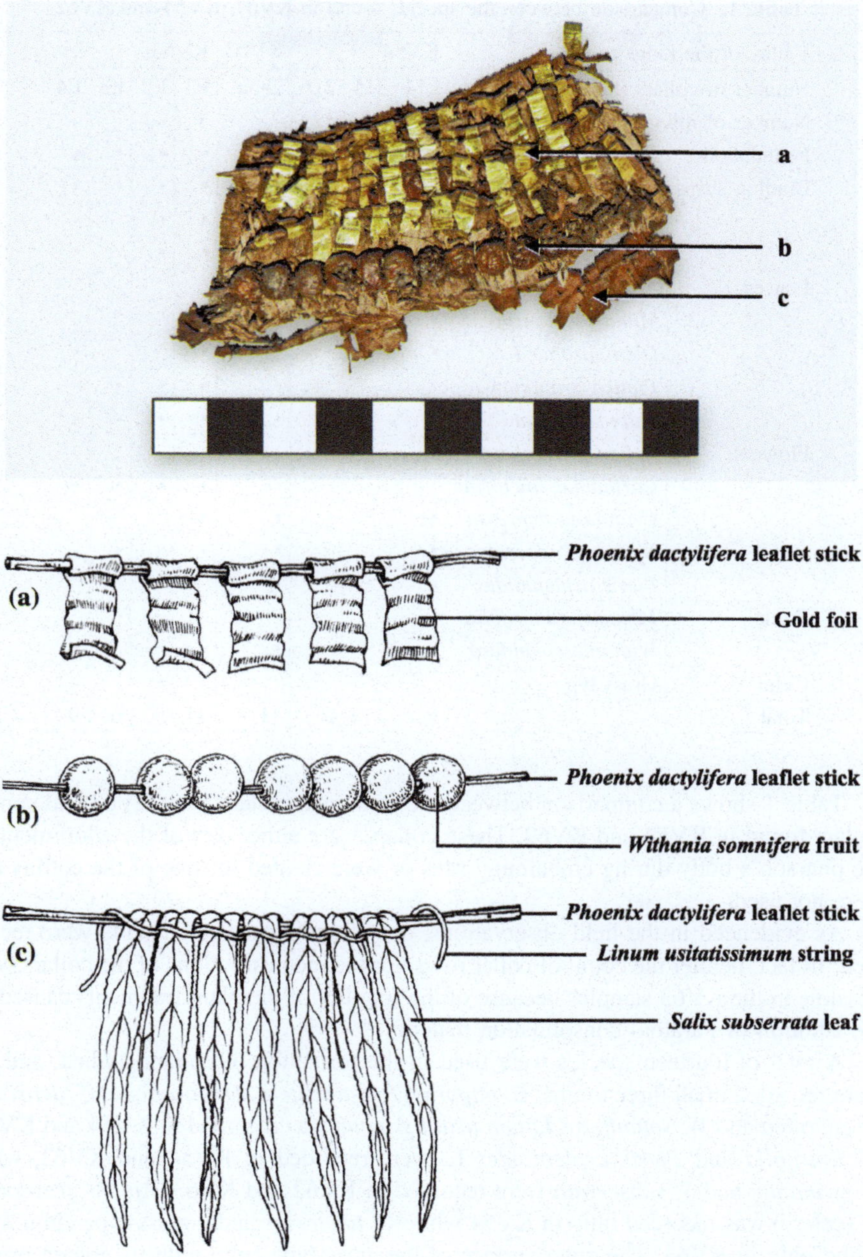

Fig. 7. Part of floral collar E4 and its diagrammatic reconstruction.

Table 1. Comparison between the species found in KV63, KV54 and KV62.

Valley of the kings tombs		KV54			KV62	KV63			
Number of collars (label)		214	215	216	255a	E1	E2	E3	E4
Number of rows		15	5	12	9	8	–	–	–
Foundation	*Cyperus papyrus*	•	•	•	•	•	•	•	•
Binding strips	*Cyperus papyrus*	•	•		•	•	•	•	•
	Linum usitatissimum		•	•		•	•		•
	Phoenix dactylifera	•	•	•	•	•	•		•
Leaves	*Apium graveolens*			•					
	Mimusops laurifolia	•	•	•					
	Olea europaea	•	•	•	•	•			•
	Punica granatum				•	•	•	•	•
	Salix subserrata				•	•	•	•	•
Flowers	*Apium graveolens*		•	•					
	Carthamus tinctorius					•	•	•	•
	Centaurea depressa	•		•	•	•			
	Nymphaea caerulea	•			•	•		•	
	Picris asplenioides	•			•				
Fruits	*Mimusops laurifolia*				•				
	Withania somnifera	•		•	•	•	•		•
Culm	Grass type						•		
Total		9	7	9	11	11	9	6	9

Table 1 shows a comparison between the examined collars from KV63 and similar collars found in KV54 and KV62. These collars were either part of the adornment on the pharaoh's body during embalming rites or were created for one of the coffins that were not used.

As evidenced in the field observations, only fragmentary collar parts were recovered. In fact, neither the count of collar rows nor the reconstructions of the collars were possible in the KV63 samples because of the fragility of the plant fragments caused by the addition of Paraloid consolidation material.

A total of fourteen species were used in the formation of collars studied, and six were recorded in all three tombs: *C. papyrus, P. dactylifera, O. europaea, C. depressa, N. caerulea* and *W. somnifera*. *Linum usitatissimum* was recorded in KV54 and KV63, *M. laurifolia* and *Picris asplènioides* L. were recorded in KV54 and KV62, while *P. granatum* and *S. subserrata* were recorded in KV62 and KV63. *Apium graveolens* L. (celery) was recorded only in KV54 while *C. tinctorius* and a grass-type culm were found only in KV63. The outer surface of the grass-type culm with its golden brown color was used instead of flowers in (E3).

During the New Kingdom and Late Period, *C. tinctorius* flower heads were rarely used in the ornamentation of garlands and collars (Hamdy 2015). Gold foils included on the gilded collar E4 were confirmed by Valentin Boissonanus, conservator at Zurich Botanical Museum, as untarnished bright yellow gold of high purity (Boissonanus personal communication). Gold in ancient Egypt represented the fire and glory of the sun, the flesh of the gods, and the idea that its luster was never lost, meaning eternal life (Axelrod 2013). Although the number of *C. tinctorius* flower heads and seeds recorded in XVIIIth dynasty collars (Germer 1989; Winlock 2010) and garlands (Hamdy 2015) is small, it is important to mention that a large number of safflowers (approximately one thousand) were recorded for the first time in all the collars examined in this study. This suggests that safflower may have been cultivated in the area during the use of KV63. This observation supports the idea that KV63 was used after the time of Tutankhamun, especially if we consider that safflower was observed on the garland of Amenophis I whose mummy was re-embalmed during the XXIst dynasty (Germer 1989) and re-examined by the authors.

Clustering Agglomerative Ward's Analysis (Henderson and Seaby 1999) was applied to the species data collected from the collars recovered in the three tombs (Fig. 8). Based on the frequency of species, the collars were segregated into two groups. Collars from KV63 are clearly separated from those discovered in KV54 which is considered to be the embalming cache of Tutankhamun (Winlock 2010) and the pectoral collar of Tutankhamun recorded from KV62 (Newberry 1927; Germer 1989).

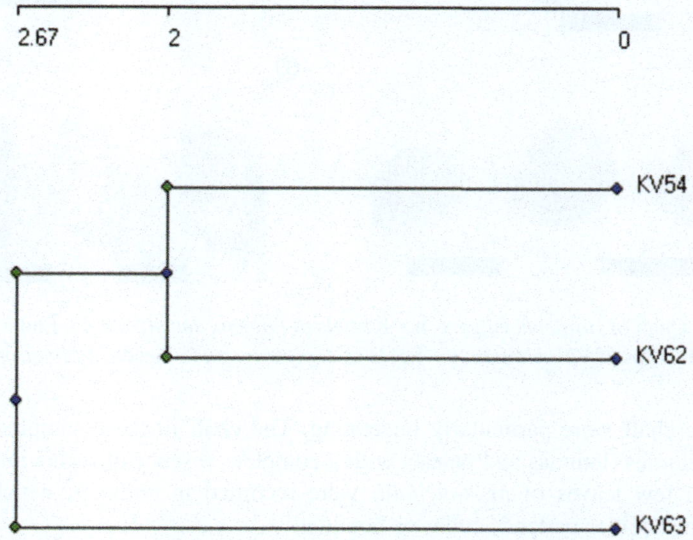

Fig. 8. Cluster analysis dendrogram for the different collars recorder from KV54, KV62 and KV63.

Other Plant Remains

A total of nine plant species were identified in KV63 in addition to those associated with the floral collars. These species included the chaff of *H. vulgare* (Fig. 9a), kernels of *P. dactylifera* (Fig. 9b), endocaprs of *O. europaea.* (Fig. 9c), and *Z. spina-christi,* nuts of *Pinus* sp., pips of *V. vinifera* (Fig. 9d), Seeds of *Juniperus* sp. (Fig. 9e), seeds of *M. laurifolia* (Fig. 9f) and *H. thebaica.* These plant remains were dispersed among the floral collars, natron, and found within sealed jars. In addition, eight bags full of

Fig. 9. **a**—Chaff of *Hordeum vulgare*, **b**—Kernels of *Phoenix dactylifera*, **c**—Endocarps of *Olea europaea*, **d**—Pips of *Vitis vinifera*, **e**—Seeds of *Juniperus* sp., **f**—Seeds of *Mimusops laurifolia*.

H. vulgare chaff were particularly interesting. The chaff in these samples consist of very fine glumes (lemmas and paleas) with a complete absence of rachis internodes. In addition, a few leaves of *M. laurifolia* were recorded in coffin F, which could be remnants of another garland, collar or bouquet.

The items present in KV63, including sealed jars, coffins, 459 kg of natron (some inside the coffins and some inside little sacks), chaff, animal bones, linen, seal impressions, sherds, and a bedstead with lion head finials suggest that the tomb was used for preparing mummies during the embalming process. In addition to many textiles, several alabaster vessels, tiny pieces of papyrus, and large swabs E5 and E6 found over the chest area of coffin E, composed of textiles wrapped around one end of a reed like stick, were all likely used to cover the coffins or mummies with resin.

Comparing the materials found in KV62 and KV54, which is considered to be the embalming cache of Tutankhamun (Winlock 2010), and the other three from Dra Abu el-Naga (Ikram and Lopez-Grande 2011), reveals that KV63 also could be an embalming cache as the associated artifacts and plant remains are similar. The diversity of plant species and the large amount of natron salt found in KV63 may indicate that it was linked to a royal burial or other royal associate as confirmed by Ikram and Lopez-Grande (2011). Likely wealthier individuals experienced more painstaking embalming processes which would result in larger embalming deposits.

Conclusion

Given the location of the tomb, it seems most likely that KV63 was initially designed as a tomb but eventually came to be used as an underground chamber, where royal bodies were embalmed, decorated and otherwise prepared for formal burial. The archaeobotanical evidence obtained from this study supports this interpretation for KV63. Morphological investigation of plant macroremains reveals the presence of a wide diversity of botanical materials, including wild and cultivated plants as well as indigenous (e.g. *Cyperus papyrus*) and non-native taxa (e.g. *Pinus* sp.). Although, there is a great similarity between the plant remains of KV63 and those identified by Germer (1989) and Hepper (1990) in the tomb of Tutankhamun, KV62, the presence of a remarkable quantity of safflower and pomegranate in KV63 suggests that this chamber was used for mummification during the Late Period or the 3rd Intermediate Period of ancient Egypt. It was during that time that these two species flourished and they were not well known during the XVIIIth dynasty (Hamdy 2015). In addition, the practice of using coffins as containers (Eaton-Krauss 2007) was previously known only from much later non-royal embalming caches unearthed outside the Valley of the Kings.

Finally after recovering and studying this unique material, we must conclude with an important recommendation. Treating botanical materials with polymers such as Paraloid should no longer continue, because it destroys the morphological and anatomical characters of macro-botanical remains, making identification difficult.

Acknowledgements. The authors would like to express their deep appreciation to Mr. Ahmad Soliman for providing descriptive illustrations for the components of floral collars.

References

Axelrod L (2013) The history and meaning of ancient Egyptian jewelry. http://www.ancientdigger.com/2013/02/the-history-and-meaning-of-ancient.html. Accessed 14 Feb 2013

Barakat HN (1986) Plant life in Doush area: Kharga Oasis. A comparative study of the present and Graeco-Roman periods. Dissertation, Master of Science, Cairo University

Baum N (1988) Arbres et arbustes de l'Egypte ancienne-la liste de la tombe Thébaine d'Ineni (no. 81), Departement Oriëntalistiek, Leuven

Baumann BB (1960) The botanical aspects of ancient Egyptian embalming and burial. Econ Bot 14(1):84–104

Boulos L (1999) Flora of Egypt, vol 1. El-Hadara Publishing, Egypt

Boulos L (2000) Flora of Egypt, vol 2. El-Hadara Publishing, Egypt

Carter H (1927) Points of interest in Egyptian burial customs. In: Carter H (ed) The tomb of Tut-Ankh-Amen, vol 2. Cassell, London, pp 92–105

Darby W, Ghallioungui P, Griffith L (1977) Food: the gift of Osiris, vol 2. Academic Press, London

Eaton-Krauss M (2007) Embalming caches in the Valley of the Kings-before and after Tutankhamun. In: Dziobek E, Höveler-Müller M, Loeben CE (eds) The mysterious tomb 63 the latest discovery in the Valley of the Kings. Verlag Marie Leidorf, Rahden, Westf, pp 70–72
El Hadidi MN (1982) The Predynastic flora of Hierakonpolis region. In: Hoffman MA (ed) The Predynastic of Hierakonpolis—an interim report. Egyptian Studies Association, vol 1. Illinois, pp 104–115
El Hadidi MN, El Fayoumi HH (1996) Catalogue of the Archaeobotanical Specimens in the Cairo University Herbarium. 1. Abu Sha'ar site, Red Sea Coast—Egypt; season 1990. Taeckholmia 16:11–30
Feucht E (2006) Die gräber des nedjemger (TT 138) und des Hori (TT 259). Verlag Philipp von Zabern, Mainz
Friis I, Hepper FN, Gasson P (1986) The Botanical Identity of the *Mimusops* in Ancient Egyptian Tombs. The Egyptian Journal of Egyptian Archaeology 72:201–204
Germer R (1989) Die pflanzenmaterialien aus dem grab des Tutanchamun. Hildesheimer ägyptologische beiträge, vol 28. Hildesheim, Gerstenberg, p 94
Hamdy RS (2015) Documentary study of floral bouquets and garlands in Ancient Egypt. Lambert Academic Publishing, Germany
Henderson PA, Seaby RMH (1999) Community analysis package, Version 1.2. http://www.irchouse.demon.co.uk
Hepper FN (1990) Pharaoh's flowers, the botanical treasures of Tutankhamun. HMSO, Royal Botanic Gardens, Kew
Ikram S, López-Grande MJ (2011) Three embalming caches from Dra Abu el-Naga. BIFAO 111:205–228
Kantor HJ (1999) The florists' trade in Egypt and its reflection in decorative work. In: Kantor HJ (ed) Plant ornament in the ancient Near East. University of Chicago, Oriental Institute
Keimer L (1931) L'Arbre TR.T: est-il réellement le saule Egyptien (*Salix* Safsaf Forsk.)? vol 28. Bulletin de l'Institut Français d'Archéologie Orientale, pp 47–54
Manniche L (1989) An ancient Egyptian herbal. British Museum Publications, Austin
Newberry PE (1889) On the vegetable remains discovered in the cemetery of Hawara. In: Petrie WF (ed) Hawara, Biahmu and Arsinoe. The Leadenhall Press, London, pp 46–53
Newberry PE (1927) Report on the floral wreaths found in the coffins of Tut-Ankh-Amen. In: Carter H (ed) The tomb of Tut-Ankh-Amen, vol 2. Cassell, London, pp 189–196
Schaden OJ (2009) KV63, 2009 season. KMT 20(3):18–29
Schaden OJ (2010) The Amenmesse project, season of 2006. ASAE 82:231–260
Schweinfurth G (1883) The flora of Ancient Egypt. Nature 28:109–114
Täckholm V, Drar M (1950) Flora of Egypt, vol2. Cairo University Press, Cairo
Vartavan de CT (1999) Hidden fields of Tutankhamun. From identification to interpretation of unknown plant material from the pharaoh's grave. Triade Exploration, London, p 220
Vartavan de CT, Amorós VA (1997) Codex of ancient Egyptian plant remains, vol 1. Triade Exploration, London, p 401
Winlock HE (1941) Materials used at the embalming of king Tut-Ankh-Amun. The Metropolitan Museum of Art, New York, p 18
Winlock HE (2010) Tutankhamun's funeral. The Metropolitan Museum of Art, New York, p 79
Zohary M, Feinbrun-Dothan N (1966) Flora Palaestina, vol 1. The Israel Academy of Sciences and Humanities, Jerusalem
Zohary M, Feinbrun-Dothan N (1972) Flora Palaestina, vol 2. The Israel Academy of Sciences and Humanities, Jerusalem
Zohary M, Hopf M (1993) Domestication of plants in the Old World, 2nd edn. Clarendon Press, Oxford

Farming and Trade in Amheida/*Trimithis* (Dakhla Oasis, Egypt): New Insights from Archaeobotanical Analysis

Valentina Caracuta[1(✉)], Girolamo Fiorentino[1], Paola Davoli[2], and Roger Bagnall[3]

[1] Laboratory of Archaeobotany and Palaeoecology, University of Salento, Lecce, Italy
valentina.caracuta@unisalento.it
[2] Department of Humanities, University of Salento, Lecce, Italy
[3] Institute for the Study of the Ancient World, New York University, New York, NY, USA

Abstract. This paper presents the results of the first archaeobotanical investigation carried out by the University of Salento archaeological team during the 2015 field season at the site of Amheida/*Trimithis* in Dakhla Oasis, Egypt. The bulk of the recovered material consists of seeds and fruits from midden deposits that lay under the foundation of an upper class fourth century AD house and the adjoining school, and similar deposits beneath streets that flanked the house. Overall, almost 600 seeds were recovered. The archaeobotanical assemblage includes nine species of fruit trees. Among these species, three belong to the local, sub-arid, vegetation of the Dakhla Oasis, such as the date palm (*Phoenix dactylifera* L.), Nile acacia (*Acacia nilotica* (L.) Delile Willd. ex Delile) and Christ´s-thorn (*Ziziphus spina-christi* (L.) Desf.), while the rest are allochthonous species that could have been locally grown, or imported as food from other areas of the Mediterranean and the Far East. The olive tree (*Olea europaea* L.), which was introduced to Egypt from the Mediterranean areas of the Levant, is quite abundant at Amheida/*Trimithis*, and its presence suggests that olives were an important source of food between the third-fourth century AD. As we know from the site of Umm Mawagir, in the nearby Kharga Oasis, olives were consumed in the oases already in the late Middle Kingdom (Cappers et al. 2013). Another species, which also comes from the Mediterranean area, is the carob tree (*Ceratonia siliqua* L.) Delile. Findings of carob seeds are recorded at Amheida/*Trimithis*, as well as at the contemporaneous site of Ismant el-Kharab/Kellis, but nowhere else in the New Valley Governorate (South-western Egypt), suggesting that this species was introduced, at the earliest, during the Roman period. The presence of black myrobalan (*Terminalia chebula* Retz.), a species that is native to South Asia, might be indicative of a network of exchanges between Amheida*Trimithis* and localities on the Red Sea coast.

Keywords: Archaeobotany · Amheida/*Trimithis* · Egypt · Farming diet Midden · Trade

A. M. Mercuri et al. (Eds.): *Plants and People in the African Past*,
https://doi.org/10.1007/978-3-319-89839-1_4

Introduction

Food habits, trade and the exchange of exotic goods are among those elements that can be addressed by studying plant remains preserved in archaeological contexts. Analyses of such material have been extensively conducted in Egypt from the end of the nineteenth century AD (Newberry 1889) and, since then, several studies have investigated a wide range of subjects, including the use and processing of plant materials. Due to the extreme environmental conditions in arid and sub-arid areas of Egypt, the preservation of organic material is such that plant parts that usually do not survive charring, such as flowers and fruit pulp, can be found in desiccated form.

The site of Amheida/*Trimithis*, in Dakhla Oasis, is among those sites where plant materials are well preserved, with an abundance of charred remains and desiccated seeds (Thanheiser and Walter 2015). The plant material discussed in this paper consists of charred and desiccated seeds collected from Area 2.1: some seeds were hand-picked from above the floors of fourth century buildings (B1 and B5), and others from beneath floors, where midden deposits were used to flatten the area on which the buildings were constructed in the first half of the fourth century AD. The midden material was intentionally spread on the area in horizontal layers and the waste had been collected from one or more dumps, probably located inside the abandoned *thermae* of the Roman period, which was then partially demolished to make space for these private buildings. Among the new buildings, there is a house (known as the house of Serenos, B1) and a school (B5) (Davoli in press; Ast and Davoli in press). The samples collected in deposits accumulated above the floors in the house of Serenos (B1), in B5, and in the streets (S2 and S3) that flanked the Serenos house are dated between the first phase of occupation (around AD 330) and the abandonment of the area (approximately 370 AD) (Bagnall et al. 2015:86–99).

This new evidence joins a rich array of data from textual sources and archaeobotanical studies on agriculture and diet in Dakhla Oasis (Bagnall 1997; Thanheiser 1999; Walter 2006; Lodwick 2013; Thanheiser and Walter 2015). These sources demonstrate an economy in the Roman period that included higher-value fruit crops, such as grapes for wine, olives largely for olive oil, dates, figs, and other fruits, together with staple cereals and legumes. In addition, it is clear that cotton and pearl millet were cultivated in the summer (Thanheiser et al. 2016) and that olive oil, dates, and cotton were largely exported (Bagnall 1997, 2008a; Gradel et al. 2012).

The Context of Study

Ecological Setting

Kharga and Dakhla Oases occupy part of a great natural depression in the southern section of the Western Desert of Egypt. This area, the New Valley Governorate, also includes the slightly elevated plain (140 m above sea level) between them (Fig. 1). The depression is open towards the south and southeast. The altitude rises gradually to the southwest, reaching 400 m in the direction of Gebel Uweinat. Dakhla Oasis is about 120 km west of Kharga Oasis. Its long axis is in a WNW-ESE direction, while length is

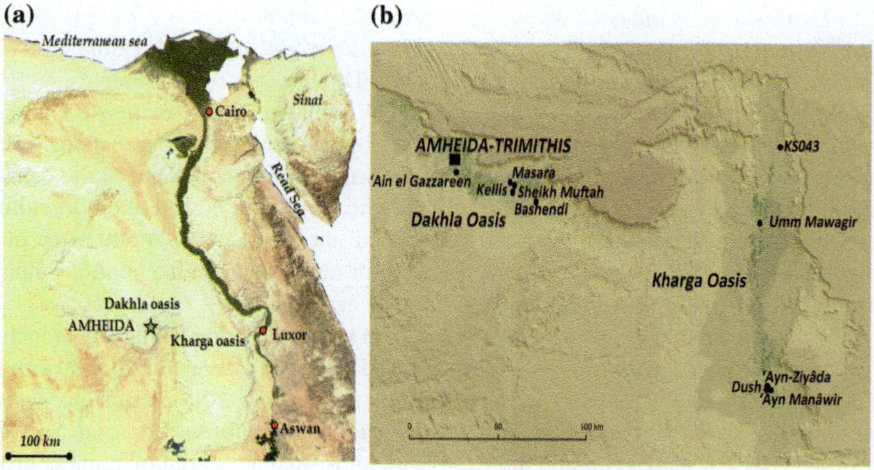

Fig. 1. The site of Amheida\Trimithis. **a** Location of Amheida\Trimithis. **b** Map of Kharga and Dakhla Oases with the archaeological sites cited in the text

about 55 km and width varies between 10 and 20 km. Elevations range between 100 and 400 m above sea level. Dakhla is bounded on the north by a precipitous escarpment running more or less irregularly for at least 250 km eastward to Kharga Oasis. The lowest point of Dakhla Oasis is 100 m above sea level, and surface rises gradually towards the rim (Zahran and Willis 2009).

The Kharga and Dakhla Oases are located in a dry, rainless part of the Great Sahara. Rainfall in recent times is below 5 mm of rain per year, although in the past decade more frequent heavy rains were observed. The temperature ranges from a minimum of 6.0–4.8 °C in winter, to a maximum of 39.2–40 °C in the summer. The water source in both oases is underground (Zahran 2010).

The plant life of the Kharga and Dakhla Oases has been studied by Oliver (1930–1931), Hassib (1951), Täckholm and Drar (1954), Migahid et al. (1960), Shalaby et al. (1975), Ritchie (1999), and Zahran and Willis (2009). Seven vegetation types have been recognized, namely: hydrophytic; reed swamp; halophytic; psammophytic; xerophytic; vegetation of cultivated lands; and vegetation of wastelands (Zahran and Willis 2009).

Hydrophytic vegetation is richly developed in the pools, ditches, wells and irrigation canals of Kharga and Dakhla Oases, which have permanent fresh or brackish waters. Reed swamp vegetation occurs in marshy areas where the water is at the surface and the soil is waterlogged throughout the greater part of the year. This vegetation, thus, occurs around ditches, swampy ground in rice fields, around wells and in pools and in drains and is dominated by *Typha domingensis* Pers. and *Phragmites australis* Adans. Halophytic vegetation is present in the salt marshes where underground water is shallow and dry salt marshes where underground water is deeper. Psammophytic vegetation is the scattered vegetation of the sand plains and the sand dunes. The xerophytic vegetation of the desert ecosystem comprises eleven communities dominated by *Citrullus colocynthis* (L.) Schrad., *Zygophyllum coccineum* L., *Salsola baryosma* Schult., *Chrozophora obliqua*

Adr., *Hyoscyamus muticus* L., *Stipagrostis scoparia* Trin. & Rupr., *Calotropis procera* Aiton, *Lagonychium farctum* Banks & Sol., *Tamarix nilotica* (Ehrenb.) Bunge, *Acacia nilotica* (L.) Delile and *Balanites aegyptiaca* (L.) Delile.

The date palm, which can be considered the dominant fruit tree in the Oases, is grown for its fruits and fibers in the dried leaves, which are used in the manufacture of baskets and other articles. In addition to date palms, the cultivable land in the oases is devoted to cereal cultivation, mainly sorghum, wheat, rice and barley, and horticulture, involving olives, grapes, citrus, pomegranate and apricot. Acacia, another tree cultivated in the oases, produces fruits that are used for tanning, while its wood is valuable for lining wells and, most importantly, it is used as timber.

The most common weeds of winter cultivation in Kharga and Dakhla Oases include *Cynodon dactylon, Melilotus indicus* (L.) All. and *Sonchus oleraceus* L. Less common winter weeds include *Chenopodium murale* (L.) Fuentes, Uotilla & Borsch, *Conyza linifolia* (L.) Less., *Eruca sativa* Mill., *Polygonum equisetiforme* Sm. and *Sorghum virgatum* (Hack.) Stapf. *Cynodon dactylon* is the most frequent weed of summer crops. Common summer weeds include *C. murale, Convolvulus arvensis* L., *Echinochloa colona* (L.) Link and *S. virgatum* (Zahran and Willis 2009). The majority of the vegetation in the oases grows in patches, in proximity to wadis or channels used for irrigation, where water is available on a seasonal basis.

The Archaeological Context

Paola Davoli

The excavation of the house of Serenos (B1), Streets 2 and 3, and building 5 was carried out between 2004 and 2011.[1] The two buildings (B1 and B5) are quite well preserved up to 2.7 m, thanks to a quick covering by sand. The roofs of the rooms, any floor above ground, or terraced roofs are missing. The structures are composed of mud brick and the floors are made of compacted mud. B1 and B5 were built at the same time, around AD 330, as a high status private house and a school of Greek (Cribiore et al. 2008). They were constructed over an imperial period bath complex (*thermae*) abandoned around the end of the third century AD. The bath was contructed in mud and baked brick and remained abandoned for a period during which it was used as a midden. Around AD 330 the area was repurposed for new buildings, and part of the bath was restored and reactivated as a smaller bath (B6) while another section was demolished and its remains leveled using materials that had accumulated inside. Some of its walls were used as foundations for B5, and the baked bricks were reused in several new buildings in the area. Serenos' house and B5 are thus built on the ruins of the *thermae* covered by midden materials and waste of different thickness. The floors of the two streets running east and west of Serenos' house also lie above a thick deposit of waste materials (Ast and Davoli in press).

A few years after the construction of B1 and B5, a natural event damaged both buildings and, as consequence, the house of Serenos was restored and the school

[1] Preliminary reports are available at: http://www.amheida.org.

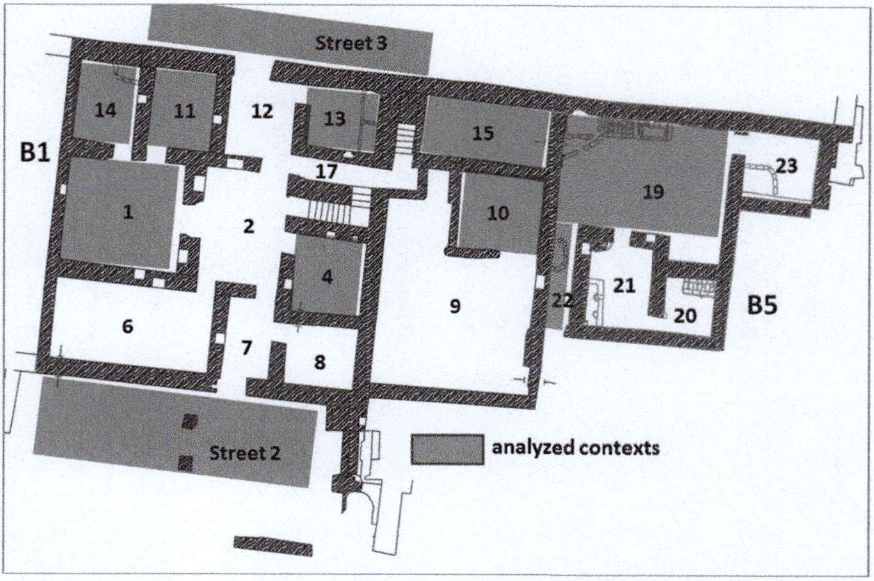

Fig. 2. Area 2.1. Serenos house, B5 and Streets 2 and 3

abandoned. The school was then annexed to the house and transformed into a storage magazine (room 15), and a stable (room 9, 10, 19–23). The last phase of habitation attested by texts (*ostraka*) and coins found in these buildings is around AD 370 (Fig. 2).

The excavation method used was by stratigraphic units: those examined in this paper belong to two main contexts, those related to the life of buildings B1 and B5, i.e., deposits found above floors in which de facto refuse is included; and those found below the floors and belonging to midden materials used to level the surface before the construction of B1 and B5. In addition, some samples were taken from floors and one from the filling of the *stibadium* (a mud-brick dining sofa) found in Street 2 and used for a short period, after which it was dismantled. The inner part of the dining sofa was filled with loose mud brick and debris (DSU 366) (Tables 1 and 2).

Materials and Methods

Macro-botanical materials were handpicked and therefore a bias against the recovery of small seeds is to be expected. The material collected included both burnt and unburnt seeds and charred wood tissue, but for the present paper, only the seeds were analyzed. The perfect state of preservation of the seeds, whether charred or desiccated, allowed an accurate study of morphological traits. The identification was carried out comparing the archaeological material to modern plants of the area with the help of an anatomical atlas (Neef et al. 2012). As result, 590 seeds were identified.

Table 1. Seed remains found above and below the floors in the house of Serenos (B1) and in B5 according to rooms

Room	US	*Acacia nilotica*	*Olea europaea*	*Phoenix dactylifera*	*Vitis vinifera fruit*	*Ziziphus spina-christi*	*Terminalia chebula*	Type of context	Chronology
4	214		1	5	1			Below floor	before ~ 330 AD
1	166			2				Below floor	before ~ 330 AD
	164		1					Below floor	before ~ 330 AD
9	122	2						Above floor	~330–370 AD
	127			4				Below floor	before ~ 330 AD
	128			2				Below floor	before ~ 330 AD
	129			4				Below floor	before ~ 330 AD
	130		1					Below floor	before ~ 330 AD
	132		5	38				Below floor	before ~ 330 AD
	137			1				Below floor	before ~ 330 AD
	151		2	58				Below floor	before ~ 330 AD
	154			1				Below floor	before ~ 330 AD
	180			1				Below floor	

(*continued*)

Table 1. (*continued*)

Room	US	*Acacia nilotica*	*Olea europaea*	*Phoenix dactylifera*	*Vitis vinifera fruit*	*Ziziphus spina-christi*	*Terminalia chebula*	Type of context	Chronology
									before ~ 330 AD
	196			1				Below floor	before ~ 330 AD
	197			1				Below floor	before ~ 330 AD
	200			1				Below floor	before ~ 330 AD
	210		1	5				Below floor	before ~ 330 AD
	212		6	54			2	Below floor	before ~ 330 AD
	215			5				Below floor	before ~ 330 AD
10	135		4	17				Above floor	~330–370 AD
	142		5	44		2		Below floor	before ~ 330 AD
	144		2	7				Below floor	before ~ 330 AD
	191		3	32				Below floor	before ~ 330 AD
	F106		8	17				floor	~370 AD
	202		1					Below floor	before ~ 330 AD

(*continued*)

Table 1. *(continued)*

Room	US	*Acacia nilotica*	*Olea europaea*	*Phoenix dactylifera*	*Vitis vinifera fruit*	*Ziziphus spina-christi*	*Terminalia chebula*	Type of context	Chronology
11	235			9				Above floor	~330–370 AD
13	207			1				Above floor	~330–370 AD
14	194		2	5				Above floor	~330–370 AD
	221		1		1			Above floor	~330–370 AD
15	157			2				Above floor	~330–370 AD
	145							Above floor	~330–370 AD
	F128			1				floor	~370 AD
19	219		76	2				Below floor	before ~ 330 AD
22	273			2				Above floor	~330–370 AD
Total		2	119	322	2	2	2		

Table 2. Seed remains found above and below the floors in Streets 2 and 3

Street	US	*Ceratonia siliqua*	*Olea europaea*	*Phoenix dactylifera*	*Prunus persica*	Type of context	Chronology
2	350				14	Above floor	~330–370 AD
	364				5	Above floor	~330–370 AD
	366				1	Above floor	~330–370 AD
	371				2	Above floor	~330–370 AD
	372				2	Above floor	~330–370 AD
	F314				3	floor	~330–360 AD
	F326			1		floor	~330–350 AD
3	396	4	18	90		Below floor	before ~ 330 AD
	F347		1			floor	~330–360 AD
Total		4	19	91	27		

Results

The study of the morphological features of recovered seeds and fruits led to the identification of nine arboreal species, a lack of small seeded species, such as staple cereals and legumes, was likely the result of recovery by handpicking during the excavation. Date palm (*Phoenix dactylifera*) and olive (*Olea europaea*) are the most ubiquitous, having been found both in the midden deposit below and above the floors in the house of Serenos, of B5 and in Streets 2 and 3. Peach (*Prunus persica* (L.) Batsch) and carob tree (*Ceratonia siliqua*), were found respectively above Street 2 and below Street 3 (Table 1). The state of preservation could raise questions about the intrusiveness of the peach in the deposit, however the stratigraphic position of the findings, the compactness of layers and the associated cultural materials suggest that the desiccated peach remains were in situ.

The midden deposit that formed the base of the foundation of the upper-class fourth century AD house and adjoining school yielded mainly refuse from edible plants such as date pits and fruits (70%), olive pits (28%), black myrobalan (*Terminalia chebula*) fruits (1%), grape pips (*Vitis vinifera* L.) (0.5%) and Christ′s-thorn fruits (*Ziziphus spina-christi*) (0.5%) (Table 1). The sediment above the floor of the house of Serenos contained plenty of date and olive seeds, and a single seed and pod of *Acacia nilotica* (Fig. 3). The botanical material found above the ancient surface of Street 2 is characterized by the predominance of peach, desiccated seeds and fruits. The assemblage, mostly below the floor of Street 3, is made up of dates, olive pits and a few carob seeds (Table 2, Fig. 4).

The data collected so far confirm that a broad spectrum of vegetal products was available to the inhabitants of Amheida/*Trimithis* in the third and early-fourth centuries AD and earlier. Some, such as the date palm, were certainly cultivated, but we can assume that some non-native fruit crops, such us grape, olive, and the plums (*Prunus*

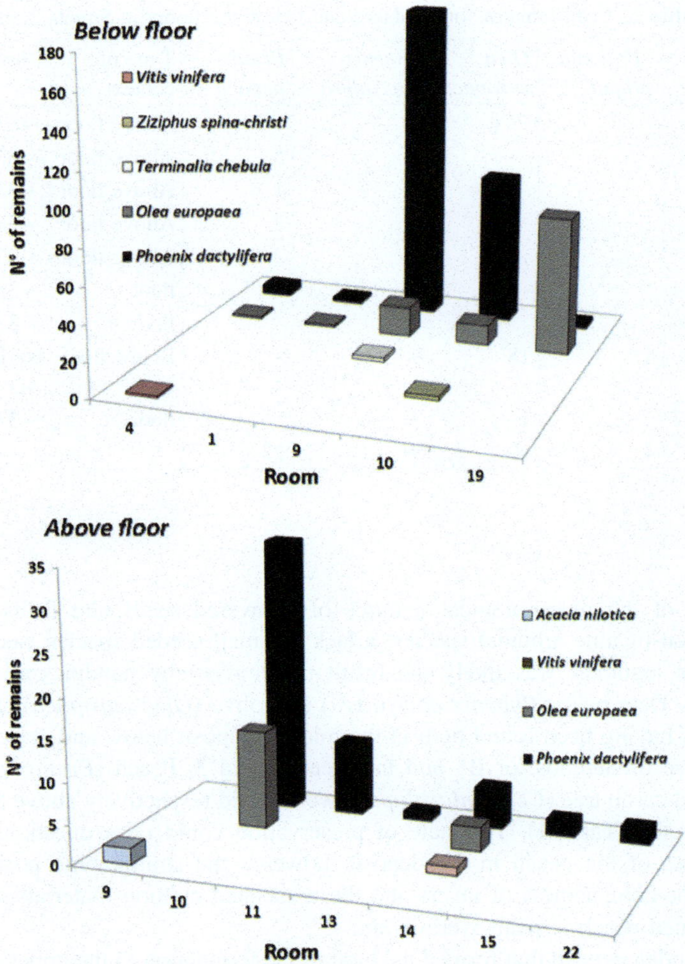

Fig. 3. Distribution of plant finds per phase. The first graph represents the archaeobotanical assemblage below the floors of the house of Serenos and of B5, datable to before the construction of the building (AD 330). The second graph represents assemblages found above the floors of B1 and B5, which accumulated between the construction and abandonment of the building (~330–370 AD)

avium L., *P. domestica/armeniaca*, *P. persica*), were also locally produced. Remains of non-native crops are plentiful in Kellis, a Roman site in Dakhla Oasis (Thanheiser et al. 2002), thus reinforcing the idea that fruit crops were an essential component of the diet of the local community and that 'exotic' plants were introduced to the oases for cultivation before the fourth century AD.

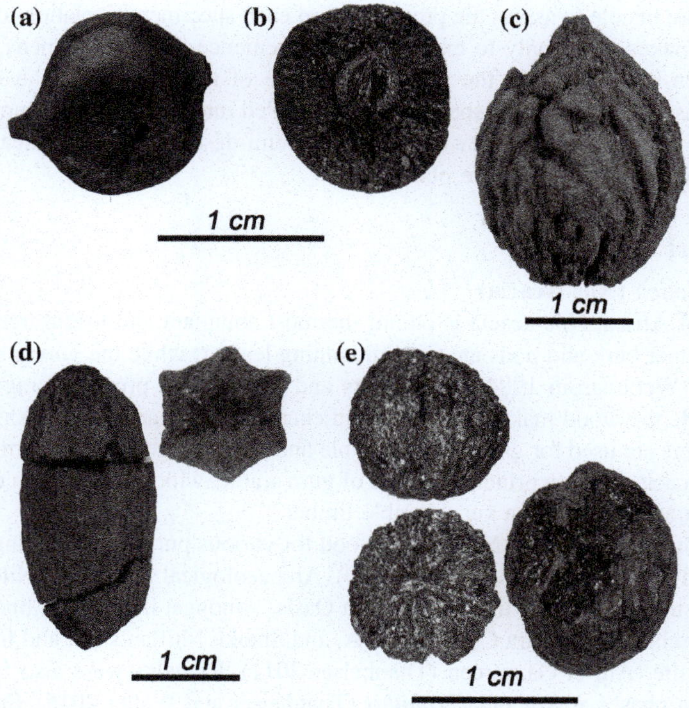

Fig. 4. Images of some species found in Amheida*Trimithis*. **a** *Acacia nilotica* (pod); **b** *Vitis vinifera* (fruit and pip); **c** *Prunus persica* (desiccated stone); **d** *Terminalia chebula* (frontal view of the stone, on the left; view of the top, on the right); **e** *Ziziphus spina-christi* (seeds, on left; fruit, on the right)

Discussion

Previous analyses at Amheida/*Trimithis* revealed the presence of a wide range of edible plants: cereals, mostly barley (*Hordeum sativum* L.) and wheat (*Triticum durum/aestivum*), lentils (*Lens culinaris* Med.) and fruit trees such as olive, grape, date palm, peach, sebesten (*Cordia* sp.) and pine (*Pinus* sp.) (Lodwick 2013). Additional findings are cotton (*Gossypium* sp.), flax (*Linum usitatissimum* L.), coriander (*Coriandrum sativum* L.), rosemary (*Rosmarinus officinalis* L.), black cumin (*Nigella sativa* L.), clover (*Trifolium* sp.), vetch (*Vicia* sp.), medick (*Medicago* sp.) along with mustard (*Brassica* sp.) and mallow (*Malva* sp.) (Thanheiser and Walter 2015). Recent archaeobotanical investigations confirm the presence of date palm, olive tree, grape, but they also identify new species of local and non-local origin (Caracuta et al. 2015; and this study). Christ's thorn, Nile acacia and carob are extensively attested in Dakhla and Kharga Oases, but black myrobalan, to date, is only found at Amheida/*Trimithis*.

The fact that all seeds presented in this study originated from the intentional accumulation of refuse in middens implies that all these remains are of mixed origins. They might have originated from different contexts and were later discarded together.

Since we are unable to assess the provenance of each specimen in relation to its context of use, it makes sense only to examine the provenience of the specimens, and in the case of non-native species, the timing and mode of introduction to the Dakhla and Kharga Oases. The identified species will be divided into 'ecological groups' based on the type of habitat to which they belong, either semi-desert or Mediterranean, and their provenance (see Asian and rare plants).

Semi-desert Vegetation

Acacia nilotica (Nile Acacia)

A native of Africa, Nile acacia is one of the most abundant and useful trees in Egypt. Both the inner bark and pods are used in tanning leather, while the young bark is used as a fiber (Wetterstrom 1984). The shoots and young pods provide forage for herbivores, while the wood makes good fuel and charcoal. In Kharga and Dakhla oases the dried flowers are used for gall bladder trouble and nasal congestion (Osborn 1968:173). The Nile acacia is also a principal source of gum arabic, which was used in making ink, and it is considered to be a very durable timber.

Ancient Egyptians used Nile acacia wood for various purposes including carpentry, dyeing and funeral offerings (Murray 2000). Archaeological remains of Nile acacia are abundant since prehistoric times in Dakhla Oasis. Amongst the earliest finds are those from the prehistoric Masara C, Bashendi A, and Sheikh Muftah sites, and from the Old Kingdom site 'Ain el Gazzareen (Thanheiser 2011). Remains were also found in the late Roman phases at Amheida/*Trimithis* (Thanheiser and Walter 2015). Specimens of *A. nilotica* are usually found associated with other species well adapted to arid and semi-arid environments, such as *Tamarix* sp. and *Z. spina-christi*. In Kharga Oasis these three species were found in the Neolithic site KS043 (Briois et al. 2012) and in the Persian site 'Ayn Manâwir (Newton et al. 2013).

Ziziphus spina-Christi (Christ's Thorn)

Christ's thorn is a tropical evergreen tree of Sudanese origin, which is found in the Sahara along main wadis, where the combination of high temperatures and available soil moisture allow trees similar to *Z. spina-christi* to grow. This species is a typical element of the natural vegetation of Dakhla Oasis, where it grows in association with *Acacia nilotica*, *P. dactylifera* and several species of *Tamarix* (Ritchie 1999).

Ziziphus spina-christi was used in ancient Egypt in Pharaonic times (de Vartavan et al. 2010) and its parts appear to have been in use in 'ancient Egyptian industry' (carpentry), diet, and medicine. The fruits were sometimes made into bread, and Egyptian farmers made similar bread as late as the beginning of the twentieth century AD (Manniche 2006).

The earliest attestation of *Z. spina-christi* in Dakhla comes from charcoal and is dated to the seventh millennium BC from the Neolithic site of Bashendi A, with later evidence found at the Old Kingdom site of 'Ain el Gazzareen and the Roman site of Kellis (Thanheiser et al. 2002; Thanheiser 2011) and *Trimithis* (this study). In the nearby Kharga Oasis, the earliest finds date to the fourth millennium BC (site KS043), while later discoveries date to the Middle Kingdom (Umm Mawagir) (Cappers et al.

2013) and Persian period ('Ayn Manâwir and 'Ayn-Ziyâda) (Newton et al. 2005; Agut-Labordère and Newton 2013).

***Phoenix dactylifera* (Date Palm)**

The date palm ranks among the first fruit trees that were brought into cultivation in the Old World. High temperatures, rainless summers, and very low humidity are particularly important conditions for fruit setting and ripening. Date palms can withstand considerable salinity and they grow well even when watered with brackish water. Date horticulture is therefore centered in the deserts south of the Mediterranean Sea and in the southern fringe of southwest Asia (Zohary and Spiegel-Roy 1975). Date palms thrived in these territories long before the initiation of agriculture. This is attested by the finds of charred fragments of *P. dactylifera* wood retrieved in Israel, at the Epi-Palaeolithic site of Ohalo II (Lipschitz and Nadel 1997). The stones of date palm reported from Egypt, Israel, Iran, and Pakistan, date to the 6th and 5th millennia BC, and probably represent fruits collected from the wild (Zohary et al. 2012).

In Egypt, remains of dates appear rarely in prehistoric sites; nonetheless, evidence of this tree is found in the Late Palaeolithic site of E-75-6 at Nabta Playa (Wasylikowa et al. 2001). Findings of date palm become more common in Pre-Dynastic sites, such as Hierakonpolis and El Omari, and prior to the Middle Kingdom much of the date remains consist of leaves, fiber, and wood, rather than fruits (see References in de Vartavan et al. 2010). Roof logs made of date palm wood were employed in the construction of the necropolis of Saqqara in tombs dated to the Proto-dynastic period (I-II dynasties) and to the Old Kingdom (V dynasty) (Nicholson and Shaw 2000).

At Kharga Oasis remains of date palm appear during the Persian period at 'Ayn Manâwir (Newton et al. 2013), and later in the Roman cemeteries of Douch (Barakat and Baum 1992). In Dakhla Oasis date palm stones are found at the Roman sites of Kellis (Thanheiser et al. 2002) and *Trimithis* (this study and Thanheiser and Walter 2015). According to the *Kellis Agricultural Account Book*, dates were used as payment for rent. Date pulp was compressed and consumed fresh, while date stones, which are also recorded in transactions, were likely crushed and milled into a flour to feed camels (Bagnall 1997).

Mediterranean Plants

***Olea europaea* (Olive Tree)**

The olive tree is a prominent feature of present-day Mediterranean vegetation. This species is adapted to cool winters and warm dry summers, but its cultivation has caused the species to surpass its natural bioclimatic limits. Today, the olive tree is extensively cultivated in northern Egypt, the Fayyum and the western Mediterranean coast and Siwa Oasis.

The distribution of olive finds in Pre-Pottery Neolithic sites indicates that the species was present, in its wild form, in a variety of habitats in the Southern Levant, ranging from the Mediterranean areas of the coast to the sub-arid highlands of the Negev (Liphschitz et al. 1991; Liphschitz 2007). However, the earliest certain evidence for olive cultivation is found in sub-arid areas of Jordan, at the Chalcolithic sites of Teleilat Ghassul and el-Khawarij (Bourke et al. 2004; Lovell et al. 2010).

Traditionally, the beginning of the cultivation of olive in Egypt is dated to the New Kingdom (Meeks 1993). The first representation of the tree on a wall painting appears in the Amarna period (Germer 1985), whereas a branch of olive is carved in a wall-relief discovered at Hermopolis, dating to the XVIIIth dynasty. The latter depicts Akenaton offering the branch to the God Aton (Cooney 1965). Information about olive orchards begin to appear in written documents dating to the XXth dynasty, such as the Harris Papyrus, which records a donation of 2735 hectares of olive orchards to temples by Ramses III (Erichsen 1933). Based on written documents, scholars have identified six major centers for the production of olives in the New Kingdom Period: Thebes, Hermopolis, Amarna, Heliopolis, Alexandria, and Qantir, but the Egyptian olive oil was of a lower quality of that imported from Syria-Palestine or Cyprus (Meeks 1993). Evidence of the import of olives and olive oil in Egypt pre-date by about five hundred years the cultivation of the plant in the countryside. The earliest archaeological remains of olive dates to the Middle Kingdom and they were found at the port-city of Memphis (Giddy and Jeffreys 1991; Murray 1993). Amongst the earliest findings of *O. europaea* in the Nile Valley are desiccated stones from a Middle Kingdom context at Umm Mawagir in Kharga Oasis (Cappers et al. 2013).

Evidence becomes more abundant at Kharga and Dakhla Oases during the Greco-Roman Period as shown by findings from the cemetery of Douch (Barakat and Baum 1992), and the sites of Kellis and *Trimithis* (Thanheiser et al. 2002; Thanheiser and Walter 2015; and this study). The textual references from Kellis point out the importance of olive oil, suggesting that this crop was one of the major products of the oasis (Bagnall 2008b).

***Ceratonia siliqua* (Carob)**

The carob tree is an evergreen species native to the eastern Mediterranean region. In Egypt today it is found along the Mediterranean coastal belt and as a cultivated garden tree (Darby et al. 1977). Carob pods are rich in protein, starch and sugar, and the medical properties of this species are commonly exploited to cure digestive diseases. Carob pods are today also used to make a syrup and sweet gum extract, which is used as a thickening agent and a substitute for coffee and chocolate (Aykroyd et al. 1982). The available archaeobotanical evidence shows that carob was part of the Levantine flora at least since early prehistoric times. The earliest finds of charred carob wood come from the prehistoric site of el-Wad terrace, in Mount Carmel, and they date to ~9500–9000 BC (Caracuta et al. 2016). In a later period, between 8000 and 5500 BC, wood and seeds of carob are found in Neolithic sites such as Nahal Oren and Atlit-Yam (Noy et al. 1973; Galili et al. 1993). Carob remains continue to appear in later Chalcolithic and Bronze Age contexts, yet many of these remains, recovered mostly in Israel, come from Roman period times or later (Zohary et al. 2012). Discoveries of carob seeds among the food offerings of a XIIth dynasty tomb seem to suggest that carob was indeed incorporated in ancient diets in Egypt as well (Täckholm 1961). Textual evidence dating to the Middle Kingdom, such as the Harris Papyrus, demonstrate that carob was included among fruits used as offerings in the oblations of festival offerings instituted by Rameses III. Carob wood was used to make chairs, tables, chests, shrines, chariots and staffs (Breasted 1906). Remains of carob seeds are attested in the Dakhla Oasis since the third-fourth century AD, as shown by the discoveries in *Trimithis* and Kellis (Thanheiser et al. 2002, and this study).

Asian and Rare Plants

Prunus persica (Peach)

The wild form of peach originated in China, where it was consumed already by the sixth-fifth millennium BC (Weisskopf and Fuller 2014). From the fourth millennium BC the tree spread outside China, first to Japan, and later to India, where it appears in its domesticated form in the third-second millennium BC (D'Andrea 2007; Boivin et al. 2012). Epigraphic sources date its introduction farther west into Persia by the fourth century BC (Hedrick 1919), but the earliest archaeological evidence is from the seventh century BC, at the site of the Heraion on Samos (Zohary et al. 2012). By the first century BC, Romans were cultivating peaches in southern Europe, Israel and Egypt (Bassi and Monet 2008; Bakels and Jacomet 2003; Sadori et al. 2009).

In Egypt, the earliest finds of *P. persica* come from el Hibeh, in the Nile valley, and date to the Late Pharaonic and Ptolemaic periods (Wetterstrom 1984). In the third-fourth centuries AD, this species appears in Dakhla Oasis, with peach being consumed in Kellis (Thanheiser et al. 2002) and *Trimithis* (Thanheiser and Walter 2015).

Terminalia chebula (Black Myrobalan)

Black myrobalan is a tree native to South and Southeast Asia and belongs to the family Combretaceae. The tree is found in forests throughout India. Black myrobalan flowers May–June, the fruits appear between November and January and reach full maturity between January and April. The fruit is 3–5 cm long, has a plum-like shape with six lines or ribs on the outer skin; it is green when unripe and yellowish grey at full maturity.

Black myrobalan is one of the most versatile plants, having a wide spectrum of pharmacological and medicinal uses. This medicinal plant is the unique source of various types of chemical compounds, which act as antioxidants, laxatives, antidiabetics and antimicrobics. The fruit paste, mixed with water, is found to be anti-inflammatory, analgesic and to have a purifying and healing capacity for wounds. Its decoction is used as a gargle to treat oral ulcers and sore throat, while its powder is a good astringent dentifrice for loose gums, bleeding and ulceration in gums (Bag et al. 2013). The fruits of *T. chebula* are rare in ancient Egypt, and prior to the finds at Amheida/*Trimithis*, the only other evidence was recovered from the port of Quseir el Qadim in layers dating to the Islamic period (Van der Veen 2011; Van der Veen and Morales 2015). Textual records report trade of black myrobalan from the Far East through Afghanistan, while medieval epigraphic sources refer to the shipment of myrobalans from India to Africa and Europe where it was used for its medical properties (Goitein and Friedman 2008).

The discovery of black myrobalan seeds in Amheida*Trimithis* in the midden deposit below the Serenos house, dated between the third and fourth century AD, represents the earliest findings of this species in Egypt and pre-dates its appearance during the Roman period, almost one thousand years before the findings of Quseir el Qadim.

Conclusion

The development of Dakhla Oasis in the Roman period is indicated by an increase in the number and extent of inhabited archaeological sites. The flourishing of settlements is likely related to the development of a successful economic system, which generated goods to be exchanged in return for commodities and rare items. A large portion of the oasis economy was probably based on the exploitation of agricultural products. The results presented herein provide insights into the local farming system and natural resources available to the inhabitants of *Trimithis*, in addition to those available through interregional trade of commodities. Species such as Nile acacia, Christ's thorn and date palm suggest that local environmental conditions around *Trimithis* did not differ from those of today, since these trees are still grown in the oasis. Textual evidence suggests that the olive was locally cultivated, and it is likely that olive oil was exported. Carob, another Mediterranean fruit tree, could have also been grown in the area surrounding *Trimithis* thanks to the groundwater system. The presence of peach is interesting, and it could have been imported or locally grown. Black myrobalan most likely was imported for its medical properties from East Asia through the ports on the Red Sea.

These results are not considered exhaustive, but rather provide a preliminary set of data which will be supplemented by further research. Additional excavations and a larger number of samples are required to understand the provenance of the macroremains deposited in the midden below the Serenos' house and adjacent streets.

References

Agut-Labordère D, Newton C (2013) L'économie végétale à 'Ayn-Manâwir à l'époque perse: archéobotanique et sources démotiques. ARTA 2013.005. http://www.achemenet.com/document/ARTA_2013.005-Agut-Newton.pdf

Ast R, Davoli P (in press) Ostraka and stratigraphy at Amheida*Trimithis* (Dakhla Oasis, Egypt): a methodological Issue. In: Proceedings of the 27th International Congress of Papyrology 29. Warsaw, Poland, July–3 August 2013

Aykroyd WR, Doughty J, Walker A (1982) Legumes in human nutrition. Paper No. 20. FAO, Rome

Bag A, Bhattacharyya SK, Chattopadhyay RR (2013) The development of *Terminalia chebula* Retz. (Combretaceae) in clinical research. Asian Pac J Trop Biomed 3(3):244–252

Bagnall RS (1997) The Kellis agricultural account book (P. Kell. IV Gr. 96) (With Contributions from Hope CA, Jenkins RG, Mills AJ, Sharp JL III, Thanheiser U, Wagner G) Dakleh Oasis Project Monograph 7. Oxbow, Oxford

Bagnall RS (2008a) SB 6.9025, cotton, and the economy of the small oasis. B Am Soc Papyrologis 45:21–30

Bagnall RS (2008b) Economy and ecology in Kellis agricultural account book. In: Wiseman MF (ed) The Oasis Papers 2. Proceedings of the 2nd International Conference of the Dakhleh Oasis Project. Oxbow Books, Oxford, p 115–117

Bagnall R, Aravecchia N, Cribiore R et al (2015) An oasis city. NYU Press, New York

Bakels C, Jacomet S (2003) Access to luxury foods in central Europe during the Roman period: the archaeobotanical evidence. World Archaeol 34(3):542–557

Barakat H, Baum N (1992) La végétation antique de Douch (Oasis de Kharga): une approache macrobotanique. IFAO, Cairo

Bassi D, Monet R (2008) Botany and taxonomy. In: Layne DR, Bassi D (eds) The peach: botany, production and uses. Wallingford, CAB International, Wallingford, pp 1–30

Boivin N, Fuller D, Crowther A (2012) Old World globalization and the Columbian exchange: comparison and contrast. World Archaeol 44(3):452–469

Bourke S, Zoppi U, Meadows J et al (2004) The end of the Chalcolithic period in the south Jordan valley: new 14C determinations from Teleilat Ghassul, Jordan. Radiocarbon 46 (1):315–323

Breasted JH (1906) Ancient records of Egypt: historical documents from the earliest times to the Persian conquest. Ancient Records, Second Series IV. University of Chicago Press, Chicago, pp 295–301

Briois F, Midant-Reynes B, Marchand S et al (2012) Neolithic occupation of an artesian spring: KS043 in the Kharga Oasis (Egypt). J Field Archaeol 37(3):178–191

Cappers R, Heinrich F, Kaaijk S et al (2013) Barley revisited: production of barley bread in Umm Mawagir (Kharga Oasis Egypt). Curr Res Egyptology 14:49–63

Caracuta V, Fiorentino G, Davoli P et al. (2015) The archaeobotanical analysis at Amheida *Trimithis* (Dakhleh Oasis–Egypt), preliminary results. In: Florenzano A, Fornaciari R (eds) Proceedings of the IWAA8. Supplemento Atti Soc Nat Mat Modena 146, pp 45–48. ISSN: 0365-7027

Caracuta V, Weinstein-Evron M, Yeshurun R et al (2016) Charred wood remains in the Natufian sequence of el-Wad terrace (Israel): new insights into the climatic, environmental and cultural changes at the end of the Pleistocene. Quaternary Sci Rev 131(1):20–32

Cooney JD (1965) Amarna reliefs from Hermopolis in American collections. Brooklyn Museum, Brooklyn

Cribiore R, Davoli P, Ratzan D (2008) A Teacher's Dipinto from *Trimithis* (Dakhleh Oasis). J Roman Archaeol 2:170–191

D'Andrea AC (2007) The dispersal of domesticated plants into northeastern Japan. In: Denham T, White P (eds) The emergence of agriculture: a global view. Routledge, London, pp 154–174

Darby WJ, Ghalioungui P, Grivetti LE (1977) Food: the gift of Osiris. Academic Press, London

Davoli P (in press) A new public bath in *Trimithis* (Amheida*Trimithis*, Dakhla Oasis). In: Redon B (ed) Le bain collectif en Égypte 2. Découvertes récentes et synthèses. Proceedings of the International Conference, Cairo, October 26, 2010, IFAO, Le Caire

de Vartavan C, Arakelyan A, Asensi-Amoròs V (2010) Codex of ancient Egyptian plants. Armenian Egyptology Centre, Yerevan

Erichsen W (1933) Papyrus Harris I. Hieroglyphische Transkcription 88, 13. Fondation Reine Elisabeth, Brussels

Galili E, Weinstein-Evron M, Hershkovitz I et al (1993) A prehistoric site on the sea floor off the Israeli Coast. J Field Archaeol 20(2):133–157

Germer R (1985) Flora des pharaonischen Aegypten. Philipp von Zabern, Mainz am Rhein

Giddy LL, Jeffreys DG (1991) Memphis, 1990. J Egypt Archaeol 77(1):1–6

Goitein SD, Friedman M (2008) India traders of the Middle Ages: documents from the Cairo Geniza 'India Book'. Brill, London

Gradel C, Letellier-Willemin F, Tallet G (2012) Une laine bien plus belle et douce que celle des moutons à El-Deir (oasis de Kharga, Égypte). In: Guédon S (ed) Entre Afrique et Égypte. Ausonius, Bordeaux, pp 119–141

Hassib M (1951) Distribution of plant communities in Egypt. Bull Fac Sci Univ Fouad I 29:59–261

Hedrick UP (1919) Sturtevant's notes on edible plants. Annual Report 27(2-II), New York Department of Agriculture

Liphschitz N (2007) Timber in ancient Israel. Dendroarchaeology and dendrochronology. Tel Aviv University Monograph Series 26

Liphschitz N, Gophna R, Hartmana M et al (1991) The beginning of olive (*Olea europaea*). cultivation in the Old World: a reassessment. J Archaeol Sci 18:441–453

Lipschitz N, Nadel D (1997) Epipalaeolithic (19,000 B.P.) Charred Wood Remains from *Ohalo II,* Sea of Galilee, Israel', Mitekufat Haeven. J Israel Prehistoric Soc 27:5–18

Lodwick L (2013) Archaeobotany. University of Reading Excavations at Amheida*Trimithis*. Preliminary Report 2013

Lovell JL, Meadows J, Jacobsen GE (2010) Upland olive domestication in the Chalcolithic period: new 14C determinations from El-Khawarij (Ajlun), Jordan. Radiocarbon 52(2–3):364–371

Manniche L (2006) An ancient Egyptian herbal. The American University in Cairo Press, Cairo

Meeks D (1993) Oléiculture et viticulture dans l'Égypte pharaonique. In: Amourreti MC, Le Brun JP (eds) Oil and wine production in the Mediterranean area. Bull Corresp Hellénique 26:3–38

Migahid AM, El-Shafei AM, Abdel Rahman AA et al (1960) An ecological study of Kharga and Dakhla Oases. Bull Soc Geogr Egypte 33:279–310

Murray MA (1993) Recent archaeobotanical research at the site of Memphis. In: Davis WV, Walker R (eds) Biological anthropology and the study of Ancient Egypt. British Museum Press, London, pp 165–168

Murray MA (2000) Fruits, vegetables, pulses and condiments. In: Nicholson PT, Shaw I (eds) Ancient Egyptian materials and technology. Cambridge University Press, Cambridge, pp 609–655

Neef R, Cappers RTJ, Bekker RM (2012) Digital atlas of economic plants in archaeology. Barkhuis & Groningen University Library, Groningen

Newberry P (1889) On the vegetable remains discovered in the cemetery of Hawara. In: Petrie WMF (ed) Hawara, Biahmu and Arsinoe. Trubner and Company, London

Newton C, Gonon T, Wuttmann M (2005) Un jardin d'oasis d'époque romaine à 'Ayn-Manâwir (Kharga-Égypte). Bull Institut Français Archéologie Orientale 105:167–196

Newton C, Whitbread T, Agut-Labordère D, Wuttmann M (2013) L'agriculture oasienne à l'époque perse dans le sud de l'Oasis de Kharga (Égypte, V^{e}-IVe s. AEC). Revue d'ethnoécologie 4

Nicholson PT, Shaw I (2000) Ancient Egyptian materials and technology. Cambridge University Press, Cambridge

Noy T, Legge AJ, Higgs ES (1973) Recent excavations at Nahal Oren, Israel. Proc Prehist Soc 39:75–99

Oliver FW (1930–1931) Oasis impression, being a visit to the Egyptian Oasis of Kharga. Trans Norfolk and Norwich Naturalists' Soc 13(2):38–52

Osborn DJ (1968) Notes on medicinal and other uses of plants in Egypt. Eco Bot 22(2):165–177

Ritchie JC (1999) Flora, vegetation and palaeobotany of the Dakhleh Oasis. In: Churcher CS, Mills AJ (eds) Reports from the survey of the Dakhleh Oasis western desert of Egypt 1977–1987. Dakhleh Oasis Project, Oxbow Books, Oxford, pp 73–80

Sadori L, Allevato E, Bosi G et al (2009) The introduction and diffusion of peach in ancient Italy. In: Morel JP, Mercuri AM (eds) Plants and culture: seeds of the cultural heritage of Europe. Edipuglia, Bari, pp 45–61

Shalaby AF, Ghanem SS, El-Habibi AM (1975) Ecological study of *Prosopis stephaniana* (Willd.) Kunth. Bull Fac Sci Mansoura Univ 3:45–63

Täckholm V (1961) Botanical identifications of the plants found at the Monastery of Phoebammon. In: Bachatly C (ed) Le Monastère de Phoebammon la Thébaidë-Tomme III. La société d'archéologie copte, Cairo, pp 1–38

Täckholm V, Drar M (1954) Flora of Egypt, vol III. Cairo University Press, Cairo

Thanheiser U (1999) Plant remains from Kellis: first results. In: Hope CA, Mills AJ (eds) The Dakhleh Oasis project: reports on the 1992–1993 and 1993–1994 field seasons. Oxbow Books, Oxford, pp 89–93

Thanheiser U (2011) Island of the blessed: 8000 years of plant exploitation in the Dakhleh Oasis, Egypt. In: Fahmy AG, Kahlheber S, D'Andrea AC (eds) Windows on the African past: current approaches to African archaeobotany. Proceedings of the 6th International workshop on african archaeobotany, Cairo. Africa Magna Verlag, Frankfurt, pp 79–90

Thanheiser U, Kahlheber S, Dupras T (2016) Pearl millet, *Pennisetum glaucum* (L.) R.Br. ssp. *glaucum*, in the Dakhleh Oasis, Egypt. In: Thanheiser U (ed) News from the past: progress in African archaeobotany. Proceedings of the 7th International workshop on african archaeobotany in Vienna, 2–5 July 2012. Barkhuis, Groningen, pp 115–125

Thanheiser U, Walter J (2015) Plant use in a Romano-Egyptian household in the third century CE. In: Boozer AL (ed) Amheida II. A Late Romano-Egyptian house in the Dakhla Oasis: Amheida House B2. NYU Press and Institute for the Study of the Ancient World, New York, pp 375–392

Thanheiser U, Walter J, Hope CA (2002) Roman agriculture and gardening in Egypt as seen from Kellis. In: Hope CA, Bowen GE (eds) Dakhleh Oasis project: preliminary reports on the 1994–5 to 1998–9 field-seasons. Oxbow Books, Oxford, pp 299–310

Van der Veen M (2011) Consumption, trade and innovation: exploring the botanical remains from the Roman and Islamic ports at Quseir al-Qadim. Africa Magna Verlag, Frankfurt, Egypt

Van der Veen M, Morales J (2015) Roman and Islamic spice trade: new archaeological evidence. J Ethnopharmacol 167:54–63

Walter J (2006) Botany and archaeobotany. In: Bagnall R (ed) Report to the supreme council of antiquities on the 2005–2006 season activities of the Dakhleh Oasis project

Wasylikowa K, Barakat HN, Boulos L et al (2001) Site E-75-6: vegetation and subsistence of the Early Neolithic at Nabta Playa, Egypt, reconstructed from charred Plant remains. In: Wendorf F, Schild R (eds) Holocene Settlement of the Egyptian Sahara. Springer, New York, pp 544–591

Weisskopf A, Fuller D (2014) Peach: origins and development. In: Smith C (ed) Encyclopedia of global archaeology. Springer, New York, pp 5840–5842

Wetterstrom W (1984) The plant remains. In: Wenke RJ (ed) Archaeological Investigations at El-Hibeh 1980: preliminary report. American Research Centre in Egypt, Reports 9. Undena Publications, Malibu, pp 50–77

Zahran MA (2010) Climate vegetation: Afro-Asian Mediterranean and Red Sea coastal lands. Springer, New York

Zahran MA, Willis AJ (2009) The vegetation of Egypt, 2nd edn. Springer, Netherlands

Zohary D, Hopf M, Weiss E (2012) Domestication of plants in the Old World. Oxford University Press, Oxford

Zohary D, Spiegel-Roy P (1975) Beginnings of fruit growing in the Old World. Science 187(4174):319–327

Archaeobotanical Studies from Hierakonpolis: Evidence for Food Processing During the Predynastic Period in Egypt

Elshafaey A. E. Attia[1], Elena Marinova[2,3(✉)], Ahmed G. Fahmy[1], and Masahiro Baba[4]

[1] Department of Botany, Faculty of Science, Helwan University, Cairo, Egypt
s.elshafaey@gmail.com

[2] Center for Archaeological Sciences, KU Leuven, Leuven, Belgium
elena_marinova@gmx.de

[3] Royal Belgian Institute of Natural Sciences, Brussels, Belgium

[4] Institute for Advanced Study, Waseda University, Tokyo, Japan

Abstract. This paper discusses recently obtained archaeobotanical evidence from locality HK11C of Predynastic Hierakonpolis, Upper Egypt, and in particular, information on plant foods and their processing. The excavations at this locality have revealed industrial food production activities dating to the Naqada II period (c. 3800–3300 BC). From one structure (Operation C) dedicated to the processing of meat and fish, plant remains were extracted through flotation of sediment from burned debris obtained from hearths. A sample from another structure, reused for refuse disposal, was also studied. The finds indicate discarded waste of cereal crop processing, fuel residues (wood and vegetative parts of herbaceous vegetation), herbivore dung and other plant remains deriving from crops (flax, melon) or representatives of the wetland and desert vegetation. The most frequent plant remains recovered from this food processing zone are crop cleaning by-products of emmer and barley, the principal cereal crops of the period (c. 75% of the total frequency of identifiable plant remains). In much smaller frequencies are remains of weeds, found as seeds and/or vegetative parts, which represent the wild growing vegetation from ruderal places, river banks or the desert. Further archaeobotanical evidence from the site HK11C comes from an installation (Operation B) where large ceramic vats were found containing charred residues. Close examination of the charred residue revealed remnants of emmer processed for food. Whole and fragmented grains were distinguishable under low magnification (10–40x). Although no longer recognisable without magnification, the presence of ground grain was demonstrated by numerous cereal pericarp fragments and aleuron cell layers visible under higher magnification (100–400x). Starch grains with perforations observed under Scanning Electron Microscope (SEM) suggest that fermentation had taken place. This processing, with subsequent heating, must have resulted in a more or less homogenous mass suggesting that this matter was wet when charred. It may represent dough for bread making or more probably, considering the

A. M. Mercuri et al. (Eds.): *Plants and People in the African Past*,
https://doi.org/10.1007/978-3-319-89839-1_5

archeological context, for initial stages of beer production. Other samples found inside the vats and their surroundings also indicate processed emmer probably for beer production.

Keywords: Archaeobotany · Ancient brewing · Emmer · Plant macroremains Predynastic Upper Egypt

Introduction

Archaeobotanical studies at Hierakonpolis, a major town in Predynastic Upper Egypt (Fig. 1), have provided crucial evidence for reconstructing the plant economy and resources used during the time of emergence of Egyptian civilization (c. 3800–3200 BC). The site includes diverse localities related to activities of daily life (settlement areas and industrial quarters where, for example, beer brewing and pottery production took place) as well as several cemeteries and a ceremonial centre (Friedman 2011). Activities that took place at the site left numerous plant remains relating to food production, processing, consumption and the dumping of refuse. The excellent preservation conditions in the arid environment at the site have allowed archaeobotanical research at Hierakonpolis to obtain detailed information on the past environment, agricultural practices, crop processing, human and animal diets, and the use of plants in architecture and ritual contexts (Fadl et al. 2013; Fahmy 2005, 2008; Fahmy et al. 2008a, b, 2011; Fahmy and Fadl 2009; Marinova et al. 2013).

In this paper we discuss recently obtained archaeobotanical evidence from locality HK11C at Hierakonpolis and in particular, new information on plant food resources and their processing. In addition to the identification of plant remains, a study of the microstructure of charred residues found in what are interpreted as brewing vats was conducted in order to observe alterations in tissues and cells of processed (cooked or/and fermented) emmer grains.

Materials and Methods

Excavations at locality HK11C, situated on the south bank of the Wadi Abu Suffian (Fig. 1), have revealed several installations dedicated to industrial food production, activities dating to the Naqada II period, c. 3800–3300 BC (Baba and Friedman 2016).

In Operation C (Square C3-4), a roughly square mudbrick structure with one rounded wall (Fig. 2a) was found. In this structure meat and fish were prepared and cooked. The interior of the structure was filled with accumulated layers of black burnt debris containing charcoal, ash, stones and potsherds overlying and surrounding numerous hearths. Two sediment samples (each with volume of 2 l) were taken from hearths (nr. 4 and 6) discovered on the floor of the structure and were processed by dry sieving with mesh sizes of 1 and 0.3 mm. Another sample of the same volume was obtained from the ash-filled debris within a larger mudbrick structure in Square C10-11, located 30 m to the south of Operation C. This material probably represents discarded waste from the industrial activities in the area and was deposited in the structure

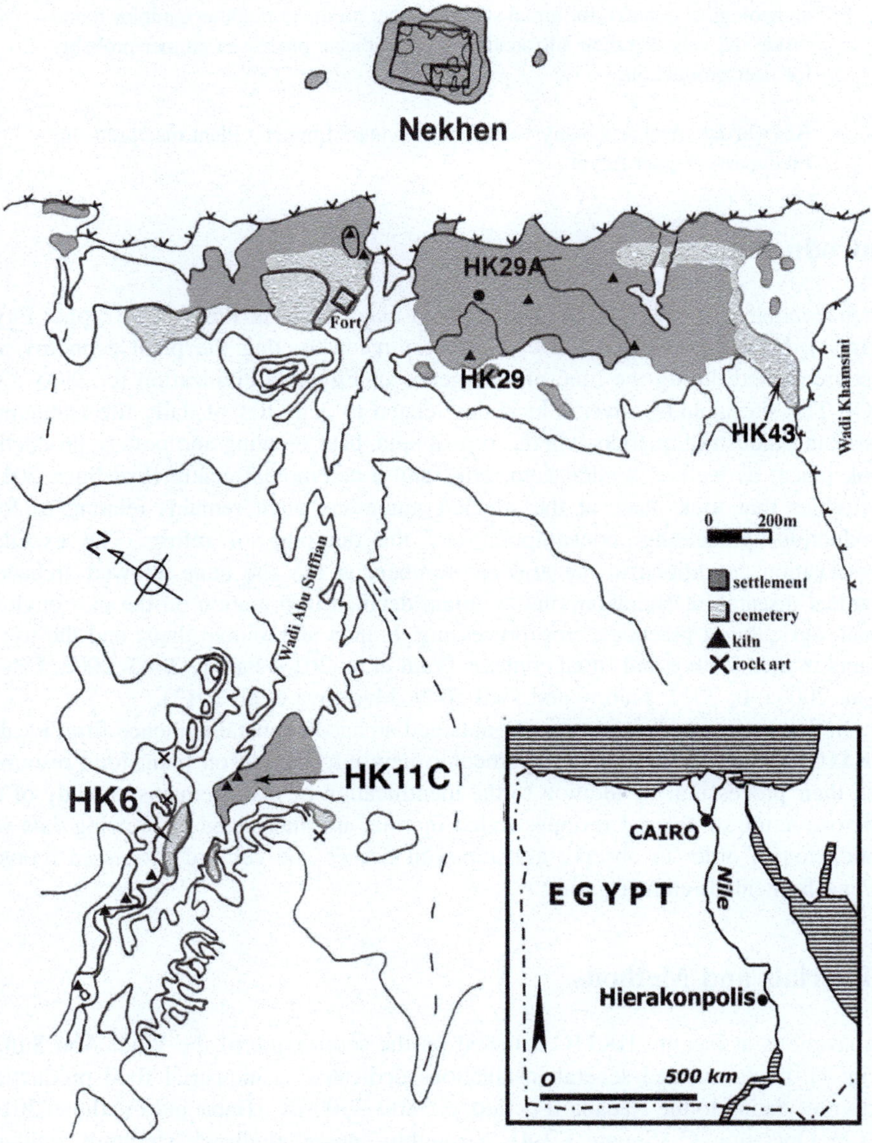

Fig. 1. Location of the study site (HK11C) within the site of Hierakonpolis

after it had fallen out of its original use. The plant remains were also extracted by dry sieving.

One industrial installation, a beer and pottery production complex dating to the early Naqada II period, was found in the lower level of Operation B (Fig. 2b). There, five large ceramic vats presumed to be for making beer were uncovered in situ (Fig. 3a). They were found within a structure composed of mud. From partly charred

(a)

(b)

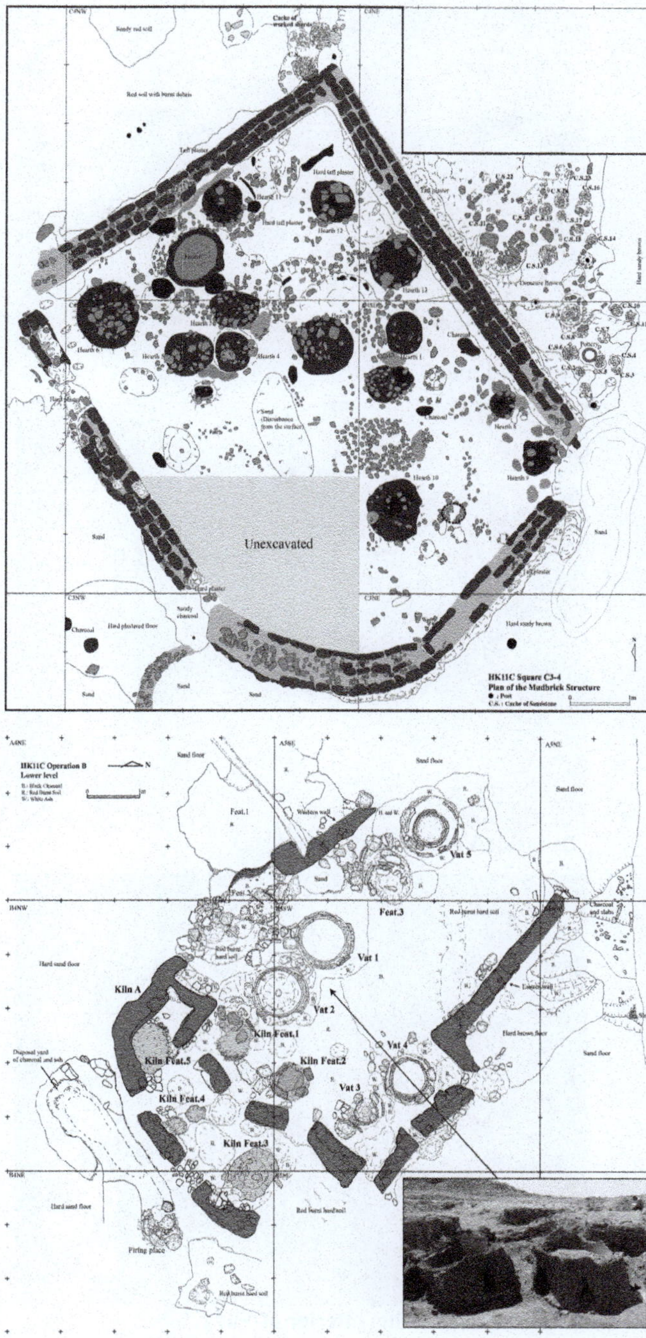

Fig. 2. Plans of the studied features: **a** Operation C (Square C3-4) **b** Operation B.

Fig. 3. **a** Location of Vats 1 and 2; **b** and **c** charred residue in Vat 2 found at Operation B.

residues found inside each of the vats (Fig. 3b, c) and adjacent deposits, six samples were taken for macro-botanical analysis. The charred residue in one of the vats (Vat nr.4) was absolutely dated to 3762–3537 BC (raw date: 4875 ± 40 BP) (Baba and Friedman 2016: 193).

All samples were analysed under low magnification stereoscope (up to 50x) while fine structures, such as the epidermal layers of plant tissues, were examined under higher magnification (up to 500x) using reflected light microscope. Fine microstructures of the residues found in vats were also documented using Scanning Electron Microscope (SEM).

Results

Well preserved plant macroremains, both desiccated and charred, were retrieved from all samples.

Dry-Sieved Samples from Squares C3-4 and C10-11

The three samples retrieved from the two mudbrick structures at locality HK11C revealed good preservation of both desiccated and charred plant macroremains (Table 1). A total of 972 of plant macroremains were identified including: 598 fragments of cereal chaff (rachis fragments of barley and glume bases of emmer); 52 cereal grains (barley, emmer or Cerealia indet.); 58 seeds of potential weeds and ruderal vegetation seeds; 18 remains of species of wetland vegetation; 3 genera of xerophytes (represented by 35 vegetative parts); 145 varia items (not identifiable below family level) seeds and fragments of vegetative parts, as well as 63 indeterminate specimens.

Cereal crops. Barley (*Hordeum vulgare* L.), an annual crop cultivated in autumn/winter, is one of the main cereals of ancient Egypt. It is present in the archaeological record from Neolithic times onward, and was favoured because it can resist drought conditions and is a suitable crop for arid regions. A total of 12 grains and 275 rachis fragments of the studied botanical assemblage at HK11C was attributed to *H. vulgare.* The majority of the identifiable fragments are chaff either charred (n = 33) or desiccated (n = 242) (Table 1). The highest number of grains was separated from the sample obtained from the structure in Square C10-11 (11 charred grains).

Emmer wheat (*Triticum dicoccum* Schrank) is an annual winter crop, which is no longer cultivated in Egypt today. Isolated grains number eight in total, and they are concentrated in the sample from the debris in Square C10-11. There is only one charred grain from the structure in Square C3-4. On the other hand, charred chaff from emmer wheat dominates (n = 282 glume bases in total) the assemblage from this area, and may represent the remains of dung cakes used as fuel for the hearths. In contrast, only 18 glume bases of emmer occur in the sample from Structure C10-11.

Other crops. One unripe desiccated capsule of flax (*Linum usitatissimum* L.) was isolated from the sample from Square C10-11 and a fragment of a charred capsule from Square C4. Flax is an annual herb that is cultivated in the winter. At present, it is

Table 1. Results of the archaeobotanical analysis of three samples from the mudbrick structures in Square C3-4 and Square C10-11.

Site			HK 11C				
Sample number			1	2	1		
Feature			C3-4	C3-4	C10-11		
Sample volume (l)			2	2	2		
Total nr. plant remains			76	526	370		
	Type of record	*Preservation*				*Total nr.*	%
Crops							
Cereal grains						**52**	**5.3**
Hordeum vulgare undiff.	Seed/fruit	Charred	1		10	11	1.1
Hordeum vulgare undiff.	Seed/fruit	Desiccated			1	1	0.1
Triticum dicoccum	Seed/fruit	Charred		1	7	8	0.8
Cerealia indet.	Seed/fruit	Charred	3		29	32	3.3
Cereal chaff						**598**	**61.5**
Hordeum vulgare undiff.	Rachis fragment	Charred	3	101	138	242	24.9
Hordeum vulgare undiff.	Rachis fragment	Desiccated			33	33	3.4
Triticum dicoccum	Glume base	Charred	2	280	18	300	30.9
Triticum dicoccum	Glume base	Desiccated			23	23	2.4
Others						**5**	**0.5**
Cucumis melo/sativus	Seed/fruit	Desiccated		3		3	0.3
Linum usitatissimum	Capsule	Charred		1	1	2	0.2
Potential weeds						**58**	**6.0**
Avena sp.	Seed/fruit	Desiccated		1		1	0.1
Lolium cf. *temulentum*	Seed/fruit	Charred		2	10	12	1.2
Phalaris cf. *minor*	Seed/fruit	Charred	1	3	38	42	4.3
Rumex cf. *dentatus*	Seed/fruit	Desiccated		1		1	0.1
Solanum sp.	Seed/fruit	Desiccated		1		1	0.1
Vicia sp.	Seed/fruit	Charred			1	1	0.1
Desert plants						**35**	**3.6**
Acacia	Wood	Charred	12			12	1.2
Pulicaria	Inflorescence	Charred			1	1	0.1
Tamarix	Branch	Charred		1		1	0.1
Tamarix	Wood	Charred		10	11	21	2.2
Riparian/Floodplain Vegetation						**18**	**1.9**
Ceruana pratensis	Seed/fruit	Charred	1		2	3	0.3
Cyperaceae	Seed/fruit	Charred			10	10	1.0
Cyperaceae	Seed/fruit	Desiccated		1		1	0.1
Cyperaceae	Culm fragment	Charred		2		2	0.2
Desmostachya bipinnata	Culm fragment	Charred			1	1	0.1
Portulaca sp.	Seed/fruit	Charred		1		1	0.1
Varia						**145**	**14.9**
Fabaceae	Seed/fruit	Charred			1	1	0.1
Lamiaceae	Seed/fruit	Desiccated			1	1	0.1
Poaceae	Seed/fruit	Charred			1	1	0.1
Poaceae	Culm fragment	Charred	20	20	24	64	6.6
Poaceae	Culm fragment	Desiccated		78		78	8.0
Indeterminated						**61**	**6.3**
	Seed/fruit	Charred		2	9	11	1.1
	Vegetative part	Charred	8	13		21	2.2
	Vegetative part	Desiccated	25	4		29	3.0

mainly cultivated in the Nile Delta and at Abu-Tig in the Assiut governorate (c. 375 km south of Cairo). It requires moderate climatic conditions for the production of a good yield.

Three desiccated seeds of melon (*Cucumis melo/sativus* L.) were found in Square C3-4. According to current knowledge (see Zohary et al. 2012) most probably the seeds represent *Cucumis melo*, a prostrate annual herb that grows in sandy loamy soils. It was a valued vegetable since Predynastic times and is attested also in the stomach contents of Predynastic humans at Hierakonpolis (Fahmy 2008).

Potential weeds and ruderal vegetation. This ecological group includes 58 seeds (55 charred and 3 desiccated) representing six species. They represent 6.2% of the recorded plant remains in the studied samples. It is observed that the weed flora of HK11C is characteristic of winter farming as both *Phalaris* cf. *minor* and *Lolium* cf. *temulentum* represent the highest percentages of weed taxa (4.3% and 1.2% respectively). Other field weeds were also recorded, for example: *Portulaca* sp., *Vicia* sp., *Rumex* cf. *dentatus, Solanum* sp., which could have originated in ruderal habitats and *Rumex* cf. *dentatus* which also may have grown along the margins of wetlands.

Reeds/wetland vegetation. The water-loving species identified in the samples are: *Desmostachya bipinnata* Stapf, *Ceruana pratensis* Forssk., and several specimens of the family Cyperaceae that cannot be identified to the genus level. These remains were concentrated mainly in Structure C10-11. The presence of such an assemblage of water loving taxa suggests the occurrence of damp soils along natural water channels that would have run through cultivated fields especially during the winter. *C. pratensis* is an annual herb which grows on the muddy banks of the Nile, major irrigation canals in cultivated fields and the inundated ground in their vicinity. Today, this taxon is an endangered species in Egypt (Boulos 2002). Its use in wattle and daub construction or incorporated in wooden architecture, such as fences, is well documented at Hierakonpolis (Fahmy et al. 2008a, b).

Xerophytes. The plant macroremains of three taxa were grouped under xerophytes and are indicative of desert habitats in the surrounding area. The remains were attributed to: *Pulicaria* sp. (represented by only one charred seed), *Acacia* sp. and *Tamarix* sp. (represented by a vegetative part).

Wood charcoals. This group includes only *Acacia* sp. and *Tamarix* sp. The studied charcoal fragments from Square C3-4 included *Acacia* sp. (12 fragments) and *Tamarix* sp. (10 fragments) while in the sample from Square C10-11 contained only *Tamarix* sp. (11 fragments). Due to the low taxonomic diversity only the half charcoals from each sample was studied.

Dung remains. Desiccated sheep/goat dung pellets were recovered in two samples from Square C3-4. They were examined for seeds and other plant remains that might have passed through the alimentary canal tracts without morphological deterioration. The analysis revealed the desiccated residues of processed cereal crops which had been used as animal fodder.

Charred Residue Samples from the Vats in Operation B

The residues found adhering to the interior side and base of three vats were examined (Fig. 3). The sample from Vat 2 was the best preserved and richest in number of identifiable plant remains. It is composed of a layer of porous charred matter up to 3 cm thick. The thickness of residues available for analysis in the other vats did not exceed 1 cm. Close examination under the microscope showed that some parts of the residue from Vat 2 are not completely charred. They are brownish in colour, indicating limited exposure to heat. The exposed surface of the residue appears rich in plant materials, especially emmer grains and chaff (Fig. 4a).

The inner surface of the residue, which adhered to the vat surface, shows a more homogenous porous structure, with few recognisable plant macrofossils (Fig. 4b). The charring on this side of the residue seems to be more intensive, as very little brownish-semi-charred plant matter was observed. This indicates greater exposure to heat from contact with the lower sides of the vat, since the fire was placed around the base (see Baba and Friedman 2016: 188). Analysis of the crust in cross section (Fig. 4c) indicates that the upper layer is rich in recognisable cereal remains (mainly chaff), while the lower part, which was adhering to the vat surface, is homogenized porous matter.

The samples from inside Vat 4 resemble those from Vat 2, but it seems that there was greater exposure to heat, as certain areas are strongly charred. On the surface some fragmented emmer grains are recognisable (Fig. 4d). There is a high frequency of partly charred and uncharred organic matter and plant remains visible in the sample (Fig. 4e, f). From one of the residue samples it was possible to isolate a single, slightly damaged grain of emmer (Fig. 5a). Most probably the damage is due to the fact that the grain was wet and softened prior to charring.

A sample was recovered from debris found around Vat 4, which was not adhered to the vat walls. It has the appearance of dried bread and has contents similar to the residue sample, but is dominated by desiccated plant remains instead of charred materials. The desiccated fragments represent most probably the same type of matter as the charred remains, but the grains inside have undergone degradation, possibly due to insect activity, to judge from the remains of insects found within this sample. Material that seems to be similar was found in a roughly contemporary brewery complex at Tel el Farkha where it was interpreted as the discarded material removed from the brewed liquid by sieving (Kubiak-Martens and Langer 2008). The uncharred matter from the Vat 4 sample was very brittle and apart from the emmer chaff it was possible, using a brush, to extract from it several seeds of possible weed plants such as *Lolium* sp., *Digitaria* sp. and other specimens not further identifiable. Especially interesting is the find of desiccated rachis of possibly free threshing wheat (cf. *Triticum durum* Desf., Fig. 5b). Free threshing wheat is rather rare in the Predynastic period of Egypt, but contemporary finds are known for Hierakonpolis (see Fahmy et al. 2011). Nevertheless, it is possible that this rachis fragment is an artefact produced by the fragmentation of emmer chaff heated in lower temperature.

The samples from Vat 5 show nearly identical consistency and composition to those from Vat 2, however the thickness of the residue is much less (c. 1 cm). The upper part of the residue contains higher amounts of coarser plant remains (mainly

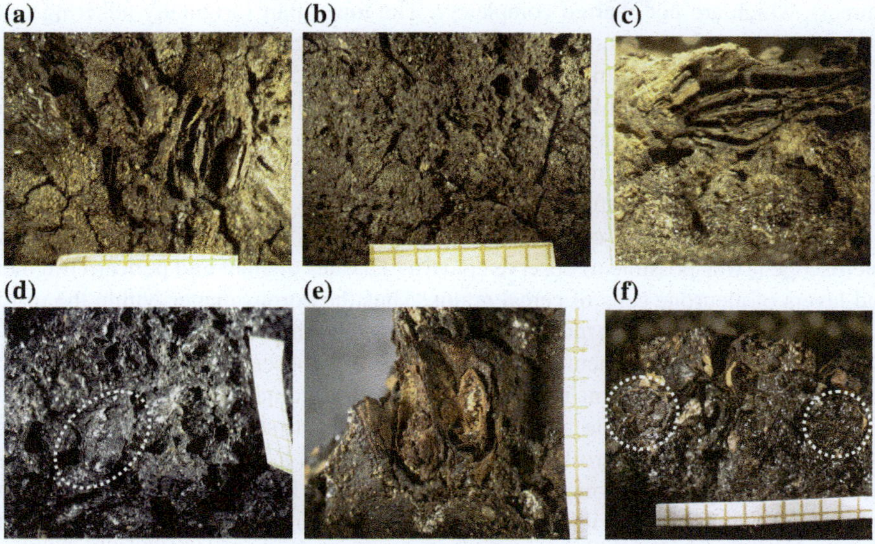

Fig. 4. Microphotographs of macro-botanical finds from the charred residues found in the vats of Operation B: **a** Surface of the charred residue, from inside Vat 2. Note the oval emmer spikelet; **b** Unexposed surface of the charred residue, attached to the wall of Vat 2; **c** View of the same residue in cross section; **d** Charred residue from Vat 4. Note an embedded grain of emmer within the dotted ellipse; **e** Partly charred emmer spikelet from Vat 4; **f** Partly charred plant remains in Vat 4, with doted ellipse are indicated broken in cross section cereal grains.

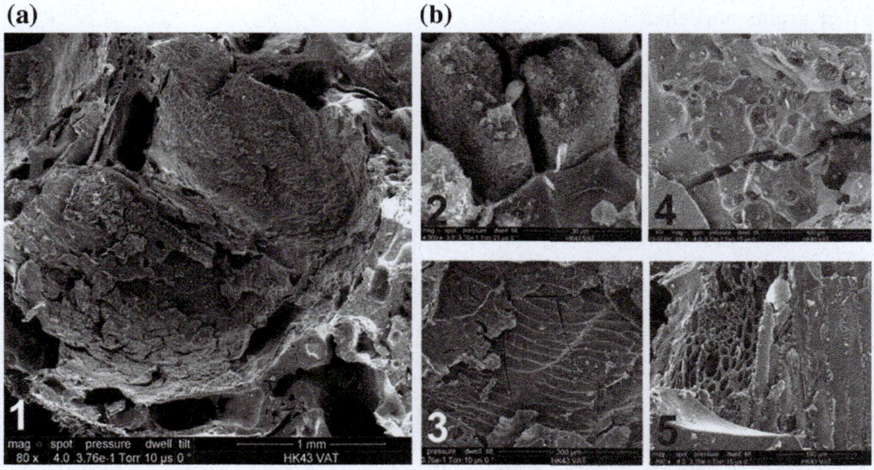

Fig. 5. **a** Charred emmer grain from the residue in Vat 4; **b** Rachis fragment, possibly free threshing wheat or artefact of the preservation of emmer chaff, from Vat 5 (scales 1 mm).

emmer), which are not charred completely. The lower part (originally adhering to the vat wall) is more homogenous and charred more strongly similar to those from Vat 4 (Fig. 4f). The sample from the surroundings of Vat 5 contains a mixture of chaff remains, but also clumps of clay that are rich in organic matter. There are also a few charred remains as well as very small fish bones and scales in the residue. This composition indicates the rather mixed character of the sample, and is probably the result of the disposal of refuse.

By close examination of the residues from all studied vats under reflected light microscope it was possible to observe epidermal tissues of chaff and pericarp fragments and layers of aleurone cells of cereal grains. Thus the porous matter within the residue consists also of cereals, but more strongly fragmented than those on the upper surface. The structures were documented with SEM (Fig. 6). Observed under higher magnification (100–400x), the burnt residues were found to contain various tissue remnants (Fig. 6a), mostly identifiable as originating from wheat (see Heiss et al. 2017). Considering the identified macro-botanical remains, this is a clear indication that emmer grains were crushed and ground together with its chaff. It is interesting to point out the presence of starch grains with perforations (Fig. 6b), suggesting, according to Samuel (2000), that they were subjected to fermentation. The presence of ground grains is also demonstrated by numerous cereal pericarp fragments and aleurone cell layers (Fig. 6c) visible under higher magnification. Closer examination of the fragmented grains and homogenous porous matter (Fig. 6d) revealed alternations (e.g. matrix of fragmented charred starchy endosperm, without any distinguishing features) similar to those observed in boiled grains by Valamoti et al. (2008). Among them also fragments of chaff epidermis and tissue (Fig. 6e) occur. This shows that the portions of the residue which appear under low magnification as porous structures are composed of crushed emmer grains and chaff.

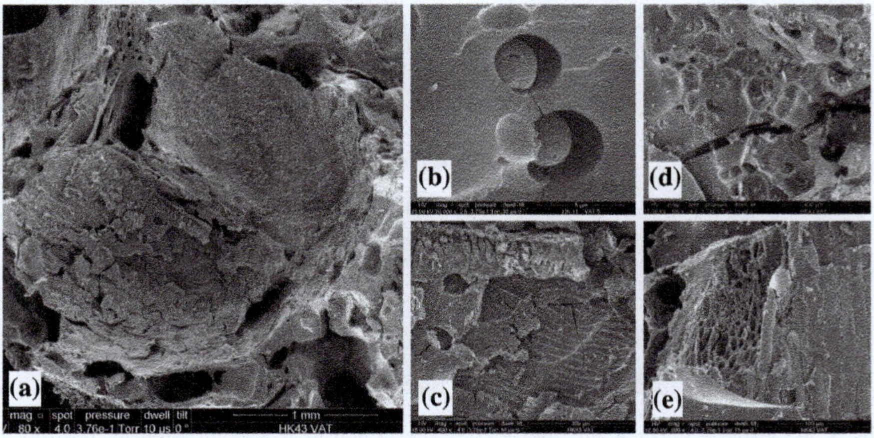

Fig. 6. SEM images of the microstructure of fragments of charred residues from the vats of Operation B: **a** overview of the residue fragment; **b** single starch grains showing predorations embeded in the charred residue; **c** transverse cell of grain pericarp; **d** matrix of fragmented charred starchy endosperm, without any distinguishing features; **e** chaff fragment with epidermal surface visible.

Discussion

The mud-brick structure of Square C3-4, within which food of animal origin was processed (Baba et al. 2017), yielded mainly by-products of cleaning emmer and barley, the principal cereal crops of the site (Fahmy et al. 2011), as well as weeds and some representatives of the wild growing vegetation. These remains probably reflect the use of crop cleaning by-products in the fuel (dung and wood) for heating during meat preparation. These findings are supplemented by the sample from the structure in Square C10-11, where burnt material, no doubt deriving from other industrial activities at the site, were dumped. The archaeobotanical analysis of those remains also provided information on weed and other wild growing flora. The field weed assemblages suggest a mono-seasonal crop regime restricted to winter. Other habitats reconstructed include marshes with saline soils surrounded by desert, probably used as pasture land for the domestic stock. Locally available *Acacia* sp. and *Tamarix* sp. were the major sources of charcoal and wood.

The archaeobotanical evidence from the partly charred residue adhering to the inside of large vats found in the food production installation at Operation B indicates that emmer grains were the main component of the residues and at least parts examined. The processing and subsequent heating must have resulted in a more or less homogenous matter (Fig. 6d) that was wet and finely ground (see Heiss 2014) when charred. It may represent either dough for bread making or, most probably, the mash for starting beer production (see Samuel 2000). Beer production is also suggested by several emmer grains showing features typical for malting (see Stika 1996, Valamoti 2017) found during the initial study of material from another installation (HK24B) at Hierakonpolis by A.G. Fahmy. The high frequencies of emmer wheat remains in archaeological sites from all over Egypt demonstrate that this crop played a very important role in the agricultural economy from Predynastic to Ptolemaic (304–30 BC) times (see Fahmy et al. 2011).

Conclusions

The botanical assemblages in the studied samples correspond well with previous findings on the Predynastic economy and environment at HK11C (Fahmy et al. 2011). The palaeoethnobotanical results from this study show that the economy of Predynastic Hierakonpolis was based on the cultivation of cereals: emmer wheat and barley. Both crops were cultivated in winter in local fields along the Nile, as shown by the weed assemblages. The fuel used in the studied cooking installation contained wood of *Acacia* and *Tamarix*, but also crop threshing residues and dung. Archaeobotanical analysis of the dung remains indicate that crop cleaning refuse and barley grains were one of the main components of the animal fodder. The archaeobotanical assemblages reflect not only cultivated plants, but also those coming from the wetlands and the desert.

Archaeobotanical finds from vat contents provide the oldest direct evidence of the cereal ingredients and procedures for beer production in ancient Egypt, since radiocarbon testing of a sample of the residue from inside Vat 4 resulted in a calibrated date of 3762–3537 BC (uncal. C^{14} 4875 ± 40 BP) (Baba and Friedman 2016). The residues are composed in part of entire unthreshed emmer grains, some preserved as such and others strongly fragmented and only recognisable through the fragments of cereal bran and chaff fragments. Observations on the microstructure of the emmer grains from the residue indicate changes in the structure of the starch grains pointing to traces of fermentation and provide strong evidence together with the archaeological context for interpreting the charred residues as the remains of beer brewing.

Acknowledgements. This archaeobotanical study was possible due to the kind scientific and financial support provided by unit "Quaternary Environments and Humans" of the Royal Belgian Institute for Natural Sciences (Brussels). The excavations at HK11C were undertaken under the auspices of the Hierakonpolis Expedition with funds provided by the Japan Society for Promotion of Science (Grant-in-Aid for Scientific Research (C) 16K03167). The authors would like also to thank Andreas Heiss for valuable comments and discussion on earlier versions of the manuscript. We are also grateful to the editor, Catherine D'Andrea, the reviewers Michelle Wollstonecroft, Lara Carretero Gonzalez and an anonymous reviewer who provided useful comments and suggestions to improve the manuscript. The research on the material presented here represents a continuation of the research begun by Ahmed G. Fahmy focussing on the investigation of the food ingredients and processing procedures (including beer production) at the site of Hierakonpolis, aiming to define the parameters involved in food production through systematic archaeobotanical analyses of the industrial facilities, ceramic vats, and burials (comparing e.g. human dental calculus, stomach contents, intestinal contents, and coprolites).

References

Baba M, Friedman RF (2016) Recent excavations at HK11C, Hierakonpolis. In: Adams MD (ed.), Midant-Reynes B, Ryan, EM, Tristant Y (coll) Egypt at its origins 4. Proceedings of the fourth international conference "Origin of the State. Predynastic and Early Dynastic Egypt", New York, 26–30th July 30, 2011. Orientalia Lovaniensia Analecta 252. Leuven/Paris/Bristol, CT:179–205

Baba M, Van Neer W, De Cupere B (2017) Industrial food production activities during the Naqada II period at HK11C, Hierakonpolis. In: Midant-Reynes B, Tristant Y (eds) Ryan EM (coll), Egypt at its origins 5. Proceedings of the fifth international conference "Origin of the State. Predynastic and Early Dynastic Egypt", Cairo, 13–18th Apr 2014. Orientalia Lovaniensia Analecta 260. Leuven/Paris/Bristol, CT:3–34

Boulos L (2002) Flora of Egypt, vol 3. El Hadara Publishing, Cairo

Fadl M, Fahmy AG, Omran W (2013) Evaluation of cultivated and wild plant macroremains from a predynastic temple in Hierakonpolis–Upper Egypt. Intern J Plant Soil Sci 2:244–262

Fahmy AG (2005) Missing plant macroremains as indicators of plant exploitation in Predynastic Egypt. Veget Hist Archaeobot 14:287–294

Fahmy AG (2008) Analysis of mummies' gut contents from Predynastic Hierakonpolis, Egypt (3750–3300 BC). In: Midant-Reynes B, Tristant, Y (eds) Rowland J, Hendrickx S (coll) Egypt and its origins 2. Proceedings of the international conference "Origin of the State. Predynastic and Early Dynastic Egypt", Toulouse (France), 5–8th September 2005. Orientalia Lovaniensia Analecta 172. Leuven/Paris/Dudley, MA:419–426

Fahmy AG, Fadl M (2009) Plant macroremains from locality HK29A at Hierakonpolis, Egypt. J Am Res Center Egypt 45:137–152

Fahmy AG, Friedman R, Fadl M (2008a) Archaeobotanical studies at Hierakonpolis locality HK6: The pre and early dynastic elite cemetery. Archéo-Nil 18:169–183
Fahmy AG, Khodary S, Fadl M et al (2008b) Plant macroremains from an elite cemetery at Predynastic Hierakonpolis, upper Egypt. Intern J Bot 4:205–212
Fahmy A, Friedman R, Fadl M (2011) Economy and ecology of Predynastic Hierakonpolis, Egypt: Archaeobotanical evidence from a trash mound at HK11C. In: Fahmy A, Kahlheber S, D`Andrea C (eds) Windows on the African past: Current approaches to African archaeobotany. Proceedings of the 6th international workshop on African archaeobotany, held June 13–15, 2009, at Helwan University, Cairo, Egypt. Reports in African Archaeology vol 3, Africa Magna, Frankfurt, pp 91–118
Friedman RF (2011) Hierakonpolis. In: Teeter E (ed) Before the pyramids. The origins of Egyptian civilization. Oriental Institute Publications 33. Chicago, pp 33–44
Heiss AG (2014) Ceremonial foodstuffs from prehistoric burnt-offering places in the alpine region. In: Chevalier A, Marinova E, Peña-Chocarro L (eds) Plants and people: choices and diversity through time. Early agricultural remnants and technical heritage (EARTH): 8,000 Years of Resilience and Innovation, vol 1. Oxbow Books, Oxford, pp 343–353
Heiss AG, Antolín F, Bleicher N, Harb C, Jacomet S, Kühn M, Marinova E, Stika H, Valamoti S (2017) State of the (t)art. Analytical approaches in the investigation of components and production traits of archaeological bread-like objects, applied to two finds from the Neolithic lakeshore settlement Parkhaus Opéra (Zürich, Switzerland). PLoS ONE 12(8):e0182401
Kubiak-Martens L, Langer JJ (2008) Predynastic beer brewing as suggested by botanical and physicochemical evidence from Tell el-Farkha, Eastern Delta. In: Midant-Reynes Tristant Y (eds), Egypt at its origins 2. Leuven, Paris & Dudley, MA, pp 427–441
Marinova E, Ryan P, Van Neer W et al (2013) Animal dung from arid environments and archaeobotanical methodologies for its analysis: an example from animal burials of the Predynastic elite cemetery HK6 at Hierakonpolis, Egypt. Environ Archaeol 18:58–71
Samuel D (2000) Brewing and baking. In: Nicholson PT, Shaw I (eds) Ancient Egyptian materials and technology. Cambridge University Press, Cambridge, pp 537–576
Stika H-P (1996) Traces of a possible Celtic brewery in Eberdingen-Hochdorf, Kreis Ludwigsburg, southwest Germany. Veg Hist Archaeobot 5:81–88
Valamoti SM (2017) Brewing beer in wine country? First archaeobotanical indications for beer making in Early and Middle Bronze Age Greece. Veg Hist Archaeobot. https://doi.org/10.1007/s00334-017-0661-8
Valamoti S, Samuel D, Bayram M et al (2008) Prehistoric cereal foods from Greece and Bulgaria: investigation of starch microstructure in experimental and archaeological charred remains. Veg Hist Archaebot 17(1):265–276
Zohary D, Hopf M, Weiss E (2012) Domestication of plants in the Old World, 4th edn. Oxford University Press, Oxford

Grapes, Raisins and Wine? Archaeobotanical Finds from an Egyptian Monastery

Mennat-Allah El Dorry(✉)

Head of the Minister's Scientific Office, Ministry of Antiquities, Cairo, Egypt
m.eldorry@gmail.com

Abstract. The study of grape (*Vitis vinifera* L.) pressing remains recovered archaeologically has been the focus of research in some areas of the Mediterranean. In Egypt, however, such remains are only seldom reported. The first substantial find of grape pressing remains in Egypt has been uncovered in the Yale Monastic Archaeology Project's excavations at the late ninth-century AD monastic settlement of John the Little in Wady al-Natrun in Egypt's Western Desert. The remains consist of pressed fruits and fruit fragments, pips, and stalks. Although grape juice can be made into a variety of products, this paper will argue that wine was a very probable end product of these grape remains. Given the history and importance of wine production in the Coptic Church, these archaeobotanical remains of grapes from John the Little provide a new and unique opportunity to investigate wine production in the Coptic Church from a different perspective. In addition to the archaeological context and the archaeobotanical remains, the textual and ethnographic records related to the manufacture and trade of wine in Egyptian monasteries will be considered.

Keywords: Coptic archaeology · Egypt · Grape-pressing · Monasticism
Vitis vinifera · Wine-making

Introduction

The by-products of pressed grapes (*Vitis vinifera* L.) have been recovered archaeobotanically in several regions of the Mediterranean basin (for example: Mangafa and Kotsakis 1996; Cartwright 2003; Margaritis and Jones 2006; Sarpaki 2012). After grapes are pressed to extract their juice, the grape-pressing by-products include pips, skins and stalks, all of which can be found archaeologically. These types of remains can have their own uses, or can be discarded as rubbish. The remains presented here were recovered and identified during excavations undertaken by the Yale Monastic Archaeology Project in Egypt's Western Desert. This assemblage presents the largest corpus of grape pressing residues recovered from archaeological sites in Egypt.

Prior to this find from John the Little, few grape pressing remains were recorded from Egypt. An example of similar material was found in the Dynasty Zero Tomb of U-j in Abydos inside imported wine jars (McGovern 2003). However, the possibility that these specimens were grape-pressing remains was deemed unlikely (Feindt 2001). Additionally, similar material has been found at Tell al-Borg (C. Malleson, pers. comm.), Middle Kingdom Saqqara Valley of the Kings (A. Fahmy, pers. comm.), and

A. M. Mercuri et al. (Eds.): *Plants and People in the African Past*,
https://doi.org/10.1007/978-3-319-89839-1_6

Persian/Roman Ayn-Ziyada Hill (C. Newton, pers. comm.). From monastic contexts, a reference is made in Monneret de Villard's publication of his excavations at the monastery of Anba Hadra in Aswan during the 1920s, informing us that squashed grapes were found dried between mats placed at a wine-pressing installation (Monneret de Villard 1927). Recent work at the site of Anba Hadra did in fact reveal a large corpus of some 1000 items of grape pressing remains from probably seventh century contexts in the monastery, identified and studied by the author.

The Monastic Settlement of Saint John the Little

Yale's archaeological investigations began at John the Little in 2007 and continued until 2012. The fieldwork aimed to understand monastic spaces and their use. The excavations have focused on a late ninth-century AD monastic residence that is part of the larger monastic settlement of Saint John the Little in the Wady al-Natrun area (Fig. 1) (Davis 2008; Brooks Hedstrom et al. 2010; Pyke and Brooks Hedstrom 2012). Wady al-Natrun is an approximately 60 × 30 km depression that lies 23 m below sea level at its lowest point (Zahran and Willis 2009). It is one of Egypt's most extensive centers for Coptic monasticism, where the earliest Coptic habitation dates to the early days of Christianity in Egypt. The excavated residential structure measures 20 × 20 m and contains 25 rooms or spaces; and it is one of the few excavated parts of the entire monastic site. The building has been cleared out of most of its mobile contents (though exactly when this happened is still not understood), leaving very little material culture that could help elucidate the lives of its residents. Remains of built platforms that could have been used for seating and/or sleeping, glass architectural elements, fragments of ceramic vessels, a few small finds (such as animal horns or a ceramic stamp), as well as both faunal and floral remains make up the corpus of finds from the residence. Additionally, 17 cooking installations have been uncovered in the residence divided over five rooms (or spaces), including bread ovens and *kanoons*, which are open-topped stove-like structures with an opening for their fuel at the bottom (Pyke and Brooks Hedstrom 2012). While several of these ovens and *kanoons* were probably used simultaneously, they are situated in different phases of use that are currently undated and are still being investigated by the team, in order to better understand their various stages of use.

Charred Grape-Pressing Remains

Archaeobotanical remains studied from the site are almost exclusively related to the use of the ovens and *kanoons*. Large ash deposits were found in and around the ovens and *kanoons*, and were sampled in full; samples ranged from 2 to 16 l. The material, connected with activities in/around the ovens and *kanoons*, is charred. The remains are concentrated in two particular ovens, which are the richest in archaeobotanical material on site. Targeted sampling was employed, where archaeologists sampled ash contexts or contexts with organic material. Samples were first sieved (with a minimum of 0.25 mm mesh) and then processed by bucket flotation. A Stereo Dissecting

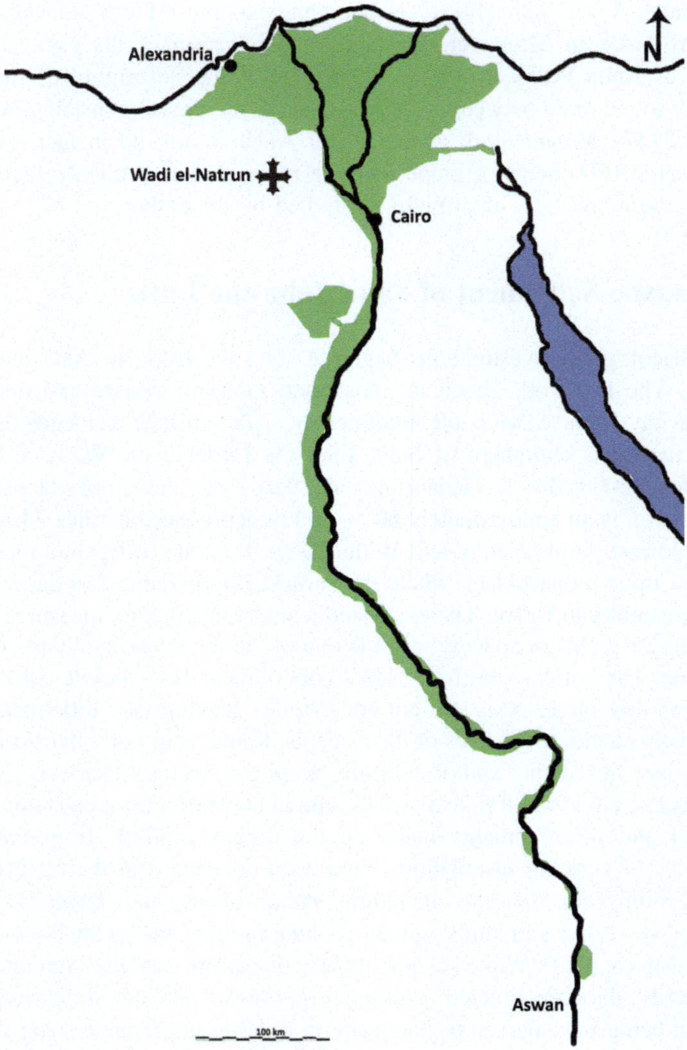

Fig. 1. Map of Egypt with location of Wady al-Natrun.

Microscope with 15–50× magnification range was used for sorting and identifying the macroremains.

Of the total assemblage of archaeobotanical material from the excavated residence (N = 12058/31 samples), a large proportion are *Vitis vinifera* remains (N = 1955), which are present in all 31 samples. The corpus of grapes is made up of mature pips

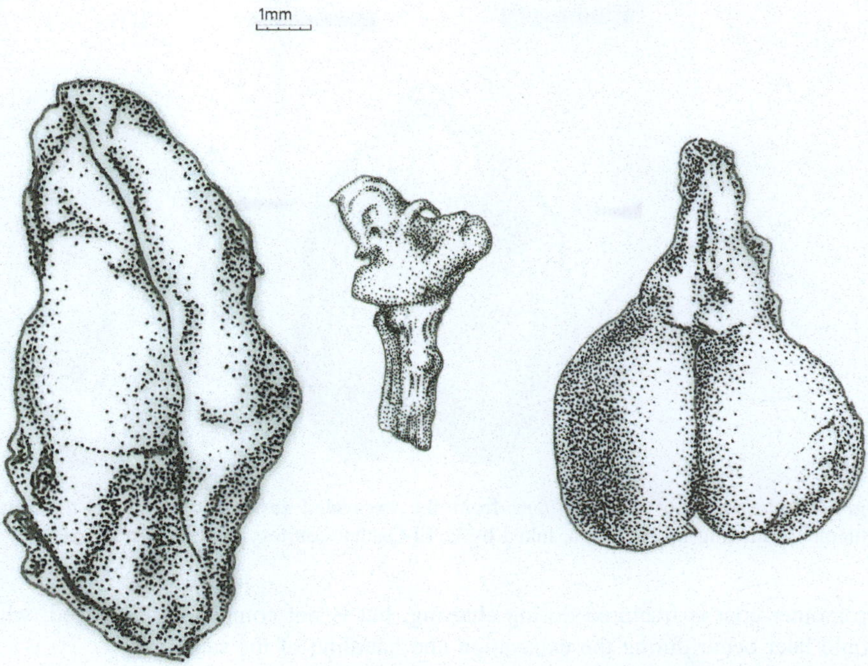

Fig. 2. Grape pressing remains, whole grapes and pips with fused skins (drawing by A. Duma, inked by E. El-Qattan, courtesy of S. Davis/YMAP)

(N = 1546), immature pips (N = 59), fruits (N = 154),[1] and pedicels (N = 150) (Fig. 2). The material is dispersed in different ovens and *kanoons*, but the largest quantity of the remains comes from one single oven. This oven provides the largest amount of items per liter, in addition to being rich in a wider variety of species.

In addition to these remains, exposed and separated grape endosperms (N = 46) were also found (Fig. 3). In some cases the pips were cracked, revealing what seems to be uncharred endosperm, thus allowing for identification. In order to check the occurrence of these remains in the samples, several intact grape pips from the samples were mechanically broken; yet none of these specimens contained such an 'endosperm'. Images of similar grape items were published by Valamoti (2015), which she identified as 'malformed' pips; however, these were desiccated. Margaritis and Jones (2006) have also recovered similar material from Hellenistic Greece, which they identified as endosperms. The Hellenistic specimens were charred. As part of Margaritis and Jones' extensive experimental charring, it was revealed that endosperms did survive in grape pips that were charred in reduced conditions at 250–300 °C. Margaritis and Jones explained that the endosperm is preserved in such a manner when a

[1] In addition to whole fruits, pip clusters are included under the count for 'fruits'. In this case, clusters of fused pips with remnants of skin adhered to them constitute one fruit.

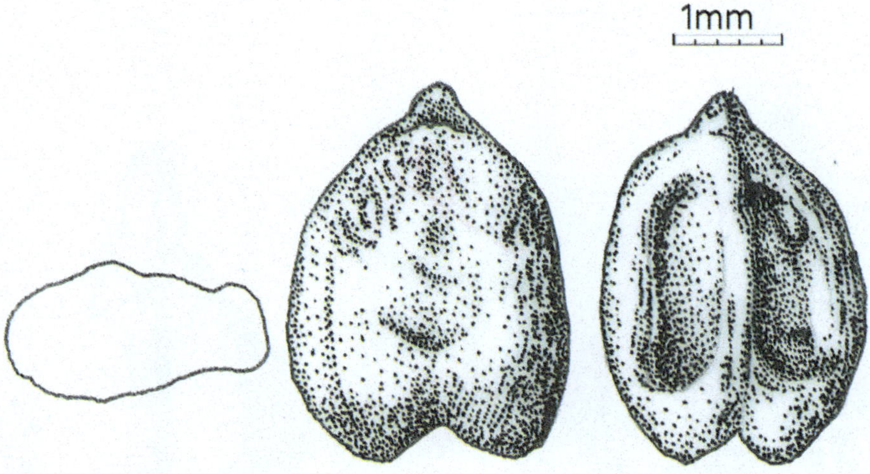

Fig. 3. Endosperms of *Vitis vinifera* from the excavated resident at St John the Little's settlement (drawing by A. Duma, inked by E. El-Qattan, courtesy of S. Davis/YMAP)

pip's outer coat is fractured during charring, but is not completely dislodged, which would later occur during the excavation and handling of the samples.

Comparing the present remains with published images enabled the identification of the Saint John the Little grape remains as most likely remains of grapes that were pressed to extract their juice (Margaritis and Jones 2006; Valamoti et al. 2007).[2] The pressed grapes in both publications have a shriveled appearance, where the skins are closely warped around the pips. This is identical to the remains from the residence at St John the Little. However, cereal processing byproducts (*Triticum durum* Desf. and *Hordeum vulgare* L. chaff) and cereal field weeds (including *Rumex* cf. *dentatus, Lolium* sp. and *Phalaris* sp.) dominated the corpus of macroremains. Identified foodstuffs include *Triticum aestivum/durum* grains, *Lens culinaris* Medick, *Vicia faba* L., *Cicer arietinum* L. and *Phoenix dactylifera* L.

Grape Pressing Remains: Evidence of Wine Production?

The presence of grape pressing remains should not automatically be regarded as evidence for wine production. Grape juice can be processed into a variety of products, including syrup, must, liquor, or vinegar (Halenko 2004; Valamoti et al. 2007). In order to argue for wine production, other data from the excavations of John the Little need to be considered, and complemented by textual, historical, and ethnographic information.

[2] I am indebted to Professor Naomi Miller who first alerted me to this kind of remain surviving archeologically and to Professor Soultana-Maria Valamoti for confirming the identification and for further discussions. Dr. Despina Moissidou generously provided me with an English translation of Greek publications I obtained from Professor Kostas Kotsakis.

First, it is important to consider the archaeological data from John the Little that can be related to wine production. So far, no wine pressing installations were found in the excavated areas of the settlement. It could be that such pressing installations have yet to be revealed, or that grapes were pressed piecemeal by hand or using small mobile presses that were moved out of the residence after abandonment. Pressing grapes by hand to make wine is known in Egypt, with special reference to wine made for the liturgy of the Coptic Church. Observations of Coptic churches in the nineteenth and twentieth centuries tell us that *Abarka (*Greek *aparxh*, meaning first fruits or offering), the thick red wine served in the liturgy of the Coptic Church, must not be tred by foot, but rather pressed by hand or small presses; and it was usually made of rehydrated raisins (Khs-Burmester 1967; Butler 1970a). Today, a Cairo-based family that makes its own *Abarka* and donates it to different churches and monasteries squeezes the grapes by hand (Nassef Family, pers. comm.). Mobile presses that are easily transported are also known (Butler 1970b). The absence of wine pressing installations is therefore not detrimental to the argument against wine production, where hand pressing or small mobile presses could have been used.

As for the ceramic evidence from the excavated residence at John the Little, all the material that has been studied so far could not be linked to wine production. Several liquid-holding type vessels were present, but none are necessarily indicative of having held wine in the past. However, the pottery from the site is still under analysis, so the above comments are preliminary. *Abarka* can be made on an as-needed basis (especially if raisins, rather than fresh grapes, were used), so that no large-scale wine pressing installations, or masses of vessels and storage space were necessary for *Abarka* production.

With the archeological data from John the Little leaving us with little to go on, historical and textual evidence can supplement the botanical and archaeological remains. Wine was a very common product of Coptic monasteries, as is attested by the large corpuses of texts related to the trade and exchange of wine, such as wine receipts and requests for wine written by or to nuns or monks (Crum and Bell 1922; Bacot 1998; Cromwell 2013). A further example of wine as a prime product of monasteries comes from the tenth century "Book of Monasteries" by Muslim writer al-Shabushty, who mentions multiple monasteries in Egypt that were popular wine drinking destinations for visitors (Atiya 1939).

The grapes at Saint John the Little could have been cultivated in Wady al-Natrun itself. There is a single mention from the ninth century AD of vineyards belonging to the neighboring Monastery of Saint Macarius in honor of the patron saint (Evelyn White 1926). The grapes could have also been sourced from the Nile Valley or Delta, which could have been transported in 24 h to Wady al-Natrun (Bagnall 1993). The grapes, whether locally sourced or imported, could have also arrived to the site as raisins, which were also used for wine production (see below).

With the popularity—and importance—of wine in Coptic monasteries, small-scale wine production seems a very likely reason for pressing the grapes. Future excavations in other parts of the monastery may furnish further evidence supporting this interpretation, or perhaps even point to large-scale wine production.

Grape-Pressing Remains: Uses and Disposal

Grape pressing wastes can be used in a variety of manners, most notably as a preservative for table grapes, as fertilizer, as fodder, and as fuel (e.g. Margaritis and Jones 2006). In the case of Saint John the Little monastery, how were grape processing byproducts used and how did they end up in the residence?

Given their recovery in the ovens and *kanoons*, the grape-pressing remains could have very well been part of the fuel. However, grape pressings do not have the same value for fuel as olive pressings or dung cakes (Foxhall 1998). In an arid environment like Wady al-Natrun with limited resources, it is a possibility that the inhabitants of the residence at the monastery of John the Little would have used the wastes of grape pressing as fuel to augment the animal dung fuel used (El Dorry 2015a; El Dorry 2015b). Alternatively, they could have been part of the animal-dung fuel due to their consumption by animals (Foxhall 1998; Miller 2008; Hentges et al. 1982). At John the Little, much of the botanical material most likely originates from fodder stuffs surviving in dung. This is a possibility, given that the remains include many species that are typical components of animal fodder during this time period, including barley, *Phalaris*, *Lolium*, and others. Additionally, many of the specimens have amorphous bits of dung adhered on to them, and several grains and seeds were found embedded in animal dung deposits. Therefore, the use of the grape wastes as fodder is a very likely suggestion. However, the good state of preservation of many of the pressed grape fruits rules out this material having passed through animal digestive systems. No experimental work has been published about the preservation of grape pressing remains in animal dung. A future element of this research would be experimental feeding of animals to study the effects of animal digestion on the remains of grape-pressing.

Although the grape residues may have had specific uses, the possibility that they were simply discarded into the ovens or *kanoons* should be entertained. Grape-pressing remains could have simply been discarded in the ovens as wastes with no further use. The frequency of wine or *Abarka* production should be considered here, but it is difficult to be certain how much wine was produced and how often. Wine for drinking purposes is more conveniently produced as a large-scale endeavor during the grape season, resulting in a large mass of waste. *Abarka* production can either be an annual large-scale event or piecemeal and made throughout the year made from stored raisins. A piecemeal production would have resulted in fewer remains that are more likely to have been discarded than used.

The larger portion of the grape pressing remains come from the ovens, rather than the *kanoons*, especially the ovens in one part of the building. This could be attributed to a preferential use in the forms of fuel employed in the *kanoons* versus the ovens. However, a more likely scenario, based on the frequency and pattern of dispersal of other species, would be that the *kanoons* were simply more frequently cleaned out than the ovens.

Alternative possibilities can be considered, despite their unlikelihood. For example, in Classical Greece, wine press-cake could have been used for preserving table grapes, by covering the grapes with the pressing remains to create a humid and cool environment (Foxhall 1998). The use of these residues as preservative could have resulted in their disposal in the ovens and *kanoons* as rubbish. However, such a form of preservation is not known from Egyptian literature. Another possibility is the use of grape pressing residues as fertilizer. Vineyards in both Egypt and Lebanon today use grape pressing wastes as fertilizer (Labib Kallas, Winery Manager/EGYBEV, pers. comm.; Winery Representative/Ksara Winery in the Bekaa Valley, Lebanon pers. comm.). This use seems unlikely for the archaeobotanical grape pressing remains from John the Little, as their use as fertilizers would not have resulted in the material being burned in the ovens.

Wine Production in the Coptic Church

Although the trade of wine in Coptic monasteries has been the topic of previous research, some aspects of the production of wine remains little understood (Crum and Bell 1922; Bacot 1998; Cromwell 2013), such as the use of raisins, rather than fresh grapes, for the production of *Abarka* for the liturgy (Drower 1956; Khs-Burmester 1967; Butler 1970a; Gabra 2009). While viticulture has a long history in Egypt, not much is known about the use of raisins in wine production. The use of raisins in wine, rather than fresh grapes, is said to have been a result of the repetitive bans set forth by Muslim rulers against the production of wine, the trade of grapes, and even the mere possession of grapes risked the bearer with legal repercussions during the eleventh to twelfth centuries AD (El Dorry 2016). Scholars have argued for raisins replacing fresh grapes for wine production during these prohibitions (Leeder 1918; Butler 1970a). Could archaeobotanical analysis help detect whether the archaeobotanical remains recovered from the excavated residence at John the Little were raisins or fresh grapes? To address this question, future analysis comparing grape-pressing wastes and raisin-pressing wastes will be necessary.

Conclusions

This find of archaeobotanical macroremains of grape pressings from an excavated residence building at the monastic settlement of John the Little is the first of its kind from a monastic context in Egypt. This is further the most substantial find of grape pressing residues from Egypt. The question of whether these remains were meant for wine (or *Abarka*) or for another grape juice-based product still remains open, but the importance of wine in Coptic liturgy and its importance as an economic commodity from monasteries, makes wine a very probable end product. However, as long as archaeological evidence confirming an economic production is missing, we may have to assume that the production may have been at least for self-supply. In order to be preserved in the ovens, the remains of the grape pressing could have been used in a variety of ways, but they could have very well just been rubbish disposed of in the fire. The grape pressing remains at Little John have brought up many questions about the origins and history of wine production in the Coptic Church; for example, how did the tradition of raisins being used in lieu of fresh grapes for *Abarka* start and why did it disappear? To answer these questions, future work will focus on experimental charring of more raisin-pressing remains, in addition to the analysis of the modern wastes of *Abarka* production. A detailed ethnographic and historical exploration of *Abarka* production is also necessary.

Acknowledgements. I am greatly indebted to Professor Stephen Davis (Yale) for allowing me to study the material. Dr. Gillian Pyke and Professor Darlene Brooks Hedstrom have provided much support during this research. I am grateful to Meredith Brand for proofreading the text. Feedback from the editors and anonymous reviewers has been of immense help in formulating this paper, and I am indebted to their input.

References

Atiya AS (1939) Some Egyptian monasteries according to the unpublished MS. of Al-Shābushtī's "Kitāb al-Diyārāt". B Soc Archéol Copte 5:1–28

Bacot S (1998) La circulation du vin dans les monastères d'Égypte à l'époque copte. In: Grimal N, Menu B (eds) Le commerce en Égypte ancienne. IFAO, Cairo, pp 269–288

Bagnall RS (1993) Egypt in late antiquity. Princeton University Press, Princeton

Brooks Hedstrom DL, Davis SJ, Herbich T et al (2010) New archaeology at ancient scetis: surveys and initial excavations at the Monastery of St. John the Little in Wady al-Natrun. Dumbart Oaks Pap 64:217–227

Butler AJ (1970a) The ancient coptic churches of Egypt II. Clarendon Press, Oxford

Butler AJ (1970b) The ancient coptic churches of Egypt I. Clarendon Press, Oxford

Cartwright CR (2003) Grapes or raisins? An early Bronze Age larder under the microscope. Antiquity 77:345–348

Cromwell J (2013) Wine and monks in Christian Egypt. In: British Museum blog. http://blog.britishmuseum.org/2013/07/10/wine-and-monks-in-christian-egypt/. Accessed 11 Jul 2013

Crum WE, Bell HI (1922) Wadi Sarga: Coptic and Greek texts from the excavations undertaken by the Byzantine Research Account. Gyldendalske Boghandel-Nordisk Forlag, Copenhagen

Davis SJ (ed) (2008) The Arabic Life of St. John the Little by Zacharias of Sakhā. Saint Mark Foundation and Saint Shenouda the Archimandrite Coptic Society, Los Angeles, California

Drower ES (1956) Water into wine: a study of ritual idiom in the Middle East. John Murray, London

El Dorry M-A (2015a) Monks and plants: a study of foodways and agricultural practices in Egyptian monastic settlements. Dissertation, Westfälische Wilhelms-Universität Münster

El Dorry M-A (2015b) Monks and plants In: Hawel M, Dopller L, Fischer-Schröter P et al (eds) Work in progress. Work on progress. Beiträge kritischer Doktorand_innen-Jahrbuch 2015 der Rosa-Luxemburg-Stiftung. Rosa-Luxemburg-Stiftung/VSA Verlag, Hamburg, p 218–227

El Dorry M-A (2016) Wine production in medieval Egypt: the case of the coptic church. In: Ayad MF (ed) Studies in coptic culture: transmission and interaction, p 92–106

Evelyn White HG (1926) The monasteries of the Wadi'n Natrûn, part III: the architecture and archaeology. The Metropolitan Museum of Art, New York

Feindt F (2001) Weintrauben-und Feigen in Gefäßen aus dem Grab U-j in Abydos (Umm el-Qaab). In: Hartung U (ed) Umm el-Qaab II: Importkeramik aus dem Friedhof U in Abydos (Umm el-Qaab) und die Beziehungen Ägyptens zu Vorderasien im 4. Jahrtausend v. Chr. Philip von Zabern, Mainz, p 391–398

Foxhall L (1998) Snapping up the unconsidered trifles: the use of agricultural residues in ancient Greek and Roman farming. Environ Archaeol 1:35–40

Gabra G (2009) The A to Z of the coptic church. Scarecrow Press, Lanham, Maryland

Halenko O (2004) Wine production, marketing and consumption in the Ottoman crimea, 1520-154. J Econ Soc Hist Orie 47:507–547

Hentges HF, Bates RP, Wasdin JG (1982) Acceptability of grape winery pomace as a cattle feed. University of Florida Report, Gainesville

Khs-Burmester OHE (1967) The Egyptian or coptic church: a detailed description of her liturgical services and the rites and ceremonies observed in the administration of her sacraments. La Société d'Archéologie Copte, Le Caire

Leeder SH (1918) Modern sons of the pharaohs: a study of the manners and customs of the copts of Egypt. Hodder and Stoughton, London

Mangafa M, Kotsakis K (1996) A new method for the identification of wild and cultivated charred grape seeds. J Archaeol Sci 23:409–418

Margaritis E, Jones M (2006) Beyond cereals: crop processing and *Vitis vinifera* L. Ethnography, experiment and charred grape remains from Hellenistic Greece. J Archaeol Sci 33:784–805

McGovern PE (2003) Ancient wine: the search for the origins of viniculture

Miller NF (2008) Sweeter than wine. Antiquity 82:937–946

Monneret de Villard U (1927) Il monastero di S. Simeone Presso Aswân. Tipografia E. Libr. Pontificia Arcivescovile S. Giuseppe, Milano

Pyke G, Brooks Hedstrom DL (2012) The afterlife of Sherds: architectural reuse strategies at the Monastery of John the Little, Wadi Natrun. In: Bader B, Ownby M (eds) Aspects of Egyptian ceramics within their archaeological context. Peeters Publishers, Leuven

Sarpaki A (2012) Re-visiting the visibility of the grape, grape products, by-products and some insights of its organization from the Prehistoric Aegean, as guided by New Evidence from Monastiraki, Crete. IANSA 382:211–220

Valamoti SM (2015) Harvesting the "wild"? Exploring the context of fruit and nut exploitation at Neolithic Dikili Tash, with special reference to wine. Veg Hist Archaeobot 24:35–46

Valamoti SM, Mangafa M, Koukouli-Chrysanthaki C et al (2007) Grape-pressings from northern Greece: the earliest wine in the Aegean? Antiquity 81:54–61

Zahran MA, Willis AJ (2009) The vegetation of Egypt. Springer, Berlin

The Role of Morphometry to Delineate Changes in the Spikelet Shape of Wild Cereals: The Case Study of Takarkori (Holocene, Central Sahara, SW Libya)

Rita Fornaciari[1,2(✉)], Laura Arru[2], Rita Terenziani[1], and Anna Maria Mercuri[1]

[1] Laboratorio di Palinologia e Paleobotanica, Dipartimento di Scienze della Vita, Università di Modena e Reggio Emilia, Modena, Italy
rita.fornaciari@unimore.it
[2] Dipartimento di Scienze della Vita, Università di Modena e Reggio Emilia, Reggio Emilia, Italy

Abstract. A morphometrical study of hundreds of spikelets recovered from archaeological deposits of Takarkori (SW Libya) provides data on the presence and size variations of wild cereals gathered by hunter-gatherers in the central Sahara during the Early and Middle Holocene (c. 10,200–c. 4600 cal yr BP). Spikelets of *Panicum laetum*, *Echinochloa colona* and *Sorghum bicolor* subsp. *verticilliflorum*, found in 18 seed/fruit concentrations, are measured using image analysis techniques. These data demonstrate that the archaeobotanical specimens have a similar typology, maturity stage and are of a uniform size, suggesting that they were selected by the human groups living in the area. Indeed, the spikelets of two samples recovered from sediments excavated elsewhere on the site compared to those from the seed concentrations, show a smaller size and greater variation in maturation status. Results are compared to metrical data obtained from modern species.

Keywords: Archaeobotany · Archaeology · Morphometry · *Echinochloa* *Panicum* · *Sorghum* · Takarkori · Libya · Central Sahara

Introduction

Many researchers consider morphometric analysis to be an effective tool to investigate the nature of plants found in archaeological sites (Dahlberg and Wasylikowa 1996; Wasylikowa 1997; Fahmy 2001; Willcox 2004; Bacchetta et al. 2008; Bandini Mazzanti et al. 2012; Chen et al. 2016). Morphometric analysis assists in species (or subspecies) identification and the characterization of plant harvesting or cultivation in a certain area (spatial pattern) during a particular period (chronological pattern) (Renfrew 1973; Pearsall 1989; Zohary and Hopf 2000; Bosi et al. 2007; Orrù et al. 2013; Mercuri et al. 2015: 252).

Morphometric studies have been carried out by several archaeobotanical projects. In North Africa, Dahlberg and Wasylikowa (1996) classified charred *Sorghum* grains from the archaeological site of Nabta Playa (Egypt) through image-analysis. A wild

A. M. Mercuri et al. (Eds.): *Plants and People in the African Past*,
https://doi.org/10.1007/978-3-319-89839-1_7

type of *Sorghum* was identified through statistical analysis of the metrical data. In addition, Fahmy led a palaeoethnobotanical study on plant remains from a Neolithic settlement in Farafra Oasis (Egypt). The analysis of morphological features and measurements of carbonised finds revealed that grasses were highly represented and mainly retrieved from hearths. Results supported the hypothesis that the subsistence strategy of Early Neolithic populations in the eastern Sahara included both animal herding and gathering of wild plants for food (Fahmy 2001). Willcox (2004) measured breadth and thickness of archaeological charred grains of wild einkorn and/or rye and barley from the Euphrates valley. Results showed an increase in spikelet size between older and younger levels of two Early Neolithic sites, demonstrating that the morphometric analysis of plant macroremains can detect morphological domestication from early farming sites. Given the importance of this methodological approach, we decided to study the large number of wild cereal spikelets collected by Holocene hunter-gatherers, excavated from the archaeological site of Takarkori, in the Libyan central Sahara. This paper reports on measurements of the maximum and minimum diameters of specimens covering a period of about 5000 years, an approach useful to observe possible trends in their morphological features. The use of a small set of parameters makes results reproducible and suitable to be used for future comparisons.

During the Early, Middle and Late Holocene of central Sahara, several archaeological sites have demonstrated that people continuously collected plants for use as food and fuel. They harvested available species in a changing environment (Wasylikowa 1997; Wasylikowa and Dahlberg 1999; Mercuri 2008a). In the Tadrart Acacus, first hunter-gatherers and then pastoral communities accumulated wild cereals in their shelters, suggesting that these plants were of critical importance in human subsistence and interrelationships with the environment (e.g. di Lernia 1999; Mercuri 2008a; Dunne et al. 2016). Indeed, wild cereals were selected for food (especially grasses with high nutritional value), fodder, to weave baskets and were sometimes used in ritual activities, as votive offerings or for the extraction of dyes (Mercuri and Garcea 2007; Mercuri 2008b, di Lernia et al. 2012).

Accumulations of large numbers of spikelets from discrete contexts enable the application of a range of statistical techniques (Fahmy 2001; Tanno and Willcox 2006; Bandini Mazzanti et al. 2012). The aim of this paper is to investigate morphometrical variations in desiccated remains from “plant accumulations” (see Olmi et al. 2012: 177) found in the archaeological site of Takarkori, during millennia of human occupation. The plant records appeared unique at first sight, because of the large number of accumulated spikelets and the excellent preservation of the macroremains. Spikelets belonging to the Poaceae family are prevalent in the record, and morphometrical data are obtained from dried spikelets of *Panicum laetum* Kunth, *Echinochloa colona* L. and *Sorghum bicolor* L. Moench subsp. *verticilliflorum* (Steud.) de Wet ex Wiersema and J. Dahlb. (syn. *Sorghum arundinaceum* (Desv.) Stapf.). Data analysis is designed to describe variation within and between the samples from Early to Middle Holocene. In addition, a comparison with other ancient and modern specimens is undertaken in this study.

The Study Area

The excavation of Takarkori was undertaken by the Italian-Libyan Archaeological Mission in the Acacus and Messak of the Sapienza University of Rome, directed by Savino di Lernia, from 2003 to 2006. The chronology of the site ranges from c. 10,200 to c. 4600 cal yr BP (Biagetti and di Lernia 2013; Cherkinsky and di Lernia 2013; Fig. 1). Takarkori is a rock shelter located on the southern side of a large valley connecting the Tanezzuft valley to the eastern fringes of the Acacus Mountains. The site is situated on a terrace (c. 2200 m^2), opening westward and sheltered to the east by a 30 m high cliff. The archaeological deposit is concentrated in the inner part of the shelter (c. 200 m^2; Biagetti et al. 2004). A total of 143 m^2 of soil was excavated using standard techniques, with contexts digitally recorded by means of an electronic total station (ETS) followed by digital post processing in a 3D environment (Biagetti and di Lernia 2013). From an archaeobotanical point of view, the shelter deposit is largely a result of plant accumulation by humans combined with micro-environmental processes (Cremaschi et al. 2014). The particular depositional setting at Takarkori reveals an extraordinary state of preservation of organic materials, which allowed preservation at the molecular level in addition to plant macroremains (Tafuri et al. 2006; di Lernia et al. 2012, 2016; Dunne et al. 2012, 2016; Olmi et al. 2012; di Lernia and Tafuri 2013; Fornaciari et al. 2016).

Fig. 1. Study area: **a** Map of the area, provided by the Italian-Libyan Mission in the Acacus and Messak (central Sahara) of the University Sapienza of Rome. The archaeological site of Takarkori is indicated by the arrow; **b** The Takarkori rock shelter (modified from Olmi et al. 2012)

At present, the area surrounding Takarkori has the typical hot-arid subtropical climate of desert areas. The region is less known from a botanical point of view, therefore even early studies remain relevant and are important for knowledge of the area (e.g. Corti 1942). Overall, the flora of Fezzan is composed of Saharo-Sindian species, with well-represented (but sporadic) Mediterranean elements, which are commonly encountered in mountainous regions of the central Sahara. Poaceae and Fabaceae are the most common plant families represented in the present day Saharan flora (Ozenda 1958, 2000). Poaceae, including wild cereals, include *Panicum*, *Setaria*, *Pennisetum* and other 21 genera which are sometimes harvested for different uses even today (Corti 1942; Table 1).

Table 1. Poaceae (Gramineae) listed by Corti (1942). Species are reported following the order in the floristic list of the flora.

Flora and vegetation of Fezzan and Ghat region (Corti 1942)				
Genus	Species	var.	Author	Ethnobotanical notes
Panicum	*repens*		(L.)	
Panicum	*turgidum*		(Forsk.)	Pasture; seeds eaten by Tuareg; ashes from the roots are blended with tobacco; ashes from the roots are blended with old dryed trunk as cicatrizant
Setaria	*verticillata*		(L.)	
Pennisetum	*dichotomum*		(Forsk.)	
Pennisetum	*dichotomum*	*subplumosum*	(Hack.)	Eaten by camels; seeds are edible also for men
Imperata	*cylindrica*		(L.)	Consolidation of the dunes; cellulose extraction; fodder
Saccharum	*ravennae*	*parviflorum*	(L.)	Animal feed; exude a sugary liquid used as sugar by Tuareg
Rottboellia	*hirsuta*		(Forsk.)	Pasture for camels, ovines and bovines
Cymbopogon	*schoenanthus*		(L.)	Basal parts of the plant are burned for fumigations or macerated with oil and used against "ahon"; eaten by camels
Eremopogon	*foveolatus*		(Del.)	
Hyparrhenia	*hirta*		(L.)	
Leersia	*hexandra*		(Sw.)	
Phalaris	*minor*	*typica*	(Retz.)	
Aristida	*coerulescens*		(Desf.)	
Aristida	*obtusa*		(Delile)	
Aristida	*plumosa*		(L.)	
Aristida	*plumosa*	*floccosa*	(Dur. et Schinz)	Excellent pasture
Aristida	*pungens*		(Desf.)	Excellent pasture, especially at blooming; Fruits are used as food for humans and camels; Leaf are used for making ropes; stabilize shifting sands
Stipa	*parviflora*		(Desf.)	

(*continued*)

Table 1. (*continued*)

Flora and vegetation of Fezzan and Ghat region (Corti 1942)				
Genus	Species	var.	Author	Ethnobotanical notes
Sporobolus	*spicatus*		(Vahl.)	
Polypogon	*monspeliensis*		(Desf.)	
Polypogon	*exilis*		(Pamp.)	
Danthonia	*forskahlii*		(Vahl.)	Excellent nutriment for camels and domestic erbivorous
Cynodon	*dactylon*	*typica*	(L.)	
Cynodon	*minor*		(Beck.)	Fodder
Phragmites	*communis*		(Trin.)	
Phragmites	*communis*	*isiacus*	(Del.)	
Phragmites	*communis*	*stenophylla*	(Boiss.)	
Koeleria	*phleoides*	*typica*	(Vill.)	
Koeleria	*phleoides*	*pseudolobulata*	(Degen et Domin)	
Eragrostis	*bipinnata*		(L.)	Not cultivated; dried and interweaved for making "zeriba"; eaten by goats, sheeps and camels
Aeluropus	*litoralis*		(Gouan.)	
Schismus	*calycinus*	*genuinus*	(L.)	Excellent fodder for rams, goats and camels
Scleropoa	*memphitica*	*genuina*	(Parl.)	
Lolium	*multiflorum*	*gaudini*	(Lam.)	
Lolium	*rigidum*		(Gaud.)	Fodder

Materials and Methods

Sample Selection

The specimens measured for this study were found in 18 "seed accumulations" (Olmi et al. 2012; Fig. 2) visible with the unaided eye on the surface of the excavation. Three species of wild cereals were selected from each accumulation: *P. laetum*, *E.colona* and *S. bicolor* subsp. *verticilliflorum*. The first 50 intact seeds for each species found in each accumulation were selected for analysis in order to include all morphological types. The taxonomic identification of specimens was completed in a previous study that coupled morphological and molecular (ancient DNA) analyses (see Olmi et al. 2012; Fornaciari et al. 2016). Depending on its chronology, each accumulation was labelled with a letter in alphabetical order (from "a" to "r" in Table 2). The spikelet record from the accumulations was compared to spikelets found in two samples (L77 O30 and L249 R22), originating from the sifting of a known volume of sediment during the excavation, labelled L77 and L249 (Table 2). These last archaeobotanical

Fig. 2. Seeds/fruits accumulation n. 8136 (Middle Holocene) found at Takarkori (Photo by the Italian-Libyan Mission, 2004, courtesy of the University Sapienza of Rome)

remains were dry sieved with stacked meshes sizes (10, 0.5, 0.2 mm) and are now conserved at the Laboratory of Palynology and Palaeobotany of Modena (Italy). Based on the cultural chronology, L249 R22 dates to the Late Acacus Phase (10,170–8180 cal yr BP), while L77 O30 belongs to the Early Pastoral Phase (8000–6890 cal yr BP; Cherkinsky and di Lernia 2013). Fertile and intact spikelets from the accumulations, as well as from the two fine meshes of samples L77 and L249, were measured. Each spikelet consists of lemma and palea including one grain, without external upper and lower glumes. In all samples, 50 specimens of each taxon were measured. In six cases (Table 3 = *P. laetum*: "f" and "g", 40 specimens; L249, 15 specimens. *E. colona*: "q", 20 specimens. *S. bicolor* subsp. *verticilliflorum*: L77, 35 specimens; L249, 15 specimens) the number of measured specimens was smaller, as a result of their availability in the sample.

Morphometrical Analysis

Samples were examined using a Leica Wild M10 stereomicroscope. Spikelets, grains and isolated glumes of *P. laetum*, *E. colona* and *S. bicolor* subsp. *verticilliflorum* were separated from soils, sorted and identified. Photos of spikelets in dorsal view were taken using a Leica MC170 HD camera connected to the stereomicroscope. All images were made with consistent parameters (default options for white background), and magnification of the stereoscope (9.0x). Maximum diameter (Dmax) and minimum diameter (dmin) of each spikelet (Fig. 3) were measured from the images using Leica Application Suite—LAS EZ version 3.0.0 software (Leica Microsystems). A further comparison between archaeological macroremains found at Takarkori and modern specimens was conducted based on comparison with data from the literature (see Results for references). Because the dmin measure is often not reported, it was calculated from drawings/photos (depending on their availability in the publication) in

Table 2. List of the plant accumulations selected for this study, their label and their chronology

Wild cereal species	Sample	Label	Chronology	
			Archaeological chronology (Cherkinsky and di Lernia 2013)	Holocene period (according to Walker et al. 2012)
Panicum laetum	1857	b	Late Acacus	Early
	6759	d	Middle Pastoral	Middle
	6713	e		
	9601	f		
	6714	g		
	8136	h		
	6715	i		
	2140	j		
	6725	k		
	9434	l		
	1291	m		
	1563	n		
	2908	p	Late Pastoral	
Echinochloa colona	6759	d	Middle Pastoral	Middle
	6713	e		
	9601	f		
	6714	g		
	8136	h		
	6715	i		
	2140	j		
	6725	k		
	9434	l		
	1291	m		
	1563	n		
	5171	o	Late Pastoral	
	2908	p		
	4556	q		
	1272	r		
Sorghum bicolor subsp. *verticilliflorum*	1423	a	Late Acacus	Early
	7040	c		

ventral view of the consulted literature. A digital image of the figures from the literature was acquired using the scanner Epson Perfection V700 Photo (Epson), and measurements were taken with the same software used for ancient spikelets.

Table 3. Samples, number of fertile spikelets measured and taken measures: average value of Dmax and dmin, standard deviation, maximum and minimum values

Sample	n. of records	Average value (mm)		Standard deviation (mm)		Maximum value (mm)		Minium value (mm)	
		Dmax	dmin	Dmax	dmin	Dmax	dmin	Dmax	dmin
Panicum laetum	**645**	**2.01**	**1.22**	**0.09**	**0.10**	**2.19**	**1.63**	**1.57**	**0.98**
1857	50	1.94	1.06	0.09	0.09	2.16	1.37	1.70	0.90
6759	50	2.02	1.24	0.07	0.11	2.20	1.86	1.41	0.93
6713	50	2.03	1.21	0.09	0.13	2.21	1.84	1.42	0.97
9601	40	2.04	1.29	0.06	0.08	2.19	1.88	1.45	1.08
6714	40	2.04	1.22	0.11	0.13	2.27	1.46	1.65	0.99
8136	50	2.03	1.26	0.06	0.06	2.19	1.38	1.85	1.07
6715	50	2.00	1.19	0.11	0.12	2.20	1.40	1.73	0.86
2140	50	2.06	1.26	0.08	0.09	2.20	1.43	1.79	1.03
6725	50	2.01	1.14	0.09	0.13	2.22	1.83	1.40	0.89
9434	50	2.04	1.26	0.07	0.08	2.19	1.83	1.40	1.03
1291	50	2.01	1.24	0.07	0.13	2.17	1.80	1.46	0.97
1563	50	1.98	1.27	0.12	0.10	2.19	1.68	1.43	1.02
2908	50	1.95	1.24	0.11	0.10	2.13	1.41	1.69	0.96
		1.92	**1.18**	**0.07**	**0.10**	**2.04**	**1.35**	**1.77**	**0.96**
L249 R22	15	1.92	1.18	0.07	0.10	2.04	1.35	1.77	0.96
Echinochloa colona	**820**	**2.11**	**1.32**	**0.13**	**0.09**	**2.40**	**1.75**	**1.60**	**1.09**
6759	50	2.11	1.34	0.12	0.09	2.43	1.90	1.51	1.17
6713	50	2.11	1.31	0.10	0.09	2.29	1.80	1.47	1.07
9601	50	2.08	1.32	0.10	0.09	2.31	1.86	1.52	1.11

(continued)

Table 3. (*continued*)

Sample	n. of records	Average value (mm)		Standard deviation (mm)		Maximum value (mm)		Minium value (mm)	
		Dmax	dmin	Dmax	dmin	Dmax	dmin	Dmax	dmin
6714	50	2.10	1.35	0.11	0.09	2.40	1.83	1.49	1.02
8136	50	2.07	1.36	0.12	0.07	2.33	1.53	1.86	1.24
6715	50	2.09	1.28	0.12	0.08	2.37	1.43	1.87	1.07
2140	50	2.11	1.34	0.11	0.09	2.33	1.49	1.84	1.12
6725	50	2.08	1.32	0.11	0.09	2.39	1.81	1.50	1.07
9434	50	2.06	1.34	0.11	0.08	2.24	1.84	1.50	1.18
1291	50	2.07	1.36	0.11	0.09	2.28	1.72	1.54	1.14
1563	50	2.09	1.30	0.19	0.12	2.62	1.69	1.51	0.99
5171	50	2.15	1.34	0.13	0.10	2.41	1.91	1.54	1.12
2908	50	2.07	1.24	0.15	0.09	2.37	1.45	1.73	0.98
4556	20	2.24	1.29	0.16	0.09	2.58	2.00	1.55	1.11
1272	50	2.26	1.30	0.15	0.12	2.64	2.00	1.55	1.00
		2.30	**1.23**	**0.23**	**0.14**	**2.72**	**1.59**	**1.85**	**0.94**
L249 R22	50	2.26	1.18	0.30	0.14	2.72	1.53	1.66	0.86
L77 O30	50	2.33	1.28	0.16	0.14	2.71	1.64	2.03	1.01
Sorghum bicolor subsp. verticilliflorum	**150**	**6.41**	**2.33**	**0.35**	**0.13**	**7.14**	**2.59**	**5.62**	**2.00**
1423	50	6.43	2.46	0.43	0.14	7.38	2.76	5.47	2.10
7040	50	6.38	2.20	0.26	0.13	6.90	2.41	5.77	1.90
		5.19	**1.83**	**0.44**	**0.25**	**6.06**	**2.24**	**4.39**	**1.39**
L249 R22	15	5.07	1.69	0.48	0.30	6.01	2.24	4.26	1.26
L77 O30	35	5.31	1.97	0.39	0.19	6.10	2.23	4.52	1.52

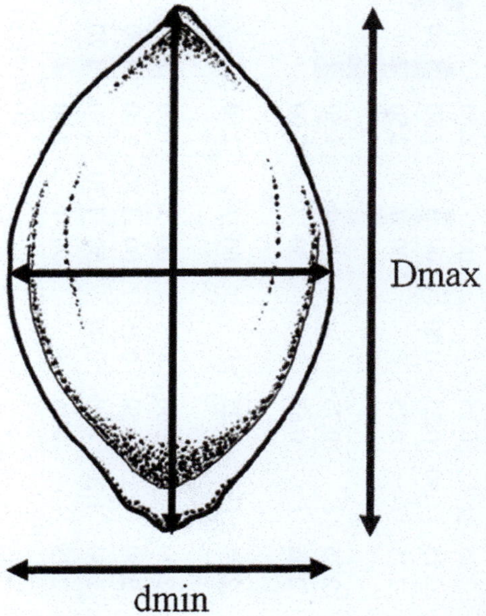

Fig. 3. Drowning of a spikelet of *Panicum laetum* showing the taken morphometrical data (Dmax = maximum diameter; dmin = minimum diameter)

Statistical Analysis

Data were imported into Microsoft Excel 2013, which was used to design the graphs. Size measurements (length and width) were statistically analysed (mean value, standard deviation, minimum and maximum value, see Liengme 2015). Data elaborations demonstrate the trend between length and width (De Leonardis et al. 2011) of seeds found in plant accumulations, ordered along a temporal scale. The linear trend has been calculated by the coefficient of determination (R^2) using linear analysis tools. Analysis of variance (one-way ANOVA) with post hoc contrast (Tukey test) was conducted to test for differences among spikelets. Differences were considered significant if $p < 0.05$ in the Tukey test. All statistical analyses were carried out with XLSTAT 2014, the software for statistical analysis in Excel. Scatter plots (Fuller et al. 2007; Kim et al. 2013) have been designed to characterise the main metrical trends in seed size.

Results

Morphometrical Analysis

Data obtained from the measurements of Dmax and dmin of the specimens from 20 samples (18 seed accumulations, L77 and L249 samples) are reported in Table 3. A total of 1615 spikelets have been examined: 645 of *P. laetum* found in 13 accumulations and in the sample L249; 820 of *E. colona* in 15 accumulations and in

Table 4. Measures of modern spikelets of *P. laetum*, *E. colona* and *S. bicolor* subsp. *verticilliflorum*: their references, label, average value of Dmax and dmin, standard deviation, maximum and minimum values

Species of wild cereal	References	Average value (mm)		Standard deviation (mm)		Maximum value (mm)		Minimum value (mm)	
		Dmax	dmin	Dmax	dmin	Dmax	dmin	Dmax	dmin
Panicum laetum	RBG Kew GrassBase (2017)	2.75	1.30[a]	0.25	0.10	3.00	1.40	2.50	1.20
	Clayton and Renvoize (1982)	2.75	/	0.25	/	3.00	/	2.50	/
	PROTA4U–Plant Resources of Tropical Africa (2017)	2.75	1.50	0.25	0.00	3.00	1.50	2.50	1.50
	Phillips (1995) Flora of Ethiopia and Eritrea, volume 7, Poaceae	2.75	/	0.25	/	3.00	/	2.50	/
Echinochloa colona	RBG Kew GrassBase (2017)	2.25	/	0.75	/	3.00	/	1.50	/
	Phillips (1995) Flora of Ethiopia and Eritrea, volume 7, Poaceae	2.65	/	0.65	/	3.30	/	2.00	/
	FAO–Food and Agriculture Organization of the United Nations (2017)	2.75	/	0.25	/	3.00	/	2.50	/
	IRRI–International Rice Research Institute (2017)	2.00	/	1.00	/	3.00	/	1.00	/
	ISTA Universal List (2017)	2.50	1.25	0.50	0.25	3.00	1.50	2.00	1.00
	Boulos (2005) Flora of Egypt, volume 4, monocotyledons	2.50	1.20[a]	1.00	0.00	3.50	1.20	1.50	1.20
	Maire (1952)	2.75	/	0.25	/	3.00	/	2.50	/

(continued)

Table 4. (*continued*)

Species of wild cereal	References	Average value (mm)		Standard deviation (mm)		Maximum value (mm)		Minimum value (mm)	
		Dmax	dmin	Dmax	dmin	Dmax	dmin	Dmax	dmin
Sorghum bicolor subsp. *verticilliflorum*	RBG Kew GrassBase (2017)[b]	6.50	/	2.50	/	9.00	/	4.00	/
	Phillips (1995) Flora of Ethiopia and Eritrea, volume 7, Poaceae[b]	7.00	/	3.00	/	10.00	/	4.00	/
	Boulos (2005) Flora of Egypt, volume 4, monocotyledons[b]	6.50	2.10[a]	2.50	0.00	9.00	2.10	4.00	2.10
	Sharp and Simon (2002)[b]	6.50	/	2.50	/	9.00	/	4.00	/
	Oliver and auct suc (1917)	6.50	/	2.50	/	9.00	/	4.00	/
	Smith and Frederiksen (2000)	6.50	/	1.50	/	8.00	/	5.00	/

[a]dmin value is obtained from drowning/pictures of the cited literature
[b]Author refers as *Sorghum arundinaceum* (syn. of *Sorghum bicolor* subsp. *verticilliflorum*)

samples L77 and L249; 150 of *S. bicolor* subsp. *verticilliflorum* from two accumulations and in samples L77 and L249.

In *P. laetum*, the average dimension is 2.01 ± 0.09 mm × 1.22 ± 0.10 mm. Dmax ranges 2.19–1.57 mm, while dmin ranges 1.63–0.98 mm in the plant accumulations. Regarding sample L249, measurements are: 1.92 ± 0.07 mm × 1.18 ± 0.10 mm. Dmax ranges 2.04–1.77 mm, while dmin ranges 1.35–0.96 mm.

E. colona spikelets found as accumulations have an average dimension of 2.11 ± 0.13 mm × 1.32 ± 0.09 mm. Dmax ranges 2.40–1.60 mm as average, while dmin ranges 1.75–1.09 mm. The samples L77 and L249 have a larger Dmax (2.30 ± 0.23 mm) and a smaller dmin (1.23 ± 0.14 mm) compared to spikelets visible during excavations, and standard deviation is higher. In fact, the length ranges 2.72–1.85 mm, while width ranges 1.59–0.94 mm.

S. bicolor subsp. *verticilliflorum* spikelets found in two accumulations have an average dimension of 6.41 ± 0.35 mm × 2.33 ± 0.13 mm. Dmax ranges 7.14–5.62 mm, while dmin ranges 2.59–2.00 mm. The specimens of samples L77 and L249 have smaller dimensions: Dmax is 5.19 ± 0.44 mm, dmin is 1.83 ± 0.25 mm, and standard deviation is higher. Length ranges 6.06–4.39 mm, while width ranges 2.24–1.39 mm.

Archaeobotanical finds of *P. laetum*, *E. colona* and *S. bicolor* subsp. *verticilliflorum* have smaller dimensions compared to modern species, with respect to both Dmax and dmin values. Modern *P. laetum* and *E. colona* reach a max length of 3 mm. All cited references indicate 2.75 mm as the average value of Dmax of *P. laetum*, while *E. colona* shows a wide range of Dmax measures (from 1 to 3.5 mm), depending on descriptions by the authors. The maximum length of the fertile spikelets of modern *S. bicolor* subsp. *verticilliflorum* approach 10 mm, but the average length is 6.5 mm (Table 4).

Statistical Analysis

Data analysis reveals the relationship between length and width of spikelets of *P. laetum* and *E. colona* found in the plant accumulations (Fig. 4). Samples are labelled with a letter according to Table 2. The average of Dmax of *P. laetum* spikelets did not change significantly (R^2 = 0.02) from Early to Middle Holocene samples, while dmax tends to increase (R^2 = 0.14) over the same period. *E. colona* shows an increasing trend in length (R^2 = 0.25) and a decreasing trend in width (R^2 = 0.17) over time, as supported by radiocarbon dating and archaeological chronology (Cherkinsky and di Lernia 2013). For *S. bicolor* subsp. *verticilliflorum*, this evaluation is not possible because the two plant accumulations date to the Early Holocene period. The statistically significant (Tukey test $p < 0.05$) size variation in between the populations (the accumulations, the samples L77 and L249, and modern species) is shown in scatter plots, which report average lengths and widths of specimens with associated standard deviations. The metrical data from *P. laetum* are clustered into two groups (Fig. 5a): the first includes 13 plant accumulations and the sample L249; the second encompasses the modern specimens. Regarding *E. colona*, three groups are distinguished (Fig. 5b): 15 plant accumulations; two samples L77 and L249; and modern samples. The statistical test clusters all specimens together when considering *S. bicolor* subsp. *verticilliflorum* (Fig. 5c).

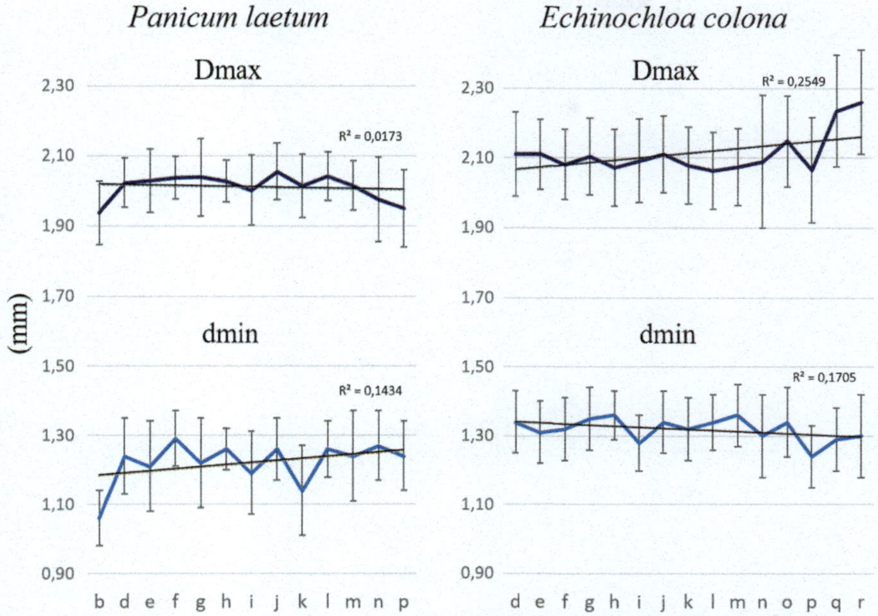

Fig. 4. Data elaboration shows the linear trend of the average length and width of the plant accumulations of *P. laetum* and *E. colona* coming from Takarkori. Samples are labelled with letters according to Table 2

Discussion

In the central Sahara, the Early and Middle Holocene savannas provided sufficient wild cereals, which were widely exploited in prehistory (Marshall and Hildebrand 2002; Kuper and Kröpelin 2006). The archaeobotanical remains recovered from Takarkori have confirmed the prevalence of wild cereals among plants collected and transported to the site.

Wild fonio—*Panicum laetum* Kunth

When comparing the 13 plant accumulations, from Late Acacus to Late Pastoral phases, is possible to observe that spikelets of *P. laetum* (Fig. 4) increased in size. The general shape became increasingly broad in specimens from the initial to the final phase of occupation, but remained elliptical in overall shape (Fig. 6), similar to modern specimens. When comparing length and width of ancient and modern samples, the scatter plot reveals two significantly different groups (Fig. 5a). We hypothesize that the well-developed spikelets found in the seed accumulations were probably collected for human consumption. In contrast, specimens found scattered in the archaeological deposits (sample L249) had smaller dimensions and included some immature spikelets. These specimens likely represent plants collected as fodder, for use in basketry (di Lernia et al. 2012) or they may have arrived to the deposit through casual and involuntary transport. The overall trend discernable in the scatter plot is that modern

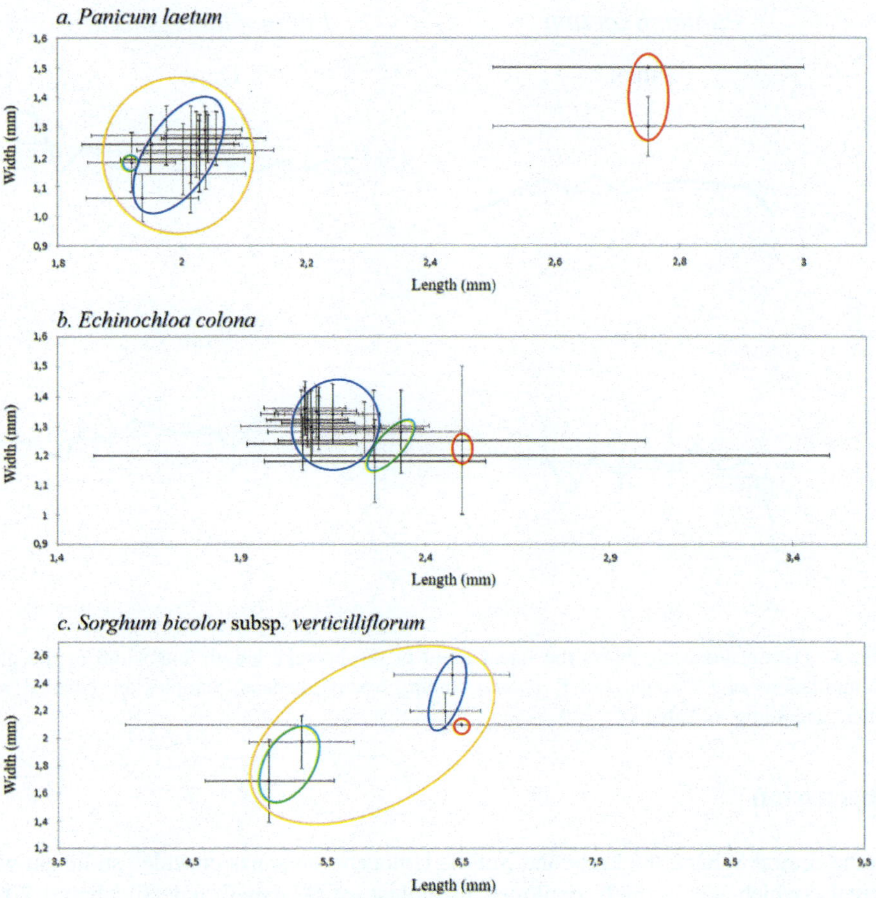

Fig. 5. Scatter plots of the average length and width of the samples with associated standard deviations. **a** *P. laetum* (blue = plant accumulations; green = L249; red = modern samples; yellow = statistical cluster); **b** *E. colona* (blue = plant accumulations; green = L77 and L249; red = modern samples); **c** *S. bicolor* subsp. *verticilliflorum* (blue = plant accumulations; green = L249; red = modern samples; yellow = statistical cluster). Clustering is supported by one-way ANOVA analysis

samples are clustered away from the ancient ones, reflecting the increase in size of contemporary species.

Jungle rice—*Echinochloa colona* L.

Specimens of *E. colona* found in the accumulations are uniform in dimensions both within the same accumulation and among the 15 other samples studied. These samples are almost totally composed of mature spikelets. A change in spikelet shape from ovate to narrow from Middle to Late Pastoral phase is evident in the metrical data (Fig. 4). Data show that spikelets from samples L77 and L249 are much more similar to modern specimens in term of shape (even if the latter are larger) than ancient spikelets in the accumulations. Moreover, samples L77 and L249 produced spikelets with higher Dmax and lower dmin compared to those found in plant accumulations (Fig. 6).

The linear trend (Fig. 4) suggests that *E. colona* was probably collected, processed and stored for food by Middle Pastoral phase human groups during their seasonal occupation of the site. This is also visible in the scatter plot comparing length and width of ancient and modern spikelets, where specimens cluster into three distinct groups (Fig. 5b). While spikelets of the Middle Pastoral phase have a homogeneous and round shape, during the Late Pastoral occupation there was a more elliptic shape similar to

Fig. 6. Spikelets in dorsal and ventral view. *P. laetum*: **a** accumulation n. 9601; **b** accumulation n. 2908. *E. colona*: **c** accumulation n. 9601; **d** sample L77 O30. *S. bicolor* subsp. *verticilliflorum*: **e** accumulation n. 1423; **f** sample L77 O30 (mature spikelet); **g** sample L77 O30 (immature spikelet)

modern specimens (Fig. 4). Decreasing availability of wild cereals, due to environmental aridification, changes in resource exploitation and probably human mobility patterns (Biagetti et al. 2004, 2009; di Lernia et al. 2012; Biagetti and di Lernia 2013) suggested changes in the use of *E. colona*.

Wild Sudan grass—*Sorghum bicolor* L. Moench subsp. *verticilliflorum* (Steud.) de Wet ex Wiersema and J. Dahlb

The *S. bicolor* subsp. *verticilliflorum* spikelets coming from two seed accumulations are uniform in dimensions and composed of only mature seeds. They are very close to the average dimension of modern spikelets, but tend to have a greater width (Tables 3 and 4). The specimens from L249 and L77 have lower values of Dmax and dmin compared to the ancient accumulations, even if the shape is elliptic. In the first two samples, mature and immature spikelets are both present (Fig. 6), and this could explain the high standard deviation of the measurements of these specimens. Instead, the high standard deviation of the accumulation n. 1423 "a" (Table 3) derives from the presence of very large spikelets, sometimes larger than average values of modern specimens. It is important to underline that *S. bicolor* subsp. *verticilliflorum* could have been selected (consciously or unconsciously) in the past, but remained a wild species, an ancestor of the cultivated crop. Samples L77 and L249 seem to reflect the natural presence of *S. bicolor* subsp. *verticilliflorum* in the area of the site, while spikelets of the accumulations were clearly selected for food purposes. This is also suggested by the dominance wild sorghum (93%) in the plant aggregation n. 1423 "a" (Table 3; Fig. 6) (Olmi et al. 2012). Statistical analysis clusters all samples together on the basis of the average measures of length and width (Fig. 5c). This suggests that modern specimens reflect the current environmental variability of a species that never became fully domesticated.

The almost uniform size and homogeneity of specimens found in the accumulations emerge from the morphometrical analysis. This underlines the critical importance of wild cereals in past human behavior, especially during the earliest occupational phases of Takarkori. It is suggested that these species grew wild in the region, but they began to be selected through collecting behaviours of the inhabitants of the rock shelter (Mercuri 2008b; di Lernia et al. 2012). In general it is assumed that this selection, which may have been done by women, favours larger and mature grains (Marshall and Hildebrand 2002; Garcea 2004).

A further consideration, suggesting the exploitation of *P. laetum*, *E. colona* and *S. bicolor* subsp. *verticilliflorum* for human consumption, is their abundance in the archaeological site. The smaller dimensions of the wild cereals of Takarkori compared to modern specimens could be related to their extinction from this area. This may not have permitted the subsequent development of larger grains in this wild species. Success in the survival of these wild cereals, which are today distributed in the southern Sahel or near wetlands, has been gradually limited by the onset of drought conditions persisting to the present day. According to archaeological chronology, *P. laetum* and *E. colona* are present in Late Acacus to Late Pastoral occupational phases of Takarkori, while *S. bicolor* subsp. *verticilliflorum* is found only in the earliest phases, coinciding with Late Acacus and Early Pastoral periods (Cherkinsky and di Lernia 2013). The Late Pastoral is a time of environmental instability, which ends with the African Humid Period (Cremaschi et al. 2014). The increasing aridity is considered to have been a main factor that led to the extinction of *Sorghum* in the area, and to the morphometrical variations in *Panicum* and *Echinochloa* spikelets observed in this study. It also underlines how species not fully adapted to drought conditions, such as *P. laetum* and *E. colona*, were able to survive climate oscillations living in mountainous refugia, thanks to a synergy between the natural geomorphology and cultural actions by hunter-gatherers and pastoralists (Mercuri 2001). As a long period of pre-domestication cultivation preceded the development of domestication traits (Tanno and Willcox 2006; Weiss et al. 2006; Willcox and Stordeur 2012; Allaby et al. 2015a, b) it will be very interesting to study ancient genomes of plants of importance in ancient human diets. It will be possible to track the steps that occurred during domestication, perhaps not visible in the phenotype but through genetic mutations.

Cereals as Key to Understanding Plant Resources in Central Sahara

It is well known that cereals are one of the most important food plants today and in the past. Despite the fact that wild cereals have been harvested worldwide for millennia (Harlan 1985; Zohary and Hopf 2000), most of the ancient valuable species that were regularly harvested in the wild are thought to have never become truly domesticated. However, some species developed domestication syndrome traits, such as non-shattering ears and an annual habit, because of adaptations to human environments (Diamond 2002; Harlan 1989; Weiss et al. 2006; Spahillari et al. 1999; Senda et al. 2006; Howard et al. 2011; Allaby et al. 2015b). Although plant domestication is not the main subject of this paper, when dealing with the exploitation of wild cereals the unquestionable links between exploitation and domestication must be considered. Wild cereals have been widely harvested by people since ancient times, and are nowadays collected in the wild or grown as crops in Saharan and sub-Saharan regions (Corti 1942; Harlan 1989).

Panicum laetum at present is not included in the flora of Fezzan (Corti 1942). It is one of the 'kreb' grasses, a group of grasses occurring in the Sahel and collected from the wild for human consumption especially in times of food shortage. The grains are ground and made into porridge and cakes, and are sometimes collected on a large enough scale to be sold in local markets. *P. laetum* is appreciated by animals for grazing and it is considered to have potential for the restoration of degraded pastures. This wild cereal is of importance in marginal areas and the selection of improved strains for grain and fodder production is recommended by the Plant Resources of Tropical Africa institution (PROTA4U 2017).

Echinochloa colona does not occur today in the Fezzan region (Corti 1942). It has been suggested that it could have been cultivated in ancient Egypt (Täckholm and Drar 1973), and, based on its use in modern times, the gathering of the wild cereals is known (Germer 1985). It is an annual plant, which grows throughout the tropics and subtropics as a weedy species of muddy or swampy places (Clayton and Renvoize 1982). It is a nitrophilous, aquatic or semiaquatic grass (Sheriff and Siddiqi 1988), a valuable grazing plant and a good pasture for cattle (Schulz and Adamou 1992) in its short season of growth. The seeds are eaten by humans in times of famine: they can be similarly cooked whole or ground into a flour, and used as a porridge. Young plants and shoots are eaten raw or cooked (Manandhar 2002, Tanaka 1976). *E. colona* can be an important weed of rice that causes substantial yield reductions (74–98%; Mukhopadhyay et al. 1972) because of its severe infestation, rapid growth and great competitive ability (FAO 2017). It is collected in the Sahel of Mali and Nigeria, where is used with other "bougrou" grasses (Harlan 1989). Zaghawa people of the Sudan and Chad harvest this grass either by beating, so that the grains fall into the basket, or by cutting with a reaping-hook (Tubiana and Tubiana 1977). In Sudan, its grains are also swept from the soil (Magid 1989). Schulz and Adamou (1992) report *E. colona* as a valued wild cereal collected in the Aïr Mountains of Niger.

The presence of *S. bicolor* subsp. *verticilliflorum* in the oldest level of the site is of great interest when framed within the discussion on *Sorghum* domestication in Africa. It is considered a wild species because of its morphometrical features, however work is in progress to examine its ancient DNA (aDNA) by means of Next Generation Sequencing (NGS). Morphological criteria on their own cannot exclude the possibility that a wild plant was cultivated (Dahlberg and Wasylikowa 1996; Wasylikowa 1997). The earliest evidence for sorghum in Africa was reported by Beldados and Costantini (2011) during the 2nd millennium BC, while domesticated forms of sorghum emerged relatively recently, around 6000 years ago (Zohary et al. 2012; Allaby et al. 2015b; Beldados et al. 2018; Fuller and Stevens 2018). *S. bicolor* subsp. *verticilliflorum* does not occur today in the Fezzan region (Corti 1942). All contemporary wild sorghum races are annual (or short-lived perennials) and are so morphologically similar that de Wet referred to them as "well defined ecotypes" (de Wet 1978). Wild sorghum grains are today harvested in the savanna, and consumed by native African peoples as famine food (Snowden 1955).

Conclusion

Studies on both pollen and macroremains point to the exploitation of wild cereals by people of the Fezzan region, extended in space and time (Cremaschi et al. 2014; Dunne et al. 2016), and to their prevalence among the panorama of plants selected and transported to shelters and caves. Those cereals, selected by type, are evidence of a deep knowledge of the plants distributed in the region and of the presence of wild cereals collected continuously in the area for millennia. Their use as medicinal or magic required much smaller quantities compared to those that have been found at Takarkori, so these cereals were likely brought to the rock shelters by humans as purposeful harvests, mainly for food, basketry and fuel (Wasylikowa 1997; Mercuri 2008b; di Lernia et al. 2012). Morphometrical analysis of the spikelets of *E. colona*, *P. laetum* and *S. bicolor* subsp. *verticilliflorum*, found as seed accumulations at Takarkori, shows that each genus has a consistent size range. Our study highlights the remarkable uniformity of grain and spikelet dimensions among the records from different seed accumulations and varying chronological and stratigraphic contexts. Collection and transport of these wild cereals must have contributed to their dissemination and growth during the changing environmental conditions during the Early and Middle Holocene. Human activity, therefore, could have prolonged the life of cereals during climatic oscillations, contributing to their survival in the mountains of Tadrart Acacus in the absence of domestication (Mercuri et al. 2018).

Acknowledgements. We thank the Italian-Libyan Archaeological Mission in the Acacus and Messak, directed by Savino di Lernia, which allowed the sampling and the analysis of the plant remains from Takarkori. The research was part of the PhD project of R.F. at the School Agri-food Science, Technologies and Bio-technologies (University of Modena and Reggio Emilia). Funds were provided by the project "SELCE—SELvatici CEreali: il futuro nella risposta delle piante ai cambiamenti climatici", sect. Scientific and Technological Research (Sime n.2015.0033), funded by the FCRMO-Fondazione Cassa di Risparmio di Modena, directed by A.M.M.

References

Allaby RG, Gutaker R, Clarke AC et al (2015a) Using archaeogenomic and computational approaches to unravel the history of local adaptation in crops. Philos T Roy Soc B 370:20130377

Allaby RG, Kistler L, Gutaker RM et al (2015b) Archaeogenomic insights into the adaptation of plants to the human environment: pushing plant-hominin co-evolution back to the Pliocene. J Hum Evol 79:150–157

Bacchetta G, Grillo O, Mattana E et al (2008) Morpho-colorimetric charachterization by image analysis to identify diaspores of wild plant species. Flora 203:669–682

Bandini Mazzanti M, Bosi G, Mercuri AM et al (2012) Analisi archeobiometriche e reperti carpologici: scopi e prospettive. In: Vezzalini G, Zannini P (eds) Atti del VII Congresso Nazionale di Archeometria (AIAr). Pàtron Editore, Bologna CD-rom

Beldados A, Costantini L (2011) Sorghum exploitation at the Kassala and its environs, North Eastern Sudan in the second and first millennia BC. Nyame Akuma 75:33–39

Beldados A, Manzo A, Murphy C, Stevens CJ, Fuller DQ (2018) Evidence of sorghum cultivation and possible pearl millet in the second millennium BC at Kassala, Eastern Sudan

Biagetti S, di Lernia S (2013) Holocene deposits of Saharan rock shelters: the case of Takarkori and other sites from the Tadrart Acacus Mountains (SW Libya). Afr Archaeol Rev 30:305–333

Biagetti S, Merighi F, di Lernia S (2004) Decoding an Early Holocene Saharan stratified site. Ceramic dispersion and site formation processes in the Takarkori rockshelter, Acacus Mountains, Libya. J Afr Archaeol 2(1):11–36

Biagetti S, Poggi G, di Lernia S (2009) Unearthing the hearths. Preliminary results on the Takarkori rockshelter fireplaces (Acacus Mts., Libya). Defining a methodological approach to interpret structural evidence. BAR Int Ser 2045:23–29

Bosi G, Costantini F, Berti P et al (2007) Applicazioni morfobiometriche in campo archeocarpologico: primi dati su *Papaver somniferum* nell'Alto Medioevo di Ferrara. Atti Soc Nat Mat Modena 137(2006):373–387

Boulos L (2005) Flora of Egypt, vol 4. Al Hadara Publishing, Cairo

Chen T, Wang X, Dai J et al (2016) Plant use in the Lop Nor region of southern Xinjiang, China: Archaeobotanical studies of the Yingpan cemetery (~25–420 AD). Quatern Int 426:166–174

Cherkinsky A, di Lernia S (2013) Bayesian Approach to 14Cdates for estimation of long-term archaeological sequences in arid environments: the Holocene site of Takarkori Rockshelter, Southwest Libya. Radiocarbon 55(2–3):771–782

Clayton WD, Renvoize SA (1982) Flora of tropical East Africa. Gramineae (Part 3). AA Balkema, Rotterdam

Corti R (1942) Flora e vegetazione del Fezzan e della Regione di Gat, Firenze

Cremaschi M, Zerboni A, Mercuri AM et al (2014) Takarkori rock shelter (SW Lybia): an archive of Holocene climate and environmental changes in the central Sahara. Quat Sci Rev 101:36–60

Dahlberg JA, Wasylikowa K (1996) Image and statistical analyses of early sorghum remains (8000 BP) from the Nabta Playa archaeological site in the Western Desert, Southern Egypt. Veg Hist Archaeobot 5(4):293–299

de Leonardis WD, Santis CD, Fichera G et al (2011) Seed morphobiometry of wild and cultivated taxa of *Phaseolus* L. (Fabaceae). Indian J Plant Genet Resour 24(3):257–264

de Wet JMJ (1978) Systematics and evolution of *Sorghum* sect. *Sorghum* (Gramineae). Am J Bot 65(4):477–484

di Lernia S (ed). (1999). The Uan Afuda cave. Hunter-gatherer societies of central Sahara, vol. 1. All'Insegna del Giglio, Florence

di Lernia S, N'siala IM, Mercuri AM (2012) Saharan prehistoric basketry. Archaeological and archaeobotanical analysis of the early-middle Holocene assemblage from Takarkori (Acacus Mts., SW Libya). J Archaeol Sci 39(6):1837–1853

di Lernia S, Tafuri MA (2013) Persistent deathplaces and mobile landmarks: the Holocene mortuary and isotopic record from Wadi Takarkori (SW Libya). J Anthropol Archaeol 32 (1):1–15

di Lernia S, Bruni S, Cislaghi I et al (2016) Colour in context. Pigments and other coloured residues from the early-middle Holocene site of Takarkori (SW Libya). Archaeol. Anthropol Sci 8(2):381–402

Diamond J (2002) Evolution, consequences and future of plant and animal domestication. Nature 418:700–707

Dunne J, Mercuri AM, Evershed RP et al (2016) Earliest direct evidence of plant processing in prehistoric Saharan pottery. Nat Plants 3(16194)

Dunne J, Eversheld RP, Salque M et al (2012) First dairying in green Saharan Africa in the fifth millennium BC. Nature 486(7403):390–394

Fahmy AG (2001) Palaeoethnobotanical studies of the Neolithic settlement in Hidden Valley, Farafra Oasis. Egypt. Veg Hist Archaeobot 10:235–246

Fornaciari R, Fornaciari S, Francia E et al (2016) *Panicum* spikelets from the Early Holocene Takarkori rockshelter (SW Libya): Archaeo-molecular and botanical investigations. Plant Biosyst 2016:1–13

Fuller DQ, Harvey E, Qin L (2007) Presumed domestication? Evidence for wild rice cultivation and domestication in the fifth millennium BC of the Lower Yangtze region. Antiquity 81 (312):316–331

Fuller DQ, Stevens CJ (2018) Sorghum Domestication and Diversification: A current archeobotanical perspective

Garcea EAA (2004) An alternative way towards food production: the perspective from the Libyan Sahara. J World Prehist 18(2):107–154

Germer R (1985) Flora des pharaonischen Ägypten, vol. 14. Verlag Philipp von Zabern, Darmstadt

Harlan JR (1985) The living fields: our agricultural heritage. Cambridge University Press, Cambridge

Harlan JR (1989) Wild grass seeds as food sources in the Sahara and Sub-Sahara. Sahara 2:69–74

Howard T, Archer JE, Turley RM (2011) Evolution, physiology and phytochemestry of the psycotoxic arable mimic weed darnel (*Lolium temulentum* L.). Prog Bot 72:73–104

Kim M, Ahn SM, Jeong Y (2013) Rice (*Oryza sativa* L.): seed-size comparison and cultivation in ancient Korea. Econ Bot 67(4):378–386

Kuper R, Kröpelin S (2006) Climate-controlled Holocene occupation in the Sahara: motor of Africa's evolution. Science 313(5788):803–807

Liengme B (2015) A guide to Microsoft Excel 2013 for scientists and engineers. Academic Press, Massachusetts

Magid AA (1989) Plant domestication in the Middle Nile Basin. British Archaeological Reports, Oxford

Maire R (1952) Flore de l'Afrique du Nord (Maroc, Algérie, Tunisie, Tripolitaine, Cyrénaïque et Sahara). Jouve, Paris

Manandhar NP (2002) Plants and people of Nepal. Timber Press, Oregon

Marshall F, Hildebrand E (2002) Cattle before crops: the beginnings of food production in Africa. J World Prehist 16(2):99–143

Mercuri AM (2001) Preliminary analyses of fruits, seeds and other few plants macrofossils from the Early Holocene sequence. In: Garcea EAA (ed) Uan Tabu in the settlement history of the Libyan Sahara. Arid Zone Archaeology, 2:161–188. All'Insegna del Giglio, Firenze

Mercuri AM (2008a) Human influence, plant landscape, evolution and climate inferences from the archaeobotanical records of the Wadi Teshuinat area (Libyan Sahara). J Arid Environ 72:1950–1967

Mercuri AM (2008b) Plant exploitation and ethnopalynological evidence from the Wadi Teshuinat area (Tadrart Acacus, Libyan Sahara). J Archaeol Sci 35(6):1619–1642

Mercuri AM, Allevato E, Arobba D et al (2015) Pollen and macroremains from Holocene archaeological sites: a dataset for the understanding of the bio-cultural diversity of the Italian landscape. Rev Palaeobot Palyno 218:250–266

Mercuri AM, Fornaciari R, Gallinaro M et al (2018) Plant behaviour through human imprints and the cultivation of wild cereals in Holocene Sahara. Nat Plants 4:71–81, https://doi.org/10.1038/s41477-017-0098-1

Mercuri AM, Garcea EAA (2007) The impact of hunter/gatherers on the vegetation in the central Sahara during the early Holocene. In: Cappers RTJ (ed) Fields of change: progress in African archaeobotany. Groningen, Barkhuis

Mukhopadhyay SK, Khara AB, Ghosh BC (1972) Nature and intensity of competition of weeds with direct-seeded upland IR8 rice crop. Int Rice Community Newsl 21(2):10–14

Oliver D and auct suc (eds) (1917) Gramineae. Fl Trop Afr 9:114

Olmi L, Mercuri AM, Gilbert MTP et al (2012) Morphological and genetic analyses of early mid-Holocene wild cereals from the Takarkori rockshelter (central Sahara, Libya): first results and prospects. In: Fahmy AG, Kahlheber S, D'Andrea AC (eds) Windows on the African past: contemporary approaches to African archaeobotany. Reports in African archaeology 3. Africa Magna Verlag, Frankfurt

Orrù M, Grillo O, Lovicu G et al (2013) Morphological charachterisation of *Vitis vinifera* L. seeds by image analysis and comparison with archaeological remains. Veg Hist Archaeobot 22:231–242

Ozenda P (1958) Flore du Sahara septentrional et Central. CNRS, Paris

Ozenda P (2000) Les végétaux: organisation et diversité biologique. Dunod, Paris

Pearsall DM (1989) Paleoethnobotany, A handbook of procedures. Academic Press, London

Phillips SM (1995) Flora of Ethiopia and Eritrea: volume 7. Poaceae (Gramineae). The National Herbarium, Biology Department, Science Faculty, Addis Ababa University, Addis Ababa

Renfrew JM (1973) Palaeoethnobotany, the prehistoric food plants of the Near East and Europe. Columbia University Press, New York

Schulz E, Adamou A (1992) Leben in der Südlichen. Die traditionelle Nutzung der Vegetation im Nord-Niger Sahara. Geographisches Institut, Würzburg

Senda T, Hiraoka Y, Tominaga T (2006) Inheritance of seed shattering in *Lolium temulentum* and *L. persicum* hybrids. Genet Resour Crop Ev 20:633–643

Sharp D, Simon BK (2002) AusGrass: grasses of Australia. CSIRO Publishing/Australian Biological Resources Study (ABRS), Melbourne

Sheriff AS, Siddiqi MA (1988) Poaceae. In: AA El-Gadi (ed) Flora of Libya 145, Tripoli

Smith CW, Frederiksen RA (2000) Sorghum: origin, history, technology, and production, vol. 2. Wiley, New Jersey

Snowden JD (1955) The wild fodder Sorghums of the section *Eu-Sorghum*. J Linn Soc Lond Bot 55(358):191–260

Spahillari M, Hammer K, Gladis T et al (1999) Weeds as part of agrobiodiversity. Outlook on Agr 28:188–199

Täckholm V, Drar M (1973) Flora of Egypt, vol II. Otto Koeltz Antiquariat, Koenigstein

Tafuri MA, Bentley RA, Manzi G et al (2006) Mobility and kinship in the prehistoric Sahara: Strontium isotope analysis of Holocene human skeletons from the Acacus Mts. (southwestern Libya). J Anthropol Archaeol 25(3):390–402

Tanaka T (1976) Tanaka's cyclopaedia of edible plants of the world. Keigaku Publishing, Tokyo

Tanno K, Willcox G (2006) How fast was wild wheat domesticated? Science 311:1886

Tubiana MJ, Tubiana J (1977) The Zaghawa from an ecological perspective; food gathering, the pastoral system, tradition and development of the Zaghawa of the Sudan and the Chad. AA Balkema, Rotterdam

Walker MJ, Berkelhammer M, Björck S et al (2012) Formal subdivision of the Holocene Series/Epoch: a Discussion Paper by a Working Group of INTIMATE (Integration of ice-core, marine and terrestrial records) and the Subcommission on Quaternary Stratigraphy (International Commission on Stratigraphy). J Quat Sci 27(7):649–659

Wasylikowa K (1997) Flora of the 8000 years old archaeological site E-75-6 at Nabta Playa, Western Desert, Southern Egypt. Acta Palaeobot 37(2):99–205

Wasylikowa K, Dahlberg J (1999) Sorghum in the economy of the early Neolithic nomadic tribes at Nabta Playa, Southern Egypt. In: The exploitation of plant resources in ancient Africa. Springer, US

Weiss E, Kislev ME, Hartmann A (2006) Autonomous cultivation before domestication. Science 5780:1608

Willcox G (2004) Measuring grain size and identifying Near Eastern cereal domestication: evidence from the Euphrates valley. J Archaeol Sci 31:145–150

Willcox G, Stordeur D (2012) Large-scale cereal processing before domestication during the tenth millennium cal BC in northern Syria. Antiquity 86:99–114

Zohary D, Hopf M (2000) Domestication of plants in the Old World, 3rd edn. Oxford University Press, Oxford

Zohary D, Hopf M, Weiss E (2012) Domestication of plants in the Old World: the origin and spread of domesticated plants in Southwest Asia, Europe, and the Mediterranean Basin. Oxford University Press on Demand

Web Sites

FAO–Food and Agriculture Organization of the United Nations (2017). http://www.fao.org/ag/agp/AGPC/doc/Gbase/data/pf000226.htm. Accessed 25 Jan 2017

IRRI–International Rice Research Institute (2017). http://www.knowledgebank.irri.org/training/fact-sheets/item/echinochloa-colona. Accessed 25 Jan 2017

ISTA–International Seed Testing Association (2017). https://seedtest.org/en/universal-list.html. Accessed 25 Jan 2017

Kew RBG–Royal Botanic Gardens (2017). http://www.kew.org/data/grasses-db.html. Accessed 25 Jan 2017

PROTA4U–Plant Resources of Tropical Africa (2017). http://www.prota4u.info/articles.asp. Accessed 25 Jan 2017

The Holocene Flora and Vegetation of Ti-n Hanakaten (Tassili n'Ajjer, Algerian Sahara)

Samira Amrani(✉)

Institute of Archaeology, University of Algiers, Beni Messous, Algiers, Algeria
amranil.samira@gmail.com

Abstract. Ti-n Hanakaten is one of the most important prehistoric sites of the Tassili n'Ajjer, in the Algerian Sahara. Since the Early Holocene, and during the Middle and Late Holocene, the occupants of Ti-n Hanakaten were hunter-gatherers, and then herders of the Bovidian culture (Aumassip et al. 2013). The Holocene stratigraphical sequence of this site consists of preserved archaeological layers with evidence of a long cultural occupation between approximately 8250 and 660 BC. During six field excavations (1974, 1975, 1976, 1978, 1980 and 1981) directed by G. Aumassip, a very large quantity of botanical remains was uncovered from the archaeological deposits of Ti-n Hanakaten. The archaeobotanical study of this site included palynological and anthracological analyses, and offered interesting palaeoenvironmental information on a fairly unknown area. Pollen data provide a broad view of the vegetation that characterized the immediate as well as the more distant surroundings of the site during the Holocene. The pollen flora of Ti-n Hanakaten belongs to different phytogeographical groups and points to the presence of a mosaic of vegetation types. Probably Mediterranean vegetation grew at 1000 m elevation during Holocene times when climatic and ecological conditions were favorable for *Cupressus* (*C. dupreziana*), *Pistacia* (*P. atlantica*) and *Olea* (*O. laperrinei*). The study of charcoal remains was useful to infer human activities and plant exploitation at the site, documenting woody plants used for fuel or other purposes. Probably small groups of inhabitants visited diverse territories in order to find wood or charcoal, and some resources may have been transported to the site from distant regions, for instance from the highest elevations of the Tassili n'Ajjer mountains where the aforementioned Mediterranean vegetation was likely found.

Keywords: Algeria · Charcoal · Ti-n Hanakaten · Central Sahara
Pollen

Introduction

Ti-n Hanakaten is a prehistoric site in the Tassili n'Ajjer which comprises archaeological deposits dating to the Late Pleistocene and Holocene. During the excavations, a large quantity of botanical remains was uncovered from layers dating to the Early, Middle and Late Holocene (Aumassip 1976; Aumassip et al. 2013). The different archaeological phases have been correlated to several rock art periods: Round Heads, Bovidian, and the more recent Horse and Camel periods (Aumassip et al. 2013).

A. M. Mercuri et al. (Eds.): *Plants and People in the African Past*,
https://doi.org/10.1007/978-3-319-89839-1_8

The archaeobotanical investigation of the continuous stratigraphic sequence, combining palynological and anthracological analyses, offers new information on the vegetation development and plant exploitation at this locality in the central Sahara during the Holocene.

Rock shelters such as Ti-n Hanakaten allow detailed reconstruction of Holocene environmental changes in the Sahara (Biagetti and di Lernia 2013; Cremaschi et al. 2014). Mercuri (2008a, b) summarised palynological research carried out on sites of hunter-gatherer and pastoralist cultures based on multidisciplinary archaeological research carried out in the Wadi Teshuinat area of the Tadrart Acacus Mts., in Fezzan, southwestern Libya. In addition, multiple archaeobotanical studies from Libyan sites have been published in combination with archaeological research (Pasa and Pasa Durante 1962; Schulz 1987; Trevisan Grandi and Mercuri 1992; Wasylikowa 1992a, b, 1993; Castelletti et al. 1998, 1999; Trevisan Grandi et al. 1998; Mercuri et al. 1998, 1999; Magid 1999; Mercuri and Trevisan Grandi 2001; Neumann and Uebel 2001; Mercuri and Garcea 2007).

In Algeria, and in particular in the Saharan regions of Ahaggar (also called Hoggar) and Tassili n'Ajjer, archaeobotanical studies including pollen and macroremains have been completed on rockshelter deposits. Early works focused on samples from non-continuous sequences which were rich in pollen grains (Quézel 1954, 1956, 1960a, b, 1963, 1967, 1978; Pons and Quézel 1957, 1958; Quézel and Thiébault 1959; Quézel and Martinez 1960). Since 1964, there have been several studies in the Ahaggar, especially in the Atakor area, studying different types of deposits within which seeds/fruits and charcoals were preserved (Table 1).

Recently, prehistoric research in the Sahara has seen much progress and gained importance in palaeoenvironmental studies (Gallin 2008; Le Quellec 2008a, b, 2012; Saliège et al. 2008; Gauthier 2009; di Lernia and Gallinaro 2010; Ehreinreich and Fuchs 2011; Gallinaro et al. 2012). In particular, impressive results were obtained through interdisciplinary research completed in southern Libya (di Lernia et al. 2012, 2016; Dunne et al. 2012, 2016; Biagetti and di Lernia 2013; Mercuri et al. 2018). Furthermore, new data have been recently made available for the Tassili n'Ajjer in Algeria by Hachid et al. (2010, 2012), Mercier et al. (2012) and Aumassip et al. (2013).

This paper presents palynological and anthracological analyses carried out at the Ti-n Hanakaten site. Much of the archaeobotanical results are qualitative and somewhat fragmentary, and as such resolution, interdisciplinary interpretations are not currently available for the site. Nevertheless, this is an interesting region for Saharan archaeology, and not well known from an archaeobotanical point of view. Therefore, the preliminary data described here constitute new baseline information which will surely be elaborated upon by future investigations.

The Study Area and the Present Vegetation

Tassili n'Ajjer is a vast plateau in southeastern Algeria at the borders with Libya and Niger, covering an area of 72,000 km^2 (Fig. 1). Since 1982 it has been classified as a UNESCO World Heritage Site for its exceptional natural and cultural legacy, and since 1986 is a biosphere reserve of UNESCO. Moreover, the valley of Ihrire, which is approximately at the center of the Tassili n'Ajjer mountain range, has been classified as

Table 1. Sites with published archaeobotanical analyses in the central Algerian Sahara in chronological order of publications.

Location	Altitude (m a.s.l.)	Notes	Chronology	Analyzed remains	References
Hoggar	1050–1700	Several sites	Pleistocene—Holocene	Pollen	Pons and Quézel (1957)
Hoggar Tassili n'Ajjer (Djanet and Sefar)	1350–2100	Several sites	Pleistocene—Holocene	Pollen	Quézel and Martinez (1960)
Atakor area: Terhananet and Illamane	2200	Several sites	Villafranchian, Pleistocene	Pollen	Van Campo (1964)
Atakor	1800	Crocodile dung	Pleistocene, Holocene	Pollen	Van Campo et al. (1965)
Hoggar	1000	Various deposits watershed	Villafranchian, Pleistocene	Pollen	Van Campo et al. (1966)
Outoul	1430	Deposits from the Wadi Outoul	Upper Pleistocene	Pollen	Van Campo et al. (1967)
Hoggar area Illamen Tahog	2100–1800	Several sites	Villafranchian, Quaternary	Pollen, charcoal, seeds	Rognon (1967)
Amekni	1200	Rock shelter deposit	Holocene	Pollen, charcoal, seeds	Guinet and Planque (1969)
Z'bib N Elias, Taessa Mountain	2150	Rock hyrax droppings	Middle Holocene	Pollen, seeds	Ballouche et al. (1995); Thinon et al. (1996)
Tan-Tartait and in Itinen caves	1500	Anthropic deposits	Holocene	Pollen	Alimen et al. (1968)
Ti-n Hanakaten site	950	Anthropic deposits	Holocene	Charcoal, seeds	Couvert (1979)
Ti-n Hanakaten site	950	Anthropic deposits	Holocene	Pollen, Charcoal	Amrani (2002, 2016)
Tassili n'Ajjer area	1180–1500	Natural and anthropic deposits	Holocene	Microcharcoal	Abdoun and Beddiaf (2002)

a "Wetland of International Importance" or "Ramsar Site" since 2001. The Ti-n Hanakaten shelter (Lat. 23° 52' N–Long. 10° 22' E; 950 m a.s.l.) is one of the most important prehistoric sites in this region (Fig. 2).

The current Saharan vegetation is limited to the north by semi-desert and to the south by a transition zone of Sahelian savanna (Monod 1957; Schulz 1986). Flora and vegetation of the Sahara were studied by Ozenda (1977, 2004), and the Tassili n'Ajjer was especially explored by Leredde (1957). Based on the geological and physical features of the region, Leredde (1957) classified the area as having Saharo-Tropical and Sahara-Mediterranean plant associations. He found twenty-one species among which only five were frequently encountered: *Artemisia herba-alba* Asso, *Erodium*

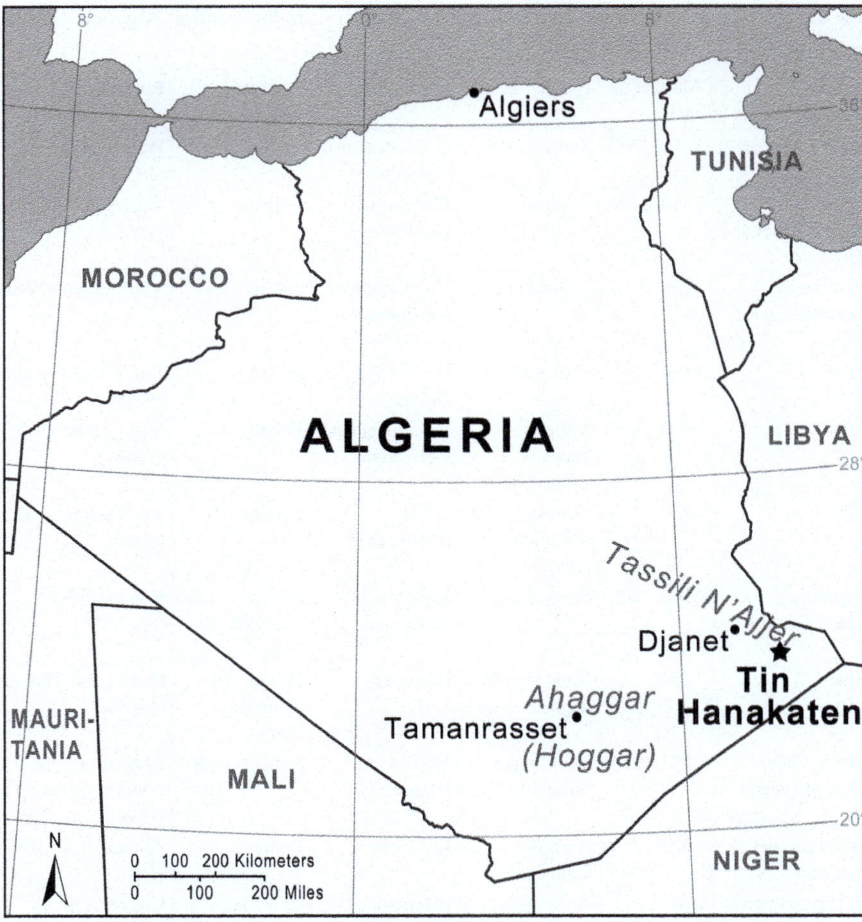

Fig. 1. Location of the Tassili n'Ajjer and Ti-n Hanakaten site.

triangulare (Cav.) Willd., *Globularia alypum* L., and the trees *Cupressus dupreziana* A.Camus and *Olea laperrinei* Batt. & Trab. The rocky wadis are refugia for the last remnants of Mediterranean vegetation. Seasonally available water and slightly more humid soils of Tassili n'Ajjer make suitable habitats for the relic flora of the Sahara-Mediterranean region which includes cypress and tamarisk (*Tamarix aphylla* (L.) Karst.). This indicates that past climate in the area was more humid, enabling the etablishment of lush vegetation accompanied by *Acacia seyal* Delile (Leredde 1957).

Today, the vegetation of the Tassili n'Ajjer is rare and often absent in vast areas, except in the wadi beds found at the bottom of the gorge, where it is denser. The Tassili mountains are extraordinarily floristically rich in herbaceous, shrubby and tree species, the latter including the famous "tarout" (cypress of Tassili) as well as the well-known olive tree of Laperrine. These two Mediterranean species are endemic and of considerable scientific interest (Cremaschi et al. 2006; Haddad et al. 2018). In terms of floristic heritage, more than 300 spontaneous plant species of diverse phytogeographical origins

Fig. 2. Ti-n Hanakaten rockshelter (photo by Aumassip)

(tropical, Mediterranean, ubiquitarian, etc.) have been recorded. In addition, the Tassili Mountains have an endemism rate approaching 50% for woody plants. Among the endemic taxa of Mediterranean origin, in addition to the above mentioned cypress and olive trees, there are: Saharan myrtle (*Myrtus nivellei* Batt. & Trab.), lavender (*Lavandula antineae* Maire), jointfir (*Ephedra altissima* Desf.) and the desert marrube (*Marrubium desertii* (Noë) Coss.). Other species of tropical, Sahelo-Sudanian and Saharo-Arabic affiliation, find favorable habitats such as: *Faidherbia albida* (Delile) A.Chev. (syn. *Acacia albida), Acacia arabica* Delile, *Acacia ehrenbergiana* Hayne, *Balanites aegyptiaca* (L.) Delile, *Calotropis procera* (Aiton) Dryand., *Maytenus senegalensis* (Lam.) Exell, *Leptadenia pyrotechnica* (Forssk.) Decne, *Maerua crassifolia* Forssk., *Artemisia judaica* subsp. *sahariensis* (L.Chevall.) Maire, *Chrozophora brocchiana* (Vis.) Schweinf., and others. Concerning physiognomy, two main characteristic plant communities can be described: extensive vegetation which covers very large areas subject to the vagaries of rainfall and possible floods, and a punctual type found in areas with permanent water sources. Only the sandy beds of the wadis offer a substrate suitable for desert savanna comprised of acacias (*Acacia raddiana* and *A. seyal*) and *Panicum turgidum* mixed with some *Maerua crassifolia* related to the influence of rare summer monsoon rains (Aumassip 1976).

Materials and Methods

The Ti-n Hanakaten sediments comprise a 5–7 m sequence of anthropic deposits, dating to the Late Pleistocene and Holocene. Figure 3 shows the general stratigraphy of the site according to Aumassip (1976, 1977, 1980–1981). The Holocene stratigraphy consists of two main formations—a lower and the upper section, each of which consists of several layers. Layer descriptions follow Aumassip et al. (2013), and each layer is labelled 'S' denoting 'stratum' from S9–10 at the bottom to S1 at the top.

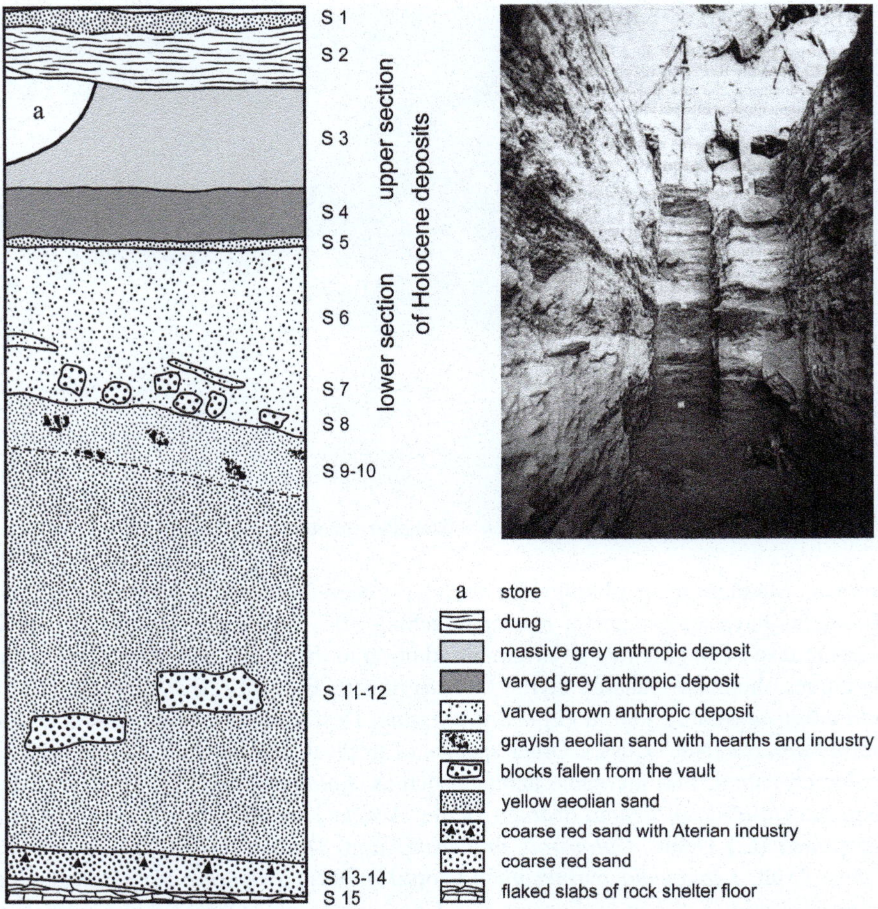

Fig. 3. Stratigraphy of the deposits at Ti-n Hanakaten (Aumassip et al. 2013, and 1982–1983, modified). The Holocene deposit (S10 to S1) is discussed in the text.

Lower section (S10–S6): The lowermost portion of the Holocene sequence, S9–10, is a 40 cm thick layer, mainly consisting of aeolian sand with grey lenses due to ash enrichment and some organic matter with microfauna. It contains both small archaeological tools and small wooden archaeological structures (Aumassip 1980–1981 and Aumassip et al. 2013). Overlying this layer, is a brown deposit 170–100 cm thick, rich in clay, ash, and organic materials that consists of three layers S8, S7 and S6. S8 is variegated, full of sandstone blocks, with a few millimeters of aeolian sand and gypsum found 10 cm from its base. S7 is a discontinuous layer of aeolian sand, a few millimeters thick, which did not reach the shelter wall. S6 is variegated, subhorizontal, with a thickness that ranges 50–70 cm. During two millennia of occupation, various archaeobotanical remains have become incorporated into this formation, among them unusually well preserved remains of plant material, such as basketry and cordage, seeds and fruits, and still colorful flowers (Aumassip 1976). Stone circles of 1.5–4.0 m diameter may have been created by the ancient inhabitants of the shelter in order to delineate space for activity areas. The most spectacular space had a floor made of small flat stones, upon which a leafy litter rested. The lower formation also contains human remains which were the subject of anthropological studies revealing the inhabitants were of tall stature (Aumassip and Heim 1989).

Upper section (S5–S1): The upper formation is almost as thick as the lower, but it is grey in colour, essentially consisting of ash. The section begins with a few millimeters of aeolian sand (S5), and is sealed by a dung deposit 30–40 cm thick (S2), which is in turn covered by another layer of 20–30 cm of aeolian sand (S1). In between S4 and S2 rested the main part of this formation, which is subdivided into sections: the basal S4 is layered and contains many plant macroremains, and the massive S3 includes many bands of aeolian sand. A discontinuous occupation is evident from lenses of 2–4 m in diameter, composed of remnant fireplaces, sometimes formerly surrounded by stones. These lenses also contain bone remains, some worked stone, and pottery fragments,

Table 2. Chronology of Ti-n Hanakaten shelter. Radiocarbon dates are reported in: uncal yr BP, cal BP and BC; dates were calibrated with OxCal 4.3 with IntCal13 (Bronk Ramsey 2009). Two more dates based on thermoluminescence on ceramics are available for S2 and S3 (2790 ± 200 and 3100 ± 240 uncal yr BP, respectively; Aumassip et al. 2013)

Samples	^{14}C on charcoal		
	(uncal yr BP)	(calBP)	(BC)
S1	No date		
S2	2555–2485	2529–2710	668 ± 127
	(2520 ± 35)	2620 ± 90	
S3	5480–5390	6210–6283	4222 ± 142
	(5435 ± 45)	6247 ± 36	
	5325–5245	5993–6166	4114 ± 122
	(5285 ± 60)	6080 ± 86	
S4	5920–5680	6478–6750	4656 ± 285
	(5800 ± 120)	6614 ± 136	
	7250–6970	7796–8090	6021 ± 306
	(7110 ± 140)	7943 ± 147	
S5	No Date		
S6	7200–7100	7945–8010	6053 ± 148
	(7150 ± 50)	7978 ± 32	
	8940–8810	9842–10124	8002 ± 241
	(8875 ± 65)	9983 ± 141	
S7	No date		
S8	8965–8845	9925–10146	8043 ± 212
	(8905 ± 60)	10036 ± 110	
S9	No Date		
S10	9620–9200	10404–11024	8772 ± 507
	(9410 ± 210)	10714 ± 310	

scattered upon floors or surfaces. Later on, the shelter was used for animals as evidenced by thick dung layers (Aumassip 1976).

The archaeological excavation brought to light layers of different cultures which, according to Aumassip (1976, 1980–1981, 2000), are attributed to the Middle Palaeolithic Aterian culture (no radiocarbon dates) and to subsequent cultures originally called Neolithic of the Saharo-Sudanese and the Bovidian period. The archaeological chronology and first dates from the site were published by Aumassip and Delibrias

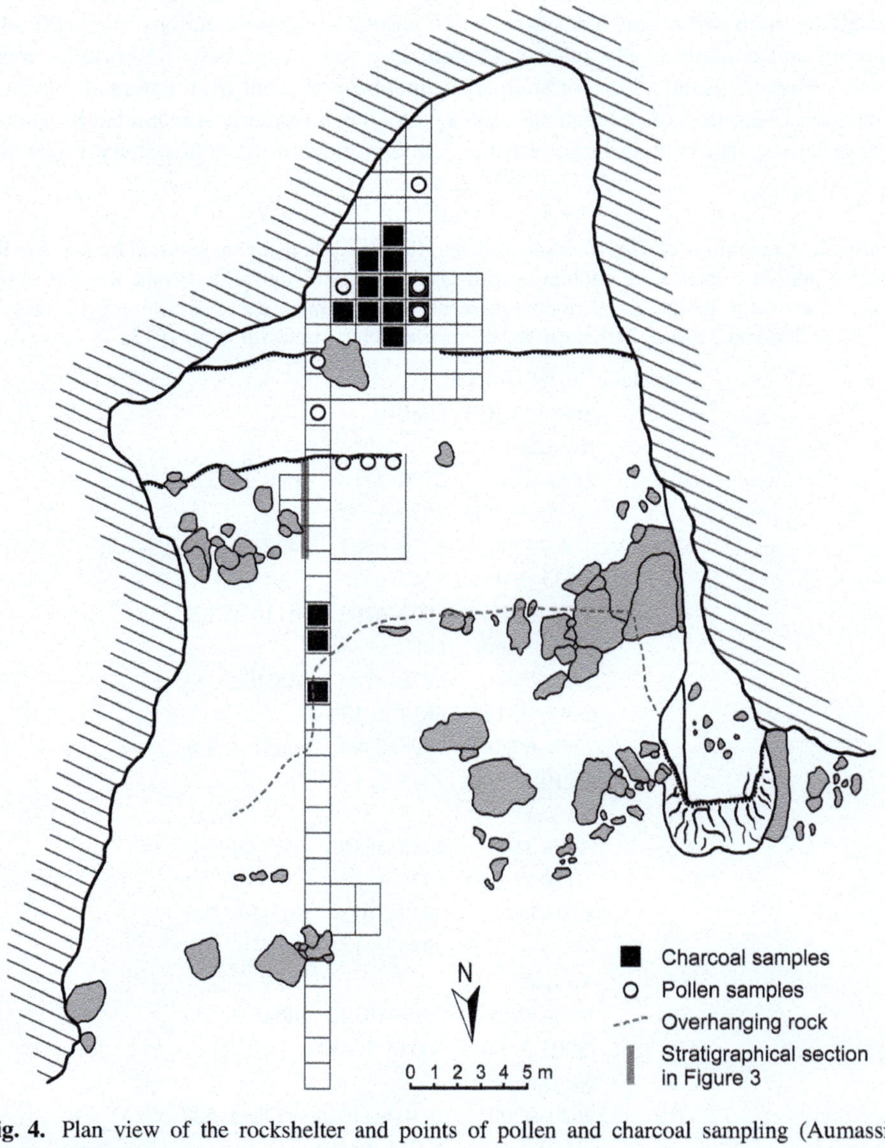

Fig. 4. Plan view of the rockshelter and points of pollen and charcoal sampling (Aumassip 1976, modified)

(1982–1983). Recently, ceramics were dated by thermoluminescence, and new radiocarbon dates were obtained from charcoal, which is abundant in all layers (Table 2; Aumassip et al. 2013). Dates have been calibrated by means of OxCal program 4.3 with IntCal13 (Bronk Ramsey 2009). They demonstrate that the depositional sequence dates from c. 8700 BC (S10 = 8772 ± 507) to c. 700 BC (S2 = 660 ± 100 BC), with most of the layers deposited between about 8000 and 4000 years BC, during the late Early (S8 = 8042 ± 212 BC) and Middle Holocene (S3; 4114 ± 122 BC).

Archaeobotanical Sampling and Laboratory Procedures

Systematic sampling of pollen and charcoal was conducted at several points in the excavated area over six seasons (1974, 1975, 1976, 1978, 1980 and 1981: Aumassip 1976, 1980–1981, 1984; Aumassip et al. 2013). Pollen samples were collected from excavation squares situated near the entrance and inside the shelter (Fig. 4). The points of sampling did not correspond to the exact positions described in Fig. 3, but there is evidence that the same layers extend over the entire site. Many pollen samples were taken from each S layer (Table 3), and a total of 47 samples were treated for pollen extraction. From S9–10, the organic lenses were selected for sampling. The laboratory processing method included enrichment treatment with zinc chloride (Cour 1974). Identification was performed using a Zeiss optical microscope (magnification: ×160, and ×1000 with immersion oil) and consulting various reference works (e.g. Maley 1981; Reille 1992, 1995, 1998).

For charcoal analysis, 92 samples were taken systematically during the archaeological excavations (hand-picked or by dry-sieving) from different levels and squares (Fig. 4; Table 3). Charcoal fragments were identified using a reflected light microscope with bright and dark fields (BF/DF), based on criteria observed in three sections (transverse, longitudinal tangential and logitundinal radial) obtained by breaking charcoal pieces by hand and/or scalpel to obtain a clear section. In some samples, the fragments were difficult to break and some crumbled during the process of fracturing

Table 3. Analyzed pollen and charcoal samples

Archaeological data		Charcoal analysis		Pollen analysis		
Cultures	*Strata*	Number of samples collected	Analyzed fragments	Depth (cm)	Samples with pollen	Pollen count
Bovidian culture	S2	3	56	120	4	1794
	S3	11	750	180	4	1575
	S4	14	950	220	6	2645
Saharo-Sudanese Neolithic	S6	37	1213	295	4	1620
	S8	15	1201	395	6	2622
	S9–10	12	1431	480	4	1440
				500	3	793
				530	2	928
Sum		**92**	**5601**		**33**	**13417**

them for identification. Identification was made with the help of keys, anatomy textbooks, digital programs, wood collections, and anatomical atlases. The main sources were Neumann et al. (2001), Thinon's numeric data base, and the Inside Wood Database (2004-onwards). The botanical nomenclature follows Ozenda (1958), in addition to the synthesis of the research by Quézel et al. (1962).

Results

Out of the 47 samples processed, 33 contained pollen. In total 13,417 microremains were counted, resulting in an average of approximately 1680 grains per sample. All charcoal samples were analysed and 5601 fragments were counted (Table 3). The integrated palynological and anthracological approach proved valuable as it revealed distinctly varied vegetation elements in the depositional history of Ti-n Hanakaten.

State of Preservation of Archaeobotanical Remains

The recovered pollen was generally in a good state of preservation; only a few grains were reddish in color, probably due to reworking and secondary deposition. In some samples, pollen clumps were observed, consisting especially of grasses, Cyperaceae and *Potamogeton*. Such clumps may indicate the presence of flowers or plant accumulation in a site (Mercuri 2008a).

Charcoals were numerous and well preserved because of the special preservation conditions of the site (Aumassip and Délibras 1982–1983) and the dry environment in general (e.g. Cremaschi et al. 2014). Furthermore, large charcoal fragments were more common than medium and small fragments. However sample size varied amongst the sampled strata.

Pollen Flora and Charcoal Species of Ti-n Hanakaten

The systematic study of the Ti-n Hanakaten archaeobotanical remains attests to the presence of mixed and varied associations which, according to several authors (Leredde 1957; Quézel et al. 1962; Ozenda 1977, 2004; Van Campo et al. 1965, 1966, 1967; Cour and Duzer 1976; Maley 1980; Vincens et al. 2007), can be grouped into different ecological groups of species living under varied palaeoenvironmental conditions. Based on this phytoecological approach, apart from the prevalent savanna/grassland taxa, three groups were distinguished in the pollen flora from the Ti-n Hanakaten deposits.

Group I: African tropical or subtropical taxa. This group includes, for example, Sapotaceae (except cf. *Argania*), Melastomataceae, cf. Meliaceae, and can be divided into Saharan (*Acacia*) and non-Saharan (Sahelian and Sudanese) taxa (Maley 1980).

Group II: Holarctic—Eurasian taxa. This group can be subdivided into sub-groups (Van Campo et al. 1967; Beucher 1975; Maley 1980) including temperate (*Pinus*, *Alnus*, *Fagus*, *Salix*, *Castanea*, *Hypericum*, *Epilobium*), and some Mediterranean taxa (*Cupressus*, *Juniperus* type, *Olea*, *Pistacia*). Some of these species currently grow in the Saharan mountains (White 1983). Other plants not living in Sahara (e.g. *Picea*), are markers of atmospheric circulation and long distance transport.

Groupe III: Hygrophilous taxa are well represented in the sampled strata, especially consisting of many Cyperaceae, *Typha*, *Potamogeton*, and *Tamarix* among the woody plants.

In addition to pollen, a number of sporomorphs have been identified such as acritarchs, algae and fern spores which also indicate the presence of wet habitats in the area.

The Pollen and Charcoal Spectra

In the following paragraphs, pollen and charcoal spectra are described for each stratum, starting from the bottom to the top of the stratigraphic sequence (Table 3, and see also Table 4). Pollen data are mainly reported at the family or genera level.

S9–10: The 9 pollen samples from S9–10 produced 60 taxa, a comparably high diversity compared to other samples. The pollen spectrum is strongly dominated by Poaceae (42%) and Cyperaceae (12%). These two families are followed by lower pollen counts of woody plants (especially *Cupressus*), herbaceous/shrubby Asteraceae (both Asteroideae and Cichorioideae) and several hygrophilous taxa (*Typha*, Juncaceae, *Potamogeton*). In addition, identified spores (monolete and trilete) are associated with humid habitats, such as the sporomorphs with aquatic affinity (acritarchs and algae).

The anthracological spectrum of 1431 identified fragments shows the dominant and abundant presence of *Tamarix* (54%). *Olea* (19%) and Fabaceae (10%) followed by 21 other taxa with low frequencies.

Overall, these data are compatible with the presence of savanna and wet habitats in the vicinity of Ti-n Hanakaten during this phase.

S8: This unit produced 40 pollen taxa in 6 samples indicating a lower species diversity. The spectrum is dominated by Poaceae (24%) and taxa of arid environments including Amaranthaceae/Chenopodiaceae (19%) and Asteraceae (15%). Wet environments are well represented by telmatophytes such as *Typha* (10%) and Juncaceae (2%), and by aquatics such as *Potamogeton* (8%) and Nymphaeaceae (4%). Indeterminate pollen grains include mainly three morphological types: inaperturates, syncolpate monads, and tetrahedral tetrads. Trilete and monolete spores were also observed.

Many charcoals (c. 1200 pieces) have been identified in this stratum: olive is abundant (44%) with a significant presence of tamarisks (24%) and Fabaceae (predominatly *Faidherbia albida*: 14%).

Overall, these data suggest that dry environments expanded during a phase that still saw the presence of wet habitats, suggesting a mosaic of habitats existed near the shelter.

S6: The 4 pollen samples obtained from S6 are more or less rich in taxa which total 42, but the same types of indeterminate pollen observed in S8 are abundant (43%). Poaceae (16%), *Potamogeton* (17%), *Typha* (13%) and Asteraceae (11%) are prevalent. Sporomorphs include acritarchs.

Table 4. Results of the analyses showing prevalent taxa/types

Strata	Pollen		Charcoal	
	Number of taxa	Prevalent taxa	Number of taxa	Prevalent taxa
S2	62	Poaceae (51%), Asteraceae (5%), Amaranthaceae/Chenopodiaceae (11%), Cyperaceae (6%); sum of hygrophilous plants (10%)	7	*Tamarix* (46%), *Ziziphus lotus* (35%), *Faidherbia albida* (11%), *Acacia* (6%), *Capparis spinosa* (2%)
S3	73	Poaceae (28%), Boraginaceae (18%), Amaranthaceae/Chenopodiaceae (6%), Cyperaceae (6%), Asteraceae (3%); sum of hygrophilous plants (13%)	30	Fabaceae (41%) including *Faidherbia albida* (26%), Moraceae (18%) and *Ficus salicifolia* (13%), *Olea laperrinei* (9%), *Zizyphus lotus* (5%), Monocotyledons (2%), *Tamarix* (< 1%)
S4	70	Poaceae (19%), Cyperaceae (10%), Asteraceae (6%), Amaranthaceae/Chenopodiaceae (5%); sum of hygrophilous plants (47%) including *Potamogeton* (15%) and *Typha* (11%); traces of *Pinus*, *Fagus* and cf. *Vitis*	28	*Olea laperrinei* (50%), *Faidherbia albida* (22%), *Acacia* (10%), *Acacia seyal* (6%), other Fabaceae (8%), *Ficus salicifolia* (6%), *F. ingens* (5%), Moraceae (9%), *Tamarix* (<1%)
S6	42	Poaceae (16%), Asteraceae (11%), *Plantago* (6%), Cyperaceae (4%); sum of hygrophilous plants (49%) including *Potamogeton* (17%), *Typha* (13%) and Nymphaeaceae (4%)	27	*Olea laperrinei* (51%), *Faidherbia albida* (8%), *Acacia* (10%), *Ficus salicifolia* (6%), *Tamarix* (3%), *Calotropis procera* (3%)
S8	40	Poaceae (24%), Amarantaceae/Chenopodiaceae (19%), Asteraceae (15%), Cyperaceae (8%); sum of hygrophilous plants (35%) including *Typha* (15%), *Potamogeton* (8%), Nymphaceae (4%) and Juncaceae (2%)	26	*Olea laperrinei* (44%), *Tamarix* (24%), *Faidherbia albida* (14%), *Ficus salicifolia* (2%), *Calligonum* sp. (3%), *Balanites aegyptiaca* (1%), *Cupressus dupreziana* (1%)
S9–10	60	Poaceae (42%), Cyperaceae (12%), Asteraceae (4%); sum of hygrophilous plants (32%) including *Typha* (7%), *Potamogeton* (6%), Juncaceae (5%); *Cupressus* (3%), *Olea* (2%) and traces of *Alnus*	29	*Tamarix* (54%), *Olea laperrinei* (19%), *Faidherbia albida* (5%), *Acacia* (5%), Amaranthaceae/Chenopodiaceae (3%), *Calotropis procera* (3%), *Helianthemum* sp. (2%)

A total of 27 taxa were distinguished during the anthracological study of more than 1200 charcoal fragments. The spectrum is dominated by *O. laperrinei* (51%) and Fabaceae (18%).

Overall, these data suggest that probably wet environments were somewhat more widespread than in the previous phase, but different habitats continued to be present near Ti-n Hanakaten.

S4: The pollen flora of 6 samples from S4 is rich, with at total of 70 taxa identified. Poaceae (19%), *Potamogeton* (15%), *Typha* (11%) and Cyperaceae (10%) are represented by comparatively high percentages. Asteraceae (6%) and Amaranthaceae/ Chenopodiaceae (5%) are also significant, with acritarchs less common.

The anthracological examination revealed 28 types. In the 950 fragments counted, the olive tree (50%) dominates, followed by Fabaceae (46%) and Moraceae (9%).

These data demonstrate that savanna prevailed in dry environments and that both were distributed in the area together with wet habitats during this phase.

S3: A rich and varied pollen flora numbering 73 taxa is present in 4 pollen samples obtained from S3, with a dominance of Poaceae (28%) and Boraginaceae (17%). Amaranthaceae/Chenopodiaceae (6%), Cyperaceae (6%) and Asteraceae (3%) have lower frequencies. A high presence of fern spores is accompanied by a few sporomorphs.

The anthracological spectrum obtained in the identification of 750 charcoal fragments is dominated by Fabaceae (41%), followed by Moraceae (18%), *Ficus salicifolia* (13%), and *O. laperrinei* (9%).

The presence of dry habitats is more evident than in the previous phase, and wet habitats are now notably reduced.

S2: Analysis of S2 sediments produced 62 taxa from 4 samples which demonstrate a slightly lower species diversity. The stratum is dominated by Poaceae (51%), followed by Amaranthaceae/Chenopodiaceae (11%) and Cyperaceae (6%).

In the anthracological analysis, a total of 56 fragments were identified, mainly belonging to *Tamarix* (46%), *Ziziphus lotus* (L.) Lam. (35%), Fabaceae (17%) and *Capparis spinosa* (2%).

Dry environments are more evident in the mosaic of vegetation types shown by these spectra.

Discussion

Floristic Elements of Ti-n Hanakaten

The samples from S3 and S4 produced the highest number of identified pollen taxa (73 and 70 types respectively). In the charcoal record, a remarkable number of taxa were identified in the samples from S3, S4 and S10 (30, 28 and 29 types respectively).

Comparing pollen and anthracological results, it is interesting to point out that most of the tree taxa were found in S3 and S4, i.e. between c. 6021 ± 306 and c. 4222 ± 142 BC, representing a crucial time of environmental and cultural dynamics during the Middle Holocene (Table 2). This evidence supports the presence of useful woody plants that were exploited by humans, both at local and regional scales, in a mosaic of wet and dry habitats. One example is the olive tree, which is represented in almost all charcoal samples (except S2) and in all pollen samples, demonstrating the local presence of this species. The same is true for *Acacia* trees which were found as pollen in many samples and charcoal in all samples. In contrast, *Tamarix* charcoal is common while its pollen is sporadic (Table 4).

Cupressus, *Fagonia*, *Ficus*, *Celtis*, *Pistacia* (*P. atlantica*), Anacardiaceae (*Rhus tripartita* (Ucria) Grande), Capparidaceae (*Capparis*) are less represented in the archaeobotanical record of Ti-n Hanakaten. Among the taxa represented only by pollen, *Quercus* (deciduous oak type), *Castanea* and *Salix* occur in almost all samples: they are allochthonous temperate trees not currently present in North Africa. The long distance transport of airborne pollen may explain their presence in the shelter together with that of *Alnus* (in S10), *Fagus*, *Pinus* and probably cf. *Vitis* (in S4). Other isolated pollen grains represent species with tropical affinity, probably distributed in southern regions such as cf. *Argania*, *Myrica*, cf. Sapotaceae, Sapindaceae. Those species represented only in the charcoal records, are *Balanites aegyptiaca*, *Calotropis procera* and *Calligonum* (in all samples except S2), and *Ziziphus lotus*.

In general, the represented trees and shrubs largely include conifers (pine, juniper, cypress) and hardwoods (oak, olive, and acacia). Among them, *Olea*, *Cupressus*, *Juniperus*, *Salix*, *Pistacia*, *Ficus*, *Tamarix*, several species of Fabaceae (*Acacia* and *F. albida*), *Rhus*, Capparidaceae, and *Celtis* may have been important resources and also representatives of the mountain vegetation of the Sahara (White 1983).

Vegetation Cover of Ti-n Hanakaten

The archaeobotanical data suggest that the vegetation around Ti-n Hanakaten was a mosaic composed of trees and associated shrubs scattered in a predominantly grassland or desert savanna habitat, with great variability during the different cultural and environmental phases. Pollen spectra demonstrate that an open landscape was present near the site. In general, grasses comprise the dominant pollen, probably reflecting a savanna-like vegetation that has been continuously present throughout the Holocene in the Ti-n Hanakaten area. Nevertheless, in some levels of the Early (S9–10) and Middle Holocene (S6, S4), there is a co-presence of hygrophilous herbs and grasses indicating savanna and wet environments. In other levels (S8, 8043 ± 212 BC), the similar proportion of grasses/Poaceae and Amaranthaceae/Chenopodiaceae suggest that xerophilous vegetation spread during relatively arid phases. At the top of the depositional sequence (S2, 668 ± 127 BC), despite the prevalence of grasses once again, the expansion of dry environments within a vegetation mosaic suggests that desert savanna spread during the Late Holocene.

In addition to Poaceae, most of the palynological spectra include Amaranthaceae/Chenopodiaceae, Cyperaceae, Brassicaceae, *Plantago* and Asteraceae (other than *Artemisia*). Taken together these taxa comprise between 75 and 95% of total pollen grains in arid to hyper-arid regions between 15 and 32 degrees northern latitude (Cour and Duzer 1978; Watrin et al. 2009). Overall, the results from Ti-n Hakanten indicate that during the Early and Middle Holocene, vegetation cover developed under relatively wetter conditions that were more favorable than those present today in the central Sahara (compare Rognon 1989–1994; Le Houerou 1997; Maley 2010; Mercuri et al. 2011; Garcea et al. 2013; Giraudi et al. 2013; Maley and Vernet 2013; Cremaschi et al. 2014).

The Ti-n Hanakaten pollen flora is similar to that observed in the rock shelters of the Wadi Teshuinat area in the Tadrart Acacus (Mercuri 2008b) where pollen analyses of anthropogenic deposits was useful to reconstruct Holocene environments that saw the development of hunter-gatherer and pastoralist cultures. For example, the pollen flora of Uan Muhuggiag is largely constituted by *Tamarix*, Poaceae and Asteroideae. The hygrophilous *Typha* has significantly higher frequencies at the bottom of the sequence while *Artemisia* characterizes the more recent samples. Poaceae, Cyperaceae, *Echium*, Fabaceae, *Pulicaria*, *Artemisia*, Cichorioideae, *Typha*, *Tamarix* and *Rhus* are well represented at Uan Afuda as well as at Uan Tabu where also *Capparis*, *Ficus* and *Tamarix* are significant. The large quantity of *Typha*, sometimes accompanied by *Potamogeton* and algae and hygrophilous trees, is consistent with more humid climates from S10 to S4, in the Early and early Middle Holocene. Dry environments, especially indicated by Amarathaceae/Chenopodiaceae and some Asteraceae, are better represented in the more recent samples of S3, S2, which date to the late Middle and Late Holocene.

Based on pollen assemblages which consist of various quantities of taxa listed above, and comparing data with the literature (e.g. Mercuri 2008b), it is evident that the pollen preserved at Ti-n Hanakaten was deposited during relatively wetter and drier climatic oscillations of the Holocene.

Wood Exploitation

The wood exploitation at Ti-n Hanakaten is well attested by the charcoal macroremains analysed. The anthracological data indicate that it is probable that the wood of several trees and shrubs, including olive, cypress, acacias, and possibly tamarisk and fig, were used during daily activities, mainly as fuel. The dwellers of the shelter extensively used the wood of the olive tree (*O. laperrinei*), as the charcoal dominates or co-dominates all samples, except for the Late Holocene unit S2. This tree is well-known for its use in cooking and heating, because its hard wood burns in a smoldering combustion process. The wood burns slowly, flameless and with a low temperature and it is also known to produce a high quality charcoal. In some other Saharo-Mediterranean species, for instance cypress, it is possible that the wood or charcoal was transported to the site from the high mountains of Tassili n'Ajjer. Wood of *F. albida*, which is known as a low quality fuel wood, is well represented in the assemblage; it was possibly burned together with *A. raddiana,* which provides excellent fuel wood with high calorific value.

During the earliest phases of the Early and early Middle Holocene (S10 and S8), it seems that shelter inhabitants extensively exploited tamarisks, mainly *Tamarix aphylla*. Although their wood is known to be a poor quality fuel, the wood and twigs of tamarisks are also common in archaeological sites of the Tadrart Acacus area (e.g. Uan Tabu: Neumann and Uebel 2001; Fewet: Zerboni et al. 2013, 2017). At Ti-n Hanakaten, the wood of the fig tree also seems to have been burned. Probably some shrubs/herbs, such as Amaranthaceae/Chenopodiaceae, Boraginaceae and indeterminated monocotyledons, were used as fuel additives, or perhaps as kindling.

Ethnobotanical analogies allow us to suggest possible uses of identified genera at Ti-n Hanakaten. The wood of *Acacia* and *Tamarix* likely served for cooking, others (*Acacia*) for heating, and others (*Olea*) have the dual-use of cooking and heating (Claudot-Hawad 2011; Zerboni et al. 2013). At Ti-n Hanakaten, some species were used despite the poor quality of their wood (such as tamarisks and figs) probably because they were readily available nearby while others were used in abundance for their good combustion proprieties such as olive and cypress.

The charcoal data confirm that Mediterranean flora has grown at 1000 m elevation which is the highest point in the region of Ti-n Hanakaten, during a large part of the Holocene. For example, *C. dupreziana* is known to have colonized these mountains (~2000 m) but also the Saharan plains during the Pleistocene and Neolithic (Ozenda 1991), and it was used in the Libyan oases until the past century (Cremaschi and Zerboni 2011).

Together with further archaeobotanical results, these charcoal investigations illustrate human activites and the use of a wide variety of plants. The uses of wood for construction or demarcation of areas in the interior of the cave (Aumassip 1976) are also attested by the presence of wooden stakes of *F. salcifolia* and *A. seyal* (Amrani 2016).

The presence of large pieces of charcoal suggests that larger branches were gathered. It is also possible that, especially for taxa not growing in the vicinity of the shelter, the wood was charred at the location of the exploited tree and the much lighter charcoal fragments transported to the shelter, instead of carrying large and heavy wooden pieces. Actually, ethnobotanical studies on traditional charcoal making technology in the central Sahara (Zerboni et al. 2013) describe a special type of charcoal production (*esed*) adopted by the Tuareg (Kel Tadrart) from the Tadrart Acacus massif (SW Libya, Central Sahara) in response to rapid changes in the environment. During difficult periods, the Kel Tadrart mainly collect wood of *Acacia* and *Tamarix* that are readily available, and incompletely burn the woods. This may result in pieces of partially carbonized wood of variable size (Zerboni et al. 2013).

Conclusion

The archaeobotanical record of the Tin-Hanakaten rock shelter covers approximately 8000 years during which several crucial environmental changes occurred under the climate oscillations of the Early, Middle and Late Holocene. The archaeobotanical results show that, at least until 1000 BC, diverse environments and natural (plant and water) resources existed in this central Saharan region. Although the resolution of these analyses will be improved in future studies, the main vegetational and environmental changes are evident based on the palynological and anthracological spectra reported in this paper. These analyses have demonstrated that from c. 8700 to c. 8100 BC (S9–10), savanna and wet habitats were already present in the area. Then, from c. 8100 to c. 8000 BC (S8), dry environments probably characterised by shrubby associations of xerophytes (Amaranthaceae/Chenopodiaceae and Asteraceae) expanded during an arid phase. Interestingly the region continued to have wet habitats with fresh water available. This mosaic of habitats made these mountains especially suitable for human survival even during the less fauvorable climatic periods. This is especially evident in a desert area because humans needed to live near water during the expansion of the desert (Garcea et al. 2013; Mercuri et al. 2016). The hygrophilous plants identified at the site were linked to neighboring wadi, while the herb cover probably spread over open spaces of rocky landscapes. The vegetation comprised a mosaic of habitats including wooded and shrubby plant associations, together with savanna-like vegetation. Tamarisks, olives and acacias were the most important representatives of the woody vegetation.

In the second part of the Early Holocene, from c. 8000 to c. 6100 BC (S6), wet environments were probably more prevalent compared to the previous phase but a variety of habitats and natural resources continued to exist near the site. During the early part of the Middle Holocene, from c. 6000 to c. 4600 BC (S4), a mosaic of savanna and dry shrubby vegetation was evident together with wet habitats with water levels probably subjected to seasonal variability (Mercuri 2008b; Cremaschi et al. 2014). In the second part of the Middle Holocene, from c. 4200 to c. 4100 BC (S3), dry environments began to spread and wet habitats were notably reduced, as occurred everywhere in the central Sahara (Mercuri et al. 2011). This process has continued for millennia that are not represented in the Ti-n Hanakaten sequence, but in the Late Holocene, c. 700 BC (S2), dry environments with the desert vegetation communities dominated the region.

The Early and Middle Holocene landscape of Ti-n Hanakaten as evidenced by our archaeobotanical studies is quite different from that of today, which is characterized by a vast open habitat with some herbaceous species (useful for grazing) and a widely spaced, stunted and degraded tree flora consisting mainly of acacias (Aumassip 1976). In general, Holocene palaeoenvironmental reconstructions completed at many sites in the central Sahara suggest that favourable environmental conditions for human occupation were offered by the mountains, particularly the Tadrart Acacus and Tassili n'Ajjer, during subsequent Holocene phases (Schulz 1987; Cremaschi 1998; Mattingly 2003; Brooks et al. 2005). *Olea laperrinei* and *Cupressus dupreziana* were the most remarkable woody plants identified at the shelter. Different subsistence practices, including hunter-gathering followed by pastoralism, took place in these areas, indicating continuous human occupation and resource exploitation at both the Ti-n Hanakaten region and the high mountains of Tassili n'Ajjer (up to ~2000 m). The plants and wood preserved at Ti-n Hanakaten represent the remnants of a number of activities such as food and craft production, in addition to having domestic uses, including cooking. These interpretations are supported by other archaeobotanical records of the central Saharan region (Mercuri 2008b; Dunne et al. 2012).

References

Abdoun F, Beddiaf M (2002) *Cupressus dupreziana* A. Camus: répartition, dépérissement et régénération au Tassili n'Ajjer, Sahara central. C R Biol 325(5):617–627

Alimen H, Beucher F, Lhote H (1968) Les gisements néolithiques de Tan-Tarteït et d'In Itinen. Tassili n'Ajjer (Sahara Central). Bull Soc Préhist Fr 65:421–458

Amrani (2002) Reconstitution des environnements holocènes de la grotte de Tin Hanakaten, (Tassili n'Ajjer, Sahara central), apport de l'archéopalynologie. Thèse obtention du diplôme de Magistère, Institut d'Archéologie, Université d'Alger

Amrani (2016) Reconstitutions des environnements holocènes du tassili n'Ajjer (Sahara central). Thèse obtention du diplôme de doctorat, Institut d'Archéologie, Université d'Alger

Aumassip G (1976) Le site préhistorique de Tin Hanakaten. Rapport préliminaire. Bull Intér CRAPE, n° 5, Ronéo

Aumassip G (1977) Campagne de fouille de Tin Hanakaten 1976. Bull CRAPE, N°7/8

Aumassip G (1980–1981) Ti-n-Hanakaten, Tassili-n-Ajjer, Algérie. Bilan de 6 compagnes de fouilles. Libyca, XXVIII–XXIX, p 115–127

Aumassip G (1984) Ti-n-Hanakaten, Tassili-n-Ajjer, Algérie. Bilan de 6 compagnes de fouilles. Libyca, p 28–29

Aumassip G (2000) Le site préhistorique de Tin Hanakaten (Tassili n, Ajjer, Algérie) et les débuts du Néolithique au Sahara, in: Premiers paysans du monde. Naissance des agricultures, J Guilaine J (ed) Paris, Errance, p 243–260

Aumassip G, Délibrias G (1982–1983) Ages de dépôts néolithique du gisement de Ti-n-Hanakaten. (Tassili n-Ajjer), Algérie. Libyca, XXX–XXXI:207–211

Aumassip G, Heim JL (1989) Les squelettes néolithiques de Tin Hanakaten, Tassili n'Ajjer, Algerie. C.R.Ac.Sc. Paris, Série III 309:187–190

Aumassip G, Jungner H, Schvoerer M (2013) Le site de Tin Hanakaten (Tassili Azjer, Algerie) et la chronologie de l'art rupestre saharien. Revue IKOSIM, N° 2: 49–60

Ballouche A, Thinon M, Reille M et al (1995) La végétation Holocène des montagnes du Sahara Central: la fin d´un mythe. Holocene 6(4):457–462

Beucher F (1975) Etude palynologique de formation Néogènes et Quaternaire au Sahara Nord-Occidental. Thèse Doc. Es. Sc., Paris, A. O., C.N.R.S., n° 5, p 796

Biagetti S, di Lernia S (2013) Holocene deposits of Saharan rock shelters: the case of Takarkori and other sites from the Tadrart Acacus Mountains (Southwest Libya). Afr Archaeol Rev 30 (3):305–338

Bronk Ramsey C (2009) Bayesian analysis of radiocarbon dates. Radiocarbon 51(1):337–360

Brooks N, Chiapello I, di Lernia S et al (2005) The climate–environment–society nexus in the Sahara from prehistoric times to the present day. J N Afr Stud 10:253–292

Castelletti L, Castiglioni E, Cottini M et al (1999) Archaeobotanical evidence at Uan Afuda: charcoal, wood and seeds from the Early Holocene sequence. In: di Lernia S (ed) The Uan Afuda Cave Hunter-gatherer Societies of Central Sahara, Arid Zone Archaeology. Monographs 1. All'Insegna del Giglio, Firenze, p 131–148

Castelletti L, Cottini M, Rottoli M (1998) Early Holocene plant remains from Uan Afuda cave (Libyan Sahara). In: di Lernia, S, Manzi G (eds) Before food production in North Africa. Questions and tools dealing with resource exploitation and population dynamics at 12,000–7000 bp, Workshop of the XIII Congress, ABACO, Forlì, p 91–102

Claudot-Hawad H (2011) 'Avoir de la cendre': ancrages nomades en territoire touareg. Techniques et culture, Editions de la Maison des sciences de l'homme 1(56):78–97

Cour P (1974) Nouvelles techniques de détection des flux et des retombées polliniques de la sédimentation actuelle au Sahara nord-occidental. «Palynology in medicine», Proceed. 3rd Intern. Palyn Conf, Moscou, p 41–58

Cour P, Duzer D (1976) Persistance d'un climat hyperaride au Sahara central et méridional au cours de l'Holocène. Rev Geogr Phys Geol Dyn 18:175–198

Cour P, Duzer D (1978) La signification climatique, edaphique et sedimentologique des rapports entre taxons en analyse pollinique. Annual Mines Belg 7–8:155–164

Couvert (1979) L'homme et l'environnement au cours du Quaternaire recent. Bull Intér CRAPE, n° 4, Ronéo

Cremaschi M (1998) Late quaternary geological evidence for environmental changes in south-western Fezzan (Libyan Sahara). In: Cremaschi M, di Lernia S (eds) Wadi Teshuinat palaeoenvironment and prehistory in South-Western Fezzan (Libyan Sahara). Quaderni Geodinamica Alpina e Quaternaria. C.N.R, Firenze, pp 13–47

Cremaschi M, Pelfini M, Santilli M (2006) *Cupressus dupreziana*: a dendroclimatic record for the middle-late Holocene in the central Sahara. Holocene 16(2):293–303

Cremaschi M, Zerboni A (2011) Human communities in a drying landscape. Holocene climate change and cultural response in the central Sahara. In: Martini IP, Chesworth, W (eds) Landscape and societies, selected cases. Springer Science, p 67–89

Cremaschi M, Zerboni A, Mercuri AM et al (2014) Takarkori rock shelter (SW Libya): an archive of Holocene climate and environmental changes in the central Sahara. Quat Sci Rev 101:36–60

di Lernia S, Bruni S, Cislaghi I et al (2016) Colour in context. Pigments and other coloured residues from the early-middle Holocene site of Takarkori (SW Libya). Archaeol Anthropol Sci 8(2):381–402

di Lernia S, Gallinaro M (2010) The date and context of Neolithic rock art in the Sahara: engravings and ceremonial monuments from Messak Settafet (South-West Libya). Antiquity 84:954–975

di Lernia S, N'siala IM, Mercuri AM (2012) Saharan prehistoric basketry. Archaeological and archaeobotanical analysis of the early-middle Holocene assemblage from Takarkori (Acacus Mts., SW Libya). J Archaeol Sci 39(6):1837–1853

Dunne J, Evershed RP, Salque M et al (2012) First dairying in green Saharan Africa in the fifth millennium BC. Nature 486(7403):390–394

Dunne J, Mercuri AM, Evershed RP et al (2016) Earliest direct evidence of plant processing in prehistoric Saharan pottery. Nat Plants 3:16194

Ehreinreich S, Fuchs G (2011) Archäologische Untersuchung eines "Goulet" (site no. 338) im Wadi Tifariti, Westsahara, Soboth, ARGIS Archäologie Service GmbH

Gallin A (2008) Peopling of the Sahel during the Late Neolithic: new data from Kobadi site (Malian Sahel, 1700-1400 BC). In: 19th Biennial conference of the society of Africanist archaeologists

Gallinaro M, Gauthier C, Gauthier Y et al (2012) The Messak project. Cultural and natural preservation and sustainable tourism (South-Western Libya). Antiquity 86:331

Garcea EAA, Mercuri AM, Giraudi C (2013) Archaeological and environmental changes between 9500 BP and 4500 BP: a contribution from the Sahara to understand expanding droughts in the "great Mediterranean". Ann Bot (Roma) 3:115–120

Gauthier Y (2009) Nouvelles réflexions sur les aires de distribution au Sahara central. Les Cahiers de l'AARS 13:121–134

Giraudi C, Mercuri AM, Esu D (2013) Holocene palaeoclimate in the Northern Sahara margin (Jefara Plain, Northwestern Libya). Holocene 23:339–352

Guinet P, Planque D (1969) "Résultats de l'analyse pollinique". In: G Camps (ed) Amekni: Néolithique ancien du Hoggar. Mém. Centre Rech Anthropol Préhist Ethnograph 10:186–188

Hachid M, Le Quellec JL, Agsous S et al (2010) Premiers résultats du projet algéro-français de datation directe et indirecte des images rupestres dans la Tasili-n-Ajjer. Sahara 21:27–58

Hachid M, Le Quellec JL, Beck L et al (2012) Quelques résultats du projet de datation directe et indirecte de l'art rupestre saharien. In: Huyge D, Van Noten F (eds) International colloquium "The signs of which times? Chronological and palaeoenvironmental issues in the rock art of Northern Africa". Royal Academy for Overseas Sciences, Brussels, pp 71–96

Haddad B, Carra A, Saadi A et al. (2018) In vitro propagation of the relict laperinne's olive (*Olea europaea* L. subsp. *Laperrinei*). Plant Biosyst 152(4):621–630

Le Houerou HN (1997) Climate, flora and fauna changes in the Sahara over the past 500 million years. J Arid Environ 37:619–647

Le Quellec JL (2008a) À propos des molettes zoomorphes du Sahara central. Sahara 19:39–60
Le Quellec JL (2008b) «What's new in the Sahara, 2000–2007». In: Bahn PG, Franklin N, Strecker M (eds) News of the world III. Oxford
Le Quellec JL (2012) Iconoclasties rupestres au Sahara. Sahara 23:29–42
Leredde C (1957) Etude écologique et phytogéographique du Tassili n'Ajjer. Inst Rech Sah, Alger II, Trav, p 455
Magid AA (1999) Preliminary study of plant impressions in pottery. The Uan Afuda Cave. Hunter-gatherer societies of Central Sahara. Arid Zone Archaeol Monogr 1:183–187
Maley J (1980) Les changements climatiques de la fin du Tertiaire en Afrique: leur conséquence sur l'apparition du Sahara et de sa végétation. In: Williams MAJ, Faure H (eds) The Sahara and the Nile. A.A. Balkema, Rotterdam, p 63–86
Maley J (1981) Etudes palynologiques dans le bassin du Tchad et paléoclimatologie de l'Afrique nord-tropicale de 30,000 ans à l'époque actuelle. Travaux et documents de l'ORSTOM 129, Paris
Maley J (2010) Climate and palaeoenvironment evolution in north tropical Africa from the end of the Tertiary to the Upper Quaternary. Palaeoecol Afr 30:227–278
Maley J, Vernet R (2013) Peuples et évolution climatique en Afrique nordtropicale, de la fin du Néolithique à l'aube de l'époque moderne. In: Vernet T, Lefebvre C, Seignobos R (eds) Histoire et archéologie du Sahel ancien: nouveaux regards, nouveaux chantiers. Afriques 04/2013
Mattingly DJ (ed.) (2003) The archaeology of Fazza^n, vol. 1. Synthesis. The Department of Antiquities/The Society for Libyan Studies, Tripoli/London
Mercier N, Le Quellec JL, Hachid M et al (2012) OSL dating of quaternary deposits associated with the parietal art of the Tassili-n-Ajjer (Central Sahara). Quat Geochronol 30:1–7
Mercuri AM (1999) Palynological analysis of the Early Holocene sequence. In: di Lernia S (ed) The Uan Afuda cave Hunter-gatherer societes of Central Sahara, Arid Zone Archaeology, Monographs 1. All'Insegna del Giglio, Firenze, pp 149–181
Mercuri AM (2008a) Human influence, plant landscape, evolution and climate inferences from the archaeobotanical records of the Wadi Teshuinat area (Libyan Sahara). J Arid Environ 72:1950–1967
Mercuri AM (2008b) Plant exploitation and ethnopalynological evidence from the Wadi Teshuinat area (Tadrart Acacus, Libyan Sahara). J Archaeol Sci 35(6):1619–1642
Mercuri AM, Florenzano A, Giraudi C et al. (2016) Humans and water in desert 'refugium' areas: palynological evidence of climate oscillations and cultural developments in early and mid-Holocene Saharan edges. Interdisciplinaria Archeologica-IANSA 6 (2)
Mercuri AM, Fornaciari R, Gallinaro M et al. (2018) Plant behaviour through human imprints and the cultivation of wild cereals in Holocene Sahara. Nat Plants 4:71–81, https://doi.org/10.1038/s41477-017-0098-1
Mercuri AM, Garcea EAA (2007) The impact of hunter/gatherers in the Central Sahara during the Early Holocene. In: Cappers RTJ (ed) Fields of Changes. Progress in African archaeobotany. Groningen Archaeological Studies 5, Groningen, p 87–104
Mercuri AM, Sadori L, Uzquiano Ollero P (2011) Mediterranean and North-African cultural adaptations to mid-Holocene environmental and climatic changes. Holocene 21:189–206
Mercuri AM, Trevisan Grandi G (2001) Palynological analyses of the late Pleistocene, early Holocene and middle Holocene layers. In: Garcea EAA (ed) Uan Tabu in the settlement history of the Libyan Sahara, Arid Zone Archaeology, Monographs 2. All'Insegna del Giglio, Firenze, pp 161–188

Mercuri AM, Trevisan Grandi G, Accorsi CA (1999) Plant exploitation during the early and middle Holocene in the Tadrart Acacus (Central Sahara, Libya): pollen evidence of changes. In: Guarino A (ed) Proceedings of the second science and technology for the safeguard of cultural heritage in the Mediterranean Basin. Elsevier, Paris, p 1081–1085

Mercuri AM, Trevisan Grandi G, Mariotti Lippi M et al (1998) New pollen data from the Uan Muhuggiag rockshelter (Libyan Sahara, VII–IV millennia Bp). In: Cremaschi M, di Lernia S (eds) Wadi Teshuinat: Palaeoenvironment and prehistory in South-Western Fezzan (Libyan Sahara). Quaderni Geodinamica Alpina e Quaternaria. C.N.R, Firenze, pp 107–124

Monod, T. (1957) Les grandes divisions Chorologiques de l'Afrique. Publ. 24. London: CCTA

Neumann K, Schoch W, Detienne P et al (2001) Woods of the Sahara & Sahel. Verlag Paul Haupt, Bern, Stuttgart, Wien, p 465

Neumann K, Uebel, D (2001) The cold Early Holocene in the Acacus: Evidence from charred wood. Uan Tabu in the Settlement History of the Libyan Sahara, All'Insegna del Giglio: Firenze, p 211–213

Ozenda P (1958) Flore du Sahara septentrional et Central. C.N.R.S, Paris

Ozenda P (1977) Flore du Sahara. C.N.R.S, Paris, p 622

Ozenda P (1991) Les relations biogéographiques des montagnes sahariennes avec la région méditerranéenne. Revue de géographie alpine, tome 79(1):43–53

Ozenda P (2004) Flore et végétation du Sahara, Paris, CNRS éditions, 3e éd., p 662

Pasa A, Pasa Durante MV (1962) Analisi paleoclimatiche nel deposito di Uan Muhuggiag, nel massiccio dell'Acacus (Fezzan meridionale). Memorie Museo Civico di Storia Naturale di Verona 10:251–255

Pons A, Quézel P (1957) Première étude palynologique de quelques paléosols sahariens. Tv Inst Rech Sah Univ Alger, XVI, pp 15–30

Pons A, Quézel P (1958) Premières remarques sur l'étude palynologique d'un guano fossile du Hoggar. C R Sc Ac Sci 244:2290–2292

Quézel P (1954) Contribution à l'étude de la flore et de la végétation du Hoggar. Inst Rech Sah, Alger, Monographie régionale, no 2:1–164

Quézel P (1956) Contribution à la flore d'Afrique de Nord: VI. Contribution à l'étude de la flore de la Tefedest (Hoggar). Bull Soc Hist Nat Afr Nord 47:131–136

Quézel P (1960a) Flore et Palynologie sahariennes: Rech Sahar, Unii, Alger. Mém. no 4:357

Quézel P (1960b) Flore et Palynologie sahariennes: quelques aspects et leur signification biogéographique et paléoclimatique. Bull Inst Fi Afr Noire 22, sér. A., n°2, p 353–360

Quézel P (1963) De l'application de techniques palynologiques à un territoire désertique. In Paleoclimatologie du Quaternaire récent au Sahara', in Arid Zone Research-Symposium XX-Changes of Climate, Paris, UNESCO p 243–249

Quézel P (1967) Signification biogéographique et paléoclimatique de quelques représentants de la flore saharienne. Palaeoecol Afr 2:62–67

Quézel P (1978) Analysis of the flora of Mediterranean and Saharan Africa. Annales Missouri Botanical Garden 65:479–534

Quézel P, Martinez C (1960) Le dernier interpluvial au Sahara central: essai de chronologie palynologique et paléo-climatique. Lybica 6–7:211–227

Quézel P, Santa S, Schotter O (1962) Nouvelle flore de l'Algerie et des regions desertiques meridionales-v. 1–2. Paris, CNRS, p 1170

Quézel P, Thébault JY (1959) Palynologie et datation du volcanisme récent de l'Ahaggar. Bull Sci Er Econ B R G M 6:59–64

Reille M (1992) Pollen et spores d'Europe et d'Afrique du Nord. URA CNRS, Marseille

Reille M (1995) Pollen et spores d'Europe et d'Afrique du Nord. (Suppl. 1), URA CNRS, Marseille

Reille M (1998) Pollen et spores d'Europe et d'Afrique du Nord. (Suppl. 2), URA CNRS, Marseille

Rognon P (1967) Le massif de l'Atakor et ses bordures (Sahara Central). Etude géomorphologique. Thèse Doctorat d'Etat ès lettres. Editions du CNRS, Paris, p 559

Rognon P (1989–1994) Biographie d'un désert. Plon, Paris, 1989, p 240 (réédition en 1994, sous le titre: Biographie d'un désert, le Sahara. L'Harmattan, Paris)

Saliège J-F, Zazzo A, Paris F (2008) Chronologie des monuments à couloir et enclos d'Emi-Lulu. Datation croisée de l'émail dentaire et des os des squelettes des humains inhumés, Les déserts d'Afrique et d'Arabie: environnement, climat et impact sur les populations. Colloque de l'académie des sciences, Institut de France, 8–9 septembre 2008, Paris, p 9–7

Schulz E (1986) Zur Vegetation der Sahara undilhrer Randgebiete. Würzburger Geogr Manuskript 16:52–64

Schulz E (1987) Holocene vegetation in the Tadrart Acacus: the pollen record of two early ceramic sites. In: Barich BE (ed), Archaeology and environment in the Libyan Sahara. The Excavations in the Tadrart Acacus, 1978–1983. British Archaeological Reports, p 313–326

Thinon M, Ballouche A, Reille M (1996) Holocene vegetation of the Central Saharan mountains: the end of a myth. Holocene 6(4):457–462

Trevisan Grandi G, Mariotti Lippi M, Mercuri AM (1998) Pollen in dung layers from rockshelters and caves of Wadi Teshuinat (Libyan Sahara). In: Cremaschi M, di Lernia S (eds), Wadi Teshuinat: palaeoenvironment and prehistory in South-Western Fezzan (Libyan Sahara). Quaderni Geodinamica Alpina e Quaternaria. C.N.R., Firenze, p 95–106

Trevisan Grandi G, Mercuri AM (1992) Ricerche geobotanico e paletnobotaniche nel Sahara libico: siti neolitici ed epipaleolitici nel Tadrart Acacus (Fezzan Sud-Occidentale). Atti Società Naturalisti e Matematici di Modena 123:53–71

Van Campo M (1964) Quelques pollens pléistocènes nouveaux pour le Hoggar. C R Acad Sci 258:1297–1299

Van Campo M, Cohen J, Guinet P et al (1965) Contribution à l'étude du peuplement végétal Quaternaire des montagnes Sahariennes II: Flore contemporaine d'un gisement de mammifères tropicaux dans L'Atakor. Pollen Spores 7(2):361–371

Van Campo M, Guinet P, Cohen J et al (1967) Contribution à l'étude du peuplement. Végétal Quaternaire des montagnes Sahariennes III: Flore de l'Oued Outoul (Hoggar). Pollen Spores 9 (1):107–120

Van Campo M, Guinet P, Dutil P (1966) Nouvelle flore pollinique des alluvions pléistocènes d'un bassin versant au sud du Hoggar. C R Acad Sci 263:487–490

Vincens A, Lezine AM, Buchet G et al (2007) African pollen database inventory of tree and shrub pollen types. Rev Palaeobot Palyno 145:135–141

Wasylikowa K (1992a) Exploitation of wild plants by prehistoric peoples in the Sahara. Würzburger Geogr Arbeiten 84:247–262

Wasylikowa K (1992b) Holocene flora of the Tadrart Acacus area, SW Libya, based on plant macrofossils from Uan Muhuggiag and Ti-n-Torha two caves archaeological sites. Origini 16:125–152

Wasylikowa K (1993) Plant macrofossils from the archaeological sites of Uan Muhuggiag and Ti-n-Thora, southwestern Libya. In: Krzyzaniak L, Kobusiewicz M, Alexander J (eds) Environmental change and human culture in the Nile Basin and Northern Africa until the 2nd Millennium BC. Poznan Archaeol Museum, Poznan, pp 25–41

Watrin J, Lézine AM, Hély C et al (2009) Plant migration and plant communities at the time of the "green Sahara". C R Geosci 341:656–670

White F (1983) The vegetation of Africa, a descriptive memoir to accompany the UNESCO/AETFAT/UNSO vegetation map of Africa (3 Plates, Northwestern Africa, Northeastern Africa, and Southern Africa, 1: 5,000,000)

Zerboni A, Biagetti S, Massamba N'Siala I et al (2013) Burning without slashing. Cultural and environmental implications of a traditional charcoal making technology in the central Sahara. J Arid Environ 98:126–131

Zerboni A, Mori L, Bosi G et al (2017) Domestic firing activities and fuel consumption in a Saharan oasis: Micromorphological and archaeobotanical evidence from the Garamantian site of Fewet (Central Sahara, SW Libya). J Arid Environ 144:123–138

The Use of Wild Plants in the Palaeolithic and Neolithic of Northwestern Africa: Preliminary Results from the PALEOPLANT Project

Yolanda Carrión Marco[1(✉)], Jacob Morales[2,3], Marta Portillo[2,4], Guillem Pérez-Jordà[5], Leonor Peña-Chocarro[5], and Lydia Zapata[2]

[1] PREMEDOC, Prehistoria Del Mediterráneo Occidental. Departamento de Prehistoria, Arqueología e Historia Antigua, Universitat de València, Valencia, Spain
Yolanda.carrion@uv.es

[2] Departamento de Geografía, Prehistoria Y Arqueología, Universidad Del País Vasco, UPV-EHU, Vitoria-Gasteiz, Spain
jacob.morales@ulpgc.es, m.portilloramirez@reading.ac.uk

[3] GI Tarha, Departamento de Ciencias Históricas, Universidad de Las Palmas de Gran Canaria, Las Palmas, Spain

[4] Department of Archaeology, University of Reading, Reading, UK

[5] GI Arqueobiología, Instituto de Historia (CCHS). Consejo Superior de Investigaciones Científicas, Madrid, Spain
{guillem.perez,leonor.chocarro}@csic.es

Abstract. This contribution presents preliminary results from PALEOPLANT, a European Research Council (ERC) multidisciplinary research project that focuses on the analyses of plant remains from Palaeolithic and Neolithic sites in North Africa. The project aims are to: (1) characterize wild plant exploitation among human communities; (2) refine current knowledge on past landscapes; and (3) improve archaeological methodology and fieldwork when studying plants. We discuss preliminary results from charred plant macroremains (wood, seeds and fruits) and plant microfossils (phytoliths and calcitic ash pseudomorphs), as well as dung spherulites recorded in several northwestern African sites spanning the Late Pleistocene to the Early-Middle Holocene (c. 33 to 2 ka BC). Charcoal assemblages indicate the presence of mixed oak/juniper woodlands during the Late Pleistocene while conifers become the dominant element of the vegetation during the transition from Late Glacial to Early Holocene. In these contexts, some of the main species of thermo- and meso-Mediterranean Holocene landscapes (*Olea europaea*, *Pistacia*, etc.) were already present but their spread occurred during the Middle Holocene, concurrently with Neolithic occupations. The combination of wood charcoal, other plant macroremains (seeds and fruits) and plant microfossils, including both plant and dung indicators, provide new insights not only into the presence of a wide range of plant resources (wild pulses, *Ziziphus*

A. M. Mercuri et al. (Eds.): *Plants and People in the African Past*,
https://doi.org/10.1007/978-3-319-89839-1_9

lotus, *Pinus halepensis*, *Quercus* sp., *Stipa tenacissima* and *Chamaerops humilis*) but also on the variability of their potential uses (including fuel, food, fiber, bedding, fodder). Archaeobotanical results suggest continuity in the use of wild plants during the Neolithic. In addition, wood charcoal assemblages reflect significant changes that occurred in plant composition including the expansion of thermo-Mediterranean species.

Keywords: Northwestern Africa · Palaeolithic · Neolithic · Charcoal Seeds · Calcitic microfossils · Phytoliths · Spherulites · Food plants Basketry

Introduction

Plants have been fundamental for human societies across the planet and many communities still depend upon gathering wild plants for their subsistence (Kelly 2013). Nevertheless, most archaeobotanical research has focused on domesticated plants from later chronological periods (Hather and Mason 2002). Consequently, there is a lack of information when it comes to archaeological evidence of plant foraging and collecting during the Palaeolithic of northern Africa. Such studies, however, are emerging in other regions (Sievers 2006; Wadley et al. 2011; Baines et al. 2015; Riehl et al. 2015; Badal García and Martínez Varea 2018, among others). Moreover, taphonomic processes generally favor the preservation of hard remains such as bone and shell while more fragile plant parts such as leaves, do not preserve as well. As a result, interpretations and syntheses of the Palaeolithic and Epipalaeolithic have most certainly overestimated the role of hunting and fishing, and underestimated the role of plants in ancient human diets. Because of these bias efforts to collect plant remains during excavations have been very limited (Kornfeld 1996; Hather and Mason 2002). The resulting outcome has been a limited picture of pre-agrarian subsistence where plants are often neglected or play a secondary role (Richards and Trinkaus 2009; Hardy 2010; Colonese et al. 2011; Mannino et al. 2012).

In this paper, we present the first results of PALEOPLANT, a multidisciplinary research project, focusing on the study of plants during the Palaeolithic and Epipalaeolithic periods. The project aims at exploring the interaction of humans with their environment, searching for new empirical discoveries which can lead to a novel understanding of plant exploitation. In order to address this urgent research question, the project and the current paper pursue the following objectives: (1) to assess wild plant exploitation among Palaeolithic groups; (2) to refine current knowledge of Palaeolithic landscapes; and (3) to improve archaeobotanical methodologies by evaluating sampling strategies and analytical procedures during fieldwork and other stages of the research. Here we present a summary of preliminary results, focusing on plant exploitation during the Middle and Later Stone Age in northwestern Africa. Some Neolithic contexts recorded in the upper layers of studied sites have been also included because during this period changes and continuities in the landscape and/or in the use of plants are detected.

Study Area

Northwestern Africa, including Morocco, Algeria and Tunisia, is the area under examination in this contribution (Fig. 1). This is a mountainous region with a Mediterranean climate that is home to a rich and varied flora (Charco 2001; Fennane et al. 2007; Le Floc'h et al. 2010). Climate conditions exert a strong influence on vegetation in North Africa: this region is isolated by barriers including Mediterranean Sea, the Atlantic Ocean and the Sahara. Mean annual temperatures range from 13 °C to 19 °C (although there are important differences within the region) and the mean minimum temperature ranges from 1 °C to 10 °C. Rainfall is restricted to the autumn to spring months, and summer is hot and dry, ranging from 20 mm in the south near the Sahara, to 2000 mm in northern Morocco (Rejdali 2004).

In terms of vegetation, the Maghreb has a rich flora, with a high level of endemism, which approaches 27% in some regions. The flora includes Mediterranean, Macaronesian, Saharo-Sindian and tropical species. The diverse climate and geology result in different forest types. The following principal forest types are found: pine (*Pinus pinaster* Aiton var. *maghrebiana* Vill., *Pinus nigra* var. *mauretanica* Maire & Peyerimh), Berber thuya (*Tetraclinis articulata* Valh.), oak (*Quercus suber* L., *Q. rotundifolia* L., *Q. faginea* Lam., *Q. coccifera* L.), wild olive (*Olea europaea* L.) and carob (*Ceratonia siliqua* L.) woodlands and maquis. Scrublands are diverse in size and coverage, ranging from shrubby formations dominated by Lamiaceae, Cistaceae, *Phillyrea* and others, to closed canopies. Steppes of *Stipa tenacissima* L. and *Artemisia* are frequent as well (Fennane 1989; Rejdali 2004).

The region is also well known for its interesting archaeological record and well-dated Middle (MSA) and Later Stone Age (LSA) sequences (Lindstädter et al. 2012). The MSA in northwest Africa is associated with anatomically modern humans and it dates as far back as c. 300,000 years ago (Barton et al. 2009; Nami and Moser 2010; Jacobs et al. 2012; Richter et al. 2017), extending to c. 27,000 years ago (Barton et al. 2016). For the LSA, chronology spans from c. 20,000 BC (22,000 cal yr BP) to 5500 BC (7500 cal yr BP; Lubell 2001; Lindstädter et al. 2012; Barton et al. 2013). The LSA is characterized by the occurrence of microlithic bladelet industries that characterize the Iberomaurusian (c. 20,000–10,000 BC; 22,000–12,000 cal yr BP) and the Capsian/Epipalaeolithic (c. 8000–5500 BC; 10,000–7500 cal yr BP). These are well known cultures distinguished by the development of rich cultural deposits and rather elaborate human burials (Camps 2000; Lubell 2001; Barton et al. 2013; Humphrey et al. 2014).

Materials and Methods

PALEOPLANT focuses on pre-agrarian plant exploitation through an interdisciplinary approach which includes the analysis of plant macroremains (seeds, fruits and non-woody plant remains), wood charcoal, and plant microfossils (phytoliths and calcitic ash pseudomorphs), as well as dung spherulites.

Given that landscapes result from the dynamic interaction between human communities and their environments, palaeoenvironmental reconstructions of these spaces require a multidisciplinary approach. Both macro- and micro-remains have been studied allowing investigations of taphonomic processes. This approach has led to an improvement in identifications which has allowed a more detailed knowledge of past plant species use.

In collaboration with various local and international institutions (Institut National des Sciences de l'Archéologie et du Patrimoine of Morocco, Centre National de Recherches Préhistorique of Algeria, Institut National du Patrimoine of Tunisia, University of Oxford, University of Cologne, University of Aix-Marseille), recovery techniques have been applied systematically to a selection of sites in northwestern Africa. Systematic sampling at the sites has led to the collection of a substantial dataset of micro and macro-botanical remains (seeds, fruits, wood charcoal, phytoliths and calcitic microfossils) spanning a long chronological sequence, with some hiatuses, across large areas of the region (Fig. 1, Table 1).

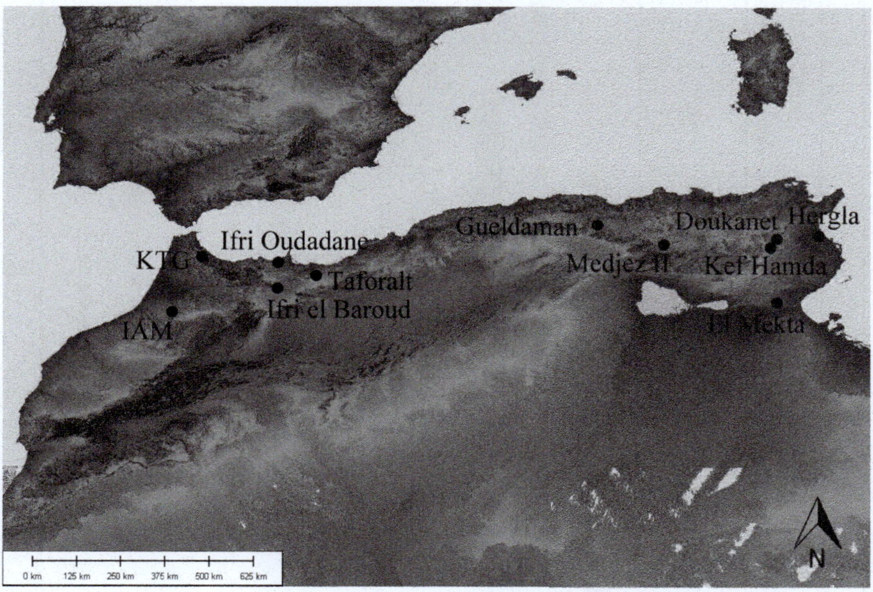

Fig. 1. Map with location of sites.

Table 1. Site location, description and volume of sediment processed.

Site	Cultures	Country	Type of site	Altitude	Volume of sediment (l.)
Taforalt	MSA, LSA	Morocco	Cave	720	561
Gueldaman	LSA, Neolithic	Algeria	Cave	500	no data
Ifri el Baroud	LSA	Morocco	Cave	535	no data
Ifri N'Amr ou Moussa	LSA, Neolithic	Morocco	Cave	200	3257
Kaf That el Ghar	LSA, Neolithic	Morocco	Cave	410	554
Medjez II	LSA	Algeria	Open-air	1000	no data
El Mekta	LSA	Tunisia	Open-air	430	133
Kef Hamda	LSA	Tunisia	Rock shelter	705	833
Ifri Oudadane	LSA, Neolithic	Morocco	Cave	50	6383
Hergla SHM-1	LSA, Neolithic	Tunisia	Open-air	70	no data
Doukanet el-Khoutifa	Neolithic	Tunisia	Open-air	672	50

Although for each site different types of plant material have been studied, depending on the availability of adequate samples (Fig. 2), the information obtained has allowed the exploration of plant resources along the sequences examined both at the local and regional scale. Data on the various materials analyzed and the methods applied are detailed below.

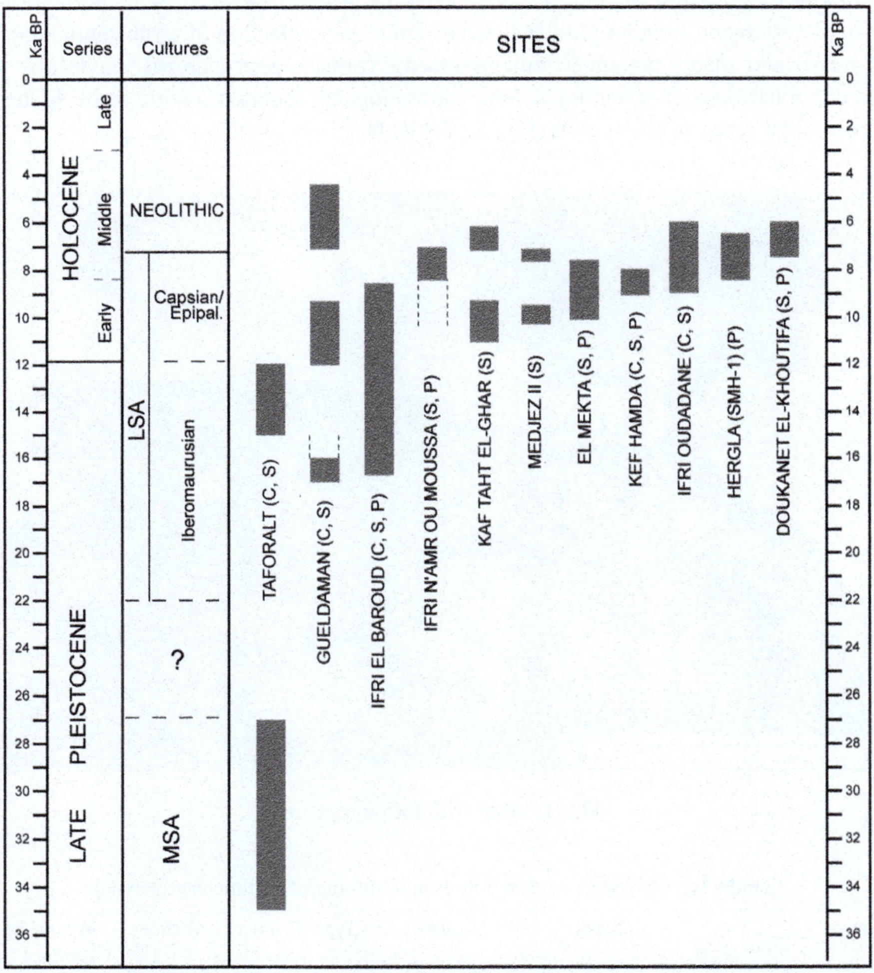

Fig. 2. Chronological assignment of each site with information of the type of remains examined. C=Charcoal; S=Seeds and fruits; P= Phytoliths and calcitic microfossils.

Seeds and Fruits

The limited application of efficient techniques for collecting plant remains in pre-agrarian contexts in comparison to later chronological periods where plant recovery methods are standardized, explains to large extent the scarcity of plant assemblages

Table 2. Presence (√)/absence (-) of seeds and fruit remains of most common taxa recorded in the sites studied.

Sites	Taforalt	Taforalt	El Mekta	Ifri Oudadane	KTG	IAM	Gueldaman	Kef Hamda	Doukanet
Period	MSA	LSA	LSA	LSA-Neolithic	LSA-Neolithic	LSA-Neolithic	LSA-Neolithic	LSA-Neolithic	Neolithic
Chamaerops humilis	–	–	–	√	√	–	–	–	–
Galium sp.	√	√	–	–	–	–	–	√	–
Juniperus sp.	–	√	–	√	–	–	√	–	–
Malva sp.	√	–	–	–	–	–	–	–	–
Olea europaea	–	–	–	√	–	√	√	–	–
Pinus spp.	–	√	√	–	–	–	–	√	√
Pistacia spp.	–	√	–	√	√	√	√	–	–
Poaceae	√	√	–	√	√	–	–	–	–
Quercus sp.	–	√	√	√	–	√	√	√	√
Stipa tenacissima	√	√	√	√	–	–	–	√	√
Taxus baccata	–	–	–	√	–	–	√	–	–
Vicia/Lathyrus	√	√	–	√	√	√	√	√	√
Vitis vinifera	–	–	–	–	–	–	√	–	–
Ziziphus lotus	√	–	–	–	–	–	–	–	–

from earlier periods. In addition, the assumption that taphonomic processes in pre-farming sites constrain plant preservation has also hampered the development of hunter-gatherer archaeobotany. Thus, one of the main goals of the project seeks to test the presence of plant macroremains in Palaeolithic sites through systematic soil sampling and sediment processing by flotation. This methodology has not been previously applied to Palaeolithic sites in this region.

Systematic sampling and recovery of plant macroremains (seeds and fruits) have been carried out at ten sites, including five sites in Morocco- Taforalt, Ifri Oudadane, Kaf Taht el-Ghar, Ifri el-Baroud and Ifri N'Amr ou Moussa (from now on IAM)-, two sites in Algeria (Gueldaman and Medjez II); and three sites in Tunisia (El Mekta, Kef Hamda, and Doukanet el Khoutifa) (Fig. 1). Their chronology spans begins in the MSA and extends to the LSA, comprising a sequence that includes the Middle-Late Pleistocene and the Early-Middle Holocene. Neolithic levels have been also sampled where available. A chronological description of the sites is summarized in Fig. 2.

Macro-plant remains were retrieved from archaeological deposits using both machine assisted and manual flotation. The number of samples and the volume of soil analyzed for each site is presented in Table 2. Flots were processed using a column of sieves with mesh sizes of 2 mm, 1 mm, 0.5 mm, and 0.25 mm. Plant remains were preserved by charring and have been identified using a binocular microscope (8–80× magnification). Identification of macroremains has been carried out by comparison to modern specimens using seed reference collections of the Archaeobiology Laboratory of the Spanish National Research Council in Madrid and the Department of Historical Sciences of the University of Las Palmas de Gran Canaria (Spain). Additionally, specialized literature on plant anatomy was also consulted (Cappers et al. 2009). Botanical nomenclature follows Charco (2001), Fennane et al. (2007) and Le Floc'h et al. (2010). Images of the most significant plant taxa are found in Fig. 3.

Wood Charcoal

Charcoal analyses have been carried out at four sites: Taforalt and Ifri el Baroud (Morocco), Gueldaman (Algeria) and Kef Hamda (Tunisia) (Figs. 1 and 2). Charcoal was recovered by flotation and analyses were completed at the Laboratory of the Department of Prehistory and Archaeology at the University of Valencia.

Charcoal analysis is based on the botanical identification of charred wood fragments based on anatomical features that sometimes enable species identification. Each charcoal fragment is observed under reflected light optical microscopy with bright and dark fields without chemical treatment. This permits the same samples to be used later for radiocarbon dating (Vernet et al. 1979). Identifications are made by comparison to a modern reference collection, specialized literature on plant anatomy (Greguss 1955, 1959; Jacquiot 1955; Jacquiot et al. 1973; Schweingruber 1990; Neumann et al. 2001) and a collection of charred woods held at the University of Valencia.

Charcoal was found to be abundant in all sites, therefore sub-sampling was undertaken. Charcoal fragments of different sizes (>4 mm and 4–2 mm) were selected

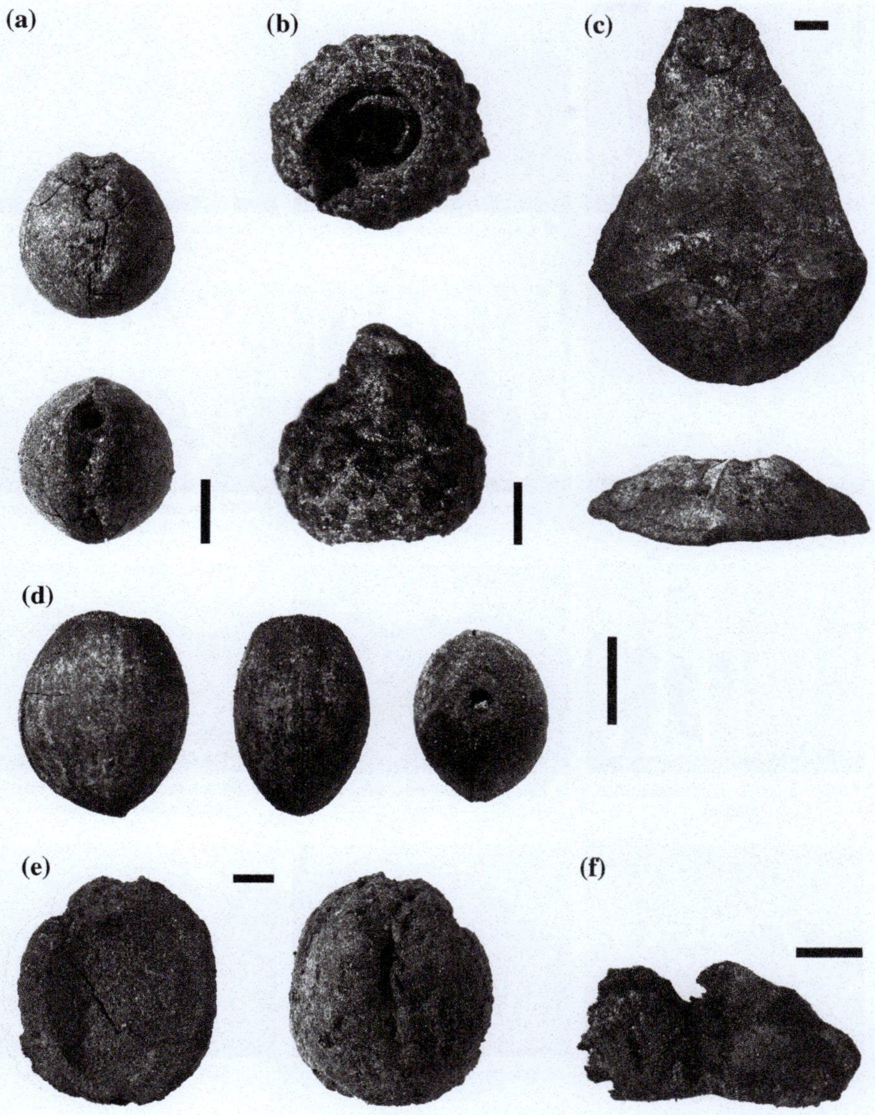

Fig. 3. Seed and fruit remains from MSA and LSA sites: (a) *Vicia/Lathyrus* sp. (MSA, Taforalt), (b) *Quercus* sp., immature cupule (LSA, Taforalt), (c) *Pinus pinaster*, seed scale (LSA, Taforalt), (d) *Pistacia lentiscus* (LSA, Ifri Oudadane), (e) *Ziziphus lotus* (MSA, Taforalt), (f) *Stipa tenacissima,* rhizome fragment (MSA, Taforalt). Scale bar = 1 mm.

for analysis but demonstrate no significant differences in species diversity between the two sieve sizes and as such, results have been combined. The identification of some taxa required the use of a scanning electron microscope (SEM) (Fig. 4). A Hitachi S-4100 MEB held at the Central Service for Experimental Research Support (SCSIE) at the University of Valencia was used for this purpose.

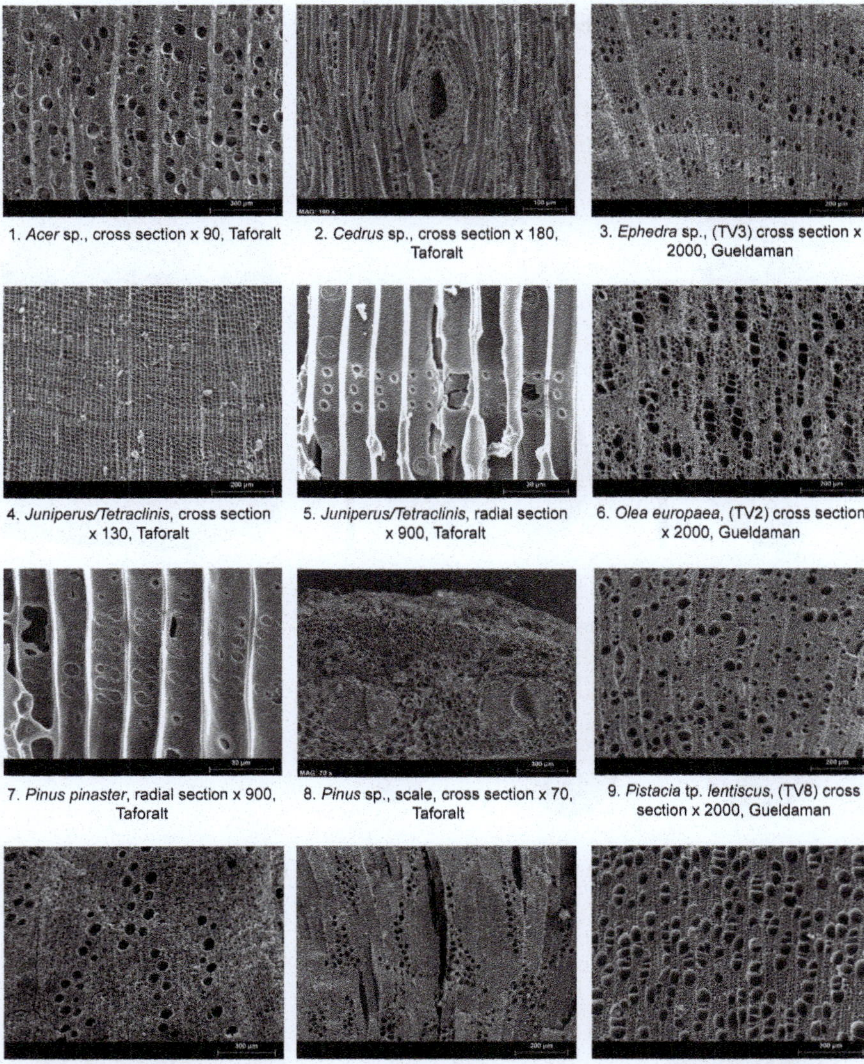

1. *Acer* sp., cross section x 90, Taforalt
2. *Cedrus* sp., cross section x 180, Taforalt
3. *Ephedra* sp., (TV3) cross section x 2000, Gueldaman
4. *Juniperus/Tetraclinis*, cross section x 130, Taforalt
5. *Juniperus/Tetraclinis*, radial section x 900, Taforalt
6. *Olea europaea*, (TV2) cross section x 2000, Gueldaman
7. *Pinus pinaster*, radial section x 900, Taforalt
8. *Pinus* sp., scale, cross section x 70, Taforalt
9. *Pistacia* tp. *lentiscus*, (TV8) cross section x 2000, Gueldaman
10. *Quercus* sp., cross section x 80, Taforalt
11. *Rhamnus-Phillyrea*, cross section x 110, Taforalt
12. *Salix* sp. (TV3), cross section x 200, Gueldaman

Fig. 4. SEM photographs of wood taxa identified in the charcoal assemblages.

The frequency of identified taxa is measured as a percentage of total fragment counts. This statistic is generally thought to reflect the range of woody species present in the local flora which were used for different purposes (Chabal 1988, 1997; Asouti and Austin 2005). Although the identified taxa represent only a portion of the existing vegetation, there are various species which might not be represented in the anthracological spectrum. This might be the case with herbaceous plants (which are not preserved) or woody species that were not collected by prehistoric groups for cultural reasons.

Some of the identifications, particularly those of species or groups of species that are anatomically very similar, were challenging. This was the case for the genus *Juniperus* which is anatomically very close to *Tetraclinis articulata* (Vahl.) Masters (both are present in the area today and are frequent in the charcoal assemblages). The basic difference between them is the presence of wedge-shaped cell walls around the bordered pits in *Tetraclinis* (Schweingruber 1990) while this feature is absent in juniper. Their presence/absence is generally confirmed by observation under SEM. Since in most cases differences between both were not significant, as observed by Zapata et al. (2013) in samples from the region, the taxon has been identified as *Juniperus/Tetraclinis*.

Phytoliths and Calcitic Microfossils

The microfossil studies reported here focused on plant exploitation using opal phytoliths (microscopic bodies composed of pure silica present in the tissues of many vegetal species, see Piperno 2006, and references therein) and calcitic ash pseudomorphs (calcite pseudomorphs remaining after heating calcium oxalate crystals to at least 450 °C), primarily originating from wood and dicotyledonous leaves (Wattez and Courty 1987; Brochier and Thinon 2003). Additionally, dung spherulites (calcitic microscopic particles that form in animal gut and found in dung (see Brochier et al. 1992; Canti 1999) were also examined dating to the transition from Epipalaeolithic hunter-gatherers to early Neolithic food producers.

In the framework of the PALEOPLANT project we have conducted phytolith and calcitic microfossil studies at six sites: Ifri el Baroud and IAM (Morocco), and El Mekta, Kef Hamda, Doukanet el Khoutifa and Hergla (SHM-1), in Tunisia (Figs. 1 and 2). Systematic sampling strategies at most sites included micromorphological analyses, which are currently in progress, and may provide detailed contextual information critical for accurately interpreting the microfossil assemblages. A full record of microfossil datasets will be published when all analyses are completed. As a case study, we present here initial results from Early Holocene IAM. Samples were taken across several contexts described in the field as fireplaces and ashy layers composed of charcoal and various artefacts and remains (i.e. lithic industries, animal bones, land snail shells, Table 3). Microfossil analyses were undertaken at the Palaeobotany Laboratory "Lydia Zapata" (Faculty of Arts) and the Department of Analytic Chemistry (Faculty of Pharmacy) of the University of the Basque Country.

Phytoliths were extracted following methods outlined by Katz et al. (2010). A weighed aliquot of ~40 mg of dried sediment was treated with 50 µl of a solution of 6 N HCl. The mineral components were then separated according to their densities in order to concentrate the phytoliths using a 450 µl 2.4 g/ml sodium polytungstate solution [$Na_6(H_2W_{12}O_{40})\cdot H_2O$]. Microscope slides were mounted with 50 µl of processed sample. A minimum of 200 phytoliths with recognizable morphologies were counted at 200X using a Nikon Eclipse 50i optical microscope. The estimated phytolith numbers per gram of sediment are related to the initial sample weight. Phytoliths that were unidentifiable because of dissolution were listed as weathered morphotypes (WM, Table 3). Multicellular structures (multi-celled or interconnected phytoliths, MC) were also recorded. These latter data may provide information regarding preservation

Table 3. Description of samples and main phytolith, ash pseudomorph and dung spherulite results obtained from Epipalaeolithic (E) and early Neolithic (EN) Ifri n'Amr ou Moussa, Morocco (2013 field season). Stratigraphical unit descriptions as recorded by the excavators in the field (J.C. Vera and R.M. Martínez Sánchez, pers. comm).

US	Phase	N. Phytoliths 1 g of sediment	% Monocotyl edonous	% Dicotyledonous leaves	% Dicotyledonous wood/bark	% WM phytoliths	% MC phytoliths	Ash pseudomorphs 1 g of sediment	Spherulites1 g of sediment	Description
2004	EN	1,060,000	68	4.6	16.1	11.3	7.4	2,560,000	150	White-gray sediment composed of ashes and charcoal fragments, hearth
2006	EN	950,000	69.1	4.5	16.5	9.9	6.7	1,990,000	52	White-gray sediment with ashes, charcoal bits, burnt clay material and faunal remains, hearth
2012	E	840,000	67.9	6.1	21	5	1.7	900,000	0	White layer with ashes, charcoal, animal bones, land snail shells and lithic artifacts

(*continued*)

Table 3. (*continued*)

US	Phase	N. Phytoliths 1 g of sediment	% Monocotyl edonous	% Dicotyledonous leaves	% Dicotyledonous wood/bark	% WM phytoliths	% MC phytoliths	Ash pseudomorphs 1 g of sediment	Spherulites1 g of sediment	Description
2014	E	790,000	60.6	8.6	19.6	11.2	2.7	1,220,000	0	White-gray sediment with ashes, charcoal, faunal bones, land snail shells and lithic artifacts
2029	E	770,000	57.9	4,2	23,5	14.4	4	1,400,000	0	Dark-gray and white-ashy layered filling with charcoal, faunal remains, burnt clay material and lithicindustry, stony hearth/oven

conditions. Morphological identification was based on modern plant reference collections from the Mediterranean region (Albert and Weiner 2001; Tsartsidou et al. 2007; Albert et al. 2008, 2011; Portillo et al. 2014) and standard literature (Twiss et al. 1969; Brown 1984; Rosen 1992; Twiss 1992; Mulholland and Rapp 1992; Madella et al. 2005; Piperno 2006).

Calcitic microfossil samples were prepared following methods developed by Canti (1999). Approximately 1 mg of dried sediment was mounted on a microscope slide with Entellan New from Merck. Slides were examined at 400X magnification under an optical microscope with crossed polarized light (XPL) for spherulites, whereas pseudomorphs where examined in plane polarized light (PPL). Similar to phytoliths, both calcitic microfossil numbers were related to the initial sediment weight. Samples were compared to a pilot modern plant reference collection, which includes selected woody species, as well as modern dung materials from the Mediterranean Maghreb (Portillo et al. 2012, 2016).

Results

Seeds and Fruits

Preliminary results indicate the presence of plant macroremains at all sites sampled. Analyses have been carried out in eight out of the ten selected sites (data from Ifri el-Baroud and Medjez II are not yet available) (Table 2). Here we present data on presence/absence of the most common taxa recorded at each site, considering the frequency and number of remains. A complete report with absolute numbers of seeds and radiocarbon dates for each site will be published when analyses are completed.

For the MSA, the only existing data come from the recent excavations at Taforalt where layers dating to c. 33,000–25,000 BC (35,000–27,000 cal yr BP; Barton et al. 2016) have yielded charred seeds of wild pulses (*Vicia/Lathyrus* sp.). Wild pulses are present throughout the sequence analyzed. Plant remains from other taxa such as alfa grass (*Stipa tenacissima*), mallow (*Malva* sp.), lotus jujube (*Ziziphus lotus* L.), and cleavers (*Galium* sp.) have been also identified. Plant remains are more abundant in LSA sites (c. 20,000–5500 BC; 22,000–7500 cal yr BP), especially in levels dated after c. 13,000 BC where a larger assemblage of seeds is available and many more species have been identified.

In Taforalt, layers belonging to the Iberomaurusian horizon and dated to c. 13,000–10,000 BC (15,000–12,000 cal yr BP) have provided abundant charred plant remains totaling 1270 (Humphrey et al. 2014). Acorns from the holm oak (*Quercus ilex* L.) and pine nuts from maritime pine (*P. pinaster*) were the most abundant taxa. Other diaspores include juniper (*Juniperus phoenicea* L.), terebinth pistachio (*Pistacia terebinthus* L.), wild pulses (*Lens* sp., *Vicia/Lathyrus* sp.) and wild oat (*Avena* sp.).

The macro-botanical assemblage from the Epipalaeolithic and Early Neolithic levels of Ifri Oudadane (c. 7000–4000 BC) is extremely rich, with more than 14,000 botanical remains recorded (Morales et al. 2013; 2016, Zapata et al. 2013). The most common taxa are *Pistacia lentiscus* L., wild pulses (*Vicia/Lathyrus* sp.), and dwarf palm (*Chamaerops humilis* L.), but other potential food plants such as acorns

(*Quercus* sp.) or wild olives (*Olea europaea*) have been also recorded. Epipalaeolithic layers (c. 9000–7000 BC) of Kaf Taht el-Ghar have yielded a few remains of *Avena* sp. and wild pulses. Seeds of wild plants are more abundant in Neolithic layers (c. 5300–4200 BC) where they appear along with emmer (*Triticum dicoccum* L.), free-threshing wheats (*Triticum aestivum/durum*) and naked barley (*Hordeum vulgare* var. *nudum* L.) (Morales et al. 2016). Wild pulses (*Vicia/Lathyrus* sp.) are the most common wild species, but other taxa such as *P. lentiscus* and dwarf palm (*C. humilis*) have been documented in this study.

At IAM, a total of 49 remains of wild plants have been identified. For the Epipalaeolithic layers, wild pulses (*Lens* sp., *Vicia/Lathyrus* sp.) and pistachio (*P. terebinthus)* are the most significant taxa, while in the Neolithic, dated to c. 5100 BC, seeds of domesticated plants such as barley (*H. vulgare*) and wheat (*Triticum* sp.) are found along with wild plants such as *O. europaea*, *Quercus* sp., and *P. terebinthus*, among others.

The plant assemblage from Iberomaurusian layers (c. 15,000–8000 BC; 17,000–10,000 cal yr BP) at Gueldaman, is composed of seeds of *Juniperus*, *P. lentiscus*, *P. terebinthus* and *Taxus baccata* L. In Neolithic layers, dated to about 4800 BC (6800 cal yr BP), wild plants such as *Vicia/Lathyrus*, *O. europaea*, *P. lentiscus*, *Vitis vinifera* L. and *Quercus* are common. No cereal crops have been recorded.

El Mekta (c. 8000–5500 BC; 10,000–7500 cal yr BP) has produced a limited number of plants and information on their use is scarce. Only 34 fragments were retrieved of which 30 were accurately identified. Three taxa including *Pinus halepensis* Mill., *Quercus* sp., and *S. tenacissima* are present (Morales et al. 2015).

At Kef Hamda (c. 7000–6000 BC; 9000–8000 cal yr BP), the most common remains are pine (*P. halepensis*), acorns (*Quercus* sp.) and a wild legume (*Lathyrus/Vicia* sp.) (Mulazzani et al. 2015). Despite the presence of pottery fragments in the upper layers no domesticated plants have been identified so far, and the only evidence comes from wild taxa.

First results from Doukanet el-Khoutifa (c. 5500–4000 BC; 7500–6000 cal yr BP) point to the presence of wild plants while domesticated species are so far absent (Mulazzani et al. 2015). In total, 22 seeds belonging to 4 taxa have been identified. The most common species are acorns (*Quercus* sp.) and wild pulses (*Lathyrus/Vicia* sp.), while pine (*P. halepensis*) is rare. Rhizome fragments of alfa grass (*S. tenacissima*) are also abundant.

Wood Charcoal

Some of the pioneering research on wood charcoal was developed in North Africa (e.g. Balout 1952; Momot 1955; Santa 1961; Couvert 1969). Among the early work, there were also attempts at interpreting wood charcoal from prehistoric sites in climatic terms (Couvert 1976). A significant contribution to the study of wood charcoal in the region is represented by the remarkable work of Wengler and Vernet (1992) on deposits from eastern Morocco. However, systematic studies with proper sampling strategies and the application of effective recovery techniques did not take place until the 21st century when several studies of seeds, fruits and wood charcoal from Moroccan Neolithic sites were undertaken by the AGRIWESTMED project (Morales et al. 2013; Zapata et al. 2013). Charcoal analyses presented here have been carried out within the

PALEOPLANT project with the aim of providing information on the mosaic of existing landscapes in northern Africa and gaining insight into their exploitation over a wide temporal sequence.

Four of the sites, Taforalt, Ifri el Baroud, Kef Hamda and Gueldaman, have provided abundant charcoal assemblages in all phases (Fig. 2). Table 4 shows the presence of identified taxa during the various chronological periods. With at least 27 different species identified, the data show significant plant diversity which includes trees (including *Acer* sp., *Cedrus*, pine, oak, and *Fraxinus*), large shrubs (such as juniper, wild olive and pistachio) and scrubs (including Cistaceae, Lamiaceae). Some charcoal pieces are of a very small size such as the 1–1.5 mm diameter monocotyledon twigs documented at several sites. Conifers (Cupressaceae and/or pines) appear systematically throughout the region with particular intensity in pre-Neolithic levels. However, the most ubiquitous taxa are *Juniperus/Tetraclinis* and legumes, the latter being represented in lower percentages. In fact, only a few taxa show significant percentages while the remaining are scarcely represented with percentages in many cases below 1%. This is the case, for instance, of carob and tamarisk at Gueldaman where only a few fragments have been recovered.

Despite the presence of common taxa at all sites, each reveals unique vegetation complexes reflecting the peculiarities of each territory. For instance, pines seem to be an important element of the Late Glacial vegetation in Taforalt and Kef Hamda, while they are almost absent at Ifri el Baroud and Gueldaman. As such the presence/absence of pines in charcoal assemblages might be a marker of regional particularities. In addition, pine wood charcoal can be identified to the species level, offering the possibility of obtaining more accurate environmental information. In Taforalt, *P. pinaster* is widely represented while the percentages of *P. halepensis* are rather low. Both species are typical of warm and temperate conditions, although in North Africa, pines can also be found up to 2000 m asl where climatic conditions are more severe. The maritime pine is associated with siliceous areas and sub-coastal sandy soils which explains its dominating presence at Taforalt. The location of the cave coincides with the area of current distribution of this species, which is restricted to the Rif, Middle Atlas and Tell Atlas. Its presence in Taforalt not only provides information on the relationship between areas of current and potential distribution of this species (Charco 1999) but it also demonstrates its presence in the region since pre-Holocene times. Likewise, Aleppo pine is dominant in Kef Hamda and is presently the most naturally widespread pine species in North Africa. Another distinctive species of the area is cedar which is only present in the Iberomaurusian layers of Taforalt. This is a water demanding taxon that can requires an annual precipitation above 500 mm. It generally prefers humid or sub-humid environments above 1000 mm. Its presence in Taforalt was probably favored by the presence of northern-exposed slopes that trapped maritime moisture where this species could thrive.

A preliminary assessment of the flora present along the long temporal sequence that characterized the sites is provided in Table 4. Taforalt presents the only available archaeobotanical example from MSA contexts. The wood charcoal identified suggests that *Quercus* sp., both evergreen and deciduous, together with *Juniperus/Tetraclinis* were the most important taxa. Other taxa, *Acer* sp., Rosaceae Maloideae, *Salix* sp. and some shrubby plants (Cistaceae, Fabaceae, Monocotyledon) were also present, forming mixed open forests.

Table 4. Presence/absence of the woody taxa identified. Shading indicates taxa with percentages over 10% (Identifications in the range of Conifer or Angiosperm group are not included).

TAXA / SITES	Taforalt MSA	Taforalt Iberomaurusian	Ifri el Baroud Iberomaurusian	Kef Hamda Epipalaeolithic	Gueldaman Epipalaeolithic	Gueldaman Neolithic
Acer sp. (maple)	*	*				
Arbutus unedo (strawberry tree)	*			*		*
Cedrus sp. (cedar)		*				
Ceratonia siliqua (carob tree)					*	
Amaranthaceae (goosefoots family)	*					
Cistaceae (rock-rose family)	*			*		*
Asteraceae (aster family)	*	*	*		*	*
Ephedra sp. (ephedra)					*	
Erica sp. (heather)					*	*
Ficus carica (fig tree)					*	*
Fraxinus sp. (ash)					*	*
Juniperus/Tetraclinis (juniper/thuja)	*	*	*	*	*	*
Lamiaceae (mint family)		*			*	*
Fabaceae (legumes)	*	*	*	*	*	*
Monocotyledon	*	*	*		*	
Olea europaea (wild olive tree)					*	*
Pinus halepensis (Aleppo pine)		*		*		
Pinus pinaster (maritime pine)	*	*				
Pinus sp.	*	*		*		
Pistacia tp. *lentiscus* (mastic)				*	*	*
Pistacia terebinthus (terebinth)					*	*
Pistacia sp.		*		*	*	*
Quercus deciduous (oak)	*	*			*	*
Quercus evergreen (holm oak/ kermes oak)	*	*	*		*	*
Quercus sp.	*	*			*	*
Rhamnus/Phillyrea (buckthorn/green olive tree)	*	*		*	*	*
Rosaceae (rose family)					*	*
Rosaceae Maloideae (juneberry subfamily)	*	*			*	*
Rosmarinus officinalis (rosemary)		*		*	*	*
Salix/Populus (willow/poplar)	*	*		*	*	*
Tamarix sp. (tamarisk)					*	*
Minimum N. of species	13	16	5	9	19	18
Charcoal fragments analysed	661	2130	540	450	1083	2360

Charcoal assemblages dating to the Iberomaurusian and Epipalaleolithic provide environmental information for the Pleistocene/Holocene transition. *Juniperus/Tetraclinis* is the most common taxon in these contexts. In Taforalt, this species is accompanied by pine (*P. pinaster*), which increases along the sequence while Cupressaceae declines (c. 13,000-10,000 BC). Conifers appear together with *Cedrus*, *Quercus* (both evergreen and deciduous), *Acer*, and low percentages of thermo-Mediterranean scrub species which gradually begin to spread (including Lamiaceae, Fabaceae, pistachio, and *Rhamnus-Phillyrea*). The Ifri el Baroud assemblage is taxonomically rather poor but it shows the predominance of *Juniperus/Tetraclinis* together with legumes. On the contrary, Epipalaeolithic levels of Kef Hamda are rich in pine (*P. halepensis*) while junipers are barely represented. Here, the charcoal assemblage shows the initial spread of scrubland Mediterranean species represented by a high percentage of Cistaceae, *Pistacia*, *Rhamnus-Phillyrea* and legumes. Some of these have been also identified in low percentages in Epipalaeolithic levels at Gueldaman where pre-Neolithic assemblages are dominated by *Juniperus/Tetraclinis* and *Pistacia* together with Lamiaceae, Asteraceae and *Ephedra*. Riparian vegetation consists of *Salix-Populus* and ash.

In Gueldaman, a radical change in the vegetation is observed in Neolithic levels. At this site, *O. europaea* becomes the dominant species reaching percentages up to 90% of the charcoal record. *Arbutus unedo* L., *Quercus* and *Pistacia*, also have been identified while the presence of conifers is minimal. Other taxa such as carob, Ericaceae (heather), legumes, *Ficus* or Rosaceae are occasionally present. This spectrum of plant species dominated by *O. europaea* may be explained by the intensive use of the cave as an animal pen during Neolithic times (Kherbouche et al. 2014), which will be further discussed below.

Phytoliths and Calcitic Microfossils

First results from transitional Epipalaeolithic- early Neolithic food producers at IAM provide both plant microfossil (phytoliths and calcitic ash pseudomorphs) as well as dung spherulite indicators. The integration of these lines of evidence both complements and supplements other sources of archaeobotanical data at the cave (i.e. charred plant macroremains, pollen) since each is influenced by different taphonomic processes. Table 3 shows the location and field descriptions of the five samples analyzed, together with the main quantitative results. Phytoliths and calcitic microfossils were noted in different amounts in the sediments, with the exception of all three Epipalaeolithic samples, where dung spherulites were absent.

Phytoliths were abundantly noted in all samples. The richest sediments were Early Neolithic hearth samples (samples 2004 and 2006, around 1 million phytoliths per gram of sediment, Table 3). The dissolution index of phytoliths does not correlate with their concentration in the samples, and the assemblages do not seem to be highly affected by post-depositional processes. The ashy-layered fill deposit from the firing installation belonging to the Epipalaeolithic, where macroscopic remains from compacted burnt clay material were observed, showed the highest dissolution rate (sample 2029, around 14%). The morphological results indicated that monocotyledonous plants and particularly grasses dominated all samples at approximately 60–70% of all counted morphotypes, whereas wood, bark and dicotyledonous leaf indicators (i.e. tracheids, globulars, and irregular morphologies) constituted about 20–30% of the assemblages. This is especially noteworthy given that woody plants are not great phytolith producers. It is also worthy of note that monocotyledonous leaves produce up to 20 times more phytoliths than wood and bark and 16 times more than dicotyledonous leaves, and they are also present as exterior contamination on woody material used as fuel (Albert and Weiner, 2001; Albert et al. 2003; Tsartsidou et al. 2007). This is consistent with abundant macro-botanical charred remains which were common in these contexts, as well as large concentrations of ash pseudomorphs (ranging between 0, 9 to 2.5 million/1 g of sediment, Table 3), indicative of wood-dominated ashes. Early Neolithic firing installations were by far the richest contexts sampled. Pseudomorphs included mostly prisms and rhomboidal morphologies, which are common in modern reference datasets, such as *Pistacia* sp. and *O. europaea* (Portillo et al. 2016, Fig. 5a). Both wild and domesticated plants were noted in the botanical record (Table 2).

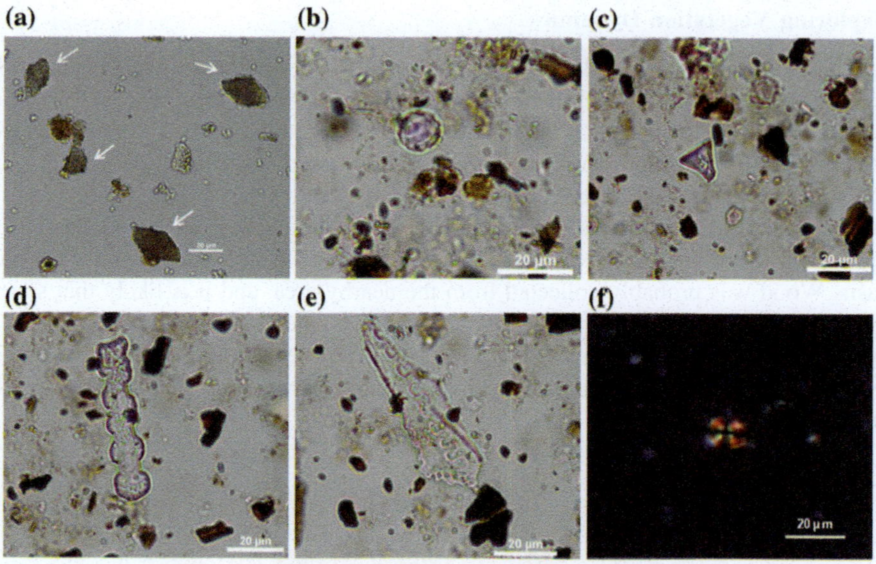

Fig. 5. Photomicrographs of phytoliths and calcitic microfossils identified in archaeological samples (at 400×) (a–e PPL; f XPL, photo A. Polo-Díaz, UPV-EHU). (a) ash pseudomorps, (b) spheroid echinate, (c) short cell rondel, (d) long cell polylobate, (e) multicellular phytoliths of dendritic long cells, (f) dung spherulite.

In addition, diagnostic morphotypes produced by the leaves of the Arecaceae family (palms) were also identified in all samples although to a lesser extent (Fig. 5b). Grass phytoliths belonged mostly to the Pooid subfamily (Fig. 5c). Grasses were in general represented by leaf/culm phytoliths (i.e. long cell polylobates, prickles, bulliforms and multi-celled phytoliths, Fig. 5d), whereas inflorescence morphotypes (i.e. papillae and decorated long cells, such as echinate and dendritics) were scarce in most assemblages, with the exception of one Neolithic sample (2004). Multi-celled phytoliths from the floral parts or husks of Pooid seeds were also noted in early Neolithic firing installations (Fig. 5e). This is consistent with findings of charred caryopses of barley (*H. vulgare*) and wheat (*Triticum* sp.) which were common in Early Neolithic deposits of the cave, as well as in other Western Maghreb sites, in addition to faunal remains from domestic animals (Ballouche and Marinval 2003; Morales et al. 2013, 2016).

Lastly, dung spherulites were noted only in Neolithic hearth samples, although in relatively lower quantities (52,000–150,000 spherulites/g, Table 3, Fig. 5f). Interestingly, spherulite abundances correlate with large numbers of grass phytoliths in both samples, possibly derived from cereal crops according to macrobotanical datasets. This suggests that these sediments were at least partly composed of livestock dung.

Discussion

Exploring Vegetation Dynamics

Archaeobotanical remains identified at the various site are spread over a rather long chronological sequence which includes some hiatuses. Such a long span of time offers the possibility of assessing not only plant diversity, and thus, landscape evolution over time, but also exploring plant uses. Indeed, most species identified in both the macro and micro remains were selected by ancient communities for a wide variety of uses. Fuel was almost certainly one of the main sources of wood charcoal at the sites. In fact, firewood collection was most certainly a non-specialized task carried out on a daily basis. Wood was probably collected from the nearby area, and it is likely that species represented in the archaeological assemblages correspond to at least part of the present-day vegetation. In addition, human groups probably collected other categories of plants (roots, rhizomes, etc.), relating to a diverse range of activities (food, drinks, crafts, bedding, fodder, etc.).

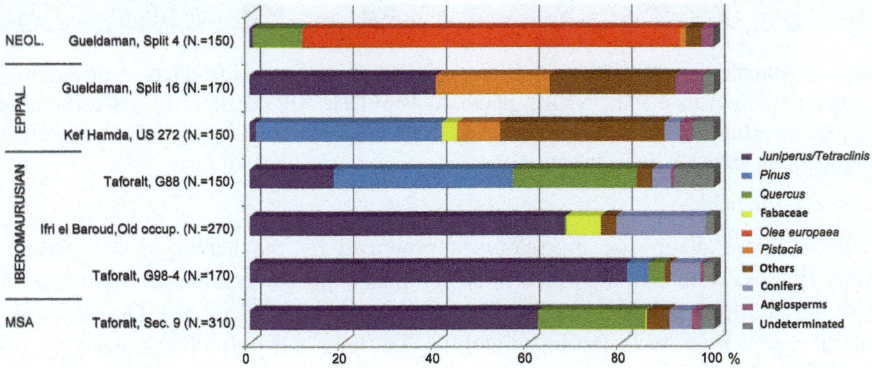

Fig. 6. Summary wood charcoal main species for the various sites analyzed.

Figure 6 presents a synthesis of the woody vegetation used for fuel at the various sites which allows the comparison of wood assemblages over time. Taforalt is the only MSA site that has offered a plant assemblage dated to c. 33,000–25,000 BC (35,000–27,000 cal yr BP; Barton et al. 2016). The samples suggest the presence of a mixed open forest dominated by junipers together with oaks (deciduous and evergreen), maple and shrubs of Cistaceae, Amaranthaceae, Asteraceae, Fabaceae, and monocotyledons. Water demanding plants such as *Salix* sp. are also present.

The Late Glacial period, and in particular the Pleistocene-Holocene transition, is well documented in all sequences (Taforalt, Ifri el Baroud, Gueldaman, Kef Hamda). Iberomaurusian and Epipalaeolithic layers (c. 13,000–5500 BC) are characterized by the ubiquity of coniferous woodlands across the whole region and the spread of juniper forests (perhaps together with *T. articulata*). *Juniperus/Tetraclinis*, species that are well-adapted to impoverished soils (Wengler and Vernet 1992), may have been pioneer species which colonized postglacial soils. These taxa are very abundant in Taforalt, Ifri el Baroud and at the base of the Gueldaman sequence dating between c. 13,000 and 8000 BC.

Wood charcoal analyses also indicate that pines were present in Late Glacial coniferous forests. Their presence is, however, more irregular as shown by the percentage values of *P. pinaster* in Taforalt and *P. halepensis* in Kef Hamda. In Taforalt, pines replace *Juniperus/Tetraclinis* forests that were dominant at the beginning of the Iberomaurusian (Fig. 6), which might indicate increasing humidity. Pines are absent from Ifri el Baroud and Gueldaman, although pollen records from marine sequences suggest this species was widespread during the Late Glacial period (Desprat et al. 2013).

The common presence, although in low percentages, of other dry-resistant taxa, such as *Ephedra* or Amaranthaceae in juniper-pine forests, points perhaps to some level of aridity. On the contrary, at Taforalt before Holocene, as mentioned above, there is a trend towards a more humid phase characterized by an increase of the proportion of pine and cedar compared to *Juniperus/Tetraclinis*.

At Gueldaman, the retreat of Cupressaceae during the Preboreal-Boreal transition occurred in parallel with a rapid development of a sclerophyllous vegetation dominated by several *Pistacia* species which is well documented both in the charcoal and seed records. Interestingly, during the Neolithic, the taxa composition of wood charcoal assemblages evidences a profound change when compared to previous phases. In fact, *O. europaea* dominates the spectrum with percentages up to 90% while *Arbutus* and evergreen *Quercus* are also present. *Pistacia* and *Juniperus* are represented in lower proportions.

From the Middle Holocene, wild olive appears to have been a key forest component of *Olea-Ceratonion* components of thermo-Mediterranean vegetation, suggesting the occurrence of warm, semi-arid conditions (Wengler and Vernet 1992). Other species such as *Pistacia*, *Juniperus*, *Tetraclinis*, *C. siliqua*, and Fabaceae were probably equally significant in these habitats as suggested by data from other Neolithic sites in the region (Wengler and Vernet 1992, Zapata et al. 2013). At Gueldaman, *O. europaea* may be overrepresented. This species is commonly used for animal fodder because of the palatability of its leaves (Badal 1999). Its dominant presence in the cave as a result of its use as animal pen may have distorted the picture and it is likely that many other species were actually present. The presence of ovicaprine coprolites in the Neolithic sediments strongly supports the use of the Gueldaman cave as livestock pen.

A similar use is proposed for the early Neolithic IAM cave where there is evidence of dung spherulites in addition to one ovicaprine charred dung pellet obtained by flotation. Phytolith results also suggest that leaves and culms of grasses, such as barley and wheat, whose seeds were common in the plant macroremain record, could have been used as bedding or matting material, and fodder. It is well known that since the Neolithic caves and rock-shelters have long been used as pens in the Mediterranean region (Brochier et al. 1992; Macphail et al. 1997; Badal 1999; Vergès et al. 2002; Karkanas 2006; Angelucci et al. 2009), and southern in the Central Sahara (Mercuri 2008; Mercuri et al. 2018; Amrani 2018).

Evidence of Food Plants

Preliminary results from Taforalt, the only site that has provided botanical evidence of wild plants from MSA layers, indicate that wild pulses (*Vicia/Lathyrus* sp.) and other species such as lotus jujube (*Z. lotus*) could have been part of the human diet during that time. Seeds from *Lathyrus* and *Vicia* species are very nutritious; they are relatively high in proteins and contain carbohydrates and also contain some toxic substances that may be removed by roasting or leaching (Aykroyd and Doughty 1982:108). However accidental burning during roasting also could account for the charred state of these remains. Charred seeds of *Lathyrus* sp. and *Vicia* sp. have been recorded in Middle Palaeolithic layers of sites in the Near East (Lev et al. 2005; Baines et al. 2015) suggesting that wild pulses were common food plants at that period.

Lotus jujube (*Z. lotus*) fruits are rich in carbohydrates and vitamins and nowadays it is a common food in North Africa and the Near East (Boulos 1983; Rivera et al. 2011). *Ziziphus* fruits are also have medicinal uses in North Africa (Boulos 1983), as such it cannot be ruled out that it was used as both food and medicine. There is no other Middle Palaeolithic evidence of *Ziziphus*, and the earliest remains of *Ziziphus* seeds have been recorded in Ohalo II, Israel (c. 21,000 BC) (Weiss 2009). Although the size of the dataset is still limited, seeds from Taforalt are one of the oldest charred plant assemblages recorded from archaeological sites in North Africa and have provided interesting new data about the use of plants during the MSA (Barton et al. 2016).

The data available for LSA layers are more extensive and have produced a richer assemblage of plant taxa. Wild pulses (*Vicia/Lathyrus* sp. and *Lens* sp.) were present, but the most abundant macrofossils were acorns (*Quercus* sp.), pine (*P. halepensis* and *P. pinaster*) and wild pistachio (*P. lentiscus* and *P. terebinthus*), recovered at most of the sites (Table 2). Although the data available from MSA contexts are still limited and the results obtained are preliminary, we propose that the increase of seeds and taxa during the LSA may be related to a higher consumption of food plants during this period. This seems to be confirmed by data from Iberomaurusian burials at Taforalt, where frequent dental caries in human remains suggest significant consumption of plants rich in starch (Humphrey et al. 2014). Charred remains of acorns, pine nuts, wild pulses and wild pistachios are also frequent in other Epipalaeolithic sites of the Mediterranean basin, confirming the widespread use of these plants (Hansen 1991;

Butler 1998; Aura et al. 2005; Vaquer and Ruas 2009). Data suggest that acorns and pine nuts may have played an important role in human subsistence. They were abundant, highly nutritious and easy to collect, process and store (Lanner 1981; Mason, 1995; Mason and Nesbitt 2009). Ethnographic data indicate that both acorns and pine nuts have been used as staple foods by hunter-gatherer populations (Barrett and Gifford 1933; Kroeber 1976; Lanner 1981; Anderson 2005). Further evidence of the importance of gathered plants in the human diet is provided by their persistent exploitation through the Middle Holocene. For instance, continuity in the use of acorns, pine nuts and wild pulses has been recorded in some Neolithic contexts studied. Crops and domestic animals have been also identified in the same contexts, especially at sites in northern Morocco (Ballouche and Marinval 2003; Morales et al. 2013, 2016; Zapata et al. 2013), suggesting a mixing foraging-farming economy where wild plants continued to play an important role. These interpretations are further supported by microfossil evidence from phytoliths.

Indirect Evidence of Basketry

Archaeological evidence of prehistoric basketry is limited due to poor preservation of organic material and deficient recovery techniques. The earliest evidence comes from the site of Dzudzuana cave in the Caucasus dated to 30,000 years ago (Kvavadze 2009). Fibers identified as wild flax or simple bast fibers as suggested by Bergfjord et al. (2010) were discovered during microscopical analyses. In addition, the sites of Dolní Věstonice and Pavlov in the Czech Republic (Adovasio et al. 1996; Soffer et al. 2000a) have provided basketry impressions in fired and unfired clay. The finds were dated to the Upper Palaeolithic, between c. 27,000 and 22,000 BC. Further evidence comes from Ohalo II (Israel) where fibers belonging to a mat were identified and dated to c. 17,000 BC (Nadel et al. 1994). Soffer et al. (2000b) collect additional data from the Upper Palaeolithic showing the presence of this technology in France, Germany, and Russia. Neolithic examples of cordage and fiber impressions appear occasionally at various sites in Europe, revealing that basketry was a technology known in the period. For the Mediterranean area, examples of basketry are found in the Neolithic Cueva de Los Murciélagos (Albuñol, Granada, Spain) (Cacho Quesada et al. 1996) where several complete baskets were exceptionally preserved due to arid conditions inside the cave. Two fragments of plant material from a sandal and a piece of plaited material using alfa grass (*S. tenacissima*) have been dated to c. 3800 BC. More recently, a textile from China dating to c. 4700–4600 BC (6775–6645 cal yr BP) and made from reeds has been identified (Zhang et al. 2016).

In Mediterranean North Africa, there is no direct evidence of baskets, mats or other products from basketry; well preserved desiccated remains of these items have only been recorded at sites from the Tadrart Acacus massifs in Central Sahara, with remains dated to the Early and Middle Holocene 'Green Sahara' (di Lernia et al. 2012). We propose that some of the carbonized rhizome fragments of alfa grass which have been recorded at most of the sites in both MSA and LSA layers, could have played role of basketry plants, although preservation conditions and inappropriate recovery techniques have not allowed the recognition of basketry items for the time being. Alfa is a native perennial grass in the western Mediterranean area which has been extensively used as raw material for basketry and other craft items in North Africa (Louis and Despois 1986; M'Handi and Anderson 2013) and the Iberian Peninsula (Kuoni 1981). The preservation of charred rhizomes suggests that alfa grass could have been collected for basketry. Leaves are the useful part of the plant, while rhizomes are a common by-product or residue of gathering and processing the fibers, which are usually discarded and re-used as fuel. Since remains of baskets or cordage have not been recorded in our analyzed sites, we cannot rule out that alfa grass was also introduced to the site as part of bedding or other subsistence activities.

In addition, seeds of the dwarf palm (*Chamaerops humilis*) has already been recorded at several Epipalaeolithic and Early Neolithic sites across the Western Maghreb (Morales et al. 2013, 2016; Zapata et al. 2013). Interestingly, phytoliths from palm leaves have been identified in fireplaces from both periods, suggesting that this plant could have been used also as fuel or discarded deliberately or accidentally into the fire. Although both seeds and leaves are potential plant foods, ethnographic research in the Rif and other Mediterranean areas (Peña-Chocarro et al. 2015, and references therein) demonstrates the extensive use of palms for basketry.

Conclusions

This project has produced important new data that fills a major gap in prehistoric knowledge, namely how plant foods and vegetal resources were used by pre-agrarian and first farming groups. Initial results of PALEOPLANT demonstrate that when properly sampled, Palaeolithic and Epipalaeolithic sites can produce direct evidence of plant use. In fact, although the analyses have not yet been completed, the available data contradicts the general assumption that plant remains do not preserve well in pre-farming sites. Moreover, the combination of multiple lines of evidence, e.g. plant macroremains, phytoliths, and other plant microfossils, has significantly improve our understanding of plant exploitation by taking into account the various pathways to archaeological record formation, each influenced by different taphonomic processes.

Our preliminary results point to the exploitation of an impressive range of plant resources, including wild pulses, acorns, pine nuts, grasses, and palms, characterized by different potential uses, such as food, fuel, basketry, bedding and animal fodder. Plant macro and micro remains have been recovered from eleven archaeological sites in our study area (Fig. 2), providing most interesting data on the range of wild plants used during the Palaeolithic and Neolithic in northwestern Africa, including wild pulses, acorns, pine nuts, alfa grass and dwarf palm. Different types of evidence (archaeological and ethnographic) suggest that wild pulses, pine nuts and acorns could have played a role as food in human subsistence while alfa grass and dwarf palm may have been used as a source of fiber for basketry or fuel.

Outstanding results have been obtained from the Middle Stone Age deposits of Taforalt (Morocco) where charred seeds of wild pulses, lotus jujube, and rhizome fragments of alfa grass in addition to other species have been identified. Continuity in the use of wild plant species is observed in Neolithic layers indicating the significant role of gathered sources in the subsistence of the first farmer communities in north-western Africa.

The large charcoal assemblages examined highlight the importance of wood gathering activities and offer a vivid picture of the existing vegetation and of its evolution throughout the sequence. Particularly interesting is the observation of the spread of thermo- and meso-Mediterranean landscapes during the Pleistocene/Holocene transition. In these transitional phases, some regional traits are distinguished in the various sites which require further work. The challenge for the future is, therefore, deepening our knowledge on regional differences observed and filling in the crucial remaining hiatuses along the chronological sequence.

Acknowledgements. Research has been carried out within the framework of the PALEOPLANT project (ERC 2013 CoG 614960). Y. Carrión and J. Morales are beneficiaries of a Ramón y Cajal research fellowship funded by the Spanish Ministry of Economy and Innovation of Spain. M. Portillo research is currently funded by the EU Horizon 2020 Marie Sklodowska-Curie action MICROARCHAEODUNG (H2020-MSCA-IF-2015-702529). G. Pérez Jordà work has been carried out within the postdoctoral contract FPDI-2013-16034 and IJCI-2015-25082 funded by the Spanish Ministerio de Economía y Competitividad. We would like to express our special gratitude to J. Bouzouggar and Y. Bokbot (Institut National des Sciences de l'Archéologie et du Patrimoine, Morocco), S. Merzoug and F. Kherbouche (Centre National de Recherches Prehistoriques, Anthropologiques et Historiques, Algeria), L. Belhouchet and N. Aouadi, (Institut National du Patrimoine, Tunisia), N. Barton (University of Oxford, UK), L. Humphrey (The Natural History Museum, UK), J. Lindstädter (Deutsches Archälogisches Institut, Germany), S. Mulazzani (University of Aix-Marseille, France), A. Polo-Díaz (UPV/EHU, Spain), J.C. Vera Rodríguez (University of Huelva, Spain) and R.M. Martínez Sánchez (University of Granada, Spain) for long-term, amicable cooperative work and for providing access to the site data.

References

Adovasio JM, Soffer O, Bohuslav K (1996) Upper Palaeolithic fibre technology: interlaced woven finds from Pavlov I, Czech Republic, *c.* 26.000 years ago. Antiquity 70:526–534

Albert RM, Weiner S (2001) Study of phytoliths in prehistoric ash layers using a quantitative approach. In: Meunier JD, Colin F (eds) Phytoliths, applications in earth sciences and human history. AA Balkema Publishers, Lisse, pp 251–266

Albert RM, Bar-Yosef O, Goldberg P et al (2003) Phytolith and mineralogical studies of hearths from the Middle Paleolithic levels of Hayonim cave (Galilee, Israel). J Archaeol Sci 30:461–480

Albert RM, Shahack-Gross R, Cabanes D et al (2008) Phytolith-rich layers from the late bronze and iron ages at tel dor (israel): mode of formation and archaeological significance. J Archaeol Sci 35:57–75

Albert RM, Esteve X, Portillo M et al (2011) Phytolith CoRe, Phytolith Reference Collection. http://phytcore.org/phytolith/index. Accessed 21 Mar 2015

Amrani S (2018) The Holocene flora and vegetation of Ti-n Hanakaten (Tassili n'Ajjer, Algerian Sahara)

Anderson MK (2005) Tending the wild: Native American knowledge and the management of California's natural resources. University of California Press, California

Angelucci DE, Boschian G, Fonatanals M et al (2009) Shepherds and karst: the use of caves and rock-shelters in the Mediterranean region during the Neolithic. World Archaeol 41(2):191–214

Asouti E, Austin P (2005) Reconstructing woodland vegetation and its exploitation by past societies, based on the analysis and interpretation of archaeological wood charcoal macro-remains. Environ Archaeol 10(1):1–18

Aura JE, Carrión Y, Estrelles E et al (2005) Plant economy of hunter-gatherer groups at the end of the last Ice Age: plant macroremains from the cave of Santa Maira (Alacant, Spain) ca. 12000–9000 B.P. Veg Hist Archaeobot 14:542–550

Aykroyd WR, Doughty J (1982) Legumes in human nutrition. FAO, Rome, Italy

Badal E (1999) El potencial pecuario de la vegetación mediterránea: las cuevas redil. Saguntum-Plav Extra 2:69–76

Badal García E, Martínez Varea C (2018) Different parts of the same plants. Charcoals and seeds from Cova de les Cendres (Alicante, Spain). Quatern Int 463B:391–400

Baines JA, Riehl S, Conard N et al (2015) Upper Palaeolithic archaeobotany of Ghar-e Boof cave, Iran: a case study in site disturbance and methodology. Archaeol Anthrop Sci 7:245–256

Balout L (1952) A propos des charbons préhistoriques. Bull Soc Hist Nat d'Afrique du Nord 43:160–163

Ballouche A, Marinval P (2003) Données palynologiques et carpologiques su la domestication des plantes et l'agriculture dans le Néolithique ancien dans le Maroc septentrional (site de Kaf Taht el-Ghar). Revue d'Archéométrie 27:49–54

Barrett SA, Gifford EW (1933) Miwok material culture. Bull Milwaukee Public Museum 2:117–376

Barton RNE, Bouzouggar A, Collcutt SN et al (2009) OSL dating of the Aterian levels at Grotte de Dar es-Soltan I (Rabat, Morocco) and possible implications for the dispersal of modern Homo sapiens. Quat Sci Rev 28:1914–1931

Barton RNE, Bouzouggar A, Hogue JT et al (2013) Origins of the Iberomaurusian in NW Africa: new AMS radiocarbon dating of the Middle and Later Stone Age deposits at Taforalt Cave Morocco. J Hum Evol 65(3):266–281

Barton RNE, Bouzouggar A, Collcutt SN et al (2016) Reconsidering the MSA to LSA transition at Taforalt Cave (Morocco) in the light of new multi-proxy dating evidence. Quatern Int 413:36–49

Bergfjord C, Karg S, Rast-Eicher A et al (2010) Comment on 30,000-Year-Old Wild Flax Fibers. Science 328(5986):1634

Boulos L (1983) Medicinal plants of North Africa. Reference Publications, Michigan

Brochier JE, Thinon M (2003) Calcite crystals, starch grains aggregates or…POCC? Comment on 'calcite crystals inside archaeological plant tissues'. J Archaeol Sci 30:1211–1214

Brochier JE, Villa P, Giacomarra M et al (1992) Shepherds and sediments: geo-ethnoarchaeology of pastoral sites. J Anthropol Archaeol 11:47–102

Brown DA (1984) Prospects and limits of a phytolith key for grasses in the central United States. J Archaeol Sci 11:345–368

Butler A (1998) Grain legumes: evidence of these important ancient food resources from early pre-agrarian and agrarian sites in Southwest Asia. In: Damania AB, Valkoun J, Willcox G and Qualset CO (eds) The Origins of Agriculture and Crop Domestication. ICARDA, Alep, pp 102–117

Cacho Quesada C, Papi Rodes C, Sánchez-Barriga Fernández A et al (1996) La cestería decorada de la Cueva de Los Murciélagos (Albuñol, Granada) Complutum 6(1):105–122

Camps G (2000) Ibéromaurusien. In: Encyclopédie Berbère, Fasc. 23, Hiempsal—Icosium, Aix-en-Provence. Edisud, pp 3579–3598

Canti MG (1999) The production and preservation of faecal spherulites: Animals, environment and taphonomy. J Archaeol Sci 26:251–258

Cappers RTJ, Neef R, Bekker RM (2009) Digital Atlas of Economic Plants. Barkhuis, Eelde
Chabal L (1988) Pourquoi et comment prélever les charbons de bois pour la période antique: les méthodes utilisées sur le site de Lattes (Hérault). Lattara 1:187–222
Chabal L (1997) Forêts et sociétés en Languedoc (Néolithique final, Antiquité tardive), L'anthracologie, méthode et paléoécologie. DAF, Paris
Charco J (1999) El bosque mediterráneo en el norte de África. Biodiversidad y lucha contra la desertificación. Ed. Mundo árabe e Islam. Agencia española de cooperación internacional, Madrid
Charco J (2001) Guía de los árboles y arbustos del norte de África. Agencia española de cooperación internacional, Madrid
Colonese AC, Mannino MA, Bar-Yosef Mayer DE et al (2011) Marine mollusk exploitation in the mediterranean prehistory: an overview. Quatern Int 239:86–103
Couvert M (1969) Identification de charbons provenant du gisement de Tamar Hat. Libyca XVII:49–52
Couvert M (1976) Traduction des éléments de la flore préhistorique en facteurs climatiques. Libyca XXIV:9–20
Desprat S, Combourieu-Nebout N, Essallami L et al (2013) Deglacial and Holocene vegetation and climatic changes in the southern Central Mediterranean from a direct land–sea correlation. Clim Past 9:767–787
di Lernia S, N'siala IM, Mercuri AM (2012) Saharan Prehistoric Basketry, Archaeological and archaeobotanical analysis of the early-middle Holocene assemblage from Takarkori (Acacus Mts., SW Libya). J Archaeol Sci 39:1837–1853
Fennane M (1989) Esquisse des séries du thuya de Bérbérie au Maroc. Bulletin de l'Institut Scientifique, Rabat 13:77–83
Fennane M, Ibn-Tattou M, Ouyahya A et al (eds) (2007) Flore Pratique du Maroc, vol 2. Institut Scientifique, Université Mohammed V, Rabat
Greguss P (1955) Identification of Living Gymnosperms on the Basis of Xylotomy. Akadémiai Kiado Budapest
Greguss P (1959) Holzanatomie der Europaïschen Laubhölzer und Straücher. Akadémiai Kiado Budapest
Hansen JM (1991) The palaeoethnobotany of Franchthi cave, Excavations at Franchthi Cave, Greece, 7. Indiana University Press, Bloomington
Hardy BL (2010) Climatic variability and plant food distribution in Pleistocene Europe: Implications for Neanderthal diet and subsistence. Quatern Sci Rev 29(5–6):662–679
Hather JG, Mason SLR (2002) Introduction: some issues in the archaeobotany of hunter-gatherers. In: Mason SLR, Hather JG (eds) Hunter-gatherer archaeobotany: Perspectives from the northern temperate zone. Institute of Archaeology, University College London, London, pp 1–14
Humphrey L, De Groote I, Morales J et al (2014) Earliest evidence for caries and exploitation of starchy plant foods in Pleistocene hunter-gatherers from Morocco. Proc Natl Acad Sci USA 111(3):954–959
Jacquiot C (1955) Atlas d´anatomie des bois des conifères. Cent Tech Bois, Paris
Jacquiot C, Trenard Y, Dirol D (1973) Atlas d´anatomie des bois des angiospermes (Essences feuillues). Cent Tech Bois, Paris
Jacobs Z, Roberts RG, Nespoulet R et al (2012) Single-grain OSL chronologies for Middle Palaeolithic deposits at El Mnasra and El Harhoura 2, Morocco: implication for Late Pleistocene human-environment interactions along the Atlantic coast of northwest Africa. J Hum Evol 62(3):377–394
Karkanas P (2006) Late Neolithic household activities in marginal areas: the micromorphological evidence from the Kouveleiki caves, Pelo- ponnese, Greece. J Archaeol Sci 33:1628–1641

Katz O, Cabanes D, Weiner S et al (2010) Rapid phytolith extraction for analysis of phytolith concentrations and assemblages during an excavation: an application at Tell es-Safi/Gath, Israel. J Archaeol Sci 37:1557–1563

Kelly RL (2013) The lifeways of hunter-gatherers: the foraging spectrum. Cambridge University Press, New York

Kherbouche F, Hachi S, Abdessadok S et al (2014) Preliminary results from excavations at Gueldaman Cave GLD1 (Akbou, Algeria). Quatern Int 320:109–124

Kornfeld M (1996) The big-game focus: reinterpreting the archaeological record of cantabrian upper paleolithic economy. Curr Anthropol 37(4):629–657

Kroeber AL (1976) Handbook of the Indians of California. Dover Publications, New York

Kuoni B (1981) Cestería tradicional Ibérica. Ediciones del Serbal, Barcelona

Kvavadze E, Bar-Yosef O, Belfer-Cohen A et al (2009) 30,000-Year-Old Wild Flax Fibers. Science 325:1359

Lanner RM (1981) The Piñon Pine: a natural and cultural history. University of Nevada Press, Reno

Le Floc'h É, Boulos L, Véla E (2010) Cataloque synonymique commenté de la Flore de Tunisie. Ministère de l'Environnement et du Développement durable, Banque Nationale de Gènes, Tunisie

Lev E, Kislev ME, Bar-Yosef O (2005) Mousterian vegetal food in Kebara cave, Mt. Carmel. J Archaeol Sci 33(3):475–484

Lindstädter J, Eiwanger J, Mikdad A et al (2012) Human occupation of Northwest Africa: a review of middle palaeolithic to epipalaeolithic sites in Morocco. Quatern Int 274:158–174

Louis A, Despois J (1986) Alfa. In: Encyclopédie Berbère, Fasc. 3, Ahaggar-Ali ben Ghaniya. Edisud, Aix-en-Provence, pp 440–442

Lubell D (2001) Late pleistoce-early Holocene Maghreb. In: Peregrine PN, Ember M (eds) Encyclopedia of prehistory, vol 1. Africa. Kluwer Academic/Plenum Publishers, New York, pp 129–149

Macphail RI, Courty MA, Hather J et al. (1997) The soil micromorphological evidence of domestic occupation and stabling activities. In: Maggi R (ed) Arene candide: a functional and environmental assessment of the holocene sequence (Excavations Bernarbo' Brea-Cardini 1940–50), Il Calamo, Roma, pp 53–88

Madella M, Alexandre A, Ball TB, ICPN Working Group (2005) International code for phytolith nomenclature 1.0. Ann Bot 96:253–260

Mannino MA, Catalano G, Talamo S et al. (2012) Origin and diet of the prehistoric hunter-gatherers on the Mediterranean island of Favignana (Ègadi Islands, Sicily). Plos One 7(11). https://doi.org/10.1371/journal.pone.0049802

Mason SLR (1995) Acorn-eating and ethnographic analogies: a reply to McCorriston. Antiquity 69:1025–1030

Mason SLR, Nesbitt M (2009) Acorns as food in southeast Turkey: implications for prehistoric subsistence in Southwest Asia. In: Fairbairn A, Weiss E (eds) From foragers to farmers. Oxbow Books, Oxford, Papers in honour of Gordon C. Hillman, pp 71–85

Mercuri AM (2008) Plant exploitation and ethnopalynological evidence from the Wadi Teshuinat area (Tadrart Acacus, Libyan Sahara). J Archaeol Sci 35(6):1619–1642

Mercuri AM, Fornaciari R, Gallinaro M et al (2018) Plant behaviour through human imprints and the cultivation of wild cereals in Holocene Sahara. Nature Plants 4:71–81. https://doi.org/10.1038/s41477-017-0098-1

M'Handi M, Anderson P (2013) Approche ethnoarchéologique d'outils et techniques de moisson de l'alfa (*Stipa tenacissima*) dans la région des Hautes Steppes en Tunisie: Premières observations et analyses. In: Anderson PC, Cheval C, Durand A (eds), Regards croisés sur les outils liés au travail des végétaux. An interdisciplinary focus on plant working tools. 23 rencontres internationales d'archéologie et d'histoire d'Antibes. Éditions APDCA, Antibes, pp 75–88

Momot J (1955) Méthode pour l'étude des charbons de bois. Bull Soc Préhist Fr 52:141–143

Morales J, Pérez-Jordà G, Peña-Chocarro L et al (2013) The origins of agriculture in North-West Africa: macro-botanical remains from Epipalaeolithic and Early Neolithic levels of Ifri Oudadane (Morocco). J Archaeol Sci 40:2659–2669

Morales J, Mulazzani S, Belhouchet L et al (2015) First preliminary evidence for basketry and nut consumption in the Capsian culture (ca. 10,000–7500 BP): archaeobotanical data from new excavation at El Mekta, Tunisia. J Anthropol Archaeol 37:128–139

Morales J, Pérez Jordà G, Peña-Chocarro L et al (2016) The introduction of South-Western Asian domesticated plants in North-Western Africa: an archaeobotanical contribution from Neolithic Morocco. Quatern Int 412:96–109

Mulazzani S, Belhouchet L, Salanova L et al (2015) The emergence of the Neolithic in North Africa: a new model for the Eastern Maghreb. Quatern Int 419:123–143

Mulholland SC, Rapp JrG (1992) A morphological clasification of grass silica-bodies. In: Rapp JrG, Mulholland SC (eds) Phytolith systematics: emerging issues. Advances in Archaeological and Museum Science. Plenum Press, New York, pp 65–89

Nadel D, Danin A, Werker E et al (1994) 9,000-year-old twisted fibers from Ohalo II. Curr Anthropol 35(4):451–458

Nami M, Moser J (2010) La Grotte d'Ifri n'Ammar; tome 2: Le Paléolithique Moyen. Zeitschrift für Archäologie Außereuropäischer Kulturen (FAAK) 9. Reichert, Wiesbaden

Neumann K, Schoch W, Détienne P et al (2001) Woods of the Sahara and the Sahel. Bern, Stuttgart, Wien

Peña-Chocarro L, Pérez Jordà G, Morales Mateos J et al (2015) Storage in traditional farming communities of the western mediterranean: ethnographic, historical and archaeological data. Environ Archaeol 20(4):379–389

Piperno DR (2006) Phytoliths: a comprehensive guide for archaeologists and paleoecologists. AltaMira Press, Lanham

Portillo M, Valenzuela S, Albert RM (2012) Domestic patterns in the Numidian site of Althiburos (northern Tunisia): the results from a combined study of animal bones, dung and plant remains. Quatern Int 275:84–96

Portillo M, Kadowaki S, Nishiaki Y, Albert RM (2014) Early Neolithic household behavior at Tell Seker al-Aheimar (Upper Khabur, Syria): a comparison to ethnoarchaeological study of phytoliths and dung spherulites. J Archaeol Sci 42:107–118

Portillo M, Belarte MC, Ramon J, Kallala N, Sanmartí J, Albert RM (2016) An ethnoarchaeological study of livestock dung fuels from cooking installations in northern Tunisia. Quatern Int 431:133–144

Rejdali M (2004) Forest cover changes in the Maghreb countries with special reference to Morocco. In: Mazzoleni S, di Pasquale G, Mulligan M, di Martino P, Rego F (eds) Recent dynamics of the Mediterranean vegetation and landscape, pp 23–32

Richards MP, Trinkaus E (2009) Isotopic evidence for the diets of European Neanderthals and early modern humans. Proc Natl Acad Sci USA 106(38):16,034–16,039

Richter D, Grün R, Joannes-Boyau R et al (2017) The age of the hominin fossils from Jebel Irhoud, Morocco, and the origins of the middle stone age. Nature 546:293–296

Riehl S, Marinova E, Deckers K et al (2015) Plant use and local vegetation patterns during the second half of the Late Pleistocene in southwestern Germany. Archaeol Anthrop Sci 7(2):151–167

Rivera D, Matilla G, Obón C et al. (2011) Plants and humans in the Near East and the Caucasus. Plants and Humans/Editum, Murcia (Spain)

Rosen AM (1992) Preliminary identification of silica skeletons from Near Eastern archaeological sites: an anatomical approach. In: Rapp JrG, Mulholland SC (eds) Phytolith Systematics:

emerging Issues. Advances in Archaeological and Museum Science. Plenum Press, New York, pp 129–147

Santa S (1961) Essai de reconstitution de paisajes végétaux quaternaires d'Afrique du Nord. Libyca 6(7):37–77

Sievers C (2006) Seeds from the middle stone age layers at Sibudu Cave. South Afr Human 18 (1):203–222

Schweingruber FH (1990) Anatomie europäischer Hölzer. Bern und Stuttgart

Soffer O, Adovasio JM, Hyland DC (2000a) The "Venus" figurines. Textiles, basketry, gender, and status in the Upper Palaeolithic. Curr Anthropol 41(4):511–537

Soffer O, Adovasio JM, Illingworth JS et al (2000b) Palaeolithic perishables made permanent. Antiquity 74:812–821

Tsartsidou G, Lev-Yadun S, Albert RM et al (2007) The phytolith archaeological record: strengths and weaknesses based on a quantitative modern reference collection from Greece. J Archaeol Sci 34:1262–1275

Twiss PC (1992) Predicted world distribution of C_3 and C_4 grass phytoliths. In: Rapp JrG, Mulholland SC (eds) Phytolith systematics: emerging issues, advances in archaeological and museum science. Plenum Press, New York, pp 113–128

Twiss PC, Suess E, Smith RM (1969) Morphological classification of grass phytoliths. Soil Sci Soc of Am Pro 33:109–115

Vaquer J, Ruas M-P (2009) La grotte de l'Abeurador Félines-Minervois (Hérault): occupations humaines et environnement du Tardiglaciaire a l'Holocène. In De Méditerranée et d'ailleurs…Mélanges offerts à Jean Guilaine. Archives d'Ecologie Préhistorique, Toulouse, pp 761–792

Vergès JM, Allué E, Angelucci DE et al (2002) La Sierra de Atapuerca durante el Holoceno: datos preliminares sobre las ocupaciones de la edad del Bronce en la Cueva de El Mirador (Ibeas de Juarros, Burgos). Trabajos de Prehistoria 59:107–126

Vernet J-L, Bazile E, Evin J (1979) Coordination des analyses anthracologiques et des datations absolues sur charbons de bois. Bull Soc Préhist Fr 76(3):76–79

Wadley L, Sievers C, Bamford M et al (2011) Middle stone age bedding construction and settlement patterns at Sibudu, South Africa. Science 334(6061):1388–1391

Wattez J, Courty MA (1987) Morphology of ash of some plant remains. In: Fédoroff N, Bresson LM, Courty MA (eds) Micromorphologie des sols—Soil Micromorphology. Association Francaise pour l'Étude du Sol, Plaisir, pp 677–682

Weiss E (2009) Glimpsing into a hut: the economy and society of Ohalo II's inhabitants. In: Fairbairn A, Weiss E (eds) From foragers to farmers: papers in honour of Gordon C. Hillman. Oxbow Books, Oxford, po 153–160

Wengler L, Vernet J-L (1992) Vegetation, sedimentary deposits and climates during the Late Pleistocene and Holocene in eastern Morocco. Palaeogeogr Palaeocol 94(1–4):141–167

Zapata L, López Sáez JA, Ruiz Alonso M et al (2013) Holocene environmental change and human impact in NE Morocco: Palaeobotanical evidence from Ifri Oudadane. Holocene 23 (9):1286–1296

Zhang J, Lu H, Sun G et al (2016) Phytoliths reveal the earliest fine reedy textile in China at the Tianluoshan site. Sci Rep UK 6:18664

Archaeology and Palaeoecology: Integrated Methods

The Translocation of Useful Trees in African Prehistory

Roger Blench(✉)

McDonald Institute for Archaeological Research, University of Cambridge, Cambridge, UK
rogerblench@yahoo.co.uk

Abstract. Agriculture in Africa is usually conceptualised as beginning in a distinct era with specific indicators of early plant domestication. However, as research on African vegetation evolves, it is increasingly clear that the identification, use and subsequent translocation of trees and other woody plants constituted a major process in the transformation of the African landscapes far earlier than agriculture proper. Indeed some authors now date this to as early as 10,000 years ago. The paper argues that tree translocation can occur through a number of inter-related processes, which are exemplified in the text. It focuses on two species in particular, the baobab and the wild date-palm, *Phoenix reclinata*, which have anthropic distributions. The paper discusses methodology of identifying such tree species, and suggests that African vegetation has been manipulated in ways comparable to early domestication in the Amazon and the Pacific. It also notes that these processes continue with the spread of fruit and timber species in the modern era.

Keywords: *Adansonia digitata* · Trees · Translocation · Wild date-palm Woody vegetation

Introduction: The 'Domesticated' Rainforest

An all-too common conceptualization of African vegetation contrasts domestic landscapes with 'wild' forest, typically the humid forest spreading along the West African coast and continuing into the equatorial regions of the Congo. NGOs and governments have a strong interest in portraying rainforests as a primeval Eden and the people who live in them as *Naturvölker*, in harmony with nature. The conclusions each institution draws are somewhat different, according to whether the goal is to protect the forests at all costs, or to log them extensively, on the presumption that the few inhabitants have no rights over the trees. However, it is becoming increasingly clear that this characterization of forests is a fantasy, that the great equatorial forests are more like gardens or farms and only our cereal-based concept of agriculture prevents us from seeing this.

Globally speaking, a key step in the identification of tree management and translocation as a distinct process was the archaeobotany of the Pacific islands (e.g. Barrau 1956; Walter and Sam 1999; Kyle Latinis 2000). Since archaeological coverage is good and increasingly reliable and direct dates relatively abundant, particular species can be identified as not part of the 'natural' flora and thus transplanted by humans

A. M. Mercuri et al. (Eds.): *Plants and People in the African Past*,
https://doi.org/10.1007/978-3-319-89839-1_10

(Blench 2005). These processes must have long preceded the horizon attributed to conventional agriculture in the Asia-Pacific (Blench 2008). It is becoming increasingly clear that similar processes occurred in the Amazon (Clement et al. 2015; Piperno et al. 2015). Trees were identified as useful as much as 11,000 years ago and were spread around the Amazon basin, notably the corozol palm, *Attalea cohune* Mart., and the peach palm, *Bactris gasipaes* Kunth. The Amazon, far from being 'untouched', is a managed resource with a modified floristic composition. Even older is evidence for the anthropic translocation of the red cabbage palm (Fig. 1), *Livistona mariae* F. Muell, into central Australia as much as 30,000 years ago (Kondo et al. 2012)[1].

Agriculture in Africa is usually conceptualized as beginning at a distinct horizon which can be delineated via the results of archaeobotany. While this can be successfully applied to cereals and pulses recovered from drier areas, evidence for tree domestication and other types of vegeculture will never be so neat. Trees may or may not show distinctive morphological changes, according to their uses and the purposes of their managers. The great majority of trees identified in early periods for food or other products may well have been translocated, protected from fire, and otherwise encouraged, but may not show morphological change. As research on African vegetation evolves, it is increasingly clear that the identification, use and subsequent translocation of trees and other woody plants constituted a major process in the transformation of African landscapes, sometimes concurrent with but often significantly earlier than seed agriculture.

As cartography of indigenous African tree species becomes increasingly available,[2] it is more evident that many species extend far beyond their natural range (Morin-Rivat et al. 2014). This is the result of a variety of processes, including both intentional translocation and facilitation by human migration as seeds of edible fruits were discarded. Well-known species in this category include baobab (*Adansonia digitata* L.) and silk-cotton (*Ceiba pentandra* (L.) Gaertn.), but there are many more, including wild date-palm (*Phoenix reclinata* Jacq.), oil-palm (*Elaeis guineensis* Jacq.), shea (*Vitellaria paradoxa* C.F. Gaertn.) and locust (*Parkia biglobosa* (Jacq.) R. Br. ex G. Don). Field guides such as Vivien and Faure (2011a, 2011b) show just how wide the range of edible fruits can be in the equatorial forest; most of these have never been recovered by archaeobotany. Most of the evidence for the exploitation and translocation of trees is biogeographical, although additional support can be gathered from palynology, ethnography and linguistics (Williamson 1993; Connell 1998; D'Andrea et al. 2006; Blench 2007a, 2007b, 2014; Bostoen 2005, 2014; Bostoen et al. 2013; Oslisly et al. 2013a, 2013b). It is worth noting that online databases such as that maintained by *Conservatoire et Jardin botaniques, Ville de Genève*, typically privilege herbarium specimens, and 'wild' specimens. Thus the map plotting the distribution of the oil palm completely omits its widespread occurrence in the semi-arid savannas, where it is a common anthropic migrant. To develop a reliable cartography depends on synthesizing both herbarium records and descriptive literature.

---

[1] Ironically this was thought to be a disappointing result, as the palm had previously been considered as a relic of Gondwanaland.

[2] See http://www.ville-ge.ch/musinfo/bd/cjb/africa/recherche.php.

Fig. 1. The red cabbage palm in central Australia (*Source* Creative Commons)

The exact circumstances and chronology of tree management in the mid-Holocene remains controversial, but in more recent times, Africa has seen the introduction and spread of a wide variety of cultivated trees. On the east coast, part of the sphere of Indian Ocean trade, important species such as coconut (*Cocos nucifera* L.), mango (*Mangifera indica* L.), orange (*Citrus sinensis* (L.) Osbeck) and areca palm (*Dypsis lutescens* (H.Wendl.) Beentje & J.Dransf.) were introduced by the medieval period or earlier. On the west coast, the Portuguese introduced the guava (*Psidium guajava* L.), the pawpaw (*Carica* papaya *L.*) and the avocado (*Persea americana* Mill.), as part of the trade triangle established with the Americas (Blench 1998). In the colonial era, trees such as the mango, and ornamentals such as the neem (*Azadirachta indica* A. Juss. 1830), Persian lilac (*Melia azedarach* L.), flame tree (*Delonix regia* (Boj. ex Hook.) Raf.) and jacaranda (*Jacaranda* spp.) were all introduced by missionaries and colonial officials, transforming the African landscape. Moreover, plants such as the *Citrus* spp. and the diploid bananas already established on the East Coast, were brought around the Cape and introduced to West Africa (Blench 2007a).

The paper discusses the process of tree translocation, the methodology for identifying such species, as well as the types of evidence we should seek in extending and testing hypotheses about individual species. It presents an extended example of

Phoenix reclinata, the wild date palm, which is not domestic in the usual sense, and which is endemic to the continent, but which has evidently been widely dispersed by humans. The broad conclusion is that current approaches to the genesis of African landscapes have consistently underestimated the contribution of managed and translocated trees and that it underwent an early domestication process in many ways comparable to the Amazon and the Pacific islands.

The Principles of Translocation

Ascertaining the motives for tree translocation hangs on the question of intentionality. Do people decide to move trees, and do they manage surrounding vegetation to encourage particular species? For example, although the distribution of the baobab is manifestly anthropic (Duvall 2007), it is highly unlikely that it was primarily transplanted intentionally. This was unnecessary as its fruit is carried as a storable snack, and when the shell is cracked open and the edible pulp is cleaned of seeds, the seeds fall and germinate. Hence baobabs often occur along the lines of former or present pastoral migration routes (Wickens and Lowe 2008). In more recent times the distribution pattern has changed, with the growth of fenced agricultural fields, as baobabs are now occasionally used as live fences. Similar debates have characterized the discussion of the spread of the oil-palm (D'Andrea et al. 2006: 199; Logan and D'Andrea 2012). They observe that '*Several authors suggest that domestication was not necessary in order to achieve optimal yields, as oil palm is capable of reproducing subspontaneously in appropriate conditions (...) The first stage is wild, followed by semi-domesticated, where oil palm becomes a "camp follower" as a result of human collection offruits, which are discarded within compounds and fields.*' Similarly, Oas et al. (2015) point to significant levels of exploitation of the incense tree, *Canarium schweinfurthii* Engl, from as early as 12,525–11,619 cal yr BP in Bosumpra Cave, in an area where it is of little importance today. They present '*an argument (...) for the management of* C. schweinfurthii *and* E. guineensis *beginning by the early Holocene*'. D'Andrea et al. (2006: 214) tabulate finds of oil-palm in West-Central Africa. The great quantity of finds in Cameroon and Central Africa date back to shortly before 3000 cal yr BP are surely indicative of a correlation with the earliest period of the Bantu expansion and indicative of some level of intentionality. By contrast, the only use of *Euphorbia* spp. is as live fences and its transplantation and management is characteristic of the late agricultural period. By examining characteristic species, we can tabulate a variety of processes responsible for anthropic tree distributions (Table 1).

The oil-palm is an example of the way multiple processes of translocation can occur over long periods. The evidence from Bosumpra cited above suggests that the oil-palm has been managed by humans for millennia. However, its presence in semi-arid regions such as Northern Nigeria is relatively recent, but prior to European presence in the region (e.g. Dalziel 1937). Nigerian oil-palms were then subject to intensive breeding in colonial breeding programs and plantations established, while the rootstock was exported to SE Asia (D'Andrea et al. 2006 has useful summary of this literature).

Table 1. Processes responsible for anthropic tree distributions

Process	Description	Example
Intentional translocation	The tree has a product so desirable that it is planted in a new location	oil palm [recently], pawpaw, orange
Unintentional translocation	The tree has a fruit or other product which is used and leads to transport of seeds to new locations	baobab
Facilitation	The tree is not translocated, but its habitat and growth prospects are improved by clearance of vegetation etc	oil-palm [in humid zones], bush-candle (incense tree)
Unintentional facilitation	The tree is not valued, but anthropic processes such a forest clearance, provide enhanced habitat	umbrella tree
Other indirect processes	Human predation of animals, insects which pollinate plants, leading to skewed distributions	all bee plants

The possible products responsible for the intentional translocation of trees are numerous and in many cases reflect economic interactions with an external trade system or broad cultural shifts within the continent. Table 2 presents a list of motives for intentionally translocating trees, including an example species.

Table 2. Reasons for intentionally translocating trees

Motivation	Example
Food and edible oil sources, fruits, roots, leaves etc.	*Vitellaria paradoxa* C.F. Gaertn.
Construction materials	*Raffia sudanica* A. Chev.
Weaving of fibres: mats, hats and others	*Phoenix reclinata* Jacq.
Boundary markers/fencing	*Newbouldia laevis* Seem.
Green manure	*Tephrosia vogelii* Hook f.
Shade	*Ficus platyphylla* Del.
Ritual importance	*Diospyros mespiliformis* Hochst. ex A. DC.
Medicinal	*Bridelia ferruginea* Benth.
Ornamental	*Melia azederach* L.
Pollen attracts bees	*Annona senegalensis* Pers.
Commercial—timber or other products sold	*Funtumia elastica* (Preuss) Stapf.
For alcohol	*Raffia hookeri* G. Mann & H. Wendl

Some of these motivations must have been already present in the pre-agricultural era, such as supplementing food supplies, edible oils, medicines or alcohol. Others are characteristic of the late agricultural period, such as large-scale fencing and construction material. Following the era of European contact, demand for products such as timber and rubber accelerated demand for particular species, while at the same time ornamentals were introduced as part of the colonial process of constructing a familiar environment. Books such as Dalziel (1937) or Busson (1965) present a conspectus of West African useful plants from a colonial perspective; the value of woods for railway sleepers, house construction and so on. Whole institutes were established to explore and value tropical products, such as the Imperial Institute in London, which became the Tropical Products Research Institute, and the Royal Museum for Central Africa, Tervuren, outside Brussels.

Our interpretation of the patterning of African tree translocation is inevitably driven by ethnobotany. We can record the uses of trees in the present, and it is tempting to read these into the past. But current tree exploitation processes are not necessarily a reliable guide to past practice. In other words, ethnobotany is not necessarily archaeobotany. Two contrastive examples illustrate these issues. The shells of the African nettle or hackberry, *Celtis integrifolia* Lam., are regularly recorded in West African sites and seem to have been eaten extensively in the past (Kahlheber 1999). But they are difficult to digest even when processed and are almost never recorded as eaten in present ethnobotany surveys, presumably because they have been displaced by more edible fruits, such as mango.

By contrast, shea, *Vitellaria paradoxa* C.F. Gaertn, is widely grown in the present from Senegal to Uganda for its oil (Hall et al. 1996; Blench 2014, 2016). Despite having potentially highly visible macroremains, shea has only a weak archaeobotanical presence. Neumann et al. (1998: 60) report a testa from the medieval village of Saouga and note that shea-butter production was recorded by Ibn Baṭṭuṭa in the 14th century. Shea is complex to process, involving roasting ovens and large-scale oil expression and purification, and it seems likely that its widespread use may well be late, reflecting an already mature agriculture.

Overview of Translocated Species

General

The examples given so far have been from trees where the motive for translocation is fairly obvious and can be attested in the ethnographic record. However, there are many species which *might* be translocated but for which the evidence is uncertain. The situation is the same in the Amazon, and only painstaking work with individual species can uncover their history. However, what we can look for is 'suspicious' distributions, i.e. those where the tree appears to reflect human presence or absence in some way.

Translocated Trees in Semi-arid Regions

Tree translocation in the semi-arid and Sahelian regions is more obvious than in the equatorial forest, since the overall floristic inventory is much smaller, and trees are hardy once established. Table 3 shows a list of common trees indigenous to the African continent, which seem to have partly anthropic distributions.

Table 3. Tree species translocated in semi-arid regions

Common name	Scientific name
African olive, incense tree	*Canarium schweinfurthii* Engl.
Baobab	*Adansonia digitata* L.
Ben oil tree	*Moringa oleifera* Lam.
Borassus palm	*Borassus aethiopum* Mart.
Dum palm	*Hyphaene thebaica* (L.) Mart.
Locust bean	*Parkia biglobosa* (Jacq.) R.Br. ex G.Don
Oil palm	*Elaeis guineensis* Jacq.
Red-flowering silk-cotton	*Bombax buonopozenze* P. Beauv.
Shea	*Vitellaria paradoxa* C.F. Gaertn.
Silk-cotton	*Ceiba pentandra* (L.) Gaertn.
Tamarind	*Tamarindus indica* L.
Whitethorn	*Faidherbia albida* (Delile) A. Chev.
Wild date palm	*Phoenix reclinata* Jacq.

Trees play a part in economic systems, and their products can change in value in various historical eras. Translocation can thus be a dynamic process, entering into feedback loops. Once a specific product of a tree is valued, familiarity leads communities to develop other uses. An example of a species translocated in semi-arid regions which has had different use-values over time is the baobab (*Adansonia digitata*). Table 4 shows the different uses of parts of the baobab with speculations about their relative antiquity.

Table 4. Changing uses of the baobab over time

Use	Antiquity	Comment
Edible fruit	Probably old	A white edible pulp surrounds the seeds[a]
Rope	Probably old	Strips of bark are rolled into rope
Religious importance	Almost certainly old	The focus of many shrines in savanna regions and trees strongly protected even in highly anthropic landscapes
Leaves	Probably recent	Making soup from leaves is a practice spread by the Hausa
Fencing	Probably recent	Young baobabs planted to fence compounds
Ice cream	Certainly recent	The white pulp surrounding the seeds used to flavour ice-cream
Tourism	Certainly recent	Old baobabs attractive sites for tourism

[a]Curiously, this pulp was only approved for food use in Europe in 2008 and in the US in 2009, despite being eaten for millennia in Sub-Saharan Africa

The value attributed to baobabs is clearly responsible for their widespread occurrence in savannas, but they also spread down the west coast of equatorial Africa, certainly not their natural environment (see discussion in Wickens and Lowe 2008). For example, Fig. 2 shows a baobab in the Kinshasa area of DRC (Democratic Republic of the Congo). It is conceivable that their presence is explained by natural spread down former savanna corridors but human intervention is the more likely explanation, since ethnographic evidence illustrates the translocation of baobabs. Rather intriguingly, evidence for the complex translocation of the baobab to India has recently emerged. It has long been assumed that the baobabs in India probably arrived in the medieval period along with the populations of former slaves of African origin, the Siddi (see discussion in Blench 2003). However, Bell et al. (2015) have mapped baobab populations around India and the evidence from genetics is that regional populations come from distinct areas on the African continent and show different levels of internal diversity, suggesting they were transported at widely varying intervals.

Fig. 2. Baobab in the Kinshasa area (*Source* Author)

Tree Translocation in the Modern Era

An intriguing feature of the African landscape is the very large number of trees that have been successfully translocated in the era of European contact. In part this reflects very different external priorities. For example, although wild fruits were eaten in pre-European Africa, they had a low prestige, indeed were often considered famine foods or suitable only for children. As a consequence, little effort was made to domesticate or improve them. This was in contrast to stimulants such as cola, or oily seeds like bush mango, *Irvingia gabonensis* (Aubry-Lecomte ex O'Rorke) Baill., which have been the subject of genetic change and anthropic dispersal. Effectively this left an open niche in the agro-ecology of the region, which was complemented by the European taste for fruit. Trees have an additional advantage compared with field crops, in that they do not need a complex agronomic infrastructure to be introduced effectively. A tree such as the pawpaw was introduced from the New World by the Portuguese in the 18th century and rapidly spread inland through farmer-to-farmer transmission (Blench 1998). By contrast, rose-apple, *Syzygium jambos* (L.) Alston, although found in a few markets along the West African coast, has never become a major part of indigenous cultivation systems, presumably because it cannot survive in a wide variety of ecosystems. Some of these fruit trees have so far escaped cultivation as to become weed species, such as mango.

The same is true with ornamentals. The Victorian enthusiasm for flowers was matched by local cultures in India and the Pacific, but flowers and the use of vegetation for decoration have never had much importance in Africa. Missionaries in particular identified a variety of trees and climbers for use in making compounds more attractive and those which were relatively hardly were transported to Africa. Neem and flame tree have become such common roadside species in many parts of the continent that many people now believe they *are* indigenous.

Table 5. Examples of tree species translocated in the modern era

Common name	Scientific name	Main use value
Orange, lemon, lime etc.	*Citrus* spp.	Fruit
Guava	*Psidium guajava* L.	Fruit
Avocado	*Persea americana* Mill.	Fruit
Coconut	*Cocos nucifera* L.	Fruit
Mango	*Mangifera indica* L.	Fruit
Pawpaw, papaya	*Carica papaya* L.	Fruit
Teak	*Tectona grandis* L.f.	Timber
Eucalyptus	*Eucalyptus grandis*[a] W. Hill ex Maiden	Firewood
Jacaranda	*Jacaranda* spp.	Ornamental
Neem	*Azadirachta indica* A. Juss.	Ornamental
Flamboyant	*Delonix regia* (Boj. ex Hook.) Raf.	Ornamental
Persian lilac	*Melia azedarach* L.	Ornamental

[a]Many other *Eucalyptus* spp. have more local distributions in Africa, for example *E. globulus*, *E. melliodora* and *E. rostrata* in Ethiopia

Table 5 is an abbreviated list of the most common translocated fruits and ornamentals in Africa which have become permanent features of the landscape and integrated into African farming systems. Many other species recorded in local handbooks of useful plants are found in restricted areas.

Open Niches and Continuing Transformation

It should not be thought that the processes behind tree translocation are static, African landscapes continue to be transformed by introductions but also by the open niches created by farming practices, the policies of development agencies and land degradation. For example, in Northern Ghana, the silk-cotton tree, *Ceiba pentandra*, was formerly a key species in the arable landscape, used for its floss, leaves and bark (Chevalier 1949). While not domesticated in the sense of morphological change, silk-cotton must have been transplanted and protected in the growing stage to account for its high incidence. The traditional uses have now been largely displaced, but the trees remain. Every dry season, the pods fall to the ground, split open and floss fills the air. The tree has thus gone from a key useful species to little more than a weed, and only its size makes it difficult to eliminate rapidly. A more modern intervention is analyzed in Awono et al. (2002) who discuss the promotion of *safoutier*, *Dacryodes edulis* H.J. Lam, an indigenous African species taken up by CIFOR (Centre for International Forestry Research). Although already widely traded in local markets of West-Central Africa, improved varieties may increase its profitability, leading to more extensive planting and sale (Leakey et al. 2002).

The Example of the Wild Date-Palm, *Phoenix reclinata* (Arecaceae)

The wild date palm, *Phoenix reclinata* (Fig. 3), is widely valued for its leaves, which are used to make hats and mats (Burkill 1997). Records from Southern Africa suggest it is also tapped for wine, although this use is not significant elsewhere in the continent. Despite its well-known relative, the fruits are virtually inedible. Ethnobotanical surveys usually record it as ‘wild’ and it is not currently transplanted. It is common in semi-arid zones, south of the date-palm proper, but is also recorded in coastal forest and disturbed habitats in equatorial forest (Fig. 4). Apart from the African mainland it is also recorded in Madagascar, the Comores and the Arabian Peninsula, and it has been naturalized in Florida, Puerto Rico, Bermuda and the Leeward Islands, presumably through the slave trade. It seems to be absent in Angola, Namibia and the dry parts of Western and Southern Africa[3].

Figure 4 shows the distribution of the wild date palm in Africa, and illustrates the hypothesis that part of its distribution at least must be anthropic. Kinnaird (1992) has shown that the fruits play an important role in the diet of mangabeys, so it is possible they may act as secondary dispersal agents. Although a tree of semi-arid savannas,

[3] According to Rodin (1985) indigenous names in Namibia translate as ‘palm of the white people’ and attempts to introduce it are within the period of historical records.

Fig. 3. Wild date-palm, *Phoenix reclinata* (*Source* Creative Commons)

there are records along the west coast of Central Africa as well as in savanna islands within the tropical moist forest. In the northwest of the equatorial forest, it has a split distribution, both along the littoral and down the riverline in the interior. This suggests anthropic colonization during forest disturbances associated with the first wave of Bantu expansion, before the opening up of the 'Sangha gap', a region of the northwest equatorial forest evident in the palynological record (Oslisly et al. 2013a; Morin-Rivat et al. 2014). This pattern of split distributions is also evident with other tree species and is attested in the linguistic record (Bostoen 2014). However, the wild date-palm also has a striking distribution on the east coast of Africa and in northwest Madagascar. Given that this part of Madagascar is where Swahili settlement was most important, it is reasonable to assume it was carried there, as to the Comores. Evidence from phytoliths shows that *Phoenix reclinata* has an early presence in East Africa, and that the lack of records in Central Africa is a relic of more recent movements in the boundary of the forest (Albert et al. 2009). As Table 6 shows, vernacular names among populations along the southeast coast of Africa, are apparently cognate with Swahili[4] and not reflected in languages of the interior. This suggests that the wild date palm was brought

[4] One reviewer suggests the Zulu form is not cognate with the others; I respectfully disagree.

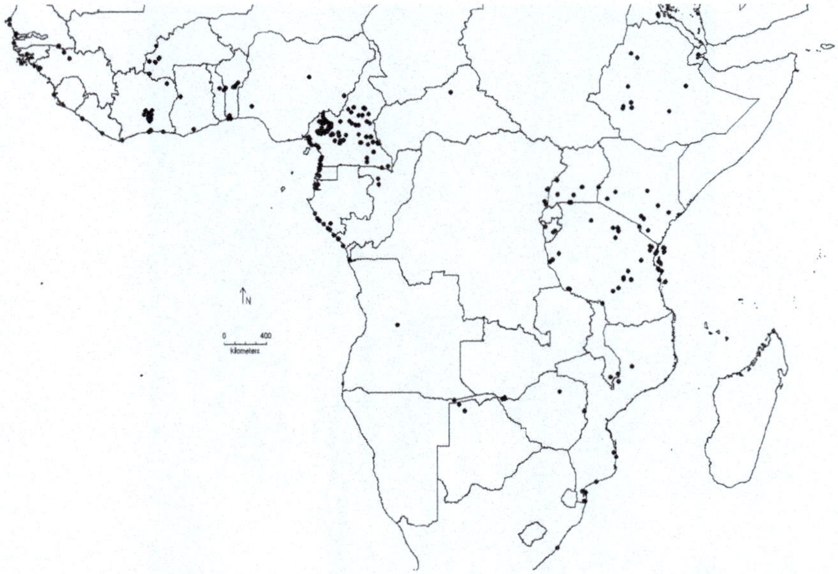

Fig. 4. Map of *Phoenix reclinata* distribution in Africa (*Source* Charles Doumenge)

there through the early coastal trade attested in an extensive archaeological literature, which left islands of Swahili in the far south of Mozambique (Blench 2012).

Is it possible to match the linguistic evidence with these hypotheses? Blench (2016) discusses some of the methodological pitfalls in using linguistic evidence. Blench (2014) is a preliminary compilation of related vernacular names for the wild date palm, and this is significantly expanded in Table 6, compiled from overviews of plant names for Bantu languages (including Wild et al. 1972; Maundu et al. 1999; Ruffo et al. 2002). West African lexemes are from my own fieldwork in Nigeria and Cameroon and Burkill (1997). Only words which I consider probably related are included. However, it is important to note these are hypotheses of cognacy; we are far from having the sort of widespread ethnobotanical data in West Africa to meet a criterion of regular phonological correspondences.

There is a widespread root in West Africa with a general shape *kVl/rV*, most commonly applied to the wild date palm, but sometimes to other species. However, in Yoruba and other southern Nigeria languages, this is shortened and then reduplicated, giving forms such as Yoruba *okũkũ*. Reflexes of this form are also attested in a variety of Bantoid languages. A similar form occurs in the Bantu languages of Eastern and Southern Africa. A Proto-Bantu form **kìndʊ́* given in the database BLR 3 (Bantu Lexical Reconstructions 3)[5] refers to Guthrie's proposal (1967–71), but the restricted distribution of supporting evidence makes their relationship uncertain. The sound correspondences between individual lexemes are not always regular, which points to

[5] http://www.africamuseum.be/collections/browsecollections/humansciences/blr.

Table 6. #-kundi-, a root for a dryzone palm in Niger-Congo languages

Family	Subgroup	Language	Attestation	Gloss as in source
Mande		Mende	kundi, keli	*Phoenix reclinata*
Mande	West	Bambara	kolo	dum palm
Atlantic	North	Wolof	koroso	*Phoenix reclinata*
Atlantic	South	Temne	a-kent/ɛ-	*Raphia* sp.
Gbaya		Ɓodoe	kò	*Borassus aethiopum*
Kwa	Ga-Dangme	Ga	kɔ̀lɔ̀	k.o. palm tree
EBC	Yoruboid	Yoruba	okũkũ	*Phoenix reclinata*
EBC	Edoid	Ẹdo	ukukõ	*Phoenix reclinata*
Bantoid	Ring	Kom	ə̄kú	palm tree
Bantoid	Momo	Ngwo	ànkɔ́n	'date palm'
Bantu		P East Bantu	*kìndʊ́	wild date palm
Bantu	[J]E15[a]	Ganda	olukindu	*Phoenix reclinata*
Bantu	E23	Zinza	mʧindu	*Phoenix reclinata*
Bantu	E51	Kikuyu	mɔkɛndɔ	*Phoenix reclinata*
Bantu	E55	Kamba	mɔkɛndɔ	*Phoenix reclinata*
Bantu	E73	Digo	makindu	*Phoenix reclinata*
Bantu	F11	Tongwe	lusanda	*Phoenix reclinata*
Bantu	F21	Sukuma	bukindu	*Phoenix reclinata*
Bantu	G42	Swahili	mkindu	*Phoenix reclinata*
Bantu	J30	Luhya	luʃindu	*Phoenix reclinata*
Bantu	L52	Lunda	makindu	palm leaves
Bantu	S10	Shona	muʧindwe	*Phoenix reclinata*
Bantu	S42	Zulu	isundu	wild date palm
Bantu	S51	Hlengwe	mʧinzu	*Phoenix reclinata*
Bantu	S53	Tsonga	m.ʧinʤu	wild date palm
Bantu	S61	Chopi	inʧindu	*Phoenix reclinata*

Bantu languages are numbered using the referential scheme of Guthrie, which may not imply anything about their genetic classification

borrowing possibly as a result of the coastal trade highlighted above. Sources for northwestern Bantu languages such as Raponda-Walker and Sillans (1961) suggest the vernacular names in Gabon are quite unrelated. However, the record of *makindu* for 'palm-leaf' in Lunda, a language of Northeast Angola may point to a possible connection with the languages of the northwest. Otherwise the connection between Bantoid and Bantu of Eastern Africa, may be found in the poorly documented Bantu languages along the northern edge of the forest savanna ecotone. Only further lexical data will clarify these intriguing connections.

Conclusions

Our understanding of the transition from foraging to farming has been strongly influenced by models deriving from the Near East, where humans discover a promising seed and gradually domesticate it by a variety of processes. Due to the density of research in the Asia-Pacific region, the origins of vegeculture are beginning to provide a competing model in terms of crops (Blench 2013). But the translocation of trees and other plants remains difficult to fit into this model. Just as there is increasing evidence for the translocation of animals long before agriculture (Drake and Blench 2017) so this is also the case with trees. Foragers can move fruits and seeds around intentionally (as we know to be the case in the Pacific Islands) or as a consequence of eating and discarding fruits. Changes in forest boundaries as palaeo-climate altered over time makes discriminating between natural and anthropic dispersals more complex. Nonetheless, although tree translocation and arboriculture are well-described processes in the Amazon and the Pacific, they hardly figure in accounts of the evolution of vegetation in Africa. Nonetheless, as the evidence mounts, both the semi-arid and equatorial landscapes must increasingly be characterized as domestic, even in periods well before the advent of agriculture.

The second aspect of this is that use-values are dynamic. One of the misleading impressions given by typical ethnobotanical characterizations is that all the uses are synchronic and by implication can be referred back to the period when humans first made use of a particular species. But it is easy to see that this is false, that a species may come into prominence for one use, others are thereby discovered and the value of the primary use may diminish or even disappear. For example, tree barks used to make rope were clearly of great importance prior to European contact, but with the importation first of stronger vegetable fibres, and then synthetic substitutes, their value has dropped to near-zero. The baobab is another example of the dynamic use of a tree species over time. Similarly, the shea, a premium oil species across much of semi-arid West Central Africa, may well have become salient relatively recently, despite being an endemic across the region (Blench 2014). Thus when reconstructing vegetation history, taking into account this type of dynamism is evidently of considerable importance.

Methodologically speaking, we are just at the beginning of developing techniques to detect anthropic translocation. Macroremains of trees do not always show the type of definitive morphological changes characteristic of cereals. Palynology can help plot the antiquity of a species in a particular region, but without associated palaeo-climatic profiles it is difficult to assess the likelihood of translocation elsewhere. The coverage of pollen cores in Africa remains patchy, with some areas well-covered, and vast regions barely known (cf. Oslisly et al. 2013b for a summary of the situation in the northwest equatorial forest). Many of the cores used to project West African vegetation patterns are from the adjacent oceans and their credibility is problematic. Bostoen et al. (2013) have demonstrated how linguistic evidence can be matched to vegetation dynamics, but this works better for well documented tree species. Descriptive ethnobotanies for individual groups remain a high priority. Genetics should also provide answers to some of these questions, especially for salient and widespread species. However, except for the baobab, the use of these techniques has barely begun.

A key part of the process, however, is simply asking the right questions. Ethnobotany can help focus archaeobotany, and archaeobotanical findings can be matched to the results from other disciplines. That landscapes are domesticated almost everywhere is beginning to be accepted as a general principle, but actually documenting particular species of useful woody plants, and identifying the patterns and mechanisms of their diffusion, remains a considerable challenge.

Acknowledgements. This paper was first presented at the 8th International Workshop for African Archaeobotany, Modena, Italy, 23–26 June 2015. However, its genesis was following an Ethnoscience Summer School in Libreville, Gabon, in 2013 hosted by Patrick Mougiama-Dauda. I would like to thank Charles Doumenge, CIRAD, Montpellier and Charles Clement, INPA, Manaus, for discussions underlying the general idea. Thanks to the editors and two referees for suggestions and corrections.

References

Albert RM, Bamford MK, Cabanes D (2009) Palaeoecological significance of palms at Olduvai Gorge, Tanzania, based on phytolith remains. Quatern Int 193(1):41–48

Awono A, Ndoye O, Schreckenberg K et al (2002) Production and marketing of safou (*Dacryodes edulis*) in Cameroon and internationally: market development issues. For Trees Livelihoods 12(1–2):125–147

Barrau J (1956) Plantes alimentaires de base des Mélanésiens. J Agri Trop Bot Appl 3(1–2):32–49

Bell KL, Rangan H, Kull CA et al (2015) The history of introduction of the African baobab (*Adansonia digitata*, Malvaceae: Bombacoideae) in the Indian subcontinent. Roy Soc Open Sci, https://doi.org/10.1098/rsos.150370. Accessed 9 Sep 2015

Blench RM (1998) The diffusion of New World Cultigens in Nigeria. In: Chastenet M (ed) Plantes et paysages d'Afrique. Karthala, Paris, pp 165–210

Blench RM (2003) The movement of cultivated plants between Africa and India in prehistory. In: Neumann K, Butler A, Kahlheber S (eds) Food, fuel and fields: progress in African Archaeobotany. Heinrich-Barth-Institut, Köln, pp 273–292

Blench RM (2005) Fruits and arboriculture in the Indo-Pacific region. Bull Indo Pac Pre Hi 24:31–50

Blench RM (2007a) Using linguistics to reconstruct African subsistence systems: comparing crop names to trees and livestock. In: Denham TP, Iriarte J, Vrydaghs L (eds) Rethinking Agriculture. Archaeological and ethnoarchaeological perspectives. Left Coast Press, Walnut Creek, p. 408-438

Blench RM (2007b) The intertwined history of the silk-cotton and baobab in West Africa. In: Cappers RTJ (ed) Fields of Change. Progress in African archaeobotany. Barkhuis, Groningen, p. 1-20

Blench RM (2008) A history of fruits on the SE Asian mainland. In: Osada T, Uesugi A (eds) Occasional Paper 4: Linguistics, archaeology and the human past. Indus Project, research institute for humanity and nature, Kyoto, pp 115–137

Blench RM (2012) Two vanished African maritime traditions and a parallel from South America. Afr Archaeol Rev 29:273. https://doi.org/10.1007/s10437-012-9115-y

Blench RM (2013) Was there once an arc of vegeculture linking Melanesia with Northeast India? In: Summerhayes GR, Buckley H (eds) Pacific Archaeology: Documenting the Past 50,000 Years: Papers from the 2011 Lapita pacific archaeology conference. University of Otago Studies in Archaeology 25. Otago University Press, Otago, pp 1–16

Blench RM (2014) New reconstructions of West African economic plants. In: Adelberger J, Leger R (eds) Language and history in the light of reconstructions. FAB 22 (2010). Rüdiger Köppe, Köln, pp 111–149

Blench RM (2016) Reconstructing African agrarian prehistory by combining different sources of evidence: methodological considerations and examples for west African economic plants. In: Thanheiser U (ed) News from the past: Progress in African archaeobotany. Proceedings of the 7th International Workshop on African Archaeobotany in Vienna, 2–5 July 2012. Barkhuis, Groningen, pp 13–26

Bostoen K (2005) A diachronic onomasiological approach to early Bantu oil palm vocabulary. Studies in African. Linguistics 34(2):143–188

Bostoen K (2014) Wild trees in the subsistence economy of early Bantu speech communities: a historical-linguistic approach. In: Stevens CJ, Nixon S, Murray MA et al (eds) African Flora, Past Cultures and Archaeo-botany. Proceedings of the fifth international workshop for African Archaeobotany, London, 3–5 July, 2006. Left Coast Press, Walnut Creek, pp 129–140

Bostoen K, Grollemund R, Koni Muluwa J (2013) Climate-induced vegetation dynamics and the Bantu expansion: evidence from Bantu names for pioneer trees (*Elaeis guineensis, Canarium schweinfurthii* and *Musanga cecropioides*). C R Geosci 345:336–349

Burkill HM (1997) The useful plants of West Tropical Africa. Families M-R Royal Botanic Gardens, Kew

Busson F (1965) Plantes alimentaires de L'Ouest Africain. Ministère de la Coopération, Paris

Chevalier A (1949) Nouvelles observations sur les arbres à kapok de l'Ouest Africain. Rev Int Bot Appl Agr Trop 321(2):377–385

Clement CR, Denevan WM, Heckenberger MJ et al (2015) The domestication of Amazonia before European conquest. Proc R Soc B 282. https://doi.org/10.1098/rspb.2015.0813

Connell BA (1998) Linguistic evidence for the development of yam and palm culture among the Delta Cross River peoples of Southeastern Nigeria. In: Blench RM, Spriggs M (eds) Archaeology and language II. Routledge, London, pp 324–365

D'Andrea AC, Logan AL, Watson DJ (2006) Oil palm and prehistoric subsistence in tropical West Africa. J Afr Archaeol 4(2):195–222

Dalziel JM (1937) The useful plants of West Tropical Africa. Crown Agents, London

Drake NA, Blench RM (2017) Reconceptualising the palaeozoogeography of the Sahara and the dispersal of modern humans. In: Boivin N, Crassard R, Petraglia, M (eds) Human dispersal and species movement: from prehistory to the present (pp 119–143). Cambridge, Cambridge University Press

Duvall CS (2007) Human settlement and baobab distribution in south-western Mali. J Biogeogr 34(11):1947–1961

Guthrie M (1967–71) Comparative Bantu: an introduction to comparative linguistics and prehistory of the Bantu languages. 4 vols. Gregg International Publishers, Farnborough

Hall JB, Aebischer DP, Tomlinson HF et al (1996) *Vitellaria paradoxa*: a monograph. School of Agricultural and Forest Sciences 8. University of Wales, Bangor

Kahlheber S (1999) Indications for Agroforestry. In: Van Der Veen M (ed) The exploitation of plant resources in ancient Africa. Kluwer Academic, New York, pp 89–100

Kinnaird MF (1992) Competition for a forest palm: use of *Phoenix reclinata* by human and nonhuman primates. Conserv Biol 6(1):101–107

Kondo T, Crisp MD, Linde C et al (2012) Not an ancient relic: the endemic *Livistona* palms of arid central Australia could have been introduced by humans. Proc R Soc B 279(1738):2652–2661

Kyle Latinis D (2000) The development of subsistence models for Island Southeast Asia and Near Oceania: the nature and role of arboriculture and arboreal-based economies. World Archaeol 32(1):41–67

Leakey RRB, Atangana AR, Kengni E et al (2002) Domestication of *Dacryodes edulis* in West and Central Africa: Characterisation of genetic variation. For Trees Livelihoods 12(1–2):57–71

Logan AL, D'Andrea AC (2012) Oil palm, arboriculture, and changing subsistence practices during Kintampo times (3600–3200 BP, Ghana). Quatern Int 249:63–71

Maundu PM, Ngugi GW, Kabuye CH (1999) Traditional food plants of Kenya. Kenya resource centre for indigenous knowledge, National Museums of Kenya, Nairobi

Morin-Rivat J, Fayolle A, Gillet JF et al (2014) New evidence of human activities during the Holocene in the lowland forests of the Northern Congo Basin. Radiocarbon 56(1):209–220

Neumann K, Kalheber S, Uebel D (1998) Remains of woody plants from Saouga, a medieval west African village. Veg Hist Archaeobot 7:57–77

Oas SE, D'Andrea AC, Watson DJ (2015) 10,000 year history of plant use at Bosumpra Cave. Ghana. Veg Hist Archaeobot 24(5):635–653

Oslisly R, Bentaleb I, Favier C et al (2013a) West Central African peoples: survey of radiocarbon dates over the past 5000 years. Radiocarbon 55(2–3):1377–1382

Oslisly R, White L, Bentaleb I et al (2013b) Climatic and cultural changes in the west Congo Basin forests over the past 5000 years. Philos T Roy Soc B 368(1625). https://doi.org/10.1098/rstb.2012.0304

Piperno DR, McMichael C, Bush MB (2015) Amazonia and the Anthropocene: What was the spatial extent and intensity of human landscape modification in the Amazon Basin at the end of prehistory? Holocene 25:1588–1597. https://doi.org/10.1177/0959683615588374

Raponda-Walker A, Sillans R (1961) Les plantes utiles du Gabon. Le Chevalier, Paris

Rodin RJ (1985) The ethnobotany of the Kwanyama Ovambos. Monographs in systematic botany from the Missouri Botanical Garden (USA). Missouri Botanical Garden Press, St. Louis

Ruffo CK, Birnie A, Tengnäs B (2002) Edible wild plants of Tanzania. Regional Land Management Unit/Sida, Nairobi

Vivien J, Faure JJ (2011a) Arbres des forêts denses d'Afrique centrale. CTA, Saint Berthevin

Vivien J, Faure JJ (2011b) Fruitiers sauvages d'Afrique (espèces du Cameroun). CTA, Saint Berthevin

Walter AE, Sam C (1999) Fruits d'Océanie. Paris: IRD Also (2002. Fruits of Oceania. [trans. P. Ferrar] Paris, ACIAR & Canberra, IRD

Wickens GE, Lowe P (2008) Baobabs—Pachycauls of Africa. Madagascar & Australia, Springer, New York

Wild HA, Biegel HM, Mavi S (1972) Rhodesian dictionary of African and English plant names. Rhodesia Government Printer, Salisbury

Williamson K (1993) Linguistic evidence for the use of some tree and tuber food plants in Southern Nigeria. In: Andah B, Okpoko A, Shaw T et al (eds) The Archaeology of Africa. Food, Metals and Towns. Routledge, London, pp 104–116

Middle Holocene Environmental Change at Mtwapa Creek, Kenya: Distinguishing Human Activity from Regional Ecological Processes

Ryan M. Szymanski(✉)

Department of Anthropology, Washington State University, Pullman, WA, USA
ryan.szymanski@wsu.edu

Abstract. Very little is known about human occupation of the East African Coast prior to the Swahili period. Only a handful of pre-Swahili archaeological sites from this region have been investigated in any detail. The Holocene archaeology of the coastal region requires research attention if the role of indigenous African populations in later ethnic mosaics is to be understood. Pollen, fungal spores, and charcoal remains from a sediment core derived from Mtwapa Creek are presented with the goal of better understanding mid-to-late Holocene (c. 7500–4000 cal yr BP) occupations of the coast. Evidence for increased local burning, herbivore activity, and overall landscape disturbance possibly derived from cultivation activities characterizes the period between c. 6200–4000 cal yr BP. Intensive exploitation of riverine habitats was practiced in the neighboring hinterland during this period, and it is proposed that similar adaptations may have been at play in coastal drainages at Mtwapa Creek and elsewhere contemporaneously. A variety of investigative methods are proposed that may yield additional information on human activities during this time period.

Keywords: Anthropogenic impact · Coastal ecosystems · Fungi
Intertidal core · Holocene · Paleoecology · Palynology · Swahili

Introduction

In contrast to the well known Swahili cultural phenomenon of the past two millennia, the archaeology of the pre-Swahili East African coast has been largely ignored. Indeed, the scope of human activities in this region prior to the Late Holocene is almost completely unstudied. Exceptions to this trend are few and far between (Omi 1982, 1984, 1986, 1991; Helm 2000; Helm et al. 2012; Shipton et al. 2018), yet this evidence suggests that the East African Coast was rife with human activity through, and perhaps previous to, the Holocene. As little archaeological detail dating to the Early and Middle Holocene is yet available for this region, a landscape-scale approach is employed here to render any significant changes to ancient coastal ecologies due to human activities visible. In an effort to clarify the nature of pre-Swahili human environmental impacts, I offer a paleoecological reconstruction of Mtwapa Creek Estuary and the surrounding environs during the Middle Holocene (c. 7500–4000 cal yr BP), using pollen, microscopic

A. M. Mercuri et al. (Eds.): *Plants and People in the African Past*,
https://doi.org/10.1007/978-3-319-89839-1_11

charcoal, and fungal spore data from a sediment core spanning this period. This data is then integrated with known archaeological knowledge of Holocene human occupation of the East African coast. First, a brief overview of integrated palynological and fungal research in East Africa is presented, followed by a review of pertinent archaeological data from the East African coast. Next, nested microbotanical data from a 5 m sediment core collected at Mtwapa Creek are discussed and interpreted. I use these interpretations to hypothesize the nature of human occupation of this area during the Middle Holocene, and advocate for the usefulness of interpreting multi-proxy environmental records in light of local archaeological evidence as a potential means of dealing with equifinal data sets. Finally, suggestions for additional topical research are put forth.

Multi-proxy paleoecological data sets have aided in defining environmental and cultural chronologies in numerous regions in East Africa (e.g. Cohen et al. 2005; Gelorini et al. 2011, 2012a, b; Lejju et al. 2005; Palacios-Fest et al. 2005a, b; Rucina et al. 2009; van Geel et al. 2011; Wooller et al. 2003; and many others). Lejju et al. (2005) have reconciled several lines of paleoecological evidence, including pollen, phytoliths, and fungal spores, with archaeological and historical evidence at Munsa, Uganda to better understand site use and occupation. Further South, past ecological change at Lake Tanganyika has been documented in a set of very detailed environmental records derived from numerous proxies by Cohen et al. (2005) in a special issue of the *Journal of Paleolimnology* (Palacios-Fest et al. 2005b). Data from fungal spores is emphasized more specifically by Gelorini et al. (2011, 2012a, b) and van Geel et al. (2011). Van Geel et al. (2011) analyzed an assemblage of fungal "non-pollen palynomorphs" (NPP's) from Lake Challa in southeastern Kenya, and found that more than two thirds of typed NPP specimens could be identified, or "linked," to known taxa of fungi. As study of microscopic African fungal bodies is still in its relative infancy, many specimens which could not be reliably assigned to taxonomic groups were "typed" according to shared morphological affinities. Nevertheless, the authors found that diagnostic spores could often be assigned to known genera or species, yielding (in these cases) a level of taxonomic resolution similar to that possible with pollen grains. Of these diagnostic types, a variety are substrate-or-environment-specific indicators. *Sporormiella,* for example, has been documented as an obligate coprophile, growing exclusively on dung substrates (Davis1987; Davis and Shafer 2006; Raper and Bush 2009; Parker and Williams 2011; Baker et al. 2013).

Local Environment and Sedimentary Context

The surrounding landscape of Mtwapa Creek today is agrarian to semi-urban in nature. Agricultural activities in the vicinity are largely coconut, mango, and maize farming. Small-scale husbandry of cattle, goats, and sheep is also practiced. The coring location is situated approximately 4 km West of the town center of Mtwapa on a tidal mudflat off the northern shore of the Mtwapa Creek estuary. A band of small agricultural homesteads shields the coring site from the small commercial area surrounding the B8 highway, the major (and only) North-South highway serving the coastal region. Moisture availability in the near-coastal plain is greater than in the more arid hinterland. Rainfall in the region varies between 800–1400 mm per annum (Oosterum 1988).

Recent coastal sediment deposits in this region, dating to the post-Pleistocene, have been described by Thompson (1956), Williams (1962), and later by Oosterum (1988), who synthesized descriptions from these and other earlier studies, and whose designations I employ here. The intertidal mudflats from which core material was extracted belong to the aptly named *recent tidal-flat deposits* identified by these authors, being derived primarily from fluvial sedimentation processes. Sedimentary material comprising these deposits originates primarily from surrounding land terraces eroded via rainfall and mechanical disturbance, as well as suspended particulate matter of small size transported into mudflats at high tide. These mudflat deposits occur frequently in mangrove environments throughout the East African littoral, and have been characterized as Thionic Fluvisols (Michieka et al. 1978; Boxem et al. 1987). These sediments are comprised of dark grey-to-olive green clays and clays with varying proportions of sand in their matrix. These deposits have been identified as Holocene and Late Pleistocene in age, with only very gentle gradation in sediment characteristics being detectable over this timespan.

Mangrove communities are by far the predominant vegetation along the estuary margins and undeveloped parts of the coast. Eight of the nine recognized species of East African mangroves are native to Kenya, including *Rhizophora mucronata* (Lam.), *Ceriops tagal* (C. B. Rob.), *Bruguiera gymnorrhiza* (L.), *Avicennia marina* (Forsk.), *Sonneratia alba* (Sm.), *Heritiera littoralis* (Aiton.), *Xylocarpus granatum* (Koenig.), and *Lumnitzera racemosa* (Willd.). The ninth species, *Avicennia officinalis*, is today found only in coastal Madagascar. Local coastal communities frequently rely on mangrove forests for building timber, craftworking (e.g. construction of fish traps), and as a fuel source (Taylor et al. 2003). According to Taylor et al. (2003) mangroves are a lynchpin of estuarine ecologies, with leaf litter from mangrove forests responsible for a large portion of organic inputs into these environments. Mangrove pollen is well represented throughout the sediments examined here.

Pre-Swahili Archaeological Research

Available data on Holocene human habitation of the East African coast is very scanty in comparison with what is known of the later Swahili occupations of the region. Linguistic and ecological evidence has been cited in argument that southwest Asian colonizers definitely encountered significant populations of indigenous African peoples occupying the coast (Schoenbrun 1993; Ehret 1998). However, only a few studies containing primary data relating to ‘stone age’ coastal occupants have been published, spanning from present day Mozambique (Morais 1988; Adamowicz 1990) north through Kenya and Tanzania (Soper 1966; Chami and Kessy 1995; Kessy 1997; Chami 1998; Omi 1982, 1984, 1986, 1991; Shipton et al. 2018), and into Somalia (Coltorti and Mussi 1984). A succinct overview of these studies is provided by Helm (2000), but the findings of Omi (1982, 1984, 1986, 1988, 1991) will be the primary sources discussed further here, as lithics pre-dating the Early Iron Age are detailed there. The very recent findings of Shipton and colleagues (2018) are not discussed here.

Omi (1982, 1984, 1986, 1991), and contributors to these reports, describe multiple coastal archaeological sites in varying degrees of detail (See Fig. 1), focusing primarily on the geomorphic and stone tool content of each. The main site of focus in these studies however, is the large, multi-component hillside site from where these volumes get their name, Mtongwe. This site, southwest of Mombasa City, across the large creek channel, is dominated by three separate hills: Sassoon Hill, Omar Hill, and the more southerly Ali Abubakar Hill. Excavations were performed at each of these hill site areas of Mtongwe. The authors provide detailed maps of the sedimentary strata encountered in various excavations at Mtongwe, but the only artifactual evidence presented is recovered lithic material.

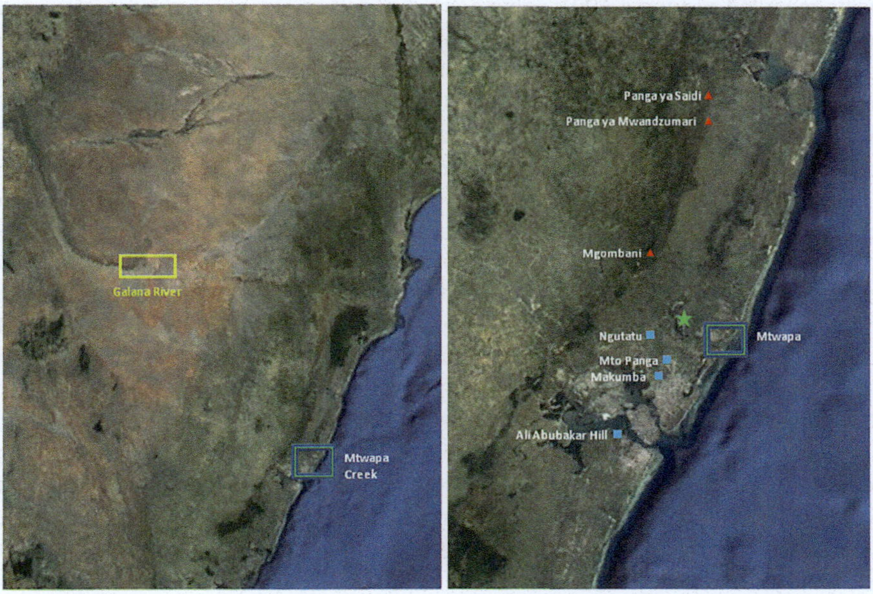

Fig. 1. Location map of the study area.

Considerable variation exists between the lithic assemblages recovered at each hill excavation. Following this, Omi (1984) argues that these assemblages reflect human activity spanning the "... *Earlier Stone Age to the Later Stone Age* (Omi 1984, p. 22; authors emphasis)." Unfortunately, Omi and colleagues do not employ lithic typologies specific to East African stone tool industries, assigning given finds to "Acheulean," "Mousterian," or "Fauresmith" industries. Nevertheless, detailed drawings and measurements, and even photographic plates in later volumes of this series, are provided of recovered lithic implements, allowing independent analysis of these assemblages. The exact identities of lithic assemblages analyzed and recovered during these excavations are beyond the scope of this chapter. It suffices to say that variation in lithic assemblages recovered from Mtongwe and other coastal sites such as Panga ya Saidi (Shipton et al. 2018) suggests that human occupation of the Kenyan coast prior to the advent of the Swahili period persisted for some time.

The few published radiocarbon dates from this project (Omi 1991), derived from materials excavated at Ali Abubakar Hill, support the notion of human coastal occupation from at least the Late Pleistocene into, and likely through, the Holocene and proto-Swahili period, slightly predating the second millennium BP. Four radiocarbon assays were made on charcoal recovered from a bed of reddish orange sand with calcareous particles in the matrix. The resultant dates varied in age considerably, ranging from 17,680 ± 270 BP [21,325 ± 367 cal BP] to 1,090 ± 90 BP [953 ± 186 cal BP]. Calibrated dates reported here were obtained via Calib 7.1, using the SHcal13 calibration curve (Stuiver et al. 2017). Alternate calibrations using the mixed Marine & NH Atmosphere curve in this program (Stuiver et al. 2017), with hypothetical marine carbon percentages of 1–5% resulted in only small departures from calibration with the SHcal13 curve. Furthermore, the youngest dates were derived from materials in deeper sedimentary levels than older dates. Excavation depth measurements recorded in Omi's 1991 volume are reported in *meters above sea level*, rather than the more traditional *centimeters below surface/datum*. Thus, higher measurement values indicate shallower sedimentary contexts from the excavation reported therein. Sediment admixture and/or the nature of recovery techniques may be responsible for this juxtaposition. A large discrete charcoal block with provenience at 41.70 masl (meters above sea level; original measurement value) yielded a date of 9,050 ± 150 BP [10103 ± 214 cal BP] (again, see Omi 1991, p. 33 for clarification of depth measurements). The authors note that the remaining three dating assays were performed on collections of small charcoal fragments collected from throughout each additional sedimentary level. Contamination by modern plant roots and insect bioturbation, they argue, resulted in influx of young carbon into lower stratigraphic levels. Therefore, the date derived from 41.70 masl likely represents the most secure date from this site. Regardless of overall stratigraphic integrity at this site, the dates indicate a long history of occupation, well preceding the last several thousand years. While Omi and colleagues posit an age for the lithic assemblages at Ali Abubakar Hill much older than these radiocarbon dates suggest, this conclusion is based upon now outdated lithic temporal associations. In fact, the lithic assemblages from numerous coastal sites documented by Omi and co-workers closely resemble Eburran lithics from the Central Rift Valley, detailed by Ambrose (1984, 1998) and others (Wilshaw 2012). Hypothesizing a cultural or evolutionary connection between coastal and Rift Valley lithics would obviously be impossible at this point, however, as lithic technologies are known to display high inter-site and inter-regional variability. Despite this, general trends in dimensional reduction over the last 10,000–15,000 years, documented by Ambrose (1984), suggest that coastal lithics from sites such as Ngutatu and Mto Panga may potentially be of Late Pleistocene or Holocene origin. Thus, coastal occupations and activity sites detailed in the works of Omi (1982, 1982, 1984, 1986, 1991) may actually be considerably younger than argued to be in these volumes. Comparison of these finds with lithic data recently available from the site of Panga ya Saidi (Shipton et al. 2018) will undoubtedly aid in contextualizing these materials.

Methods

A five meter sediment core was extracted from the intertidal zone of Mtwapa Creek using a modified piston corer. Sediments were extruded and sampled every 5 cm along the entire length of the core. Approximately 15–20 ml of sediment were excised for each sample. Sampled sediments were sealed in whirl-paks© and returned to the Department of Anthropology at Washington State University for processing and analysis in the Palynology and Paleoenvironments Laboratory (now the Archaeobotany Laboratory) housed there. Nineteen sediment samples were selected for analysis. Sediment processing was performed using the following procedure: quantification of 2 ml of sediment for each sample, addition of two commercially produced *Lycopodium* tablets to allow calculation of pollen and spore concentrations, carbonate removal with 10% Hydrochloric Acid, screening through 150 μm wire mesh, overnight immersion in 65% Hydrofluoric Acid to remove silicates, 20" treatment with 1% Potassium Hydroxide to dissolve humates, dehydration in 99% Glacial Acetic Acid, acetolysis treatment as outlined by Erdtman (1960), heavy density separation with Sodium Polytungstate at 2.34 specific gravity, curation with glycerin in glass vials. Each chemical treatment step was followed by repeated rinsing in deionized water and centrifugation at >2,000 rpm.

Samples material was mounted on glass microscope slides to undergo microscopy. Counting of pollen, spores, and charcoal was performed under a light microscope at 100–400x magnification, with higher magnifications of up to 1000x being used when necessary. Vertical traverses of each slide were made until at least 400 pollen grains were counted for each sample. This procedure was repeated for counting of fungal spores and of charcoal fragments. Identifications were made using published micrographs and available reginal pollen (Caratini and Guinet 1974; Bonnefille and Riollet 1980) and spore (Jarzen and Elsik 1986; Gelorini et al. 2011; van Geel et al. 2003, 2011) keys, as well as online resources such as the African Pollen Database. Charcoal was categorized into size ranges [small-sized 3–25 μm, medium-sized 26–75 μm, large-sized 76–150 μm], and counted following the procedure outlined by Rucina et al. (2009), with identifications proceeding according to the recommendations of Waddington (1969) and other authors (Patterson et al. 1987; Pitkanen and Huttunen 1999; Clark 1988). Graphical rendering was performed on Tilia 2.0 (Grimm 1991). Pollen, spore types, and charcoal size classes are rendered as percentages of the overall assemblage in each sample.

Results

Radiocarbon Dates

Six bulk carbon dates were obtained via Direct AMS, which indicate that this sediment core captures the majority of the Middle Holocene (See Table 1). All dates were calibrated using CALIB 7.1 (Stuiver et al. 2017) under the MARINE13 calibration curve to account for marine reservoir effects. Results confirmed that the period between c. 7500–4000 cal yr BP is reflected in these sediments, comprising the majority of the Middle Holocene (8200–4200 cal yr BP), and likely verging into the earliest part of the Late Holocene (4200 cal yr BP-forward) as defined by Walker et al. (2012).

Table 1. Radiocarbon dates and associated depths from Mtwapa Creek.

Depth (cm)	uncal yr BP (BP)	cal yr BP (BP)	Age range (1 sigma) (BP)	Laboratory code
20	4247 ± 30	4355	4307–4406	D-AMS 010033
60	5191 ± 29	5555	5516–5590	D-AMS 007685
155	5785 ± 31	6215	6183–6253	D-AMS 007686
265	5901 ± 50	6319	6271–6372	D-AMS 017968
380	6875 ± 34	7391	7351–7427	D-AMS 017969
492	6940 ± 34	7448	7414–7477	D-AMS 005585

After graphical rendering of results, sum of squares analysis suggested the existence of two major analytical zones: a younger zone defined by sediments above 175 cm below ground level, and an older zone comprised of all deeper sediments. The older zone is from here on called Zone 1, and the younger zone, Zone 2. This phenomenon was captured independently in pollen, fungi, and charcoal datasets. Association between dates suggests rapid sediment accumulation during the mid-8th millennium BP and during the mid-to-late 7th millennium BP. During each of these periods, over 100 cm of sediment accumulated over very short timeframes. During the earlier of these two periods, diversity of pollen and fungal spores is relatively high (See Figs. 2 and 3). High rates of sediment accumulation such as witnessed at these times are likely the result of precipitation and/or erosion rates. As will be discussed later, climate models indicate that wetter conditions than present predominated during and through these timeframes. Especially wet periods during this extended humid phase may be responsible for these brief periods of extreme sediment deposition. Erosion due to landscape disturbance is another possibility which may account for high sedimentation, but charcoal frequencies and other palynological indicators of disturbance present conflicting signals with regard to forces affecting landscape alteration, leaving this interpretation dubious, but not able to be ruled out.

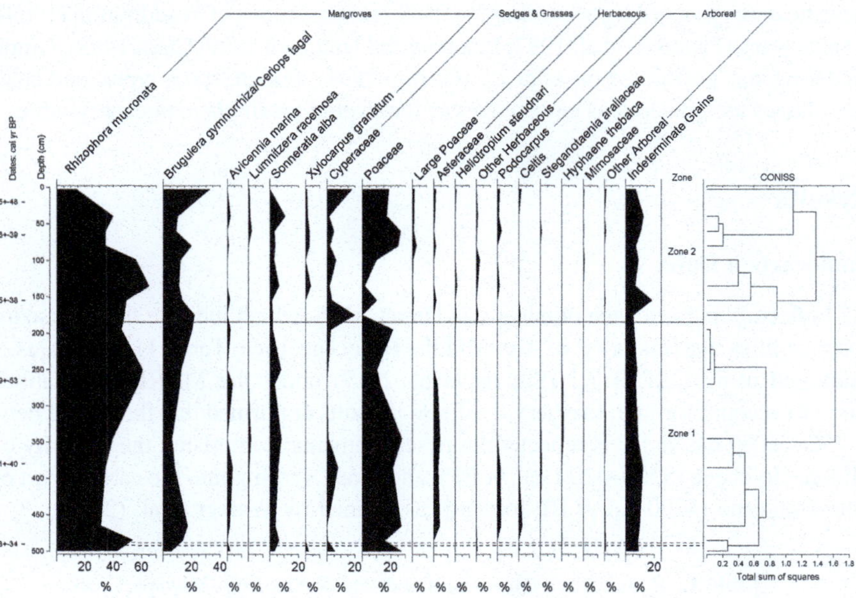

Fig. 2. Pollen representation at Mtwapa Creek.

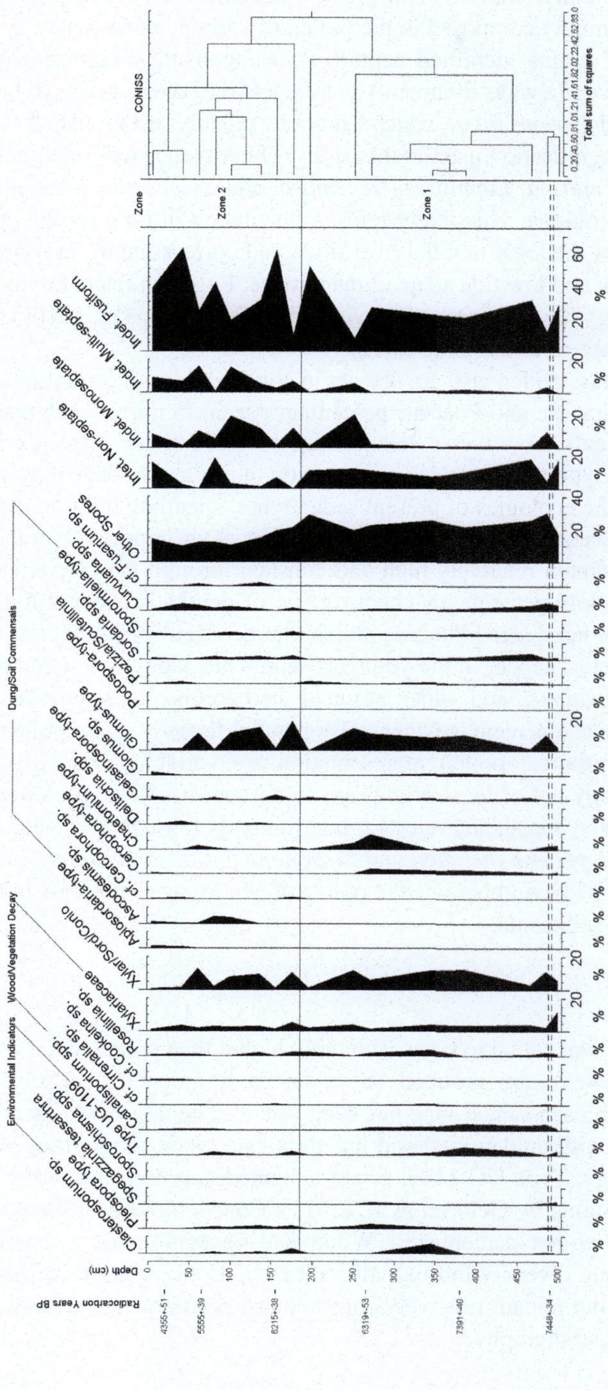

Fig. 3. Fungal representation at Mtwapa Creek.

Pollen

The pollen assemblage from Mtwapa Creek was found to be of relatively low diversity, being overwhelmingly composed of the pollen of various mangrove types (See Fig. 2). The majority of grains identified capture an antagonistic relationship between *Rhizophora mucronata* (always dominant) and the less diagnostic grains of both *Bruguiera gymnorrhiza* and *Ceriops tagal*, which cannot be reliably distinguished at the relatively low levels of magnification used in this study. Other mangroves, including *Sonneratia alba*, *Avicennia marina*, *Lumnitzera racemosa*, and *Xylocarpus granatum* are present in lower concentrations. This likely reflects fluctuation in tidal heights and/or salinity conditions within the local mudflat over time, and corresponding movement of groves of high, medium, and low tide adapted mangroves. Research related to the dynamics of mangrove ecology has been recently reviewed by Punwong et al. (2013a, b), and is not rehashed here (see also, Gastaldo 2012).

Sedge and grass pollen also are present in considerable concentrations. In the oldest sediments, Cyperaceae and Poaceae percentages seem to mirror each other. Yet in the youngest sediments, an inverse relationship between these types develops. Closer examination of Cyperaceae species composition in this assemblage may yield pertinent information on the ecologies of present sedge types, shedding light on the nature of the relationship between grasses and sedges in this environment. Overall grass pollen concentrations remain relatively high and constant through the core stratigraphy, with fluctuations and low percentages characteristic of depths between 100–200 cm. Low concentrations of large-sized Poaceae pollen ("large-sized" being defined as >37 μm in maximum diameter) appear in the youngest sediments, along with trace percentages of *Heliotropium steudneri* and other assorted herbaceous taxa, present in very low amounts. Asteraceae percentages generally parallel those of grass pollen.

Patterns in arboreal pollen representation are variable. *Podocarpus* and *Celtis* pollen (presumably not of local origin) are a constant, low-level presence in the oldest sediments, but then vanish and reappear intermittently through the youngest sediments. Low levels of *Hyphaene thebaica* and *Steganotaenia arialiaceae* are present in older sediments in very low numbers, while pollen of Mimosaceae presents in trace amounts in the youngest sediments.

Fungi

Fungal diversity was found to be considerably higher than pollen diversity (See Fig. 3). Encountered spores were grouped based on environmental and substrate-mediated conditions which each fungal type has been found to be associated (see Fig. 3; fungal representation), with unidentified and indeterminate types being grouped according to spore morphology. Type UG 1109, as-yet unnamed, has been posited to be associated with xeric conditions by Gelorini et al. (2011). Concentrations of these spores closely follow Poaceae pollen percentages. Wood and vegetation decay commensals were found in moderate diversity through all sediments. Concentrations of individual types vary over time, but remain relatively constant as a percentage of overall spore content throughout core stratigraphy.

Spores commensal with dung and soil isolates were very diverse in all levels. The strongest presence is that of *Glomus*-type spores, a soil erosion indicator with concentrations which vary through time (especially in the youngest sediments), yet remain a relative constant. Many saphrotrophic and coprophilous fungal types present in low concentrations throughout the core, while others, including *Apiosordaria*, *Ascodesmis*, and *Podospora* appear only in the top 1–2 m of sediment.

Spores which are often found in association with particular plants or types of vegetation were also recovered. *Curvularia* spp. have been found to grow on grasses, maize, wheats, and rice (Ellis 1971). Various species of *Fusarium* have also been reported on grasses, but more frequently on stored sorghum and millet (Onyike and Nelson 1992; Penugonda et al. 2011). Overall spore diversity, to include unidentified and non-diagnostic spores, increased in sediments shallower than 200 cm below ground level.

Discussion

The often equifinal nature of anthropogenic and climatic environmental signatures poses obvious problems in the investigation of ancient ecologies. This is particularly true for archaeologists interested in human landscape impacts. As discussed earlier, multiple lines of proxy evidence have been used to approach this issue with great affect in East Africa and elsewhere. Paleoenvironmental records from East Africa suggest that persistent humid conditions progressively gave way to a more xeric climate between the Early Holocene and c. 5700 cal yr BP, after which dry conditions seem to have persisted until well after 4000 cal yr BP. A warming trend seems to have ensued after c. 4500 cal yr BP, although conditions remained dry (Richardson 1972; Nicholson 1996; Bergner et al. 2003).

Zone 1: c. 7500–6300 cal yr BP

If human impacts indeed shaped the Middle Holocene ecology of the Mtwapa Creek catchment, data patterns should differ from what would be predicted by climate models for the region. Of particular importance is the period after c. 6200 cal yr BP, which seems to capture much environmental fluctuation. If ecological changes captured in this core are solely climatologically mediated, proxy data should point toward a relatively humid, yet drying, period between c. 7000–5700 BP, and a dry period between c. 5700–4000 BP, when core chronology terminates.

Climate models predict a relatively humid phase prior to 5700 cal yr BP in this region (van Zinderen Bakker 1966; Thompson et al. 2002; Kiage and Liu 2006). Mangrove pollen ratios remain relatively stable during this timeframe, while grass pollen percentages are relatively high, as are Asteraceae and *Podocarpus* percentages. While these patterns would not necessarily be expected under conditions of relative humidity, moisture availability may not have been great enough to favor the development of forest cover that could outcompete grassland environments. Based on floristic analysis of numerous grid squares surrounding a mangrove coring site, Punwong et al. (2013a) contend that Poaceae representation in sediment cores from mangrove environments may not accurately reflect the actual presence of these environs in the immediate landscape. However, based on the presence of considerable amounts of grass-dominated landscape near the Mtwapa Creek coring site, there does not seem to be sufficient reason to discount the importance of grasses in this analysis. *Podocarpus* presence is likely a relic of regional pollen rain, which may have been preferentially introduced during precipitation events. Sedges, many of which favor less saline environments, also experience an extended period of stasis during this early period. Based on these pollen data, there does not seem to be much justification for discounting climatic patterns as a primary mediator of environmental conditions during the first approximately 1000 years captured in sediments from Mtwapa Creek.

Fungal data from the oldest sediments unfortunately do not much aid in further clarifying the source(s) of ecological mediation during this period. Spores of both *Glomus* sp. and *Glomus*-type, which are taken as indicators of soil erosion, display elevated concentrations in sediments dating between c. 7500–6200 cal yr BP. The means by which erosion may have been affected, however, are still unclear. Precipitation and/or disturbance potentially related to dessication, mechanical disturbance by human activity or animal traffic, or grazing by herbivores may all result in similar erosional signals. *Cercophora* sp. and *Cercophora*-type spores also show slightly elevated concentrations in these sediments. While often found on dung, *Cercophora* also occurs on decayed wood and/or plant material (Bell 1983). *Canalisporium* spp. is commensal with decaying and inundated wood, and is found in elevated percentages in the oldest sediments. Wood decay likely would be favored by moist conditions, as predicted by climate data. Gelorini et al. (2011) identified Type UG-1109 as a possible indicator of locally dry conditions based on the similarity of this type to a European spore type which indicated such conditions. Elevated percentages of Type UG-1109 during the likely relatively wet period between 7500–6000 BP cast this relationship into doubt. The existence of human-created dry microhabitats in this vicinity which might have played host to xeric-commensal fungi should not be ruled out, however. Thusfar though, the most likely scenario favors the prevailing climatic regime as the primary shaper of the Mtwapa Creek ecology prior to c. 6200 BP.

Zone 2: c. 6200–4000 BP

In sediments shallower than approximately 175 cm below ground level, relatively rapid fluctuations in pollen and fungal representation occur. Climate models predict the existence of very dry conditions during this period, with a notable warming trend beginning slightly prior to 4500 BP. Punwong et al. (2013a) appear to capture the predicted cool and dry period in the Rufiji Delta of Tanzania between c. 5600 and 4600 cal yr BP as a seaward movement of mangroves during this period, indicating a sea level regression. After this time, sea levels appear to rise again, enabling a coincident mangrove resurgence. If this phenomenon occurred at Mtwapa Creek, it does not appear readily apparent in the pollen data. Indeed, both *Rhizophora mucronata* and *Sonneratia alba* present with relatively high values during this period. This would not be expected if sea levels were subsiding. It may be that sea levels dropped in this vicinity, yet that tidal levels did not drop sufficiently to render the immediate mudflat environment unsuitable for *Rhizophora mucronata*. Tidal levels may have dropped

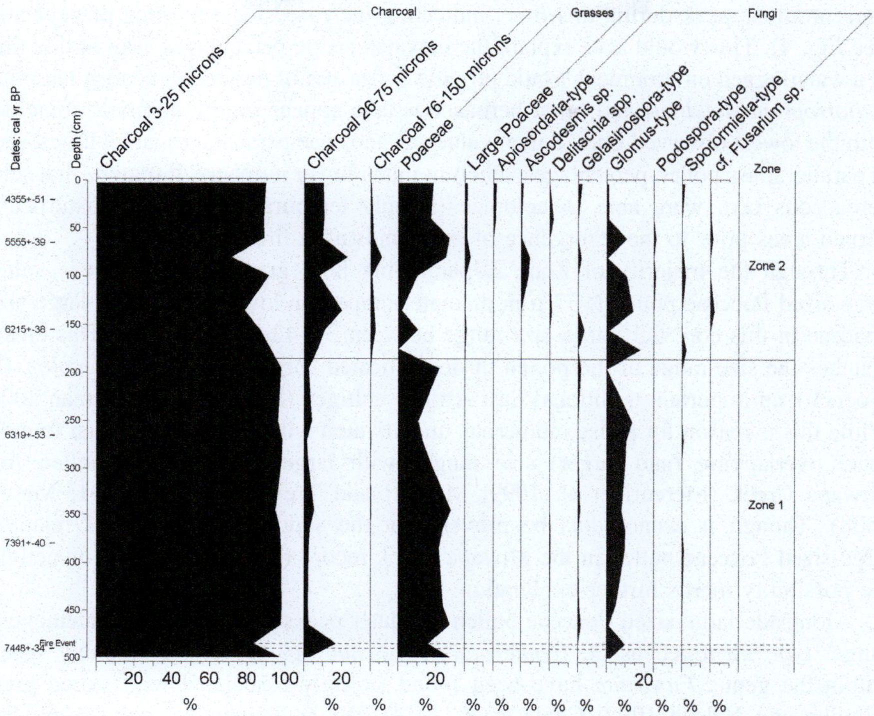

Fig. 4. Integrated charcoal and palynomorph indicator representation at Mtwapa Creek.

enough, however, to create favorable nearby conditions for *Xylocarpus granatum*, which prefers much less frequent inundation.

In the terminal sediments of Zone 2, *Rhizophora* percentages drop, while percentages of *Ceriops tagal* and *Bruguiera gymnorrhiza* increase. This may reflect selective mangrove harvesting for building materials and other uses, as has been documented in later Swahili communities (Kusimba 1999). This could indicate sea level subsidence during the period shortly preceding 4000 cal yr BP. Sedge percentages also increase during this latest period. These patterns, as well as the fact that no further significant sediment accumulation apparently took place, may be due to the absence of sufficient suspended sediment in tidal events to affect a depositional environment. Fluvial transport inputs from landward freshwater runoff likely were not powerful enough to reach mid-mudflat contexts, leading to the effective termination of sedimentation at the coring location.

The near disappearance (and subsequent reappearance) of grass pollen during the earliest part of Zone 2 is anomalous. *Rhizophora* pollen may have simply been too numerous for many grass pollen grains to be captured during microscopy. Cool and dry conditions are not a likely culprit responsible for very low grass pollen values, as sedge percentages are elevated at this point in the Mtwapa Creek chronology. A potential scenario which could account for both low Poaceae and high mangrove percentages would be burning of grasslands. Large-sized (>75μm) charcoal fragment concentrations peak during this time, indicating increased local burning of vegetation (see Fig. 4). This would also explain the elevated sedge percentages seen at this time, as a waterlogged environment would provide a measure of protection against land fires. *Heliotropium steudneri* and other herbaceous taxa appear in low numbers coincident with the lowest observed grass pollen values. Asteraceae percentages, at all times, seem to parallel grass pollen percentages, albeit in much lower numbers. It may be that select herbaceous taxa were able to colonize (if only temporarily) recently disturbed or burned areas prior to the resurgence of grasslands after fire events.

Through the majority of Zone 2, paralleling both grass and Asteraceae values, large-sized Poaceae pollen (>37 μ in diameter) appear in low numbers. All large-sized Poaceae in this core fall into a size range between 37–42 μ in diameter. This closely matches the size range of the pollen of domesticated sorghum (*Sorghum bicolor,* (L.) Moench), an important traditional East African cultigen (Schmidt and Bothman 2005). While this is reason for pause, numerous uncultivated wild grasses have been found to share overlapping (and larger) size ranges with large-sized Poaceae pollen from Mtwapa Creek (Mercuri et al. 1999; Mercuri and Trevisan Grandi 2001; Mercuri 2008). Though it cannot thus be proven that the sudden appearance of relatively large-sized Poaceae pollen in the Mtwapa Creek record represents cultivation activity, the possibility merits further exploration.

Alongside large-sized Poaceae pollen and later peaks in Asteraceae percentages, a fungal type identified as cf. (*conferatur*) *Fusarium* sp. appears. Numerous species within the genus *Fusarium* have been found strongly associated with stored grains (Onyike and Nelson 1992; Chala et al. 2014). While filamentous grass-commensal fungi are often found in association with grassland environments, *Fusarium* has been found to predominate in stored-grain contexts. While, again, this does not represent an evidentiary "smoking gun" for grain storage or domesticated grains, *cultivation* and *collection* of wild grains by foraging populations may be responsible for this particular fungal signature. Also a possibility is that a particular type(s) of wild grass (as yet unidentified specifically) producing relatively large pollen grains found an amenable habitat niche during the increasingly arid phase captured in Zone 2. Concurrently, fungal types including species of *Fusarium* may have found the same.

Also of particular note, unique increases in the representation of coprophilous fungi occur within Zone 2. *Ascodesmis* sp. and *Delitschia* spp. fungal spores appear only during Zone 2, concurrent with, or slightly later than evidence for large-sized Poaceae, *Fusarium,* and increased percentages of large-sized charcoal, indicating more localized burning. *Ascodesmis* spores have been reported on various types of dung, while *Delitschia* spp. have been identified as distinct indicators of large herbivore activity by Gelorini et al. (2012b). *Sporormiella*-type spores, while present in very low concentrations in Zone 1, experience peak percentages in Zone 2. Non-obligate coprophiles including *Apiosordaria*-type and *Podospora*-type also only appear in Zone 2.

The representation of *Gelasinospora*-type spores also deserves attention. The presence of *Gelasinospora* in sediments has been shown to be strongly associated with burning. Indeed, the initial appearance of *Gelasinospora*-type in Mtwapa Creek sediments occurs immediately after a posited fire event captured in the oldest sediments of Zone 1, evidenced by increased occurrence of charcoal in the size range of 26–75 μ. While a long period of relative stasis characterizes both larger sized charcoal and

Gelasinospora-type concentrations, in Zone 2 sediments percentages of both are elevated. Furthermore, concentration increases in both of these (and other) proxies are episodic and irregular, rather than progressively increasing. An already dry environment experiencing progressive warming after c. 4500 cal yr BP, as predicted by climate models (Karlen et al. 1999; Kiage and Liu 2006), might be expected to likewise undergo progressively more (and more frequent) burning events. This is apparently not the case, as large-sized charcoal concentrations level off during the period of the expected heatwave. Overall charcoal data patterns support the idea that episodic burning was characteristic of the period between c. 6200–4000 BP. With the exception of the previously mentioned small spike in larger-sized charcoal fragment concentrations after a putative burning event early in Zone 1 sediments, charcoal in the largest two size classes only appears in any number during the period represented by Zone 2.

Middle Holocene Landscape Use in the Coastal Hinter

Multi-proxy research is an important analytical strategy for investigating Middle Holocene coastal occupations in East Africa, as archaeological information dating to this period is chronically underrepresented, not only in the coastal region, but in hinterland regions as well (Kusimba et al. 2005; Wright 2007; Wright et al. 2007). Kusimba et al. (2005) surveyed and excavated numerous sites from the Tsavo region, with a minority of sites from the Bungule complex containing occupations dating to the mid-and-earliest Holocene. Wright (2005), and later, Wright et al. (2007) provided optically stimulated luminescence (OSL) dates from several sites along the Galana River in Tsavo which placed early floodplain occupations in the region at c. 6000 BP. Wright (2007) and Wright et al. (2007) argue that the Galana River floodplain provided an attractive living environment to hunter-gatherers exploiting mostly local resources for over 5000 years, from roughly 6000–900 BP. This more intensive exploitation of riverine environments would have been a potential means of dealing with increasingly xeric conditions. Wright et al. (2007) also contend that hunter-gatherers living along the Galana River adopted intensive local resource exploitation strategies after c. 6000 BP, relying on locally available game and lithic materials, mainly quartz. A similar lithic material procurement strategy seems to be characteristic of coastal occupations described by Omi. Local slate was utilized primarily, with quartzite being of secondary importance.

Evidence for human activity on the East African Coast during this period also appears at inland sites in the form of marine shell (Leakey 1966; Nelson 1993; Wright 2005), indicating the existence of geographically extensive resource exchange networks better evidenced by lithic source materials (Kusimba 2003). Unfortunately, these occupations are located hundreds of kilometers away from Mtwapa Creek, and thus cannot be considered part of the “local” coastal archaeological record. Lithic evidence described by Omi (1982, 1984, 1986, 1988, 1991), attributed to Early-to-Late Stone Age peoples, provides the only described record of potential Middle Holocene human occupation of actual coastal environments in present-day Kenya.

Of the sites and materials described by Omi, the most proximal to the coring site is the site if Ngutatu, roughly seven kilometers southwest of the mudflat coring location at Mtwapa Creek. According to Omi (1982):

> At Ngutatu A, seven crescent-type stone tools, four micro-blades, seven cores, fifty-three flakes and one small chopper were collected. All implements are made of shale, except for one chopper and one flake made of quartzite. At Ngutatu B, four side-scrapers, one end-scraper, one micro-blade, thirty-one cores and seventy-one flakes and chips were collected, all implements being made of shale. (Omi 1982, p. 145).

Ambrose (1984) identifies progressive reduction in microlith sizes as a useful means of inferring the age of lithic assemblages from the Central Rift Valley. While exact numerical metrics for lithics from Ngutatu are not given, microlith dimensions reconstructed from scaled illustrations of Ngutatu microliths indicate mean length and width measurements that fall within the range of Ambrose's Eburran 4 and 5, corresponding to a time range between c. 8500 BP and c. 2000 BP. A thorough overview of Eburran chronology is provided by Wilshaw (2012). As this estimate uses comparison with a non-local lithic typology, this is obviously a very unsatisfactory means of inferring the age of the Ngutatu occupation. However, considering metric and technological similarities between the Ngutatu lithic assemblage and lithic types from the Holocene of central Kenya (including Eburran assemblages), as well as marine shell remains found in these contexts, it is not outside the realm of possibility that microlith size reduction trends identified by Ambrose (1984) might be in evidence during the Middle Holocene in the coastal region of Kenya as well. Although the existence of a relationship between coastal and Rift Valley lithic technologies remains pure speculation at this point, I argue that it is worthwhile to consider the possibility that similar lithic technologies and manufacturing trends were practiced during the Holocene in these two very different regions.

Conclusion

Even taken together, microbotanical, settlement, and lithic evidence presented here in support of human modification of East African coastal landscapes during the Middle Holocene is hardly irrefutable, and based at times on little more than speculation. Having said this, the list of coincident data patterns with uncertain origins presented here should not be ignored. Fungal and palynological data suggesting that large herbivore activity may have increased between c. 6200–4000 BP co-occur with increased evidence for intermittent localized vegetation burning at Mtwapa Creek. Novel pollen types including large-sized Poaceae, and fungal types found associated with grain storage (e.g. cf. *Fusarium* sp.) also appear and persist during this period, suggesting that cultivation of plant foods may have occurred at an early date. Macro-botanical evidence will obviously be needed to confirm this. Even if these patterns can potentially be interpreted as merely the result of fluctuations in the local climatic and/or precipitation regimes and floral and faunal adaptations to them, human adaptations in the bordering Tsavo region after c. 6000 BP reflect settlement and resource use strategies designed to exploit virtually the same conditions that environmental models predict existed at Mtwapa Creek during this period. Less extensive and more intensive occupation and use of more circumscribed environments would presumably lead to less ephemeral exploitation of such landscapes, leading to more distinctive environmental signatures of land and resource use. Even if human occupation of the East African Coast was of relatively low intensity prior to the last few millennia comprising the Swahili and proto-Swahili periods, these ancient human populations merit the archaeological attention that has historically been denied them. It is rarely demonstrated directly or indirectly (Morris et al. 2015) that colonizing populations from southwest Asia and elsewhere around the Indian Ocean littoral encountered indigenous African populations upon arrival. These populations are often given little more than lip-service acknowledgement as undoubtedly an "important" part of the multi-ethnic mosaic that came to comprise later East African coastal populations. Cultural traditions documented to have been introduced from afar are often emphasized in the archaeological literature, while it is implicitly assumed that indigenous cultural adaptations must have filled the "vacuum" left unexplained in the creation of the Swahili mosaic. I argue that these indigenous populations deserve our renewed future attention.

Several researchers, in addition to Omi and colleagues, have attempted to address these pre-Swahili contexts in recent years (Crowther 2012; Helm et al. 2012; Shipton et al. 2018). Building on this work, several lines of inquiry may be useful in future investigation of the history of coastal occupation in East Africa. More finely resolved records of sea levels through the Holocene will shed light on the significance of mangrove pollen fluctuations over time. Marine terrace records detailed by Oosterum (1988) and others are effectively geologic in time scale resolution. Century, or at least millennial scale knowledge of ancient sea levels would be more appropriate for making inferences concerning past environments. Some research has already been performed in this area (Woodroffe et al. 2015). Mangrove pollen data will undoubtedly be part of this investigation.

Metric data on pollen of coastal grasses will shed light on the nature of relatively large-sized Poaceae pollen recovered in Zone 2. Ecological information on endemic grasses producing pollen of this size will render a clearer picture of grassland dynamics through the Holocene. Sampling of fungal communities commensal with specific grassland contexts will further enlighten this investigation, allowing for comparison of modern and ancient grassland fungal communities. Additionally, modern fungal samples gathered from coastal livestock grazing plots will elucidate the potential value of coprophilous fungi as indicators of herbivore density. In sum, the constellation of proxy signals presented here is sufficient to suggest that human occupation of the East African coast during the Middle Holocene was sufficient to impact coastal ecologies considerably.

Finally, renewed archaeological investigation of known archaeological sites predating the Swahili period will undoubtedly provide new insight into the nature of Holocene coastal occupations. Focus on other aspects of human behavior, aside from lithic technology, that may be preserved at these sites will be a valuable source of information. Additional survey of coastal riparian and hinterland environments for evidence of pre-Swahili sites should provide a clearer picture of the task ahead.

References

Adamowicz L (1990) Newly discovered Stone Age cultures in Nampula province, Northern Mozambique. In: Sinclair P, Pwiti G (eds) Urban origins in eastern Africa: project proposals and workshop summaries. Paper No. 6. Stockholm, Sweden. The Swedish Central Board of National Antiquities, pp 137–197

Ambrose SH (1984) The introduction of pastoral adaptations to the highlands of East Africa. In: Clark D, Brandt SA (eds) From hunters to farmers: the causes and consequences of food production in Africa. University of California Press, Berkeley

Ambrose SH (1998) Chronology of the Later Stone Age and food production in East Africa. J Arch Sci 25(4):377–39

Ambrose SH, DeNiro MJ (1989) Climate and habitat reconstruction using stable carbon and nitrogen isotope ratios of collagen in prehistoric herbivore teeth from Kenya. Quat Res:407–422

Ambrose SH, Sikes NE (1991) Soil carbon isotope evidence for Holocene habitat change in the Kenya Rift Valley. Science 253:1402–1405

Baker AG, Bhagwat SA, Willis KJ (2013) Do dung fungal spores make a good proxy for past distribution of large herbivores? Quat Sci Rev 62:21–31

Bell A (1983) Dung fungi, an illustrated guide to the coprophilous fungi in New Zealand. Victoria University Press, Wellington

Bergner AGN, Trauth MH, Bookhagen B (2003) Paleoprecipitation estimates for the Lake Naivasha basin (Kenya) during the last 175 k.y. using a lake-balance model. Global Planet Change 36:117–136

Bonnefille R, Riollet G (1980) Pollens des Savanes d'Afrique Orientale. Centre National de la Recherche Sientifique, Paris, France

Boxem HW, de Meester T, Smaling EMA (eds) (1987) Soils of the Kilifi area. Reconnaissance soil survey report No. R11, Kenya Soil Survey, Nairobi and Department of Soil Science and Geology, Agricultural University of Wageningen, Pudoc, Wageningen, The Netherlands

Caratini C, Guinet P (1974) Pollen et Spores d'Afrique Tropicale. Association des Palynologues de Langue Francaise. Ministere de l'Education Nationale, Talence, France

Chala A, Taye W, Ayalew A et al (2014) Multimycotoxin analysis of sorghum (*Sorghum bicolor* L. Moench) and finger millet (*Eleusine coracana* L. Garten) from Ethiopia. Food Control 45:29–35

Chami FA (1998) A review of Swahili archaeology. Afr Archaeol Rev 15:199–218

Chami FA, Kessy E (1995) Archaeological work at Kisiju, Tanzania. Nyame Akuma 43:38–45

Clark JS (1988) Particle motion and the theory of stratigraphic charcoal analysis: source area, transport, deposition and sampling. Quat Research 30:67–80

Cohen AS, Palacios-Fest MR, Msaky ES (2005) Paleolimnological investigations of anthropogenic environmental change in Lake Tanganyika: IX. summary of paleorecords of environmental change and catchment deforestation at Lake Tanganyika and impacts on the Lake Tanganyika ecosystem. J Paleolimnol 34(1):125–145

Coltorti M, Mussi M (1984) Late stone age hunter-gatherers of the Juba Valley. Nyame Akuma 28:32–33

Crowther A (2012) Agriculture, subsistence and trade in coastal eastern Africa: archaeobotanical evidence from Kenya, Zanzibar and Tanzania. Paper presented at the 21st Biennial conference of the Society of Africanist Archaeologists, University of Toronto, Ontario, Canada: 20–23 June 2012

Davis OK (1987) Spores of the dung fungus *Sporormiella*: increased abundance in historic sediments and before Pleistocene megafaunal extinction. Quat Research 28:290–294

Davis OK, Shafer DS (2006) *Sporormiella* fungal spores, a palynological means of detecting herbivore density. Palaeogeogr Palaeocl 237:40–50

Ehret C (1998) An African classical age. University of Virginia Press, Charlottesville

Ellis MB (1971) Demaiaceous Hyphomycetes. Kew Publishers, Surrey, Commonwealth Mycological Institute

Erdtman G (1960) The Acetolysis method: A revised description. Sven Bot Tidskr 54:561–564

Gastaldo RA (2012) Taphonomic controls on the distribution of palynomorphs in tidally influenced coastal deltaic settings. Palaios 27:798–810

Gelorini V, Verbeken A, van Geel BJ et al (2011) Modern non-pollen palynomorphs from East African lake sediments. Rev Palaeobot Palyno 164:143–173

Gelorini V, Ssemmanda I, Verschuren D (2012a) Validation of non-pollen palynomorphs as paleoenvironmental indicators in tropical Africa: contrasting ~200-year paleolimnological records of climate change and human impact. Rev Palaeobot Palyno 186:90–101

Gelorini V, Verbeken A, Lens L et al (2012b) Effects of land use on the fungal spore richness in small crater-lake basins of western Uganda. Fungal Divers 55:125–142

Grimm E (1991) Tilia 1.7. Illinois State Museum, Research and Collection Center, Springfield

Helm R (2000) Conflicting histories: the archaeology of the iron-working, farming communities in the central and southern coast region of Kenya. Dissertation, University of Bristol

Helm R, Crowther A, Shipton C et al (2012) Exploring agriculture, interaction and trade on the eastern African littoral: preliminary results from Kenya. Azania 47:39–63

Jarzen DM, Elsik WC (1986) Fungal palynomorphs recovered from recent river deposits, Luangwa Valley, Zambia. Palynology 10:35–60

Karlen W, Fastook JL, Holmgren K et al (1999) Glacier fluctuations on Mount Kenya since ~6000 cal. years BP: implications for Holocene climatic change in Africa. Ambio 28:409–418

Kessy ET (1997) Archaeological sites survey from Kisiju to Dar-es-Salaam. Nyame Akuma 48:57–69

Kiage LM, Liu K (2006) Late Quaternary paleoenvironmental changes in East Africa: a review of multiproxy evidence from palynology, lake sediments, and associated records. Progr Phys Geog 30:633–658

Kusimba CM (1999) The rise and fall of Swahili States. Altamira Press, Walnut Creek

Kusimba SB (2003) African foragers: environment, technology, interactions. Altamira Press, Walnut Creek

Kusimba CM, Kusimba SB, Wright DK (2005) The development and collapse of precolonial ethnic mosaics in Tsavo, Kenya. J Afr Arch 3:345–365

Leakey MD (1966) Excavation of burial mounds in Ngorongoro Crater. Tanzania Notes and Records 66:23–135

Lejju BJ, Taylor D, Robertshaw P (2005) Late-Holocene environmental variability at Munsa archaeological site, Uganda: a multicore, multiproxy approach. Holocene 15(7):1044–1061

Mercuri AM, Trevisan Grandi G, Accorsi CA (1999) Plant exploitation during the early and middle Holocene in the Tadrart Acacus (Central Sahara, Libya): pollen evidence of changes. In: Guarino A (ed) Proceedings of the second science and technology for the safeguard of cultural heritage in the Mediterranean Basin. Elsevier, Paris, pp 1081–1085

Mercuri AM, Trevisan Grandi G (2001) Palynological analyses of the Late Pleistocene, Early Holocene and Middle Holocene layers. In: Garcea EAA (ed) Uan Tabu in the settlement history of the Libyan Sahara. Arid Zone Archaeology, Monographs 2. All'Insegna del Giglio, Firenze, pp 161–188

Mercuri AM (2008) Plant exploitation and ethnopalynological evidence from the Wadi Teshuinat area (Tadrart Acacus, Libyan Sahara). J Archaeol Sci 35:1619–1642

Michieka DO, van der Pouw BJA, Vleeshouwer JJ (1978) Soils of the Kwale-Mombasa-Lungalunga area. Reconnaissance Soil Survey Report No. R3, vol I: Main Report, Draft Edicion, Kenya Soil Survey, Nairobi

Morais J (1988) The early farming communities of southern Mozambique. Maputo, Eduardo Mondulane University and Stockholm, The Swedish Central Board of National Antiquities, Studies in African Archaeology, 3

Morris A, Monge J, Kusimba C et al (2015) Decoding the Swahili: ancient DNA studies on the Kenyan Coast. Paper presented at the 80th meeting of the society for American archaeology, 15–19 April, San Francisco, CA

Msaky ES, Livingstone D, Davis OK (2005) Paleolimnological investigations of anthropogenic environmental change in Lake Tanganyika: V. Palynological Evidence for Deforestation and Increased Erosion. J Paleolimnol 34(1):73–83

Nelson CM (1993) Evidence for early trade between the coast and interior of East Africa. Paper presented at WAC Mombassa intercongress conference volume

Nicholson SE (1996) A review of climate dynamics and climate variability in Eastern Africa. In: Johnson AI (ed) Limnology, climatology and paleoclimatology of the East African Lakes. CRC Press, Boca Raton, pp 25–56

Nirenberg HL, O'Donnell K (1998) New *Fusarium* species and combinations within the Gibberella fujikuroi species complex. Mycologia 90:434–452

Omi G (1982) Mtongwe 1980: an interim report of the East and Northeast African prehistory research project 1980. Nagoya University, Nagoya

Omi G (1984) Mtongwe 1982: an interim report of the East and Northeast African prehistory research project 1982. Nagoya University, Nagoya

Omi G (1986) Mtongwe 1984: an interim report of the East and Northeast African prehistory research project 1984. Nagoya University, Nagoya

Omi G (1988) Mtongwe and Mgonga 1986: an interim report of the East and Northeast African prehistory research project 1986. Nagoya University, Nagoya

Omi G (1991) An interim report of the East and Northeast African prehistory research project, 1988/1989. Nagoya University, Nagoya

Onyike NBN, Nelson PE (1992) *Fusarium* species associated with *Sorghum* grain from Nigeria, Lesotho and Zimbabwe. Mycologia 90:452–458

Oosterum AP (1988) The geomorphology of Southeast Kenya. Dissertation. Agricultural University, Wageningen

Palacios-Fest MR, Cohen AS, Lezzar K (2005a) Paleolimnological investigations of anthropogenic environmental change in Lake Tanganyika: III. Physical Statigraphy and Charcoal Analysis. J Paleolimnol 34(1):31–49

Palacios-Fest MR, Alin SR, Cohen AS (2005b) Paleolimnological investigations of anthropogenic environmental change in Lake Tanganyika: IV. Lacustrine Paleoecology. J Paleolimnol 34(1):51–71

Parker NE, Williams JW (2011) Influences of climate, cattle density, and lake morphology on *Sporormiella* abundances in modern lake sediments in the US Great Plains. Holocene 22:475–483

Patterson WA III, Edwards KJ, MacGuire DJ (1987) Microscopic charcoal as a fossil indicator of fire. Quat Sci Rev 6:3–23

Penugonda S, Girisham S, Reddy SM (2011) Elaboration of mycotoxins by seed-borne fungi of finger millet (*Eleusine coracana* L.). Int J Biotechnol 2:198–200

Pitkanen A, Huttunen P (1999) A 1300 year forest-fire history at a site in eastern Finland based on charcoal and pollen records in laminated lake sediment. Holocene 9:311–320

Punwong P, Marchant R, Selby K (2013a) Holocene mangrove dynamics and environmental change in the Rufiji Delta, Tanzania. Veg Hist Archaeobot 22:381–396

Punwong P, Marchant R, Selby K (2013b) Holocene mangrove dynamics in Makoba Bay, Zanzibar. Palaeogeogr Palaeocl 379–380:54–67

Raper D, Bush M (2009) A test of *Sporormiella* representation as a predictor of megaherbivore presence and abundance. Quat Res 71:490–496

Richardson JL (1972) Paleolimnological records from Rift lakes in central Kenya. Palaeoecol Afr 6:131–136

Rucina SM, Muiruri VM, Kinyanjui RN et al (2009) Late Quaternary vegetation and fire dynamics on Mount Kenya. Palaeogeogr Palaeocl 283:1–14

Schmidt MR, Bothman G (2005) Indications of bee pollination in sorghum and its implications in transgenic biosafety. Int Sorghum Millets Newsl 46:72–75

Schoenbrun DL (1993) We are what we eat: ancient agriculture between the Great Lakes. J Afr Hist 34:1–31

Shipton C, Roberts P, Archer W et al (2018) 78,000-year-old record of Middle and Later stone age innovation in an East African tropical forest. Nat Commun 9(1832):1–8

Soper RC (1966) Archaeological evidence from the coastal hinterland of Kenya and north-east Tanzania. Unpublished paper, Fort Jesus, Mombasa

Stuiver M, Reimer PJ, Reimer RW (2017) CALIB 7.1 [WWW program]. http://calib.org. Accessed 23 Jan 2017

Taylor M, Ravilious C, Green EP (2003) Mangroves of East Africa. UNEP WCMC, Banson, Cambridge

Thompson AO (1956) Geology of the Malindi Area. Report No. 36, Geological Survey of Kenya, Government printer, Nairobi

Thompson LG, Mosley-Thompson E, Davis ME et al (2002) Kilimanjaro ice core records: evidence of Holocene climate change in tropical Africa. Science 298:589–593

van Geel BJ, Buurman O, Brinkkemper J (2003) Environmental Reconstruction of a Romap Period settlement site in Uitgeest (The Netherlands), with special reference to coprophilous fungi. J Arch Sci 30:873–883

van Geel B, Gelorini V, Lyaruu A et al (2011) Diversity and ecology of tropical African fungal spores from a 25,000-year palaeoenvironmental record in southeastern Kenya. Rev Palaeobot Palyno 164:174–190

van Zinderen Bakker EM (ed) (1966) Palaeoecology of Africa and of the surrounding islands and Antarctica. Balkema, Rotterdam

Waddington JCB (1969) A stratigraphic record of the pollen influx to a lake in the Big Woods of Minnesota. Geol S America S 123:263–283

Walker MJC, Berkelhammer M, Bjorck S et al (2012) Formal subdivision of the Holocene series/epoch: a discussion paper by a working group of INTIMATE (Integration of ice-core, marine and terrestrial records) and the Subcommission on Quaternary Stratigraphy (International Commission on Stratigraphy). J Quat Sci 27:649–659

Williams LAJ (1962) Geology of the Hadu-Fundi Isa area North of Malindi. Report No. 52, Geological Survey of Kenya, Government printer, Nairobi

Wilshaw A (2012) An investigation into the LSA of the Nakuru-Naivasha Basin and surrounding, Central Rift Valley, Kenya: technological classifications and population considerations. Dissertation, St. John's College, Cambridge

Woodroffe SA, Long AJ, Punwong PA et al (2015) Radiocarbon dating of mangrove sediments to constrain Holocene relative sea-level change on Zanzibar in the southwest Indian Ocean. Holocene 25:820–831

Wooller MJ, Swain DL, Ficken KJ et al (2003) Late Quaternary vegetation changes around Lake Rutundu, Mount Kenya, East Africa: evidence from grass cuticles, pollen and stable carbon isotopes. J of Quat Sci 18(1):3–15

Wright DK (2005) New perspectives on early regional interaction networks of East African trade: a view from Tsavo National Park. Kenya. Afr Arch Rev 22(3):111–140
Wright DK (2007) Tethered mobility and riparian resource exploitation among Neolithic hunters and herders in the Galana River basin, Kenyan coastal lowlands. Environ Arch 12:25–47
Wright DK, Forman SL, Kusimba CM et al (2007) Stratigraphic and geochronological context of human habitation along the Galana River. Kenya. Geoarchaeol 22(7):709–728

Multiscalar Perspectives on Holocene Climatic and Environmental Changes in the Sahara and Nile Corridor, with Special Consideration of Archaeological Sites on Sai Island, Sudan

Elisabeth Hildebrand[1(✉)], Elena A. A. Garcea[2], Assunta Florenzano[3], and Anna Maria Mercuri[3]

[1] Anthropology Department, Stony Brook University, Stony Brook, NY, USA
elisabeth.hildebrand@stonybrook.edu
[2] Dipartimento di Lettere e Filosofia, Unversità di Cassino e del Lazio Meridionale, Cassino, Italy
[3] Laboratorio di Palinologia e Paleobotanica, Dipartimento di Scienze della Vita, Università di Modena e Reggio Emilia, Modena, Italy

Abstract. This multiscalar study explores Holocene environmental changes across the Sahara, within the eastern Sahara, and along the Nile in northern Sudan. The Early Holocene saw increased moisture across most parts of northern Africa after c. 10,000 BC, with peak humid conditions 7800–7000 BC. A short but significant dry interval after c. 7000 BC was followed by wetter conditions around 6000 BC, and then a gradual aridification from 5000 BC. The latter dry phase has continued until present times. The exceptional environments near the Nile are known to have seen impressive variations as climate oscillations and flora left traces in the palaeobotanical record. Multidisciplinary archaeological studies in this area—including analysis of plant macroremains—have focused on the transitions from hunting-fishing-gathering (Khartoum Variant) to pastoralism (Abkan) and later to agro-pastoralism (Pre-Kerma). The palynological data from four Sai Island sites (8-B-10C, 8-B-76, 8-B-81, and 8-B-10A) and the nearby mainland site of Amara West (2-R-66) provide new perspectives on local environmental shifts during this time of profound economic and social change. Despite poor pollen preservation, the high number of samples enables comparisons that show both diachronic changes and synchronic variation. Since the earliest phases, pollen spectra reflect mixed flora from various habitats and some seasonal variability. During the Early Holocene and the initial part of the Middle Holocene, dramatic floods on Sai's east side and seasonal desiccation on Sai's west side together created an ecological mosaic that exposed people to several different habitat types within a short distance. These included swamps and marshes, wooded savannas, grasslands and desert savanna, providing access to plants used for food, medicine, and other purposes. Documenting localized patterns of vegetation variation and change can lay important groundwork for explaining changes in subsistence and social organization.

Keywords: Eastern Sahara · Nile · Palaeoenvironment · Pollen
Sudan · Upper Nubia · Sai Island

A. M. Mercuri et al. (Eds.): *Plants and People in the African Past*,
https://doi.org/10.1007/978-3-319-89839-1_12

Introduction

As Saharan climates fluctuated dramatically through the past 12,000 years, Northern Sudan has seen important innovations in subsistence and social life: food producing economies emerged, use of the Nile Valley intensified, storage and mortuary practices transformed, and eventually powerful polities arose (Bonnet 1992, 2006; Reinold 2001, 2006; Honegger 2004a, 2007; Garcea and Hildebrand 2009; Garcea 2011–2012; Garcea et al. 2013, 2016a, b; Honegger and Williams 2015; Hildebrand and Schilling 2016). Interactions between humans and local habitats were crucial to processes of social and economic change; these have been clarified by geoarchaeological, zooarchaeological and macro-botanical investigations of datasets that directly reflect human activities (Chaix and Grant 1993; Lario et al. 1997; Salvatori et al. 2011). Palynological research complements these efforts by establishing ecological baselines using datasets that reflect natural regional conditions rather than specific human actions. Although poor preservation has hindered palynological studies in North Sudan (Watrin et al. 2009; Florenzano et al. 2016), archaeological deposits represent an exciting data source for examining diachronic environmental changes, and evaluating possible natural and anthropic causes in arid lands (Horowitz 1992; Mercuri 2008).

Throughout Holocene changes in climate, and their profound effects on Saharan environments and plant migrations, the Nile has always offered unusual sets of resources (Maley 2010). When arid conditions reduced inland vegetation to desert types, waterplaces were refugia, maintaining wet habitats that sustained plants, animals and people (Garcea et al. 2013; Mercuri et al. 2015). Cultural trajectories were often concurrent with climate oscillations; these changes eventually affected expansions or contractions of past human populations influencing mobility and land use in the Sahara (Kuper and Kröpelin 2006, Kröpelin et al. 2008; Mercuri et al. 2011; deMenocal and Tierney 2012; Garcea et al. 2013; Manning and Timpson 2014; Macklin and Lewin 2015; Maley and Vernet 2015). However, not all arid phases triggered crisis or abandonment; some created opportunities for new subsistence strategies and social networks (Zerboni et al. 2016).

The Northern Sudanese Nile is a fascinating setting to study human/climate interaction during dramatic environmental changes (flood to drought) and human responses (periodic environmental reassessment driven by unstable ecosystems and land degradation). We report a synthesis of pollen analyses from four sites (8-B-10C, 8-B-10A: Florenzano et al. 2018; 8-B-76, 8-B-81: Florenzano et al. 2016) on Sai Island, in Northern Sudan (Fig. 1). In this synthesis we discuss the results of interdisciplinary investigations on human adaptations and subsistence strategies under Holocene climate changes in light of paleoclimate and paleoethnobotanical information obtained from the literature on this region.

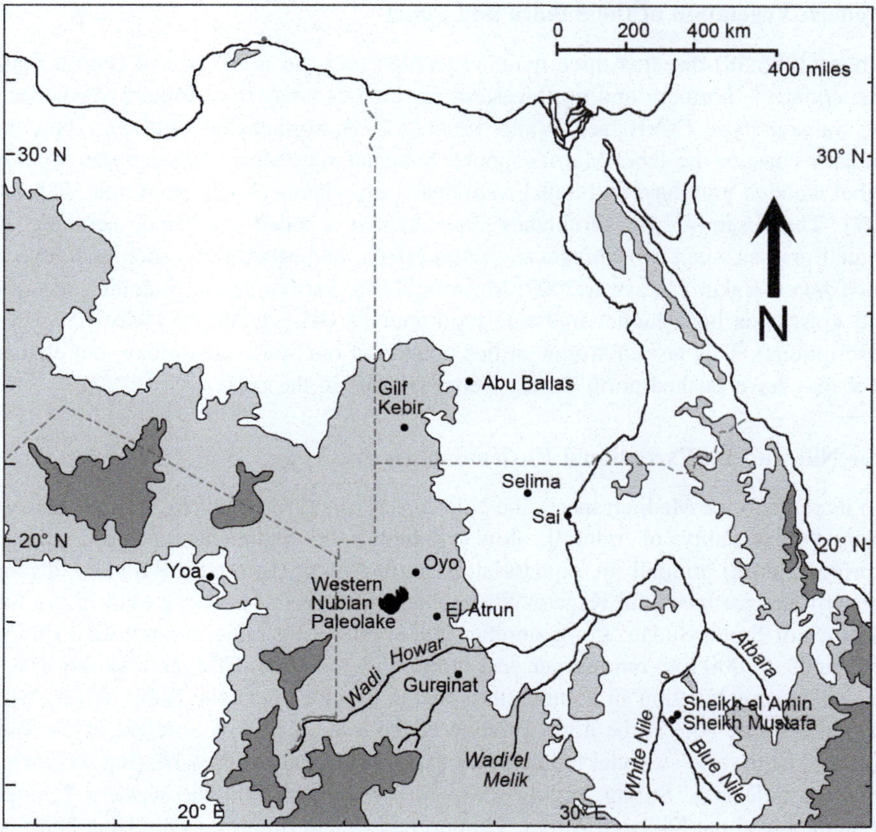

Fig. 1. Paleoenvironmental archives in the eastern Sahara (information from Hoelzman et al. 2001, 2010; Neumann 1989). Lightly shaded areas are at least 500 m asl; heavily shaded areas are highlands with significantly variable topography.

Environmental Setting: Modern Climate, Vegetation, and Land Use

Today Central and Northern Sudan encompass an incredible range of environments, from semi-arid areas near Khartoum to arid full desert farther north, crossed by a 'green corridor' of the riparian Nile. Near Khartoum, the rainy season (July to September) delivers an average 125 mm per year. In far northern Sudan at Abri near Sai Island, average annual rainfall is 2 mm (Climate Data n.d.). At this latitude ($\sim$20.5° N), areas away from the river have numerous sand dunes west of the Nile, or shift to a rocky substrate east of the Nile. Today, many areas >200 m from the river have scant vegetation, and some animals rely on the Nile and its floodplains for critical resources and water.

Modern Vegetation of the Sahara and Sahel

White (1986:38) discerns three main vegetation areas in northern and central Sudan. The country's northern and northwestern extremities areas have limited vegetation of the Saharan type (XVII; see White 1986:223). A northeast-to-southwest belt from Sudan's coast to the Jebel Marra supports Sahelian vegetation of Khordofan type, e.g. Sahel wooded grassland and Sahel semi-desert grasslands (XVI; see White 1986:206–207). These semi-desert shrublands have *Acacia* wooded grassland, perennial and annual grasses, such as *Panicum turgidum* Forsk., and other herbs such as *Indigofera hochstetteri* Baker (Abuswar 2007; Mustafa 2007). Farther south, Sudanian woodland and grasslands have higher moisture requirements (III; see White 1986:105–107 for descriptions). This last environment lies outside of our study area today, but elements of it may have pushed north during wetter periods in the past.

The Nile and Its Exceptional Environments

On its route to the Mediterranean, the Nile carries runoff from several source areas with distinct seasonality of rainfall, flow gradient, and sediment transport. Bimodal (spring/autumn) rainfall in equatorial eastern Africa (Burundi, Rwanda, Uganda, western Kenya) feeds the Kagera-White Nile; much of this water is held in the Sudd marshes of South Sudan. Long summer and short spring rains in southwest Ethiopia highlands (>2000 mm rainfall per year drained by the Baro, Gilo, and Akobo Rivers), as well as precipitation in southeastern Sudan together feed the Sobat River, which joins the White Nile at the northern edge of the Sudd. The low gradient of the White Nile and its diverse seasonal inputs result in a perennial, slow flow bearing sediment of small particle size. Strong, highly seasonal summer rains in the western Ethiopian highlands feed the Didessa, Abbay, Dinder, and Rahad (uniting in the Blue Nile), and the northwestern Ethiopian highlands and Ethio-Eritrean borderlands (Takezze/Atbara). Due to their seasonality, steeper gradients, and passage through arid areas more vulnerable to erosion, these rivers carry higher volumes of sediment than the White Nile, and coarser particles (Garzanti et al. 2015). Today, the largest overall contributor to the main Nile is the Blue Nile. However, tributary flows have varied in the past, when the eastern Saharan drainage systems of Wadi Howar and Wadi el Melik were also active.

In Northern and Central Sudan, riparian and arable lands are richer in plant species than other habitats (Abdelkreem and Ibrahim 2016). Today, land use is practically limited to riverine environments or areas that can be irrigated. Along the Nile, farming activity is well-developed thanks to irrigation technologies: the historic shaduf (wherein a pole mounted in an upright frame and balanced with a counterweight is used to raise water from the Nile channel to field level) and modern pumps. In places where the Nile follows a deep channel, well-watered riverine habitats are separated from many mainland and island surfaces. Elsewhere, more shallow riverine margins sustain swamp communities rich in sedges and grasses (Ali 2006; Madani et al. 2015). Islands sometimes have at least one shallow margin that is periodically inundated, receiving both flood moisture and enriched waterborne silts; thus, they may sustain higher plant species diversity.

Modern conditions in northern Sudan offer insights into environmental parameters that affected people throughout the Holocene. These include strong distinctions between riparian and interior habitats, latitudinal differences in rain-fed vegetation away from the river that responded to amplified or weakened monsoons, changes in channel depth and flood level, and fluctuating contributions from different Nile tributaries that would capture rainfall at different times of the year. Together these suggest a complex mosaic of factors which interacted to produce highly variable environments along the Nile and nearby hinterlands over the last 12,000 years.

Paleoclimate and Paleovegetation

Northern Africa has witnessed several dramatic climatic changes. During the Last Glacial Maximum (24,000–17,000 BC[1]; Tierney and deMenocal 2013) and for many millennia thereafter, extremely dry conditions made the Sahara hardly habitable (see Cancellieri et al. 2016) and substantially reduced the flow of the Nile, although several Late Pleistocene sites were recorded in Lower Nubia and farther north (Schild and Wendorf 2010). Beginning c. 10,000 BC, the Early Holocene saw increased moisture across northern Africa, with peak humid conditions 7800–7000 BC. A short but significant dry interval began just after 6500 BC, and affected many parts of the Sahara. Wetter conditions resumed 6000 BC, but a gradual aridification began after 5000 BC and has continued until present times.

Although enhanced moisture during the Early and Middle Holocene has been termed the 'African Humid Period' (deMenocal et al. 2000), the exact manifestations of this climate event varied across Africa (Costa et al. 2014; Williams et al. 2015a). This is especially true north of the equator, where different circulation systems govern different regions. For example, during periods of exceptional monsoon strength, the ITCZ extends farther north during the northern hemisphere summer, delivering more moisture to the Horn of Africa, the northern Sahel, and the southern Sahara. There is complex interplay between tropical moisture-bearing systems from the Indian Ocean (SE Africa) versus Atlantic/Gulf of Guinea (SW Africa), and between these and higher-latitude circulation systems (Gasse 2000, 2002).

Global climatic events and their local manifestations exerted a profound impact on environments and resources, altering the relative utility of riverine versus hinterland contexts for human livelihoods. To contextualize our new data from the Middle Nile, we review current perspectives on climatic and environmental changes at three geographic scales (Fig. 1). *Saharan* denotes the region now encompassing the Sahara desert, which for much of the Holocene sustained Sahelian and Sudanic environments. The *Eastern Sahara* encompasses the plains between rain-capturing mountains farther west (Tibesti, Ennedi, and Jebel Marra) and the Nile; it includes Saharan tributaries of the Nile that were active during moist periods (e.g. Wadi Howar) and their watersheds, as well as closed lake basins. *Nile Corridor* denotes areas adjoining the Nile in its current active channels and tributaries.

[1] All dates are given in cal BC. In cases where dates were originally published in uncalibrated form, we have calibrated them using Calpal (www.calpal-online.de). Please note that authors of some dates kept them uncalibrated in recognition of potential reservoir effects that may have varied over time, making calibration less accurate.

Regional Context: Environmental Changes Across the Sahara

Onset of Wet Conditions

The onset of the moist conditions occurred in two episodes within the Sahara and Sahel: water levels rose somewhat 13,000–12,500 BC, retreated during the Younger Dryas (10,900–9600 BC), and then resurged 9500–9000 BC (Gasse 2000). High water tables fostered the development of lakes, ponds and wetland areas in many parts of the Sahara by c. 9000 BC, with watercourses linking vast areas (Fig. 2; see also Drake et al. 2011). Many parts of the Sahara show multiple lines of geomorphological, climatic and environmental evidence for increased moisture (Fontes et al. 1985; Haynes et al. 1989; Schulz 1991; Cremaschi and di Lernia 1998, 1999, 2001; Cremaschi and Trombino 1998; Cremaschi 2002; Gasse 2000; Kröpelin et al. 2008; Mercuri 2008; Cole et al. 2009; Cremaschi and Zerboni 2009; Cremaschi et al. 2010, 2014; Zerboni et al. 2011; Giraudi et al. 2013). In the southern Saharan massifs, floods from upland areas created permanent lakes in lower depressions c. 9000–8500 BC (Gasse 2000).

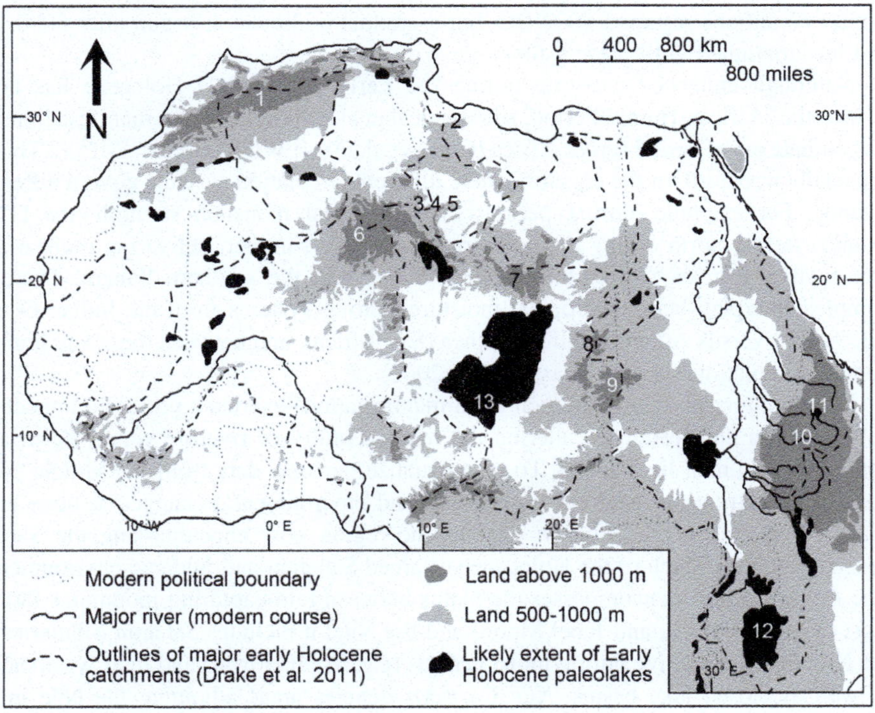

Fig. 2. Major geographical features and paleoenvironmental study localities in the Sahara and the Nile. *1*–Atlas Mountains. *2*–Jebel Gharbi. *3*–Tadrart Acacus. *4*–Erg Uan Kasa. *5*–Edeyen of Murzuq. *6*–Ahaggar Mountains. *7*–Tibesti Mountains. *8*–Ennedi Mountains. *9*–Jebel Marra. *10*–Northwestern Ethiopian Highlands. *11*–Lake Tana. *12*–Lake Victoria (probable early Holocene extent). *13*–(Paleo) Lake Chad (probable early Holocene extent). Approximate catchment boundaries and lake maxima from Drake et al. 2011.

Vegetation responded: tropical taxa probably followed riverine corridors north into greening parts of the Sahara, reaching their northernmost extent by c. 8500 BC. Because species responded individually to climate oscillations, and riverine and hinterland areas existed side by side, pollen spectra record a mosaic of plant taxa that does not have any analogues in modern environments (Watrin et al. 2009). Many animals—including fish and molluscs—also recolonized the Sahara, but some species requiring deep water (hippo, Nile perch) or other particular conditions stayed south of the Ahaggar (Drake et al. 2011).

Peak Moisture Delivery

By c. 7800 BC the InterTropical Convergence Zone (ITCZ) shifted as far north as 24° N in some parts of Africa, bringing tropical rainfall to much of the Sahara (Gasse and Roberts 2004). The exact timing and duration of this "hydrological optimum" varied geographically (Gasse 2000; Lézine et al. 2011; see also Costa et al. 2014; Williams et al. 2015a). In the Libyan Sahara's Tadrart Acacus, tufa development shows that heavy monsoon rains saturated the Tanezzuft Formation and created freshwater springs 7600–5600 BC (U-series and ^{14}C dates; Cremaschi et al. 2010). Lacustrine sedimentation indicates rainfall sufficient to sustain open bodies of water in the erg Uan Kasa 7200–6100 BC and the Edeyen of Murzuq 7500–5600 BC (Cremaschi 1998; Cremaschi and Zerboni 2009). The southern Sahara saw highest lake levels 7300–6600 BC (Gasse 2000), but the Chad Basin had a short regression c. 7500–7000 BC (Maley 1981, 2010).

Pioneer tropical taxa that pushed north during the earliest Holocene continued growing in Saharan areas during this time of peak moisture (Watrin et al. 2009). In addition, Sahelian plants such as *Acacia* migrated north into central Saharan mountain ranges sometime between 6300–5800 BC, where plants associated with permanent freshwater environments (e.g., *Juncus*, *Potamogeton*) peaked early in this phase (Mercuri 2008).

Brief Arid Event, and Renewed Moisture

Northern Saharan records show an arid episode c. 6000–5500 BC (Gasse 2000). Reduced tufa sedimentation in the Tadrart Acacus (c. 6100 BC) suggests decreased local rainfall, though spring activity continued until c. 5600 BC (Cremaschi et al. 2010).

Although regional records suggest aridity 6340–6210 BC (Cremaschi and di Lernia 1999), pollen spectra do not intercept the arid event in the Tadrart Acacus (Mercuri 2008). Local effects may have varied. Archaeological surveys suggest people were more mobile and had more ephemeral sites (Cremaschi and di Lernia 1998; Cremaschi and Zerboni 2009). The last major transgression of lakes in the southern Sahara occurred c. 5000 BC (Gasse 2000).

Late Holocene Arid Phase

By 4000–3000 BC, lower summer insolation weakened monsoons (Lézine 1989; Williams 2009). Progressively, the 100-mm isohyet moved 250 km south and rainfall declined as the ITCZ retreated south. Many lakes regressed; those fed by groundwater persisted longer than rain-fed playas (Nicoll 2004). The northern Sahara saw lacustrine regressions and fluctuations and increasing salinity 5400–3400 BC; lakes dropped to present levels after 3000 BC (Fontes et al. 1985). The southern Sahara saw regressions after 5000 BC, with a brief, modest increase in moisture 1800 BC (Gasse 2000).

Diminished moisture eventually affected flora. Dendroclimatic research in Wadi Tannezuft, SW Libya, shows the growth of *Cupressus dupreziana* was affected by droughts 3900–3760 BC and again at 3120 BC, and extremely arid conditions that prevailed during the past 500 years (Cremaschi et al. 2006). In the central Sahara, tropical plant communities were replaced by herbaceous-dominated desert ecosystems (Damnati 2000; Lézine et al. 2011; Mercuri et al. 2011). Some central Saharan areas had fairly sudden vegetation changes: a shift toward permanent aridity began 4000 BC (Mercuri et al. 2011; Cremaschi et al. 2014), and by 3400 BC hyperarid habitats were present, as indicated by psammophilous vegetation (Mercuri et al. 2015). In the northern Sahara aridity increased gradually in the Middle Atlas, Morocco and in Libya (Nourelbait et al. 2015; Giraudi et al. 2013). There, at Jebel Gharbi, Zygnemataceae (algae) HdV-Type E indicate fresh but stagnant water, possibly due to drier conditions. As aridity increased between 3400 and 2500 BC grassland-like vegetation first expanded and then contracted in favor of even more xeric-adapted plants (Giraudi et al. 2013).

The Eastern Sahara

The Eastern Sahara is commonly regarded as more arid than other parts of northern Africa, and encompasses inland environments closest to the Nile (Fig. 1). Few data sources exist for the initial stages of the AHP, but several lines of evidence become available starting c. 8500 BC. Here we review environmental inferences drawn from sediment and pollen sequences from paleolakes in this part of the Sahara, and additional terrestrial evidence. Our review begins with Lake Gureinat on the southern edge of the eastern Sahara, then follows a series of archives eastward across northern Sudan from El Atrun to Selima, and finally concludes with Lake Yoa in the northwest corner of the eastern Sahara. We then explore spatial variation in precipitation and vegetation across the eastern Sahara region.

Gureinat Paleolake

The southernmost well-studied paleolake in the eastern Sahara is Gureinat, south of Wadi Howar in western Sudan. Gureinat is thought to have been mainly rain-fed, and not linked to other paleodrainage or groundwater systems, although the presence of hippos suggest connections to other deep-water habitats existed at some point. Still, Gureinat deposits are regarded as excellent indicators of local precipitation, providing an important benchmark of monsoon strength. Its sequence begins before 9000 BC when aeolian sands underwent fluvial redeposition and groundwater levels rose and fluctuated. Around 9000 BC, the water table reached the basin's lowest point and a shallow lake formed. Sediments, diatoms, and chemical records suggest lake levels and conditions continued to fluctuate, with regressions likely at c. 8500 and 7000 BC and high stands likely at c. 8100 and 6700 BC. Then Gureinat's levels became more stable. After a brief (possibly century-long) major regression between 6200 and 5600 BC, the lake refilled to its maximum. Although the latest clear evidence for Lake Gureinat occurs at 5100 BC, hippos as late as 2000 BC indicate continuation of local conditions wetter than today (Hoelzman et al. 2010).

El Atrun

Lying c. 70 km north of Wadi Howar, the large El Atrun depression contains a sequence of lacustrine sediments covered by sands. Pollen data show tropical arboreal taxa as continuously present c. 9300–7500 BC, increasing 7500 BC, and declining c. 5700 BC (Ritchie and Haynes 1987).

Jahns (1995) divided El Atrun's sequence into four local pollen assemblage zones (LPAZ). Zone A (9100–8900 BC) sediments show initial lake formation and high biogenic activity; pollen suggests abundant sedge and steppic vegetation and few trees, probably due to the underrepresentation of pollen of Sudanian arboreal taxa in spectra. Zone B's (8900–8000 BC) carbonate mud sediments point to decreased salinity and alkalinity and a more humid climate; pollen contains more Sudano-Sahelian and Saharo-Sahelian elements suggestive of wooded savanna. Zone C (8000–7400 BC) has multiple lines of evidence for more arid conditions: higher sedimentation rates, chemical indicators of increased alkalinity and reduced open water, overall reduction in pollen concentration, diminished representation of pollen from moist environments, and heightened frequency of pollen from steppic vegetation. Zone D (7400–5900 BC) records a further increase in salinity and alkalinity, but other chemical indicators suggest open water returns. Overall pollen concentration increases again, and suggests a wooded savanna with the sequence's highest representation of Sudanian pollen such as *Celtis*, *Piliostygma*, *Hyphaene*, *Phyllanthus*, and *Salvadora persica* (Jahns 1995).

The Western Nubian Paleolake

A large paleolake area ~100 km NW of El Atrun has yielded five stratigraphic sequences from four profiles. These suggest lake formation began by 8900 BC. A large, coherent freshwater lake existed by 6300 BC, becoming segmented or discontinuous by 4000 BC. Progressive desiccation reduced lakes, but water may have lingered in some areas as late as c. 2300 BC. Human occupation of the area is documented between c. 6000 and 2000 BC, when the lake dried up (Hoelzman et al. 2001).

Oyo

About 50 km NE of the Western Nubian Paleolake's eastern shore, Paleolake Oyo began forming before 7500 BC. Oyo received seasonal input of freshwater until c. 5000 BC. Pollen from these fully lacustrine deposits include tropical Sudano-Sahelian taxa, so the area probably supported a savanna woodland/scrub mosaic similar to that found today in Darfur and Ennedi uplands, and savanna-woodlands in Khordofan ~500 km south of Oyo (Ritchie 1994).

At c. 5000 BC, Oyo sediments show downslope resedimentation and slumping, possible indicators of regression. Increasing amounts of windblown sands suggest lake levels dropped further by c. 4750 BC. Vegetation incorporated more Sahelian (savanna) and Saharan (thorn scrub) taxa and fewer deciduous savanna-woodland elements. Final desiccation occurred by 3700 BC, and vegetation was reduced to desert scrubs and grasses near other remaining oases (Ritchie et al. 1985; Ritchie 1994). After 3500 BC, remnants of Sudanian and Sahelian savannas around the lake were replaced by *Acacia*-thorn savanna and shrub-grasslands with increasing Cyperaceae until 2500 BC, when aeolian sands covered the lacustrine sediments (Ritchie 1994).

Selima

Located >400 km NE of Oyo and El Atrun, Selima lies in the heart of the eastern Sahara's driest areas, ~100 km west of the Nile's Second Cataract. Around 7700 BC the water table rose, creating a 14-m deep lake at Selima and other lakes in the area (Haynes et al. 1989). Pollen data from a Selima core c. 9200–5100 BC indicate a drier set of surrounding vegetation than contemporary deposits at Oyo and El Atrun. After initial development of *Typha* marshes, fully lacustrine conditions emerged by 9200 BC and Poaceae and Cyperaceae dominate the pollen spectrum. A sparsely wooded savanna was comparable to modern Sahelian-Sudanian vegetation, and featured *Acacia*, *Commiphora* and *Maerua*, Chenopodiaceae/Amaranthaceae shrubs, and perennial herbs (*Tribulus*, *Blepharis*) (Ritchie et al. 1985; Ritchie and Haynes 1987; Haynes et al. 1989; Ritchie 1994).

Pollen of Sudanian taxa typically associated with moister areas were never abundant at Selima, but extrapolations from core dates suggest slight peaks c. 9200, 6900, and 6500 BC (Ritchie and Haynes 1987). Later research suggested a narrower moist interval: c. 7500–5900 BC (Haynes et al. 1989). At c. 6100 BC (2.4 m core depth), Chenopodiaceae/Amaranthaceae increased, and Saharo-Sahelian taxa present throughout Selima's sequence reached their peak frequency c. 6100–5550 BC. Vegetation changed to a semi-arid steppe or desert with chenopods 5000–4000 BC (Ritchie 1994), and the lake dried up by 2000 BC (Hoelzman et al. 2001).

Charcoal from contexts 5250–4560 BC complements pollen data, recording Sahelian taxa such as *Acacia* sp., *Acacia albida*, *Maerua crassifolia*, *Leptadenia pyrotechnica*, *Ziziphus* sp., *Boscia senegalensis*, *Balanites aegyptica*, cf. *Cassia senna*, and Chenopodiaceae (Neumann 1989).

Lake Yoa

In the NW corner of the eastern Sahara region, northern Chad's Ounianga lakes (including Yoa) drain the Tibesti mountains. Lake Yoa deposits cover the last 6000 years. Low surface water conductivity and magnetic susceptibility, coupled with high organic content, suggest a productive freshwater lake that formed c. 3600 BC. Yoa changed to a more shallow, saline water body 2200–1900 BC (Kröpelin et al. 2008).

Wooded communities shifted gradually to desert in this large basin. Tropical and montane trees and shrubs and fern spores were important elements of pollen assemblages 4000–2300 BC, reflecting local humid conditions and continuing freshwater runoff from the Tibesti mountains. Xeric taxa, such as *Acacia*, *Boerhavia*, and *Tribulus*, increased c. 2800 BC. *Commiphora* and *Acacia* became significant locally c. 2700–2100 BC. Steppic vegetation (Chenopodiaceae/Amaranthaceae, *Artemisia*, *Ephedra*, *Salvadora persica*) became dominant after 700 BC, along with *Typha* (Lézine 2009; Lézine et al. 2011; see also Kröpelin et al. 2008).

Other Terrestrial Environmental Archives

Wood charcoal from terrestrial sources records environmental changes in Egyptian portions of the eastern Sahara (Neumann 1989). Areas north of 25°N had contracted desert vegetation of the northern type (dwarf shrubs, likely sustained by winter rainfall) 8150–5100 BC.

At Abu Ballas (~24° N), flora was still limited c. 7200 BC, with acacias and tamarisks dominating and isolated examples of *Maerua*, *Leptadenia*, and

Chenopodiaceae. By 5900–5500 BC the floral assemblage was more diverse, adding *Grewia*, *Calotropis*, *Ziziphus* and cf. *Cassia senna*. These probably represent the limits of expansion for these Sahelian taxa, and Neumann (1989) notes that runoff capture made the Ballas Mudpans exceptionally moist for that latitude.

The Gilf Kebir, slightly farther south and 1000 m asl., has charcoal assemblages dominated by *Tamarix* and (to a lesser degree) *Ziziphus* 6500–2900 BC. Slightly wetter conditions may have occurred at 5550, 4530, and 3900–3600 BC given the appearance of *Acacia*, *Maerua*, and/or *Balanites*, and 5000 BC (*Acacia albida*).

Wadi Howar (the 'Yellow Nile' drainage system, North Sudan) drained portions of the Ennedi Mountains and Jebal Marra. Analogies with other parts of the Sahara (based on pollen comparisons) have led Neumann (1989) to propose that Sudan Zone vegetation would have been present c. 5900 BC in the dunes near Wadi Howar. Such vegetation might have included taxa such as *Terminalia* cf. *macroptera*, *Annona senegalensis*, cf. *Rhus* sp., *Crateva adansonii*, *Celtis integrifolia*, *Ximenia americana*, *Ficus* sp., *Boscia* cf. *salicifolia*, *Balanites aegyptiaca*, *Cadaba farinosa*, *Ziziphus* sp., and *Acacia albida*. By 4150 BC more arid conditions fostered a more restricted set of Sahelian taxa, recovered in charcoal assemblages around Wadi Howar itself: besides the dominant *Acacia* sp., charcoal samples have only *Acacia nilotica, Ziziphus* sp., cf. *Grewia tenax* and Capparidaceae (Neumann 1989).

In central Sudan, charcoal analysis suggests a thorn savanna vegetation 4900–3800 BC near the archaeological site of Kadero (Barakat 1995). Various *Acacia* species (including *A. nilotica*), *Commiphora* sp., Capparidaceae (*Boscia senegalensis*), *Grewia* sp., *Balanites aegyptiaca*, and *Cadaba* sp. are all present. Barakat notes that Kadero's charcoal assemblage may reflect proximity to the Nile, human selection of taxa for burning and use, and environmental modification by humans such as landscape burning and grazing.

Regional Patterns Within the Eastern Sahara

Several eastern Saharan records belie the notion of sudden climate changes during the Middle Holocene, or quick, dramatic oscillations between arid and wet conditions (Kröpelin et al. 2008). Tropical taxa may have diminished gradually, depending on local sediments' propensity to trap or shed runoff, and the dependability of precipitation entering local drainage systems.

A west-east transect of paleoenvironmental data sources at $\sim$23° N shows that freshwater lakes existed to the west (closer to the central Sahara) at the same time as mudpans existed to the east (closer to the Nile), thereby suggesting that western areas received more moisture and sustained a more generous complement of vegetation. Exceptions to this trend are found in two settings. First, where uplifted basement rocks created higher groundwater pressure, freshwater surface ponds could form (e.g., near Bir Tarfawi). Second, montane locations experiencing orographic effects (e.g., Meidob Hills) may have captured moisture earlier and in greater quantity, and contributed to lowland parts of the eastern Sahara through distributaries (Pachur and Hoelzmann 2000).

A south-north transect reveals a similar gradient of moisture from high (south) to low (north). As one moves north, freshwater lakes give way to playas and pans, Mg and Sr ions become more concentrated in lake carbonates, and ^{18}O values become more depleted (Pachur and Hoelzmann 2000).

The Nile Corridor

The Nile and areas immediately adjacent are environments that vary independently from inland areas. Because fluctuations in Nile flow are mainly influenced by precipitation in the western Ethiopian highlands, Uganda, and South Sudan (Fig. 2), they are not necessarily in step with broader environmental changes seen in the eastern Sahara or the greater Saharan region. This is partly due to the fact the SE African monsoons from the Indian Ocean govern precipitation in the Nile headwaters, Nile Valley and eastern Sahara, whereas SW African monsoons from the Atlantic/Gulf of Guinea govern moisture delivery in central and western parts of the Sahara. It also reflects the topography, gradient, and physiographic contexts of the various Nile tributaries (reviewed above in "The Nile and Its Exceptional Environments").

Here we review major changes in Nile flow as seen through deposits along the Blue Nile, White Nile, and sapropel activity indicative of outflow into the Mediterranean Sea. We also consider pollen archives from areas near to Nile sources and from riparian environments adjacent to the middle Nile (from the confluence of the Blue and White Niles at Khartoum, to Aswan near the First Cataract).

The AHP Onset: Resumption of Constant Flow c. 13,000–10,000 BC

Nile flow was intermittent and seasonal during the Late Glacial. Returning monsoons raised the levels of Lake Victoria and Lake Albert to overflow c. 13,000 BC (OSL dating), restoring White Nile flow and pushing it to extremely high levels c. 12,700–11,100 BC (Williams et al. 2006; Williams 2009). Alluvial plains near Ed Dueim (Fig. 3) show elevated White Nile flow c. 11,300 BC (OSL dating) (Williams et al. 2010). Lake Tana overflowed c. 12,750 BC and restored heavy flow to the Blue Nile 11,900–11,200 BC, but underwent a minor, temporary regression at 10,000 BC (Lamb et al. 2007; Williams 2009).

Nile Flow Patterns During Peak Saharan Moisture c. 10,000–7000 BC

Alluvation resumed along the White Nile at Ed Dueim 7800 BC (OSL) (Williams et al. 2010). C. 7700–7000 BC, White Nile floods were higher (3 m above present levels, but not as high as during the initial overflow), and other western Nile tributaries became active (Gatto and Zerboni 2015). The Blue Nile does not show a high river level during this time interval, but its sedimentation record is more fragmentary than the White Nile's (Williams 2009).

Wadi Howar (the "Yellow Nile") was an important Nile tributary by 7300 BC. For the remainder of the Early Holocene, it brought water from the Ennedi and northern Jebel Marra to join the main Nile at Dongola Reach (Nicoll 2004). Downstream near Kerma on the Middle Nile's eastern bank, a large, swampy plain existed 8300–5300 BC (Honegger and Williams 2015).

A sharp increase in Nile freshwater outflow caused the deposition of Mediterranean sapropel S1. Starting dates for sapropel deposition range from c. 11,000 BC to 7000 BC, and ending dates range from c. 4800 to 4000 BC (Aritzegui et al. 2000; deMenocal and Tierney 2012; reviewed by Williams et al. 2015b:104). Depleted ^{18}O signatures in sapropel S1 point to strengthened contributions from the Indian Ocean monsoon system, rather than Atlantic monsoons (Hennekam et al. 2014). Thus, Indian

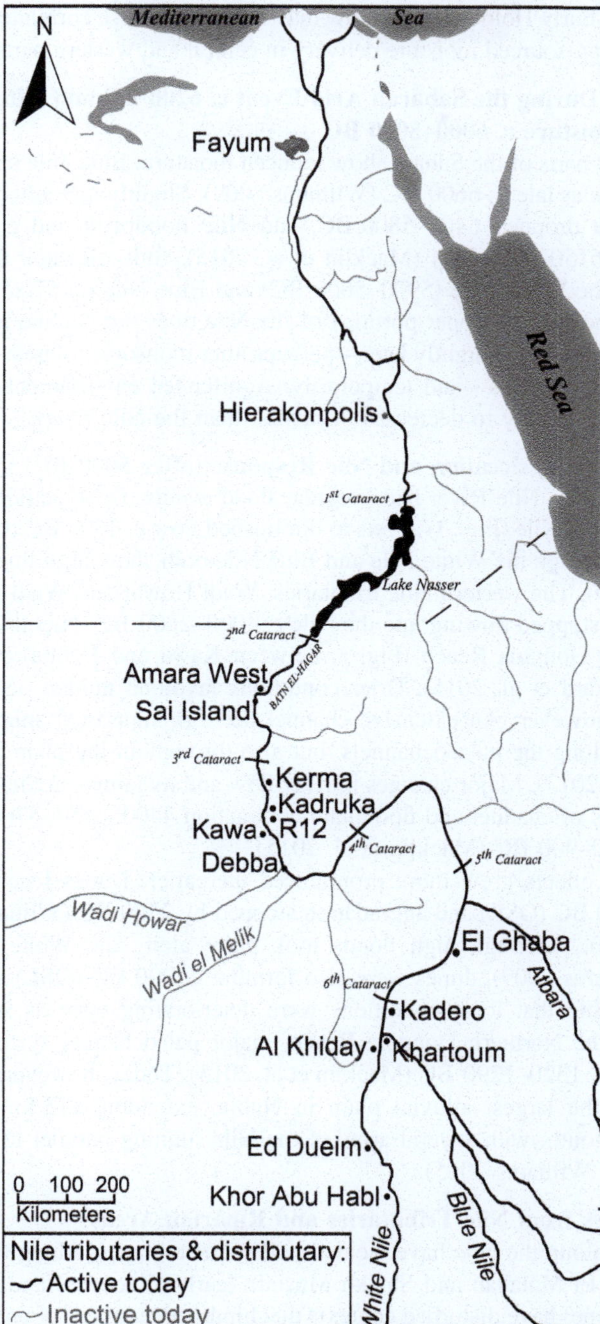

Fig. 3. Archaeological sites along the Nile mentioned in the text.

Ocean-sourced Early Holocene Nile flow may not always have correlated with changes in Atlantic Ocean-sourced moisture delivery in central and western parts of the Sahara.

Nile Behavior During the Saharan Arid Event c. 6200 BC and After Resumption of Saharan Moisture c. 6000–5000 BC
Although many parts of the Sahara show reduced moisture, Blue Nile sediments record high river levels as late as 6600 BC (Williams 2009). Slightly thereafter, Lake Victoria and Lake Tana dropped 6400–5800 BC, and Nile floodplain and channel systems contracted (c. 6150–5750 BC) (Macklin et al. 2015). Still, elevated flood levels are evident along the White Nile (5900–5600 BC) and Blue Nile (c. 5700 BC) (Williams 2009). Thus, the flow of upper portions of the Nile does not decline in step with the Saharan aridity event but slightly later—perhaps after monsoonal retreat was felt farther south in Nile headwaters—and temporarily. Aquifer-fed environments in the Sahara responded more quickly to decreases in rainfall than the Nile rivers.

Middle Holocene Desiccation and Nile Response (after 5000 BC)
By c. 5000 BC, the Nile River had irregular flood events. Dune activity at Khor Abu Habl by the White Nile (Fig. 3) attests to diminished flow c. 4600 BC (OSL) (Williams et al. 2010), although the White Nile and Blue Nile both show high floods at 4300 BC (Williams 2009). The western Nile tributaries, Wadi Howar and Wadi Melik, lowered and eventually stopped flowing into the Nile c. 4000–2800 BC (Nicoll 2004), but local wadis near the Dongola Reach (Fig. 3, between Kawa and Debba) continued to be active (Woodward et al. 2015). Drier conditions favoured human occupation of the Kerma plain, however: An extensive channel network supported animal and vegetal life, not only along the palaeochannels, but also throughout the plain 5300–3500 BC (Macklin et al. 2013). Major changes in river flow and dynamics in the Nile catchment included phases of channel and floodplain contraction 4400–4150, 3700–3450, 2700–2250, and 1350–900 BC (Macklin et al. 2015).

Desiccation effects grew more pronounced thereafter. Dunes reactivated at Khor Abu Habl 2800 BC (OSL) and aggradation stopped by 2500 BC (Williams et al. 2006; Williams 2009). Although high floods took place along the White Nile at 1200–800 BC (Williams 2009), dunes were also forming c. 900 BC (OSL) (Williams et al. 2010), suggesting that local conditions were deteriorating even as some Nile flow continued. In the Northern Dongola Reach, major paleochannel systems east of the Nile dried out c. 1320–1290 BC (Macklin et al. 2013). Today, however, their remnants still represent the largest alluvial plain in Nubia, extending >13 km east-west and 100 km north-south, with several arms of the Nile running parallel to the main river (Honegger and Williams 2015).

Pollen Archives from Nile Tributaries and Riparian Areas
Few locations along the Nile have yielded pollen assemblages. Along the Blue Nile in central Sudan, El Malahab and Sheikh Mustafa (early Holocene) and Sheik el Amin (middle Holocene) have disturbed contexts that hinder assessment of precise diachronic changes. Pollen from all three sites suggests open savanna, with *Acacia* only present at El Malahab. Arid elements such as Chenopodiaceae/Amaranthaceae, *Capparis*, *Boscia*, and *Maerua* are present. Sheikh Mustafa has evidence for periodic landscape clearance

via burning, and Sheikh el Amin has coprophilous ascospores suggestive of pastoral land use (López-Sáez and López-García 2003).

In northern Sudan several archaeological sites on and near Sai Island have yielded pollen assemblages (Florenzano et al. 2016, 2018). These spectra, which together comprise a sequence covering much of the Holocene, will be discussed in greater detail below; see "Pollen records from archaeological sites on Sai Island"

Implications for Human Behavior and Resource Access Issues

Reviewing paleoenvironmental data at these three scales suggests a few important points and unresolved issues. The general Saharan and eastern Saharan records are distinct in two ways that have important implications for resource distributions and human economic activities.

First, some sudden trends in the main Sahara seem muted in the eastern Sahara: vegetation shifts, and consequent changes in human behavior, may have been more gradual. Thus, although the eastern Sahara has the reputation of being the driest part of the Sahara, its fluctuations may have been less extreme.

Second, within the eastern Sahara, aridity increases in an easterly direction. Areas near the Nile, but away from its floodplain and riparian edge, would have been among the most arid environments in the Sahara at any given time. They may have offered significantly less vegetation and resources than the northwest Sudan paleolakes reviewed above, and more closely resembled environments documented at Selima. Eastern Saharan data support broader impressions that "greening" of the Sahara was neither universal nor homogeneous; rather, Saharan environments changed in a patchy fashion depending on localized conditions of circulation, topography, and substrate.

However, the Nile and its riparian environments fluctuated in distinct ways. While the Sahara depended on Atlantic circulation for monsoon rains, Nile flow depended on low-latitude precipitation influenced by Indian Ocean monsoons. Changes in Nile flow and floodplain size may therefore have been out of step—delayed—in comparison to Saharan patterns. In addition, the Nile would have served as a linear refugium, maintaining taxa that could no longer survive farther west during arid phases, and potentially allowing their more speedy introduction inland when moist conditions resumed.

Holocene Plant Use Along the Nile

The foregoing multiscalar review demonstrates the diverse forces that influenced environmental changes throughout the Holocene in the Sahara and Nile. To understand their localized effects on resources and human livelihoods requires explicit examination of archaeobotanical records from sites near the Nile itself. In this section, we review prior studies of plant use at archaeological sites near the Sudanese and Egyptian Nile (Fig. 3).

Plant Use Along the Nile During Peak Moisture Delivery

Early Holocene sites in Sudan (c. 8000–5000 BC) have many material cultural commonalities, such as pottery decorated with 'wavy-line' incisions and 'dotted-wavy line' and rocker-stamp impressions. Slight differences in decorative techniques and motifs distinguish those in central Sudan ('Early Khartoum') from those downstream from Debba ('Khartoum Variant') (Garcea and Hildebrand 2009). However, archaeobotanical records from northern and central Sudan are not comparable.

While northern Sudanese sites from this period have yielded little botanical evidence, macrobotanical remains and plant impressions in pottery at Early Khartoum sites suggest that fisher-gatherer-hunters ate plants that did not need special processing and/or preparation (e.g. fruits and berries such as *Celtis integrifolia*, *Ziziphus spina-christi* and *Grewia tenax*), as well as grains of wild cereals that do (e.g., *Sorghum*, *Setaria*, *Echinochloa* and *Panicum* species) (Magid 2003). This type of edible plant exploitation indicates an economic strategy based on a complex and broad spectrum of vegetal foods.

Farther south in Central Sudan, consumption of wild tubers, including *Cyperus rotundus* L., was documented via dental calculus from foraging and farming/herding individuals buried in the cemetery of Al Khiday, on the White Nile in central Sudan (Buckley et al. 2014). This sedge is appreciated for its high carbohydrate content and other non-dietary values (as a perfume and/or medicine).

Plant Use Along the Nile During the Onset of Food Production

Herding preceded the onset of village agriculture in most areas. In Sudan broad spectrum exploitation of plant foods continued without significant changes even after the arrival of domestic livestock at Shaheinab, Sai Island, and other sites dating to around 5000–4000 BC. Macro-botanical evidence includes fruits of *Celtis* and *Ziziphus*, as well as cereals such as *Sorghum*, *Setaria*, *Echinochloa* and *Panicum*. The primary activities continued to be focused on the consumption of plants that did not require processing and preparation (Magid 2003).

Currently, the earliest evidence for domestic plants along the Nile is at cemetery R12 in northern Sudan (Salvatori and Usai 2008), where domestic cereals of southwest Asian origin are present by c. 5311–5066 BC (Out et al. 2016). Phytoliths in human burial deposits are attributed to *Hordeum* sp. and/or *Triticum* sp., which are thought to be domesticated because there are no wild ancestors in the region (Madella et al. 2014). It has been suggested that these plants were used in the burials for ritual purposes as they represent rare or exotic (luxury?) plants. They may have been used as pillows or mats (Out et al. 2016) or as funerary offerings (Garcea et al. 2016a). In addition, dental calculus with starch of *Hordeum* sp./*Triticum* sp. from human burials farther south at El Ghaba (4750–3650 cal BC) may suggest that these plants were also used as food (Out et al. 2016).

Farther south up the Nile, phytoliths and starch in human dental calculus at the cemetery of El Ghaba (4750–3650 cal BC) show that *Hordeum* sp. and/or *Triticum* sp. were used for both ritual purposes and food (Madella et al. 2014; Out et al. 2016). On the other hand, wild C_4 grasses were much more common than domestic C_3 cereals, indicating continuing regular wild plant exploitation and even management for centuries. Bucrania (domestic cattle skulls) were found together with phytoliths of wild panicoid grasses. While these finds document consumption of crops, it is not yet clear whether wheat and barley were locally farmed at this time.

The earliest large-scale evidence for domestic plants is at Kom W and K in the Fayum, where the southwest Asian domesticates emmer wheat (*Triticum turgidum* L. ssp. *dicoccum* (Desf.) Husn.) and hulled barley (*Hordeum vulgare* L.) are stored in bulk by 4650–4350 BC (Wetterstrom 1993; Wendrich et al. 2010). Away from the Nile, emmer has been found in deposits dating to 4450 BC at site KS043 at Kharga Oasis (Briois et al. 2012; Wuttmann et al. 2012). Because only eight grains of emmer were found, chaff was absent, and grains had been cleaned before entering the site, one cannot build a systematic case for local farming. It is possible that grains were brought in from elsewhere: KS043 also at this time has large catfish, and bivalves such as *Spathosis rubens* and *Corbicula consobrina*. Because these taxa could not have lived in the oasis, but only in the Nile River, they argue for long-distance exchange networks between Kharga and the Nile (Briois et al. 2012).

Full agriculture was practiced in the Egyptian Nile Valley at Hierakonpolis c. 3750–3300 BC (Fahmy 2003). Emmer wheat provided food and drink, but bread wheat (*Triticum durum/aestivum*) was also present. Plants were also grown for non-alimentary purposes. Six-rowed barley (*H. vulgare*) and two-rowed barley (*H. distichon* L.), which appeared in higher quantities than wheat, were seemingly grown as fodder for livestock, indicating that herding was important. Linen bandages in tombs show that flax (*Linum usitatissimum* L.) was cultivated for weaving. Wild plants such as *Cyperus esculentus* L., *Ziziphus spina-christi* (L.) Desf., and *Balanites aegyptiaca* (L.) Del. continued to provide food (Fahmy 2003).

Between the 1st and 2nd Cataracts near the Egypt/Sudan border, A-Group sites (c. 3700–2400 BC) have evidence for farming of domestic wheat, barley, and legumes; herding cattle, sheep, and goats; and hunting and fishing. Sites occur both by the Nile and away from the river, and inland herding may have yielded to farming near the Nile banks as rainfall diminished. Some A-Group sites have evidence for small- or large-scale storage, and exchange with Egypt 3700–3300 BC; A-Group pottery is also found as an exotic ceramic at Pre-Kerma sites farther south. Food production, physical storage, and long-distance trade all appear to take hold together (Lal 1967; Nordström 1972; Gatto 2006; Lange 2007; Garcea and Hildebrand 2009).

Farther upstream in northern Sudan, despite early dates for (presumably imported) domestic cereals at R12 and Ghaba cemeteries, regular crop cultivation is not documented until c. 2700 BC at site 8-B-52A on Sai Island (Hildebrand 2006–2007; Hildebrand and Schilling 2016). Agriculture was well-developed during the Kerma period, dating from 2500 BC (Iacumin et al. 1998).

In central Sudan, farming of wheat and barley is recorded at Meroe, which is only dated between around 1000 BC and 1000 AD (Out et al. 2016). The earliest clear evidence for farming of indigenous crops (*Sorghum*) occurs c. 780–400 BC at Kawa (Fuller and Smith 2004), and by 0 AD sorghum was more widely cultivated (Rowley-Conwy 1989). Delayed adoption of southwest Asian taxa may relate to local environmental conditions. Consequently, native wild species continued to play a greater weight in the diet than in Egypt. In fact, the intensive consumption of wild grains adopted during the Early Khartoum period became widespread and lasted a long period of time (Stemler 1990; Fuller and Smith 2004; Iacumin et al. 1998).

Pollen Records from Archaeological Sites on Sai Island

Four archaeological sites together span economic transitions from fishing-hunting-gathering to herding to farming at Sai. They also reflect regional changes in material culture, from pre-pastoral Khartoum Variant (c. 7600–4800 BC), pastoral Abkan (c. 5000–4300 BC) and agricultural pre-Kerma/Kerma (c. 1800–1700 BC). Ceramic assemblages reveal significant cultural shifts at Sai during this time interval, involving cross-cultural interactions (D'Ercole et al. 2015; Garcea 2006, 2006–2007, 2011–2012, 2016a, b, in press; Garcea and Hildebrand 2009; Garcea et al. 2016a, b; D'Ercole et al. 2017a, b).

Pollen from Sai sites can clarify the environmental contexts of these economic and social changes. The following sites and contexts have yielded pollen samples: 8-B-10C Level 2 (7600–7200 BC, 2 samples); 8-B-10C Level 1 (5000–4800 BC, 10 samples); 8-B-76 (6400–6250 BC, 5 samples), 8-B-81 (5000–4300 BC, 5 samples) and 8-B-10A (1700 BC, 6 samples) (Fig. 4). Florenzano et al. (2016, 2018) describe pollen recovery, preparation, analysis methods and results for all Sai samples. Florenzano et al. (2016) also report on pollen analyses from the Amara West district (not described here). Pollen samples show poor preservation and low concentrations, presenting taxa from local plant cover or long-distance arrival of pollen. State of preservation and phytogeographical attributions help to distinguish local from regional or extra-regional pollen composition. Steady alteration of seasonal wet/dry conditions may have impeded the full preservation of exines throughout the Holocene sequence, and Nile flood events may have further diluted representation of organic matter.

Site 8-B-10C

This Khartoum Variant site consists of gravel bars with archaeological deposits atop an early Holocene fluvial terrace. Seasonal Nile floods created a marshy environment with stagnating water pools; gravel bars occasionally became separated islets (Hesse 1996; Geus 1998, 2002). The 105-m^2 excavation area had two superimposed Khartoum Variant occupation phases: Level 2 (7600–7200 BC), an occupational phase with 75 post holes; and Level 1 (5000–4800 BC), an occupational phase c. 2500 years after Level 2, consisting of a large (semi-) permanent settlement with seven hut floors, 100 post holes, three hearths, and other structural features such as a rubbish pit.

Level 2 has a tiny concentration of <260 pollen grains per gram (p/g) probably reflecting the increased inputs of sediments with poor organic content from Wadi Howar, and the White and Blue Niles (Williams 2009; Gatto and Zerboni 2015). Pollen spectra show low diversity with prevalence of Poaceae (31%) and low percentage of woody plants (8%). Different vegetation types are represented: desert savanna, just out of the green borders of the river (*Capparis* 3%, *Salvadora persica* 1%, Chenopodiaceae 4%, Asteraceae 4% excl. *Artemisia*), wet environments (Cyperaceae 4%), and tropical forest (*Celtis*, *Myrica*). The latter type of forest probably did not grow locally but likely

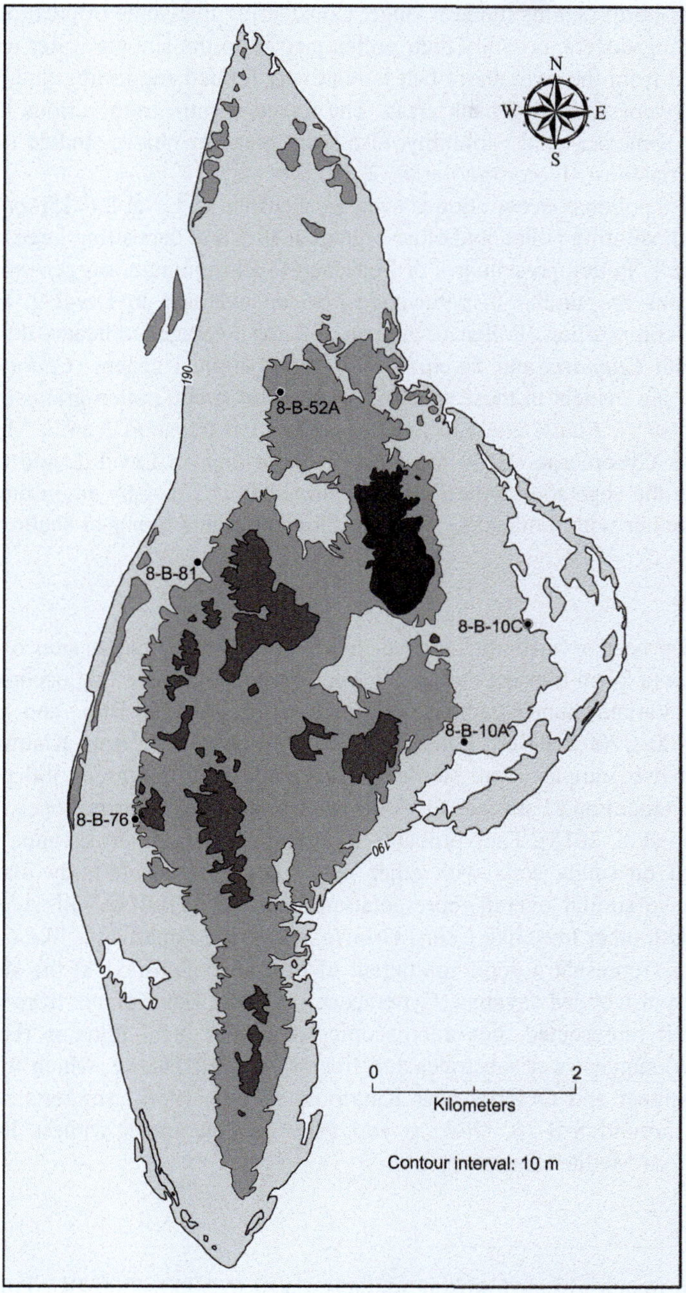

Fig. 4. Topographic map of Sai Island with archaeological sites mentioned in text. Base map by T. Schilling, with contour data from Despagne n.d. Topographic data are not available for mainland areas east of the main channel and west of the minor channel; these areas are the lightest shade of grey.

lived farther north than its modern range. Considering that some tropical plants are not high-pollen producers, possibly their pollen arrived to the site via water with the river flow. Pollen from the river shore belt is relatively limited suggesting that flood events may have swept away riverbank areas. The mixed spectra from various habitats may also reflect some seasonal variability, also seen in earlier phases studied in the nearby site of Amara West (Florenzano et al. 2016).

Level 1's pollen concentration is even weaker than in Level 2 (<150 p/g), probably due to floods diluting pollen and other organic matter and depositing large quantities of silt or gravels. Pollen percentages of Poaceae (44%) dominate, suggesting grasslands. Woody plants are similar in percentage (7% on average) to Level 2, but differ in taxonomic composition. Pollen of *Acacia* (3% on average) indicates desert wooded savanna with *Capparis* and *Tamarix*. Also other tropical genera (*Cadaba*, *Barleria*, *Indigofera*) are evident in these spectra. Several Holoarctic pollen grains (broadleaved *Quercus*, *Ostrya*, *Pinus*) attest to long-distance wind transport from a Mediterranean source area. Cyperaceae (12%) are more frequent than in Level 2, and indicate that remnants of the vegetation typical of the riverine belt and marshy environments remain visible, together with tamarisks, reeds and floating plants living in shallow water.

Site 8-B-76

The site lies near a currently inactive floodplain on the western side of Sai Island. Excavations (a 27-m transect and a 2 × 1-m test unit) revealed two occupations: early Khartoum Variant hunter-gatherer occupations (6400–6250 BC), and then Abkan herders (5400–3700 cal BC). Pollen samples were only taken from Khartoum Variant levels, and two samples were sterile. Pollen concentration was c. 300 p/g. Poaceae (18% on average) and Fabaceae (17%) prevail, suggesting their presence near the site (Florenzano et al. 2018). They probably represent shrubby desert savanna where these plants lived on sandy soils with other xerophilous species. Woody plants (7% on average) have similar overall representation to that at 8-B-10C with *Acacia* (0.6%). Together with other trees like *Celtis*, *Grewia*, and shrubs/small trees like *Capparis* and *Myrica*, they represent a good contingent of tropical pollen taxa at the site, probably produced by a wooded savanna. Cyperaceae (3%) and other plants from wet habitats are not well represented, but microscopic freshwater algal remains (HdV-181 cf., *Fritschiella*) suggest wet substrates and freshwater. *Fritschiella*, which today blooms between August and October near Khartoum (Brook 1956), suggests seasonal dry conditions around 8-B-76. *Quercus* and *Pinus* pollen grains witness long-distance transport from Mediterranean regions.

Site 8-B-81

A 4 × 1-m trench revealed Abkan artefacts, dated to 5000–4300 BC. Pollen concentration was poor (<150 p/g). High values of Poaceae (52% on average) and low Cyperaceae (2%) likely represent local grassland or wetland plant cover. According to the phytogeographical affinity (Watrin et al. 2009; Lézine et al. 2011; Hely et al. 2014), pollen from woody plants (11%) represents both long-distance wind transport (e.g., *Quercus* and *Pinus*) and a complex indicating local wooded grassland or savanna woodland along rivers (this is, for example, the plant association suggested by pollen by *Acacia*, *Grewia*, *Morus* cf. and *Myrica*, 5% altogether). These values are consistent with northern penetration of tropical flora seen elsewhere (Lézine 2007).

Comparing our two c. 5000 BC contexts shows differences between the west (8-B-81) and east (8-B-10C, Level 1) sides of Sai Island. *Fritschiella*, seen at 8-B-81 and earlier at the west side site of 8-B-76, indicates that for millennia, this side of Sai experienced a distinct set of seasonal oscillations: less availability of water overall, and probable drought crises during dry seasons. Less extreme flooding on the eastern side may have allowed more permanent vegetation to grow, such as the riverine wooded grassland mentioned above.

Site 8-B-10A

This stratified site is located in the eastern part of the island about 2 km south of 8-B-10C (Geus 1998; Garcea and Hildebrand 2009). Its ceramic assemblage parallels Pre-Kerma sites elsewhere (D'Ercole et al. 2015, 2017a, b), but its chronology appears to overlap with the Kerma period (2500–1500 BC; Hildebrand and Schilling 2016). Archaeological deposits yielded numerous animal remains, the majority of which were domestic livestock (Chaix 2011). Caprines were much more numerous than cattle (270 vs. 14 remains, or 93.1% vs. 4.8%), but the weight of cattle bones was considerable (94.3 g, or 19.3% of assemblage weight). Small livestock that could be identified as sheep versus goat exhibited a prevalence of sheep (*Ovis aries* L.), including lambs and young individuals, over goats (96.7% vs. 3.3%). Birds and fish were insignificant. High representation of sheep contrasts with the contemporary site of Kerma, where goats make up 21.8% of the faunal assemblage. Sai sheep appear to have been much smaller in size than Kerma sheep, whereas goats did not show a size difference between the two sites. Because vertebrae and ribs were underrepresented with respect to the other bone segments, Chaix (2011) has suggested slaughtering probably occurred elsewhere, away from 8-B-10A's habitation area. In pollen spectra, the desert environment is evident from the presence of *Capparis* (3%), *Salvadora persica* (0.2%), Chenopodiaceae (3%) and Asteraceae (6%). *Celtis*, cf. *Hagenia abyssinica* and *Rhynchosia* are tropical taxa that probably arrived via river flow from southeastern regions. Poaceae (20%) and Cyperaceae (5%), indeed, represent grassland and riverine vegetation. *Plantago*, Cichorieae and other Asteraceae may be also connected to browsing and trampling by domestic animals. Although no cereal pollen was found, this is the only site where pollen of palms (*Phoenix*, and possibly *Elaeis*?) are recorded. This indicates palms were growing either along the river or inland at oases.

Relations Between Environmental and Economic Changes

Since the beginning of the Holocene, areas in and near to Sai Island have showed a mosaic of habitats with open-vegetation and wet environments (Fig. 5). Palynology reveals that vegetative changes along the northern Sudanese Nile paralleled those documented in the eastern Sahara, albeit with some interesting variations and details.

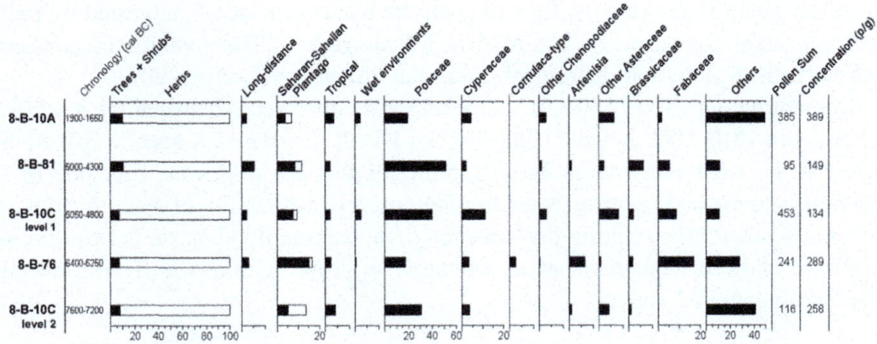

Fig. 5. Percentage pollen diagram of the Sai Island sites (redrawn from Florenzano et al. 2018)

Local Environmental Changes

8700–8500 BC: The palaeoenvironmental picture emerges from site 2-R-66, in the Amara West district dated to the Arkinian period (Florenzano et al. 2016; see also Woodward et al. 2015). The mixed pollen flora including species from southeastern regions testifies that some plant displacement near this area had occurred during the Early Holocene, paralleling Watrin et al. (2009)'s documentation that tropical trees extended their northernmost expansion from 12° to 19° N at c. 9800–5600 BC. Grasses were part of the ephemeral or perennial desert grassland vegetation, as well as the savanna vegetation. The environment was characterised by an open vegetation with a Saharan gravel desert or shrubby vegetation and swamp communities with hygrophilous trees.

7600–7200 BC: Grass cover was again distributed in several habitats including desert savanna and swamp or riverine communities. The xerophilous communities and dry habitats would likely have been positioned just beyond the riparian belt. Even at this time of maximum moisture delivery to the Sahara, we do not see evidence for the local growth of tropical forest species at this time. Vegetation signalling different habitats suggests that there may have been seasonal vegetative growth in different locations—some in response to Nile floods, and some in response to seasonal rainfall.

6400–6250 BC: Wooded savannas appear to have shifted north, such that a good contingent of tropical plants could grow near Sai. A steppe-like environment rich of xerophilous taxa spread during the dry oscillation that characterised this phase.

5000–4800 BC: A desert wooded savanna spread locally while grass cover expanded in grasslands and in the communities typical of the riverine belt and marshy habitats. Floods became less dramatic and even irregular, but still supported well-developed wet environments. The eastern banks of Sai Island had a more open environment, while the western side of the island sustained a wooded grassland or savanna woodland.

2500–1500 BC: At Sai, the spreading of date palms 1800–1700 BC is the first clear evidence of the establishment of anthropogenic environments. Although the desert environment continued to spread as rainfall diminished, the local riverine vegetation continued to grow along the Nile.

Push and Pull: Hinterland and Refugium

Sai area pollen spectra, but especially those from Amara West, suggest that tropical species spread quickly to this area during the Early Holocene. This is in contrast with other archives from the eastern Sahara (e.g., charcoal from Abu Ballas) that suggest gradual ecological changes. Spatial trends of greater aridity to the east and north do not hold for the Nile; the riverine corridor may have abetted the extension of tropical taxa north from their rain-fed distribution areas. Although Sai Island's Early Holocene record has not yielded well-dated evidence for sustained human use until 7600–7200 BC (8-B-10C Level 2), it is possible that such sites underlie the more recent deposits on Sai. Also, Amara West has evidence for human occupation 8700–8500 BC, suggesting that hunter-gatherer-fishers used this stretch of the Nile during the wettest phases of the AHP, even though Saharan environments away from the Nile may have offered attractive sets of resources which people may have seasonally exploited. Subsequent Khartoum Variant horizons (8-B-10C Level 1 and 8-B-76) suggest occupation of Sai intensified as time went on.

Around 5000 BC, the wooded grasslands and grassy woodland environments suggested by the Sai pollen spectra are evidence for increased aridity (Fig. 5), but not as marked as effects seen in other parts of the Eastern Sahara, from the Selima Sandsheet west to Lake Oyo, where lakes shrank dramatically and desert vegetation expanded. Continuing activity of Nile tributaries near Dongola suggests that some eastern Saharan areas still received rainfall or were sustained by large aquifers whose Early Holocene recharges allowed them to continue contributing water even after local precipitation decline; it also suggests that Nile headwaters were still capturing abundant rainfall from the Indian Ocean monsoon. The delayed deterioration of the Sai flora may be due to either of these factors, or simple maintenance of mesic vegetation near a still-active river. Distinguishing among these possibilities would be important to understanding how the last of the Khartoum Variant hunter-gatherers, and the early Abkan pastoral populations, might have juggled their use of riverine versus inland areas.

By 2500 BC, desert environments are clearly expanding at Sai, although grassland and riparian habitats remain. With most Eastern Saharan lake basins drying up by c. 2000 BC, focusing on intensive exploitation of resources near the Nile is the only viable strategy. Sai shows the use of subterranean pits by 3600 BC and the start of an agricultural economy as early as 2700 BC at the specialized storage facility of 8-B-52A (Fig. 4), and presumably human-made stands of date palms by 1800 BC via pollen at 8-B-10A. However, we are lacking knowledge of a critical chronological interval (4500–2800 BC) that might show the exact nature of local environmental changes and human responses leading up to farming and a full focus on Nile resources. Analysis of macroremains from 8-B-52A (dated to 3600–2800 BC) can fill part of this gap, but sites covering earlier parts of this interval (4500–3600 BC), such as 8-B-76 and 8-B-81 have so far not yielded sufficient macroremains for a secure analysis.

Final Remarks

Our knowledge of changing plant uses along the Nile is far from complete, but a few interesting trends and avenues for further research are clear. First, generalized exploitation of grasses and riparian resources continues even after herding takes hold. Second, there seems to be a clear contrast between the Egyptian Nile—where domestic plants occur in contexts and quantities that argue for increasing emphasis on agricultural activities—and the Sudanese Nile, where southwest Asian cereals arrive early in mortuary contexts but economic shifts from mobile herding to sedentary farming and village life take place much later (Hildebrand and Schilling 2016). The first appearances of *Hordeum* and *Triticum* at R12 and El-Ghaba are not accompanied by clear signs of an agricultural economy, such as well-defined storage facilities (subterranean storage pits, above-ground granaries or storage bins) and more permanent village architecture. Although scholars have noted strong deflation in northern Sudan (Reinold 2001), such features *do* appear in the archaeological record in Pre-Kerma times, where they are correctly interpreted as harbingers of a more sedentary lifestyle with a new economic focus on farming (Hildebrand and Schilling 2016). Other material indications of local agricultural production, such as the presence of chaff impressions from domestic plants in pottery, are so far not known from R12 or El-Ghaba, again suggesting that long-distance exchange may have brought wheat and/or barley up the Nile for local consumption at these sites.

What might have caused the delay from the *presence* of domestic plants to a *commitment* to agriculture in Sudan? The paleoenvironmental contexts outlined here can help lay a firmer foundation for future practice-based considerations of agricultural adoption: knowledge of a resource or crop, or occasional consumption, does not necessarily guarantee immediate sweeping economic changes (e.g., Denham and Haberle 2004). One set of events may stimulate initial introduction of a domesticate, followed by decades, centuries, or millennia of occasional, sporadic, or small-scale use; and a completely different set of environmental or social factors may subsequently trigger a profound economic reorientation in which local production and consumption of a crop becomes pervasive. Non-economic factors should also be taken into consideration with regard to the delayed shift to village-based farming along the Sudanese Nile, namely environmental and technical ones. Unlike Egypt, Sudan is located in the summer rain belt, which is not naturally favorable to the cultivation of southwest Asian crops; this would have hindered the adoption of rain-fed (non-irrigated) farming. Well-dated benchmarks for specific environmental changes along the Nile can help evaluate whether environmental events contributed to (or delayed) economic changes.

Multi-scalar examination of paleoenvironmental data from the eastern Sahara and Nile gives important insights about changes in environments, resources, and likely strategies for human livelihoods across the Holocene. As more and more stratigraphic archives yield more precise data, it becomes clear that specific parts of the Sahara experienced distinct changes that were influenced by localized patterns of rainfall and runoff, and substrates and groundwater. Moving beyond the generalized concept of a "greening" and "browning" Sahara to more nuanced readings of local circumstances is essential to providing environmental contexts for specific archaeological sequences, and explaining particular changes in economy and society.

Establishing a detailed palynological record for Sai Island, comparing it to records from other eastern Saharan localities, and exploring its implications for local economic practices is one small step in a massive endeavour. Ideally, such pollen sampling should be applied throughout the Middle Nile to achieve coverage of numerous sites. Riverine sites at different latitudes could be compared to track the advances—and limits—of tropical taxa, and eventual resurgence of desert flora. Such data could provide a much more nuanced picture of the challenges faced by past peoples of the Nile and Eastern Sahara, and the motivations that spurred important innovations in subsistence and social life.

Acknowledgements. Archaeological excavations at Sai Island and Amara West were supported by the National Geographic Society, Committee for Research and Exploration under Grant # 9201-12 and by the University of Cassino and Southern Latium, Italy (grants to EAAG), and the US National Science Foundation (NSF BCS-0519434 to EH). EAAG and EH wish to thank the Sai Island Archaeological Mission (successively directed by Francis Geus, Didier Devauchelle, and Vincent Francigny), and Sudan's National Corporation for Antiquities and Museums, formerly directed by Hassan Hussein Idris and currently directed by Abdelrahman Ali Mohamed, for having granted permissions to conduct fieldwork. In addition, EAAG thanks the Amara West Research Project (directed by Neal Spencer). We thank Martin Williams and an anonymous reviewer for their insights, which were extremely thoughtful and helpful. Lastly, we are grateful to the people of Sai Island and Amara West for their continuous interest and involvement in our field research.

References

Abdelkreem MIM, Ibrahim DA (2016) Check list of Flora and vegetation of an archeological habitat in North Sudan. P J B C 1:1–9

Abuswar AO (2007) Range management. UNESCO chair on desertification. University of Khartoum, Sudan

Ali MM (2006) Shoreline vegetation of Lake Nubia, Sudan. Hydrobiol 570:101–105

Aritzegui D, Asioli A, Lowe JJ et al (2000) Paleoclimate and the formation of sapropel S1: inferences from Late Quaternary lacustrine and marine sequences in the central Mediterranean region. Palaeogeogr Palaeoclimatol Palaeoecol 158:215–240

Barakat HN (1995) Middle Holocene vegetation and human impact in central Sudan: charcoal from the Neolithic site at Kadero. Veg Hist Archaeobot 4:101–108

Bonnet C (1992) Excavations at the Nubian royal town of Kerma: 1975–91. Antiquity 66:611–625

Bonnet C (2006) Les établisments des cultures Kerma. In: Caneva I, Roccati A (eds) Acta Nubica: proceedings of the X international conference of Nubian studies. Istituto Polygrafico e Zecca dello Stato, Roma, pp 15–20

Briois F, Midant-Reynes B, Marchand S et al (2012) Neolithic occupation of an artesian spring: KS043 in the Kharga Oasis. Egypt J Field Archaeol 37(3):178–191

Brook AJ (1956) A note on the ecology of the terrestrial alga *Fritschiella tuberosa* in the Sudan. New Phytol 55:130–132

Buckley S, Usai D, Jakob T et al (2014) Dental calculus reveals unique insights into food items, cooking and plant processing in prehistoric Central Sudan. PLoS ONE 9(7):e100808

Cancellieri E, Cremaschi M, Zerboni A et al (2016) Climate, environment, and population dynamics in Pleistocene Sahara. In: Jones S, Stewart B (eds) Africa from 6-2: population dynamics and palaeoenvironments. Springer, Dordrecht, pp 123–145

Chaix L (2011) Contribution à l'étude de l'économie de la période Pré-Kerma. Premiers résultats sur la faune du site 8-B-10A sur l'Ile de Saï (nord Soudan). In: Rondot V, Alpi F, Villeneuve F (eds) La pioche et la plume–Autour du Soudan, du Liban et de la Jordanie: Hommages archéologiques à Patrice Lenoble. Presses de l'université Paris-Sorbonne, Paris, pp 207–223

Chaix L, Grant A (1993) Palaeoenvironment and economy at Kerma, Northern Sudan, suring the third millennium B.C.: archaeozoological and botanical evidence. In: Krzyżaniak L, Kobusiewicz M, Alexander J (eds) Environmental change and human culture in the Nile Basin and Northern Africa. Poznan, Poznan Archaeological Museum, pp 399–404

Cole JM, Goldstein SL, deMenocal PB et al (2009) Contrasting compositions of Saharan dust in the eastern Atlantic Ocean during the last deglaciation and African Humid Period. Earth Planet Sci Lett 278:257–266

Costa K, Russell J, Konecky B et al (2014) Isotopic reconstruction of the African Humid Period and Congo Air Boundary migration at Lake Tana, Ethiopia. Quat Sci Rev 83:58–67

Cremaschi M (1998) Late Quaternary geological evidence for environmental changes in south-western Fezzan (Libyan Sahara). In: Cremaschi M, di Lernia S (eds) Wadi Teshuinat—palaeoenvironment and prehistory in South-western Fezzan (Libyan Sahara). Quaderni di Geodinamica Alpina e Quaternaria, vol 7. CNR, Roma-Milano, pp 13–47

Cremaschi M (2002) Late Pleistocene and Holocene climatic changes in the central Sahara: the case study of the southwestern Fezzan Libya. In: Hassan, FA (ed) Droughts. Food and culture. Kluwer Academic, New York, pp 65–82

Cremaschi M, di Lernia S (1998) Thegeoarchaeological survey in central Tadrart Acacus and surroundings (Libyan Sahara): environment and cultures. In: Cremaschi M, di Lernia S (eds) Wadi Teshuinat—palaeoenvironment and prehistory in South-western Fezzan (Libyan Sahara). Quaderni di Geodinamica Alpina e Quaternaria, vol 7. CNR, Roma-Milano, pp 243–296

Cremaschi M, di Lernia S (1999) Holocene climatic changes and cultural dynamics in the Libyan Sahara. Afr Archaeol Rev 16:211–238

Cremaschi M, di Lernia S (2001) Environment and settlements in the Mid-Holocene palaeo-oasis of Wadi Tanezzuft (Libyan Sahara). Antiquity 75:815–825

Cremaschi M, Trombino L (1998) The palaeoclimatic significance of paleosols in Southern Fezzan (Libyan Sahara): morphological and micromorphological aspects. CATENA 34:131–156

Cremaschi M, Zerboni A (2009) Early to Middle Holocene landscape exploitation in a drying environment: two case studies compared from the central Sahara (SW Fezzan, Libya). C R Géosci 341:689–702

Cremaschi M, Pelfini M, Santilli M (2006) Cupressus dupreziana: a dendroclimatic record for the middle-late Holocene in the central Sahara. Holocene 16(2):293–303

Cremaschi M, Zerboni A, Spötl C et al (2010) The calcareous tufa in the Tadrart Acacus Mt. (SW Fezzan, Libya): An early Holocene palaeoclimate archive in the central Sahara. Palaeogeogr Palaeoclim Palaeoecol 287:81–94

Cremaschi M, Zerboni A, Mercuri AM, Olmi L, Biagetti S, di Lernia S (2014) Takarkori rock shelter (SW Libya): an archive of Holocene climate and environmental changes in the central Sahara. Quat Sci Rev 101:36–60

Damnati B (2000) Holocene lake records in the Northern Hemisphere of Africa. J Afr Earth Sci 31:253–262

deMenocal PB, Tierney JE (2012) Green Sahara: African humid periods paced by Earth's orbital changes. Nat Educ Knowl 3(10):12

deMenocal P, Ortiz J, Guilderson T et al (2000) Abrupt onset and termination of the African humid period: rapid climate response to gradual insolation forcing. Quat Sci Rev 19:347–361

Denham T, Haberle S (2004) Agricultural emergence and transformation in the Upper Wahgi valley, Papua New Guinea, during the Holocene: theory, method and practice. Holocene 2008; 18:481–496

D'Ercole G, Budka J, Sterba JH et al (2017a) The successful 'recipe' for a long-lasting tradition: Nubian ceramic assemblages from Sai Island (northern Sudan) from prehistoric times to the New Kingdom period. Antiquity 91:24–42

D'Ercole G, Garcea EAA, Eramo G et al. (2017b) Variability and continuity of ceramic manufacturing of ethnographic productions since prehistory in Upper Nubia, Sudan. J Archaeol Sci Rep 16:553–563. http://dx.doi.org/10.1016/j.jasrep.2017.04.012

D'Ercole G, Eramo G, Garcea EAA et al (2015) Raw material and technological changes in ceramic productions at Sai Island, Northern Sudan from the 7th to the 3rd Millennium BC. Archaeometry 57(4):597–616

Despagne R (nd) Topographic data files for Sai Island. Geus archives, Lille

Drake NA, Blench RM, Armitage SJ et al (2011) Ancient watercourses and biogeography of the Sahara explain the peopling of the desert. Proc Nat Acad Sci 108:458–462

Fahmy AG (2003) Palaeoethnobotanical studies of Egyptian Predynastic cemeteries: new dimensions and contributions. In: Neumann K, Butler A, Kahlheber S (eds) Food, fuel and fields: progress in African archaeobotany. Heinrich-Barth-Institut, Köln, pp 95–106

Florenzano A, Mercuri AM, Altunoz M et al (2016) Palynological evidence of cultural and environmental connections in Sudanese Nubia during the Early and Middle Holocene. Quatern Int 412B:65–80. https://doi.org/10.1016/j.quatint.2016.01.001

Florenzano A, Mercuri AM, Fornaciari R et al (2018) Plants, water and humans: pollen analysis from Holocene archaeological sites on Sai Island, northern Sudan. Palynology (online). https://doi.org/10.1080/01916122.2017.1384411

Fontes JCh, Gasse F, Callot Y et al (1985) Freshwater to marine-like environments from Holocene lakes in northern Sahara. Nature 317:608–610

Fuller DQ, Smith L (2004) The prehistory of the Bayuda: new evidence from the Wadi Muqaddam. In: T Kendall (ed) Nubian studies 1998: proceedings of the ninth conference of the international society of Nubian Studies Northwestern University, Boston, pp 265–281

Garcea EAA (2006a) Semi-permanent foragers in semi-arid environments of North Africa. World Archaeol 38(2):197–219

Garcea EAA (2006–2007) The Holocene prehistory at Sai Island, Sudan. In: Gratien, B (ed) Mélanges offerts à Francis Geus. Cahiers de Recherches de l'Institut de Papyrologie et d'Egyptologie de Lille 26:107–113

Garcea EAA (2011–2012) Revisiting the Khartoum Variant in its landscape. Cahiers de Recherches de l'Institut de Papyrologie et d'Egyptologie de Lille 29:139–150

Garcea EAA (2016a) Multi-stage dispersal of Southwest Asian domestic livestock and the path of pastoralism in the Middle Nile Valley. Quatern Int 412B:54–64

Garcea EAA (2016b) Semi-permanent foragers in North Africa–An archaeological perspective. In: Tvedt T, Oestigaard T (eds) A history of water, Series 3. Water and Food From Hunter-Gatherers to Global Production in Africa, vol 3. I.B. Tauris, London, pp 29–54

Garcea EAA (in press) The Southern periphery of Egypt in the Predynastic period: Nubia in the 5th and 4th Millennium BC. In: Knabb K, Boswell A (eds) Life at the margins of the state: comparative landscapes from the Old and New World. University Press of Colorado, Boulder

Garcea EAA, Hildebrand EA (2009) Shifting social networks along the Nile: middle Holocene ceramic assemblages from Sai Island, Sudan. J Anthropol Archaeol 28:304–322

Garcea EAA, Karul N, D'Ercole G (2016a) Southwest Asian domestic animals and plants in Africa: routes, timing and cultural implications. Quatern Int 412B:1–10

Garcea EAA, Mercuri AM, Giraudi C (2013) Archaeological and environmental changes between 9500 BP and 4500 BP: a contribution from the Sahara for the understanding of expanding droughts in the "Great Mediterranean". Annali di Botanica (Roma) 3:115–120

Garcea EAA, Wang H, Chaix L (2016b) High-precision radiocarbon dating application to multi-proxy organic materials from late foraging to early pastoral sites in Upper Nubia. Sudan J Afr Archaeol 14(1):83–98

Garzanti E, Andò S, Padoan M et al (2015) The modern Nile sediment system: processes and products. Quat Sci Rev 130:9–56

Gasse F (2000) Hydrological changes in the African tropics since the last glacial maximum. Quat Sci Rev 19:189–211

Gasse F (2002) Diatom-inferred salinity and carbonate oxygen isotopes in Holocene waterbodies of the western Sahara and Sahel. Quat Sci Revi 21:737–767

Gasse F, Roberts CN (2004) Late Quaternary hydrologic changes in the arid and semiarid belt of Northern Africa. Implications for past atmospheric circulation. In: Diaz HF, Bradley RS (eds) The Hadley circulation: present, past and future. Kluwer Academic, Dordrecht, pp 313–345

Gatto MC (2006) The Nubian A-Group: a reassessment. Archéo-Nil 16:61–76

Gatto MC, Zerboni A (2015) Holocene supra-regional environmental changes as trigger for major socio-cultural processes in northeastern Africa and the Sahara. Afr Archaeol Rev 32:301–333

Geus F (1998) Saï 1996-1997. Archéologie du Nil Moyen 8:85–126

Geus F (2002) Saï 1998–1999. Archéologie du Nil Moyen 9:95–134

Giraudi C, Mercuri AM, Esu D (2013) Holocene palaeoclimate in the northern Sahara margin (Jefara Plain, Northwestern Libya). Holocene 23:339–352

Haynes CV, Eyles CH, Pavlish LA et al (1989) Holocene palaeoecology of the Eastern Sahara; Selima Oasis. Quat Sci Rev 8:109–136

Hely C, Lézine A-M, APD contributors (2014) Holocene changes in African vegetation: tradeoff between climate and water availability. Climate of the Past 10:681–686

Hennekam R, Jilbert T, Schnetger B et al (2014) Solar forcing of Nile discharge and sapropel S1 formation in the early to middle Holocene eastern Mediterranean. Paleoceanography 29:343–356

Hesse A (1996) Reconnaissance archéologique sur les rives du Nil: le cas du site SKP1 de l'île de Saï. Préhistoire Archéologie Méditerranéennes 5:105–110

Hildebrand EA (2006–2007) The significance of Sai Island for early food production in Sudan. Cahiers de Recherches de L'Institut de Papyrologie et d'Egyptologie de Lille 26:173–182

Hildebrand EA, Schilling TM (2016) Storage amidst early agriculture along the Nile: perspectives from Sai Island, Sudan. Quatern Int 412B:81–95

Hoelzman P, Keding B, Berke H et al (2001) Environmental change and archaeology: lake evolution and human occupation in the Eastern Sahara during the Holocene. Palaeogeogr Palaeoclimatol Palaeoecol 169:193–217

Hoelzman P, Schwald A, Roberts N et al (2010) Hydrological response of an east-Saharan palaeolake (NW Sudan) to early-Holocene climate. Holocene 20:537–549

Honegger M (2004a) The Pre-Kerma: a cultural group from Upper Nubia prior to the Kerma civilisation. Sudan Nubia 8:38–46

Honegger M (2007) Aux origines de Kerma. Genava, n.s. 55:201–212

Honegger M, Williams M (2015) Human occupations and environmental changes in the Nile valley during the Holocene: the case of Kerma in Upper Nubia (northern Sudan). Quat Sci Rev 130:141–154

Horowitz A (1992) Palynology of Arid Lands. Elsevier, Amsterdam

Iacumin P, Bocherens H, Chaix L et al (1998) Stable carbon and nitrogen isotopes as dietary indicators of ancient Nubian populations (Northern Sudan). J Archaeol Sci 25:293–301

Jahns S (1995) A Holocene pollen diagram from El Atrun, northern Sudan. Veg Hist Archaeobot 4:23–30

Kröpelin S, Verschuren D, Lézine A-M et al (2008) Climate-driven ecosystem succession in the Sahara: the past 6000 years. Science 320:765–768

Kuper R, Kröpelin S (2006) Climate-controlled holocene occupation in the Sahara: motor of Africa's evolution. Science 313:803–807

Lal BB (1967) Indian archaeological expedition to Nubia, 1962: a preliminary report. Fouilles en Nubie 1961–1963:99–118

Lamb HF, Bates CR, Coombes PV et al (2007) Late Pleistocene desiccation of Lake Tana, source of the Blue Nile. Quat Sci Rev 26:187–299

Lange M (2007) Development of pottery production in the Laqiya region, Eastern Sahara. CRIPEL (Cahiers de Recherches de L'Institut de Papyrologie et d'Egyptologie de Lille) 26:243–251

Lario J, Sanchez-Moral S, Fernandez V et al (1997) Palaeoenvironmental evolution of the blue Nile (Central Sudan) during the early and mid-Holocene (Mesolithic-Neolithic transition). Quat Sci Rev 16:583–588

Lézine A-M (1989) Late Quaternary vegetation and climate of the Sahel. Quat Res 32:317–334

Lézine A-M (2007) Pollen records, postglacial. In Scott, EA (ed) Encyclopaedia Quaternary Science, vol. 4, pp 2682–2698

Lézine A-M (2009) Timing of vegetation changes at the end of the Holocene humid period in desert areas at the northern edge of the Atlantic and Indian monsoon systems. CR Geosci 341:750–759

Lézine A-M, Hely C, Grenier C et al (2011) Sahara and Sahel vulnerability to climate changes, lessons from Holocene hydrological data. Quat Sci Rev 30:3001–3012

López-Sáez JA, López-Garcia P (2003) Pollen analysis from Early and Middle Holocene archaeological sites in the Blue Nile area, Central Sudan. Complutum 14:397–400

Macklin MG, Lewin J (2015) The rivers of civilization. Quat Sci Rev 114:228–244

Macklin MG, Toonen WHJ, Woodward JC et al (2015) A new model of river dynamics, hydroclimatic change and human settlement in the Nile Valley derived from meta-analysis of the Holocene fluvial archive. Quat Sci Rev 130:109–123

Macklin MG, Woodward JC, Welsby DA et al (2013) Reach-scale river dynamics moderate the impact of rapid Holocene climate change on floodwater farming in the desert Nile. Geology 41:695–698

Madani I, Tahir IF, Hamdeen HM et al (2015) Vegetation ecology and taxonomy of El-Ga'ab area, north-western Sudan. Europ Acad Res 3:2927–2943

Madella M, García-Granero JJ, Out WAD et al (2014) Microbotanical evidence of domestic cereals in Africa 7000 years ago. PLoS ONE 9:e110177

Magid AA (2003) Exploitation of food-plants in the Early and Middle Holocene Blue Nile area, Sudan and neighbouring areas. Complutum 14:345–372

Maley J (1981) Etudes palynologiques dans le bassin du Tchad et paléoclimatologie de l'Afrique nord-tropicale de 30 000 ans à l'époque actuelle. Travaux et documents de l'ORSTOM 129, Paris

Maley J (2010) Climate and Palaeoenvironment evolution in north tropical Africa from the end of the Tertiary to the Upper Quaternary. Palaeoecol Afr 30:227–278

Maley J, Vernet R (2015) Populations and climatic evolution in north tropical Africa from the end of the Neolithic to the dawn of the modern era. Afr Archaeol Rev 32:179–232

Manning K, Timpson A (2014) The demographic response to Holocene climate change in the Sahara. Quat Sci Rev 101:28–35

Mercuri AM (2008) Human influence, plant landscape, evolution and climate inferences from the archaeobotanical records of the Wadi Teshuinat area (Libyan Sahara). J Arid Environ 72:1950–1967

Mercuri AM, Florenzano A, Giraudi C et al (2015) Humans and water in desert 'refugium' areas: palynological evidence of climate oscillations and cultural developments in early and mid-Holocene Saharan edges. IANSA–Interdisciplinaria Archaeologica 6:151–160

Mercuri AM, Sadori L, Uzquiano Ollero P (2011) Mediterranean and north-African cultural adaptations to mid-Holocene environmental and climatic change. Holocene 21:189–206

Mustafa MA (2007) Desertification processes. UNESCO chair on desertification. University of Khartoum, Sudan

Neumann K (1989) Holocene vegetation of eastern the Sahara: Charcoal from prehistoric sites. Afr Archaeol Rev 7:97–116

Nicoll K (2004) Recent environmental change and prehistoric human activity in Egypt and Northern Sudan. Quat Sci Rev 23:561–580

Nordström H-A (1972) Neolithic and A group sites. Scandinavian University, Uppsala

Nourelbait M, Rhoujjati A, Benkaddour A et al (2015) Climate changes since the mid-Holocene in the Middle Atlas, Morocco. Clim Past Discuss 11:4097–4121

Out WA, Ryan P, García-Granero JJ et al (2016) Plant exploitation in Neolithic Sudan: a review in the light of new data from the cemeteries R12 and Ghaba. Quatern Int 412B:36–53

Pachur HJ, Hoelzmann P (2000) Paleoclimatic implications of late Quaternary lacustrine sediments in Western Nubia, Sudan. Quat Res 36:257–277

Reinold J (2001) Kadruka and the Neolithic in the northern Dongola Reach. Sudan & Nubia 5:2–10

Reinold J (2006) Les cimetières préhistoriques au Soudan – cotoumes funéraires et systemes sociaux. In: Caneva I, Roccati A (eds) Acta Nubica: proceedings of the X international conference of Nubian Studies. Istituto Polygrafico e Zecca dello Stato, Roma, pp 139–162

Ritchie JC (1994) Holocene pollen spectra from Oyo, northwestern Sudan: problems of interpretation in a hyperarid environment. Holocene 4:9–15

Ritchie JC, Haynes CV (1987) Holocene vegetation zonation in the eastern Sahara. Nature 330:645–647

Ritchie JC, Eyles CH, Haynes CV (1985) Sediment and pollen evidence for an Early to Mid-Holocene humid period in the Eastern Sahara. Nature 314:352–355

Rowley-Conwy P (1989) Nubia AD 0–550 and the "Islamic" agricultural revolution: preliminary botanical evidence from Qasr Ibrim, Egyptian Nubia. Archéologie du Nil Moyen 3:131–138

Salvatori S, Usai D (eds) (2008) A Neolithic cemetery in the Northern Dongola Reach (Sudan): Excavation at Site R12. Sudan Archaeological Research Society, London

Salvatori S, Usai D, Zerboni A (2011) Mesolithic site formation and palaeoenvironment along the White Nile (Central Sudan). Afr Archaeol Rev 28(3):177–211

Schild R, Wendorf F (2010) Late Palaeolithic Hunter-Gatherers in the Nile Valley of Nubia and Upper Egypt. In: Garcea EAA (ed) South-Eastern Mediterranean peoples between 130,000 and 10,000 years ago. Oxbow Books, Oxford, pp 89–125

Schulz E (1991) Holocene environments in the central Sahara. Hydrobiol 214:359–365

Stemler ABL (1990) A Scanning Electron Microscopic Analysis of Plant Impressions in Pottery from the Sites of Kadero, El Zakiab, Um Direiwa and El Kadada. Archéologie du Nil Moyen 4:87–106

Tierney JE, deMenocal PB (2013) Abrupt shirts in the Horn of Africa hydroclimate since the last glacial maximum. Science 342:843–846

Watrin J, Lézine A-M, Hély C et al (2009) Plant migration and ecosystems at the time of the "green Sahara". CR Geosci 341:656–670

Wendrich W, Taylor RE, Southon J (2010) Dating stratified sites at Kom K and Kom W: fifth millennium BCE radiocarbon ages for the Fayum Neolithic. Nucl Instrum Methods 268(7–8):999–1002

Wetterstrom W (1993) Foraging and farming in Egypt. In: Shaw T, Sinclair PJJ, Andah B, Okpoko A (eds) The archaeology of Africa: food, metals and towns. Routledge, London, pp 165–226

White F (1986) La végétation de l'Afrique. UNESCO, ORSTOM

Williams MAJ (2009) Late Pleistocene and Holocene environments in the Nile basin. Global Planet Change 69:1–15

Williams MAJ, Talbot M, Aharon P et al (2006) Abrupt return of the summer monsoon 15,000 years ago: new supporting evidence from the lowe White Nile valley and Lake Albert. Quat Sci Rev 25:2651–2665

Williams MAJ, Williams FM, Duller GAT et al (2010) Late Quaternary floods and droughts in the Nile valley, Sudan: new evidence from optically stimulated luminescence and AMS radiocarbon dating. Quat Sci Rev 29:1116–1137

Williams MAJ, Usai D, Salvatori S et al (2015a) Late Quaternary environments and prehistoric occupation in the lower White Nile valley, central Sudan. Quat Sci Rev 130:72–88

Williams MAJ, Duller GAT, Williams FM et al (2015b) Causal links between Nile floods and eastern Mediterranean sapropel formation during the past 125 kyr confirmed by OSL and radiocarbon dating of Blue and White Nile sediments. Quat Sci Rev 130:89–108

Woodward J, Macklin M, Fielding L et al (2015) Shifting sediment sources in the world's longest river: a strontium isotope record for the Holocene Nile. Quat Sci Rev 130:124–140

Wuttmann M, Briois F, Midant-Reynes B et al (2012) Dating the end of the Neolithic in an Eastern Sahara Oasis: modeling absolute chronology. Radiocarbon 54(3–4):305–318

Zerboni A, Trombino L, Cremaschi M (2011) Micromorphological approach to polycyclic pedogenesis on the Messak Settafet plateau (central Sahara): Formative processes and palaeoenvironmental significance. Geomorphol 125:319–335

Zerboni A, Biagetti S, Lancelotti C, Madella M (2016) The end of the Holocene Humid Period in the central Sahara and Thar deserts: Societal collapses or new opportunities? PAGES Mag 24 (2):60–61. https://doi.org/10.22498/pages.24.2.60

Archaeological Deposit in Motako, Southwest Nigeria, Investigated by Pollen Analysis

Kingsley C. Daraojimba[1(✉)], Philip A. Oyelaran[1], Marcia A. de Barros[2], Jeanne Cordeiro[3], and Cynthia F. Pinto da Luz[4]

[1] Department of Archaeology and Anthropology, University of Ibadan, Ibadan, Nigeria
kingsleyjohnbosco@yahoo.com, philipoyelaran04@yahoo.co.uk

[2] Laboratório de Palinologia, Department of Geology, Universidade Federal do Rio de Janeiro, Rio de Janeiro, Brazil
marcabarros@yahoo.com

[3] Laboratório de Arquelogia Brasileira, Rio de Janeiro, Brazil
jeanne.cordeiro@hotmail.com

[4] Núcleo de Pesquisa em Palinologia, Instituto de Botânica, São Paulo, Brazil
cyluz@yahoo.com.br

Abstract. Pollen, in association with phosphate and archaeological data from a 120 cm deep midden deposit provide insight into site formation processes at Motako. Twelve samples were retrieved from the five distinct stratigraphic layers and subjected to pH and phosphate analyses, and pollen with some archaeobotanical analyses. Results showed that environmental factors influenced the oxidation of the deposit in addition to fungal activity, hence the low pollen content recorded. Five pollen zones were established according to the pollen, phosphate, lithological and archaeological data. The identified palynomorphs were dominated by fungal spores and Amaranthaceae pollen. Moreover, ruderals, including species of ethnobotanical relevance, and *Elaeis guineensis* pollen were identified, indicating open vegetation initiated possibly by human activity. Zone 1 (110–85 cm) was characterised by low phosphate content, coupled with archaeological materials, broken palm kernels, and low pollen diversity suggesting the area was likely used for activities related to palm oil production. At Zone II (85–70 cm), we suspected that the deposit had gradually been turned into a midden heap as indicated by increase in organic matter content which corresponded to the increased concentration of phosphate and cultural materials, in addition to the increase of pollen of ruderal species. In Zone III (70–45 cm) the midden deposit was fully established with high activity as reflected in the peak in phosphate content, increased diversity of kitchen wares, organic remains and pollen of ethnobotanical significance. The midden was probably beginning to be deactivated in Zone IV (45–25 cm) and/or used for another activity such as gardening for the cultivation of plants with several domestic purposes. Zone V (25–0 cm) shows a reduction of both organic and inorganic materials. The varying concentrations of archaeological material, phosphate and pollen content suggest changing site activities across the

A. M. Mercuri et al. (Eds.): *Plants and People in the African Past*,
https://doi.org/10.1007/978-3-319-89839-1_13

sedimentary units, depicting changes in land use and management over time for refuse disposal. The occurrence of decreasing concentration of organic waste towards the top of the sediment profile in tandem with phosphate concentration was considered to be an indicator of the gradual abandonment of the midden.

Keywords: Pollen · Midden · Charcoal particles · Motako · Ruderal species Economic plants · Nigeria

Introduction

Studies involving the investigation of archaeobotanical microremains (pollen, spores and other microfossils) or macroremains (seed, leaf, wood, flower, fruit, bark, root and more) from archaeological sites have been useful for understanding environmental settings of human cultures and the role played by humans in modifying their environment. In addition, such plant remains have the potential of providing vital information on diet, subsistence, warfare, technology, trading activities and even religious aspects of life. These information equip the archaeologist to speak more confidently about many aspects of past human cultures (e.g. Bryant and Holloway 1983; Mercuri et al. 2015). One of the many areas of archaeobotanical studies now attracting attention is the extraction, examination and interpretation of on-site fossil pollen and spores through pollen analytical procedures. This procedure has been used to generate data not only for reconstructing past vegetation, but also in providing insight into the impact of humans on the vegetation (Shackley 1981; Faegri and Iversen 1989; Mercuri 2014). In this paper, we explore the potential use of pollen analysis in the study of a historic-era midden feature at a local community, Motako, situated in the southwestern part of Nigeria. This approach is rare for the area and the quite recent times investigated. We adopted it to understand different aspects of site formation processes and human activities that feed into it.

Motako is a place of a Yoruba speaking society, one of the many communities in Southwest Nigeria occupied by the people of Owu. Amongst other Yoruba speaking groups in Nigeria, Owu is one of the earliest and important (Mabogunje and Omer-Cooper 1971). However, this formidable town was sacked during the internecine war in the early part of the nineteenth century by the allied forces of Ife, Oyo and Ijebu (Akintoye 2010). The war outbreak in primordial Owu town (which arguably is claimed to be Motako) led to the dispersal of its inhabitants into communities in different parts of Nigeria. The resettlement of the Motako space had no significant effect on their settlement patterns except for the introduction of foreign settlers, such as people of Ibadan origin, making up the population. Economically speaking, Motako is not as affluent as its neighbor, Orile-Owu, where a portion of its inhabitants resettled after the invasion. This difference in economic status is attributed mainly to the larger space and population density of Orile-Owu which supports diverse commercial activities. Presently, not much has been done archaeologically or archaeobotanically to investigate these people with the intention of understanding aspects of their cultural and socio-economic life in regard to the use of space. Although archaeobotanical studies in Nigeria are still a growing field of research, several archaeobotanical studies have been conducted in the northern part of Nigeria (Breunig and Neumann 1999; Klee et al. 2000, 2004; Zach and Klee 2003), in the Southeast (Umeji et al. 2012) and Southwest (Orijemie et al. 2010; Orijemie and Sowunmi 2014; Orijemie 2015), as well as in the central region (Kahlheber et al. 2009; Höhn and Neumann 2014, 2016). One of the earliest published works was part of a multidisciplinary study of the Kariya area in Northeast Nigeria (Sowunmi 1991). The aforementioned works have studied archaeobotanical remains to determine the origin of plant remains through morphological criteria; investigate subsistence economy, dry phases, Late Holocene climatic changes, paleovegetation and growing densities of human population and their environmental implications. Our investigations at Motako seek to understand behaviors surrounding site formation through a study of fossil palynomorphs, archaeological remnants and phosphate contained in the midden feature. Identified pollen taxa are also examined in the light of extant ethnobotanical data.

Environmental Setting, Economic Activity and Land Use in Motako

The Motako site (Lat. 7°16′15′′N–7°18′15′′N, and Long. 4°17′20′′E–4°18′20′′E) is located inside the Ayedaade local government area of Osun State, southwestern part of Nigeria (Fig. 1). Motako is close to another Owu town, Orile-Owu, having an average annual temperature of 26.5 °C and 1405 mm of rainfall annually with double maxima in June and September (https://en.climate-data.org/location/371279/). It has been pointed out by Swaine and Hall (1986) that forest vegetation in West Africa varies in relation to rainfall. Given the region as mapped by White (1983), Motako area is floristically located in the Guineo-Congolian center of endemism (Fig. 1) and the vegetation type is characterized by drier peripheral semi-evergreen Guineo-Congolian rain forest. The ecology of identified taxa are attributed after Lebamba et al. (2009), Ngomanda et al. (2009), and White (1983). Also, there are areas of cultivation with *Cola gigantea* A. Chev, *Cola nitida* (Vent.) Schott & Endl., *Theobroma cacao* L., *Carica papaya* L. and herbaceous stands of *Musa* x *paradisiaca* L. and *Musa* x *sapientum* L., and secondary forests of different ages with pioneer species such as *Elaeis guineensis* Jacq., *Trema orientalis* (L.) Blume, *Rauvolfia vomitoria* Afzel, *Asystasia gangetica* (L.) T. Anderson, *Cyathula prostrata* (L.) Blume, *Luffa cylindrica* (L.) M. Roem, and *Petiveria alliacea* L. Notwithstanding that population density in Motako is relatively sparse, portions of the forest have been modified by slash-and-burn agriculture and moderate logging activities. The comparative openness of the forest canopy initiated by these anthropogenic activities, encourages the growth of secondary forest species. Abundance of stands of oil palm trees in the study area is associated with its socio-economic relevance to the people of the area. Major products of the oil palm exploited by people of the area are its fruit and sap which are rich sources of edible oil and wine, respectively usually used as part of their livelihood (Sowunmi 1999; Corley and Tinker 2003; Daraojimba 2016). Palm oil production is a common activity in the local area and close to the excavated site a local palm oil production centre is located (Fig. 2) with its surroundings characterized by heaps of palm kernel shells from the oil extraction exercise.

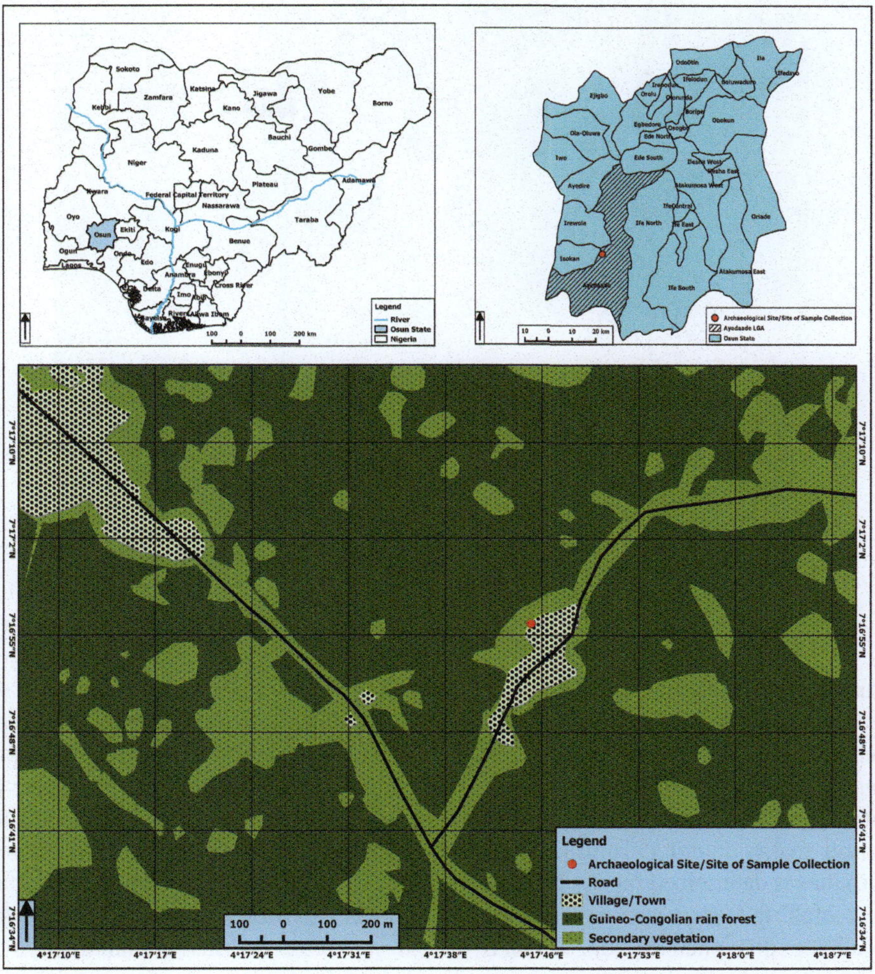

Fig. 1. Map of modern vegetation with archaeological site at Motako

The cause for the strong rise of pollen of oil palm during the Late Holocene is still an on-going debate. Sowunmi (1999) regards this increase as an indicator of human activities, because humans, through slash-and-burn agriculture and other mechanisms of forest disturbance, create favorable conditions for oil palm to thrive. On the other hand, Maley and Chepstow-Lusty (2001) have strongly contested this standpoint. They reasonably argued that the marked increase in oil palm pollen cannot be used as an index of human activities nor plant cultivation, to them climatic disturbance of the vegetation and drier conditions are the contributors to the oil palm expansion.

In Motako, farming is usually complemented with hunting and it is noteworthy that the farmers are also hunters; however, most farmers practice hunting as pastime. Farmers concentrate on intense farming activities during the wet season (March through to November) but switch over to hunting during the dry season. This could be

Fig. 2. Local palm oil production industry

attributed to availability of water supply through rainfall for farming during the wet season and the increased thickness of the forest which makes penetration difficult hence making hunting for games near impossible. On the contrary, inadequate water supply during the dry season (December to March) discourages intensive farming and the burning of the forest vegetation to drive out games from cover associated with this period makes hunting fruitful. The lifestyle of the people with regard to farming activities is determined mostly by the availability of land and partly because of shortage in rainfall. In comparison with towns, village areas like Motako receive more rainfall. Realizing this phenomenon, Motako people capitalize on this rainfall pattern by commencing farming activity in the second week of the third month of every year when rainfall should have set in. Land acquisition is usually either through purchase, gift or inheritance. Just like every other part of Nigeria, land ownership in Motako is serious matter and so, unauthorized land encroachment would be seen as a serious crime. Farmers therefore would only farm on their land (regardless of location or distance from place of residence) or, with appropriate permission from concerned authority, on another person's land. This has serious implications for farming activities in the area because farming is limited by availability of space.

Materials and Methods

Twelve sediment samples were collected from the northern wall of a 2 m × 2 m excavated midden feature of 120 cm depth at Motako. The stratigraphy of the excavated pit was delineated into five different strata based on color, texture, structure and consistency (Fig. 3). Macroremains were observed under a stereomicroscope. The samples were collected from varying depth intervals such that each of the delineated strata was represented in the samples, with the help of hand trowels washed with clean water at intervals. It is usually advised to start sample collection from the bottom layer so as to avoid contamination. Twenty grams of each sample were submitted to standard treatment (Faegri and Iversen 1989) (60% HF, 36% HCl, $ZnCl_2$/HCl [specific gravity: 2.0], 0.5% HCl) for pollen and spore recovery. A large amount of sediment was used to offset the poor preservation of archaeobotanical materials in the samples studied. 100% glycerin was added to the residue and 10 μl volume of the resulting mixture was mounted on duplicates of temporary slides. Pollen grains and spores were counted besides microcharcoal particles (Kangur 2002).

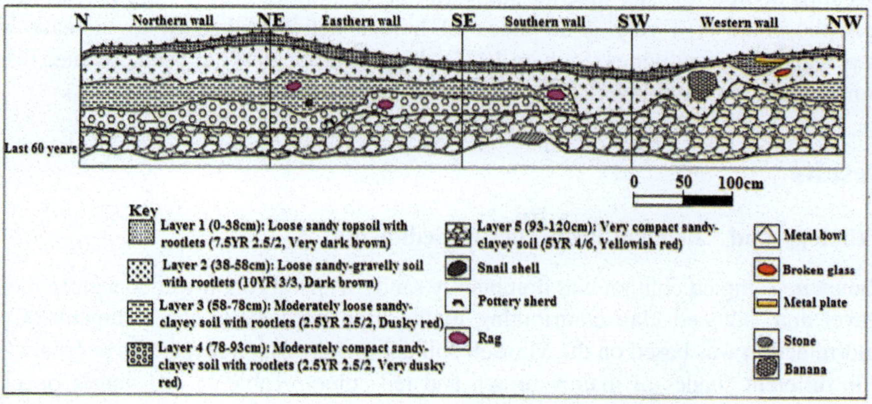

Fig. 3. Stratigraphy of Motako excavated pit

Identification of palynomorphs was done using light microscopy at 400x magnification based on the 3600 reference pollen slides collection of 187 families of extant species and unpublished photomicrographs housed at the Palynology Laboratory in the Department of Archaeology and Anthropology at the University of Ibadan, and published literatures (APLF 1974; Salard-Cheboldaeff and Locquin 1980; van Geel et al. 2011; Gosling et al. 2013). The term "spore/pollen type" was used and refers to a single plant species or a group of species, or higher taxa, presenting similar pollen morphology. Plant taxa represented in the pollen assemblages were assigned to the biomes to which they have the greatest resemblance (Lézine et al. 2009). Pollen grains whose ecological significance could not be determined because its plant sources are or seemingly are everywhere were categorized as 'ubiquitous'. Similarly, pollen grains that were not known were grouped as 'unidentified', while those whose morphological features could not be discerned due to their poor state of preservation were classified as 'unidentifiable'. Selected taxa such as *Elaeis guineensis*, *Alchornea*, Poaceae, fungal and fern spores were considered separately in the pollen diagram due to their ecological importance. *E. guineensis* is noted for its association with forest clearance (Sowunmi

1981). Both *Alchornea* species and oil palm are recognized as pioneer forest taxa (Elenga et al. 2004; Sowunmi 2004).

Pollen sum includes all pollen grains counted. Charcoal particles, fungal spores and fern spores were excluded from the pollen sum but expressed separately as a percentage of the pollen sum. TILIA and TILIAGRAPH software were used for demonstration of the pollen grains, fern spores, fungal spores and charcoal particles occurrence (Grimm 1987). Ethnobotanical and economic data of plant usage in the area were collected mainly from literature (Ebana et al. 1991; Verger 1995; Weniger et al. 2004; Chukwuma et al. 2015; Daraojimba 2016) and anecdotal sources of oral informants.

Additionally, pH analysis of the samples was carried out with a Jenway 3520 electronic pH meter, whilst the phosphate content of the samples was determined through the Mehlich-3 extraction procedure and the colorimetric method. pH analysis of the samples was done in order to determine the suitability of samples for palynomorph preservation. Soil phosphate is a very good index of human activities known to increase with increasing human activities (Sjoberg 1976). A radiocarbon date has been obtained from a sample from the basal layer (110–120 cm) of the midden deposit (see results).

Results

Lithology and Chemical Properties of Sediments

The entire sampled column was dominantly sandy with varying lower concentrations of gravel and, silt and clay contributing to the high porosity of the sediments. Color determination was based on the Munsell soil color chart which revealed five color types with different shades of mainly brown and red colors (Table 1). pH values occurred within a range of acidic (6.7) to alkaline (7.2), while phosphate content decreased in concentration from top to mid-way of the sequence, followed by an increase and a subsequent decrease to the lowest values. This pattern of phosphate concentration runs parallel to the pattern with the concentration of archaeological materials in each of the sampled levels, especially the organic remains (Table 1).

Radiocarbon Date

The sample collected from the basal layer (110–120 cm) of the midden deposit was radiocarbon dated at Beta Analytic, Florida, U.S.A. The date obtained from the sample is represented in pMC (percent modern carbon) indicating that the materials dated were not older than the last 60 years in age (Table 2).

Palynomorph Types

A total of 31 pollen types and three spore types: monolete, trilete (*Pteris*) and fungal spores, were recognized (Table 3). Pollen frequency was low compared to fungal spores (Fig. 4). Palynomorphs counted ranged from 215 (80 cm) to 3462 (23 cm), while charcoal particles counted comprised from 280 (110 cm) to 17 particles (4 cm). Some of the collected samples showed oxidation as assigned to their yellowish and reddish colors (Hassan 1978), perhaps due to high porosity of the sediments.

Table 1. Lithological, chemical, palynomorph and archaeological characteristics of Motako sediments

Depth (cm)	Color	Soil type	pH	Phosphate content (mg/kg P)	Archaeological details	Palynomorphs
0–25	Very dark brown (10 YR 3/3)	Sandy	7.2–6.8	936.3	**Organic**: broken palm kernel shell, intact palm kernel shell (15), charcoal, snail shell (9) **Inorganic**: Potsherds (253), metal scraps, broken bottles, plastic scraps, small plastic pipe (1), bead (1)	Amaranthaceae, ***Elaeis guineensis***, *Tetrorchidium didymostemon*, Cyperaceae, ***Alchornea***, Poaceae, fern spore, fungal spore and charcoal.
25–45	Dark brown (10 YR 3/3)	Sandy-gravelly	6.7	835.0	**Organic**: indeterminate animal bones (8), broken palm kernel shells, intact palm kernel shells (10), charcoal, snail shells (7) **Inorganic**: potsherds (193), metal scraps, metal bowl (1), beads (2), cowrie (1), broken bottles, plastic scraps, dry cell battery (1), bead strand (1), plastic button of cloth (1), shoe ornament (1), hair comb (1),	***Elaeis guineensis***, Amaranthaceae, ***Allophylus africanus***, ***Paullinia pinnata***, ***Rauvolfia vomitoria***, ***Alstonia boonei***, ***Alternanthera sp***, Asteraceae, ***Chromolaena odorata***, ***Talinum triangulare***, Cyperaceae, Anacardiaceae, cf. ***Canthium setosum***, Combretaceae-Melastomataceae, *Justicia cf. glabra*, ***Luffa cylindrica***, Myrtaceae, *Myrcia* type, Poaceae, fern spore, fungal spore and charcoal.
45–70	Dusky red (2.5 YR 2.5/2)	Sandy-clayey	6.8	1135.0	**Organic**: indeterminate animal tooth (4), indeterminate animal bones (11) and lower jaw bone (1), thoracic vertebrae of rodent (1), broken palm kernel shell, intact palm kernel shell (4) charcoal, snail shell (1) **Inorganic**: potsherds (448), plastic button of cloth (4), metal bowl (1), ear-ring (2), beads (15), aluminum cooking spoon (1), Muller (1), bowl of smoking pipe (1), stem of smoking	***Elaeis guineensis***, Amaranthaceae, *Podocarpus*, ***Adenia cissampeloides***, *Tetrorchidium didymostemon*, Asteraceae, ***Oldenlandia corymbosa***, ***Talinum triangulare***, cf. ***Canthium setosum***, ***Sida***, ***Triplochiton scleroxylon***, Meliaceae, Combretaceae-Melastomataceae, *Justicia sp.*, ***Alchornea***, Poaceae, Cyperaceae, fern spore, fungal spore and charcoal.

(*continued*)

Table 1. (*continued*)

Depth (cm)	Color	Soil type	pH	Phosphate content (mg/kg P)	Archaeological details	Palynomorphs
					pipe (1), porcelain (6), cowries (3), coin currency (1), broken bottles, rubber sandal (1), hair comb (1)	
70–85	Very dusky red (2.5 YR 2.5/2)	Sandy-clayey	6.7	532.3	**Organic**: Indeterminate animal bones (1), cowrie (1), broken palm kernel shell, charcoal **Inorganic**: potsherds (193), metal scraps, bead (1), ear-ring (1), broken bottles	***Elaeis guineensis***, Amaranthaceae, *Allophyllus* cf. *nigericus*, *Mallotus*/*Macaranga*, ***Borreria ocymoides***, Cyperaceae, ***Alchornea***, fern spores, charcoal, fungal spores
85–120	Yellowish red (5 YR 4/6)	Sandy-clayey	6.9	385.0	**Organic**: broken palm kernel shell, charcoal **Inorganic**: potsherds (543), local lamp pedestal (2), metal scraps, broken bottles, goethite (7), hammer stones (3)	***Elaeis guineensis***, *Tetrorchidium didymostemon*, Combretaceae/Melastomataceae, Amaranthaceae, ***Alchornea***, Cyperaceae, Poaceae, fern spores, fungal spores and charcoal.

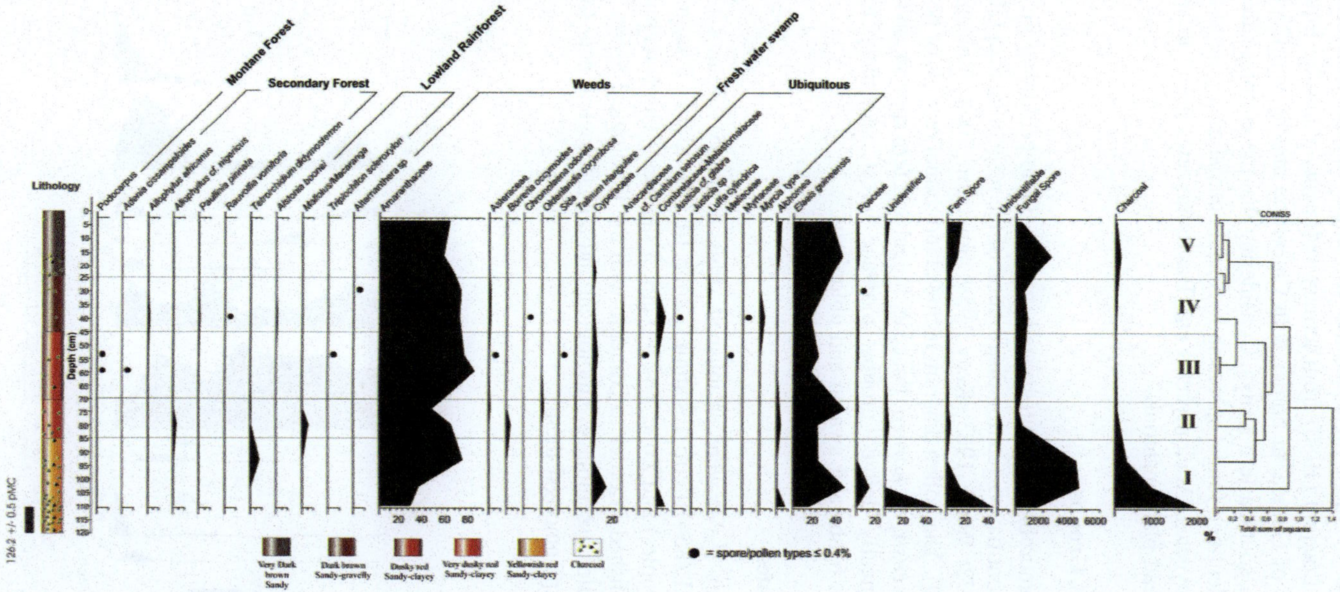

Fig. 4. Pollen diagram of Motako excavated pit (7º16′56′′N, 4º17′44′′E). Bullets means spore/pollen types ≤ 0.4%.

Several amounts of unicellular, bicellular and multicellular fungal spore types were recognized. Some of them were similar to fungal spores recovered by Salard-Cheboldaeff and Locquin (1980) from tertiary sediments along the Equatorial Africa coast of the Gulf of Guinea, at border with Cameroon and by van Geel et al. (2011) from Lake Challa, southeastern Kenya. Considering that the samples studied were from a midden heap, identified fungal spores showed characteristics similar to those of extant species of Ascomycetes, Adelomycetes and Basidiomycetes found in wood of deciduous trees and dead plant materials.

Phyto-Ecological Groups

Identified palynomorphs were categorized into six phyto-ecological groups (Table 3) based on the present-day vegetation distribution of the respective taxa (Hutchinson and Dalziel 1958–1972; White 1983; Keay 1959; Sowunmi 2004; Vincens et al. 2007; Lézine et al. 2009). The pollen diagram included individual records of pollen and spore taxa (*Elaeis guineensis*, *Alchornea*, Poaceae, fungal and fern spores), and records of the following ecologically defined groups: montane forest, secondary forest, lowland rainforest, weeds, freshwater swamp and ubiquitous.

Local Pollen Zones

Five pollen zones (I–V) were established (Figs. 4 and 5): Zone I (110–85 cm: 3 samples), Zone II (85–70 cm: 1 sample), Zone III (70–45 cm: 3 samples), Zone IV (45–25 cm: 2 samples) and Zone V (25–0 cm: 3 samples). The zonation of the pollen record was established in order to coincide with the lithological layers enhancing the integration of pollen assemblage, lithological and archaeological data for a discussion of the relationships amongst them.

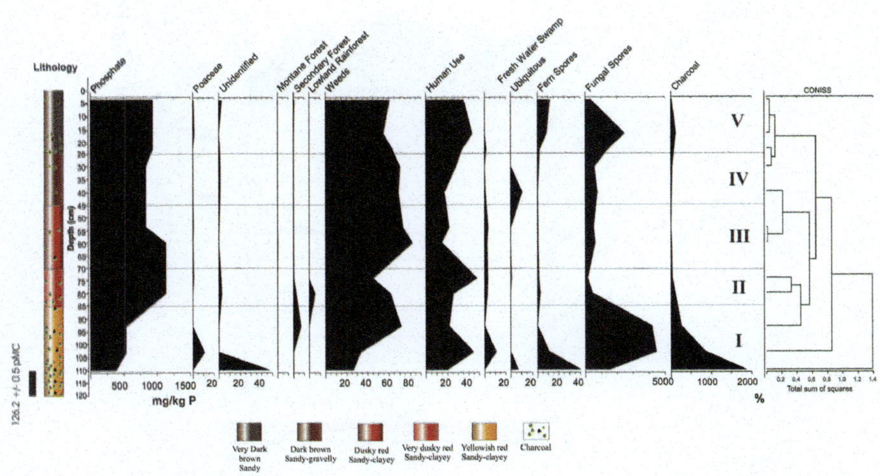

Fig. 5. Pollen diagram highlighting plants of economic use and the total amount of phosphate on Motako excavated pit

Table 2. AMS ^{14}C date for Motako deposit

Depth (cm)	Sample type	Lab. No.	Measured radiocarbon age	Conventional radiocarbon age
110–120	Charcoal + charred palm kernel endocarp	Beta-403756	126.2 ± 0.3 pMC	126.5 ± 0.3 pMC

Zone I (110–85 cm: 3 Samples; 126.5 ± 0.3 pMC)

This zone is characterized by low amounts of pollen and spores. Lowland rain forest species are absent. Meanwhile, secondary forest is sparsely represented by *Tetrorchidium didymostemon* (Baill.) Pax & K. Hoffm. (7.1%) at the upper part of the zone (93 cm). Carbonized particles are frequent in the organic deposits, especially in the lower part of the zone (110 cm) where they occur at their peak representing the 1866.7% of the pollen sum. Charcoal percentage decreases to the upper part of the zone (93 cm) where values decline to 250.0%. *Elaeis guineensis* fluctuates within the range of 13.3–44.4% of the pollen sum. *Alchornea* occurs in low percentages (6.7%) at the base of the deposit and Poaceae appear only at 103 cm with 11.1%. Weeds, represented by Amaranthaceae, occur with increasing percentages (26.7–71.4%), from the bottom to the top. Monolete fern spores remain concentrated at the bottom of the zone (13.3%) while the trilete (*Pteris*) spores decrease upwards (26.7–11.1%). Fresh water swamp is represented by Cyperaceae (11.1%). Ubiquitous species are solely represented by Combretaceae-Melastomataceae which contributes 6.7% to the pollen sum at the lower part of the zone (110 cm). Fungal spores shows increasing values within the range of 1533.3–4466.7%.

Zone II (85–70 cm: 1 Sample)

Lowland rain forest species appears for the first time but in low frequency represented mainly by pollen grains of *Mallotus/Macaranga* (5.1%). Similarly, secondary forest (*Allophylus* cf. *nigericus*) continues to occur in low values (2.6%). *Alchornea* occurs in low percentages (6.7%). In relation to the previous zone, weeds increases in diversity represented by the pollen of *Borreria ocymoides* (Burm f.) DC., a plant with medicinal properties, and Amaranthaceae. However, Amaranthaceae reduces to 61.5% likewise *Elaeis guineensis* (20.5%). Freshwater swamp (Cyperaceae), *Alchornea* and monolete spores occur in low frequencies (2.6%). Charcoal particles decrease further in values (100.0%) as well as fungal spores (446.15%).

Zone III (70–45 cm: 3 Samples)

In this zone charcoal particles continue to decrease, from 32.7% to 6.8%, while pollen diversity is higher than in the previous Zone II. Secondary forest species occur in low frequencies and are represented by the pollen of *Tetrorchidium didymostemon* (0.5%) and *Adenia cissampeloides* (Planch. ex Hook.) Harms. (0.4%), as well as lowland rainforest species (*Triplochiton scleroxylon* K. Schum, with pollen 0.2% at the top). Montane forest is represented by *Podocarpus* (0.2–0.4%) at the upper part of the zone (0.2%). Weeds increase in diversity, dominated by Amaranthaceae (47.1–82.1%) in

Table 3. Allocation of identified pollen taxa to plant forms, phyto-geographical zones, phyto-ecological groups and economic significance. Bolded taxa are economically useful in Nigeria and/or on the inventory of plants with ethnobotanical significance in Motako. Underlined taxa are present or closely related to the modern vegetation in Motako.

Family	Pollen types (local name)	Characteristic plant forms/main occurring environment/economical use	Main vegetation type	Phyto-ecological group
Acanthaceae	*Justicia* undiff.	Herbs or shrubs/occur in lowland-upland (edges) forest, woodland, wooded grassland, bushland and grassland	Tropical Africa	Ubiquitous (UB)
Acanthaceae	*Justicia cf. glabra*	Herb/growing in riverine woodland, also in drier montane forest, usually on the margins, riverine forest, scrub and woodland and on termite mounds; at low to medium altitudes	Tropical Africa	Ubiquitous (UB)
Amaranthaceae	Amaranthaceae/Chenopodiaceae undiff.	Herbs or subshrubs, sometimes climbers/all habitats	Tropical Africa	Weeds (W)
Amaranthaceae	***Alternanthera*** undiff.	Herb/weed/supplies food and used with medicinal and ritualistic purposes in Nigeria	Tropical Africa	Weeds (W)
Anacardiaceae	Anacardiaceae undiff.	Trees or shrubs/all habitats	Tropical Africa	Ubiquitous (UB)
Apocynaceae	***Alstonia boonei***	Tree/lowland (swamp/gallery) rain forest/used with medicinal purposes in Nigeria	Guineo-Sudanian/Guineo-Congolian/ Lake Victoria/Old secondary Guineo-Congolian rain forest	Lowland rainforest (LRF)
Apocynaceae	***Rauvolfia vomitoria*** (Wada)	Tree or shrub/lowland (gallery) rain forest edges/pioneer secondary forest/used with medicinal purposes in Nigeria	Secondary Guineo-Congolian rain forest/Sudanian/Guineo-Sudanian/ Guineo-Congolian transition woodland/ Guineo-Zambezian/Zambezian/Lake Victoria	Secondary forest (SF)
Arecaceae	***Elaeis guineensis*** (Ope, Igi-Ope)	Tree palm/lowland secondary rain forest/cultivated. Supplies food and used with medicinal and ritualistic purposes in Motako. Exploited for the local palm oil production industry	Guineo-Congolian (now found throughout the moister parts of Africa)	Stands separately on the pollen diagram
Asteraceae	Asteraceae undiff.	Herbs or shrubs, sometimes trees or lianas/all habitats	Tropical Africa	Weeds (W)
Asteraceae	***Chromolaena odorata*** (Akintola, Awolowo)	Shrub/invasive weed of disturbed areas. Used for the treatment of malaria in Motako	Planted or introduced plant	Weeds (W)
Combretaceae/Melastomataceae	Combretaceae/Melastomataceae undiff.	Trees, shrubs or lianas/all habitats	Tropical Africa	Ubiquitous (UB)
Cucurbitaceae	***Luffa cylindrica***		Planted or introduced plant/Naturalized	Ubiquitous (UB)

(*continued*)

Table 3. (*continued*)

Family	Pollen types (local name)	Characteristic plant forms/main occurring environment/economical use	Main vegetation type	Phyto-ecological group
		Climber/on forest gaps and edges, also on agricultural lands and other disturbed areas/used for medicinal purposes in Nigeria		
Cyperaceae	Cyperaceae undiff.	Herb/hygrophytes or occurs on moist habitats	Herbaceous fresh-water swamp and aquatic vegetation	Fresh water swamp (FWS)
Euphorbiaceae	***Alchornea cordifolia*** (Esinsin, Ipa, Eepa, Weere)	Trees or shrubs/lowland-upland forest (secondary/edges/gallery/swamp)/woodland/coastal bushland. Used for medicinal and ritualistic purposes in Nigeria. Used for the treatment of headache in Motako.	Sudanian/Guineo-Sudanian/Guineo-Congolian/Guineo-Zambezian/Zambezian/Zanzibar-Inhambane/Afromontane/Lake Victoria/Tongaland-Pondoland	Stands separately on the pollen diagram
Euphorbiaceae	*Tetrorchidium didymostemon*	Tree or shrub/pioneer secondary forest	Secondary Guineo-Congolian rain forest	Secondary forest (SF)
Euphorbiaceae	*Mallotus/Macaranga* undiff.	Tree/lowland (gallery/secondary) (rain/dry) forest/bushland/thicket	Guineo-Sudanian/Guineo-Congolian/Guineo-Zambezian/Zambezian/Zanzibar-Inhambane/Lake Victoria	Lowland rainforest (LRF)
Malvaceae	*Sida* undiff.	Herbs, rarely shrubs/all habitats except rain forest/*Sida acuta* and *S. veronicifolia* are medicinal plants in Nigeria	Tropical Africa	Weeds (W)
Meliaceae	Meliaceae undiff.	Trees or shrubs/all habitats except grassland and desert	Tropical Africa, except Saharan	Ubiquitous (UB)
Myrtaceae	Myrtaceae undiff.	Trees or shrubs/all habitats except semi-desert/desert scrub	Tropical Africa	Ubiquitous (UB)
Myrtaceae	*Myrcia/Psidium* undiff.	Trees or shrubs/secondary species/some types (*Psidium guajava* and *P. cattleianum*) were planted for food and occurs in areas of agricultural crops, plantations and sub-humid pastures	Guineo-Congolian transition woodland	Ubiquitous (UB)
Passifloraceae	***Adenia cissampeloides*** (Godogbo, Arokeke)	Climber/popular medicinal and ritualist plant, also supplies food and various commodities for the local population in Nigeria/cultivated or at least retained when land is cleared/the crushed plant is used as a fish poison in Nigeria	West Malagasy deciduous thicket/Tropical Africa except Saharan	Secondary forest (SF)
Poaceae	Poaceae undiff.	Herb/open nature of the vegetation	Guineo–Congolian vegetation mosaic/savannah forest/Sudanian zone/Semi-desert grassland and shrubland of the Sahel/absolute desert and	Stands separately on the pollen diagram

(*continued*)

Table 3. (*continued*)

Family	Pollen types (local name)	Characteristic plant forms/main occurring environment/economical use	Main vegetation type	Phyto-ecological group
			sandy to rocky environments of the Sahara region	
Podocarpaceae	*Podocarpus* undiff.	Trees or shrubs/upland forest/mid-altitude (swamp/gallery) forest/transitional bushland	Afromontane/Lake Victoria/Tongaland-Pondoland/Kalahari-Highveld	Montane forest (MF)
Pteridaceae	*Pteris* undiff.	Herbaceous perennial or annual fern/on river banks, also in forest and palm groves and in areas of agricultural crops, plantations and sub-humid pastures	Guineo-Congolian rain forest	Stands separately on the pollen diagram as Fern Spores
Rubiaceae	**cf. *Canthium setosum***	Woody climber/in riparian forest, on river banks/regenerates under forest canopy, but usually found in small gaps/largely used as traditional antimalarial plant in Nigeria	Zambezian thicket/Zambezian termite-mound thicket/scrub forest/Zambezian dry deciduous forest and scrub forest/Zambezian rupicolous bushland and thicket/Guineo-Congolian elfin thicket/East African evergreen and semi-evergreen bushland and thicket/Scrub forest/Evergreen and semi-evergreen bushland and thicket and derived communities	Ubiquitous (UB)
Rubiaceae	***Borreria ocymoides*** (Pepe, Ijan)	Herb/weed of arable crops, lawns or waste areas/in Nigeria, supplies food and the juice of the leaves is applied for ring worm and eczema and the sap is squeezed on to the wound or lesion. Used with medicinal purposes and as food in Motako	Tropical Africa	Weeds (W)
Rubiaceae	***Oldenlandia corymbosa*** (Oyigi)	Herb/grassland with long or short grass, bushland, montane scrub, shallow soil on rocks, sandy river ridges, furrows and dry ponds on black-cotton soil, cultivated and disturbed ground/supplies food and used with medicinal and ritualistic purposes; also it is a source of dye in Nigeria	Tropical and subtropical Africa	Weeds (W)

(*continued*)

Table 3. (*continued*)

Family	Pollen types (local name)	Characteristic plant forms/main occurring environment/economical use	Main vegetation type	Phyto-ecological group
Sapindaceae	***Allophylus africanus*** (Odu-oko)	Shrub or small tree/in riverine thickets, open woodland and forest edges, often associated with termite mounds/lowland (secondary/gallery/edges) forest/woodland/wooded grassland/bushland/river banks/used for medicinal purposes in Nigeria. Used for the improvement of libido and food preservation in Motako	Evergreen and semi-evergreen bushland and thicket and derived communities/Sahelian/Sudanian/Guineo-Sudanian/Guineo-Congolian/Guineo-Zambezian/Zambezian/Tongaland-Pondoland	Secondary forest (SF)
Sapindaceae	*Allophylus* cf. *nigericus*	Shrub/forest	Tropical Africa except Saharan	Secondary forest (SF)
Sapindaceae	***Paullinia pinnata*** (Obi, Isu-Omode)	Climber/lowland (secondary/gallery/edges) rain forest/wooded grassland/used with medicinal purposes in Nigeria	Guineo-Congolian transition woodland/Afromontane/Tropical Africa except Saharan-Sahelian	Secondary forest (SF)
Sterculiaceae	***Triplochiton scleroxylon*** (Obeche)	Tree/rapidly growing, light-demanding, valuable timber trees, often occur gregariously, and can regenerate abundantly on abandoned farmland/often in the transition zone between rainforest and savannah/used occasionally as a vegetable and host of the African silk moth to the silk cloth manufacture	Old secondary Guineo-Congolian rain forest/Drier peripheral semi-evergreen Guineo-Congolian rainforest and similar forest in the transition	Lowland rainforest (LRF)
Talinaceae	***Talinum triangulare*** (Gbure)	Herb/grown along the banks of rivers. Supplies food and used with medicinal and ritualistic purposes in Nigeria. Commonly used as vegetable in Motako	Planted or introduced plant	Weeds (W)

association with other weed taxa which occur in lower frequencies (Asteraceae, *Talinum triangulare, Sida* and pollen of *Oldenlandia corymbosa* L.). *Talinum triangulare* (Jacq.) Willd., also known as water leaf (Grubben and Denton 2004), is a vegetable of the secondary forest and is used as food. An increase in pollen diversity of ubiquitous species (such as *Canthium setosum* Hiern, Combretaceae/Melastomataceae, *Justicia* and Meliaceae) is noted in this zone. Fresh water swamp species represented by Cyperaceae occur in all samples (2.2–3.4%). *Elaeis guineensis* occurs in high quantities at the base of the zone (44.2%) but decreases upwards (21.1%). *Alchornea* is present only at the base (1.4%). Poaceae (1.0–0.8%) and monolete fern spores (1.0–0.2%) occur in decreasing frequencies while trilete fern spores increases to the top (0.5–0.8%). Fungal spores increase gradually to the top (from 146.2 to 533.7%).

Zone IV (45–25 cm: 2 Samples)

There is an increase in the diversity of secondary forest species at the bottom, represented by the pollen of *Paullinia pinnata* L. (medicinal plant), *Rauvolfia vomitoria,* and *Allophylus africanus* P. Beauv. (0.2–1.7%). Rainforest elements (*Alstonia boonei* De Wild.) continue to be poorly represented (0.5%). Weeds maintain high diversity, dominated by Amaranthaceae (c. 69%). Amongst the other weeds, there are pollen of *Alternanthera*, Asteraceae, *Chromolaena odorata* (L.) R. M. King & H. Rob. and *Talinum triangulare*. The pollen of the exotic and medicinal plant, *Chromolaena odorata*, has the least frequency (0.2%). Fresh water swamp pollen (Cyperaceae) decreases in frequency (1%). A further increase is noted in the diversity of ubiquitous species (Anacardiaceae, Combretaceae/Melastomataceae, *Justicia glabra* K. D. Koenig Roxb., Myrtaceae and *Myrcia*) at the bottom; only *Luffa cylindrica* (1.2%) occurs at 30 cm. *Elaeis guineensis* increases slightly at the top (25.8%). Poaceae, though present in all the levels, decreases further in this zone (0.5–0.4%). Fern spores constituting both mono- and trilete (*Pteris*) forms occur with low percentages (0.2–1.2%). Fungal spores decrease from 765.3% to 562.9% in the zone while charcoal particles increase to the top (from 14.6% to 39.9%).

Zone V (25–0 cm: 3 Samples)

This zone is marked by a slight increase in charcoal particles that reach their peak value (111.3%) at 17 cm depth along with fungal spores (2452.1%); they both decline to 11.2% and 347.4% at 4 cm, respectively. Secondary forest is singly represented by the pollen of *Tetrorchidium didymostemon* (0.7%) at the topmost 4 cm sample. *Elaeis guineensis* occurs in higher frequency (32.2–42.3%). Weeds continue to be dominated by Amaranthaceae, the only taxon in this category (54.9–63.4%). *Alchornea* reappears with increasing values (1.4–3.3%), likewise Poaceae (0.7–1.4%). Lowland rain forest elements are not present in this zone. Fresh water swamp, Cyperaceae, are low (1.8%) at the base of the zone (23 cm; Table 4).

Discussion

For a better understanding of the interplay among palynomorphs, chemical, lithological and archaeological data that characterize the midden feature, the deposit is divided into "phases" or horizons similar to the pollen and lithological zones in order to integrate

Table 4. Frequency (%) of spore/pollen types, fungal spores and charcoal in the sediments.

Spore/pollen types	Phyto-ecological groups	04 cm	17 cm	23 cm	30 cm	40 cm	54 cm	60 cm	74 cm	80 cm	93 cm	103 cm	110 cm
Podocarpus undiff.	MF	0	0	0	0	0	0.2	0.4	0	0	0	0	0
Adenia cissampeloides	SF	0	0	0	0	0	0	0.4	0	0	0	0	0
Allophylus africanus	SF	0	0	0	0	1.7	0	0	0	0	0	0	0
Allophyllus cf. *nigericus*	SF	0	0	0	0	0	0	0	0	2.6	0	0	0
Paullinia pinnata	SF	0	0	0	0	0.5	0	0	0	0	0	0	0
Rauvolfia vomitoria	SF	0	0	0	0	0.2	0	0	0	0	0	0	0
Tetrorchidium didymostemon	SF	0.7	0	0	0	0	0	0	0.5	0	7.1	0	0
Alstonia boonei	LRF	0	0	0	0	0.5	0	0	0	0	0	0	0
Mallotus/Macaranga undiff.	LRF	0	0	0	0	0	0	0	0	5.1	0	0	0
Triplochiton scleroxylon	LRF	0	0	0	0	0	0.2	0	0	0	0	0	0
Alternanthera undiff.	W	0	0	0	0.4	0	0	0	0	0	0	0	0
Amaranthaceae undiff.	W	60.5	54.9	63.4	69.8	68.7	73.1	82.1	44.7	61.5	71.4	33.3	26.7
Asteraceae undiff.	W	0	0	0	0.8	0.5	0.2	0	0.5	0	0	0	0
Borreria occymoides	W	0	0	0	0	0	0	0	0	2.6	0	0	0
Chromolaena odorata	W	0	0	0	0	0.2	0	0	0	0	0	0	0
Oldenlandia corymbosa	W	0	0	0	0	0	0	0	1.4	0	0	0	0
Sida undiff.	W	0	0	0	0	0	0.2	0	0	0	0	0	0
Talinum triangulare	W	0	0	0	0.4	0	0	0	0.5	0	0	0	0
Cyperaceae undiff.	FWS	0	0	1.8	0	1	3.4	2.2	2.9	2.6	0	11.1	0
Anacardiaceae undiff.	UB	0	0	0	0	0.5	0	0	0	0	0	0	0
cf. *Canthium setosum*	UB	0	0	0	0	0	0.2	0	0	0	0	0	0
Combretaceae-Melastomataceae undiff.	UB	0	0	0	0	6.5	0.2	0	0.5	0	0	0	6.7
Justicia cf. *glabra*	UB	0	0	0	0	0.2	0	0	0	0	0	0	0
Justicia undiff.	UB	0	0	0	0	0	0	0	0.5	0	0	0	0
Luffa cylindrica	UB	0	0	0	1.2	0	0	0	0	0	0	0	0
Meliaceae undiff.	UB	0	0	0	0	0	0.2	0	0	0	0	0	0
Myrtaceae undiff.	UB	0	0	0	0	0.2	0	0	0	0	0	0	0
Myrcia undiff.	UB	0	0	0	0	3.3	0	0	0	0	0	0	0

(*continued*)

Table 4. (*continued*)

Spore/pollen types	Phyto-ecological groups	04 cm	17 cm	23 cm	30 cm	40 cm	54 cm	60 cm	74 cm	80 cm	93 cm	103 cm	110 cm
Alchornea undiff.	POL	3.3	1.4	0	0	0	0	0	1.4	2.6	0	0	6.7
Elaeis guineensis	POL	32.2	42.3	33.9	25.8	14.4	21.1	14.4	44.2	20.5	21.4	44.4	13.3
Poaceae undiff.	POL	0.7	1.4	0	0.4	0.5	0.8	0.4	1	0	0	11.1	0
Unidentified	POL	2.6	0	0.9	1,2	1.2	0.2	0	1.9	2.6	0	0	46.7
Monolete spore undiff.	FS	9.9	2.8	0.9	0,4	0.2	0.2	0.9	1	2.6	0	0	13.3
Trilete spore undiff.	FS	2	5.6	1.3	1,2	0.2	0.8	0	0.5	0	0	11.1	26.7
Unidentifiable	PAL	0	0	0	0	0.2	0	0	0	2.6	0	0	0
Fungal spores	FUNG	347.4	2452.1	1422.9	562.9	765.3	533.7	621.4	146.2	446.2	4214.3	4466.7	1533.3
Charcoal	CH	11.2	111.3	71.8	39.9	14.6	6.8	10.5	32.7	100	250	844.4	1866.7
Montane forest	PS	0	0	0	0	0	0.2	0.4	0	0	0	0	0
Secondary forest	PS	0.7	0	0	0	2.4	0	0.4	0.5	2.6	7.1	0	0
Lowland rainforest	PS	0	0	0	0	0.5	0.2	0	0	5.1	0	0	0
Weeds	PS	60.5	54.9	63.4	71.4	69.4	73.5	82.1	47.1	64.1	71.4	33.3	26.7
Fresh water swamp	PS	0	0	1.8	0	1	3.4	2.2	2.9	2.6	0	11.1	0
Ubiquitous	PS	0	0	0	1.2	10.8	0.6	0	1	0	0	0	6.7

and establish relationships amongst the different data recovered. It should be noted that ordinarily, micro and macro charcoal particles from the deposit should indicate burning activity. However, charcoal particles recorded in the samples are not interpreted to necessarily suggest fire activity in the area for the following reasons. First, putting into consideration the nature of the deposit being investigated, the charcoal may have been swept into the midden feature from hearths and other features in the archaeological site, and its association with fire activity is very unlikely. Second, the taxa from which the charcoal was derived were not identified nor was the size of the charcoal measured in detail. Third, the sieving and decanting procedure adopted during the pollen preparation can lead to loss of some charcoal particles, hence altering significantly the original size distribution of charcoal particles (Tinner and Hu 2003).

Motako archaeological deposit exhibits low pollen content which may be attributed to oxidation, fungal activity and infiltration of pollen grain through the sandy deposit. Samples show signs of oxidation as depicted from reddish and yellowish colours of the deposit. Similarly, fungal spores occur in all the levels of the midden feature in high frequencies within a range of 146.2–4466.7%. However, Amaranthaceae dominate the pollen assemblage with peaks of 73.1% and 82.1% (at 54 cm and 60 cm, respectively). There was no observable evidence of wet/dry cycles which could also have been a contributory factor to the low amount of pollen in the deposit.

Fourteen of the thirty-one pollen types identified are associated with medicinal, food and ritualistic usage among the present-day people of Motako and by extension Nigeria, according to ethnobotanical data (Table 3). These plants were widely used by the over two hundred inhabitants of Motako as confirmed from oral sources. The people of Motako believe in the use of traditional therapy for treatment of ailments rather than orthodox medicine, owing partly to their low income.

In the region, it is common that native forest areas are devastated by fire to enhance growth performance of the oil palm. Field observations made at Motako show that palm oil and palm wine are sold as sources of income. Moreover, an ethnographic study showed that oil palm is one of the essential ingredients during *Ogun-Ija* festival in the area (Daraojimba 2016). The festival is marked by the use of *Vigna unguiculata* (L.) Walp. (cowpea), *Dioscorea cayenensis* subsp. *rotundata* (Poir.) J. Miège (yam) and *Cola nitida* (kola nut) in addition to palm oil, during the ritual exercise. Oil palm which constitutes an important element in the domestic economy of the Motako people had its pollen ubiquitously represented in the samples. The significance of the oil palm to the people of Motako and the abundance of cultivated stands of this versatile tree crop dotting the landscape of the area gives a good explanation for the pollen making its way into the midden deposit.

Phase I (110–85 cm): This bottom zone is characterized by the highest concentration of fungal spores but low diversity of archaeological materials and pollen content. Ordinarily, high values of fungal spores would expectedly suggest increased concentration of organic matter. However, poor organic matter constituent was recorded which synchronizes with low phosphate content. This anomaly may be a result of the possible reworking of the sediment. *Elaeis guineensis* pollen increases but near the top of the zone suffers a reduction, whereas Amaranthaceae pollen, which comprise ruderal species, increase to the top. Broken palm kernel shell and charcoal are very sparse at the base but only appear appreciably upwards from level 110 cm of the zone indicating a possibility that the deposit was not yet a midden prior to this time. The mottled yellowish colour of the sediment suggests poor drainage (Hassan 1978), and low content of pollen and archaeological materials recorded also support poor activity on the midden. However, increasing activity is observed towards the top of the zone where high concentration of potsherds in association with hammer stones, local lamp pedestal and cowrie are noted. Hammer stones are heavy stone tools that were possibly used in breaking the palm kernels to extract its nuts, a common activity in the local area. Knowledge of the local population on the use of goethite as a source of iron and pigment is in doubt as this is not currently in operation in the Motako area.

Phase II (85–70 cm): Here, it is suspected that the deposit is gradually turning into a midden feature as indicated by increased activity at the top of the previous Zone I. This suspicion is further strengthened by the increase in organic matter content (bones, cowrie, broken palm kernel shell and charcoal) which corresponds with the increased concentration of phosphate and other cultural materials (ear-ring, potsherds, metal scraps, bead and broken bottles), but not still in tandem with fungal spore occurrence which reduces drastically. Pollen of *Elaeis guineensis* and Amaranthaceae maintain similar high values in this zone associated with an increase in taxa with ethnobotanical significance (*Borreria ocymoides* and *Alchornea*).

Phase III (70–45 cm): At this phase, the midden feature is fully established with high activity as reflected in the peak in phosphate content, increased diversity of kitchen wares (aluminium cooking spoon, muller, porcelain, metal bowl), organic remains (indeterminate animal bones and tooth, thoracic vertebrae of rodent, charcoal, snail shell, kernel shell) and pollen of plants with ethnobotanical significance (*Elaeis guineensis*, *Adenia cissampeloides*, *Oldenlandia corymbosa*, *Talinum triangulare*, *Sida*, *Triplochiton scleroxylon* and *Alchornea*). Also, there is a peak in the pollen of *Elaeis guineensis* which later declines at the top of this zone, whilst Amaranthaceae pollen shows an inverse behavior. The presence of *Podocarpus*, whose dispersion is anemophilous and which does not occur in the present-day vegetation of the area, indicates the existence of air currents enabling the transport of the pollen grains from long-distance to the archaeological site.

Phase IV (45–25 cm): As reflected in the reduced phosphate content and the increasing frequency in unbroken palm kernels, the midden at this phase is beginning to be deactivated and/or used for other activites, perhaps gardening for the cultivation of plants with several domestic purposes. The location of the study area just behind the residential building of the community head makes it a perfect site for gardening. Presently, the area is characterized by stands of *Musa* sp. (plantain) and other herbs (*Alternanthera sessilis* (L.) R. Br. ex DC., *Heliotropium indicum* L., *Amaranthus spinosus* L., *Synedrella nodiflora* L. (Gaertn), *Ageratum conyzoides* (L.) L., and others). In this zone the greatest richness of pollen types is present and, for the most part representing, plants that are used for different purposes by the population (*Elaeis guineensis*, *Allophylus africanus*, *Paullinia pinnata*, *Rauvolfia vomitoria*, *Alstonia boonei*, *Alternanthera* sp., *Chromolaena odorata*, *Talinum triangulare*, *Canthium setosum* and *Luffa cylindrica*). There is a reduction of archaeological material compared to the previous zone, which is in consonance with reduction of the phosphate content and probably implies less use for the disposal of organic material. Identification of diversity of weeds of cultivated fields, such as *Talinum triangulare*, Asteraceae,

Amaranthaceae and *Chromolaena odorata* (Akobundu and Agyakwa 1998), suggests an increasing accumulation of their pollen resulting from reduced activity on the midden feature.

Phase V (25–0 cm): Although the phosphate content increases in this zone, there is a reduction of both organic and inorganic materials. However, a further increase is noted in the frequency of unbroken palm kernel which gets to its peak in this zone, a possible indication of decreasing palm kernel breaking activity in the area and a signal of abandonment of the midden as a refuse dump site. It is worth mentioning that as at the time of excavation of the midden and subsequent sample collection, the midden was already inactive, that is to say, was no longer in use. However, the length of abandonement is not known, perhaps recently, considering the recent radiocarbon date of not later than 60 years from the basal layer of the deposit. The richness of pollen types is reduced, with anthropogenic weed dominance towards the top of the layer suggesting human activities associated with cultivation of plants in the area.

Conclusion

From the palynological analysis of a recent anthropogenic deposit associated with archaeological data and phosphate analysis, it was possible to verify the management of the deposit by the local population. Amaranthaceae pollen associated with other ruderal species, including several species of ethnobotanical uses, was dominant in the deposit, indicating open vegetation possibly modified by anthropic action. The presence of *Elaeis guineenses* pollen can be justified by the management of the palm tree for oil production within the local space. The pattern of occurrence of anthropogenic pollen taxa may be a reflection of the gradual stabilization, especially from 45 cm to the top, of the deposit. Perhaps, also considering the phosphate values, it reflects site activities changing across the sedimentary units. This hypothesis is supported by the changing concentrations of archaeological materials which followed an identical pattern with phosphate concentrations across the sediment profile. The pattern of phosphate content distribution was consistent with pollen content in the various phases. Therefore, we suppose that different types of land management and land use over a period of c. 60 years were caught in infill of the sampled pit at Motako.

Acknowledgements. We thank the anonymous reviewers whose insightful comments guided us in preparing the final version of this paper. Dr Tubosun is appreciated for his relevant contributions to the first draft of this paper. Our sincere gratitude goes to Ms. Elizabeth Adeyemo, Messrs. Emmanuel Nwagbara and Moyib Segun for their assistance during the fieldwork. Mr. Nwagbara was very helpful in interpretation during all oral interviews conducted in the local community. Thanks are also offered to Mr. P. C. Opara for his assistance with sample preparation. Our inestimable friends, Messrs. Macham Mangut and Abiodun Areola, are appreciated for being instrumental in producing the satellite image vegetation map of the study area. In addition, we are greatly indebted to the National Council for Scientific and Technological Development (CNPq) for the financial support received by the third and fifth authors (research fellows). The funding facilitated the purchase of the Tilia software used in the production of pollen diagrams for this study.

References

Akintoye SA (2010) A history of the Yoruba people. Amalion, Dakar

Akobundu IO, Agyakwa CW (1998) A handbook of West African weeds, 2nd edn. African Book Builders, Ibadan

A.P.L.F (1974) Pollen et spores d'Afrique tropicale. Centre d'Etudes de geographie tropicale, Talence

Breunig P, Neumann K (1999) Archäologische und archäobotanische Forschungen in Westafrika. Arch Nachrichtenbl 4:336–357

Bryant VM, Holloway RG (1983) The role of palynology in archaeology. Adv Archaeol Method Theory 6:191–224

Chukwuma EC, Soladoye MO, Feyisola RT (2015) Traditional medicine and the future of medicinal Plants in Nigeria. Med Plants Stud 3:23–29

Corley RHV, Tinker PB (2003) The oil palm, 4th edn. Blackwell, Oxford

Daraojimba KC (2016) Humans and oil palm (*Elaeis guineensis* Jacq.) exploitation in Orile-Owu, southwest Nigeria, ca 1450–1640A.D: Archaeo-botanical evidence. Dig It 3:14–23

Ebana RU, Madunagu BE, Ekpe ED et al (1991) Microbiological exploitation of cardiac glycoside and alkaloids from *Garcinia kola*, *Borreria ocymoides*, *Kola nitida* and *Citrus aurantifolia*. J Appl Bacteriol 71:398–401

Elenga H, Maley J, Vincens A et al (2004) Paleoenvironments, Paleoclimates and landscape development in Atlantic Equatorial Africa: a review of key sites covering the last 25 kyrs. In: Battarbee RW, Gasse F, Stickley CE (eds) Past climate variability through Europe and Africa. Springer, Dordrecht, pp 181–198

Faegri K, Iversen J (1989) Textbook of pollen analysis, 4th edn. The Blackburn Press, Caldwell

Gosling WD, Miller CS, Livingstone DA (2013) Atlas of the tropical West African pollen flora. Rev Palaeobot Palyno 199:1–135

Grimm EC (1987) CONISS: a 77 program for stratigraphically constrained cluster analysis by the method of incremental sum of squares. Comput Geosci 2:13–35

Grubben GJH, Denton OA (eds) (2004) Plant resources of Tropical Africa 2. Vegetables, PROTA Foundation, Wageningen

Hassan FA (1978) Sediments in archaeology: methods and implications for palaeoenvironmental and cultural analysis. Field Archaeol 5:197–213

Höhn A, Neumann K (2014) Millet and more: farming and food in the time of the Nok Culture. In: Breunig P (ed) Nok. African sculpture in archaeological context. Africa Magna, Frankfurt, pp 178–185

Höhn A, Neumann K (2016) The palaeovegetation of Janruwa (Nigeria) and its implications for the decline of the Nok culture. African Archaeol 14:331–353

Hutchinson J, Dalziel JM (1958–1972) Flora of West Tropical Africa, vol II and III. Crown Agents for Oversea Governments and Administrations, London

Kahlheber S, Höhn A, Rupp A (2009) Archaeobotanical studies at Nok sites: an interim report. Nyame Akuma 71:2–17

Kangur M (2002) Methodological and practical aspects of the presentation and interpretation of microscopic charcoal date from sediments. Veg Hist Archaeobot 11:289–294

Keay RWJ (1959) An outline of Nigerian vegetation, 3rd edn. Federal Government of Nigeria, Lagos

Klee M, Zach B, Neumann K (2000) Four thousand years of plant exploitation in the Chad Basin of northeast Nigeria I: The Archaeobotany of Kursakata. Veg Hist Archaeobot 9:223–237

Klee M, Zach B, Stika H (2004) Four thousand years of plant exploitation in the Lake Chad Basin (Nigeria), part III: plant impressions in potsherds from the final Stone Age Gajigama culture. Veg Hist Archaeobot 13:131–142

Lebamba, J, Ngomanda A, Vincens A et al. (2009) Central African biomes and forest succession stages derived from modern pollen data and plant functional types. Clim Past:403–429

Lézine A-M, Julie W, Vincens A et al (2009) Are modern pollen data representative of West African vegetation? Rev Palaeobot Palyno 156:265–276

Mabogunje A, Omer-Cooper JD (1971) Owu in Yoruba History. Ibadan University Press, Ibadan

Maley J, Chepstow-Lusty A (2001) *Elaeis guineensis* Jacq. (Oil palm) fluctuations in Central Africa during the late Holocene: climate or human driving forces for this pioneering species? Veg Hist Archaeobot 10:117–120

Mercuri AM (2014) Genesis and evolution of the cultural landscape in central Mediterranean: the 'where, when and how' through the palynological approach. Landscape Ecol 29(10):1799

Mercuri AM, Montecchi MC, Pellacani G et al (2015) Environment, human impact and the role of trees on the Po plain during the Middle and Recent Bronze Age: pollen evidence from the local influence of the terramare of Baggiovara and Casinalbo. Rev Palaeobot Palyno 218: 231–249

Ngomanda A, Neumann K, Schweizer A et al (2009) Seasonality change and the third millennium BP rainforest crisis in southern Cameroon (Central Africa). Quat Res 71:307–318

Orijemie EA (2015) Archaeobotanical study of Late Stone Age rock shelter AHI, Abeokuta, south-western Nigeria. In: Florenzano A, Fornaciari R (eds) Proceedings of IWAA8 Conference. Supplemento Atti Soc Nat Mat Modena, vol 146, pp 27–30

Orijemie EA, Sowunmi MA (2014) Humans and the mangrove in southern Nigeria. In: Stevens CJ, Nixon S, Murray MA, Fuller DQ (eds) The archaeology of African plant use. Left Coast Press, Walnut Creek, pp 103–112

Orijemie EA, Ogunfolakan A, Aleru JO et al. (2010) The archaeology and palynology of Ajaba, a late iron-age settlement in north-east Yoruba Land, Nigeria: some preliminary results. In: Allsworth-Jones P (ed) West African Archaeology: new developments, new perspectives. BAR International Series 2164. Archaeopress, Oxford, pp 103–115

Salard-Chelbodaeff M, Locquin MV (1980) Champignons présents au Tertiaire le long du littoral de l'Afrique equatoriale. C R Congr Nat Soc Savantes 105:183–195

Shackley MC (1981) Environmental Archaeology. Allen and Unwin, London

Sjoberg A (1976) Phosphate analysis of anthropic soils. Field Archaeol 3:447–454

Sowunmi MA (1981) Aspects of late quaternary vegetation changes in West Africa. Biogeogr 8:457–474

Sowunmi MA (1991) Pollen analysis of Kariya Wuro Rock Shelter, Bauchi State. Nigerian Field 56:163–170

Sowunmi MA (1999) The Significance of the Oil Palm (*Elaeis guineensis* Jacq.) in the Late Holocene environments of West and West Central Africa: a further consideration. Veg Hist Archaeobot 8:199–210

Sowunmi MA (2004) Aspects of Nigerian coastal vegetation in the Holocene: some recent insights. In: Battarbee RW, Gasse F, Stickley CE (eds) Past climate variability through Europe and Africa. Springer, Dordrecht, pp 199–215

Swaine MD, Hall JB (1986) Forest structure and dynamics. In: Lawson GW (ed) Plant ecology in West Africa. Wiley, New York, pp 47–93

Tinner W, Hu FS (2003) Size parameters, size-class distribution and area-number relationship of microscopic charcoal: relevance for fire reconstruction. Holocene 13:499–505

Umeji OP, Ibeanu AM, Agwu COC (2012) Holocene human occupation of the eastern Nigeria scarplands: an impact assessment study. Quatern Int 4:2–13 van Geel B, Gelorini V, Lyaruu A et al (2011) Diversity and ecology of tropical African fungal spores from a 25,000-year palaeoenvironmental record in south-eastern Kenya. Rev Palaeobot Palyno 164:174–180

Verger PF (1995) Ewé- O uso das plantas na sociedade Iorubá. Companhia das Letras, São Paulo

Vincens A, Lézine A-M, Buchet G et al (2007) African pollen database inventory of tree and shrub pollen types. Rev Palaeobot Palyno 145:135–141

Weniger B, Lagnika L, Vonthron-Sénécheau C et al (2004) Evaluation of ethnobotanically selected Benin medicinal plants for their in vitro antiplasmodial activity. Ethnopharmacol 90:279–284

White F (1983) The vegetation of Africa: a descriptive memoir to accompany the UNESCO/AETFAT/UNSO vegetation map of Africa. UNESCO, Paris

Zach B, Klee M (2003) Four thousand years of plant exploitation in the Chad Basin of NE Nigeria II: discussion of the morphology of caryopses of domesticated *Pennisetum* and complete catalogue of the fruits and seeds of Kursakata. Veg Hist Archaeobot 12:187–204

Pits at Pangwari: Charcoal Taphonomy at a Multi-phased Nok Site, Central Nigeria

Alexa Höhn(✉), Gabriele Franke, and Annika Schmidt

Archaeology and Archaeobotany of Africa, Institute for Archaeological Sciences, Goethe University, Norbert-Wollheim-Platz 1, 60629 Frankfurt am Main, Germany

{a.hoehn,g.franke,a.schmidt}@em.uni-frankfurt.de

Abstract. Pits are common features at sites of the Nok Culture, which is known for its masterful terracotta sculptures and for early iron metallurgy in sub-Saharan Africa. If pits are stratified and well-dated, contained archaeological charcoal may serve as a proxy for vegetation change—especially for changes mediated by human impact. Recent investigations into charcoal samples from several pits at Pangwari, a Nok site that has been—not necessarily continuously—in use during the last 1.5 millennia BC, revealed that the taphonomy of these assemblages is not as straightforward as expected. Find distribution analysis documents that regularly only the lowermost parts of the pits are still in their original state. Later re-working of the original pit filling, erosion, and recent illicit digging for terracotta figurines challenge the interpretation of the charcoal data. Moreover, the distribution of ^{14}C dates and of potsherds assigned to different pottery groups, as well as differences in the charcoal assemblages of some pits, hint towards divergent backfill histories. Some pits seem to have been refilled quite quickly, possibly even in a single event, while other pits might have remained open for a longer time. Additionally, pits likely were reused in later phases and the infill became mixed. For charcoal analysis, it is hence crucial to consider the backfill history of each pit based on the archaeological evidence.

Keywords: Archaeological charcoal · Taphonomy · Find distribution Pits · Nok Culture · Later stone age · Iron age · Nigeria

Introduction

The Nok Culture in Central Nigeria has been known since the middle of the last century for its masterful terracotta sculptures, representing the earliest sophisticated figural sculptures in sub-Saharan Africa, as well as for its iron metallurgy, which is among the earliest securely dated evidence for iron working in Africa (Fagg 1977; Eggert 2014; Junius 2016). Since 2005, aspects of the Nok Culture reaching beyond the terracotta figurines, such as chronology, settlement structures, regional distribution, subsistence, and vegetation have been investigated (Breunig and Rupp 2010). More than 300 new sites have been recorded and about 80 were excavated. Often the Nok sites consist of a shallow layer of scattered finds and features are only recognizable as find concentrations and/or darker colorations. Among the features undifferentiated pits are

A. M. Mercuri et al. (Eds.): *Plants and People in the African Past*,
https://doi.org/10.1007/978-3-319-89839-1_14

distinguished from terracotta depositions and graves, though definite proof for the latter is difficult as bones are not preserved in the acid soils of the region (Breunig and Rupp 2016). Human activities during the 1500 years that the Nok Culture lasted, from 1500 BC to the turn of the Common Era, must have influenced the landscape and left their traces in the woody vegetation. Archaeobotanical investigations revealed subsistence strategies including mixed cropping of pearl millet (*Pennisetum glaucum* (L.) R. Br.) and cowpea (*Vigna unguiculata* (L.) Walp.); the development of a cultural landscape with fields and fallows in different states of regeneration seems conceivable (Kahlheber et al. 2009).

Anthracological investigations, aimed at tracing this development in the woody vegetation, were carried out at the Nok site Pangwari for several reasons: (a) the large excavation furnished a wide range of ^{14}C dates covering different phases of the Nok Culture; (b) material for radiocarbon dating, mostly pearl millet, was, whenever possible, taken from archaeobotanical samples instead of handpicked material—making it possible to analyse the charcoal from the same samples that many dates had been retrieved from; (c) during the excavation the positions of all finds, including archaeobotanical samples, were individually measured by total station. This allows relating the samples from which archaeological charcoal was analysed to the distribution of other finds by creating three-dimensional distribution diagrams; and (d) when the charcoal analysis began, archaeological investigations into find distribution and ceramic chronology on the material from Pangwari had started as well.

In Sudanian and Sahelian savannas of West Africa the development of a cultural landscape has successfully been traced through the analysis of archaeological charcoal (Höhn and Neumann 2012; Huysecom et al. 2015). There, mounds of debris, accumulated over several centuries and containing comparably large amounts of charcoal fragments, offered good conditions for environmental reconstructions. Long-term deposition of scattered charcoal from domestic household origin, as necessary for sound environmental reconstructions (Chabal 1992; Théry-Parisot et al. 2010), guaranteed a representative cross-section through the exploited woody vegetation and allowed to trace its change over time. At Pangwari and other Nok sites the archaeological situation is entirely different: Sufficient amounts of charcoal for environmental reconstruction only came from the deeper pits, which were assumed to be refuse pits, filled with debris, ashes and organic household rubbish as indicated by the frequent finds of pearl millet (*Pennisetum glaucum*) and grinding stone fragments. However, when finally merging the results of radiocarbon dating, find distribution, ceramic chronology and charcoal analysis, it became clear that the taphonomy of the pits was more complicated than expected. The taphonomy of pits is determined by their function and thus the history of their infill. They determine the context in which the charcoal enters the pit and thus the story it can tell. Generally, pits found in archaeological sites may have had different functions: primary functions, for which they were dug in the first place, and often secondary functions, forms of reutilization of the pits after the first function had been served. In general primary functions might have been the storage of harvested crops (Gronenborn 1997), processing of organic products (e.g. tanning of leather), and uses as water reservoirs or as (animal) traps (Mbida Mindzie et al. 2000). Pits may also have been the result of clay or laterite mining (Wotzka 1993). Rituals connected to mortuary practices or burials are conceivable as well, especially in pits characterized by a thin coating of organic material and carefully deposited ceramics (Wotzka 1993). The most obvious secondary function is that as a refuse pit for domestic waste or, in the presence of slag, tuyère fragments and iron objects, for residues of smithing activities (Meister 2008). If initially dug with the intention of using the pit for refuse, this would then be the primary function. But it is difficult, if not impossible, to distinguish archaeologically between primary and secondary and even a possible tertiary use of a pit (Meister and Eggert 2008). This is the first study which combines archaeological and archaeobotanical information to capture the taphonomy of pits at a Nok site, focusing on the following questions: how did the charcoal assemblages develop and how representative are they for the woody vegetation in the vicinity of the site?

The Nok Culture

The distribution area of the Nok Culture, based on Nok terracotta figurines found between 1943 and 1976, encompasses an oval-shaped area of about 50,000 km^2 west of the Jos Plateau, stretching about 480 km from Kagara in the northwest to Katsina Ala in the southeast, and 160 km from Abuja to the foot of the Jos Plateau in the east (Fig. 1). The majority of the early terracotta finds, however, came from the area around

Fig. 1. Map of Nigeria with areas, sites and locations mentioned in the text.

Nok in the northeast-central section of the distribution area. It is in the same region, about 40 km west of Nok, in which the key study area of Goethe University Frankfurt's project lies.

Originally, the duration of the Nok Culture had been set between about 500 BC and 200 AD (Fagg 1977). More than 170 radiocarbon dates and almost 30 luminescence results from the project's excavations have amended this estimate. Together with a comprehensive pottery analysis which concentrated on decoration and form elements, a chronology for the Nok Culture has been established (Franke 2016, 2017). On an absolute chronological basis, the duration of the Nok Culture is set between c. 1500 BC and the turn of the Common Era. After an early phase of about 500 years (Early Nok, c. 1500–900 BC), settlement activities increased significantly and terracotta figurines are present in almost every site (Middle Nok, c. 900–400 BC). Sometime after the 8th century BC first traces of iron metallurgy appear, mostly in form of furnaces. The last centuries BC witness a considerable decrease in sites within the key study area, but the few sites found still contain terracotta figurines and typical Nok pottery (Late Nok, c. 400–1 BC). With the beginning of the Common Era, archaeological sites with Nok Culture material disappear completely and it is suggested that the Nok Culture ended at some point between the last centuries BC and the first centuries AD.

Especially the development within the Middle Nok phase with its terracotta figurines and iron technology remains unclear. The radiocarbon calibration curve plateau between c. 800 and 400 BC does not allow accurate calendar year age estimates. Here, pottery analysis has been able to provide a more detailed succession based on different pottery styles (Franke 2017). It also has become clear that many Nok sites were occupied or visited at different times during or after the Nok Culture period. In total, seven Nok pottery groups and two Common Era pottery groups have been defined. The pottery of the second millennium BC sites is grouped together in the Puntun Dutse pottery group. The Ido group forms a small transitional group in the 10th and 9th centuries BC. It is from the subsequent Ifana and Pangwari groups onwards that terracotta figurines start appearing regularly and in large numbers. While the Ifana group consists of only a few sites that date between c. 900 and 750 BC, sites with Pangwari pottery fall within the calibration curve plateau, mostly between 800 and 550 BC. Possibly slightly younger is the Tsaunim Gurara pottery group that also appears at some point between 800 and 550 BC and lasts no longer than to 300 BC. Often, evidence of ironworking in form of furnaces is found in association with Tsaunim Gurara pottery. Two more Nok pottery groups—Pandauke and Ungwar Kura—are attributed to the last 400 years BC, and mainly define the Late Nok phase. The Common Era groups are the older Janruwa group, which features carved wooden roulette decorations, and the younger Gimba group, which also includes fibre roulette decorations.

Environmental Settings

The key study area is located in the northern High Plains, the hilly foreland of the Jos Plateau. The relief is undulating with striking granite outcrops and elevations between 300 and 900 m. At Kafanchan, a town about 75 km to the east of Pangwari, mean annual rainfall is about 1800 mm, mean monthly temperature is 25 °C and relative humidity averages about 63% (Abaje et al. 2010). The rainy season lasts from April to October. The key study area belongs to the northern part of the Southern Guinea vegetation zone of Nigeria (Keay 1953). Floristically, it is part of the Guinea-Congolia/Sudanian regional transition zone (White 1983), with a mosaic of secondary (wooded) grassland, semi-evergreen rain forest and woodland. Woodlands are open stands of trees at least 8 m tall with a canopy cover of 40% or more and a field layer usually dominated by high grasses; they are not true forests (White 1983). They are annually swept by fire and the degree of canopy closure may vary, resulting in more and less open patches of savanna woodland. At the border of rivers and in deeply incised gorges denser gallery forests are present. Where cultivation is possible, the natural vegetation has been profoundly modified to wooded farmland, with fields including useful trees and bush fallows, i.e. woodland in various stages of regeneration (Keay 1953).

Archaeology of the Site

Pangwari stretches along the slope of a hill and thus belongs to the "hilltop sites" which are considered to represent the most favorite Nok site location (Breunig and Rupp 2016). The upper part of the site lies on a saddle between two low-rise summits in an

Fig. 2. Excavation at Pangwari, view from the saddle downslope to the West (Photograph: P. Breunig).

advantageous position with even ground and distant views to the east and west (Fig. 2). To capture its spatial extension, ten trenches with a total area of 2617 m^2 were excavated. Altogether almost 13,000 single measurement points were taken by total station and 20 features were recorded (Fig. 3). Shallow bedrock and stone agglomerations, mostly of granite and gneiss, characterize the pattern of the site. Due to the hillside situation, erosional processes partly changed its original appearance and relocated finds, as indicated by their dispersed distribution in the upper layer of the site and the bad preservation of potsherds at the foot of the slope (Schmidt 2014).

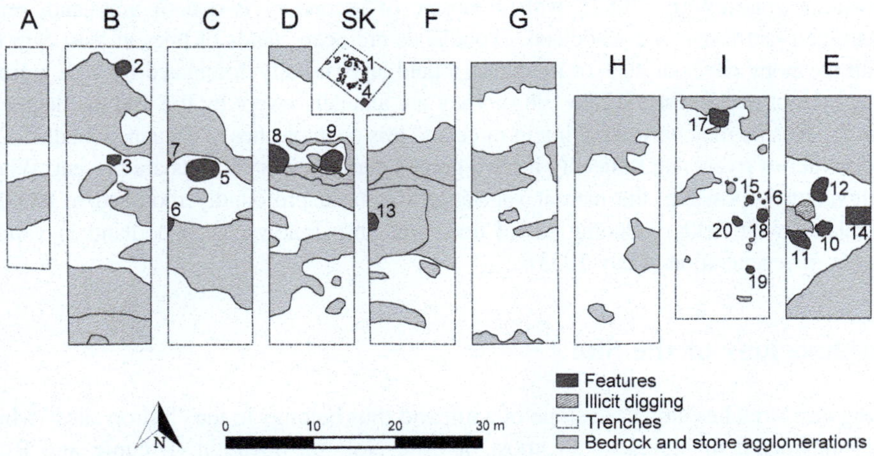

Fig. 3. Excavated trenches and features at Pangwari. Feature outlines measured by total station in different levels or reconstructed on the basis of pictures and drawings. Deep pits: 2, 4, 8, 9, 11, 12, 17; graves: 18 & 20; terracotta deposit: 10; shallow pits/find accumulations: 5, 6, 7, 13, 14, 15, 16, 19.

37 radiocarbon dates were obtained on charred plant remains, mostly pearl millet (*Pennisetum glaucum*), from different trenches, features and depths (Franke 2016). The calibrated calendar dates for the whole site range from 1500 BC to modern times as evidenced by one bomb ^{14}C result (after 1955 AD) for trench SK (Fig. 3). The summed probability density graph (Fig. 4) depicts the 31 radiocarbon dates for the Nok Culture period and clearly shows the recurring use of the site for about 1500 years, with the highest density around 800 BC. Seven dates attest the first occupation episode during the Early Nok phase between c. 1500 and 900 BC. The majority of dates (n = 21) range between the 9th and 5th century BC. Almost half of them (all on pearl millet) date into a roughly 60-year window between 826 and 763 BC, giving a quite accurate time frame for an occupation horizon. Due to the calibration curve plateau between c. 800 and 400 BC the other age estimates are less precise. Thus, the absolute dates cannot provide information whether the site was settled continuously for a longer period or whether there were several, rather short occupation episodes. In any case, the main occupation at the site was during the Middle Nok phase (900–400 BC). Only three dates may indicate a continued site use in the last centuries BC (Late Nok phase). After the end of

the Nok Culture period around the turn of the Common Era, the site was unoccupied until the beginning of the second millennium AD from which time onward a few radiocarbon dates (n = 5) prove occasional activities between the 11–13th and the 15–17th century AD. Several pottery groups are present in the ceramic inventory and evidence the different occupation episodes that took place between the second half of the second millennium BC and the second millennium AD: Puntun Dutse, Ifana, Pangwari, and Gimba groups. The morphology of the site allows distinguishing three different layers (Fig. 5). The 30–50 cm thick upper layer A extends over the whole excavated area and contains finds and dates from all occupation episodes. Disturbed areas affected by erosional processes as well as probably undisturbed sections occur. In some places pits were dug into this layer A. The filling of the pits is designated as layer B. For its finds an in situ position can be assumed, although often the upper part of the pit seems mixed with layer A. The transition between layer A and B is hardly visible and in most cases it can only be differentiated by the find distribution. Only in the upper part of the site (trench E) an additional top layer C is present (Fig. 5b). Material from the elevated areas to the north and south of trench E has accumulated in the center of the saddle after the Nok period has ended. This may be the reason why it only contains a few finds, mostly undecorated potsherds.

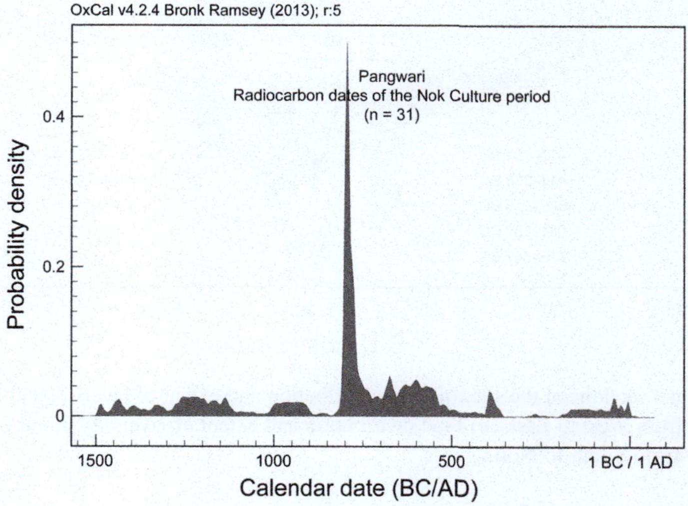

Fig. 4. Summed probability density graph of the 31 radiocarbon dates from Pangwari falling within the Nok Culture period (OxCal 4.2, IntCal13, 95.4% probability; using the "sum" command). The program produces the sum of the ^{14}C dates' individual probability distributions for each year (Bronk Ramsey 2009; Reimer et al. 2013). The high peak around 800 BC is caused by the large number of ^{14}C dates that have their highest probability at the beginning of the calibration curve plateau.

Different types of features occur at Pangwari: eight shallow pits or find accumulations, seven deep pits, two assumed graves, one terracotta deposition, one AD stone circle, and one AD hearth (Schmidt 2014). The analyzed charcoal samples come from

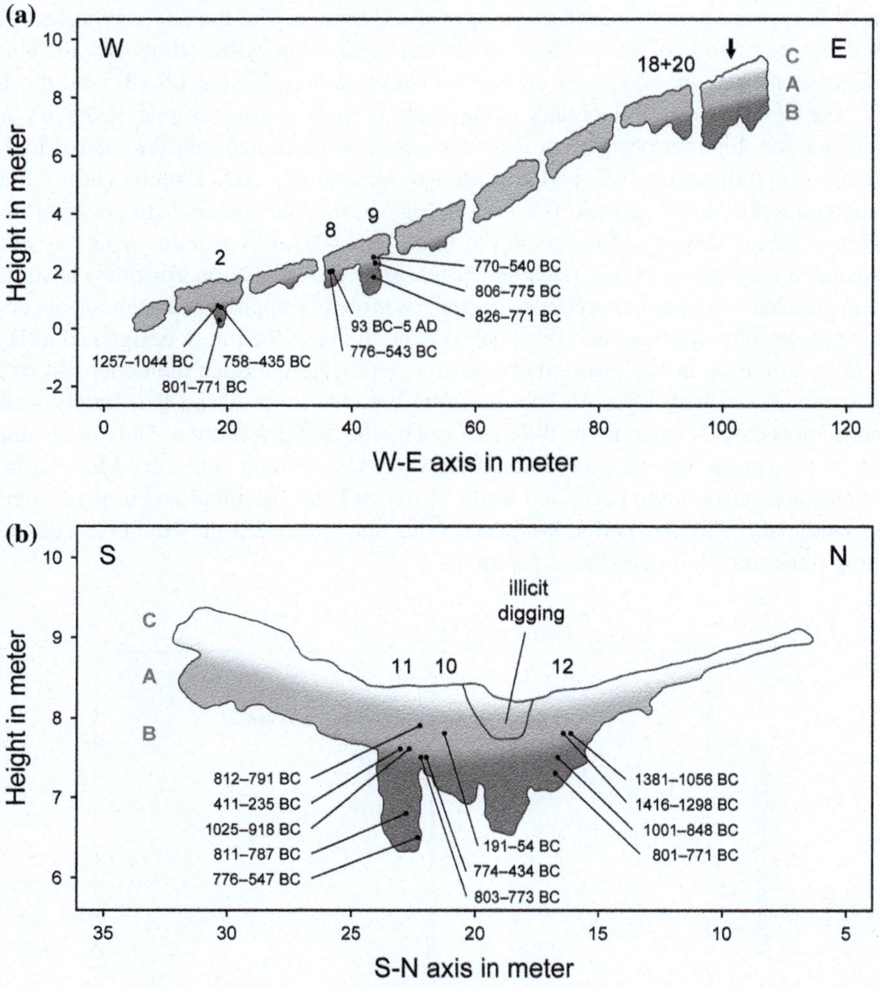

Fig. 5. Virtual sections of the site. (a) Find distribution and extent of layers A and B per trench, lateral view from West to East. (b) Find distribution and extent of layers A, B and C in trench E, lateral view from South to North.

several of the deep pits, which are up to 1.5 m deep and mostly identifiable by a distinct change in color. While the shallow find accumulations were most likely formed as a result of hillside erosion at natural barriers, as material transported by erosion accumulated at the rocky formations, the deep pits have remained in situ.

Charcoal Analyses—Material and Method

Twelve charcoal samples from the deep pits have been analyzed so far (Table 1). The charcoal fragments derive from the large fraction (2.5 mm^2 mesh size) of the flotated sediment (=archaeobotanical) samples. Between one and twelve liters of sediment were processed. Five samples came from feature 9 (find no. 2910, 3105, 3199, 3406, 3753), two samples each from feature 2 (1105, 1352), feature 8 (3004, 3246) and feature 11 (6549, 6636) and one from feature 12 (5224).

The find numbers roughly reflect the depth of the sample in so far as they have been assigned to all finds consecutively during the excavation. Higher numbers thus are from lower parts of the pits. For all finds, regardless of find category (e.g. pottery, terracotta, stone, etc.), three-dimensional coordinates have been recorded by total station allowing the spatial reconstruction of find distributions and thus comprehending the context of the analyzed samples.

Between 30 and 80 fragments per sample, adding up to a total of 709 fragments, were analyzed with a reflected-light microscope (Leica DM 4000 M) after manually fracturing them along the three planes of wood (transverse, longitudinal tangential and longitudinal radial). Fragments with similar wood anatomical characters were grouped into types and the types were identified. Each type was assigned a name using the InsideWood database (InsideWood 2004-onwards), the reference collection of microscopic wood slides of the Frankfurt Archaeobotanical Laboratory and the Frankfurt DELTA anatomical database of African woods (Dallwitz 1980; Neumann et al. 2001). The differences between charred and fresh wood (e.g. Prior and Gasson 1993) were taken into consideration during the identification process. The names of the charcoal types are given in small capitals to discriminate the anatomical charcoal types from the botanical taxa, with which they are not identical (Joosten and de Klerk 2002; Höhn and Neumann 2018). The suffix spp. indicates that several species of this genus share the characters of the charcoal type. If more than two genera have the same set of characters, tribe, subfamily or family names are assigned; if several subgroups can be distinguished within this taxonomical unit, it is indicated by attached Roman numbers (e.g. *Caesalpinioideae* I).

Ecological interpretations rely mainly on Arbonnier (2002), additional references are cited where appropriate.

Results

Charcoal Types

Charcoal types that are present in three or more of the analyzed samples (Table 1) are characterized in alphabetical order of the families they are assigned to. Because the focus is on taphonomy, types present in only one or two samples with not more than one to two fragments are omitted from the descriptions but given at the end of this list. Only diagnostic and few general characters of the types are given; otherwise reference is made to published descriptions. Most of the types comprise several species, of which

Table 1. Results of charcoal analysis. Find numbers roughly reflect stratigraphy within pits: lower numbers indicate location in upper parts of pits, higher numbers in lower parts.

Trench	B	B	D	D	D	D	D	D	D	E	E	E		
Feature	2	2	8	8	9	9	9	9	9	11	11	12		
Find no.	1105	1352	3004	3246	2910	3105	3199	3406	3753	6549	6636	5224		
Sample volume [l]	10	11	6	9	12	6	1	5.5	5	9	7	12		
Associated pottery group	Middle Nok undiff.		Pangwari		Pangwari/Ifana				Ifana	Pangwari		Puntun Dutse		
14C-date	BC 754-416	BC 799-676	-	BC 93-17 AD	BC 766-525	-	-	BC 804-771	BC 810-770	-	BC 809-798	BC 1380-1057	Sum per type	Ubiquity of type
Anogeissus leiocarpa	13	1	5	9	24	16	16	26	20	49	15	35	229	100%
Caesalpinioideae	4	6	10	25	16	4	34	4	14	4	9	6	136	100%
Phyllanthaceae I	9	1	4	1	.	2	1	1	18	1	.	10	48	83%
Parkia spp.	2		5	4	3	10	3	3	1	.	.	.	31	67%
Terminalia spp.	5	.	.	.	3	3	5	3	3	.	1	1	24	67%
Uapaca spp.	.	.	3	2	3	3	4	4	.	.	1	1	21	67%
Khaya spp.	.	.	.	1	7	1	1	4	.	.	.	6	20	50%
Lophira spp.	.	.	.	.	7	7	2	13	7	.	.	.	36	42%
Flueggea/Hymenocardia	6	.	5	1	.	.	.	.	.	.	6	6	24	42%
Sapindaceae I	1	.	.	1	.	1	.	1	.	.	.	5	9	42%
Syzygium spp.	.	.	.	.	6	.	2	1	3	.	.	.	12	33%
Sapotaceae	1	.	2	.	.	.	.	.	1	.	4	.	8	33%
Pericopsis spp.	.	.	.	.	.	.	1	1	.	3	2	.	7	33%
Chrysobalanaceae I	.	.	.	.	.	.	1	1	1	.	.	1	4	33%
Rubiaceae I	1	.	.	.	1	.	1	1	.	.	.	.	4	33%
Detarieae I	.	21	.	1	.	.	.	.	.	.	.	4	26	25%
Combretum glutinosum	.	.	.	.	.	.	1	.	1	.	1	.	3	25%
Others (no. types)	1(1)	.	2(2)	.	2(1)	2(2)	4(3)	3(3)	1(1)	1(2)	5(2)			
Indet.	7	1	4	4	8	1	4	4		1	6	4		
sum of analyzed fragments	50	30	40	49	80	50	80	70	70	60	50	79	708	
number of types per sample*	10	4	9	9	10	11	16	16	11	6	10	10		

* Indet. (non-identifiable fragments) excluded

[a]Indet. (non-identifiable fragments) excluded

those typically present in the Sudanian and Guinean zones of Nigeria according to Keay (1989) are listed.

***Chrysobalanaceae* I**

Solitary and large vessels, apotracheal parenchyma, mostly uniseriate homocellular rays of procumbent cells with abundant silica characterize this type (Höhn and Neumann 2016; Höhn 2005, as *Parinari curatellifolia* type). Species with matching wood anatomy are *Maranthes polyandra* (Benth.) Prance, *M. kerstingii* (Engl.) Prance ex F. White, *Parinari curatellifolia* Planch. ex Benth., *P. congensis* Didr. and *Neocarya macrophylla* (Sabine) Prance ex F. White.

***Anogeissus leiocarpa*, Combretaceae**

This type is characterized by small vessel groups, scanty to vasicentric parenchyma, uni- to biseriate rays with rows of square cells (often with prismatic crystals) within rays of mostly procumbent cells (Neumann et al. 1998; Höhn 2005; Höhn and Neumann 2016).

***Combretum glutinosum*, Combretaceae**

The only type in the assemblage with included phloem. Vessels are solitary, tend to be of two different sizes, with larger diameters around 100 μm, and have vestured pits. The rays are slightly heterocellular and mostly uniseriate (Neumann et al. 1998; Höhn 2005).

***Terminalia* spp., Combretaceae**

Solitary and grouped vessels, aliform parenchyma of 4 and more cells per strand and uni- to biseriate heterocellular mixed rays are main characters of this charcoal type (Neumann et al. 1998; Höhn 2005). Probable matches are *T. glaucescens* Planch. ex Benth., *T. laxiflora* Engl., *T. macroptera* Guill. & Perr. and *T. mollis* M. A. Lawson.

***Caesalpinioideae* I, Fabaceae, Caesalpinioideae**

The rays are mostly biseriate and exclusively homocellular and thus different from *Detarieae* I. Otherwise the type is similar: solitary and grouped (2–3) vessels with vestured pits, aliform and confluent parenchyma, and crystals in chambered axial parenchyma are main characters (Höhn and Neumann 2018). While many Detarieae and some other Caesalpinioideae have this kind of wood anatomy, some of the more probable matches are *Afzelia africana* Pers., *Tamarindus indica* L. and *Erythrophleum africanum* (Benth.) Harms.

***Detarieae* I, Fabaceae, Caesalpinioideae**

Depending on the preservation and assemblage this type often is part of *Caesalpinioideae* I (Höhn and Neumann 2016). In Pangwari as in some other sites (Kahlheber et al. 2009), the occurrence of fragments with mostly uniseriate homocellular rays, but occasional occurrence of rays with distinctly enlarged cells in the marginal rows, and strictly aliform parenchyma allowed to differentiate between the two types. Probable matches are *Isoberlinia doka* Craib & Stapf, *I. tomentosa* (Harms) Craib & Stapf and *Berlinia grandiflora* (Vahl) Hutch. & Dalziel.

***Pericopsis* spp., Fabaceae, Faboideae**
Axial parenchyma and rays are storied. Other characters are solitary and small groups of vessels with vestured pits, aliform and confluent parenchyma of mostly 2-celled strands, 2- to 3-seriate rays with exclusively procumbent cells and crystals in chambered axial parenchyma (Höhn and Neumann 2016). The most probable match is *Pericopsis laxiflora* (Baker) Meeuwen.

***Parkia* spp., Fabaceae, Mimosoideae**
Solitary and grouped (2–3) vessels with vestured pits, aliform and confluent parenchyma, fibres in radial alignment, mostly 5-seriate rays of procumbent cells, together with long chains of crystals in chambered parenchyma characterize this type (Höhn 2005, as cf. *Parkia biglobosa* type; Höhn and Neumann 2016). The general wood anatomy is quite similar to that of *Caesalpinioideae* I but the larger rays separate the types. *Parkia biglobosa* (Jacq.) G. Don is widespread in Sudanian and Guinean savannas.

***Khaya* spp., Meliaceae**
Typical characters are septate fibres, solitary and grouped (2–3) vessels with small pits, scanty to vasicentric parenchyma, mostly 5–6-seriate rays, with 1–2 rows of upright/square marginal cells that may contain prismatic crystals (Neumann et al. 1998 as *Khaya senegalensis* type). *Khaya senegalensis* (Desv.) A. Juss. and *K. grandifoliola* C. DC. are possible matches.

***Syzygium* spp., Myrtaceae**
Combination of vestured pits and vessel-ray pitting with much reduced borders is characteristic. Vessels are mostly in radial rows of 2–3 cells; parenchyma is mostly aliform and confluent; heterocellular, often mixed, mostly 3-seriate rays are present together with uniseriate rays with predominately square/upright cells (Neumann et al. 2009). *Syzygium guineense* (Willd.) DC. is the most widespread and abundant *Syzygium* species in Nigeria.

***Lophira* spp., Ochnaceae**
Large vessels with minute to small intervessel pits, wide (3–5 cells) parenchyma bands, often touching the vessels unilaterally, parenchyma strands of 7–10 cells and 2-3-seriate rays of procumbent cells are typical for this charcoal type (Höhn and Neumann 2018). In humid savannas *Lophira lanceolata* Tiegh. ex Keay is widespread and abundant.

***Phyllanthaceae* I**
The type is very similar to *Flueggea/Hymenocardia* and is only distinguished by its larger vessels (>100 μm). The other characters are identical: Vessels solitary and in groups of 3 and more, vessel-ray pits with reduced borders, scanty parenchyma, septate fibres, heterocellular mixed 4-5-seriate rays, uniseriate rays of square/upright cells, and crystals in square/upright ray cells (Höhn and Neumann 2016). Probable matches are *Antidesma venosum* E. Mey. ex Tul., *Bridelia* spp. and *Margaritaria discoidea* (Baill.) G. L. Webster.

***Flueggea/Hymenocardia*, Phyllanthaceae**

While otherwise similar to *Phyllanthaceae* I, the diameter of the vessels is very small with less than <50 µm (Höhn 2005 as *Flueggea virosa/Hymenocardia acida*, Höhn and Neumann 2016). *Flueggea virosa* (Roxb. ex Willd.) Royle, *Hymenocardia acida* Tul. and *H. heudelotii* Müll.Arg. are probable matches.

***Uapaca* spp., Phyllanthaceae**

This charcoal type differs from *Phyllanthaceae* I because of exclusively non-septate fibers, silica in ray cells and some scalariform perforations. Vessels are solitary and in groups of 2–3, vessel-ray pits are gash-like with much reduced borders, parenchyma is scanty, and rays are heterocellular and mostly 4-5-seriate (Höhn and Neumann 2016). Possible matches are *Uapaca togoensis* Pax and *U. heudelotii* Baill.

***Rubiaceae* I**

Small (<50 µm), mostly solitary vessels with vestured pits, scanty parenchyma and 3-4-seriate rays with several rows of upright/square cells and low procumbent cells characterize this charcoal type (Neumann et al. 1998 as *Feretia* type; Höhn 2005; Höhn and Neumann 2016). Numerous species of the family share a similar wood anatomy, among them *Feretia apodanthera* Delile and several *Gardenia* species.

***Sapindaceae* I**

Characteristic is the combination of scanty, hardly visible parenchyma, septate fibres, mostly uniseriate, mostly homocellular rays of procumbent cells occasionally containing crystals in radial alignment and vessels which are either solitary or in radial multiples of up to 4 (Höhn and Neumann 2016). In some of the Pangwari fragments helical thickenings on the vessel walls are visible; they are not coded for *Lecaniodiscus cupanoides* in Inside Wood (2004-onwards) but still described for the genus in Metcalfe and Chalk (1950). For the Pangwari assemblage probable matches are thus *Blighia sapida* K. D. Koenig and *L. cupanoides* Planch.

Sapotaceae

The type is characterized by groups of mostly two vessels, often with tyloses, irregular and narrow (2-celled) banded parenchyma, 2-4-seriate, heterocellular rays with several rows of upright/square cells, some vessel-ray pits with reduced borders and silica in ray cells and parenchyma (Neumann et al. 1998, Höhn 2005 as *Vitellaria paradoxa* syn. *Butyrospermum paradoxum* types, Höhn and Neumann 2016). Probable matches are *Vitellaria paradoxa* C. F. Gaertn., *Manilkara obovata* (Sabine & G. Don) J. H. Hemsl., *Malacantha alnifolia* (Baker) Pierre and *Pachystela pobeguiniana* (Dubard) Pierre ex Lecomte.

14 types are present in only one or two samples. These are: *Millettia* spp., *Petrocarpus* spp., *Celtis* spp. (see Höhn and Neumann 2016), *Rubiaceae* II, *Ziziphus* spp., *Detarium* spp. (see Höhn 2005), *Oncoba* spp., *Anisophyllea* spp., *Mareya* spp. (see Höhn and Neumann 2018) and a monocotyledon type. The remaining four types have only been tentatively assigned to resemble wood from *Acacia*, *Vitex*, *Zanthoxylum* and the Apocynaceae family; definite wood types have not been established yet due to their scarcity in the assemblage so far, but investigations into the charcoal of Pangwari are ongoing.

Charcoal Assemblage

A. Leiocarpa and *Caesalpinioideae* I are ubiquitous charcoal types and present in every sample. The 229 fragments of *A. Leiocarpa* make up one third of all fragments; together *A. Leiocarpa* and *Caesalpinioideae* I (136 fragments) account for about half of them (Table 1). *Phyllanthaceae* I is also present in every feature but missing in two samples. *Terminalia* spp., *Parkia* spp. and *Uapaca* spp. are present in more than half of the samples, but each type is totally absent from one of the investigated. The presence of *Lophira* spp. is striking; it is exclusively present in all samples from feature 9 and, with the exception of sample 3199, always with at least 7 fragments, equaling about 10%. *Syzygium* spp. shows a similar distribution, but with lower fragments counts, and is absent in sample 3105. *Sapindaceae* I, *Sapotaceae*, *Pericopsis* spp., *Chrysobalanaceae* I, *Rubiaceae* I and *Combretum glutinosum* are present with a single or very few fragments in 25–33% of the samples. The fragment counts in the different samples are very variable, indicating that neither absence nor frequency represent much information. Possible exceptions are samples 1352 (feature 2) and 6459 (feature 11). In both samples one type represents more than 70% of the fragments and accordingly diversity is low with only four, respectively six types (Table 1). In sample 1352 *Detarieae* I dominates with 21 out of 30 analyzed fragments and in sample 6549 it is *A. Leiocarpa* with 49 out of 60 fragments. The highest diversity is present in samples 3406 and 3199 from feature 9. They contain 16 types, while the average is around 10 types. A possibly higher concentration of charcoals in feature 9 is indicated by the higher number of analyzable fragments, even in a soil sample of just one liter (3199, Table 1). Saturation curves have been compiled for all analyzed samples and substantiate that 50 fragments are sufficient to catch the most ubiquitous and well represented types in this context, as the curve flattens between 30 and 40 fragments. Thus, they represent the samples diversity, which is not at all to equate with the diversity of the environment.

The Features and Their Charcoal Assemblages

Feature 2 is situated in trench B and consists of a convex-shaped pit with a depth of c. 70 cm. The pit filling, assigned to layer B, is considered in situ. It is overlain by layer A which is about 30 cm thick (Fig. 5a) and is present throughout trench B. Three radiocarbon dates were obtained—two in the pit's upper part (1257–1044 BC and 758–435 BC) and one at the bottom of the pit (801–771 BC). While the upper dates confirm the mixture of material, which is typical of the upper parts of pits at Pangwari, the one at the bottom dates the in situ finds to the Middle Nok period. The finds consist of a few poorly preserved potsherds and terracotta fragments as well as charcoal and burnt clay. Due to the potsherds' poor preservation, no classification into a specific pottery group was possible; they can only be addressed as Middle Nok pottery. Their distribution is almost exclusively restricted to feature 2, whereas terracotta fragments are found throughout trench B (Fig. 6). Charcoal from two archaeobotanical samples (1105,

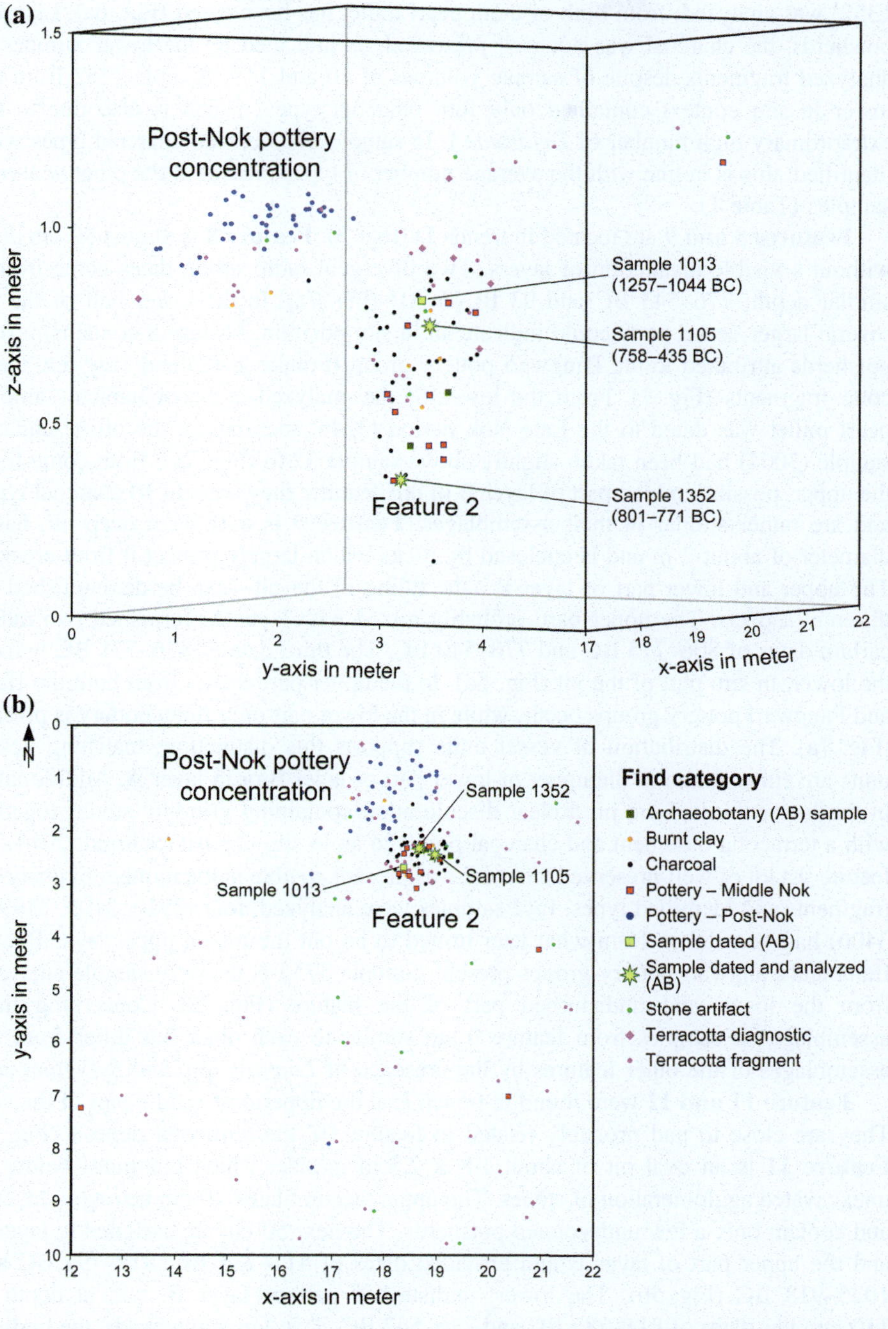

Fig. 6. Trench B. (a) Three-dimensional plot of significant finds, archaeobotanical samples and dated samples in the 5 × 5 m section containing feature 2. The Middle Nok pottery cannot be assigned to specific pottery groups. (b) Two-dimensional plot of significant finds and samples in the 10 × 10 m section containing feature 2.

1352) was analyzed; from both of them pearl millet has been dated (Fig. 6a). Like the potsherds, the charcoal was not well preserved, as indicated by the lower number of analyzed fragments despite of sample volumes of 10 and 11 l. Sample 1352 from the lower in situ context contained only four charcoal types, which is also due to the extraordinary high number of *Detarieae* I. In sample 1105 eleven charcoal types were identified; this is in line with the average number of types present in the other analyzed samples (Table 1).

Features 8 and 9 are located in trench D (Fig. 3). **Feature 8** is almost 80 cm deep without a visible distinction of layers. Two divergent radiocarbon dates come from a similar depth: 776–543 BC and 93 BC–5 AD (Fig. 7a). In the lower half of the pit several larger vessel parts could indicate an in situ position. Feature 8 contains mostly potsherds attributed to the Pangwari pottery group (Franke 2017) and only few terracotta fragments (Fig. 7). From the lower of the analyzed archaeobotanical samples pearl millet was dated to the Late Nok period (3246, see above); the other, undated sample (3004) had been taken slightly above sample 3246 (Fig. 7a). Both come from the upper, possibly mixed part of layer B in this feature; they contain 10 charcoal types and are rather similar in their assemblages. **Feature 9** is a 100 cm deep pit, has a diameter of about 2 m and is enclosed by rocks which largely protect it from erosion. The upper and lower part of layer B—the filling of the pit—can be distinguished by different aspects. The upper part, probably mixed with layer A, furnished two radiocarbon dates of 806–775 BC and 770–540 BC. The third date of 826–771 BC is from the lower, in situ part of the pit (Fig. 5a). In the upper part sherds from both the Ifana and Pangwari pottery groups occur, while in the lower part only Ifana pottery is present (Fig. 7a). The distribution of vessel units supports this distinction; matching vessel units are either found in the upper or lower part of layer B or in layer A. Additionally, in the lower third of the pit a clear discoloration containing grinding stones together with a terracotta basement and charcoal points to an in situ deposit (Schmidt 2014). In feature 9 a lot of well-preserved charcoal was present, visible in the number of analyzed fragments and identified types. Five samples were analyzed, four (2910, 3105, 3199 & 3406) had been taken from what later turned to be out the mixed upper layer B with Ifana and Pangwari pottery groups present. Sample 3753 is the only sample analyzed from the lower and undisturbed part of the feature (Fig. 7a). Concerning their assemblage all samples from feature 9 are similar to each other but differ from the assemblages of the other features by the presence of *Lophira* spp. and *Syzygium* spp.

Feature 11 and 12 were found in trench E at the uppermost saddle part of the site. They are close to and probably related to feature 10, the terracotta deposit (Fig. 8). **Feature 11** is an oval pit of about 1.8 × 2.5 m in size, which continues below an unexcavated agglomeration of stones. The upper, accumulated 40 cm belong to layer C and contain only a few undecorated potsherds. The next 50 cm are assigned to layer A and the upper part of layer B and provided dates of 411–235 BC, 812–791 BC and 1025–918 BC (Fig. 5b). The lower, undisturbed part of layer B, with a depth of 110 cm, has dates of 811–787 BC and 776–547 BC. Two terracotta heads, the body of a kneeling person holding a quadruped (Männel and Breunig 2016) as well as a grinding stone and fragments of two pots were excavated from feature 11. The majority

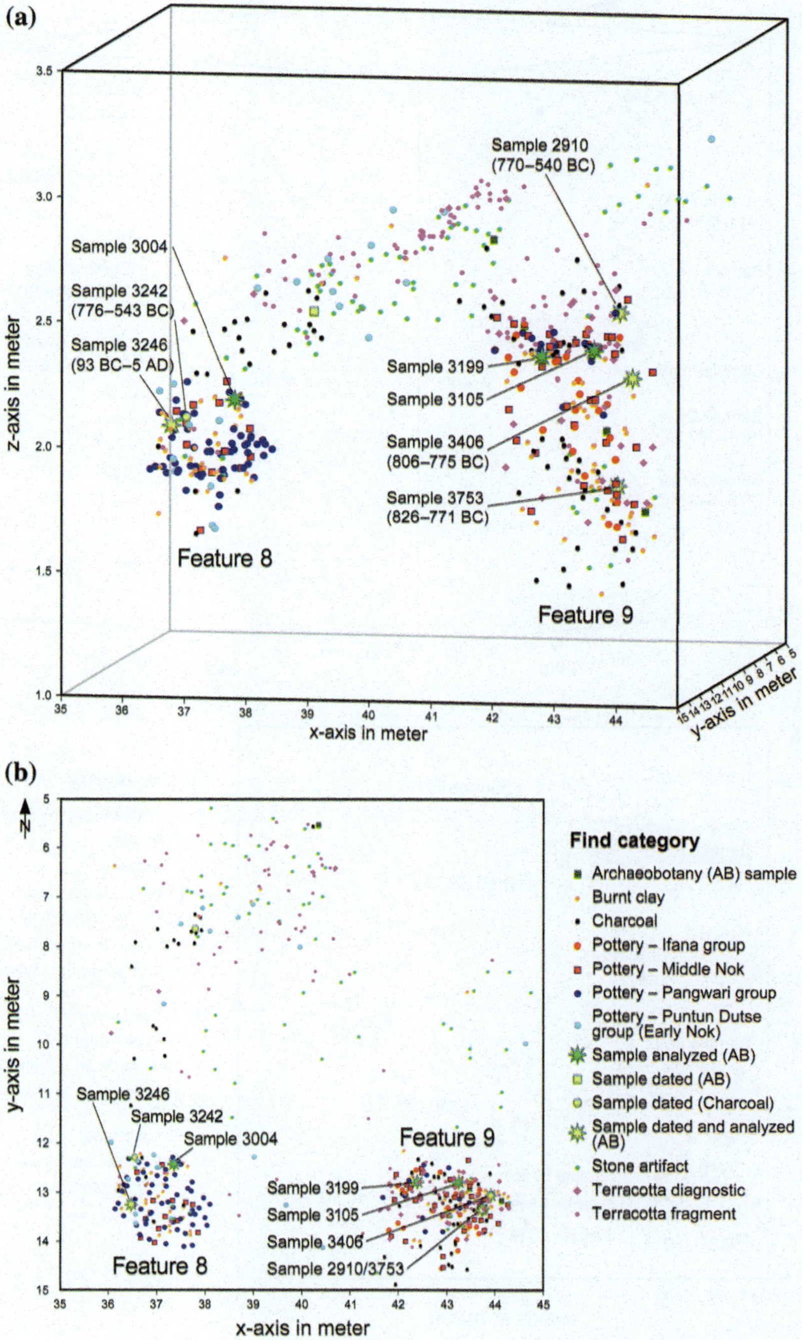

Fig. 7. Trench D. (a) Three-dimensional plot of significant finds, archaeobotanical samples and dated samples in the 10 × 10 m section containing features 8 & 9. (b) Two-dimensional plot of significant finds and samples in the 10 × 10 m section containing features 8 & 9.

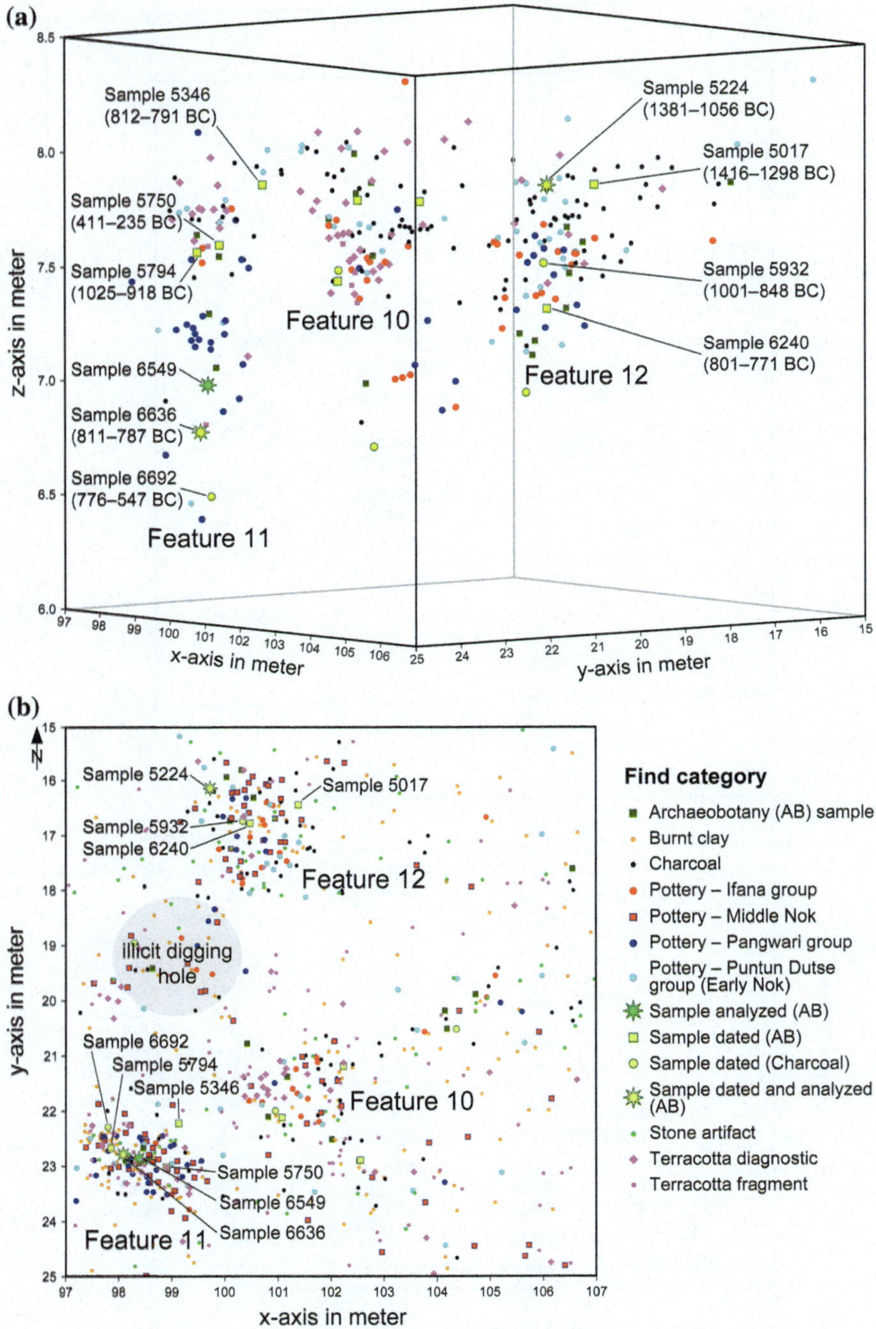

Fig. 8. Trench E. (a) Three-dimensional plot of significant finds, archaeobotanical samples and dated samples in the 10 × 10 m section containing features 10–12. (b) Two-dimensional plot of significant finds and samples in the 10 × 10 m section containing features 10–12. The outline of the illicit digging hole is approximated.

of the sherds belong to the Pangwari group (Fig. 8). Two charcoal samples were analyzed. While the lower, dated sample 6636 is not different from the general picture concerning diversity and presence of types, sample 6549 is of low diversity: 49 of the analyzed 60 fragments belong to *A. Leiocarpa*. This is not only the highest percentage (81%) but also the highest absolute number of this type in any of the samples analyzed from Pangwari. **Feature 12** is a 90 cm deep pit starting about 30 cm beneath layer C (Fig. 5b). Layer A and B cannot be separated by the find distribution, but the dispersion of vessel units allows a distinction of two horizons, of which the upper one furnished older dates (1381–1056 BC and 1416–1298 BC) than the lower one (1001–848 BC and 801–771 BC). The lower horizon contains a grinding stone and a terracotta basement as well as Early Nok (Puntun Dutse group) and Middle Nok potsherds (Pangwari and Ifana group) (Fig. 8a). Terracotta fragments, a mixture of potsherds, and stone artifacts were found in the upper horizon; at least some finds may derive from the spoil of an illicit digging hole adjacent to feature 12 (Fig. 8b). Only one sample from the upper part of feature 12 has been analyzed (Fig. 8a). Like in feature 11 *A. Leiocarpa* is well represented with 35 fragments, corresponding to 44% of the analyzed fragments. But in general the sample does not differ from the others.

Other features, essential for understanding the function of the site, are **feature 10** in trench E and **features 18 and 20**, two possible graves, in trench I (Fig. 3). Charcoal assemblages of these and other features, especially from trench I, are currently investigated and will be published elsewhere. **Feature 10** contained several larger terracotta parts covered by a stone pavement at a depth of 90 cm, beneath layer A and C (Fig. 9c). Unlike the nearby feature 11, no discolorations to identify its extension were recognized. The terracotta parts include a well-preserved large torso, broken up in two pieces, a second torso (Fig. 9a), a single hand with a bracelet, the head and upper body of a man with a shell decoration on his forehead (Fig. 9a), two shattered heads (not matching any of the other fragments) as well as the head of a therianthropic being. An interpretation as a ritual act of destruction seems a reasonable explanation, considering that the fragments cannot be assembled to form complete figures (Männel and Breunig 2016). **Features 18 and 20** are located in trench I and consist of unmodified and grinding stones arranged upright and one, respectively two complete pots (Fig. 9b). This type of arrangement is known from other Nok sites; at Ido, the additional presence of stone beads arranged as if on a string (forming a necklace) allows the interpretation as a grave (Rupp 2014). The two features are 7–10 m away from the terracotta deposition, still on the saddle. They emerge 70 cm beneath layer A, isolated in a layer of finer texture and void of any finds other than the complete pots, which supports the interpretation as burials. Charcoal from below the pot in feature 18 was just recently dated to 780–541 BC; this puts the graves into the Middle Nok phase and, based on the pot's style, into the Pangwari pottery group.

Fig. 9. (a) Nok terracotta figurines depicting a man with a shell on his head and a male torso (feature 10, trench E). (b) Stone-pot-arrangements (features 18 and 20, trench I) interpreted as graves. (c) View of the terracotta deposition (feature 10, trench E), back of the male torso visible.

Discussion

Site History

The site of Pangwari has been used several times during the last 3500 years—with different intensities. The earliest traces of human occupation date back into the second half of the second millennium BC (Early Nok). Some concentrations of the characteristic Puntun Dutse pottery are present in trenches E and I, both located on the saddle, indicating that Early Nok activities were concentrated there. From there, erosion has probably washed potsherds down the slope, since Puntun Dutse pottery is found in almost all trenches in layer A (Figs. 7 and 8). Some Early Nok ^{14}C dates and potsherds are also present in some of the Middle Nok pits; they probably entered the pits as part of the backfill material or were washed into the pits. Since no Early Nok features are identified, no conclusion can be drawn as to the character of the site during this period. In any case, the period between the late 9th and possibly 5th century BC (Middle Nok) is the main phase of activity at the site (see Fig. 4 for a summed probability density graph). The first pits were dug around 800 BC (stage1) and filled with Ifana group pottery and other finds. Only the lower parts of feature 9 and possibly feature 2 are in situ remains of these activities. Somewhat later (stage 2), new pits were added and filled with Pangwari group pottery, and obviously older pits were reused as indicated by the mixture of material and dates in the upper parts of layer B (Figs. 7a, 8a); whether this happened accidentally or intentionally remains unclear. Most probably the older pottery and organic material entered the stage 2 pits as backfill or by erosion from the surface—causing a mixture of Early Nok and Middle Nok stage 1 and 2. The obvious reuse of older pits is also an argument for a brief span between the two stages.

After this main phase of activity that supposedly took place in the 9th and 8th century BC, occupation of the site is hardly traceable. There is no evidence of activity in the later part of the Middle Nok phase until c. 400 BC. No dates fall into this period and the typical pottery groups (Tsaunim Gurara, Pandauke) are also absent. Evidence of the Lake Nok phase is also weak: only three ^{14}C dates on pearl millet grains (411–235 BC, 191–54 BC, 93 BC–5 AD; Fig. 5) may indicate some occasional stays. While there are at least some dates for the final centuries of the Nok Culture, there is no evidence for any occupation in the first millennium AD. Only after 1000 AD several ^{14}C dates attest various occupation episodes (1000–1200 AD, 1450–1650 AD, after 1700 AD). The concentration of AD potsherds close to the surface near feature 2 probably belongs to the period around the middle of the second millennium BC. The stone circle (Fig. 3, SK) to the north of trenches D and F also belongs to the Common Era, but has not been dated specifically.

Feature History. Can We Rely on the Dates?

The temporal correlation between charcoal assemblages and ^{14}C dates retrieved from macroremains of the same archaeobotanical samples is disturbed by the human activities in stage 2, which resulted in a severe mixture of older and younger materials. Only the charcoal assemblages located in the lower undisturbed in situ pit fills of feature 2 and 9, which are dated to stage 1, allow a temporal correlation between dates and assemblages. Accordingly, two charcoal assemblages can be considered representative for the exploitation of woody vegetation in the vicinity of the site around 800 BC: Sample 3753 from feature 9 and sample 1352 from feature 2. All other samples come from sediments that have been disturbed at some point of their history, either by recurrent human activities or through erosion, the influence of which is not to be neglected, especially for layer A.

No representative charcoal assemblages are present for stage 2; all samples associated with stage 2 dates come from mixed contexts. The best choice for tracing a stage 2 assemblage could have been feature 8. The pit probably was newly dug during stage 2 and filled with the correspondent Pangwari group pottery. But, whether as part of the backfill or washed into the pit from the sides, some Puntun Dutse potsherds found their way into the pit as well (Fig. 7a). Thus, a mixture of Early Nok and stage 2 organic material also has to be expected even at the bottom of the pit. The same is true for feature 11; it also consists mainly of Pangwari group pottery, but the few Puntun Dutse potsherds suggest that the infill of this pit also contains surface or layer A material (Fig. 8a).

Initially, two radiocarbon dates on pearl millet suggested samples originating from before (sample 5224, feature 12, 1381–1056 BC) and after (sample 3246, feature 8, 93 BC–5 AD) the Middle Nok peak activities at Pangwari. But the final merging of all information from find distribution and ceramic chronology revealed that these samples also derive from mixed sediments. The Early Nok date comes from the upper part of feature 12, from material that had been disturbed during the activities in Middle Nok stage 1 and 2. However, the relatively large number of three Early Nok dates in the upper part of pit 12 (Fig. 8a) might indicate that an Early Nok feature had been present in the environs of the feature before the Middle Nok pit was dug. How the Late Nok date found its way into feature 8 is unclear; bioturbation seems the best explanation since there is no other evidence for Late Nok material in this pit and in this depth.

Summing up, the different ^{14}C dates of Pangwari on pearl millet grains just prove that pearl millet was prepared at the site at the dated times. But due to the mixed character of layer A and of the upper parts of layer B, most of the radiocarbon dates do not necessarily represent the age of all archaeobotanical (including charcoal) remains retrieved in the respective samples. Though it is quite probable that the dated grains are not the only organic material dating to these periods, it is not possible to separate the remains from one period or event from those dating to other events—unless submitting every single macro-remain to radiocarbon dating. As this is no practical solution, answers have to be found within the analyzed material: The different assemblages from mixed contexts do resemble each other, especially concerning the ubiquitous and better represented charcoal types. Moreover, the assemblages do not differ in general composition from the in situ sample 3753 from feature 9 (except for *Lophira* spp. and *Syzygium* spp. which will be discussed in the next paragraph). Also, considering that most pottery sherds as well as radiocarbon dates belong to stage 1 and 2, we can assume that most of the organic material entered the site during these peak activity stages. Many Middle Nok charcoal fragments should thus be mixed with a few earlier or later organic remains. Consequently, the information from the majority of the Middle Nok organic material should override the information from the little organic material from previous, and subsequent events and the mixed samples should predominantly represent Middle Nok vegetation exploitation.

Sample History. How Representative Are the in situ Assemblages?

Only the two samples from in situ contexts certainly represent exploitation of the woody vegetation precisely in stage 1. The types present in the two assemblages must have been present in the vicinity; however, these samples are not representative for the woody vegetation at Pangwari in general. There is no direct relation between the assemblages of the two samples and the woody vegetation as indicated by the considerable differences between the two assemblages. Each assemblage represents just one small subsample of the woody vegetation probably collected during one event at that time. In sample 3753 from feature 9 *Lophira* spp. and *Syzygium* spp. are present (as in the other samples from feature 9), but both types are absent in the other features. Sample 1352 from feature 2 is dominated by *Detarieae* I, which is only present with few fragments in two more samples from other pits (Table 1). These salient assemblages do not result from several collection events and several fire events, which slowly accumulated in the pits, but rather are derived from one fire event, which contained wood of *Lophira* spp. and *Syzygium* spp. in the case of feature 9 and *Detarieae* I in the case of feature 2. While these samples do not match the requirements for environmental reconstructions, in particular that to be derived from several collection events (Chabal 1992), they allow conclusions regarding the taphonomy of the two pits and support the already outlined history of the features.

As explained, **feature 9** was dug in stage 1. It was filled with the Ifana pottery and the remains of a fire, which contained, amongst others, charcoal from *Lophira* spp. and *Syzygium* spp. as indicated by the two distinct charcoal types only present in this feature. When the other—later—pits at Pangwari were filled, the remains of fires did not include these two types, whether by chance or by intention cannot be told. The amount of charred material from *Lophira* spp. and *Syzygium* spp. must have been large enough to survive the possible reuse of the pit during stage 2. During the activities in this location in stage 2, parts of the charred large branches deposited in stage 1 were removed and later reused as backfill—together with Pangwari group pottery. Fragments of *Lophira* spp. and *Syzygium* spp. thus were redeposited into the upper part of the feature. The two types indicate that feature 9 was created by two consecutive events of digging and reusing the pit and that it was quickly refilled. Archaeological evidence supports this interpretation. Strong discoloration and the good preservation in feature 9 are indications of quick fills, both in stage 1 and stage 2. Moreover, neither Early Nok Puntun Dutse pottery nor an Early Nok date is present in this feature. This implies that the material used for filling the pit could not have been taken from layer A, which usually contains some Early Nok material, and that the pit must have been closed or protected before the rainy season began as otherwise Early Nok material would have been washed into the pit. The clear discoloration containing a grinding stone, a terracotta basement and charcoal in the lower, in situ part of the pit is another fact supporting the distinction between the lower and upper part of feature 9, as is the fact that matching vessel units are only found in either the upper or lower part (Schmidt 2014).

The in situ sample 1352 from the undisturbed part of **feature 2** is striking as well. It contains only four charcoal types, one of them *Detarieae* I, which dominates the assemblage. Like in feature 9 remnants of a fire, in this case containing branches of the *Detarieae* I type, were discarded at the bottom of the pit, together with not specifically classified Middle Nok pottery. However, in this case the branch or branches came to lie below the area of the pit that was reworked during stage 2 and thus did not blend with the material in the upper part that was added in stage 2 and led to the deposition of organic material dating to Early Nok and stage 2 times. The other charcoal assemblage from feature 2 comes from this mixed, upper context and is similar to those from the other features (Table 1). The lack of discoloration and the bad preservation of the pottery suggest, however, a slow refilling of feature 2, at least for the upper part containing the Early Nok date. On the other hand, the pottery could already have been in a bad condition before it entered the pit.

The Function of the Pits

The pits clearly date to the Middle Nok activities at the site and have to be seen together with the Middle Nok terracotta deposition and the graves, putting the function of the site in a ritual and funeral context (Schmidt 2014; Männel and Breunig 2016). The recurring presence of large vessel fragments and grinding stones associated with terracotta fragments indicate a temporal and functional correlation with the terracotta deposition and graves. Also, the quick nature of the refill of at least several features rather points to ritual context than to a gradual filling of a refuse pit. However, the presence of settlement inventory within the pits raises questions. Grinding stones and finds of pearl millet point to food preparation, which, however, must not necessarily have taken place in a household context. A ritual context like some kind of feasting is also conceivable.

Deep pits and the speculation about their function are not restricted to the Central Nigerian Nok Culture. They are present in many sites of West and Central Africa and several possible functions have been addressed (e.g. Wotzka 1993; Gronenborn 1997; Mbida Mindzie et al. 2000; Magnavita and Magnavita 2001; Meister and Eggert 2008; Seidensticker 2010, 2013). The different characteristics and functions of pits in relation to ethnographic surveys were treated seminally by Wotzka (1993) and an attribute catalogue was developed. A comparison shows that some characteristics of the pits at Pangwari contradict the interpretation as profane waste pits within a settlement: (a) The location of features 11 and 12 is too close to the terracotta deposition and the assumed graves to be unrelated to them; (b) The pits with depths of up to 1.5 m seem excessive for simple waste pits; and (c) The find inventories contain finds untypical of settlement waste, such as a terracotta statuette and a terracotta basement together with grinding stones (feature 9), mostly terracotta fragments and charcoal (feature 2), or large vessel fragments and complete pots associated with well-preserved terracotta pieces (feature 11). It cannot be ruled out that the deep pits initially served other purposes. They might have been used for the removal of material or for storage purposes, and were re-used during Middle Nok times, when the terracotta deposition and the graves were laid out. However, no clear functions have yet been determined for Nok pits excavated since traces of usage are rare and difficult to interpret. As the example of Kintampo pits in northern Ghana has shown, people not necessarily stored grain in pits but preferred above-ground facilities to allow for air circulation (Casey 2000), which also today is the case in our key study area. Although the assumption that ordinary storage or waste pits from Early Nok times were re-used during the Middle Nok phase for ritual purposes would fit our hypothesis, the state of preservation of such pits does not allow any such conclusion.

The Environment

In spite of the problems of taphonomy, erosion and disturbance, the charcoal assemblages are indicative of the woody vegetation in the vicinity of Pangwari around 800–700 BC. It is reasonable to assume that the majority of charcoal fragments entered the site during this main activity phase. At that time the vegetation at Pangwari was not heavily impacted by human influence as species typical for young fallows or open savanna are missing or rare in the assemblages. In the immediate surrounding of the site, on the slope itself, there were probably no fields and fallows; the charcoal types rather point to a dry forest or woodland. At the foot of the slope or on the less inclined slopes behind the hilltops of Pangwari, fields and older fallows might have been present, and some of the charcoal types might have been derived from this vegetation.

Dry forest or woodland as well as regrowth in old fallows are possible sources for *A. Leiocarpa*, which dominates the assemblages in ubiquity as well as in frequency. Because the type is well recognizable and the genus is mono-specific, we can be sure that all fragments are derived from *A. Leiocarpa*. The tree is present in most of the savanna areas from the semi-arid Sahel to the forest-savanna boundary. In the Southern Guinea savanna zone it occurs in several different vegetation types from the margins of forest outliers (Jones 1963) and of gallery forests to dry forests and old fallows. The tree tolerates temporary inundations, regenerates on rather shallow soils, resprouts after cutting, is moderately fire-resistant, and kills out grass effectively (Irvine 1961). The presence of *Caesalpinioideae* I in the assemblages may hint towards the exploitation of dry forests. *Afzelia africana* is a very common tree of West African savannas and its wood is a probable match for *Caesalpinioideae* I in the Pangwari assemblages. It is also typical for the tree stratum of dry forests as described for southern Togo (Ern 1979). When cut down, this dry forest is quickly replaced by humid Guinean savannas with *Lophira lanceolata* and *Daniellia oliveri* (Rolfe) Hutch. & Dalziel. The latter is missing from Pangwari, but *Lophira* spp. was regularly present in feature 9 and must have been present in stage 1 of the Middle Nok activities at Pangwari. *A. leiocarpa* and *Caesalpinioideae* I, possibly *Afzelia africana,* are probably overrepresented in the assemblage. Both species furnish high-quality firewood and charcoal; they might have been actively preferred when collecting firewood, whether it was for household fires, feasting or ritual purposes at the site.

Phyllanthaceae I, *Uapaca* spp. and *Terminalia* spp., which are regularly present in the samples, also might have been collected in the immediate vicinity of the site. *Antidesma venosum*, a possible match for *Phyllanthaceae* I, grows on shallow soils as present at Pangwari today, as does *Uapaca togoensis*. Some *Terminalia* species also might have been part of the wooded vegetation on the slope. But the presence of *Phyllanthaceae* I together with *Parkia* spp. might also indicate human impact. *Bridelia ferruginea* and *B. micrantha*, which are possible matches for the first type, are typical in fallows, and *Parkia biglobosa*, a fire-resistant heliophyte, is also conspicuous and abundant in bush fallow and wooded farmland. Several *Terminalia* species also grow in fallows. Even though several charcoal types might represent species of gallery forests, like *Uapaca* spp. (*U. heudelotii*), *Terminalia* spp. (several *Terminalia* species) and *Phyllanthaceae* I (*Margaritaria discoidea*), the distance of 2 km to the Gurara River renders it unlikely that the wood was transported from there. In Janruwa, where gallery forest exploitation was postulated (Höhn and Neumann 2016), a different set of charcoal types was co-dominant with *A. Leiocarpa*.

Conclusions

The taphonomy of charcoal samples at Pangwari is intricate. Besides obvious challenges like erosion and destruction through illicit digging, the Middle Nok activities themselves, especially the disturbance of the original pit fillings during stage 2, impede the chronological interpretation of the charcoal data. The radiocarbon dates on single macro-remains are not necessarily representative for all macro-remains from the same archaeobotanical sample. Only two in situ samples allow a glimpse into the exploitation of woody vegetation at a defined period of time. However, the quick refilling of the pits does not allow reconstructing the woody vegetation from them alone. Being derived from single events, taphonomic biases have to be considered: single types are present or dominate by chance and not because of their importance in the surrounding woody vegetation.

The charcoal assemblages of Pangwari do not—as initially hoped for—allow tracking the development of a cultural landscape, in spite of recurring human activities in the last 3500 years, because the organic remains in the archaeobotanical samples are mainly derived from the early Middle Nok period of the 9th and 8th century BC. Earlier and later dates on single pearl millet grains are present, but the assemblages they are derived from are mixed and do not represent the dated earlier or later periods. Considering the evidence from all assemblages together, a prudent sketch of the exploited woody vegetation has been drafted for the peak activity period: woodland, whether natural or secondary, was present and exploited, but human impact seems to still have been low.

Pangwari is an excellent example of how the large number of individual measurements of find positions during excavation allows evaluating the distribution of all find categories, such as charcoal, dated and undated archaeobotanical samples, sherds of different pottery groups, terracotta fragments, and more, in overall context and in relation to each other. Only this holistic view renders it possible to trace the history of the individual features, develop ideas about their use, and reflect on the history and function of the site in general. Based on 3-D distribution diagrams the taphonomy of the charcoal assemblages was established, which is the prerequisite for their interpretation. The joint efforts of archaeology and archaeobotany demonstrate that for palaeoenvironmental reconstructions based on organic material from pits it is absolutely necessary to consider the taphonomy of the infill. Radiocarbon dates alone may not reflect the original history of pits and thus provide wrong context information. The combination of archaeological information, in form of absolute dates, pottery chronology, and find distributions, and of archaeobotanical analysis will be applied to other Nok sites as well as to trench I at Pangwari, where several pit features and possible graves were excavated in a subsequent excavation campaign.

Acknowledgements. Funding was provided by the Deutsche Forschungsgemeinschaft (DFG) within the frame of the long-term project on the Nok Culture (NE 408/5-3, BR 1459/7-3). We thank the archaeological director P. Breunig, our Nigerian partners, especially the National Commission for Museums and Monuments, and all colleagues and workers involved in the excavation and documentation, as well as two anonymous reviewers for their comments and suggestions. We appreciate sample processing by F. Nijimi and technical assistance by J. Markwirth and M. Ruppel. E. Eyub provided geographical information and maps and G. Försterling the graphic design.

References

Abaje IB, Ishaya S, Usman SU (2010) An analysis of rainfall trends in Kafanchan, Kaduna State, Nigeria. Res J Environ Earth Sci 2:9–96

Arbonnier M (2002) Arbres, arbustes et lianes des zones sèches d'Afrique de l'Ouest. Editions Quae, Versailles

Breunig P, Rupp N (2010) Outline of a New Research Project on the Nok Culture of Central Nigeria, West Africa. Nyame Akuma 73:46–54

Breunig P, Rupp N (2016) An outline of recent studies on the Nigerian Nok Culture. J Afr Archaeol 14(3):237–255

Bronk Ramsey C (2009) Bayesian analysis of radiocarbon dates. Radiocarbon 51(1):337–360

Casey J (2000) The Kintampo complex. The late Holocene on the Gambaga Escarpment, Northern Ghana. Archaeopress, Oxford

Chabal L (1992) La représentativité paléo-écologique des charbons de bois archéologiques issus du bois de feu. Bull Soc Bot Fr Actual Bot 139(2–4):213–236. https://doi.org/10.1080/01811789.1992.10827101

Dallwitz MJ (1980) A general system for coding taxonomic descriptions. Taxon 29:41–46

Eggert MKH (2014) Early iron in West and Central Africa. In: Breunig P (ed) Nok—African sculpture in archaeological context. Africa Magna Verlag, Frankfurt, pp 50–59

Ern H (1979) Die Vegetation Togos. Gliederung, Gefährdung, Erhaltung. Willdenowia 9(2):295–312

Fagg B (1977) Nok Terracottas. The National Museum, Lagos, and Ethnographica Ltd., London

Franke G (2016) A chronology of the Central Nigerian Nok Culture—1500 BC to the beginning of the Common Era. J Afr Archaeol 14(3):257–289

Franke G (2017) Potsherds in time—the Pottery of the Nigerian Nok Culture and its chronology. Dr. Hut Verlag, München

Gronenborn D (1997) An ancient storage pit in the SW Chad Basin, Nigeria. J Field Archaeol 24(4):431–439

Höhn A (2005) Zur eisenzeitlichen Entwicklung der Kulturlandschaft im Sahel von Burkina Faso. Untersuchungen von archäologischen Holzkohlen. Dissertation, Goethe Universität Frankfurt. http://publikationen.ub.uni-frankfurt.de/frontdoor/index/index/docId/3159. Accessed Feb 2016

Höhn A, Neumann K (2012) Shifting cultivation and cultural landscape development during the Iron Age (0–1500 AD) in the Sahel of Burkina Faso, West Africa-Insights from archaeological charcoal. Quatern Int 249:72–83. https://doi.org/10.1016/j.quaint.2011.04.012

Höhn A, Neumann K (2016) The palaeovegetation of Janruwa (Nigeria) and its implications for the decline of the Nok Culture. J Afr Archaeol 14(3):331–353

Höhn A, Neumann K (2018) Charcoal identification in a species-rich environment–the example of Dibamba, Cameroon. IAWA J 39(1):87–113

Huysecom E, Ozainne S, Jeanbourquin C et al (2015) Towards a better understanding of Sub-Saharan settlement mounds before AD 1400: The tells of Sadia on the Seno Plain (Dogon Country, Mali). J Afr Archaeol 13(1):7–38

InsideWood (2004-onwards) Inside Wood database. http://insidewood.lib.ncsu.edu/search. Accessed Dec 2015

Irvine FR (1961) Woody plants of Ghana. Oxford University Press, Oxford

Jones EW (1963) The forest outliers in the Guinea zone of northern Nigeria. J Ecol 51:415–434

Joosten H, De Klerk P (2002) What's in a name? Some thoughts on pollen classification, identification, and nomenclature in Quaternary palynology. Rev Palaeobot Palyno 122:29–45

Junius H (2016) Nok early iron production in Central Nigeria—new finds and features. J Afr Archaeol 14(3):291–311

Kahlheber S, Höhn A, Rupp N (2009) Archaeobotanical studies at Nok sites: an interim report. Nyame Akuma 71:2–14

Keay RWJ (1953) An outline of Nigerian vegetation. Government Printer, Lagos

Keay RWJ (1989) Trees of Nigeria. Clarendon, Oxford

Männel TM, Breunig P (2016) The Nok terracotta sculptures of Pangwari. J Afr Archaeol 14 (3):313–329

Magnavita C, Magnavita S (2001) New evidence of proto-urban settlements in the Lake Chad area. Nyame Akuma 55:46–50

Mbida Mindzie C, van Neer W, Doutrelepont H et al (2000) Evidence for banana cultivation and animal husbandry during the first millennium BC in the forest of Southern Cameroon. J Archaeol Sci 27:151–162

Meister C (2008) Recent archaeological investigations in the tropical rain forest of South-West Cameroon. In: Runge J (ed) Dynamics of forest ecosystems in Central Africa during the Holocene: Past–Present–Future. Palaeoecology of Africa 28, Taylor & Francis, London, pp 43–57

Meister C, Eggert MKH (2008) On the early Iron Age in Southern Cameroon: the sites of Akonétye. J Afr Archaeol 6:183–202

Metcalfe CR, Chalk L (1950) Anatomy of the dicotyledons, vol 1 and 2. Clarendon Press, Oxford

Neumann K, Kahlheber S, Uebel D (1998) Remains of woody plants from Saouga, a medieval West African village. Veg Hist Archaeobot 7:57–77

Neumann K, Schoch W, Détienne P et al (2001) Woods of the Sahara and the Sahel: an Anatomical Atlas. Paul Haupt, Bern, Stuttgart, Wien

Neumann K, Fahmy A, Lespez L et al (2009) The Early Holocene palaeoenvironment of Ounjougou (Mali): Phytoliths in a multiproxy context. Palaeogeogr Palaeoclimatol Palaeoecol 276:87–106

Prior J, Gasson P (1993) Anatomical changes on charring six African hardwoods. IAWA J 14:77–86

Reimer PJ, Bard E, Bayliss A et al (2013) IntCal 13 and Marine13 radiocarbon age calibration curves 0–50,000 Years Cal BP. Radiocarbon 55(4):1869–1887

Rupp N (2014) Tradition and innovation. In: Breunig P (ed) Nok–African sculpture in archaeological context. Africa Magna, Frankfurt, pp 157–166

Schmidt A (2014) Studien zur Struktur der Nok-Fundstelle Pangwari in Zentralnigeria anhand der Befunde und der Fundverteilung. MA thesis, Goethe Universität Frankfurt. http://publikationen.ub.uni-frankfurt.de/frontdoor/index/index/docId/41376. Accessed Feb 2016

Seidensticker D (2010) Grubenbefunde in Campo (Südkamerun). AInf 33(1):159–164

Seidensticker D (2013) Die Nutzung und Funktion von Gruben in Zentralafrika. Ethnologische Analogien und geochemische Analysen. In: Karl R, Leskovar J (eds) Interpretierte Eisenzeiten. Fallstudien, Methoden, Theorie. Tagungsbeiträge der 5. Linzer Gespräche zur interpretativen Eisenzeitarchäologie. Studien zur Kulturgeschichte von Oberösterreich 37. Landesmuseum, Linz, pp 51–60

Théry-Parisot I, Chabal L, Chrzavzez J (2010) Anthracology and taphonomy, from wood gathering to charcoal analysis. A review of the taphonomic processes modifying charcoal assemblages, in archaeological contexts. Palaeogeogr Palaeoclimatol Palaeoecol 291:142–153

White F (1983) The vegetation of Africa. A descriptive memoir to accompany the Unesco/AETFAT/UNSO vegetation map of Africa, Unesco, Paris

Wotzka HP (1993) Zur Tradition der Keramikdeponierung im äquatorialen Regenwald Zaïres: Ein Bekenntnis zur allgemein-vergleichenden Analogie. Ethnogr-Arch Zeitschr 34:251–283

Plant Use, Agricultural History and Ethnoarchaeology: Foods and Fields

Mapping Global Agricultural History: A Map and Gazetteer for Sub-Saharan Africa, c.1800 AD

Mats Widgren(✉)

Department of Human Geography, Stockholm University, Stockholm, Sweden
mats.widgren@humangeo.su.se

Abstract. Recent global historical cropland modelling grossly underestimates the pre-colonial development of agriculture in the Americas and many parts of sub-Saharan Africa. Such models are usually developed by back casting from recent land cover, combined with environmentally deterministic algorithms. Historical geographers have been slow in responding to a new demand for a global synthesis. In this paper, a preliminary map of African agricultural systems dating to AD 1800 is presented. It forms a component of the project Mapping Global Agriculture and is based on the existing historical literature, observations by early travelers, archaeology and archaeobotany. It should be emphasized that the generated map should be considered preliminary.

Keywords: Land cover · Land use · Agricultural systems
Historical croplands

Introduction

The 1992 special issue of Annals of the Association of American Geographers *The Americas before and after 1492* (Butzer 1992) formed an important platform in the synthesis of pre-Columbian landscapes of America. With the completion of the three volume work on the cultivated landscapes of native America (Doolittle 2000; Denevan 2001; Whitmore and Turner 2001) and the subsequent dissemination of these syntheses to a broader audience in map form (Mann 2005), a new state of was research established, which had consequences more broadly for globally oriented historical geography. Because similar syntheses are lacking for the rest of the world, these research efforts represent a major challenge to historians, archaeologists and geographers working in other continents. Asia, with its long history of agriculture has some national agrarian histories, but no broader overview or any attempts at expressing agricultural history in map form. Similarly for Africa no synthesis of precolonial farming systems and agrarian landscapes has been published. Such empirically based global overviews of past land use and agriculture are badly needed to deal with some of the scientific challenges facing our world today. Within the humanities and the social sciences, the quest for understanding the globalized state we live in, fuels an inquiry into global production systems in existence prior to industrialization and western economic and political domination. Within climate science and global change studies, modelers use

A. M. Mercuri et al. (Eds.): *Plants and People in the African Past*,
https://doi.org/10.1007/978-3-319-89839-1_15

reconstructions of past land use in order to explain early human influence on the global climatic system. Both fields of research use old data—or in the case of climate modeling—replace data with assumptions. The work presented in this paper forms part of an effort towards answering these challenges. It presents a preliminary map of agricultural systems of sub-Saharan Africa for the cross-section AD 1800. It will be followed by cross-sections for AD 1000 and AD 1500 and forms part of a larger endeavor in producing a series of global maps for these three cross-sections.

PART I: Previous Research

The now well established relationship between vegetation, carbon cycles and climate has made knowledge of past agriculture a central focus in climate change research. It is increasingly recognized that early developments in land use have implications for the present day global climatic system. Since the publication of a seminal article by Ruddiman (2003), which addresses the role of early human land use on climate, historical aspects of land use have shifted increasingly into the focus of climate research. Ruddiman proposed that Neolithic farming as well as continued agricultural expansion well before the industrial revolution, had an impact on the emissions of greenhouse gases and the global climate system.

In the literature on long term carbon cycles and historical land use, different global models of past land use have been utilized. These are all based on back casting from recent land cover. The SAGE (Center for Sustainability and the Global Environment) dataset was initially published in 1999 and covered the period AD 1700–1999 (Ramankutty and Foley 1999). At the same time HYDE (History Database of the Global Environment) was established and initially also covered the past 300 years (Goldewijk 2001). Pongratz et al. (2008) published a series of maps covering AD 800–AD 2000. All are based on recent global distribution of croplands, from which cropland distribution in previous centuries are calculated based on historical population estimates.

HYDE is now the most commonly used and quoted model of past croplands. The recent version covers the last 12,000 years (Goldewijk et al. 2011). The basic assumption behind the HYDE model is that past agriculture developed within the confines of the recent distribution of croplands and successively expanded to fill these areas. Allocation of historical cropland between recent state territories (and in some cases between regions within states) is made in relation to historical population estimates (McEvedy and Jones 1978, with some recent updates) and an assumption on cropland per capita. The distribution within these geographical territories is based on algorithms and weighting maps which take into consideration among other things soil suitability for crops according to recent FAO maps of global agro-ecological zones. In addition, coastal areas and river plains are given positive weight and steep terrain is given negative weight. Historical data on agricultural history and cropland distribution are not entered into this weighting model. The location and extent of historical cropland reconstructions can only be changed by different population estimates, different per capita assumptions or different assumptions on how environmental variables impact on location of croplands. The results of such modeling are presented as "datasets" and used widely in climate modeling. But their connection to actual empirical data of settlement and land use in the past is weak. When they, as Goldewijk et al. (2011), claim to depict the distribution of croplands and pasture in distant archaeological past periods, they are profoundly misleading.

However, there are alternate approaches that are empirically based. Three different methods of land cover reconstructions based on data, instead of modeling, can be identified. First, in areas with reliable historical sources, this information can be used to reconstruct detailed gridded cropland maps for different historical periods. Chinese historical geographers are at the forefront of this approach and have published gridded maps of croplands of the mid-northern Song Dynasty (AD 1004–1085, He et al. 2012). Similar methods have been developed for Scandinavian countries, but they will not be able to reach further back than the 16th century (Li et al. 2013).

A second approach is through the quantitative treatment of pollen data (e.g. Gaillard et al. 2010). Pollen percentages over time are recalculated to classes of land cover. There are in principle no restrictions in the regional or chronological reach of such pollen based approaches, but the work necessary to achieve a global synthesis is extensive.

A third approach is represented by the mapping of global agrarian systems rather than precise land use and land cover. During the first part of the 20th century, several attempts were made by geographers to move beyond the mere mapping of agro-climatic zones, towards a more substantial understanding of different types of agriculture, which takes into account not only various environmental conditions, but also different historical pathways that agriculture has taken in different regions (see e.g. Whittlesey 1936). Grigg's (1969) theoretically informed review and his reflections on previous work on agricultural regions, along with his subsequent book (Grigg 1974) on the agricultural systems of the world, are still fundamental to any work on the history of global agriculture. Neither Whittlesey nor Grigg formulated their respective regionalizations based on one main simple classificatory characteristic, such as crops, tools or nutrient circulation. In Griggs words we are thus dealing with *multi-purpose regions* rather than *single purpose* regions. Although Grigg in a previous paper discussed regionalization as a formal and rigorous process (Grigg 1965), in the case of global agrarian systems he realized the process had to be intuitive (Grigg 1969). Grigg's work was mainly based on and developed further the map presented by Whittlesey (1936). With different updates and alterations it remains to this day one of the most reproduced maps of global agricultural systems. The Whittlesey map, representing the beginning of the 20th century is a natural conceptual and empirical starting point for a retrogressive historical analysis of global agriculture. Whittlesey emphasized five aspects in his classification of global agricultural regions: "1. The crop and livestock association. 2. The methods used to grow the crops and produce the stock. 3. The intensity of application to the land of labor, capital, and organization, and the outturn of product which results. 4. The disposal of products for consumption. 5. The ensemble of structures used to house and facilitate the farming operations." (Whittlesey 1936:209)

Based on Whittlesey's map and the modifications made to it by subsequent writers, Grigg (1974) identified the following categories: 1. shifting agriculture; 2. wet rice cultivation in Asia; 3. pastoral nomadism; 4. Mediterranean agriculture; 5. mixed farming in Western Europe and North America; 6. dairying; 7. the plantation system; 8. ranching; and 9. large scale grain production.

Mapping Global Agricultural History

The approach taken by Whittlesey and Grigg forms the basis for the work reported in this paper. The goals, categories and methods were developed by a small group of historians and historical geographers in Sweden and the US in the collaborative project *Mapping Global Agricultural History*. The project aims at summarizing the existing knowledge of land use and agricultural regions on a global scale during the last millennium (Widgren 2010a). The design, as well as the proposition of global categories of agricultural regions, is the result of a collective effort by this group.[1] The map I present here is the first in a series of three maps AD 1000, AD 1500 and AD 1800, which will be included in a forthcoming global map of agricultural regions. The *Mapping Global Agricultural History* project now also forms part of a wider program studying land cover changes during the last 6000 years on a global scale (LANDCOVER6K, Gaillard 2015) supported by the PAGES (Past Global Changes) network and involving all approaches mentioned above including a large group of paleoecologists, archaeologists, historians and historical geographers as well as modelers active in this field.

A central aim of the *Mapping Global Agricultural History* project is to express current knowledge of global agrarian history on maps, in a format that can be compared with and challenge maps based on back-casting. The project will produce a series of maps of global agricultural systems during the last millennium at the same level of detail and scale as the map of Whittlesey (1936). Three cross-sections have been selected: (1) AD 1000 which was a time when especially African and American polities and landscapes were distinctly different from those of the late fifteenth century; (2) AD 1500, on the eve of European oceanic expansion and before the Columbian exchange; and (3) AD 1800, before the nineteenth-century wave of globalization that drew large parts of the global South into commercial agriculture.

There is one important difference between our approach and the above mentioned parallel efforts to reconstruct land cover. We do not map land cover, but a series of qualitatively different agricultural regions. We map the presence of agriculture in a region during a specific time-window, and characterize the dominant type of agricultural system in that region. It is therefore not possible to directly translate these maps to areas of different land cover or land use. However, as will be shown below, we characterize agricultural systems on a scale of intensity which facilitates comparison with other methods.

[1] Ulf Jonsson (Stockholm University), Janken Myrdal (Swedish University Agricultural Sciences), William E. Doolittle (University of Texas, Austin), Mats Widgren (Stockholm University) and William I. Woods (1947–2015, University of Kansas).

Global Categories

The selection of categories of agricultural regions to be displayed on a global map is, as is always the case with regionalization, a compromise. In our categories, we have drawn on both Whittlesey (1936) and Grigg (1974), with the three first criteria used by Whittlesey as the starting point. Criterion four was excluded, because in the longer time perspective that we take into consideration, the degree of commercialization is considered to have been low. Before the mid-19th century AD global land use was not characterized to same extent by global flows of food and fibers. Moreover, recent research in historical and economic anthropology has come to question the idea of pure subsistence farming, as most agricultural communities produced a surplus and were involved in exchange with other groups (Håkansson 1994:249–251, 2008:240–242, 248–252). Similarly we have not considered Whittlesey′s fifth aspect, the structures to house and facilitate farming operations, because we do not have global data on this subject for earlier periods. The categories in Grigg (1974) that refer to commercial agriculture typical of the late 19th and the 20th century, including specialized dairying (6), plantations (7) and large scale grain production (9), have been excluded and instead we have added categories more relevant for the preindustrial period.

Agricultural systems can be considered as qualitatively different ways of utilizing the land and they do not immediately translate to any quantitative measure, or to any simple evolutionary sequence. For that reason, most maps of agricultural regions use a nominal scale, with map signatures that are qualitatively different but not in any quantitative order. However, systems do differ in the degree of intensity (for different aspects of intensity see below). In order to make our maps more legible and comparable with other maps of historical croplands, we have used a combination of nominal and ordinal scales.

We have therefore made an attempt at ranking degrees of intensity from husbandry of non-domesticated plants and pastoralism, to extensive farming, permanent fields and to more intensive forms of agriculture involving modification of soils through manuring and major landscape transformations, such as terracing, canal irrigation and dykes. The measure of intensity in Table 1 is of course a very rough estimate. In reality, we are dealing with at least three different ways of expressing intensity. First, of interest for agrarian history and for understanding the possibilities of producing a surplus, input and output per hectare are central measures. In our categories, we first consider labor intensity, i.e. input of labor per defined areal unit. Secondly, taking a landscape perspective, the degree to which the land has been modified (anthropogenic soils, irrigation, terracing etc.) can also be classified into varying intensities. Thirdly, for comparison with other land cover maps, the density of cropland (for example hectares of cropland per square kilometer) would have been the best measure of intensity. However, the map basically describes dominant agricultural systems in each region and this measure of intensity cannot always be read from the maps. The intensity of fields on the ground can differ significantly within otherwise homogenous agricultural regions. To take an example from the forthcoming global map: mixed farming existed historically in the Paris basin, as well as in the central forested parts of Scandinavia. Differences in the density of croplands versus meadows, grazing land and forests, would have been obvious in this vast region. In our very cursory indication of intensity in Table 1 and as reflected in the color sequence on the global map we have ranked, on a very general level, the combined intensity, taking into consideration the three aspects mentioned above.

A specific region is often characterized by a mosaic of different forms of agriculture and pastoralism. Especially at the intermediate level of intensity (rank 5 in Table 1) agriculture has been characterized by regionally distinct forms of farming, which combined extensive and intensive methods in the same area. The Mediterranean agriculture with its intensive irrigation and horticulture, grain-fields with winter crops, and at the same time extensive transhumance exemplifies such a mosaic. In order not to reduce agricultural history to just a linear development towards more intensive agriculture, but to give weight to regional diversity, we allowed for a limited number of regionally distinct forms of such mosaics or agricultural complexes to be represented on the map.

In the global map that we have prepared, eleven broad categories are used on a preliminary basis. Some of them are not represented in sub-Saharan Africa. However, some of the global categories can, for the purpose of the African map, be further subdivided. Table 1 presents the global categories and subdivisions that are possible and relevant for Africa.

Table 1. Classification of agricultural systems.

Intensity rank	Global categories	Categories in sub-Saharan Africa	
1	Pastoralism, ranching	1	Pastoralism
2	Husbandry of non-domesticated plants		NA
3	Extensive or undifferentiated farming	3.1	Grain
		3.2	Grain and roots
		3.3	Rice
		3.4	Vegeculture
4	Permanent fields	4.1	Permanent fields
		4.2	Flood retreat and other wetland cultivation
5	Mediterranean complex		NA
	Mesoamerican complex		NA
	Taro complex		NA
	Mixed farming	5.1	Infield-outfield
		5.2	Mixed farming, general
		5.3	Mixed farming with terracing
6	Intensive farming	6.1	Banana gardens
		6.2	Canal irrigation
7	Irrigated rice	7	Irrigated rice
	Oases		NA

Sources and Methods

Syntheses of African farming taking a long term perspective have, to a large extent, focused on the introduction and spread of farming, and to a lesser degree on the changes in tools, crops, field systems and farming practices during subsequent periods (see Sutton 1996; Fuller and Hildebrand 2013; Gifford-Gonzales and Hanotte 2011, 2013; Russell et al. 2014). For a general understanding of the diversity of pre-colonial agriculture on the continent the works of William Allan (1965) and Marvin Miracle (1967) do provide an important starting point. Since then, several studies have demonstrated how innovative and ecologically aware many local agricultural systems were and often remain so in the present. But as the contrasting perspectives of Kjekshus (1977) and Koponen (1988) do show, there is indeed an intellectual and empirical endeavor to move from a general appreciation of the diversity and ingenuity of pre-colonial agriculture towards a chronologically rich and precise articulation of agrarian history. The special issue of Azania *Towards a history of cultivating the fields* is an important milestone in the study of precolonial African field systems (Sutton 1989). Some works take a broad view of the agricultural history of particular areas (David 1976; Blench 1997; McCann 1995) but for the whole continent, no grand syntheses are published. Blench (2006) contributes important syntheses towards an agrarian history of Africa. Studies that document a clear regional sequence of agricultural change during the last millennium, however, remain few in number.

Earlier writings sometimes confuse the 19th and 20th century ethnographically-documented African farming practices with what existed in the more distant past, and assume an almost essentialist connection between ethnic groups and their systems of agriculture (see, for example, Beck 1943). But agriculture developed in Africa as it did elsewhere and there are indeed regional works on agrarian history based on oral history and early travelers, which focus directly on this change and challenge and go beyond simplistic assumptions of inertia in agricultural development (see e.g. von Oppen 1993; Cochet 2001; Hawthorne 2003). Such regional studies of agricultural change permit the production of a credible delimitation on a map.

When previous authors have expressed their research on agrarian systems in map form, as is the case e.g. for the permanent fields in the Congo basin (Vansina 1990), this information has been included in our AD 1800 map. Also when the distribution of ancient and abandoned field system is known and archeologically dated, as is the case for Bokoni and Nyanga in Southern Africa, delimitation of these features on a map is uncontroversial. In other cases the characterizations and delimitations of the regions on a map are based on a combination of more recent ethnographic observation of agricultural systems, supported by scant contemporary evidence from travelers' reports from the early to late 19th century AD. Such reports seldom include general characterizations of types of farming, but they can nevertheless, in their detailed observations of crops and work on the fields, be used to lend support to the conclusion that a similar agricultural system existed already by 1800 AD. In a few cases analogous reasoning and informed guesses have also been used. This has been done in the case of dating the Mossi infield-outfield system (see below) and terraced highland areas in Western Africa (cf. Widgren 2010b).

A problem with the map of AD 1800 for Africa is that the 19th century was a dynamic period. In some case what was observed on the ground in the mid-19th century may have been developed shortly before that as the result of, for example, expanding caravan trade in eastern Africa (e.g. the Baringo canal irrigation system—see below under 6.1). In other cases, such as in the abandonment of field systems at Engaruka in Tanzania and Nyanga in Zimbabwe, AD 1800 represents the latest possible, but not yet fully proven, date of its abandonment. As such the dating of the map to AD 1800 should be considered an estimate and not exact. Moreover, it must be said that in Africa many non-tangible components of agricultural systems have evaded documentation, aspects which have been well recorded in Europe, India and China, with their extensive literature on agrarian history. Such information that is lacking for Africa includes yield ratios, productivity, cropping sequence and knowledge systems. There is always a danger that spectacular landscape modification, visible as archaeological features and notable farming methods observed by travelers, take a more important role than daily activities of the knowledgeable women and men that shaped the agriculture of Africa (cf. the "tyranny of monuments" Sutton 1989:112).

PART II: Agricultural Systems by AD 1800

At a general level there was a close relation between climatic zones and agricultural regions (Fig. 1). In the mainly semi-arid savanna zones ("arid steppe") grain-based farming of sorghum and millets dominated and was interspersed with pastoralism. In contrast, cattle was absent in equatorial climatic zones and cultivation was based either on a combination of grain and roots/tubers or (especially in the fully humid or monsoonal equatorial climate zones) a vegeculture complex based on roots, tubers and bananas. In temperate highland zones ("warm temperate") we find many different examples of more intensive agriculture with mixed farming, terracing and, as in the highlands of Madagascar, paddy rice cultivation. However, the map (Fig. 2) will also show many exceptions, where areas of mixed farming or irrigation existed as islands in more intensive agriculture within the arid steppes.

The map presented here (Fig. 2) is an attempt at visualizing our present knowledge of late precolonial African agricultural systems. The starting point has been the global categories as described in Table 1. The following categories can be inferred from the sources described above.

Pastoralism and Ranching

Areas dominated by specialized pastoralism, where cultivation was of minor importance, have fluctuated in extent over time in Africa. It is a difficult task to clearly delimit areas where such specialized pastoralism was prevalent. The boundary between extensive cultivation and pastoralism has been fluid over time and, moreover, most areas of extensive cultivation in the savanna zones were also interspersed with pastoralism. Especially difficult is eastern Africa, where rainfall is highly variable within short distances and the result is a complicated mosaic of semiarid, temperate and equatorial climates (cf. Fig 1, or the more detailed map by Pratt and Gwynne

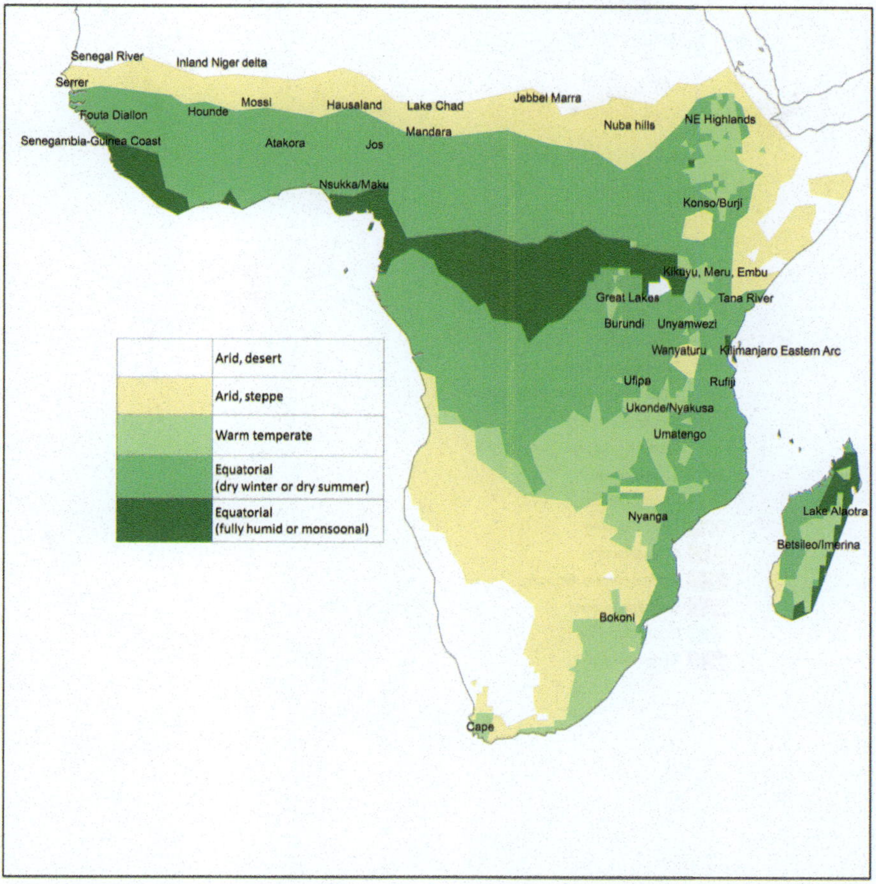

Fig. 1. Main climatic zones with a selection of places and regions mentioned in the text. The climate zones follows the updated Köppen-Geiger classification (Kottek et al. 2006) and is for the purpose of this map divided into five classes: Desert (BW), steppe (BS), warm temperate (C), equatorial with dry winter or dry summer (As and Aw) and equatorial fully humid or monsoonal (Af and Am).

reproduced in Håkansson 2004:579). Specialized pastoralism expanded in this area during the 18th and 19th centuries AD (Bollig et al. 2013), but there have been only a few attempts to map these changes (Sutton 1993; Håkansson 2012). The boundary between pastoralism and farming as depicted on this map (Fig. 2) must therefore be used with caution, especially in eastern Africa. More work is needed to make a sharper delimitation of the extent of pastoralism in AD 1800.

Husbandry of Non-domesticated Plants

On global maps, the category *husbandry of non-domesticated plants* will be used in areas of the world where the protection, encouragement and planting of

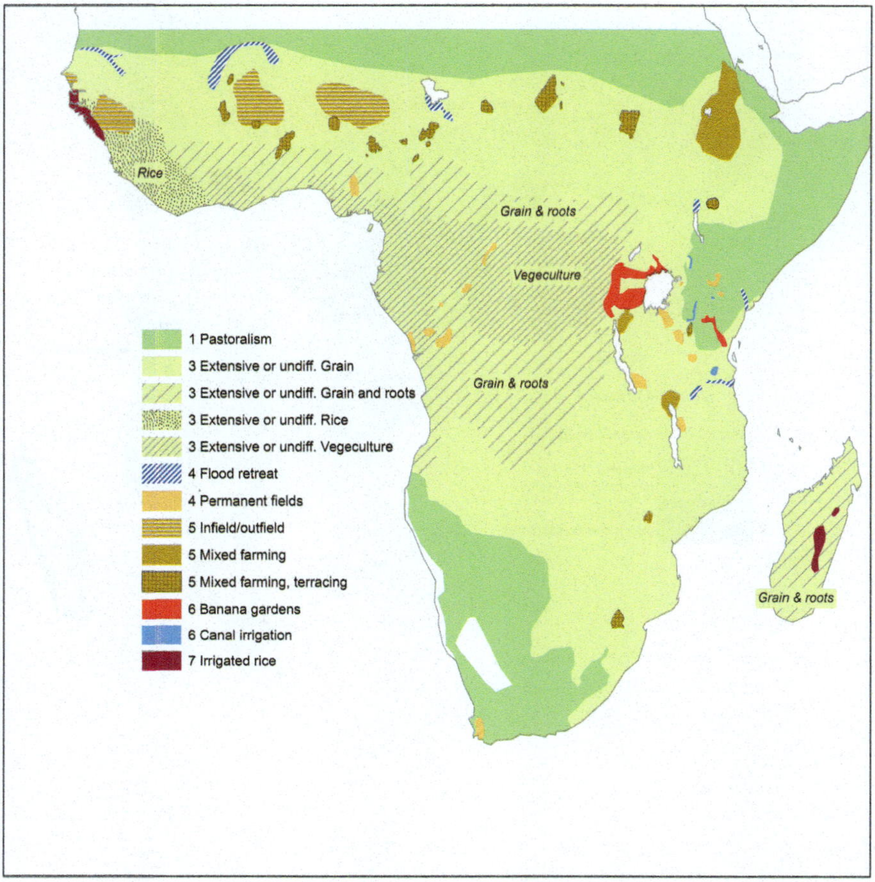

Fig. 2. Agricultural systems in sub-Saharan Africa c. 1800 AD.

non-domesticated plants constituted the main plant-based activities and where pastoralism or cultivation of domesticated plants were not practiced. Such activities probably were present among many hunter-gatherers. In Africa, husbandry of non-domesticated plants is documented, e.g. in the Congo rain forest (Yasuoka 2013), but because this was practiced in pockets within larger regions dominated by extensive agriculture, no attempt has been made to map them at this scale.

Extensive or Undifferentiated Agriculture

This somewhat clumsily named category refers to, on the one hand, areas where agriculture is known to have been extensive (e.g. slash-and-burn), and on the other hand to areas where agriculture existed, but its character is unknown. It was important in this context and in comparison with non-historical modeling to show the full extent of farming on the AD 1800 map, even when its nature is not fully understood, rather than to mark these areas as “no data”.

There is tendency among researchers to assume that shifting cultivation was the norm for all early agriculture. In this project, a more conservative standpoint is taken and hence the same degree of proof for shifting cultivation has been required as for intensive farming systems. With better evidence from paleoecology there are now some regions in Africa that have provided direct evidence that fields were shifting, fire was used regularly, and the vegetation was characterized by secondary woodlands, but such direct evidence is still scarce (Höhn and Neumann 2012).

Within the area delimited as extensive or undifferentiated agriculture there is scattered evidence from the 18th and 19th centuries AD for intensive farming methods. This includes the case of mounding and ridging in the upper Zambezi valley (Almeida et al. 1873:92; von Oppen 1993:108ff) and the use of manure in many parts of sub-Saharan Africa (Beck 1943). But in these and other cases, neither the full character of the agricultural system nor its definite delimitation has been possible to document at this stage. Another important aspect of this category is the role of tree crops and the specific form of agricultural landscapes characterized by parklands with extensive grain cultivation. This type of agrarian system and its resulting clearly anthropogenic landscape has not been mapped here. Overviews of present parklands exist (Boffa 1999), but to my knowledge, their historical depth has not yet been mapped on regional scale.

Evidence from the 20th century AD shows that the term shifting cultivation in sub-Saharan Africa comprises a large array of different types of farming (cf. the map by Morgan 1969:252–253), but at this point such evidence for the precolonial period is not at hand. Another important distinction within this larger region is the division into areas dominated by different staples: vegeculture (roots and bananas) mainly in regions of equatorial climate and grain in savanna regions. Large areas where grain and roots were intermixed also existed. We know now that this division was not fixed, and not fully determined by climate, but that it changed over time (Kahlheber et al. 2013). The arrival of American crops to Africa expanded the crop repertoire within the vegeculture complex with cassava and sweet potatoes. The introduction of maize may have opened up more humid environments for grain cultivation. The effects of these new arrivals on the zoning of vegeculture versus grain cultivation have not yet been documented.

In the AD 1800 map presented here (Fig. 2), the regions of main staples as described by Morgan for the 20th century (1969:252–253) are shown stippled and hatched on top of the main category of extensive or undifferentiated farming. Morgan's zones roughly correspond to maps presented by Harris on West African traditional crop zones (Harris 1976: Fig. 4) and the map for the Congo basin by Marvin Miracle on principal food crops in the Congo basin 1900–1915AD (Miracle 1967:281). Extending the mid-20th century AD distribution back in time to AD 1800 is hypothetical but mapping that distribution may form a basis for further research.

The outer limits of this agricultural region also correspond to the limits of agriculture at this time. The northern boundary is based on the more recent evidence from Dixon et al. (2001) with some consideration also taken of the limits of agriculture in different countries as documented in *L'atlas de l'Afrique* (Ben Yahmed and Sablayrolles 2000). The southern limit roughly follows Russell et al. (2014) and for South Africa the more precise limits of African farming settlements as documented by Maggs (1980, 1984).

Permanent Fields on Dry Lands

In many parts of Africa, fields were more or less permanent. This does not exclude periods of fallowing, but farming was permanent enough for regular field systems to develop. There is also archaeological evidence of permanent field systems in arid regions as in Tagant in Mauretania and in the Niamey region in Niger (Khattar et al. 1994; Khattar 1995; Guillon et al. 2013), but it is not clear whether they survived into the 19th century AD and therefore they are not represented on this map.

Several areas in present day Tanzania, often at the junction of important 19th century AD caravan routes, were described by Gillman (1936:361) as 'cultivation steppes' (see also Håkansson and Widgren 2007). The evidence from Stanley's journey through Ugogo and neighboring areas is in line with this characterization (Stanley 1872). Some of the areas in the Kenyan highlands, inhabited by the Kikuyu, the Meru and the Embu, do belong to this category (Table 2, Fig. 3). Their role as specialized grain producers is evidenced in the regional economic history of Ambler (1988) and the travel accounts of Gedge (1892) give an impression of some densely cultivated landscapes, albeit in a later period. These areas are marked as permanent fields on the AD 1800 map. Most likely many other areas in sub-Saharan Africa shared this characteristic, but the evidence has not yet come to light.

Table 2. Gazetter, referring to the numbers in Fig. 3.

ID	Name	Present country	Category	Sources
1	Sahara/Sahel		Pastoralism	
2	Northern Savannah		Extensive or undiff. Grain	Morgan (1969:252)
3	Senegal River	Senegal, Mauretania	Flood retreat	Caillié (1830a:170), Ben Yahmed and Sablayrolles (2000:160), Le Roy et al. (2005)
4	Inland Niger delta	Mali	Flood retreat	Caillié (1830b:136), Barth (1858:150, 157, 224), Carney (2001:39)
5	Lake Chad	Nigeria, Cameroon, Chad	Flood retreat	Denham et al. (1826:251, 259)
6	Serrer	Senegal	Infield/outfield	Pélissier (1966), Widgren (2017)
7	Fouta Diallon	Senegal, Guinea Bissau, Guinea	Infield/outfield	Caillié (1830a), Ajayi et al. (1988:85), Widgren (2012)
8	Mossi	Burkina Faso	Infield/outfield	Sautter (1962)
9	Hausaland	Nigeria, Niger	Infield/outfield	Sutton (1979), revised
10	Rice Coast	Gambia, Senegal, Guinea Bissau, Guinea	Irrigated rice	Carney (2001:39), Hawthorne (2003), Fields-Black (2008)
11	Hounde	Burkina Faso	Mixed farming, terracing	Widgren (2010b)
12	Dogon	Mali	Mixed farming, terracing	Widgren (2010b)
13	Kassena	Burkina Faso, Ghana	Mixed farming, terracing	Widgren (2010b)
14	Atakora	Togo, Benin		Widgren (2010b)

(*continued*)

Table 2. (*continued*)

ID	Name	Present country	Category	Sources
			Mixed farming, terracing	
15	Kamuku	Nigeria	Mixed farming, terracing	Widgren (2010b)
16	Jos	Nigeria	Mixed farming, terracing	Widgren (2010b)
17	Eggon	Nigeria	Mixed farming, terracing	Widgren (2010b)
18	Tangale Waja	Nigeria	Mixed farming, terracing	Widgren (2010b)
19	Mandara	Nigeria/Cameroon	Mixed farming, terracing	Widgren (2010b)
20		Cameroon	Mixed farming, terracing	Widgren (2010b)
21		Cameroon	Mixed farming, terracing	Widgren (2010b)
22	Guere	Chad	Mixed farming, terracing	Widgren (2010b)
23	Jebbel Marra	Sudan	Mixed farming, terracing	Widgren (2010b)
24	Jebbel Marra	Sudan	Mixed farming, terracing	Widgren (2010b)
25	Jebbel Marra	Sudan	Mixed farming, terracing	Widgren (2010b)
26	Nuba hills	Sudan	Mixed farming, terracing	Widgren (2010b)
27	NE Highlands	Ethiopia	Mixed farming	McCann (1995:42)
28	Omo River	Ethiopia	Flood retreat	Bassi (2011)
29	Konso/Burji	Ethiopia	Mixed farming, terracing	Amborn (2009), Watson (2009)
30	West African rice zone		Extensive or undiff. Rice	Morgan (1969:252f), Harris (1976:319)
31	West Africa		Extensive or undiff. Grain and roots	Morgan (1969:252f), Harris (1976:319)
32	West Africa		Extensive or undiff. Vegeculture	Morgan (1969:252f), Harris (1976:319)
33	West Africa		Extensive or undiff. Vegeculture	Morgan (1969:252f), Harris (1976:319)
34	Nsukka/Maku	Nigeria	Permanent fields	Widgren (2010b)
35	Central Africa		Extensive or undiff. Grain and roots	Morgan (1969:252f)
36	Cantral Africa		Extensive or undiff. Vegeculture	Morgan (1969:252f)
37	Central Africa			Morgan (1969:252f)

(*continued*)

Table 2. (*continued*)

ID	Name	Present country	Category	Sources
			Extensive or undiff. Grain and roots	
38		Congo, Congo DRC	Permanent fields	Vansina (1990:212f)
39		Congo	Permanent fields	Vansina (1990:212f)
40		Congo, Congo DRC	Permanent fields	Vansina (1990:212f)
41		Congo	Permanent fields	Vansina (1990:212f)
42		Congo	Permanent fields	Vansina (1990:212f)
43		Congo DRC	Permanent fields	Vansina (1990:212f)
44	Great Lakes	Congo DRC, Uganda Rwanda	Banana gardens	Schoenbrun (1998:79ff), Ben Yahmed and Sablayrolles (2000:95), Reid (2001)
45	Burundi	Burundi	Mixed farming	Cochet (2001)
46	Gusi	Kenya	Permanent fields	Håkansson (1994)
47	Marakwet	Kenya	Irrigation	Östberg (2004:20), Davies et al. (2014)
48	Kikuyu, Meru, Embu	Kenya	Permanent fields	Gedge (1892), Ambler (1988)
49	Mbooni Hills	Kenya	Irrigation	Jackson (1976)
50	Kitui	Kenya	Permanent fields	Ambler (1988)
51	Tana River	Kenya	Flood retreat	Ylvisaker (1979:55f), Gedge (1892)
52	Eastern Africa		Pastoralism	
53	Unyamwezi	Tanzania	Permanent fields	Gillman (1936), Kjekshus (1977:42)
54	Wanyaturu	Tanzania	Permanent fields	Gillman (1936), Schneider (1970)
55	Engaruka, Sonjo, Pagasi	Tanzania	Irrigation	Berntsen (1976), Adams et al. (1994), Stump (2006)
56	Iraqw	Tanzania	Mixed farming, terracing	Börjeson (2004)
57	Meru, Kilimanjaro, Pare, Usambara	Tanzania	Banana gardens	Stump and Tagseth (2009)
58	Ugogo	Tanzania	Permanent fields	Stanley (1872), Gillman (1936)
59	Uluguru	Tanzania	Irrigation	Kjekshus (1977:36), Paul (2003:171)
60	Ufipa	Tanzania	Permanent fields	Kjekshus (1977:41)
61	Ukonde/Nyakusa	Tanzania, Malawi	Mixed farming	Kjekshus (1977:36f)
62	Kilombero	Tanzania	Flood retreat	Monson (1991:56ff), Beez (2005:173ff)
63	Rufiji	Tanzania	Flood retreat	Havnevik (1993:68f), Beez (2005:171f)
64	Umatengo	Tanzania	Permanent fields	Kjekshus (1977:37f)
65	Southern Savannah		Extensive or undiff. Grain	Morgan (1969:262), Maggs (1984)
66	Nyanga	Zimbabwe	Mixed farming, terracing	Soper (2002)

(*continued*)

Table 2. (*continued*)

ID	Name	Present country	Category	Sources
67	Madagascar	Madagascar	Extensive or undiff. Grain and roots	Le Bourdiec (1974:53ff), Campbell (2005:24)
68	Betsileo/Imerina	Madagascar	Irrigated rice	Le Bourdiec (1974:53ff), Campbell (2005:24)
69	Lake Alaotra	Madagascar	Irrigated rice	Le Bourdiec (1974:53ff), Campbell (2005:24)
70	Southern Africa		Pastoralism	
71	Bokoni	South Africa	Mixed farming, terracing	Delius et al. (2014)
72	Cape	South Africa	Permanent fields	Guelke (1976), Christopher (1982)

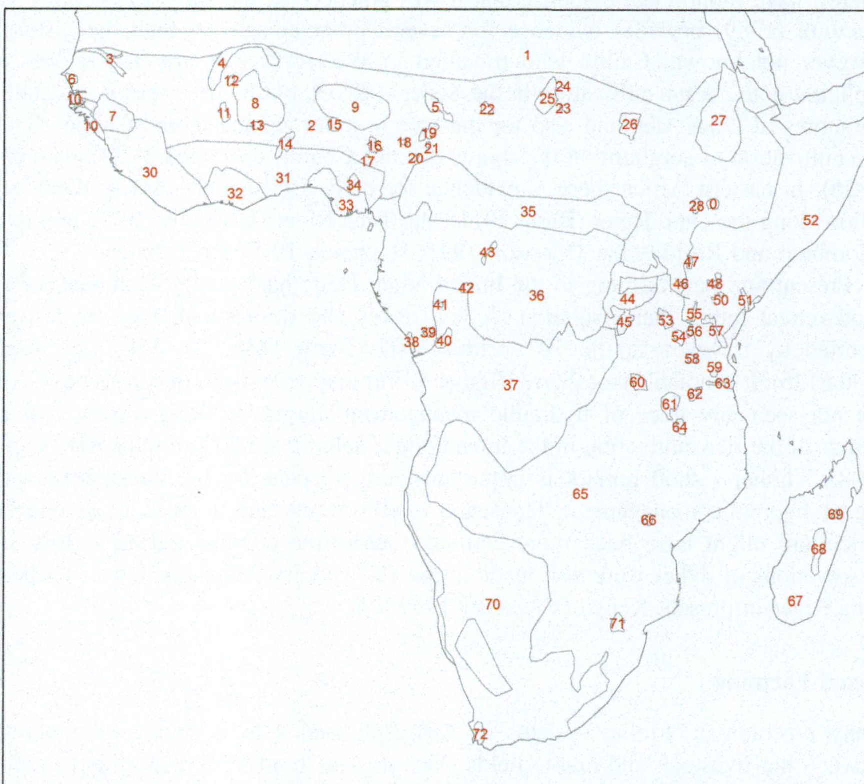

Fig. 3. Agricultural systems. Red figures refer to Table 2.

Flood Retreat and Other Wetland Cultivation

In the global map, flood retreat farming and other types of wetland cultivation will be classified as a subcategory of permanent fields. Wetland cultivation forms an important part of African agricultural systems (for details and distinctions between *crue*, *décrue* and other wetland cultivation techniques see Marzouk 1989 and Adams 1991:76ff). The reason for making a clear difference between this type of wetland cultivation and irrigated agriculture is that the degree of landscape modification was low or non-existent in ordinary flood retreat farming in contrast to irrigated agriculture, where the building and repair of dykes and canals are central activities and accordingly labor intensity is higher.

The Inland Niger Delta forms the largest concentrated area for flood recession agriculture in Africa which is evidenced in the 19th century AD by Caillié (1830a) and Barth (1858). Similar farming was practiced along many rivers and lake shores in other parts of Africa. I have restricted the cases presented on the map to the few instances where I have found clear evidence that it was practiced in the early 19th century AD. Marzouk (1989) provides evidence for several other areas, in which the history is however not known. Caillié who travelled in West Africa in the 1820s does not explicitly note *décrue* cultivation in the Senegal River, but he describes it in detail on the shores of Lake Aleg and also for the area in more general terms (Caillié 1830b). The cultivation of sorghum on the shores of Lake Chad is described by Denham et al. (1826). In eastern Africa there is evidence for early 19th century AD wetland cultivation along the Omo River (Bassi 2011), the Tana River (Ylvisaker 1979) and in the Kilombero and Rufiji deltas (Monson 1991; Havnevik 1993; Beez 2005).

Present and past farming in the Inland Niger Delta has usually been described as flood retreat rather than irrigation. A few dykes and drainage ditches are however reported by travelers in the 19th century AD (Barth 1858:150, 157, 224; Caillié 1830a:136). It is notable that Pierre Viguier writes that he himself, in contrast to Caillié, has not seen any trace of hydraulic management during his long observation and studies of the rice cultivation in the Inland Niger delta in the 20th century AD (Viguier 2008). Viguier's short remark is important since it opens for the question to which degree hydraulic management, landscape modifications and a more labor intensive agriculture might have been more common before the colonial period in this area. Observations of dykes were also made in the 19th century AD in the lower reaches of Tana River in present Kenya (Ylvisaker 1979:56).

Mixed Farming

Many precolonial farming systems in Africa showed a high degree of integration between the livestock and arable fields. Manure was used as fertilizer, and products from the fields were used as fodder for livestock. In the AD 1800 map, the term mixed farming is used for these types of systems. A technical definition of mixed farming formulated in the context of the propagation of mixed farming to Africa is the following: "… mixed farming is the cultivation of crops and the raising of cattle, sheep and/or goats by the same economic entity, such as a household or a 'concession', with animal inputs (e.g. manure, draft power) being used in crop production and crop inputs (e.g. residues, fodder) being used in livestock production" (Powell and Williams 1993).

In the African context the term mixed is farming is often connected to a development agenda in which the "modern" methods of farming that were developed during the agricultural revolution in Western Europe were exported to Africa in the 20th century. As with many other terms for agricultural systems the concept is thus tainted by its specific usage in development discourse (cf. the discussion of the concept as used in Zimbabwe in Wolmer and Scoones 2000). It can therefore be questioned if such a term based on a European perspective should be used as a generic and globally applicable concept. Except for the issue of draft power, which during this period in sub-Saharan Africa only existed in highlands of the Horn of Africa, mixed farming in the sense of the quoted definition was already present on the continent long before it became a part of part of a development agenda. I cannot find a better term than *mixed farming* to describe this farming system. On the map I have identified the following different subcategories of mixed farming in sub-Saharan Africa.

Mixed Farming: Infield-Outfield Systems

Sautter has provided a comprehensive overview of infield-outfield systems in West Africa (Sautter 1962). They combined an intensively cultivated and manured infield, immediately surrounding the village or town, with an extensively cultivated outfield, which received no or little manure. In some cases the infield only consists of small gardens such as in Fouta Djallon. These systems have also been termed ring-cultivation systems (see Widgren 2012, 2017 and references cited there). The argument that such systems, which are documented among the Serer in Senegal, in Fouta Djallon, among Mossi in Burkina Faso and in Hausaland, were already well developed in the early 1800s AD is based on observations by Caillié (1830a, b), Denham et al. (1826) and other travelers as well as the historical analysis of Watts (1983) for Hausaland and the retrogressive analyses of Pélissier (1966) on the Serer. The close integration between crop and livestock in that area may go back to the thirteenth century when the Serer first settled in the area (Reinwald 1997). In the Mossi area such farming systems have been documented mainly in the 20th century AD by early travelers and later researchers (Prudencio 1993; Stamm 1994 and references therein). The conclusion on Mossi is thus based on analogy, for lack of direct evidence.

Mixed Farming, General

The mixed farming systems of northern Europe (and in the development agendas) included the use of oxen as draft power and the use of ploughs. But in sub-Saharan Africa plough agriculture was by AD 1800 more or less restricted to the highlands in the Horn of Africa and as McCann (1995:42) has shown it was by then mainly restricted to areas north of 9 degrees latitude. Only later did it spread to almost all highland areas in Ethiopia. The integration of livestock and agriculture was in this area mainly restricted to draft power, and to a lesser degree based on the use of manure as fertilizer on agricultural fields (McCann 1995:57). In large areas of the highlands terracing was common, but I have not been able to find a good overview of the historical situation (for later periods see Tadesse et al. 2006:86).

Beck (1943) provides a map of areas where manure was used according to different sources from the 19th and 20th century AD, but it is difficult to pinpoint the delimitation of specific agrarian systems and their dates. Outside the areas of infield-outfield farming and the areas where manuring was combined with extensive terracing (see below), it is mainly from the evidence of Kjekshus (1977) for Tanzania and Cochet (2001) for Burundi that it can be argued that a fully-fledged mixed farming existed in the early 19th century AD.

Mixed Farming with Terracing

A detailed overview of intensive and terraced agriculture in West Africa and Sudan and a discussion of the chronology of these systems can be found in Widgren (2010b). For eastern Africa Widgren and Sutton (2004) provide an overview. The mapped occurrences in eastern Africa are based on the Iraqw of Tanzania (Börjeson 2004) and the Konso-Burji complex in Ethiopia (Amborn 2009; Watson 2009). The occurrences of abandoned terracing in Southern Africa are based on recent archaeological research in Nyanga and Bokoni (Soper 2002; Delius et al. 2014; Widgren et al. 2016).

Intensive Systems

Whittlesey (1936) identified a category termed *Intensive subsistence tillage without paddy rice*. In Africa for AD 1800 *intensive systems* is used for systems that had clearly higher productivity per area than most other categories, except the irrigated rice (see below).

Intensive Systems: Banana Gardens

By AD 1800 the banana gardens of interlacustrine Africa had not yet reached their full extent, with, for example, Burundi still dominated by a maize system with mixed farming (Cochet 2001). The expansion of banana cultivation in the interlacustrine area and in Uganda has been described by Schoenbrun (1998:79ff) and Reid (2001). In the highlands of Meru, Kilimanjaro and in the eastern arc mountains of Tanzania, home gardens with bananas were already established at this time and based on canal irrigation (Stump and Tagseth 2009).

Intensive Systems: Canal Irrigation

Specialized canal irrigation complexes focusing on grain were found in the Marakwet and Engaruka-Sonjo-Pagasi complex (Berntsen 1976; Adams et al. 1994; Östberg 2004; Stump 2006; Davies et al. 2014), in the Mbooni hills in Kenya (Jackson 1976) and the Uluguru mountains in Tanzania (Kjekshus 1977:36; Paul 2003:71). By AD 1800 it is possible that the now defunct system at Engaruka was still in active use (Westerberg et al. 2010), and although questions have recently been raised on the antiquity of the Marakwet, it remains probable that the irrigation in Marakwet was already developed (Davies et al. 2014). The canal irrigation system at Baringo in Kenya developed later in the 19th century AD and is therefore not included in this map (Anderson 1988, 1989; Petek and Lane 2017).

Irrigated Rice

The expansion of the irrigated paddy rice complex of West Africa is described and discussed in the works of Carney (2001), Hawthorne (2003) and Fields-Black (2008). The conclusions there are partly contradictory as to the early history of irrigated rice in the mangrove areas, but the distribution by 1800 AD seems uncontroversial. The irrigated rice cultivation in Madagascar saw its start in the 1600s AD and its growth was closely connected to the Merina kingdoms (Campbell 2005). The detailed reconstruction of the transformation from extensive swidden rice cultivation to the establishment of irrigated rice cultivation of the highlands, with extensive landscape modification in the form of dykes, canals and terracing is mapped by Le Bourdiec (1974) based on early travelers accounts.

Summary and Discussion

The purpose of this paper has been to map the distribution and character of agriculture in sub-Saharan Africa at approximately 1800 AD. As can be seen from the map (Fig. 2), it has been possible to delimit a series of qualitatively different agricultural systems for that period mainly in Western and Eastern Africa. Large parts of Central and Southern Africa provide a much more homogenous picture. With few exceptions, this vast region of Africa exhibits variation only in the different staple crops under cultivation. It must be noted that these boundaries are hypothetical and based mainly on data which are some 150 years later than 1800 AD. It is of course unlikely that this large region was so homogenous. Evidence exists for differentiated, and sometimes more intensive farming methods, but it has not been possible to delimit larger regions. The evidence from early travelers is not systematically used all over Africa and it is likely that more work based on these sources will improve the map, especially in Central and Southern Africa. The systematic use of archaeological databases may also substantially improve the data for this vast region. Such work for Africa is now planned within the framework LANDUSE6K (http://landuse.uchicago.edu).

It is both a weakness and strength to summarize scanty evidence in map form. While a written synthesis can omit areas with no or little evidence, the map exhibits openly its weaknesses which become more transparent. Many readers will immediately see problems in the map, based on their specific regional knowledge, which will hopefully provoke responses and critique. In order to facilitate critique and scrutiny of the inference from the literature to the polygons as depicted on the maps, a downloadable GIS-file lists the main sources for each region. Table 2 is based on the GIS file and should be read together with Fig. 3. Comments, criticism and additions are more than welcome and may be incorporated into later versions of the map.

Acknowledgements. A map like this is by nature always in progress. This version benefitted greatly from important critical remarks and suggestions from the participants of 8th International Workshop for African Archaeobotany in Modena 2015 and of the PAGES LandCover6k workshop in Utrecht 2016, two anonymous reviewers, from the editors of this volume and many other colleagues. But as a work in progress much more can be done to improve it. The GIS-file can be critically scrutinized by downloading it from https://doi.org/10.17045/sthlmuni.5173477. Please quote this chapter if the GIS file is used in further work.

References

Adams WM (1991) Wasting the rain: the ecology and politics of water resources in Africa. Earthscan, London

Adams WM, Potkanski T, Sutton JEG (1994) Indigenous farmer-managed irrigation in Sonjo, Tanzania. Geogr J 160:17–32

Ajayi JFA, Crowder M, Coquery-Vidrovitch C, Laclavère G (1988) Atlas historique de l'Afrique. Les Editions du Jaguar, Paris

Allan W (1965) The African husbandman. Oliver & Boyd, Edinburgh

Almeida FJM, de Lacerda E, Burton RF et al (1873) Lacerda's journey to Cazembe in 1798. John Murray, London

Ambler CH (1988) Kenyan communities in the age of imperialism: the central region in the late nineteenth century. Yale historical publications. Miscellany, vol 36. Yale University Press, New Haven

Amborn H (2009) Flexibel aus Tradition: Burji in Äthiopien und Kenia. Unter Verwendung der Aufzeichnungen von Helmut Straube. With explanation of some cultural items in English. Aethiopistische Forschungen, vol 71. Harrassowitz, Wiesbaden

Anderson DM (1988) Cultivating pastoralists: Ecology and economy among the Il Chamus of Baringo, 1840–1980. In: Johnson D, Anderson DM (eds) The ecology of survival: case studies from Northeast African history. Crook, London, pp 241–260

Anderson DM (1989) Agriculture and irrigation technology at Lake Baringo in the nineteenth century. Azania 24:89–97

Barth H (1858) Travels and discoveries in North and Central Africa: being a journal of an expedition undertaken under the auspices of H. B. M.'s government, in the years 1849–1855, vol. 5. Longman, Brown, Green, Longmans, Roberts, London

Bassi M (2011) Primary identities in the lower Omo valley: migration, cataclysm, conflict and amalgamation, 1750–1910. J East Afr Stud 5(1):129–157

Beck WG (1943 (reprint 1968)) Beiträge zur Kulturgeschichte der afrikanischen Feldarbeit, Stuttgart 1943. Repr. Johnson Reprint, New York, NY, London (tr. Meisenheim (Glan))

Beez J (2005) Die Ahnen essen keinen Reis: Vom lokalen Umgang mit einem Bewässerungsprojekt am Fuße des Kilimanjaro in Tansania. Universität Bayreuth, Bayreuth

Ben Yahmed D, Sablayrolles J (2000) L'atlas de l'Afrique. Jaguar, Paris

Berntsen JL (1976) Maasai and their neighbors: variables of interaction. Afr Econ Hist 2:1–11

Blench RM (1997) The history of agriculture in northeastern Nigeria. In: Barreteau D, Dognin R, von Graffenried C (eds) L'Homme et le milieu végétal dans le Bassin du Lac Tchad. Orstom, Paris, pp 69–112

Blench RM (2006) Archaeology, language, and the African past. AltaMira Press, Lanham

Boffa J-M (1999) Agroforestry parklands in Sub-Saharan Africa. FAO, Rome

Bollig M, Schnegg M, Wotzka H-P (2013) Pastoralism in Africa: past, present and future. Berghahn Books, New York

Börjeson L (2004) A History under Siege: intensive agriculture in the Mbulu Highlands, Tanzania, 19th century to the present. Acta Universitatis Stockholmiensis, Stockholm

Butzer KW (1992) The Americas before and after 1492: an introduction to current geographical research. Ann Assoc Am Geogr 82(3):345–368

Caillié R (1830a) Journal d'un voyage a Temboctou et a Jenné dans l'Afrique centrale précéde observations faites chez les maures braknas, les nalous et d'autres peuples; pendant les années 1826–1828 T. 2. Imprimerie royale, Paris

Caillié R (1830b) Journal d'un voyage a Temboctou et a Jenné dans l'Afrique centrale précéde observations faites chez les maures braknas, les nalous et d'autres peuples; pendant les années 1826–1828 T. 1. Imprimerie royale, Paris

Campbell G (2005) An economic history of Imperial Madagascar 1750–1895: the rise and fall of an island empire. African studies series, vol 106. Cambridge University Press, New York

Carney JA (2001) Black rice: the African origins of rice cultivation in the Americas. Harvard University Press, Cambridge

Christopher AJ (1982) Towards a definition of the nineteenth century South African Frontier. S Afr Geogr J 64(2):97–113

Cochet H (2001) Crises et révolutions agricoles au Burundi. Karthala, Paris

David N (1976) History of crops and peoples in North Cameroon to A.D. 1900. In: Harlan JR, De Wet JMJ, Stemler ABL (eds) Origins of African plant domestication. Mouton, The Hauge, pp 223–268

Davies MIJ, Kipruto TK, Moore HL (2014) Revisiting the irrigated agricultural landscape of the Marakwet, Kenya: tracing local technology and knowledge over the recent past. Azania 49 (4):486–523

Delius P, Maggs T, Schoeman A (2014) Forgotten World: the stone walled settlements of the Mpumalanga Escarpment. Wits University Press, Johannesburg

Denevan WM (2001) Cultivated landscapes of Native Amazonia and the Andes: triumph over the soil, Oxford geographical and environmental studies. Oxford University Press, Oxford

Denham D, Oudney W, Clapperton H (1826) Narrative of travels and discoveries in Northern and Central Africa, in the years 1822, 1823 and 1824, by Major Denham, F.R.S., Captain Clapperton, and the late Doctor Oudney, vol I, 2nd edn. John Murray, London

Dixon J, Gulliver A, Gibbon D (2001) Farming systems and poverty: improving farmers' livelihoods in a changing world. FAO, Rome

Doolittle WE (2000) Cultivated landscapes of native North America. Oxford geographical and environmental studies. Oxford University Press, Oxford

Fields-Black EL (2008) Deep roots: rice farmers in West Africa and the African diaspora. Indiana University Press, Bloomington

Fuller DQ, Hildebrand E (2013) Domesticating plants in Africa. In: Mitchell P, Lane P (eds) The Oxford handbook of African archaeology. Oxford University Press, Oxford, pp 507–526

Gaillard MJ, Sugita S, Mazier F et al (2010) Holocene land-cover reconstructions for studies on land cover-climate feedbacks. Clim Past 6(4):483–499

Gaillard M-J (2015) LandCover6k: global anthropogenic land-cover change and its role in past climate. Pages Mag 23(1):38–39

Gedge E (1892) A recent exploration, under Captain F. G. Dundas, R. N., up the River Tana to Mount Kenia. Proc R Geogr Soc Monthly Rec Geogr 14(8):513–533

Gifford-Gonzalez D, Hanotte O (2011) Domesticating animals in Africa: implications of genetic and archaeological findings. J World Prehist 24:1–23

Gifford-Gonzalez D, Hanotte O (2013) Domesticating animals in Africa. In: Mitchell P, Lane P (eds) The Oxford handbook of African archaeology. Oxford University Press, Oxford, pp 491–506

Gillman CA (1936) A population map of Tanganyika territory. Geogr Rev 26(3):353–373

Goldewijk KK (2001) Estimating global land use change over the past 300 years: The HYDE database. Global Biogeochem Cy 15(2):417–433
Goldewijk K, Beusen A, van Drecht G, de Vos M (2011) The HYDE 3.1 spatially explicit database of human-induced global land-use change over the past 12,000 years. Global Ecol Biogeogr 20(1):73–86
Grigg D (1965) The logic of regional systems. Ann Assoc Am Geogr 55(3):465–491
Grigg DB (1969) The agricultural regions of the world: review and reflections. Econ Geogr 45 (2):95–132
Grigg DB (1974) The agricultural systems of the world: an evolutionary approach. Cambridge Geographical Studies, vol 5. Cambridge University Press, Cambridge
Guelke L (1976) Frontier settlement in early Dutch South Africa. Ann Assoc Am Geogr 66 (1):25–42
Guillon R, Quiquérez A, Petit C et al (2013) Stone lines and heaps on South-Western Niger plateaus as remains of ancient agricultural land. In: Djindjian F, Sandrine (eds) Understanding landscapes, from land discovery to their spatial organization. Archaeopress, Oxford, pp 87–96
Håkansson NT (1994) Grain, cattle, and power: social processes of intensive cultivation and exchange in precolonial Western Kenya. J Anthropol Res 50(3):249–276
Håkansson NT (2004) The human ecology of world systems in East Africa: the impact of the ivory trade. Hum Ecol 32:561–591
Håkansson NT (2008) The decentralized landscape: Regional wealth and the expansion of production in Northern Tanzania before the eve of colonialism. In: Cliggett L, Pool CA (eds) Economies and the transformation of landscape. Society for Economic Anthropology monographs, vol 25. Altamira Press, Lanham, pp 239–266
Håkansson NT (2012) Ivory: Socio-ecological consequences of the East African ivory trade. In: Hornborg A, Clark B, Hermele K (eds) Ecology and power: struggles over land and material resources in the past, present, and future. Taylor & Francis, Hoboken
Håkansson NT, Widgren M (2007) Labour and landscapes: the political economy of landesque capital in nineteenth century Tanganyika. Geogr Ann B 3:233–248
Harris DR (1976) Traditional systems of plant food production and the origins of agriculture in West Africa. In: Harlan JR, De Wet JMJ, Stemler ABLA (eds) Origins of African plant domestication. Mouton, The Hague, pp 311–356
Havnevik K (1993) Tanzania: the limits to development from above. Scandinavian Institute of African Studies (Nordiska Afrikainstitutet) in cooperation with Mkuki na Nyota Publ., Tanzania, Uppsala
Hawthorne W (2003) Planting rice and harvesting slaves: transformations along the Guinea-Bissau coast, 1400–1900. Heinemann, Portsmouth
He FN, Li SC, Zhang XZ (2012) Reconstruction of cropland area and spatial distribution in the mid-Northern Song Dynasty (AD1004–1085). J Geogr Sci 22(2):359–370
Höhn A, Neumann K (2012) Shifting cultivation and the development of a cultural landscape during the Iron Age (0–1500 AD) in the northern Sahel of Burkina Faso, West Africa: insights from archaeological charcoal. Quatern Int 249:72–83
Jackson K (1976) The dimensions of Kamba pre-Colonial history. In: Ogot BA (ed) Kenya before 1900: eight regional studies. East African Publishing House, Nairobi, pp 174–261
Kahlheber S, Höhn A, Neumann K (2013) Plant and land use in southern Cameroon 400 B.C.E.–400 B.C.E. In: Stevens CJ, Nixon S, Murray M-A, Fuller DQ (eds) Archaeology of African plant use. Left Coast Press, Walnut Creek
Khattar MO (1995) Les sites Gangara, la fin de la culture de Tichitt et l'origine de Ghana. J Africanistes 65(2):31–41
Khattar MO, Grébénart D, Vernet R (1994) Les parcellaires préhistoriques du Tagant (Mauritanie). Bull Soc Préhist Fr 91(6):457–458

Kjekshus H (1977) Ecology control and economic development in East African history: the case of Tanganyika, 1850–1950. Heinemann, London

Koponen J (1988) People and production in late precolonial Tanzania: history and structures. Monographs of the Finnish Society for Development Studies, 0284–4818, vol 2. Finnish Society for Development Studies, Helsinki

Kottek M, Grieser J, Beck C et al (2006) World map of the Koppen-Geiger climate classification updated. Meteorol Z 15(3):259–263

Le Bourdiec F (1974) Hommes et paysages du riz à Madagascar: étude de géographie humaine. s. n., Antananarivo

Le Roy X, Kane M, Marcellin C et al (2005) Pauvreté et accès à l'eau dans la vallée du Sénégal. Paper presented at the international Symposium on Pauvreté hydraulique et crises sociales, Agadir, 12–15 December 2005

Li BB, Jansson U, Ye Y et al (2013) The spatial and temporal change of cropland in the Scandinavian Peninsula during 1875-1999. Reg Environ Change 13(6):1325–1336

Maggs T (1980) The Iron-Age sequence South of the Vaal and Pongola Rivers-some historical implications. J Afr Hist 21(1):1–15

Maggs T (1984) The Iron Age south of the Zambezi. In: Klein RG (ed) Southern African prehistory and paleoenvironments. Balkema, Rotterdam, pp 329–360

Mann CC (2005) 1491: new revelations of the Americas before Columbus, 1st edn. Knopf, New York

Marzouk Y (1989) Societes rurales et techniques hydrauliques en Afrique. Étud rurales 115–116:9–36

McCann J (1995) People of the plow: an agricultural history of Ethiopia, 1800–1990. University of Wisconsin Press, Madison

McEvedy C, Jones R (1978) Atlas of world population history. Penguin, London

Miracle MP (1967) Agriculture in the Congo basin: tradition and change in African rural economies. Madison

Monson J (1991) Agricultural transformation in the Inner Kilombero Valley of Tanzania, 1840–1940. University of California, Los Angeles

Morgan WB (1969) Peasant agriculture in tropical Africa. In: Thomas MF, Whittington GW (eds) Environment and land use in Africa. Methuen, London, pp 241–272

Östberg W (2004) The expansion of Marakwet hill-furrow irrigation in Kenya. In: Widgren M, Sutton JEG (eds) Islands of intensive agriculture in Eastern Africa. James Currey, Oxford, pp 19–48

Paul J-L (2003) Anthropologie historique des Hautes Terres de Tanzanie orientale: stratégies de peuplement et reproduction sociale chez les Luguru matrilinéaires. IFRA, Paris; Nairobi, Karthala

Pélissier P (1966) Les paysans du Sénégal: les civilisations agraires du Cayor à la Casamance. Fabrègue, Saint-Yrieix

Petek N, Lane P (2017) Ethnogenesis and surplus food production: communitas and identity building among nineteenth-and early twentieth-century Ilchamus, Lake Baringo, Kenya. World Archaeol:1–21

Pongratz J, Reick C, Raddatz T et al. (2008) A reconstruction of global agricultural areas and land cover for the last millennium. Global Biogeochem Cy 22(3)

Powell JM, Williams TO (1993) An overview of mixed farming systems in sub-Saharan Africa. In: Powell JM, Fernandez-Rivera S, Williams TO, Renard C (eds) Livestock and sustainable nutrient cycling in mixed farming systems of sub-Saharan Africa, vol 2, Technical Papers. International Livestock Centre for Africa, Addis Ababa

Prudencio CY (1993) Ring management of soils and crops in the West-African semiarid tropics:-the case of the Mossi farming system in Burkina-Faso. Agr Ecosyst Environ 47(3):237–264

Ramankutty N, Foley JA (1999) Estimating historical changes in global land cover: Croplands from 1700 to 1992. Global Biogeochem Cy 13(4):997–1027

Reid A (2001) Bananas and the archaeology of Buganda. Antiquity 75(290):811–812

Reinwald B (1997) 'Though the earth does not lie': Agricultural transitions in Siin (Senegal) under colonial rule. Paideuma 43:143–169

Ruddiman WF (2003) The anthropogenic greenhouse era began thousands of years ago. Clim Change 61(3):261–293

Russell T, Silva F, Steele J (2014) Modelling the spread of farming in the Bantu-speaking regions of Africa: An archaeology-based phylogeography. Plos One 9(1)

Sautter G (1962) A propos de quelques terroirs d'Afrique Occidentale: Essai comparatif. Études rurales 4:24–86

Schneider HK (1970) The Wahi Wanyaturu: economics in an African society. Viking Fund Publications in Anthropology, vol 48. Wenner-Gren Foundation, New York

Schoenbrun DL (1998) A green place, a good place: agrarian change, gender, and social identity in the Great Lakes region to the 15th century. Heinemann, Portsmouth

Soper R (2002) Nyanga: ancient fields, settlements and agricultural history in Zimbabwe. Memoirs of the British Institute in Eastern Africa, vol 16. The British Institute in Eastern Africa, London

Stamm V (1994) Anbausysteme und Bodenrecht in Burkina Faso. Afr Spectr:247–264

Stanley HM (1872) How I found Livingstone: travels, adventures, and discoveries in Central Africa; including four months' residence with Dr. Livingstone. Sampson Low, Marston, Searle and Rivington, London

Stump D (2006) The development and expansion of the field and irrigation systems at Engaruka, Tanzania. Azania 41:69–94

Stump D, Tagseth M (2009) The history of pre-colonial and early colonial agriculture on Mount Kilimanjaro: A review. In: Clack TAR (ed) Culture, history and identity: landscapes of inhabitation in the Mount Kilimanjaro Area, Tanzania. BAR international series, vol 1966. Archaeopress, Oxford, pp 107–124

Sutton JEG (1979) Towards a less orthodox history of Hausaland. J Afr Hist 20(2):179–201

Sutton JEG (1989) Towards a history of cultivating the fields. Azania 24:99–113

Sutton JEG (ed) (1996) The growth of farming communities in Africa from the Equator southwards. Azania 29(30). British Institute in Eastern Africa, Nairobi

Sutton JEG (1993) Becoming Maasailand. In: Spear T, Waller R (eds) Being Maasai: ethnicity and identity in East Africa. James Currey, Oxford, pp 38–68

Tadesse M, Alemu B, Bekele G et al (2006) Atlas of the Ethiopian rural economy. International Food Policy Research Institute (IFPRI); Central Statistical Agency; Ethiopian Development Research Institute

Vansina J (1990) Paths in the rainforests: toward a history of political tradition in equatorial Africa. University of Wisconsin Press, Madison

Viguier P (2008) Sur les traces de René Caillié: le Mali de 1828 revisité. Éditions Quae, Versailles

von Oppen Av (1993) Terms of trade and terms of trust: the history and context of pre-colonial market production around the Upper Zambezi and Kasai. Studien zur afrikanischen Geschichte, vol 6. Lit, Münster

Watson EE (2009) Living terraces in Ethiopia: Konso landscape, culture and development. Eastern Africa series. James Currey, Oxford

Watts M (1983) Silent violence: food, famine, and peasantry in northern Nigeria. University of California Press, Berkeley

Westerberg L-O, Holmgren K, Börjeson L et al (2010) The development of the ancient irrigation system at Engaruka, northern Tanzania: physical and societal factors. Geogr J 176:304–318

Whitmore TM, Turner BL (2001) Cultivated landscapes of middle America on the eve of conquest. Oxford geographical and environmental studies. Oxford University Press, Oxford

Whittlesey D (1936) Major agricultural regions of the earth. Ann Assoc Am Geogr 26(4):199–240

Widgren M (2010a) Mapping Global Agricultural History. In: Kinda A, Komeie T, Mnamide S, Mizoguchi T, Uesugi K (eds) Proceedings of the 14th international conference of historical geographers, Kyoto 2009. Kyoto University Press, Kyoto, pp 211–212

Widgren M (2010b) Besieged palaeonegritics or innovative farmers: historical political ecology of intensive and terraced agriculture in West Africa and Sudan. Afr Stud 69(2):323–343

Widgren M (2012) Slaves: Inequality and sustainable agriculture in pre-colonial West Africa. In: Hornborg A, Clark B, Hermele K (eds) Ecology and power: struggles over land and material resources in the past, present, and future. Routledge, Abingdon, Oxon, pp 97–107

Widgren M (2017) Agricultural intensification in Sub-Saharan Africa, 1500–1800. In: Austin G (ed) Economic development and environmental history in the Anthropocene: perspectives on Asia and Africa. Bloomsbury Academic, London, pp 51–67

Widgren M, Maggs T, Plikk A et al (2016) Precolonial agricultural terracing in Bokoni, South Africa: typology and an exploratory excavation. J Afr Archaeol 14(1)

Wolmer W, Scoones I (2000) The science of 'civilized' agriculture: the mixed farming discourse in Zimbabwe 99(397):575–600

Yasuoka H (2013) Dense wild yam patches established by hunter-gatherer camps: beyond the wild yam question; toward the historical ecology of rainforests. Hum Ecol 41(3):465–475

Ylvisaker M (1979) Lamu in the nineteenth century: land, trade, and politics. African research studies, vol 13. African Center, Boston University, Boston

Agriculture and Wild Plant Use in the Middle Senegal River Valley, c. 800 BC—1000 AD

Daphne E. Gallagher[1(✉)], Susan K. McIntosh[2], and Shawn S. Murray[2]

[1] Department of Anthropology, University of Oregon, Eugene, USA
daphne@uoregon.edu
[2] Department of Anthropology, Rice University, Houston, USA

Abstract. The histories of the diverse agricultural strategies used in the Middle Senegal River Valley (MSV) are poorly understood. The MSV is a productive inland floodplain environment located within the highly unpredictable West African Sahel, and is capable of reliable flood recession agriculture augmented by rain-fed strategies. Archaeological research from 1990–1993, directed by R. McIntosh, S. McIntosh, and H. Bocoum at Sincu Bara, Siwré and Cubalel sites along with additional excavations at Waladé by A. Déme in 1998–1999 produced diverse archaeological materials spanning almost three millennia of occupation. These included an extensive archaeobotanical record comprising 315 flotation samples from 14 excavation units in which the period 800 BC–1000 AD is particularly well represented. This paper presents a synthesis of the archaeobotanical seed and fruit sequence from these sites, drawing upon previous work by several scholars, including Capezza, Fuller, Gallagher, M.A. Murray and S. Murray, to explore regional patterns in plant use. Results indicate that pearl millet (*Pennisetum glaucum* (L.) R.Br.) was cultivated in the MSV for almost 3000 years, and during this time residents also exploited a range of wild plants, including jujube (*Ziziphus* sp.). In contrast, although common in the ethnohistoric record, sorghum (*Sorghum bicolor* L. Moench) was extremely rare and cotton (*Gossypium* sp.) was not conclusively identified at any of the excavated sites. The absence and/or rarity of these important plants in the archaeobotanical record of the MSV may be a result of preservation biases or it may be that they were of lesser significance prior to 1000 AD.

Keywords: Middle Senegal valley · Senegal river · Archaeobotany Archaeology · Décrue agriculture · Pearl millet · Sorghum · *Ziziphus*

Introduction

Located in the arid Sahel, the Middle Senegal Valley (MSV) is a rich environmental zone for agricultural production. Historically, farmers took advantage of the local ecology to practice intensive, year-round agriculture, growing dry-land crops such as millet during the summer rainy season, and utilizing flood-recession techniques to grow sorghum during the winter, in addition to collecting wild plant resources. Little is known of the antiquity of these practices, as few archaeological projects have incorporated systematic flotation collection for macro-botanical remains. This paper

A. M. Mercuri et al. (Eds.): *Plants and People in the African Past*,
https://doi.org/10.1007/978-3-319-89839-1_16

synthesizes the results of paleoethnobotanical research at the MSV sites of Walaldé, Siwré, Sincu Bara, and Cubalel, spanning the 1st millennium BC to the end of the 1st millennium AD (Table 1). Results indicate that 1st millennium AD societies in the MSV may have utilized different agricultural strategies than those common in the region historically.

Agroecology of the Middle Senegal Valley

The agroecology of the MSV is shaped by the complex geomorphology of the river valley and the unpredictable rainfall and seasonal floods that combine to create numerous microenvironments suitable for different forms of agriculture. The region consists of two major zones: seasonally flooded areas (*waalo*) near the primary river channels, which include both deep basins (*hollalde*) and high levees (*fonde*); and uplands outside maximal flood areas (*jeeri*), which are connected to the *waalo* through a transitional slope (*jejeengol*) (Boutillier et al. 1962; McIntosh and McIntosh 2016a) (Fig. 1). Within our study region, *fonde* extend along the Senegal River in bands that are typically 1–2 km wide (Fig. 2). They are much more limited in extent along the Doué River, which forms the southern margin of the *waalo*. These zones are today exploited by the Tukolor – sedentary agriculturalists who practice décrue cultivation in the basins – and Fulbe or Peul, who enter the MSV for dry season pasture in the basins after the sorghum harvest. Both groups grow pearl millet (*Pennisetum glaucum* (L.) R. Br.) on the *jeeri*. In the mid-late twentieth century, researchers conducted several detailed assessments of land use and agricultural practice in the region during a period when "traditional" practices were being impacted by the construction of dams and expansion of irrigation. These studies provide a snapshot of subsistence strategies and land use at that time.

The MSV is located in the Sahelian zone, which is characterized by a short summer rainy season (typically July-October) and long dry winters that become increasingly hot through the spring. While on average the region receives 300–500 mm of rainfall annually, interannual variability in both the quantity and timing of precipitation is very high. As a result, both the vegetation and agricultural strategies on the rainfed *jeeri* uplands favor resilient, drought resistant species. Wild vegetation is dominated by drought resistant grasses (e.g., *Aristida* spp., *Brachiaria* spp., *Cenchrus biflorus* Roxb., *Dactyloctenium aegyptium* (L.) Willd., *Pennisetum pedicellatum* Trin., and *Panicum turgidum* Forssk.) and other herbaceous annuals (e.g., *Cleome* sp., *Commelina* spp., *Gisekia pharnaceoides* L., *Indigofera senegalensis* Lam.). Small trees, including *Acacia raddiana* Savi, *Acacia senegal* (L.) Britton, *Balanites aegyptiaca* (L.) Del. (desert date), *Combretum aculeatum* Vent., and *Ziziphus mauritiana* Lam. (jujube) are distributed sparsely throughout the landscape, and herbaceous perennials such as *Tephrosia purpurea* (L.) Pers. are found only occasionally (Trochain 1940; Adam 1962). In the mid-twentieth century, farmers planted primarily pearl millet, which thrives in poor, well-drained sandy soils and can grow with as little as 200 mm of precipitation (NRC 1996), and pearl millet was found in over 95% of *jeeri* fields (Boutillier et al. 1962). These fields were frequently intercropped or rotated with cowpea (*Vigna unguiculata* (L.) Walp.) and the introduced crops peanuts (*Arachis*

Table 1. Summary of archaeological sites incorporated in this synthesis.

Village	Archaeological site	Unit size	Mound height	Location	Site dimensions	Site area
Cubalel	C-1	3 × 4 m	3 m	9–10 m levee	115 × 90 m	0.7 ha
	C-3	A: 6 × 5 m	6 m	9–10 m levee	135 × 110 m	1.5 ha
		B: 6 × 5 m				
	C-6	3 × 4 m	3.6 m	9–10 m levee	140 × 160 m	2.0 ha
	C-8	3 × 3 m	0.8 m	9–10 m levee	70 × 55 m	
	C-20	1.5 × 1.5 m	5.5 m	9–10 m levee	90 × 55 m	
Siwré	S-1	1S: 4 × 3 m	6 m	9–10 m levee	411 × 300 m	10 ha
		2S: 4 × 3 m				
		3S: 2 × 2 m				
Walaldé	WAL1	4 × 3 m	4.5 m	9–10 m levee	550 × 100 m	5 ha
Sincu Bara	SB	SB1: 3 × 3 m	non-mound	*jeeri*		67 ha
		SB2: 3 × 3 m				
		SB3: 3 × 2 m				
		SB4: 2.5 × 2.5 m				

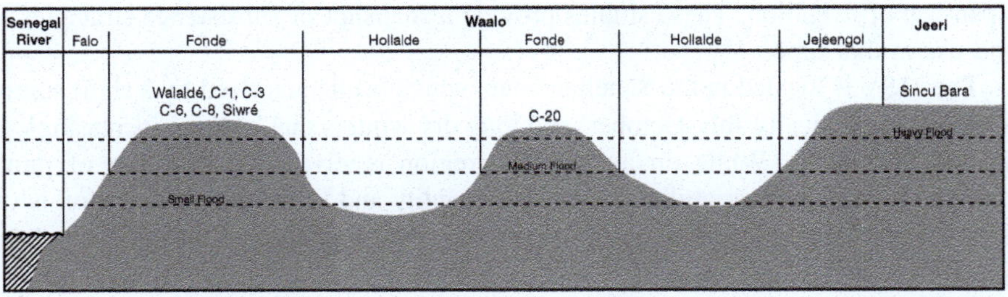

Fig. 1. Schematic of the Middle Senegal Valley landscape (based on Boutillier et al. 1962).

hypogaea L.) and maize (*Zea mays* L.), and farmers sometimes cultivated a few fields of sorghum (*Sorghum bicolor* L. Moench), although the crop was more likely to fail in a dry year. *Jeeri* fields typically required regular fallows to replenish their nutrients (Boutillier et al. 1962).

In contrast to the *jeeri*, the vegetation and agricultural strategies in the *waalo* were shaped as much if not more by the river and its seasonal floods as by rainfall. The Senegal River originates in the highland Fuuta Jallon region to the south, where much higher annual rainfalls (average 1500–2000 mm) feed the flow of the river (McIntosh and McIntosh 2016a). The timing and quantity of rainfall in the Fuuta Jallon and

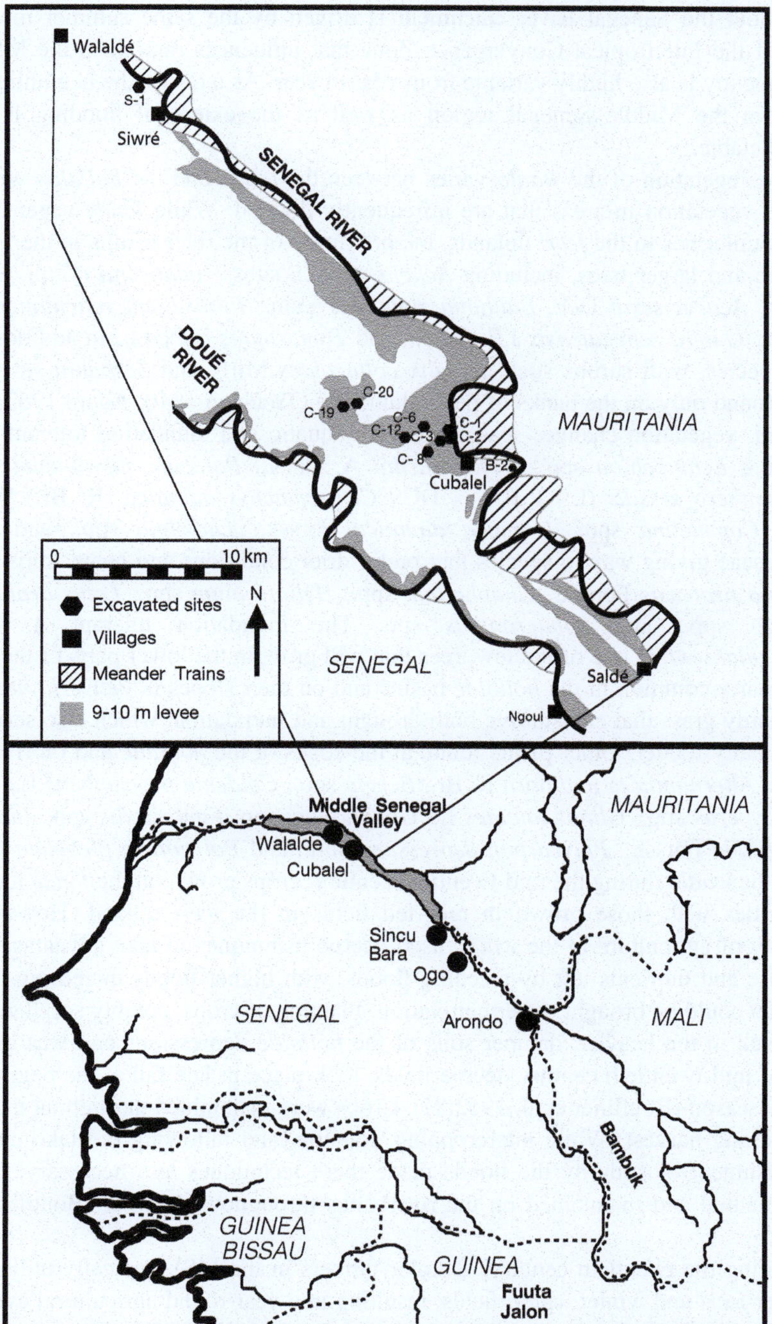

Fig. 2. The Middle Senegal Valley, showing sites mentioned in the text.

throughout the Senegal River catchment is driven by the same summer movement north of the Intertropical Convergence Zone that influences rainfall in the MSV and consequently is also highly variable from year to year. As a result, the beginning of the floods in the Middle Senegal region as well as the extent of flooding is highly unpredictable.

The vegetation of the *waalo* varies between the *fonde* and the *hollalde* with transitional vegetation in areas that are infrequently flooded. While *fonde* vegetation has some similarities to the *jeeri* uplands, the proximity of the river results in the presence of more and larger trees, including *Acacia nilotica* subsp. *adansonii* (Guill. & Perr.) Brenan, *Acacia seyal* Del., *Bauhinia rufescens* Lam., *Piliostigma reticulatum* (DC.) Hochst., *Guiera senegalensis* J.F. Gmel., and *Tamarindus indica* L. in addition to the *jeeri* species, with shrubs such as *Salix coluteoides* Mirb. and *Sesbania sesban* (L.) Merr. found only on the banks of perennial rivers (Trochain 1940; Adam 1962). In the *hollalde,* vegetation changes seasonally with aquatic and inundation tolerant species including *Echinochloa* spp., *Oryza barthii* A. Chev., *Panicum subalbidum* Kunth., *Alternanthera sessilis* (L.) R.Br. ex DC., *Centrostachys aquatica* (R. Br.) Wall. ex Moq., *Commelina* spp., *Ipomoea aquatica* Forssk., *Ludwigia* spp., and diverse Cyperaceae giving way to species that prefer drier conditions and heavy soils such as *Ceruana pratensis* Forssk., *Gnaphalium* spp., *Heliotropium* spp., *Lotus arabicus* L., *Mollugo* spp., and *Sphaeranthus* spp. The inundation tolerant *A. nilotica* subsp. *adansonii* is one of the few trees that will grow in the lower parts of the basins. Particularly common in the *hollalde* basins and on their slopes is *Vetiveria negritiana* L., a hardy grass that can survive both drought and inundation, which also serves as a high quality fodder. Other plants found at the edges of the *hollalde* and on riverbanks include *Alternanthera nodiflora* R. Br., *Bergia* spp., *Coldenia procumbens* L., *Crypsis* spp., *Cyperus* spp., *Glinus lotoides* L., *Gnaphalium* spp, *Fimbristylis* spp., *Indigofera oblongifolia* Forssk., *Polycarpon depressum* Nutt., and *Polygonum plebeium* R. Br.

In the *waalo* during the mid-twentieth century, crops grown on the *fonde* had many similarities with those grown in rain fed fields in the *jeeri* upland. However, the majority of agriculture in the *waalo* used décrue techniques to take advantage of soil moisture and nutrients left by receding floods, with higher floods increasing the land area that could be brought under cultivation. Planting of crops, notably sorghum which does well in the heavier, damper soils of the *hollalde* depressions and was grown in 98% of mid-twentieth century décrue fields, took place in late fall at the beginning of the dry season (Boutillier et al. 1962:87). Crops were maintained throughout the winter for a spring harvest. While intercropping, rotation, and fallowing did take place, the replenishment of soils by the floods made these techniques less necessary. Gardens were planted and maintained on the riverbanks throughout the year (Boutillier et al. 1962).

During the twentieth century, Tukolor farmers in the MSV typically utilized both summer *jeeri* and winter *waalo* fields, resulting in a year-round agricultural cycle with both a pearl millet and a sorghum harvest (Delolme 1937; Boutillier et al. 1962; Lericollais and Schmitz 1984; Schmitz 1986; Boutillier and Schmitz 1987). While both field systems are ultimately dependent on rainfall from the same weather system, the spatially patchy nature of precipitation is such that a bad year for local rainfall does not necessary result in a bad year for floods or vice versa. Mid-twentieth century studies indicated a slightly more important role for the *waalo* fields, although diets reflected seasonal shifts from a reliance on sorghum to pearl millet depending on the most recent harvest (Boutillier et al. 1962). Also noted as important in the diet were vegetables grown in gardens; while many American domesticates played an important role, foods such as okra (*Abelmoschus esculentus* (L.) Moench), melons (*Citrullus* spp.), and roselle (*Hibiscus sabdariffa* L.) likely had greater antiquity in the region (Boutillier et al. 1962). Several wild plants were also described as regularly consumed: *Panicum laetum* Kunth. grains, leaves from baobab and *Senna tora* (L.) Roxb., water lily roots and fruits (*Nymphaea* spp.), and the fruits of tamarind, desert date, and jujube (Boutillier et al. 1962:175).

Currently available paleoclimate data suggest that the first two centuries of the 1st millennium BC were arid and apparently coincided with initial colonization of the MSV by northern agropastoralists (Déme and McIntosh 2006; McIntosh and McIntosh 2016a). The remainder of the 1st millennium BC was marked by a more humid climate with arid episodes, although different climate proxies provide conflicting interpretations and chronologies (Maley and Vernet 2015; McIntosh and Déme 2016). Sometime between 200 BC and 200–300 AD, Sudanian taxa were replaced by Sahelian species in pollen cores near the Senegal Valley, and sand dunes were reactivated along the Niger Bend during a period marked by considerable climatic instability. After 300 AD, the climate became more stable. The early 2nd millennium is regionally heterogeneous, as numerous proxy records from the western Sahel show increased humidity during the Medieval Climate Anomaly (950–1300 AD), while equatorial West Africa was drier (Azzoug et al. 2012:60). Arid episodes, some severe, were common in the MSV from the 2nd millennium AD into the historic period (McIntosh and McIntosh 2016b:406). While in general rain fed agriculture was likely more reliable and floods more extensive in the MSV during humid periods, interannual variability would have remained high.

Archaeological Research in the Middle Senegal Valley

During the colonial period, French administrators, historians, and interested amateurs discovered, dug, and surface collected a variety of sites in the MSV, but in most instances these individuals lacked the training to properly record or interpret their efforts. They were content to fit their discoveries into the broadest possible culture historical frameworks: Neolithic – Iron Age – Sub-recent. Time-space systematics were non-existent well into the 1980s, despite widespread application of radiocarbon dating after 1960. Excavations were routinely plagued by a lack of stratigraphic integrity, and only a handful involved collection and analysis of archaeobotanical and faunal material and systematic description of material culture (e.g., Chavane 1985; Thiaw 2012; McIntosh and McIntosh 2016a). Consequently, a primary goal of the Middle Senegal Valley Archaeological Project (MSVAP) in the 1990s was the development of a reliable chronological framework for human settlement using data from excavations at multiple sites to identify change through time and space in material culture, subsistence, technology, trade activity, and environment (Déme and McIntosh 2006; McIntosh et al. 2016). Excavation was followed by intensive, pedestrian surface survey to

Table 2. Chronology of sites incorporated in this analysis. To avoid confusion, the equivalent Cubalel/Siwré phase names are used throughout this chapter when referring to specimens from Sincu Bara.

			Walaldé	Phase I	800–550 BC
			Walaldé	Phase II	550–200 BC
Sincu Bara	Phase I	≈	Cubalel/Siwré	Phase IA	0–200 AD
			Cubalel/Siwré	Phase IB	200–400 AD
Sincu Bara	Phase II	≈	Cubalel/Siwré	Phase II	400–550/600 AD
Sincu Bara	Phase IIIA	≈	Cubalel/Siwré	Phase IIIA	550/600–750 AD
Sincu Bara	Phase IIIB/IV	≈	Cubalel/Siwré	Phase IIIB	750–850/900 AD
			Cubalel/Siwré	Phase IV	900–1600 AD
Sincu Bara	Phase V	≈	Cubalel/Siwré	Phase V	1600–1900 AD

locate and map the full range of sites present in the study area, with regard to location, size, access to basic resources, surface features and artifacts, and chronology.

The settlement mound sites in the core study area around Cubalel are up to 6 m high, and allow the recovery of centuries-long sequences from individual excavation units. The key mounds in this study are all located on levees 9–10 m above mean sea level, which places them above the level of annual flooding in most years. Sincu Bara, located 100 km upriver from the Walaldé, Siwré and Cubalel mound sites, is a vast, non-mound site on the *jeeri* that revealed 3.5 m of subsurface deposits (McIntosh and Bocoum 2000; Bocoum and McIntosh 2002) (Table 1).

A detailed chronological framework for the 1st millennium BC and AD has been established for the study region, based on pottery sequences that define different temporal phases, and anchored by more than four dozen calibrated radiocarbon dates, although there are possibilities for a longer or shorter chronology for any of the approximately 200-year long ceramic phases and subphases, given the calibrated date ranges (Table 2). The 2nd millennium AD remains poorly defined, as excavated sequences provided scant evidence of undisturbed deposits from this period.

Over the two or more millennia documented by MSVAP research, a variety of factors have impacted human societies living in the MSV, offering possibilities for change and regional transformation. Our goal is to identify the archaeological evidence that informs us on the nature and magnitude of these changes in subsistence, technology, settlement structure, and social organization. Interannual variability in rainfall and flooding as well as longer-term climate fluctuations that may involve sustained periods of drought are a perennial challenge for the pastoralists, agropastoralists, and farmers who exploit the MSV. The impact of historically documented climate change in the Sahel, involving famine, political instability and large-scale movements of people and herds, is well documented (e.g., Cissoko 1968; Webb 1995; Millàn and Rodrigo 2014). Initial colonization of the MSV by northern agropastoralists in the early 1st millennium BC was almost certainly provoked by severe droughts.

The data presented here inform us on plant use during the 1500 or more years preceding the economic and social transformations that attended the growing participation of the MSV in long-distance trade. Excavation data indicate that northern trade goods such as copper alloys and glass began reaching the MSV in the 9th or 10th century AD, presaging the growth of Takrur as a major trade town and polity by the 12th century (al-Idrisi, in Levtzion and Hopkins (1981:106–107), McIntosh et al. (2016)). From that point forward, the MSV was a key node in the trans-Saharan trade for gold, slaves, horses and other goods. Further political and economic developments throughout the 2nd millennium AD had significant impact on agriculture and subsistence systems and resulted in the elaboration of specialist producers, traders, warriors, clerics, and servile populations in the developing political economy. The most important of these developments include the arrival of Europeans on the coast, beginning with the Portuguese in the 15th century, and the settlement reorganization effected by the theocratic imamate of Futa Toro in the late 18th century (Robinson et al. 1972; Robinson 1975; Kane 1986; Déme and Guèye 2007). The imamate ended in the later 19th century with the colonization of the MSV by France. In this paper, we investigate domesticated and wild plant use in the MSV through 1000 AD in order to assess the subsistence base prior to these transformations. In order to understand the ways in which these later developments impacted societies, it is essential to understand the diversity of crop plants, use of landscape and extent of agricultural intensification employed by early farmers in the region.

Archaeobotanical Research in the MSV

Prior to the MSVAP, archaeobotanical sampling (aside from the collection of wood charcoal samples for dating) was not widely employed on archaeological projects in the MSV. The exception is Chavane's (1985) research at Ogo, a site on the north bank of the Senegal River occupied from c. 900–1200 AD. One hermetically sealed jar near the surface contained "*grains ovales de mil*" which quickly blew away in the wind during excavation (Chavane 1985:87). In this case, the general term "*mil*" (millet) is ambiguous. While most commonly used to describe pearl millet, the grains of that plant are typically club-shaped rather than oval. Sorghum, fonio (*Digitaria exilis* (Kippist) Stapf) and wild Paniceae grasses (e.g., *Panicum laetum*) all have a more oval grain shape. Targeted flotation samples were taken from hearths in one excavation unit, although analysis yielded only 11 carbonized jujube nuts (Chavane 1985: 85).

While Chavane's macro-botanical sampling was largely opportunistic, he did sample systematically for pollen in one excavation unit (Chavane 1985:109–112). The results indicate that over the course of the site's occupation, tree cover decreased as agricultural fields and fallow vegetation became more common (although human favored tree species such as *Adansonia digitata* L. appear, and agriculturally useful trees including the nitrogen-fixing *Acacia* spp. remain frequent). Most interesting is the possible identification of cotton pollen (cf. *Gossypium* sp.) beginning in the early 2nd millennium AD. Cotton cultivation at this site is supported by the presence of spindle whorls and other material culture associated with spinning.

For sites excavated as part of MSVAP (Cubalel, Siwré, and Sincu Bara), a systematic macrobotanical sampling program was designed and carried out by M.A. Murray in the 1990–1991 season, and the sampling (4–12 l systematic) and bucket flotation protocols she developed were followed by Ndiaye in subsequent seasons (see

Murray et al. 2007:63). To date, all analyses have focused on seeds and fruits from the sites. M.A. Murray and Capezza carried out initial sorting and analyses on all Sincu Bara samples and a preliminary subset of samples from Cubalel and Siwré. The Sincu Bara sample identifications were checked and expanded on by Gallagher and initially interpreted by S. McIntosh (McIntosh and Bocoum 2000) while M.A. Murray and Capezza's preliminary results from Cubalel were written up in collaboration with Fuller (Murray et al. 2007). The samples were then sent to S. Murray, who sorted the remaining samples under 10–40× magnification and compiled the comprehensive tables of identified specimens (Murray 2008; Murray and Gallagher 2016).[1] Gallagher, working in consultation with S. McIntosh and S. Murray, is primarily responsible for the interpretation and write-up of results from Cubalel and Siwré, including a more detailed analysis of samples from the richest excavation unit (Cubalel C3-A) (Gallagher and Murray 2016). At Waladé, Alioune Déme followed the flotation protocols used by MSVAP, but collected only a single one-liter sample from most excavation levels. S. Murray sorted the samples under binocular magnification, identified seeds and fruits, and analyzed the results (Murray and Déme 2014). This paper synthesizes these dispersed results in a regional analysis encompassing all currently available data.

Characterizing the Archaeobotanical Assemblage

This paper incorporates analyses of the light fractions of 315 flotation samples from eight archaeological sites (Waladé, Siwré (S-1), Sincu Bara (SB1, SB2, SB3, and SB4), and Cubalel sites C-1, C-3, C-6, C-8, and C-20). Of these, 269 samples included seed or fruit specimens that could be identified. Overall, 25,431 individual specimens were identified, although their distribution is uneven in both space and time, with 59.7% of identified specimens from site C-3 and 49.5% of specimens from Phase II (over 98% of which were from sites C-3 and C-6) (Table 3). Even within these, a few particularly large samples frequently accounted for the majority of identified specimens. For example, in Unit C-3A, Phase II (33 samples), over 75% of identified specimens were recovered from just 5 samples. Rich samples were primarily from hearths, charcoal scatters, and small ashy pits. All these features were commonly identified in many levels, and their richness varied significantly. High densities of seeds and fruits were not consistently associated with any particular characteristics of these features, suggesting primarily circumstantial carbonization of seeds and fruits in the region.

Across the eight sites, plants from 82 taxonomic designations were identified (Tables 4, 5). Of these, 28 taxa are identified to species, 39 to genus or tribe, and 15 to family (Fig. 3). While this analysis maintains continuity in the taxa present at each site, some of these distinctions may be functionally equivalent. For example, cf.

[1] At least two extremely rich samples from C3-A documented by Murray et al. 2007 are not included in the tables compiled by S. Murray and are not incorporated in this analysis. While these samples would significantly increase the absolute numbers of pearl millet grain and chaff for this unit, they minimally affect ubiquity calculations and do not broadly affect the overall patterns discussed in this paper.

Table 3. Distribution of Flotation Samples and Identified Specimens. S (Samples): N samples with carbonized seeds/fruits (N samples analyzed); ID (Identifications): Number Identified Specimens (Number Identified Taxa). C-20 samples are from Phase III.

Phase	Walaldé	Siwré	Sincu Bara	C-1	C-3	C-6	C-8	C-20	Total
N Units	1	3	4	1	2	1	1	1	
WI	S: 6 (10)								S: 6 (10)
	ID: 88 (13)								ID: 88 (13)
WII	S: 5 (10)								S: 5 (10)
	ID: 13 (6)								ID: 13 (6)
IA			S: 1 (1)		S: 6 (6)				S: 7 (7)
			ID: 26 (4)		ID: 1430 (15)				ID: 1456 (16)
IB		S: 11 (12)			S: 7 (8)				S: 18 (20)
		ID: 1765 (19)			ID: 269 (11)				ID: 2034 (23)
II			S: 10 (11)	S: 4 (5)	S: 39 (41)	S: 29 (31)			S: 82 (88)
			ID: 200 (17)	ID: 18 (8)	ID: 6629 (39)	ID: 5747 (32)			ID: 12594 (57)
IIIA			S: 5 (5)	S: 8 (17)	S: 39 (42)	S: 5 (5)	S: 6 (7)		S: 63 (76)
			ID: 630 (20)	ID: 19 (8)	ID: 3798 (37)	ID: 521 (23)	ID: 182 (14)	S: 19 (23)	ID: 5150 (51)
IIIB					S: 27 (29)			ID: 59 (8)	S: 27 (29)
					ID: 3066 (24)				ID: 3066 (24)
IIIB/IV			S: 10 (13)			S: 13 (15)			S: 23 (28)
			ID: 305 (27)			ID: 321 (18)			ID: 626 (37)
V		S: 4 (4)	S: 1 (3)						S: 4 (6)
		ID: 38 (4)	ID: 1 (1)						ID 38 (3)
IB/IV		S: 14 (17)							S: 14 (17)
		ID: 306 (13)							ID: 306 (13)
Total	S: 11 (20)	S: 29 (33)	S: 27 (33)	S: 12 (22)	S: 118 (126)	S: 47 (51)	S: 6 (7)	S: 19 (23)	S: 269 (315)
	ID: 101 (16)	ID: 2109 (24)	ID: 1162 (39)	ID: 37 (15)	ID: 15192 (53)	ID: 6589 (40)	ID: 182 (15)	ID: 59 (8)	ID: 25431 (82)

Eragrostidae is recorded at Sincu Bara, while cf. *Eragrostis/Sporobolus* spp. is recorded at Cubalel sites. While samples with larger numbers of identified specimens typically include greater species richness, in general most taxonomic designations occur in both large and small samples, and rare specimens are as likely to be found in samples with only a few identifications as in a large rich sample.

Staple grains (N = 16,390, 64.4% of assemblage)

Pearl millet (*P. glaucum*) is by far the most common species (n = 15,727, 99.7% of staple grains), and is present at every site and in every occupational phase (65.1% ubiquity). Pearl millet occurs most frequently as carbonized caryopses (n = 12,380, 59.5% ubiquity), which are present in every phase at every site. In contrast, threshing

Table 4. Taxa identified at Middle Senegal Valley sites summarized by occupation phase. xxxx = > 500 specimens; xxx = 100 − 499 specimens; xx = 20 − 99 specimens; x = 1 − 19 specimens. a = achenes, b = biomineralized drupe, c = caryopses, i = involucres, n = nutshell, s = seeds, g = glumes.

Family	Taxon	Part	NISP	W	IA	IB	II	IIIA	IIIB	IIIB/IV	V
POACEAE	*Pennisetum glaucum* (includes partial and damaged specimens)	C	12380	xx	xxxx	xxxx	xxxx	xxxx	xxxx	xxx	xx
POACEAE	*Pennisetum* cf. *glaucum*	I	3963			xxx	xxxx	xxxx	xxxx	x	
POACEAE	*Pennisetum glaucum* vel. *Sorghum bicolor*	C	21			x	x				
POACEAE	*Digitaria* sp., short broad type (possible fonio)	C	26				xx	x	x	x	
POACEAE	Unidentified	C	1251	x	xx	xxx	xxx	xxx	x	xx	x
POACEAE	Unidentified	G	117			xxx					
POACEAE	*Digitaria* spp.	C	308		xx	xx	xxx	xx	xx	x	
POACEAE	*Brachiaria/Echinochloa* spp.	C	238		x	x	xxx	xx	x		x
POACEAE	*Echinochloa* cf. *colona*	C	1							x	
POACEAE	*Setaria* spp.	C, G	194	x	x	x	xxx	xx	xx	x	
POACEAE	*Cenchrus biflorus*	I	109		xxx		x				
POACEAE	cf. *Cenchrus biflorus*	C	2		x		x				
POACEAE	cf. *Panicum* spp.	C	87		x	x	xx	x	x	x	
POACEAE	*Panicum turgidum*	C	39		x		xx	x	x	x	
POACEAE	cf. *Panicum laetum*	C	17			x		x			
POACEAE	Paniceae	C	82	x	x	x	xx	x		x	
POACEAE	*Pennisetum* spp. (wild)	C	49					xx		xx	
POACEAE	cf. *Pennisetum violaceum*	C	1				x				
POACEAE	*Pennisetum* cf. *purpureum*	C	1				x				

(continued)

Table 4. (*continued*)

Family	Taxon	Part	NISP	W	IA	IB	II	IIIA	IIIB	IIIB/IV	V
POACEAE	*Paspalum scrobiculatum*	C	31	x			xx	x	x	x	
POACEAE	cf. *Paspalum* spp.	C	8				x	x			
POACEAE	cf. *Acroceras* spp.	C	2							x	
POACEAE	*Dactyloctenium aegyptium*	C	48			x	xx				
POACEAE	cf. *Eragrostis/Sporobolus* spp.	C	33				x	x			
POACEAE	cf. Eragrostidae	C	4				x			x	
POACEAE	cf. *Sporobolus* sp.	C	1				x				
POACEAE	Chloridoideae	C	8							x	
POACEAE	*Eleusine indica*	C	4				x			x	
POACEAE	cf. *Chloris pilosa*	C	2					x			
POACEAE	cf. Andropogoneae	C	22			x	x	x		x	
POACEAE	cf. *Bothriochloa bladhii*	C	9		x	x		x			
POACEAE	cf. *Bothriochloa* spp.	C	2				x				
POACEAE	cf. *Aristida* spp.	C	3							x	
POACEAE	cf. *Rottboelia* spp.	C	2				x	x			
POACEAE	cf. *Hackelochloa* sp.	C	1				x				
POACEAE	cf. *Rhytachne* sp.	C	1				x				
RHAMNACEAE	*Ziziphus* spp.	N, S	4071	x	xxxx	xxx	xxxx	xxxx	xxxx	x	
BOMBACACEAE	*Adansonia digitata*	N, S	305		x	x	xxx	xx		x	
cf. MIMOSACEAE	Unidentified	S	163			x	xx	xx	x	x	
MIMOSACEAE	cf. *Prosopis africana*	S	17		x	x	x	x	x		
PAPILIONACEAE	cf. *Tephrosia* spp.	S	74		x	x	xx	xx	x	x	x

(*continued*)

Table 4. *(continued)*

Family	Taxon	Part	NISP	W	IA	IB	II	IIIA	IIIB	IIIB/IV	V
CAESALPINACEAE	cf. *Piliostigma reticulatum*	S	6					x			
MALVACEAE	cf. *Abutilon* spp.	S	38				xx	x	x		
ANACARDIACEAE	*Sclerocarya birrea*	N	12	x			x	x	x		
TILIACEAE	*Grewia* spp.	N	12	x		x	x	x			
ULMACEAE	*Celtis* spp.	S	10				x				
ULMACEAE	*Celtis integrifolia*	B	5	x							
VITACEAE	*Vitex* spp.	S	7				x				
AMARANTHACEAE/CHENOPODIACEAE	Unidentified	S	529				xx	xx	xxx	x	
AIZOACEAE	*Zaleya pentandra*	S	520				xx	xx	xxx	x	
AIZOACEAE	*Trianthema portulacastrum*	S	4	x				x			
AIZOACEAE	cf. *Gisekia pharnaceoides*	S	2	x						x	
SOLANACEAE	*Physalis/Solanum* spp.	S	115	x			x	x	xx	x	
SOLANACEAE	*Solanum* spp.	S	39					x	xx		
SOLANACEAE	*Solanum* cf. *nigrum*	S	15							x	
SOLANACEAE	cf. *Physalis* spp.	S	2	x							
SOLANACEAE	*Physalis* spp.	S	2					x			
cf. SOLANACEAE	Unidentified	S	20				x	x		x	
cf. FABACEAE	Unidentified	S	91	x	x	x	xx	xx	x	x	
PAPILIONACEAE	cf. *Indigofera* spp.	S	32				x	x		xx	
PAPILIONACEAE	cf. *Crotalaria* spp.	S	10			x	x	x			
PAPILIONACEAE	cf. *Stylosanthes* sp.	S	1					x			
RUBIACEAE	*Borreria* spp.	S	30				x	x		xx	
PORTULACACEAE	*Portulaca oleracea*	S	25			x	x	x	x		
PORTULACACEAE	*Portulaca foliosa*	S	7				x		x	x	

(continued)

Table 4. *(continued)*

Family	Taxon	Part	NISP	W	IA	IB	II	IIIA	IIIB	IIIB/IV	V
CYPERACEAE	Unidentified	A	16			x	x	x			
CYPERACEAE	*Fimbristylis* spp.	A	6				x	x			
CYPERACEAE	cf. *Cyperus* sp.	A	1							x	
CYPERACEAE	cf. *Scirpus* sp.	A	1							x	
cf. ASTERACEAE	Unidentified	A	11				x	x		x	
CARYOPHYLLACEAE	Unidentified	S	10				x	x		x	
LAMIACEAE	cf. *Hyptis lanceolata*	S	10				x				
cf. LAMIACEAE	Unidentified, *Teucrium* type	S	7				x				
cf. LAMIACEAE	Unidentified	S	6				x	x		x	
cf. MALVACEAE	Unidentified	S	8				x	x		x	
CAPPARACEAE	cf. *Cleome* spp.	S	5	x							
COMMELINACEAE	*Commelina* spp.	S	5				x	x	x		
COMMELINACEAE	*Commelina* cf. *benghalensis*	S	1					x			
cf. APIACEAE	Unidentified	S	4				x				
CAPPARACEAE	Unidentified	S	3				x		x		
cf. BRASSICACEAE	Unidentified	S	2			x		x			
cf. CONVOLVULACEAE	Unidentified	S	2	x							
BORAGINACEAE	*Heliotropium* sp.	S	1					x			
cf. CUCURBITACEAE	Unidentified	S	1					x			

Table 5. Taxa identified at Middle Senegal Valley sites summarized by site. xxxx = > 500 specimens; xxx = 100 − 499 specimens; xx = 20 − 99 specimens; x = 1 − 19 specimens. a = achenes, b = biomineralized drupe, c = caryopses, i = involucres, n = nutshell, s = seeds, g = glumes.

Family	Taxa	Part	NISP	Wal	Siwré	C-1	C-3	C-6	C-8	C-20	SB
POACEAE	*Pennisetum glaucum* (includes partial and damaged specimens)	C	12380	xx	xxxx	x	xxxx	xxxx	x	x	xxxx
POACEAE	*Pennisetum* cf. *glaucum*	I	3963		xxx		xxxx	xxxx			
POACEAE	*Pennisetum glaucum* vel. *Sorghum bicolor*	C	21		x		x	x			x
POACEAE	*Digitaria* sp., short broad type (possible fonio)	C	26				x	x	x		
POACEAE	Unidentified	C	1251	x	xxx	x	xxxx	xxx	x	x	xxx
POACEAE	Unidentified	G	117		xxx						
POACEAE	*Digitaria* spp.	C	308		x	x	xxx	xx			x
POACEAE	*Brachiaria/Echinochloa* spp.	C	238		x	x	xx	xxx			
POACEAE	*Echinochloa* cf. *colona*	C	1								x
POACEAE	*Setaria* spp.	C, G	194	x	xx		xxx	xx			x
POACEAE	*Cenchrus biflorus*	I	109			x	xxx				
POACEAE	cf. *Cenchrus biflorus*	C	2				x				
POACEAE	cf. *Panicum* spp.	C	87				xx	x			x
POACEAE	*Panicum turgidum*	C	39		x		xx	x			
POACEAE	cf. *Panicum laetum*	C	17		x	x					x
POACEAE	Paniceae	C	82	x	x		xx	xx			x
POACEAE	*Pennisetum* spp. (wild)	C	49								xx
POACEAE	cf. *Pennisetum violaceum*	C	1				x				
POACEAE	*Pennisetum* cf. *purpureum*	C	1				x				
POACEAE	*Paspalum scrobiculatum*	C	31	x		x	x	x			x

(*continued*)

Table 5. (*continued*)

Family	Taxa	Part	NISP	Wal	Siwré	C-1	C-3	C-6	C-8	C-20	SB
POACEAE	cf. *Paspalum* spp.	C	8				x	x			
POACEAE	cf. *Acroceras* spp.	C	2								x
POACEAE	*Dactyloctenium aegyptium*	C	48		x	x	xx				
POACEAE	cf. *Eragrostis/Sporobolus* spp.	C	33				x	x	x	x	
POACEAE	cf. Eragrostidae	C	4								x
POACEAE	cf. *Sporobolus* sp.	C	1								x
POACEAE	Chloridoideae	C	8								x
POACEAE	*Eleusine indica*	C	4				x				x
POACEAE	cf. *Chloris pilosa*	C	2					x			
POACEAE	cf. Andropogoneae	C	22		x				x		xx
POACEAE	cf. *Bothriochloa bladhii*	C	9				x				
POACEAE	cf. *Bothriochloa* spp.	C	2					x			
POACEAE	cf. *Aristida* spp.	C	3								x
POACEAE	cf. *Rottboelia* spp.	C	2				x				
POACEAE	cf. *Hackelochloa* sp.	C	1								x
POACEAE	cf. *Rhytachne* sp.	C	1				x				
RHAMNACEAE	*Ziziphus* spp.	N, S	4071	x	xxx	x	xxxx	xxx	xxx	x	x
BOMBACACEAE	*Adansonia digitata*	N, S	305		x		xx	xx			xxx
cf. MIMOSACEAE	Unidentified	S	163		xx		xx	xx	xx		
MIMOSACEAE	cf. *Prosopis africana*	S	17		x		x				
PAPILIONACEAE	cf. *Tephrosia* spp.	S	74		x	x	xx	x	x		
CAESALPINACEAE	cf. *Piliostigma reticulatum*	S	6				x				

(*continued*)

Table 5. (*continued*)

Family	Taxa	Part	NISP	Wal	Siwré	C-1	C-3	C-6	C-8	C-20	SB
MALVACEAE	cf. *Abutilon* spp.	S	38				x	xx			
ANACARDIACEAE	*Sclerocarya birrea*	N	12	x			x	x			x
TILIACEAE	*Grewia* spp.	N	12	x	x		x	x			
ULMACEAE	*Celtis* spp.	S	10					x			
ULMACEAE	*Celtis integrifolia*	B	5	x							
VITACEAE	*Vitex* spp.	S	7								x
AMARANTHACEAE/CHENOPODIACEAE	Unidentified	S	529				xxx	xx			xx
AIZOACEAE	*Zaleya pentandra*	S	520			x	xxx	x	x	x	xx
AIZOACEAE	*Trianthema portulacastrum*	S	4	x	x				x		
AIZOACEAE	cf. *Gisekia pharnaceoides*	S	2	x				x			
SOLANACEAE	*Physalis/Solanum* spp.	S	115	x		x	xxx	x			x
SOLANACEAE	*Solanum* spp.	S	39				xx				x
SOLANACEAE	*Solanum* cf. *nigrum*	S	15								x
SOLANACEAE	cf. *Physalis* spp.	S	2	x							
SOLANACEAE	*Physalis* spp.	S	2								x
cf. SOLANACEAE	Unidentified	S	20				x	x	x		
cf. FABACEAE	Unidentified	S	91	x	x	x	xx	xx		x	xx
PAPILIONACEAE	cf. *Indigofera* spp.	S	32				x	x			xx
PAPILIONACEAE	cf. *Crotalaria* spp.	S	10		x		x		x	x	x
PAPILIONACEAE	cf. *Stylosanthes* sp.	S	1					x			
RUBIACEAE	*Borreria* spp.	S	30				x				xx
PORTULACACEAE	*Portulaca oleracea*	S	25		x		xx	x			
PORTULACACEAE	*Portulaca foliosa*	S	7		x		x	x		x	x
CYPERACEAE	Unidentified	A	16				x	x			

(*continued*)

Table 5. (*continued*)

Family	Taxa	Part	NISP	Wal	Siwré	C-1	C-3	C-6	C-8	C-20	SB
CYPERACEAE	*Fimbristylis* spp.	A	6				x	x			
CYPERACEAE	cf. *Cyperus* sp.	A	1								x
CYPERACEAE	cf. *Scirpus* sp.	A	1								x
cf. ASTERACEAE	Unidentified	A	11				x				x
CARYOPHYLLACEAE	Unidentified	S	10				x	x			
LAMIACEAE	cf. *Hyptis lanceolata*	S	10				x				
cf. LAMIACEAE	Unidentified, *Teucrium* type	S	7			x		x			
cf. LAMIACEAE	Unidentified	S	6				x				x
cf. MALVACEAE	Unidentified	S	8				x	x	x		
CAPPARACEAE	cf. *Cleome* spp.	S	5	x							
COMMELINACEAE	*Commelina* spp.	S	5			x	x		x		
COMMELINACEAE	*Commelina* cf. *benghalensis*	S	1								x
cf. APIACEAE	Unidentified	S	4				x				
CAPPARACEAE	Unidentified	S	3				x	x			
cf. BRASSICACEAE	Unidentified	S	2		x		x				
cf. CONVOLVULACEAE	Unidentified	S	2	x	x						
BORAGINACEAE	*Heliotropium* sp.	S	1						x		
cf. CUCURBITACEAE	Unidentified	S	1					x			

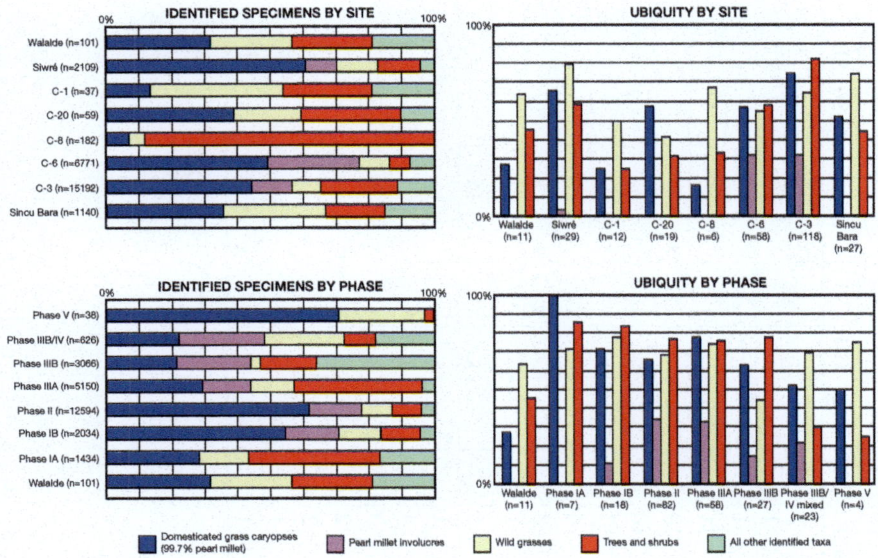

Fig. 3. Summary of the occurrence of major plant categories in the Middle Senegal Valley

waste in the form of involucres (n = 3963) occurs exclusively at C-3 (Phases IB-IIIB), C-6 (Phases II-IIIB/IV), and Siwré (Phase IB), and is present in only 28.2% of samples at these sites.

Evidence for the presence of other staple grains is minimal at best. Only 21 seeds could be tentatively identified to sorghum, and given their rarity, poor preservation and the possibility of morphological overlap with pearl millet outliers, they cannot be considered confirmed identifications and were recorded as *S. bicolor* vel. *Pennisetum glaucum* (see photographs in Gallagher and Murray 2016:304). The largest assemblage of possible sorghum consists of 16 caryopses from a single sample dating to Phase IB at Siwré, a sample that also contains 440 caryopses of pearl millet. The remaining five seeds are all from Phase II at Sincu Bara, C-3 and C-6. Similarly, 26 *Digitaria* sp. seeds were categorized as a "short, fat" type that could fall within the morphological range of contemporary samples of fonio (*D. exilis*), but the rarity of their occurrence makes their identification difficult to confirm, particularly since individual seeds are spread out over Phases II, IIIA, IIIB, and IIIB/IV at sites C-3, C-6, and C-8. *Digitaria* spp. caryopses with a more elongated morphology consistent with contemporary wild varieties are much more common. Rice *(Oryza* sp.) is not present in any samples.

Wild grasses (N = 2677, 10.5% of assemblage)

Diverse wild grasses occur in virtually all phases at all sites (65.4% ubiquity). Although grasses in the assemblage are assigned to 31 taxa, most specimens (n = 1368, 51.1% of wild grasses) were not identified beyond the family level. Among the remaining specimens, the vast majority are members of Paniceae (n = 1169, 43.6% of wild grasses). Within individual samples, caryopses from a single genus or species are usually present in small numbers (<10) with higher concentrations (>35) occurring in only a few exceptional cases: 44 *Dactyloctenium aegyptium* caryopses in a C-3 Phase II sample, and 135 *Brachiaria* vel. *Echinochloa* caryopses in a C-6 Phase II sample. Involucres, rachis, and glumes from wild grasses were rarely identified. When they are, they occur in high concentrations in single samples in Phase IB: 100 *Cenchrus biflorus* involucres in a C-3 sample and 105 mixed Poaceae parts in a Siwré sample. Both samples also contain over 250 pearl millet caryopses and ~40 wild grass caryopses.

***Trees and Shrubs* (N = 4720, 18.6% of assemblage)**

Jujube (*Ziziphus* sp.) (n = 4071, 86.3% of trees and shrubs) constitutes the majority of the tree fruit assemblage and overall is the second most common taxon identified at sites in the MSV. It is a significant component of the assemblage in every phase at every site with the notable exception of Sincu Bara, where only 3 nutshell fragments were identified in the assemblage. Like pearl millet, jujube is also highly ubiquitous (53.5% of samples, 59.5% if Sincu Bara is excluded), and occurs in large concentrations of >100 shell fragments in 8 samples, most of which are from C-3. Complete stones (n = 43) and interior seeds (n = 168) are comparatively rare, with fragmented shell comprising the vast majority of identifications. Baobab (*Adansonia digitata*) is the second most common tree fruit (n = 305), and was most frequently identified at Sincu Bara, where it is present in all phases (n = 204, 33% ubiquity). The remaining specimens are from C-3 and C-6, where baobab occurs in small quantities throughout the sequence (5.6% ubiquity).

Other commonly identified trees and shrubs include cf. Mimosaceae (n = 163) and cf. *Tephrosia* sp. (n = 74) which occur primarily in small quantities, and cf. *Abutilon* sp. which was primarily recovered from a C-6 Phase II sample (26 of 38). The significantly more rare (n < 20) *Piliostigma reticulatum* (DC.) Hochst., cf. *Prosopis africana* (Guill. &Perr.) Taub., cf. *Celtis* spp., *Grewia* spp., *Vitex* spp. and *Sclerocarya birrea* Hochst. (marula) all occur across multiple sites and phases.

***Other Taxa* (N = 1544, 6.1% of assemblage)**

This category encompasses 36 taxa from 19 different families. Many of these taxa are at the family level or genera containing several hundred diverse species (e.g., *Borreria* spp.). Most occur in small numbers (<20) across multiple samples and sites. Of the nine taxa with more than 20 identified specimens, most are likewise broadly distributed across multiple samples. However, four are concentrated in two Phase IIIB ash pits from the same level of unit C-3A (*Zaleya pentandra* (L.) C. Jeffrey, 402 of 518; Amaranthaceae/Chenopodiaceae, 440 of 526; *Physalis/Solanum* sp., 87 of 113; *Solanum* sp., 37 of 39), and cf. *Indigofera* sp. was primarily recovered from a Phase IIIB sample at Sincu Bara (25 of 32). However, in all of these cases the remaining examples are widely distributed.

Temporal Patterns

The majority of taxa identified at archaeological sites in the MSV occur throughout the sequence, particularly once the smaller samples from Walaldé and Phase I are taken into account. Pearl millet, wild grasses (in particular Paniceae) from dry and wet habitats, diverse legumes (e.g., shrubs *Indigofera* sp., *Tephrosia* sp. and Mimosaceae trees), common herbaceous plants (e.g., *Commelina* spp., Solanaceae), and trees with edible fruits (baobab, jujube, marula, *Grewia* sp., *Vitex* sp.) together reflect the remarkable consistency found throughout the sequence. While there are some indications of changing relative importance of these plants (e.g., decreasing pearl millet in the latter half of the 1st millennium AD), most of these patterns do not hold up when assessed using multiple quantification methods (e.g., NISP, ubiquity, ratios) and are equally likely to be artifacts of sampling.

The temporal patterns in the presence/absence data are difficult to interpret as most taxa are either identified throughout the sequence or are present in low numbers and/or with low ubiquity. *Celtis* spp., *Cenchrus biflorus, Dactyloctenium aegyptium,* sorghum, and several plants identified at only the family level are present only at Walaldé and/or in Phases I and II. In contrast, a few additional sedge species and *Chloris* spp. are the only notable additions to the assemblage from Phase III onwards. Of the 30 taxa with at least 20 identified specimens, only four have temporal patterning and three of those could be contained within a different designation (e.g., *Solanum* spp. occurs beginning in Phase III, but *Physalis/Solanum* spp. is present from Phase II).

While sample sizes likely account for much of the variability, it is interesting that Phases IA, IB, and IIIB include not only a lower diversity of taxa, but also no unique taxa: everything identified also occurs in Phases II and/or IIIA (Table 6). Phase IIIB/IV appears to be the exception with 7 unique taxa, however this difference is due primarily to the influence of samples from Sincu Bara (see discussion below), as all but one unique taxon is from that site. In contrast, the large samples from Phases II and IIIA include not only unique specimens from Sincu Bara (n = 7) but also from other sites (n = 11). Once again though, similarity is more significant than difference. When the unique taxa are not included, there is 75% overlap in taxa in Phases II and IIIA, and these shared taxa account for 99.0% of the identified specimens from these two phases.

Site Variability

The majority of identified taxa are found at multiple sites and reflect the same consistent combination of pearl millet and wild plants described above. While only the very common pearl millet, jujube, and wild grass (Poaceae) were identified at every site, of the 30 taxa with more than 20 identified specimens, one (*Dactyloctenium aegyptium*) was identified at only one site and 25 were identified at three or more sites. Even rare taxa are frequently distributed across multiple sites; 15 of 24 taxa with 5–19 identified specimens were identified at two or more sites, and three taxa in this group were recovered from five sites.

Most major taxa seem to have played an equivalent role in the assemblages at different sites, particularly when the much smaller samples from Walaldé, C-1, C-8, and C-20 are taken into account. However, there were some significant patterns in the

Table 6. Summary data on distribution of taxa.

Phase	WAL	IA	IB	II	IIIA	IIIB	IIIB/IV	V
N Samples	11	7	18	82	63	27	23	4
N Taxa	16	16	23	57	51	24	37	4
Mean N Taxa/Sample	2.6	4.6	3.6	4.8	5	3.3	3.8	1.4
Median N Taxa/Sample	2	4	3	4	4	2	3	1
Maximum N Taxa/Sample	9	8	13	17	30	13	18	3
N Taxa > 500 Specimens	0	2	1	2	2	2	0	0
N Taxa 100–499 Specimens	0	1	2	5	1	2	1	0
N Taxa 20–-99 Specimens	1	2	1	12	10	4	4	1
N Taxa 1–19 Specimens	15	11	19	38	38	16	32	3
N Taxa Unique to Phase	4	0	0	11	7	0	7	0
N Taxa >50% Ubiquity	0	2	2	2	2	2	0	1
N Taxa 6–49% Ubiquity	6	5	7	17	14	12	14	0
N Taxa <5% Ubiquity (or in 1 sample if <20 samples)	10	9	14	38	35	10	23	3
Site	**WAL**	**Siwré**	**C-1**	**C-3**	**C-6**	**C-8**	**C-20**	**SB**
N Samples	11	29	12	118	47	6	19	27
N Taxa	16	24	15	53	40	15	8	39
Mean N Taxa/Sample	2.6	3.2	2.2	4.7	4.6	3.2	1.7	4.4
Median N Taxa/Sample	2	3	1	4	3	1	1	4
Maximum N Taxa/Sample	9	13	7	30	14	11	4	18
N Taxa >500 Specimens	0	1	0	3	1	0	0	1
N Taxa 100–499 Specimens	0	2	0	6	3	1	0	2
N Taxa 20–99 Specimens	1	2	0	11	8	1	0	7
N Taxa 1–19 Specimens	15	19	15	33	28	13	8	29
N Taxa Unique to Site	3	0	0	9	5	1	0	14
N Taxa >50% Ubiquity	0	1	0	2	2	0	0	0
N Taxa 6–49% Ubiquity	6	13	4	15	22	2	5	20
N Taxa <5% Ubiquity (or in 1 sample if <20 samples)	10	10	11	36	16	13	3	19

distribution and relative importance of a few taxa. Despite the high numbers and ubiquity of carbonized pearl millet caryopses at all sites, involucres (threshing waste) are present in large numbers and with high ubiquity at C-3 and C-6, high numbers but low ubiquity at Siwré, and are absent from the other sites. Jujube is one of the most numerous plants in the assemblage at every site except Sincu Bara, where it plays only a minor role (n = 3), while baobab follows the reverse pattern with significant numbers at Sincu Bara and only a few finds at other sites. Finally, Sincu Bara has a particularly high proportion of wild grasses, whether measured by NISP, ubiquity, or ratio of wild grasses to either pearl millet or trees and shrubs. Unfortunately, over 75% of the grasses at the site are identified only to the family level.

In general, sites with larger samples included not only more taxa, but also more unique taxa (Table 6). For example, several unique taxa were found at C-3 (n = 9) and C-6 (n = 5) while C-8 had only one unique taxon and none were found at Siwré, C-1, or C-20. The two exceptions are Walaldé, with three unique taxa despite a very small sample size, and Sincu Bara, which has 14 unique taxa. The unique taxa at Sincu Bara include eight wild grasses, two Solanaceae taxa, two Cyperaceae taxa, *Vitex* spp. and *Commelina* cf. *benghalensis* L.

As Cubalel, Siwré, and Walaldé are all located on the *fonde*, it was expected that their assemblages could differ from the assemblage at Sincu Bara, located outside the *waalo* and significantly to the east of the other sites. While the patterns described above do indicate some key differences at Sincu Bara, notably in the tree fruits and grasses, overall the assemblage was more similar than different, as 91.1% of identified specimens are from taxa found at other sites. Among the 33 taxa occurring at three or more sites, their likelihood of appearing at a given site is largely related to sample size, and these taxa are as common at Sincu Bara (n = 20) as at Siwré (n = 21). Even the high numbers of taxa unique to Sincu Bara may not be particularly significant, as similar Solanaceae, Cyperaceae, *Commelina* spp., and wild grass taxa with higher or lower levels of taxonomic specificity occur at other sites.

It was likewise expected that the assemblage from Walaldé, as a 1st millennium BC site, might also differ significantly from later assemblages. While the low sample size makes it difficult to draw conclusions, from the beginning of its occupation Walaldé seems to have the same mix of pearl millet, wild grasses, and tree fruits (including jujube) found at later sites (Murray and Déme 2014). As with Sincu Bara, its unique species are similar to plants identified at higher or lower levels of taxonomic specificity at other sites.

Human Use of Wild Plants in the Middle Senegal Valley

Wild plants were likely used by societies along the Middle Senegal River during the 1st millennium AD for a variety of purposes including food, fuel wood, material culture, fodder, and medicines.

Trees and Shrubs

While there are few trees present in the *hollalde,* the *fonde* supports a wide diversity of species, including many not found in the *jeeri.* In contrast, the sparse trees of the *jeeri* are almost all species that also grow in *fonde* environments, making it difficult to identify spatial patterns of resource use in the tree and shrub assemblage. With the exception of the Mimosaceae taxa and the shrubs cf. *Tephrosia* spp., and cf. *Abutilon* spp., the identified tree fruits were from edible species.

The most commonly identified wild fruit at MSV sites is jujube, which grows throughout the region and would likely have been found near all sites where it was identified. Jujube fruits have a very thin pericarp, such that their caloric contribution is likely to be low even when present in large numbers. However, jujube fruits are high in sugars and in vitamin C, and could make an important nutritional contribution to the diet (Burkill 1997). While jujube ripen in the early spring (February–April), they are easily dried and the dried flesh is often powdered and made into small sweet cakes (Chevalier 1947; Burkill 1997). Although jujube are consumed throughout the region, at most archaeological sites they are found at much lower rates than is the case in the MSV; the only other site at which they have been documented in large carbonized concentrations is Arondo, located on the Upper Senegal River (Gallagher 1999). This could suggest that residents of the Senegal River Valley during the 1st millennium AD utilized a processing method that enabled frequent carbonization of stones such as parching over a fire rather than in the sun or the discard of stones on a fire after defleshing. The rarity of jujube at Sincu Bara, where jujube plants would be just as likely to grow, could indicate a lesser role for the fruit at the site or the use of a different processing method. Alternatively, it could be a sampling artifact.

Baobab has an inverse distribution to jujube, as it is common at Sincu Bara and rare elsewhere. Ethnoecological research has demonstrated that baobab trees in particular are highly associated with human settlement throughout West Africa (e.g., Duvall 2007) and this is the case in the MSV as well. In addition to their edible fruits and leaves, baobabs are used as a source of cordage and have diverse medicinal uses (Burkill 1985). Similarly, the more rarely identified marula (which is found at several sites) likely grew exclusively in the rich soils in or near human settlements. *Vitex* spp. is also more likely to be found in settlements or along the river (Burkill 2000).

The unidentified Mimosaceae could be from an edible species or from one of the more common non-edible species (Burkill 1995). Mimosaceae (including *Acacia* spp.) grow in both *waalo* and *jeeri* environments, and tend to be good fodders. As nitrogen-fixing legumes, Mimosaceae (including *Prosopis africana*) and *Piliostigma reticulatum* are beneficial in agricultural fields, and also provide a source of other products including gums, tannins, firewood, and medicines. *Grewia* spp. and *Celtis* spp. are more likely to grow in the *jeeri* environment, but can be found on some *fonde.* Interestingly, desert date, which like Mimosaceae and jujube is found throughout the landscape and would likely have been widely consumed, is not present in the samples.

The two semi-woody shrubs included in this category both tend to favor well-drained environments. *Tephrosia* spp. seeds are commonly used as fish poisons (Burkill 1995). These poisons work best in a confined area with low water flow, and could be very effective in *hollalde* depressions (Neuwinger 2004). The stems of *Abutilon* spp. produce strong fibers (Burkill 1997).

Finally, it should be noted that most flotation samples from MSV sites were composed primarily of wood charcoal. While many of the species listed above can be used as firewood, to date, no wood charcoal assemblages from the region have been analyzed, other than samples from iron furnaces, which were almost exclusively *Acacia nilotica* and *A. raddiana*, with one example of *Prosopis africana* (Killick 2016). In other parts of West Africa where wood charcoal analyses have been completed (e.g., northern Burkina Faso, eastern Mali), the charcoal analyses typically identify a broader range of taxa than those represented by their fruits (e.g., Höhn and Neumann 2012; Eichhorn and Neumann 2014).

Wild grasses

The interpretation of the wild grass assemblage is challenging. Mixed wild grasses, in particular those from the Paniceae tribe, are known to be collected as food by groups throughout the Sahel (Harlan 1989; Burkill 1994). Ethnographically, wild grass harvesting can take several forms, including beating, scything, sweeping, and collection from termite nests (Harlan 1989). In this case, the assemblage includes (often within single samples) grasses that grow in *hollalde* and *jeeri/fonde* environments as well as grasses that grow in large homogenous stands and those that favor more mixed communities.

In the MSV, Boutillier et al. (1962) have noted that *Panicum laetum* was regularly consumed as a minor component of the diet although they did not describe the collection method. *P. laetum* does grow in large stands that can be extremely productive (and in ideal conditions even out-produce pearl millet fields), and in other parts of West Africa is collected using a swinging basket technique (Harlan 1989; van Damme 1998). While only 17 *P. laetum* caryopses were identified at MSV sites, more could be among the 82 Paniceae and 1251 unidentified Poaceae specimens. Even if wild grasses were not primarily used as food, their collection could still be significant. Mueller and Almedom (2008) have described the importance of continuous use of wild foods that may be relied on in times of scarcity in order to ensure that knowledge related to collection, processing, and consumption is maintained within the community.

At the same time, there are numerous other pathways by which these grasses could have been brought into sites. Ruderal and weedy species such as *Dactyloctenium aegyptium* frequently grow in and near settlements and produce as many as 16000 seeds per plant. Consequently, these taxa easily enter the archaeological record through both natural (wind distribution, presence in soil, etc.) and cultural (sweeping, weeding, burning waste, etc.) processes. Wild grasses may also be brought from greater distances, either intentionally (e.g., species with strong culms such as *Aristida* sp. may be collected and used as thatch) or unintentionally (e.g., the glumes of *Cenchrus biflorus* cling easily to clothing and animals) (Burkill 1994). Wild grasses are consumed by animals as fodder, and may be collected and brought to sites for this purpose (Burkill 1994). As a result, it is not uncommon for wild grass grains to carbonize when dung is used as fuel (Miller and Smart 1984; Gallagher 2014).

The only ethnoarchaeological study of dung fuel in West Africa was carried out in the Inland Niger Delta by Murray (2005). In her examination of cow, horse, and goat/sheep dung, she found that the seed populations (which were highly variable in density and diversity from sample to sample) typically reflected the grasses, sedges, and other herbaceous plants growing in and around near-by rice fields, and even included occasional seeds from species known to be toxic as fodders (e.g., *Paspalum scrobiculatum* L.). Aside from the challenges of distinguishing seeds originating in dung fuel from those accidentally harvested as weeds, many of the identified species have other ethnobotanical uses, including being collected as food or for use in medicines. Murray found no individual species or combinations of species that could be used as unambiguous indicators for dung fuel burning. Across the MSV assemblage, numerous samples contain diverse assemblages of wild grasses and other herbaceous species that could be equally indicative of both dung fuel use and other events.

Another possible line of evidence for dung fuel is the presence of carbonized pearl millet chaff. Murray et al. (2007:66) noted that carbonized chaff from C-3A was recovered from a charred dung matrix and speculated that the dung was used as fuel. While this may be the case in some individual samples, the correlation was not always consistent as at the level of both the individual sample and the occupation phase at a particular site, high densities of carbonized chaff did not regularly occur in association with high densities of wild grasses and diverse herbaceous fodders, nor did they regularly occur in association with high densities of carbonized pearl millet caryopses. Likewise, the proportion of wild grasses was not dependent on the quantity of pearl millet in a sample. Interestingly, chaff is most common at sites with higher mean and median number of taxa per sample. Perhaps some of this diversity is associated with the carbonization of threshing waste, in which case the lower numbers of taxa per sample at C-1, C-8, and C-20 could result from the excavation of areas where threshing was not taking place (Table 6). However, this pattern does not hold for Sincu Bara, where high mean and median number of taxa per sample were identified despite the absence of threshing waste.

Other Taxa

Several other taxa have possible ethnobotanical uses, although in many cases the high level of the taxonomic identification makes specificity difficult. For example, while the leaves of *Indigofera* sp. are notably used to make dark blue indigo dyes, not every species of the genus produces a good color (Burkill 1995). Similarly, many Amaranthaceae/Chenopodiaceae and *Portulaca* species produce edible leaves, and Solanaceae include both edible species and hallucinogens, among other properties (Burkill 1985, 1997, 2000). In addition, many of these may be ruderal and/or weedy species that can enter the archaeological record through the diverse pathways discussed for wild grasses above.

Discussion

Aside from jujube, the role of wild plants in the 1st millennium AD diet of the MSV is ambiguous, as several of the most important wild plants documented by Boutillier et al. (1962), were not identified at MSV sites. This could be in part preservation bias, although water lily and desert date (as well as baobab and jujube) were recovered from much smaller samples at Arondo, a 1st millennium AD site further upstream along the

Senegal River (Gallagher 1999). Nonetheless, there is clear evidence for the consumption of diverse wild fruits, and it is probable that people were eating other wild plant foods including leaves (from trees and herbs) and wild grasses. Wild plants were almost certainly used for a variety of other tasks: the assemblage includes species with medicinal properties, those used for construction and thatch, fiber plants, dye plants, and firewood. Particularly interesting is the broad distribution of the fish poison *Tephrosia* spp. Fish bones are present in all excavated units, though not in the numbers one might expect at sites located directly on the river (van Neer 2016)

The wild plant assemblage includes taxa that collectively produce seeds/fruit year round, although in general most of the herbaceous species fruit at the end of summer, while many of the trees bear fruit in the winter or spring. There are very few plants in the local environment that produce their fruits exclusively from November to April, which means that winter movements of the population would be difficult to identify. However, in the twentieth century, populations in the region were more likely to cluster in the *waalo* during the winter and move out to the *jeeri* during the summer rainy season, such that most sites in the study region would be more likely to experience a decrease in population during the summer months if seasonal mobility was significant during the 1st millennium AD.

Agriculture in the Middle Senegal Valley

From the early occupation at Waladé, pearl millet has been an important crop in the MSV. Found in all phases at all sites, it was likely grown on both the *fonde* levees where sites like Cubalel are located, as well as on the *jeeri* uplands. While millet is clearly the primary crop throughout the sequence, it may have become both less common and less ubiquitous in Phases IIIB and IV. However, there is no evidence at that time for another crop becoming more significant, and this pattern most likely derives from sampling bias. There is a small possibility that fonio is also present throughout the sequence, although in very low numbers. Like pearl millet, fonio grows well in poor soils. However, most varieties prefer at least 400 mm of rainfall, making it slightly less drought resistant (NRC 1996).

In contrast, there is little evidence for crops that would grow in the *hollalde* basins, in particular, rice and sorghum. Since the wild *Oryza barthii* is historically common, the Senegal River Valley had been considered a possible early center of African rice cultivation (*O. glaberrima* Steud.), although the main centers of rice cultivation in the Senegambia are further to the south (Linares 2002). Delolme's (1937) survey of the Senegal River does occasionally note the presence of rice fields, and Boutillier et al. (1962) mention rice as a minor component of the diet, although not of the agricultural system. However, while rice is regularly recovered from archaeological sites in other West African rice-growing regions (e.g., McIntosh 1995; Murray 2005), rice is completely absent from MSV samples. Based on current data, rice was not grown in the MSV during the 1st millennium AD.

In the twentieth century MSV sorghum was the primary crop cultivated in *hollalde* basins yet the species is only tentatively identified in very small quantities in the archaeological assemblage. This result is consistent with its apparently minor role at other 1st millennium AD archaeological sites in West Africa (Kahlheber and Neumann 2007:332). It is possible that sorghum was grown in the 1st millennium AD but did not carbonize at the same rate as pearl millet. In the MSV today, sorghum is prepared and consumed in the same way as pearl millet and would presumably have equal opportunities for carbonization. However, in other parts of West Africa, sorghum is particularly favored for beer brewing, and prior to the introduction of Islam in the MSV, sorghum could have been used for this purpose. If this were the case, a portion of the sorghum harvest could have followed a different processing path with perhaps fewer opportunities for carbonization.

More likely, sorghum was not grown or grown only on a very limited basis during the 1st millennium AD in the MSV which, combined with the absence of rice, suggests that the winter *hollalde* crop was not an important part of the agricultural system during the 1st millennium AD. While the planting of both winter and summer crops mitigates risk of crop failure, there are disadvantages. With two crops year-round agricultural labor is required, and depending on the timing of rainfall and floods, planting and harvest labor bottlenecks overlap in some years. *Hollalde* crops also must be protected from livestock and wild animals that congregate in the *waalo* seeking water and fresh vegetation as the *jeeri* dries out during the winter.

Interestingly, the few possible grains of sorghum are found in deposits dating to the early 1st millennium AD dry period. It is certainly possible that during this period, when rainfall was likely more erratic, communities experimented with growing sorghum in the *hollalde*. It is also possible that sorghum was traded in small quantities from the south to supplement food stores, as has been suggested for sites in northern Burkina Faso (Kahlheber 2004; Kahlheber and Neumann 2007). No sorghum threshing waste was recovered, although given the rarity of the species its absence is not necessarily significant.

Wild grasses and herbaceous species can provide a secondary line of evidence regarding the spatial structure of the agricultural system. While many specimens could not be identified to the degree of taxonomic specificity necessary for environmental reconstruction, several taxa were identified that would be likely associated with well-drained *jeeri*/high *fonde* (e.g., *Cenchrus biflorus, Dactyloctenium aegyptium, Panicum turgidum, Zaleya pentandra*) or seasonally inundated *hollalde* environments (e.g., *Bothriochloa bladhii* (Retz.) S.T.Blake, *Hyptis lanceolate* Poir., *Paspalum scrobiculatum, Scirpus* sp.). Unfortunately, these taxa tended to be rare in the assemblage, and given the locations of the sites, would be expected to occur in small quantities regardless of the agricultural strategies in use. Most phases at individual sites included samples with *jeeri/fonde* indicators, samples with *hollalde* indicators and samples that contained a mix of the two. As described above, several of the dry species occur early in the sequence, while a broader diversity of species associated with the *hollalde* are present later in the sequence. Of the five samples with possible sorghum, three contained *hollalde* and *fonde* indicators, one contained only *fonde* indicators, and one had no environmentally sensitive species. At the same time, in the 41 samples with more than 20 pearl millet caryopses, 8 contained only *jeeri/fonde* indicators, 2 contained only *hollalde* indicators, and the remainder contained both or neither.

There is very little evidence in the assemblage for the cultivated legumes and vegetables that would likely have been grown in the 1st millennium AD. The unidentified Cucurbitaceae seed (from C-3 Phase IIIA) could be from a domesticated melon and eight unidentified Malvaceae seeds (from C-3, C-6, and C-8) could derive from okra or roselle (the unidentified Fabaceae are too small for cowpea). This lack of vegetables is fairly common in Sahelian and savanna sites in West Africa (e.g., Kahlheber 2004; Murray 2005; Kahlheber and Neumann 2007) and is likely the result of preservation biases.

The absence of cotton (*Gossypium* sp.) is consistent with temporal patterns observed elsewhere in West Africa. While there are a few cotton seeds and spindle whorls that may date to the late first millennium AD at Dia, both botanical finds and associated material culture are significantly more common beginning in the 2nd millennium AD (Murray 2005, 2007, unpublished). In the MSV in particular, cotton is absent from the botanical samples, 98% of which pre-date the 10th century AD, and spindle whorls are present only in Phase IV and V deposits at Siwré. At Ogo, Chavane (1985) documented a similar pattern, with spindle whorls and possible cotton pollen recovered only from upper deposits dating to the early 2nd millennium AD.

Discussion

Pearl millet farming has a deep history in the greater region, as some of the earliest finds of domestic pearl millet are from southeastern Mauritania (e.g., Fuller et al. 2007; Person et al. 2012; Ozainne et al. 2014). While we know little about the initial residents of the MSV, data from Walaldé indicate that cultivation of pearl millet was a component of the subsistence economy from the occupation of the region by semi-sedentary pastoralists by c. 800 BC.

The continuity of the agricultural system, with its focus on rain fed cultivation of *fonde* and/or *jeeri* environments is all the more remarkable when placed in the context of the dynamic climatic events of the 1st millennium AD. While there is a slim possibility that there was some experimentation with *hollalde* cultivation during an arid phase in the early 1st millennium AD, farmers appear to have taken advantage of the drought resistant properties of pearl millet and were able to reliably produce a sufficient staple crop using primarily rain fed agriculture. Given the extensive, deep root systems of pearl millet, it is possible that by planting it on the *fonde, jejeengol*, or even *jeeri* near the edge of the *waalo*, farmers were able to take advantage of the greater soil moisture near the Senegal River. This approach may have mitigated some of the risk related to interannual rainfall variability and allowed farmers to use consistent agricultural strategies throughout the 1st millennium AD despite periods of climate change. This use of a less intensive strategy could be indicative of lower population densities, less need for the production of agricultural surplus, or of an annual cycle that placed emphasis on different activities during the winter dry season. Regardless, farming would likely have articulated in different ways with other subsistence activities. For example, dry season pasturage would have been less constrained as herds of cattle, sheep, and goats could have grazed more freely in the *waalo* if sorghum was not being grown, perhaps limiting the degree to which residents needed to transhume away from the river.

Conclusions

Paleoethnobotanical results indicate that agricultural practices in the MSV during the 1st millennium AD differed significantly from those described in the ethnohistoric record. Still unknown is when the more intensive year-round agricultural cycle became standard in the region, and what factors spurred its adoption. If there were experiments with sorghum in the mid-1st millennium AD, it is possible the crop was not widely adopted because the available varieties were not considered productive. McIntosh (McIntosh and Scheinfeld 2012:156; McIntosh and McIntosh 2016a) has suggested that the appearance of sorghum farther upriver at Arondo in the late 1st millennium AD hints at the chronology for its wider incorporation into the MSV economy, and it is possible that at this time, a new variety more suited for décrue cultivation was introduced from the east. Evidence for sorghum cultivation in West Africa is more common beginning in the late 1st millennium AD, although it remains significantly less abundant than pearl millet (Kahlheber and Neumann 2007). If this proves to be the case, the contemporaneous appearance of exotic goods, sorghum, and possibly cotton would not be coincidental. Alternatively, appropriate sorghum varieties may have been available throughout the 1st millennium AD and not employed until the added production capacity of a year-round intensive agricultural cycle was required.

Regardless, year-round agricultural productivity could have reduced the proportion of the population engaged in agriculture and/or produced a surplus that would have supported the expansion of specializations such as spinning/weaving, and trading that are documented by the 11th century in connection with Takrur and other towns along the MSV (al-Bakri, in Levtzion and Hopkins 1981:78). In this scenario, the previously diversified agropastoral economy of the prior two millennia became more sedentary, with an expanded focus on farming, thus taking on the characteristics of the Tukolor economy historically. If al-Bakri's and al-Idrisi's accounts are reliable, population densities were high in and around trade towns on the MSV. With the development of trade networks northward, agricultural surplus was needed to provision entrepots on the desert edge. Historically, surpluses for export were made possible by servile labor working for "noble" and warrior lineages (Robinson 1975), a practice that may have considerable time depth. In order to understand the contexts within which the historic agricultural program developed, more archaeological work from the 2nd millennium AD in the MSV is needed.

Acknowledgements. Many thanks to all of the archaeologists who contributed to the collection and analysis of the MSVAP botanical samples, in particular Cecilia Capezza, Dorian Fuller, Mary Anne Murray, and Mamadou Ndiaye. Funding for the MSVAP was provided by the National Science Foundation (grant BNS90-11811 and 9820919) and the National Geographic Society (grant 4397-90). The Dean of Social Sciences at Rice University, Lynn Ragsdale, generously funded the analysis of the archaeobotanical materials. Co-directors of the MSVAP were R.J. McIntosh, S.K. McIntosh, and H. Bocoum. We are grateful to all our key excavation and supervisory personnel: E. Davies, Alioune Déme, Sigrid Gabler, the late Ndèye Sokhna Guèye, Mary Anne Murray, Ibrahima Ndiaye, Gavin Rees, Moustapha Sall, the late Mansour Samb, Karol Stoker, and Ibrahima Thiaw. Thank you also to Anna Maria Mercuri and the Laboratorio di Palinologia e Paleobotanica for organizing IWAA8, to the volume editors for their encouragement and support and to two anonymous reviewers for their thoughtful comments on this manuscript.

References

Adam J (1962) Itinéraires botaniques en Afrique occidentale. J Agr Trop Bot App 9:297–416

Azzoug M, Carré M, Chase BM et al (2012) Positive precipitation–evaporation budget from AD 460 to 1090 in the Saloum Delta (Senegal) indicated by mollusk oxygen isotopes. Glob Plan Change 98–99:54–62

Bocoum H, McIntosh SK (2002) Excavations at Sincu Bara, middle Senegal valley (Senegal). CRIAA-Université de Nouakchott, Nouakchott

Boutillier J, Schmitz J (1987) Gestion traditionelle des terres (système de décrue/système pluvial) et transition vers l'irrigation. Cah Sci Hum 23(3–4):533–554

Boutillier J, Cantrelle P, Causse J et al (1962) La moyenne vallée du Sénégal. Presse Universitaire du France, Paris

Burkill HM (1985) The useful plants of West Tropical Africa, vol 1, 2E. Royal Botanical Gardens, Kew

Burkill HM (1994) The useful plants of West Tropical Africa, vol 2, 2E. Royal Botanical Gardens, Kew

Burkill HM (1995) The useful plants of West Tropical Africa, vol 3, 2E. Royal Botanical Gardens, Kew

Burkill HM (1997) The useful plants of West Tropical Africa, vol 4, 2E. Royal Botanical Gardens, Kew

Burkill HM (2000) The useful plants of West Tropical Africa, vol 5, 2E. Royal Botanical Gardens, Kew

Chavane B (1985) *Villages de l'ancien Tekrour*. Karthala, Paris

Chevalier A (1947) Les jujubiers ou *Ziziphus* de l'ancien monde et l'utilisation de leurs fruits. Rev Int Bot App Agr Trop 27:470–483

Cissoko SM (1968) Famines et épidemies à Tombouctou et dans las Boucle du Niger XVIe au XVIII siècle. Bull IFAN 30:806–821

Delolme M (1937) Rapport sur les renseignements divers, sur les terres, les cultures, les biens et les coutumes dans la Vallée du Sénégal. Unpublished manuscript

Déme A, Guèye NS (2007) Enslavement in the Middle Senegal Valley: historical and archaeological perspectives. In: Ogundiran A, Falola T (eds) The archaeology of Atlantic Africa and the African diaspora. Indiana University Press, Bloomington, pp 122–139

Déme A, McIntosh SK (2006) Excavations at Walaldé: new light on the settlement of the Middle Senegal Valley by iron-using peoples. J Afr Arch 4:317–347

Duvall C (2007) Human settlement and baobab distribution in south-western Mali. J Biog 34:1947–1961

Eichhorn B, Neumann K (2014) Holocene vegetation change and land use at Ounjougou, Mali. In: Stevens CJ, Nixon S, Murray MA, Fuller DQ (eds) Archaeology of African plant use. Left Coast Press, Walnut Creek, pp 83–96

Fuller DQ, Macdonald KC, Vernet R (2007) Early domesticated pearl millet in Dhar Nema (Mauritania). In: Cappers RTJ (ed) Fields of Change. Barkhuis Publishing, Groningen, pp 71–76

Gallagher D (1999) Analysis of seeds and fruits from the sites of Arondo and Ft. Senedebu, Senegal. In Thiaw I, Archaeological investigation of long term culture change in the Lower Falemmé (Upper Senegal Region). Dissertation, Rice University, pp 385–431

Gallagher D (2014) Formation processes of the macrobotanical record. In: Marston JM, d'Alpoim Guedes J, Warinner C (eds) Method and theory in paleoethnobotany. University of Colorado Press, Boulder, pp 19–34

Gallagher D, Murray S (2016) Paleoethnobotanical analysis of seeds and fruits from unit C3A. In: McIntosh et al. (ed) The search for Takrur, pp 299–310

Harlan JR (1989) Wild-grass seed harvesting in the Sahara and sub-Sahara of Africa. In: Harris DR, Hillman GC (eds) Foraging and farming. Routledge, New York, pp 79–98

Höhn A, Neumann K (2012) Shifting cultivation and the development of a cultural landscape during the Iron Age (0–1500 AD) in the northern Sahel of Burkina Faso, West Africa. Quatern Int 249:72–83

Kahlheber S (2004) Perlhirse und Baobab—Archäobotanische Untersuchungen im Norden Burkina Fasos. Dissertation, Johann Wolfgang Goethe-Universität

Kahlheber S, Neumann K (2007) Plant cultivation in west Africa. In: Denham T, Iriarte J, Vrydaghs L (eds) Rethinking agriculture. Left Coast Press, Walnut Creek CA, pp 320–346

Kane O (1986) Le Fuuta-Tooro des Satigi aux Almaami. Université Cheikh Anta Diop, Thesis

Killick D (2016) Iron-working in the middle Senegal valley. In: McIntosh et al. (ed) The search for Takrur, pp 191–280

Lericollais A, Schmitz J (1984) La calabasse et la houe. Cah ORSTOM sèr Sci Hum 20:427–452

Levtzion N, Hopkins JFP (eds) (1981) Corpus of early Arabic sources for west African history. Cambridge University Press, Cambridge

Linares O (2002) African rice (*Oryza glaberrima*): history and future potential. Proc Nat Acad Sci 99(25):16360–16365

Maley J, Vernet R (2015) Populations and climatic evolution in north tropical Africa from the end of the Neolithic to the dawn of the modern era. Afr Arch Rev 32:179–232

McIntosh RJ, McIntosh SK (2016a) Background to the research. In: McIntosh et al. (ed) The search for Takrur, pp 9–32
McIntosh RJ, Déme A (2016) Reconnaissance of the Cubalel study region. In: McIntosh et al. (ed) The search for Takrur, pp 341–380
McIntosh RJ, McIntosh SK, Bocoum H (eds) (2016) The search for Takrur: archaeological excavations and reconnaissance in the Middle Senegal Valley. Yale University Publications in Anthropology, New Haven
McIntosh SK (ed) (1995) Excavations at Jenné-Jeno, Hambarketolo, and Kaniana (Inland Niger Delta, Mali), the 1981 season. University of California Press, Berkeley
McIntosh SK, Bocoum H (2000) New perspectives in Sincu Bara. Afr Arch Rev 17:1–41
McIntosh SK, McIntosh RJ (2016) Overview and conclusions. In: McIntosh et al. (ed) The search for Takrur, pp 381–426
McIntosh SK, Scheinfeldt LB (2012) It's getting better all the time: comparative perspectives from Oceania and West Africa on genetic analysis and archaeology. Afr Arch Rev 29:131–170
Millàn V, Rodrigo FS (2014) Climatic information of Western Sahel (1535–1793 AD) in original documentary sources. Clim Past Disc 10:3877–3900
Miller N, Smart T (1984) Intentional burning of dung as fuel. J Ethnob 4:15–28
Mueller J, Almedom A (2008) What is "famine food"? Hum Eco 36(4):599–607
Murray MA, Fuller DQ, Capezza C (2007) Crop production on the Senegal river in the early first millennium AD. In: Cappers RTJ (ed) Fields of change. Barkhuis Publishing, Groningen, pp 63–69
Murray SS (2005) The rise of African rice farming and the economic use of plants in the Upper Middle Niger Delta (Mali). Dissertation, University of Wisconsin
Murray SS (2007) Medieval cotton and wheat finds in the middle Niger Delta (Mali). In: Cappers RTJ (ed) Fields of change. Barkhuis Publishing, Groningen, pp 43–52
Murray SS (2008) A report on the charred botanical remains from Sincu Bara. Nyame Akuma 69:56–63
Murray SS (unpublished) Report on the botanical remains from Djenné Museum Site excavations. On file, Rice University
Murray SS, Déme A (2014) Early agro-pastoralism in the middle Senegal Valley. In: Stevens CJ, Nixon S, Murray MA, Fuller DQ (eds) Archaeology of African plant use. Left Coast Press, Walnut Creek, pp 97–102
Murray SS, Gallagher D (2016) Appendix F: macrobotanical data. In: McIntosh et al. (ed) The search for Takrur, pp 485–524
National Research Council (1996) Lost crops of Africa, vol 1. National Academy Press, Washington
Neuwinger HD (2004) Plants used for poison fishing in tropical Africa. Toxicon 44(4):417–430
Ozainne S, Lespez L, Garnier A et al (2014) A question of timing. J Arch Sci 50:359–368
Person A, Amblard-Pison S, Jousse H et al (2012) Influence de l'environnement sur la gestion des ressources au Néolithique. J Afr Arch 10:133–164
Robinson D (1975) Chiefs and cleric. Clarendon Press, Oxford
Robinson D, Curtin P, Johnson J (1972) A tentative chronology of Futa Toro from the sixteenth through the nineteenth centuries. Cah Étud Afr 48:555–592
Schmitz J (1986) Agriculture de décrue. Cah Rech Dev 12:65–77
Trochain J (1940) Contribution à l'étude de la vegetation du Senegal. Dakar, IFAN

Thiaw I (2012) Colonization and the development of archaeology in Senegal. In: Van der Linde SJ, van de Dries MH, Schlanger N, Slappendel CG (eds) European archaeology abroad. Sidestone Press, Leiden, pp 349–374

Webb JL (1995) Desert frontier. University of Wisconsin Press, Madison

van Damme P (1998) Wild plants as food security in Namibia and Senegal. In: Bruins H, Lithwick H (eds) The Arid Frontier. Springer, Netherlands, pp 229–247

van Neer W (2016) Fish remains from Cubalel and Siwré. In: McIntosh et al. (ed) The search for Takrur, pp 335–340

The Archaeobotany of the Later Stone Age (LSA) in Nigeria: A Review

Emuobosa A. Orijemie[1,2](✉)

[1] Department of Archaeology and Anthropology, University of Ibadan, Ibadan, Nigeria
Orijemie5@yahoo.com,
eo317@cam.ac.uk
[2] McDonald Institute for Archaeological Research, University of Cambridge, Cambridge, UK

Abstract. A review of the evidence for plant use during the Later Stone Age (LSA), a period which spanned c. 10,000-0 BC (12,000–2000 yr BP), in Nigeria is presented. Available data indicate that although there was contemporaneity in the use of artefacts (pottery, trapezoids and ground-stone axes) associated with plant use at LSA sites, there were marked regional differences in the plant species exploited. Linguistic evidence indicates the exploitation of *Elaeis guineensis* Jacq., *Cola acuminata* (P.Beauv.) Schott & Endl., *Dioscorea* spp. and *Raphia hookeri* G.Mann & H.Wendl. in the Proto Atlantic-Congo period in southern Nigeria, and subsequent exploitation of *E. guineensis* at c. 1000 BC (3000 yr BP). In the northeast, wild grasses, rice and fruits were exploited from c. 2000 BC (4000 yr BP), and domesticated *Pennisetum glaucum* (L.) R.Br. by c. 1200–1000 BC (3200–3000 yr BP). In the north-central areas, from c. 800–450 BC *P. glaucum, Eleusine* cf. *coracana* L. Gaertn., *Vigna unguiculata* (L.) Walp., *Canarium schweinfurthii* Engl., *Vitex* spp. and *Isoberlinia* were exploited in the Nok area of north-central Nigeria. At Kariya Wuro, *Bridelia scleroneura* Müll.Arg., *Pavetta* sp., *E. guineensis* and *Sarcocephalus latifolius* Sm. were gathered prior to c. AD 800. Incontrovertible evidence of the utilisation of culturally significant crops such as *Dioscorea* spp. is lacking due partly to the paucity of archaeobotanical research and to ineffective recovery methods. The combination of ethnographic and palaeoethnobotanical data is necessary to refine the understanding of human-plant interactions in the LSA, the archaeobotanical picture of which is still unfolding.

Keywords: Later stone age · Human-plant interactions · Microliths
Palaeoenvironment · Nigeria

Introduction

Plants are crucial to the survival and sustenance of human life. Archaeobotany, a sub-discipline of archaeology, deals with human-plant relationships and deciphers plants and the uses to which they were put in the remote and recent past. Owing to taphonomic challenges, only a fraction of plants utilised by humans is usually recovered from archaeological contexts. Hence, the analyses of both micro- (phytoliths, pollen, spores and starch granules) and macro-botanical remains (wood charcoal, fruits, leaves,

A. M. Mercuri et al. (Eds.): *Plants and People in the African Past*,
https://doi.org/10.1007/978-3-319-89839-1_17

roots and seeds) are needed to provide a fuller picture of human-plant relationships (Neumann 1999). The origins of agriculture and its associated crops have, and still dominate archaeobotanical research globally. The academic patronage such studies have received is enormous and skewed in favour of selected crops most notably cereals. This tends to create the impression that plants were only useful to early humans with respect to diet on the one hand, and that plants other than cereal cultivars were less significant in human history before and after the emergence of agriculture, on the other hand. Agriculture encompasses just a fraction of human history, and in some parts of the world outside the Levant, China and the Americas, did not begin until c. 2000–1000 BC (4000–3000 yr BP; McDonald 1998).

The methods by which plants are processed into useable products, their disposal and burial contribute to what plants are preserved and eventually retrieved from archaeological contexts. An understanding of these dynamic processes is central to the success of archaeobotanical research. In practice, areas such as savannas and deserts where conditions are almost always dry, as well as waterlogged places preserve plants well. Because of the high level of iron-oxide in tropical soils such as those present in Nigeria, it has often been assumed, sometimes unsubstantiated, that plant remains might not preserve well in such archaeological sites. Indeed pollen usually does not preserve well in archaeological deposits but charred plant remains are known to survive in such conditions hence should be targeted. Ecological approaches with respect to anthropogenic plants/weeds remain pivotal to archaeobotany particularly where remains of "useful/economic" plants are absent. Although it has been difficult to establish weed-host specificity, it is known that certain weeds accompany different farming strategies and land-use management practices (Jones 2002; Bogaard et al. 2016).

The Stone Age and its associated cultures present opportunities for investigating plant use before the beginning of farming. In Nigeria, the Later Stone Age (LSA) is the best documented and dated period of the Stone Age, and is closest to the food-producing period. The current state of knowledge of human-plant interactions during this period shall be the main thrust of this paper. It reviews the archaeobotany of LSA sites in Nigeria, considering the methods employed in such studies, and highlighting the successes and obstacles to archaeobotanical research. It shall conclude by underscoring the importance of archaeobotany to archaeological research, to promote the use of archaeobotanical methods that are suitable to the region, and advocate archaeobotanical research that looks at a broader spectrum of the ways humans use plants as against the current over-emphasis on particular plant groups and their use as food and/or fuel.

Palaeoenvironment in the last 12,000 years in Nigeria

The major environmental events during the Holocene are summarised in Table 1. For the purpose of this review, attention shall be on the last 12,000 years a special focus on c. 2000 BC – 0 (4000–2000 yr BP) because most LSA sites and archaeobotanical finds date to this period. In most parts of Nigeria during the Last Glacial Maximum

(LGM) and prior to c. 10,000 BC (12,000 yr BP) the environment was dry and characterised by Sudan-Sahel savannah. During this time, the Sahara desert encroached into northern Nigeria. In the south, there were southward shifts in savanna zones such that rainforests dispersed southwards towards the coast. Humid conditions returned from c. 10,000–8900 BC. Then, in the period c. 8900–8500 BC, conditions were dry followed by an extremely humid phase during the Middle Holocene c. 6500–5000 BC. During this wet phase, *Typha* sp. became abundant around lakes while Guinea savanna elements, and in some areas rainforest, expanded northwards. In the south, rainforests and mangroves became diverse and dominated regions especially during c. 7000–3500 BC. The humid climate in the north existed until c. 4800 BC after which conditions became extremely dry at c. 1800 BC with marked increases in Sudanian-Sahel elements and a concomitant decrease in Guinea savanna (Salzmann 2000; Salzmann et al. 2002). In north-central Nigeria, areas around the Katsina Ala floodplain experienced wet conditions from c. 300 BC – 0 (2300–2000 yr BP; Tubosun 1995). In the south, drier conditions did not begin until c. 2000 BC. By c. 1000 BC Guinean savanna elements dominated while rainforests and mangroves declined drastically (Sowunmi 2004; Orijemie and Sowunmi 2014). Subsequently favourable conditions were restored and forests recovered but did not attain the same extent and floristic diversity of the Middle Holocene.

The Archaeobotany of the Later Stone Age in Nigeria

The LSA is known as a period (c. 10,000 BC – 0; 12,000–2000 yr BP) during which small sized-stone tools referred to as microliths replaced larger core tools. These microliths which are usually one to three cm in length, include crescents, points, blades, chips, chisels, flints, micro-blades and scrapers (Soper 1965) and they have "at least one edge trimmed by steep backing or blunting" (Allsworth-Jones et al. 2010: 33). The locations of some important Nigeria LSA sites are shown in Fig. 1 while a summary of plants used during this period from several archaeological sites is presented in Table 2.

Evidence of Plant Exploitation in the Nigerian LSA

Linguistics and Ethnography

Evidence for the earliest known use of plants in the LSA of Nigeria can be found in linguistics studies. Williamson (1993) suggested that oil palm (*Elaeis guineensis* Jacq.), kolanuts (*Cola acuminata* (P.Beauv.) Schott and Endl.), yams (*Dioscorea* spp.) and wine palm (*Raphia hookeri* G.Mann and H.Wendl.) constitute some of the earliest plants utilised by humans in southern Nigeria. Similarities in the names of these plants in several languages spread over a wide geographical area of southern Nigeria enabled the reconstruction of their root words to the Proto-Atlantic-Congo period. In addition, the inferred antiquity of these plants is supported by ethnographic data which reveal their use in rituals and similar traditional practices (Andah 1993; Williamson 1993).

Table 1. The major environmental events during the Holocene based on reference studies carried out in Nigeria

Chronology (approximate range)	Major environmental events	Proxy data	Author(s)
12,000–11,000 BC (*c.* 14,000–13,000 BP)	Wet and warm conditions; fluvial flow; rejuvenation of rivers; flow of marine sediments in the Niger Delta area	Sedimentology	Pastouret et al. (1979)
Prior to 10,000 BC (*c.* 12,000 BP)	Prevalence of Sudanian savanna; significant increased amount of charcoal indicative of constant fire frequencies in northern Nigeria	Palynology	Salzmann et al. (2002)
Prior to 10,000 BC (*c.* 12,000 BP)	Southward shifts in the savannas such that rainforests were pushed to refugia along the coast in the south.	Palynology	Sowunmi (1981a, b; 1985)
10,000–8900 BC (*c.* 12,000–10,900 BP)	Humid conditions in the north with dominance of wetter savannas	Palynology	Salzmann et al. (2002)
10,000–8000 BC (*c.* 12,000–10,000 BP)	Wet conditions in the south; rainforests regenerated and became established in most parts of the south of the region	Palynology and geology	Sowunmi (1981a, b; 1985)
8900–8500 BC (*c.* 10,900–10,500 BP)	After dry conditions, there was onset of humid and wetter conditions as well as increased lake levels (abundance of *Typha* sp.)	Palynology and sedimentology	Salzmann (2000); Salzmann et al. (2002)
8000–6500 BC (*c.* 10,000–8500 BP)	Wet and warm period dominated; rainforest and mangroves regenerated in the south of the region	Palynology	Sowunmi (1981a, b; 1985); Dupont et al. (2000)
8500–4800 BC (*c.* 10,500–6800 BP)	Initially humid and cool, and wetter period from about 6500 BC	Palynology and sedimentology	Salzmann (2000); Salzmann et al. (2002).
6500-2500 BC (*c.* 8500–4500 BP)	The wettest period is at around 6500-3000 BC, and maximum extension of the rainforests in the south of the region	Palynology	Sowunmi (1981a, b); Sowunmi (2004)
4800–1800 BC (*c.* 6800–3800 BP)	Dry conditions began to be established, and by around 1800 BC conditions became extremely dry resulting in increased Sudanian-Sahel elements and concomitant decrease in Guinean taxa	Palynology and sedimentology	Salzmann (2000); Salzmann et al. (2002)
2500–1100 BC (*c.* 4500–3100 BP)	Fluctuating climate in the south; disturbances in lowland rainforest; gradual appearance of savanna elements	Palynology	Sowunmi (2004)
1800 BC (*c.* 3800 onwards)	Dry conditions were established and continued till the present	Palynology	Salzmann (2000); Salzmann et al. (2002)
1100–500 BC (*c.* 3100–2500 BP)	Dry conditions established in the south; rainforest degraded; mangroves declined and disappeared in some areas	Palynology	Sowunmi (2004); Orijemie and Sowunmi (2014)
300 BC – 0 (*c.* 2300–2000 BP)	Wet conditions in the Benue Valley; in the hinterland and along floodplain of Katsina Ala River	Geomorphology	Tubosun and Andah (1983); Tubosun (1995)

Furthermore, the recovery of kernels of *E. guineensis* from deposits dated to 7580–7350 BC (8410 ± 40 yr BP) in Bosumpra Cave, Ghana (Oas et al. 2015) suggests the deep antiquity of oil palm use in the southern parts of West Africa.

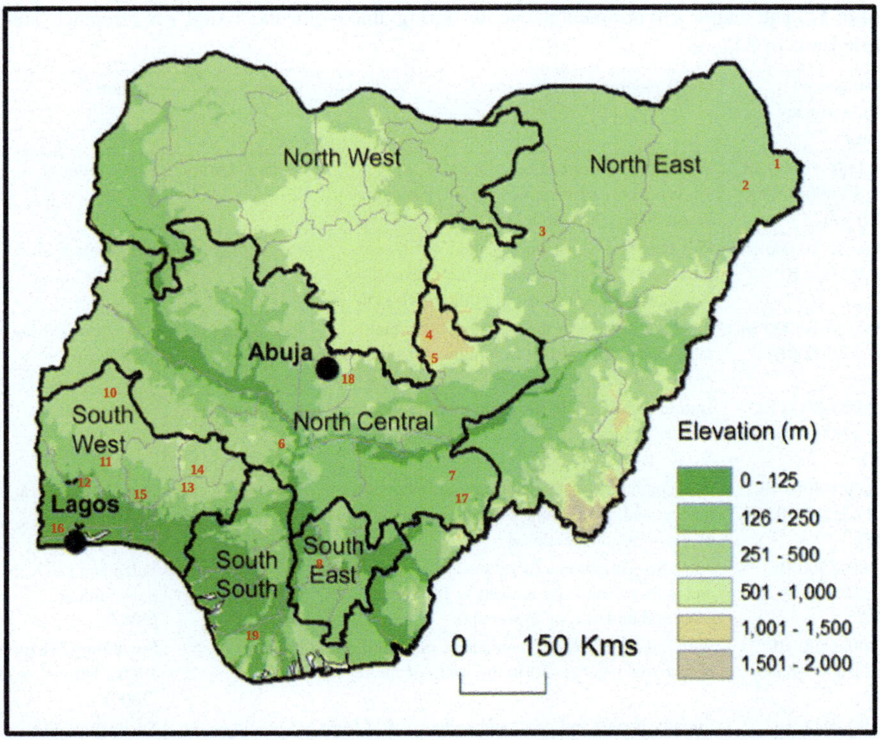

Fig. 1. Map of Nigeria, showing Later Stone Age (LSA) sites mentioned in the text. 1. Kursakata, 2. Gajiganna, 3. Kariya Wuro, 4. Dutsen Kongba, 5. Rop, 6. Itaakpa, 7. Katsina Ala River, 8. Udi-Okigwe Plateau, 9. Afikpo, 10. Mejiro Cave (Old Oyo), 11. Atamora, 12. Abeokuta, 13. Iwo Eleru, 14. Ita-Ogbolu, 15. Ifetedo, 16. Apa, 17. Tse Dura, 18. Nok, 19. Niger Delta

Palynology

The antiquity of oil palm exploitation in Nigeria is supported by palynological evidence from eastern (Umeji et al. 2012) and southern regions of the country (Sowunmi 1981a, b; 1985). Umeji et al. (2012) reported the occurrence of oil palm pollen from three rock shelters in the Udi-Okigwe plateau, one of which was dated to c. 3000 BC. However it is unclear what sampling methods were employed, and moreover the archaeological contexts from which the samples were collected and period of deposition are unknown. Therefore, understanding the utilisation of the oil palm in a chronological sequence and how its occurrence relates to other LSA sites is difficult. Exploitation of the oil palm was noted in pollen records of the Ofuabo creek core at 950–850 BC based on an unprecedented abundance of *E. guineensis* pollen during that period. Although it has been argued that distinguishing between cultivated and wild pollen of *E. guineensis* is almost impossible, Sowunmi (1985) indicated that this unprecedented increase in oil palm pollen was accompanied by a drastic reduction in tree pollen as well as increases in weed pollen. These phenomena indicate that the oil palms were protected while other forest trees were cleared in the typical slash-and-burn agriculture, which is still practised today. At Kariya Wuro

Table 2. Review of archaeobotanical researches and some plants used during the LSA from archaeological sites of Nigeria

Site	Time span	Dates	Lithics	Pottery	Plants	Identification	Uses	Author (s)
Southern Nigeria	PAC	LSA	Not Applicable	Present	*Elaeis guineensis*	Linguistics	Food, building; rituals	Andah (1993); Williamson (1993)
					Raphia hookeri		Wine, fish poison, building	
					Cola acuminata		Food, stimulant	
					Dioscorea cayenensis, D. rotundata, D. bulbifera		Food; rituals	
Iwo Eleru	LSA	3500 BC (*c.* 5500 BP)	Microliths and ground-stone axes	Present	Tubers (?)	Indirect inference from ground-stone axe and trapezoid use	Food	Shaw and Daniels (1984)
		1500 BC (*c.* 3500 BP)	Trapezoids	Present	Wild grasses (?)			
Udi-Okigwe plateau	LSA	[*c.* 3000 BC uncal.]	Unreported	Present	*Elaeis guineensis*	Pollen	Food	Umeji et al. (2012)
Gajiganna	LSA	1800–1500 BC (*c.* 3800–3500 BP)	Microliths	Present	Suite of wild Poaceae (including *Oryza* sp., *Paspalum, Brachiaria, Cenchrus, Echinochloa, Eleusine, Sorghum*), *Ziziphus* sp., *Celtis integrifolia, Grewia* sp. and *Balanites aegyptiaca*	Impressions on pottery; charred kernels, charcoal	Food and fuel	Klee et al. (2004); Breunig and Neumann (2002)
		1500–800 BC (*c.* 3500–2800 BP)			*Pennisetum glaucum, Ziziphus* sp., *Celtis integrifolia, Grewia* sp. and *Balanites aegyptiaca*	Fruits and seeds	Food	
Niger Delta	LSA	950–850 BC (2950–2850 cal. BP)	Not Applicable		*Elaeis guineensis*	Pollen	Food	Sowunmi (1981a, b)

(*continued*)

Table 2. (*continued*)

Site	Time span	Dates	Lithics	Pottery	Plants	Identification	Uses	Author (s)
Apa	LSA	1060–540 BC (*c.* 3060–2540 BP)	Ground-stone axe	Present	*Elaeis guineensis*	Charred palm kernels	Food; Fuel (?)	Alabi (1999, 2000, 2002)
Kursakata	LSA	1000 BC–AD 100 (*c.* 3000–1900 BP)	Microliths	Present	*Pennisetum*, Paniceae, *Oryza* sp., *Vitex* sp., *Celtis integrifolia*	Charred seeds and fruits	Food	Klee et al. (2000)
					Kigelia africana, Mitragyna type, *Acacia nilotica, Prosopis africana, Vitex* sp., *Ziziphus* sp., *Kigelia africana, Capparis tomentosa, Celtis* cf. *integrifolia, Securinega/Hymenocardia, Khaya* sp., *Acacia raddiana* type, *Balanites aegyptiaca, Acacia albida, Tamarindus* sp., *Maerua crassifolia,* cf. Bauhinieae.	Charcoal	Fuel	
Nok sites (Iddah, Janruwa and Ungwar Kura)	LSA	800–450 BC (c. 2800–2450 BP)	Unreported	Present	cf. *Cynometra vogelli,* cf. *Pterocarpus erinaceus, Dialium* type, *Faidherbia albida, Antifdesma, Drypetes, Flaeggea virosa/Hymenocardia, Uapaca, Anogeissus leiocarpus, Khaya* sp., *Parinari* sp.	Charcoal	Fuel	Kahlheber et al. (2009)
					Pennisetum glaucum, Eleusine cf. *coracana, Vigna unguiculata, Canarium schweinfurthii* and *Vitex* spp.	Charred seeds and fruits	Food	
Tse Dura	LSA	[2330 ± 70–2200 ± 150 BP uncal.]	Microliths, hammer and grinding stones	Present	*Elaeis guineensis* and unidentified seeds	Kernels, seeds	Food	Andah (1983)
Itaakpa	LSA	410–50 BC (*c.* 2400–2050 BP)	Microliths	Present	*Elaeis guineensis*	Palm kernels, pollen	Food	Allsworth-Jones et al. (2012)
Rop	LSA	0 (*c.* 2000 BP)	Microliths	Present	Starchy cultivars (?millets)	Dental caries	Food	Eyo (1972)

(*continued*)

Table 2. (*continued*)

Site	Time span	Dates	Lithics	Pottery	Plants	Identification	Uses	Author (s)
Kariya Wuro	LSA	AD 1020–1160 (*c.* 930–790 BP)	Microliths	Present	*Elaeis guineensis, Sarcocephalus latifolius* (syn. *Nauclea latifolia*)	Pollen	Food	Sowunmi and Awosina (1991)
					Bridelia scleroneura, Pavetta sp.		Medicine	
Itaogbolu	LSA	[Not dated]	Microliths; ground-stone axes	Present	Tubers (?)	Indirect inference from ground-stone axe use	Food	Alabi et al. (2009)

rock shelter, Bauchi State, a date of AD 1020-1160 (950 ± 30 yr BP) was obtained from 20–30 cm level of a 50 cm deep trench (Switsur et al. 1994). The pollen of *Bridelia scleroneura* Müll.Arg., *Pavetta* sp., *E. guineensis* and *Sarcocephalus latifolius* Sm. (syn. *Nauclea latifolia* Sm.) were recovered from inside the rock shelter (Sowunmi and Awosina 1991). The plants from which the pollen was derived are thought to have been deliberately brought into the rock shelter for medicinal and dietary purposes. In addition, the recovery of their pollen, baring wind and insect transport, indicate their occurrence within the region, hence the prevalence of wetter conditions than at present. Both *E. guineensis* and *S. latifolius* are no longer present in Kariya Wuro. It is possible that human occupation of this rock shelter predates AD 1020–1160 as indicated by the shallow depth from where the radiocarbon dated charcoal was obtained and the origin of Bauchi languages, which has been put at c. 1000 BC – 0; Blench 2015.

Pottery, Stone Tools and Physical Anthropology

Four sets of cultural innovations which relate to plant use in the LSA included the appearance of artefacts namely: pottery and ground-stone axes at c. 5000 BC; trapezoids at c. 1500 BC Shaw and Daniels 1984); and grinding stones prior to c. 2000 years ago. Although it is not certain which plants were exploited by the inhabitants of Iwo Eleru rock shelter, the occurrence of pottery along with ground-stone axes were thought to be linked to the beginning of farming, however direct evidence for this conclusion is lacking (Shaw 1984; Shaw and Daniels 1984; Phillipson 2005). In fact, pottery appeared at least 2000–600 years before recognisable signs of farming occurred in Nigeria (Connah 1981; Breunig and Neumann 2002). Ground-stone axes are thought to have been used for felling trees, characteristic of slash-and-burn which is a common practice in the forest zone of southern Nigeria (Alabi 1999, 2000), as well as in harvesting tubers (Alabi et al. 2009).

That ground-stone axes occurred during the period when climate was humid and rainforest was moderately diverse suggests they could have been used in forest clearance. Similar explanations have been given for the occurrence of ground-stone axes at Apa c. 1060–540 BC (2670 ± 90 BP, Beta-89266) (Alabi 1999, 2000) and might also be applicable to the site of Itaakpa c. 410–50 BC (2210 ± 80 yr BP, OxA-1150) (Allsworth-Jones et al. 2010; Oyelaran 2002). The subsequent change in tool kit, i.e., the appearance of trapezoids from c. 3500 years ago onwards at Iwo Eleru (Shaw and Daniels 1984) seems to reflect an exploitation of wild grasses (Phillipson 2005; Allsworth-Jones et al. 2010). This use is not unexpected when one considers that it was a response to an abundance of grasslands under the prevailing dry climate of the Late Holocene (Sowunmi 2004; Orijemie and Sowunmi in prep.). The possibility that the trapezoids were used in harvesting wild *Sorghum* as suggested by Shiner (1973) based on sheen observed on their edges, is unlikely due to the shattering nature of wild sorghum during harvesting (Harlan 1993). In addition, there are complications with this suggestion in that both sorghum and pearl millet (*Pennisetum glaucum* (L.) R. Br.) are harvested the same way (Pelling 2014). The recovery of grinding stones, along with microliths such as projectile points, pointed tools, cores and flakes, from the lowest levels (75–90 cm) of the Tse Dura rock shelter (Andah 1983) may indicate the exploitation of plant resources and game hunting. Oil palm and unidentified seeds are reported to have been recovered but their stratigraphic context was not indicated.

The Rop rock shelter is similar to Iwo Eleru in context but with more recent human skeletal remains dated to c. 2000 years ago. There are no reports of archaeobotanical finds however, wear on teeth, in addition to the occurrence of dental caries, led to the conclusion that the individual consumed starch-rich cultivated plants, a diet in contrast to typical hunters-gatherers (Gaherty 1968). Generalised studies indicate that dental caries are rare among hunter-gatherers (0–5.3%), but common (2.2–48.1%) among farmers in the Americas and Europe (Lanfranco and Eggers 2012). Dental caries have also been associated with the consumption of cultivated foods high in sugars (sucrose) such as cereals (Schollmeyer and Turner 2004), and are usually used as a proxy for identifying dietary habits and patterns among prehistoric humans (Bryant 1995). Given the relative antiquity of what appears to be domesticated pearl millet in north-eastern Nigeria (c. 1200–1000 BC) there is a high probability that the Rop individual consumed millet.

At the Itaakpa rock shelter, near Ajaokuta in present-day Kogi State, kernels of oil palm were recovered along with human remains (Allsworth-Jones et al. 2012), both of which were associated with a calibrated date of 410–50 BC (OxA-1150). In addition, the nature and level of wear on human teeth from Itaakpa, in contrast to those from Europe whose diets were dominated by cereals, have been described as "typical of hunter-gatherers" (Allsworth-Jones et al. 2012: 11), but no specific plant foods were suggested. Pollen analysis of sediments from an undated pond near the rock shelter revealed an abundance of forest taxa, and *E. guineensis* pollen corresponding to the aceramic and ceramic phases respectively in the rock shelter (Oyelaran 2002). Hence, it is known that the Itaakpa human community exploited oil palm, a practice that persists for over two thousand years. Yams (*Dioscorea* spp.) and melon (*Citrullus lanatus* (Thunb.) Matsum. and Nakai) the macroremains of which are relatively rare in the archaeological record might have equally been exploited in the Itaakpa rock shelter.

The most important festival of ritual initiation rites among the Ososo people of northern Edo State known as *Itakpo*, involve traditional meals of yams (*ovwhila*) and melon soup (*osomo-okoshi*) cooked with palm oil (*owhiri*). Activities of this festival culminate at some rock shelters in Ososo from where the individuals performing the rites emerge as Men thus re-enacting an ancient passage rite connected with the rock shelters. It is interesting that part of the oral traditions of Ososo cite Ajaokuta, a neighbouring town of Itaakpa, as one of their ancestral homes. Hence, this festival possibly has its origins in Itaakpa. Migration and distance may have separated the two peoples but their shared memory is re-enacted in the *Itakpo* rites. Pottery and charred oil palm kernels at Apa (Alabi 1999, 2000, 2002), near Badagry in southern Nigeria, dated to 1060–540 BC (2670 ± 90 yr BP, Beta-89266), suggests the exploitation of oil palm took place at a period earlier than that at Itaakpa.

Ifetedo (Fatunsin 1996) and Ita-Ogbolu rock shelters, two LSA sites in Osun and Ekiti States in south-western Nigeria yielded lithics and a type of subsistence which revealed environmental change from forest to savanna similar to the sequence at Iwo Eleru (Alabi et al. 2009). One major difference however lies in the shallow nature of the deposits. In addition, the current lack of radiocarbon dates prevents a direct correlation to be made between these sites and Iwo Eleru. Sedimentological (Alabi and Abraham 2013) and preliminary pollen analyses of the Ita-Ogbolu sediments seem to support the archaeological interpretation of the finds. The recovery of ground-stone axes from the Ita-Ogbolu excavation was interpreted as evidence of the likely use of the tool for harvesting tubers (Alabi et al. 2009). The Atamora rock shelter, located on Ibadan-Ife Express Way, south-western Nigeria is yet to be dated. It too was compared with Iwo Eleru on the basis of similarities and pattern of occurrence of lithics (Akpobasa 2009). Remains of oil palm and gamhar (*Gmelina arborea* Roxb.) were recovered from the uppermost sections of one of the trenches indicating a very recent association. Gamhar, a native of southern Asia was introduced to Nigeria in the 20th century.

Macrobotanical Remains (Charcoal, Seeds and Fruits) and Plant Impressions

In the Gajiganna complex in northeastern Nigeria, plant impressions on excavated pottery and charred caryopses showed that pearl millet, wild Paniceae, wild rice species (*Oryza* spp.) and tree fruits constituted the earliest plant food resources exploited there c. 2000–1000 BC (Klee et al. 2000, 2004). Other plants exploited included *Ziziphus* sp., *Celtis integrifolia* Lam., *Grewia* sp., and *Balanites aegyptiaca* (L.) Delile (Klee et al. 2004). Pearl millet subsequently became the dominant crop from 1200-1000 BC (Klee et al. 2000, 2004; Breunig and Neumann 2002). Breunig and Neumann (2002) argue that domesticated pearl millet and a unique mat impression on pottery, appear to have diffused from an area farther north of Borno in the Sahara, both appearing at about the same time (c. 3200 years ago) in the archaeological record of northeastern Nigeria.

Pearl millet, finger millet (*Eleusine* cf. *coracana* L. Gaertn.), cow pea (*Vigna unguiculata* (L.) Walp.), *Canarium schweinfurthii* Engl.and *Vitex* spp. were exploited in the Nok area of north-central Nigeria by 800–450 BC (Kahlheber et al. 2009). Based on charcoal remains, the habitat surrounding the Nok sites at that period is thought to have been *Isoberlina*-woodland with some riparian forest possibly nearby. The authors claim that, in comparison with areas further north (e.g., Lake Tilla), these sites were slightly wetter during the Late Holocene, and that the riparian forests probably constituted the likely source from where wood was harvested for fuel.

Discussion

LSA Groups and Plant Exploitation

A survey of LSA sites in Nigeria shows that similarities existed in the pattern of occurrence and chronological sequence of microliths, pottery and ground-stone axes in particular, of LSA groups in northern and southern Nigeria. Given these similarities, a likely result of several interactions over a period (MacDonald 1998), it is apparent that both ecological zones i.e., forest and savanna, were occupied at the same time. However, despite this contemporaneity, there are marked differences in the plants exploited by the various groups. These differences largely reflect cultural and ecological and/or climatic variations in the environments of both areas. Therefore, the notion that LSA groups migrated southwards from the north c. 1000 BC spreading agricultural knowledge suggests a unidirectional movement of ideas, which is unlikely given the current evidence. What may have occurred is that first given the earlier dates in Iwo Eleru (and southwestern Cameroon), LSA groups initially occupied southern Nigeria from c. 10,000 BC (12,000 yr BP) and migrated northwards probably forming separate but fairly homogeneous groups. Secondly, from c. 2000–1500 BC onwards, these groups were met by incoming migrants from the Sahara and cultural interaction and exchange of ideas took place. People selected plants which were "foreign" but most suited their needs and local environment. This accounted for the southward spread of cereals such as pearl millet and *Sorghum* spp. from the north beginning c. 1000-800 BC, and of oil palm and kolanuts (*C. acuminata*) in the opposite direction. For instance oil palm was present in Kariya Wuro at c. 800 BC, while kolanuts reached the north from Yorubaland via Nupe at an unknown time (Williamson 1993).

Pearl millet, *Sorghum* spp., *Oryza* spp., wild Paniceae and a host of wild fruits including *C. integrifolia, S. latifolius, Vitex* and *Ziziphus* spp., which constituted plants consumed in northern Nigeria were recovered from archaeological contexts. In contrast, plants except oil palm, thought to have been exploited in the south (Andah 1993) are yet to be recovered from archaeological contexts. In fact that their use was inferred mainly from linguistics and ethnographic data is not without problems. This presents a challenge to archaeobotanical research in southern Nigeria. Therefore, it is pertinent to identify archaeological features and/or areas from where these remains can be retrieved including hearths, kitchens, store houses and granaries which are known to present the best opportunities for plant recovery. Ethnographic data (Folorunso 1998), where available, are crucial and should be explored and integrated with archaeological data (Hather and Mason 2002). There is also a need to devise effective methods for recovering and identifying plant remains such as tubers and roots (Hather and Mason 2002). Upon recovery it will also be necessary to deal with the questions of what constitutes domesticated plants on the one hand, and "autochthonous" and "introduced" plants on the other hand. While it has been possible to identify domesticated plants and perhaps the processes involved in northern Nigeria, the same cannot be said for the south, prompting the use of the term "incipient vegeculture" to describe subsistence in areas where evidence for agriculture is lacking (McDonald 1998). Studies of hunter-gatherers have shown that they have complex social systems where plants are utilised in ways other than as foods (Brooks et al. 1984). For instance there is linguistic evidence of the use of several plants as fuel, building materials and in rituals in southern Nigeria (Williamson 1993; Sowunmi 2002), and as wood fuel in Nok (Kahlheber et al. 2009) and northeastern Nigeria (Ballouche and Neumann 1995). Hence, it is by investigating these other uses to which plants were put that archaeobotanists can have a fuller sense of past human-plant relationships.

Transition to Farming and Beyond

What factors led to the beginning of agriculture specifically the dominance of domesticated pearl millet over gathering of wild grasses and fruits in Gajiganna, northeastern Nigeria? It is believed that climatic variability in the Sahara contributed to the diffusion of farming cultures into the Gajiganna areas (Breunig and Neumann 2002). However, Muzzolini (1993) indicates that the change to food production occurred during a wet phase, and therefore there is no need to invoke a dry phase cause-and-effect phenomenon. Instead, the need for people occupying environmentally-marginal areas to have predictable, regular and higher protein contents in plants (and animals) may have compelled individuals to adopt farming. These views have been supported by Barker (2006), who went further to add that technological advancement contributed to the adoption of farming. The richness of the forest, and forest-Guinea savanna ecotones in terms of biotic diversity provided LSA groups in southern Nigeria access to a wider variety of plant resources. Hence, until the need arose for higher availability and/or predictability of resources in these areas, adopting farming that was practised in the north was perhaps unnecessary (Wetterstrom 1993). Thus, while the diet of the Rop individual seemed to have been dominated by Paniceae at c. 2000 years ago, that of the Itaakpa individual, who existed at about the same time as the Rop group, was different. The Itaakpa individual possibly consumed roots and tubers which were reflected in very large molars (Allsworth-Jones et al. 2012). Similar kinds of plant foods may also have been exploited by the Iwo Eleru individual whose molars were not significantly different from those of the Itaakpa individual. Harvati et al. (2011:3) stated that the Iwo Eleru individual had large teeth with his "browridge… slightly more forward projecting than the average modern human shape". Unfortunately the preservation of plants in archaeological contexts in these sites is poor. In addition, it is also worth investigating the factors responsible for the delay in accepting agriculture in these areas. While the Saharan emigrant populations brought along farming ideas to the northeast, farming did not commence there until after a substantial amount of time (Neumann 1999; Breunig and Neumann 2002). Moreover, it is known that the people still continued to gather wild fruits after farming had been introduced. A parallel situation but of earlier occurrence is known in the Sahara (Lucarini et al. 2015).

In the majority of LSA sites discussed above, microliths disappeared after c. 2000 years ago marking the end of that technology. What happened to the LSA peoples; were they ultimately assimilated by Iron Age people or did they go into extinction? In their studies of Dobe! Kung of southern Africa Brooks et al. (1984) stated that hunter-gatherers embraced plant domestication as an addition to their already broad food base, subsequently becoming independent food producers. Unsuccessful groups became isolated and this partly accounted for why modern hunter-gatherers such as those in southern Africa are located in environmentally-marginal areas. Those who could neither sustain nor continue their hunter-gatherer status due to socio-economic pressures eventually became assimilated into food-producing communities. This is reflected in the continued use of stone tools in hunting, grinding, as hammers for cracking hard nuts, sharpening blunt knives and igniting fires, among several cultures in Nigeria—a true testament to a lasting legacy bequeathed to succeeding generations by their LSA ancestors.

Archaeobotany in Nigeria: Looking to the Future

The lack of systematic archaeobotanical research in Nigeria significantly hampers our knowledge of plant resources exploited by LSA groups. It is also daunting to accurately decipher the period and factors responsible for the choice of plants and under what circumstances. Traditional archaeobotanical techniques employed so far include pollen, spores and macro-botanical remains (charcoal, fruits, seed and wood). Starch granules, phytoliths and parenchyma cells are yet to be fully investigated at Nigerian LSA sites. Although there are challenges in identifying the specific plants that produced starch granules (Barton et al. Barton 1998), there has been some success (Piperno and Holst 1998). Residues obtained from LSA tools could also provide additional information on plant-use especially harvesting and felling of trees (Allen et al. 1995; Yates et al. 2014). Attention should also be focused on phytoliths which can detect remains of plants otherwise not visible as macroremains (Dominguez-Rodrigo et al. 2001).

Apart from pollen, and spores of Pteridophytes, fungal spores have not been given adequate attention. In some cases, fungal spores are better preserved than pollen and as such can provide additional information on plant-use and in some cases animal husbandry. Some fungi provide evidence of certain plants based on fungi-host relationships. For example the most common fungal pathogens of pearl millets, *Aspergillus flavus* Link. and *Fusarium* spp. are specific to both *Pennisetum* and *Sorghum* (Abdulsalaam and Shenge 2011; González et al. 1997; Zida et al. 2008). Hence, their occurrence in an archaeological context could provide indirect evidence for cultivation or use of these grasses. This accounts for why fungal spores are now recognised as important constituents of non-pollen palynomorphs, and their data are often complementary to those of pollen (Cook et al. 2011; Gelorini et al. 2011; van Geel et al. 2011). As these complement one another, where possible, a combination of these techniques should be employed in archaeological investigations. There is also an urgent need to conduct archaeobotanical investigations of important LSA sites such as those of Afikpo (Andah and Anozie 1980) and Mejiro (Old Oyo) (Willett 1962) through re-excavation and flotation for archaeobotanical remains, as was successfully done at Tse Dura recently. The results will add to the growing knowledge of plants exploited during the LSA of southern Nigeria, a region with comparatively fewer archaeobotanical data.

The Department of Archaeology and Anthropology, University of Ibadan has, for a very long time, remained the only Archaeology Department that conducts archaeobotanical research in Nigeria. Sadly, the lack of experienced personnel and training laboratories has prevented other Archaeology Departments in the country from incorporating archaeobotanical evidence to understand past human-plant relations. It is however gratifying that two young Nigerian archaeologists are currently undergoing archaeobotanical training in Canada and the UK. Their return is highly anticipated, and will no doubt expand the frontiers of archaeobotanical research in Nigeria.

References

Abdulsalaam S, Shenge KC (2011) Seed borne pathogens on farmer-saved Sorghum. J Stored Prod Postharvest Res 2(2):24–28

Akpobasa JJ (2009) Stone age studies in Southwest Nigeria with particular emphasis on the assessment of material from Atamora Rock Shelter. Unpublished PhD Dissertation, Eberhard-Karls-University Tübingen

Alabi RA (1999) Human-environment relationship: a synthesis of ethnoarchalogical evidence of human impact on the Badagry coastal area of South Western Nigeria. J Sci Res 5(1):25–31

Alabi RA (2000) Apa: a later stone age occupation layer in the Southwestern coast of Nigeria. Nyame Akuma 53:29–34

Alabi RA (2002) Environment and subsistence of the early inhabitants of coastal southwestern Nigeria. Afr Archaeol Rev 19(4):183–201

Alabi RA, Abraham R (2013) Sedimentological analysis of late stone age environment of Itaogbolu rock shelter, Ondo State, Nigeria. J Environ Culture 10(1,2):1–17

Alabi RA, Aleru JO, Opadeji AO et al (2009) A hunter-gatherer community in the rain forest belt of southwestern Nigeria. Nyame Akuma 72:74–77

Allen J, Newman ME, Riford M et al (1995) Blood and plant residues on Hawaiian stone tools from two archaeological sites in upland Kāne'ohe, Ko'olau Poko District. O'ahu Island. Asian Perspect 3(2):283–302

Allsworth-Jones P, Harvati K, Stringer C (2010) The archaeological context of the Iwo Eleru cranium from Nigeria and preliminary results of new morphometric studies. In: Allsworth-Jones P (ed) West African archaeology, new developments, new perspectives BAR S2124. Archaeopress, Oxford, pp 29–42

Allsworth-Jones P, Oyelaran PA, Stringer C et al (2012) Itaakpa, a Late stone age site in southwestern Nigeria. J Field Archaeol 37(3):163–177

Andah BW (1983) The Bantu "Homeland" Project: ethnoarchaeological investigations in parts of the Benue valley region. W Afr J Archaeol 13:23–60

Andah BW (1993) Identifying early farming traditions in West Africa. In: Shaw T, Sinclair P, Andah B, Okpoko A (eds) The archaeology of Africa: food, metals and towns. Routledge, London, pp 240–254

Andah BW, Anozie F (1980) Preliminary report on the prehistoric site of Afikpo (Nigeria). W Afr J Archaeol 10:83–102

Ballouche A, Neumann K (1995) A new contribution to the Holocene vegetation history of the West African Sahel: pollen from Oursi, Burkina Faso and charcoal from three sites in northeast Nigeria. Veg Hist Archaeobot 4:31–39

Barker G (2006) The agricultural revolution in prehistory: why did foragers become farmers?. Oxford University Press, Oxford

Barton H (1998) Clues to stone tool function re-examined: comparing starch grain frequencies on used and unused obsidian artefacts. J Archaeol Sci 25:1231–1238

Blench R (2015) Linguistic and cultural background to the North Bauchi region and the Wiihə people. In: Allsworth-Jones P (ed) Kariya Wuro: a late stone age site in northern Nigeria. Reports in African Archaeology 7. Africa Magna Verlag, Frankfurt

Bogaard A, Hodgson J, Nitsch E et al (2016) Combining functional weed ecology and crop stable isotope ratios to identify cultivation intensity: a comparison of cereal production regimes in Haute Provence, France and Asturias, Spain. Veg Hist Archaeobot 25:57–73

Breunig P, Neumann K (2002) From hunters and gatherers to food producers: new archaeological and archaeobotanical evidence from the West African Sahel. In: Hassan FA (ed) Droughts, food and culture ecological change and food security in Africa's later prehistory. Kluwer Academic Publishers, New York, pp 123–156

Brooks AS, Diane EG, Yellen JE (1984) Food production and culture change among the !Kung San: implications for prehistoric research. In: Clark JD, Brandt SA (eds) From hunters to farmers. University of California Press, Berkley, pp 293–310

Bryant VM (1995) Eating right is an ancient rite. The World & I. January 216–221

Connah G (1981) Three thousand years in Africa. Cambridge University Press, Cambridge

Cook EJ, van Geel B, van der Kaars S et al (2011) A review of the use of non-pollen palynomorphs in palaeoecology with examples from Australia. Palynology 35(2):155–178

Dominguez-Rodrigo M, Serrallonga J, Juan-Tresserras J et al. (2001) Woodworking activities by early humans: a plant residue analysis on Acheulian stone tools from Peninj (Tanzania). J Hum Evol 40:289–299

Dupont LM, Jahns S, Marret, F et al. (2000). Vegetation change in equatorial West Africa; time-slices for the last 150ka. Palaeogeogr Palaeoclimatol Palaeoecol 155:95–122

Fatunsin FK (1996) Ifetedo: a late stone age site in the forest region of southwestern Nigeria. W Afr J Archaeol 26(1):71–87

Folorunso CA (1998) The compound of the Tiv in Benue State Nigeria: the reality of ethnoarchaeology In: Andah BW, Sowunmi M.A, Okpoko AI, Folorunso CA (eds) Africa: the challenge of archaeology. Heinemann Educational Books, Nigeria, pp 235–255

Gaherty G (1968) The human skeleton from Rop Rock Shelter, Nigeria. West Afr Archaeological Newsl 9:18–19

Gelorini V, Verbeken A, van Geel B et al (2011) Modern non-pollen palynomorphs from East African lake sediments. Rev Palaeobot Palyno 164:143–173

González HHL, Martinez EJ, Resnik SL (1997) Fungi associated with *Sorghum* grain from Argentina. Mycopathologia 139(1):35–41

Harlan JR (1993) The tropical African cereals. In: Shaw T, Sinclair P, Andah B, Okpoko A (eds) The archaeology of Africa: food, metals and towns. Routledge, London, pp 53–60

Harvati K, Stringer C, Grun R et al (2011) The later stone age calvaria from Iwo Eleru, Nigeria: morphology and chronology. PLoS ONE 6(9):1–8

Hather JG, Mason SLR (2002) Introduction: some issues in the archaeobotany of hunter-gatherers. In: Mason SLR, Hather JG (eds) Hunter-gatherer archaeobotany: perspectives from the northern temperate zone. University College London, Institute of Archaeology, pp 1–14

Jones G (2002) Weed ecology as a method for the archaeobotanical recognition of crop husbandry practices. Acta Palaeobot 42(2):185–193

Kahlheber S, Höhn A, Rupp N (2009) Archaeobotanical studies at Nok sites: an interim report. Nyame Akuma 71:2–15

Klee M, Zach B, Neumann K (2000) Four thousand years of plant exploitation in the Chad Basin of northeast Nigeria I: the archaeobotany of Kursakata. Veg Hist Archaeobot 9:223–237

Klee M, Zach B, Stika H (2004) Four thousand years of plant exploitation in the Lake Chad Basin (Nigeria), part III: plant impressions in potsherds from the final stone age Gajiganna culture. Veg Hist Archaeobot 13:131–142

Lanfranco LP, Eggers S (2012) Caries through time: an anthropological overview. In: Ming-Yu Li (ed) Contemporary approach to dental caries. http://www.intechopen.com/books/contemporary-approach-to-dental-caries/caries-archaeological-andhistorical-record. In Tech

Lucarini G, Radini A, Barton H et al. (2015) The exploitation of wild plants in Neolithic North Africa: use-wear and residue analysis on non-knapped stone tools from the Haua Fteah cave, Cyrenaica, Libya. Quatern Int http://dx.doi.org/10.1016/j.quaint.2015.11.109

MacDonald KC (1998) Archaeology, language and the peopling of West Africa: a consideration of the evidence. In: Blench R, Sriggs M (eds) Archaeology and language II archaeological data and linguistic hypotheses. Rouledge, London, pp 33–66

Muzzolini A (1993) The emergence of a food-producing economy in the Sahara. In: Shaw T, Sinclair P, Andah B, Okpoko A (eds) The archaeology of Africa: food, metals and towns. Routledge, London, pp 227–239
Neumann K (1999) Early plant food production in the West Africa Sahel: new evidence. In: van der Veen M (ed) The exploitation of plant resources in ancient Africa. Kluwer Academic Publishers, NewYork, pp 73–80
Oas SE, D'Andrea AC, Watson DJ (2015) 10,000 year history of plant use at Bosumpra Cave, Ghana. Veg Hist Archaeobot 24:635–653
Orijemie EA, Sowunmi MA (2014) Humans and the mangrove in southern Nigeria. In: Stevens CJ, Nixon S, Murray MA, Fuller DQ (eds) The archaeology of African plant use. Left Coast Press, Walnut Creek, California, pp 103–112
Oyelaran PA (2002) Vegetation change in the pollen assemblage of Osaru Pond. Environmental Issues 24:32–37
Pastouret L, Chamley H, Delibrias G et al. (1979) Late Quaternary climate changes in western tropical Africa deduced from Deep-sea sedimentation off Niger Delta. Oceanologica Acta 1 (2):212–232.
Pelling R (2014) Patterns in the archaeobotany of Africa: developing database for North Africa, the Sahara and the Sahel. In: Stevens CJ, Nixon S, Murray MA, Fuller DQ (eds) Archaeology of African plant use. Left Coast Press, Walnut Creek, California, pp 205–224
Piperno DR, Holst I (1998) The presence of starch grains on prehistoric stone tools from the humid neotropics: indications of early tuber use and agriculture in Panama. J Archaeol Sci 25:765–776
Phillipson DW (2005) African archaeology, 3rd edn. Cambridge University Press, Cambridge
Salzmann U (2000) Are modern savanna degraded forests? – a Holocene pollen record from the Sudanian vegetation zone of NE Nigeria. Veg Hist Archaeobot 9:1–15
Salzmann U, Hoelzmann P, Morczinek I (2002) Late quaternary climate and vegetation of the Sudanian zone of northeast Nigeria. Quat Res 58:73–83
Schollmeyer KG, Turner CG (2004) Dental caries, prehistoric diet, and the pithouse-to pueblo transition in southwestern Colorado. Am Antiquity 69(3):569–582
Shaw T (1984) Archaeological evidence and effects of food-producing in Nigeria. In: Clark JD, Brandt SA (eds) From hunters to farmers. University of California Press, Berkeley, pp 152–157
Shaw T, Daniels SGH (1984) Excavations at Iwo-Eleru. W Afr J Archaeol 14:1–269
Shiner JL (1973) Wear patterns on stone tools. Nyame Akuma 3:32
Soper RC (1965) The Stone Age in northern Nigeria. J Historical Soc Nigeria 3:175–194
Sowunmi MA (1981a) Aspects of late Quaternary vegetational changes in West Africa. J Biogeogr 8:457–474
Sowunmi MA (1981b) The late Quaternary environmental changes in Nigeria. Pollen Spores 23 (1):125–148
Sowunmi MA (1985) The beginnings of agriculture in West Africa: botanical evidence. Curr Anthropol 26(1):127–129
Sowunmi MA (2002) Environmental and human responses to climatic events in west and west central Africa during the late Holocene. In: Hassan FA (ed) Droughts, food and culture: ecological change and food security in Africa's later prehistory. Kluwer Academic, New York, pp 95–104
Sowunmi MA (2004) Aspects of Nigerian coastal vegetation in the Holocene: some recent insights. In: Battarbee RW, Gasse F, Sticklay CF (eds) Past climate variability through Europe and Africa. Springer, Dordrecht, The Netherlands, pp 199–218
Sowunmi MA, Awosina EO (1991) Pollen analyses of Kariya Wuro Rock Shelter, Bauchi State. Nigerian Field 56:163–170

Switsur R, Otto T, Allsworth-Jones P (1994) New dating evidence and wood identification of wood charcoal from Kariya Wuro Rock shelter, Bauchi State. The Nigerian Field 59:135–145

Tubosun BJ (1995) Geo-archaeological investigations in Adikpo area of Katsina-Ala River basin, Benue State, Nigeria. Unpublished PhD Dissertation, University of Ibadan

Umeji OP, Ibeanu AM, Agwu COC (2012) Holocene human occupation of the eastern Nigerian scarp lands: an impact assessment study. Quatern Int 4:2–13

van Geel B, Gelorini V, Lyaruu A et al (2011) Diversity and ecology of tropical African fungal spores from a 25,000-year palaeoenvironmental record in southeastern Kenya. Rev Palaeobot and Palyno 164:174–190

Wetterstrom W (1993) Foraging and farming in Egypt: the transition from hunting and gathering to horticulture in the Nile valley. In: Shaw T, Sinclair P, Andah B, Okpoko A (eds) The archaeology of Africa: food, metals and towns. Routledge, London, pp 165–226

Willett F (1962) The microlithic industry from Old Oyo, western Nigeria. Actes du Ive Congrès Panafrician de Préhistoire et de l'étude du Quaternaire 3:261–271

Williamson K (1993) Linguistic evidence for the use of some tree and tuber food plants in southern Nigeria. In: Shaw T, Sinclai P, Anda B, Okpoko A (eds) The archaeology of Africa: food, metals and towns. Routledge, London, pp 139–153

Yates A, Smith AM, Parr J et al (2014) AMS dating of ancient plant residues from experimental stone tools: a pilot study. J Archaeol Sci 49:595–602

Eyo EO (1972) Rop rock shelter excavations 1964. W Afr J Archaeol 2:13–16

Zida PE, Sereme P, Leth V, Sankara P, Somda I, Neya A (2008) Importance of seed-borne fungi of *Sorghum* and pearl millet in Burkina Faso and their control using plants extracts. Pak J Biol Sci 11:321–331

Cottoning on to Cotton (*Gossypium* spp.) in Arabia and Africa During Antiquity

Charlène Bouchaud[1(✉)], Alan Clapham[2], Claire Newton[3], Gaëlle Tallet[4], and Ursula Thanheiser[5]

[1] AASPE-UMR 7209, CNRS-MNHN, Universités Paris-Sorbonne, Paris, France
charlene.bouchaud@mnhn.fr
[2] Ely, Cambridgeshire, UK
a.j.clapham@hotmail.co.uk
[3] Laboratoire D'Archéologie et de Patrimoine, Université Du Québec à Rimouski, Rimouski, Canada
clairennewton@gmail.com
[4] CRIHAM-EA 4270, University of Limoges, Limoges, France
gaelle.tallet@gmail.com
[5] Vienna Institute for Archaeological Science, University of Vienna, Vienna, Austria
ursula.thanheiser@univie.ac.at

Abstract. The occurrences of cotton in texts and in the archaeological record (seeds, fibres and textiles) demonstrate the emergence of cotton production centres in north-eastern Africa and western Arabia during the 1st–4th centuries AD, which is concurrent with an increase of cotton trade. These finds could correspond to any of the two Old World domestic cotton species: *Gossypium arboreum* L., probably domesticated in the Indus valley and traded since the 3rd millennium BC, or *Gossypium herbaceum* L., an African species about which very little is known, beside its presence in Nubia during Antiquity. Our paper reviews the archaeobotanical, textile and textual data from north-eastern Africa and western Arabia, with specific attention to several sites located in Central Sudan (Muweis), Lower Nubia (Qasr Ibrim), western Egypt (Kellis, Amheida) and north-western Arabia (Madâ'in Sâlih/Hegra). The intention of this review is to a) document how cotton production was integrated into agrarian and trade economies and b) examine current hypotheses regarding the diachronic distribution of the two species. The results highlight the importance of cotton in different agrosystems from the 1st–2nd centuries AD. In Central Sudan, Nubia and Dakhleh oasis, cotton cultivation appeared together with other new tropical/sub-tropical crops, such as sorghum (*Sorghum bicolor*) and pearl millet (*Pennisetum glaucum* subsp. *glaucum*). This was not the case in north-western Arabia. It seems that cotton production occurred at first as small-scale experiments before scaling up during the 3rd century AD, in conjunction with the spread of the water-wheel in the Nile valley. Cotton in Nubia, and possibly in other neighbouring areas, probably belonged to the African species *G. herbaceum*, which was in all likelihood domesticated in southern regions, perhaps

A. M. Mercuri et al. (Eds.): *Plants and People in the African Past*,
https://doi.org/10.1007/978-3-319-89839-1_18

Ethiopia. We suggest that the increase of exchanges across the Indian Ocean during Antiquity created a favourable context for the emergence of cotton production and its relative expansion before the Islamic period.

Keywords: Antiquity · Arabia · Cotton · *Gossypium arboreum/herbaceum* Northeastern Africa

Introduction

Two main hypotheses concerning cotton distribution in the Old World prevail to date. The first is that commercial expansion of cotton production and textile in south-west Asia, northern Africa and the Mediterranean basin was concomitant with the Arab conquest. The second suggests that until the 19th century and the spread of the American cotton, cotton textiles mainly came from India where they had been manufactured and transformed for many years (Mazzaoui 1981; Watson 1983; Wild 1997; Lombard 2002; Bulliet 2009). Recent archaeological discoveries of cotton seeds and textile are significantly modifying our understanding, demonstrating the existence of cotton production centres in north-western Arabia and north-eastern Africa already at the turn of the 1st–3rd centuries AD (Clapham and Rowley-Conwy 2009; Bouchaud et al. 2011; Tallet et al. 2012; Fuller 2014). Some of these data include cotton plants domesticated in Africa (Palmer et al. 2012). The present collaborative study is based on archaeobotanical and textual corpuses, and to a lesser extent, on textile remains. We aim to review the diachronic distribution of cotton in western Arabia and north-eastern Africa during the first centuries AD. In addition we investigate the impact of cotton production on agrarian dynamics by examining five archaeological sites located in Central Sudan (Muweis), Lower Nubia (Qasr Ibrim), western Egypt (Kellis, Amheida) and north-western Arabia (Madâ'in Sâlih).

The paper begins with a presentation of biological aspects of cotton and presents the methodological framework of the study, followed by a description of archaeobotanical results from the five archaeological sites mentioned above. We then review general results on seed, textile and text discoveries of cotton in western Asia and north-eastern Africa. These data allow us to address several issues concerning the diffusion of cotton products and its cultivation in these regions, including management practices, the significance of plant macroremains of cotton in archaeological contexts, the biological origin of cotton in Africa and economic and social aspects associated with the spread of cotton cultivation.

Botanical and Genetic Background

Cotton belongs to the Malvaceae family and the genus *Gossypium*, which comprises approximately 45 herbaceous and woody perennial wild species growing at tropical and sub-tropical latitudes (Wendel and Cronn 2003; Page et al. 2013). The fruit of cotton is a rigid capsule that opens at maturity in form of a boll composed of oleaginous seeds surrounded by fine fibres used for textile production. The bolls are divided into three to five loculi, each containing six to nine seeds. Two kinds of epidermal hairs (trichomes) cover the seed coat. The longest constitute the lint that can be easily pulled off at maturity to be spun and woven. The shortest trichomes or "fuzz" (2–7 mm long) are more difficult to remove and are used today as cotton wool (Reis et al. 2006:47).

Cotton relies for growth and maturation on hot and humid short days. Today, it is grown between latitudes of 37° north (45° north in China) and 30° south in temperate, subtropical and tropical regions. The cotton plant is a perennial small tree but has been progressively domesticated and selected to be grown as a pseudo-annual shrub. Present-day varieties grow 25 cm to over 2 m high, depending on how they are cultivated. Cotton still grows as perennial bushes only in a few regions, including South America (Peru, Brazil), Africa (Ethiopia), Central Asia and Northern India. The basic conditions required for its successful production include a long frost-free period, a temperature range of 18°–32 °C and a minimum of 500 mm of water over the long growing cycle, which lasts 125–225 days depending on varieties and local climate. Whereas winter crops, such as barley and wheat, are sown in the autumn with lengthening days after the winter solstice leading to a spring harvest, cotton requires shortening day lengths after the mid-summer solstice to stimulate flowering. However, most modern cultivars are day-length neutral. In the northern hemisphere, sowing occurs in March to April and harvest lasts from the end of July to December–January. Cotton exhibits a certain degree of tolerance to salt and drought (Stephens 1976; Smith and Cothren 1999; Chaudhry and Guitchounts 2003; Reis et al. 2006:49).

Four different species of cotton were independently domesticated, two of the tetraploid AD-group in the New World and two others of the diploid A-group in the Old World (Paterson et al. 2012). Of the two species from America, *Gossypium hirsutum* L. was domesticated in Central America and *G. barbadense* L. in Peru (Smith and Stephens 1971; Stephens 1975; Dillehay et al. 2007). *Gossypium hirsutum* is the most frequently cultivated cotton plant in the world today and has often replaced the other species. The two Old World species, *G. herbaceum* L. and *G. arboreum* L., are more resistant to water stress and pests and are adapted to shallow and sandy soils, but have a shorter staple length and are now cultivated less than the American species. The staple length of cotton fibres varies from 10 to 60 mm according to the four different species but the fibres of *G. arboreum* and *G. herbaceum* do not exceed 20 mm on average (Benedict et al. 1999). The Old World cottons are still exploited by small farmers in West Africa, western Egypt, India, Bangladesh, Pakistan and Iran (Brubaker et al. 1999; Wendel et al. 1999; Boulos 2000:111; Eyhorn et al. 2005).

There are still important issues concerning the domestication processes of *G. arboreum* and *G. herbaceum*. Genetic and morphological arguments indicate that *G. herbaceum* was derived from its wild ancestor *G. herbaceum* var. *africanum* (G. Watt) Vollesen, which grows naturally in southern Africa today. *Gossypium arboreum* has a primitive perennial form *G. arboreum* race '*indicum*' growing in western India, but its wild ancestor is unknown (Stanton et al. 1994; Kulkarni et al. 2009). Archaeological evidence raises contrasting and supplementary views. First, the extant distribution of *G. herbaceum* var. *africanum* is much more southern than the most ancient and reliable archaeological evidence of the use of cotton, known to date to the 1st century BC–1st

century AD in Lower Nubia (Clapham and Rowley-Conwy 2009). Secondly, the discovery of textiles and seeds dating to the 6th–4th millennium BC point to a possible centre of domestication of *G. arboreum* in Pakistan (Moulhérat et al. 2002).

Nature of the Available Data

This review is based on archaeological plant macroremains, textiles, and epigraphic evidence. The possibility of recovering each type of evidence is the result of several complex taphonomic, processing (*chaîne opératoire*) and preservation processes. As a result the information obtained may not be consistent or directly comparable to one type of evidence to another.

Production and Preservation of the Archaeological and Textual Data

The harvest of cotton bolls is traditionally done by hand, leaving behind most of the bracts that surround the bolls as well as stems and leaves, which can then be harvested separately and used as fuel (Abasaeed 1992; Gomes et al. 1997) or fodder (Suttie 2004, 168). To release the fibres from the seeds, the ginning process was carried out until modern times by hand or by rolling a cylinder or wooden stick on the bolls laid on a smooth flat stone. Within the domestic sphere, women generally undertook this task (Nicholson 1960:14; Leroi-Gourhan 1971:247). The carding process, using a comb or a bow, allows the fibres to be fluffed. This last step is however not compulsory. Until recently in Sudan, the fibre was spun immediately after gently removing the lint from the seed in order to keep the fibres in their original order, lying as parallel to each other as possible, so as to avoid the soft fluffs to be mixed up and to make the fibres more durable, soft and warm (Crowfoot 1924; Nicholson 1960:14). In Egypt and Nubia, modern and archaeological examples show the occurrence of different spinning methods with a hand-held or suspended spindle. The spindle whorls are made of unfired clay, stone, wood, bone, ceramic or potsherds, and placed at the top of the spindle, often secured by the insertion of an iron hook between the shaft and the whorl's central perforation (Crowfoot 1931; Yvanez 2016). After spinning, cotton threads are then ready to be woven using a horizontal loom or a warp-weighted loom (Leroi-Gourhan 1971).

The discovery of cotton seeds, spun or unspun fibres and textiles in archaeological contexts is thus a direct link to specific processing steps of the *chaîne opératoire* (Leroi-Gorhan 1971) as well as evidence of household and commercial activities (through the transport of raw or manufactured cotton). However, their preservation heavily depends on taphonomic processes, as organic materials such as seeds, fibres, papyri, wooden books, etc. are generally destroyed through biological decay. Thankfully, some regions of north-eastern Africa and the Arabian Peninsula offer a hyper-arid climate and/or contexts without much oxygen, such as tombs, which permit a partial preservation of organic remains. In other, more "humid" areas, only charred material tends to be preserved. This study also takes into consideration textual sources that are mostly Greek texts found in Egypt in the form of ostraca, papyri and wooden books. In Sudan and Nubia, documents written in the Meroitic script could possibly reveal new data, but their translation still remains difficult and requires further study and analysis (Rilly 2007).

Interpretive Potential

Plant remains, textiles and texts can provide details on the location of production centres, distribution and trade activities, original uses, and the taxonomic status of cotton.

Textiles are often found in tombs of high-ranking individuals, providing selective information on elite production, and a few, more recently discovered specimens have arisen from household contexts. Textiles were essential commodities and often the subject of exchange and trade. As such, their presence in a context is not necessarily directly linked to their place of production. On the contrary, spinning and weaving techniques are strongly embedded in social identity and can potentially be good spatiotemporal markers (Wild 1997). The yarn itself for example, the direction of the fibre twist, which depends on the rotation direction of the spindle, varies from one region to another, based on long-lived cultural traditions (Barber 1993:39–50; Wilkinson 2014:249–50). The substantial textile corpus from the Nile valley shows that S-spun yarns—created by an anti-clockwise rotation of the spindle—are dominant in both Egypt and Nubia (Crowfoot 1931; Wild 1997). This specificity seems relates to the pharaonic tradition of flax-spinning, for this fibre naturally curls to the left (S) when dampened (Barber 1993). Z-spun threads—clockwise rotation of the spindle—are however dominant in India and Central Asia (Wild and Wild 2014a; Wilkinson 2014). The spinning tradition of the ancient Arabian Peninsula is not well defined to date. When they are sufficiently well preserved, the type of fabrics, weaves, patterns, and colours are all very good indicators of the origin of textiles.

Charred and desiccated non-manufactured elements, such as bolls, isolated fibres, wads of raw cotton, empty capsules (i.e., pericarp), whole and broken seeds likely lie within agricultural production centres as they result from the cleaning of the cotton plant prior to any use of the fibres. However, we need to consider that the harvests of different places (fields, owners, villages) could have been centralised in domestic or in specialised (urban) workshops for further processing. In this way, remains of cotton seeds would indicate that cotton likely grew in the neighbouring villages, but perhaps not around the town itself (see section IX.B.). Raw fibres with seeds still attached could also circulate by long-distance trade (see section IV.E.). Identifying local production only from carpological clues such as seeds thus remains problematic, except when other harvesting by-products are also found, such as empty capsules (Clapham and Rowley-Conwy 2009). Whether it be seeds or textiles, the strongest element indicating local production is the statistical domination of cotton in both the archaeobotanical samples and textile assemblage of the site under study. Cotton wood, which would be relevant to attest local production, has never been identified to date (Bouchaud et al. 2011), with the exception of twigs found at Douch, Egypt (Newton unpublished data; Tallet et al. 2012:126). Finally, the analysis of textile implements, especially spindle whorls, could also benefit our understanding of fibre and thread production. The weight of the whorl is particularly interesting, as different fibres require different force to be effectively spun: light to medium weight (10–30 g) for short fibres such as cotton and medium length wool, and heavier tools for long length wool and linen (see Barber 1993; Yvanez 2016:163).

Seeds and textiles provide the same taxonomic signal at the genus level (*Gossypium* sp.). In other words, there is no possibility to distinguish *G. herbaceum* from *G. arboreum* on a morphological basis. Several identification proposals have already been made, notably based on the thread diameter (Chowdhury and Buth 1971; Coombs et al. 2002), but the results are ambiguous and should be reassessed through new morphometric studies. On well-preserved material, archaeogenomic analyses also have provided interesting perspectives and show promise (Palmer et al. 2012).

The majority of texts dealing with cotton are private letters mentioning garments or threads, official agricultural surveys citing cotton plants, military accounts and orders of delivery, as well as private book accounts and receipts listing quantities of cotton. In the Egyptian written corpus, cotton is generally designated by the Greek term ἐρεόξυλον (*ereoxylon*, literally the wool tree or the wool wood). This term is virtually absent from the other Greek sources. During the 4th century BC, Theophrastus for instance uses the periphrasis δένδρα ἐριοφόρα (*dendra eriophora*, the wool-bearing tree) to designate the cotton tree in Bahrein[1] (*Historia Plantarum* 4.7.7). The anonymous author of the *Periplus Maris Erythraei*, a Greek handbook compiled by an Alexandrian sailor as a guide to merchants engaged in long-distance trade in the Red Sea and Indian Ocean, uses a transcription of the Sanskrit term *kârpâsa* into Greek κάρπασο (*karpaso*), by which he specifically designates the material Indian clothes are made of (Casson 1989). Several Latin texts refer to *gossypinus* and *xylon*, showing that the plant was known, at least to initiated persons, throughout the Mediterranean basin. The texts give information about the location of agricultural and craft production centres, the various uses and shape of the plant, as well as trade. Despite their scarcity and their focus on Indian and Egyptian cultures, literary accounts provide detailed historical descriptions that usefully complete archaeobotanical or textile studies.

Archaeobotanical Finds in Nubia, Egypt and Saudi Arabia

At least fifteen archaeological settlements located in north-eastern Africa and western Arabia have produced cotton archaeobotanical remains more or less securely dated to the Antique and Late Antique periods (Fig. 1, Table 1). We exclude from this list the

[1] For a complete view of cotton production in Bahrein during Achaemenid period, see Bouchaud et al. (2011); Tengberg and Lombard (2001).

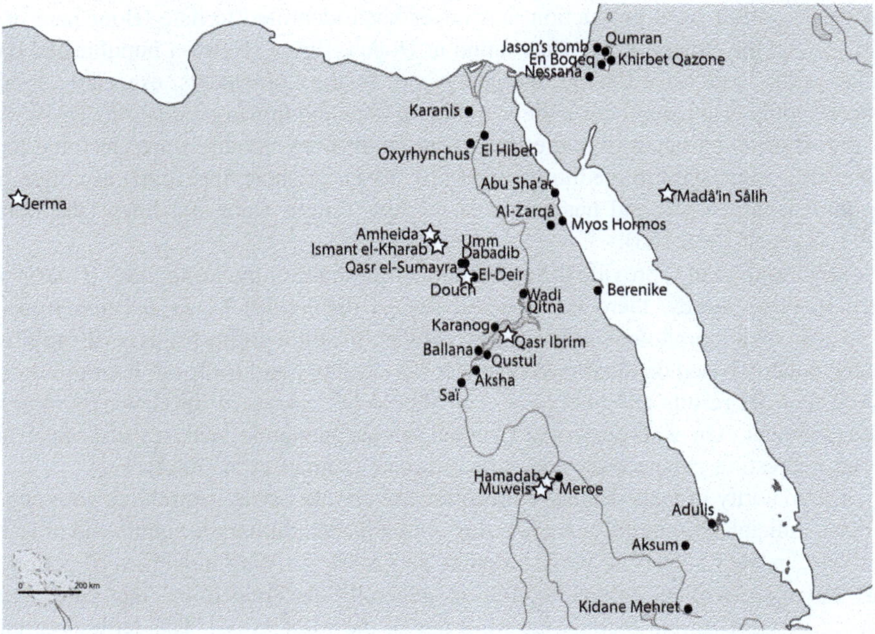

Fig. 1. Mapping the evidence for cotton in Northeastern Africa, Western Arabia and the Near East, 1st century BC—7th century AD. For more details, see Table 1. The names refer to either ancient or modern names. Stars designate the sites for which cotton production is assumed, either locally or in the region.

unique discovery of desiccated cotton seeds at Afyeh (Nubia) insecurely dated to the 3rd millennium BC because of the context of the finds (in goat faeces) and the absence of direct radiocarbon dating (Chowdhury and Buth 1971; see criticisms in Clapham and Rowley-Conwy 2009, 249; Fuller 2015:14). Five of these sites, Qasr Ibrim (Lower Nubia, Egypt), Muweis (Central Sudan), Madâ'in Sâlih (north-western Arabia), Kellis and Amheida (Dakhleh Oasis in the Western Desert of Egypt) and their cotton discoveries are detailed below to consider the representation of cotton items and their potential significance in terms of use and production in various environments. The quantitative description of the cotton finds is based on the occurrence of cotton remains, i.e., the number of samples where the taxon is recognized, and in some cases on the proportion of each type of remain (raw number of remains). For two sites, the occurrence of cotton is compared with that of other economic plants. Qualitative descriptions are used when the quantitative ones are not yet available. A list of other significant cotton finds is also added to these detailed descriptions.

Table 1. Archaeobotanical, textile and textual evidence for cotton in north-eastern Africa, western Arabia and the southern Near East, 1st c. BC—7th c. AD

Region	Site	Present Country	Context	Type of cotton finds	Preservation of archaeological material	Chronology[a]	Reference
Central Sudan	Muweis	Sudan	Domestic (town/village)	Seeds	Charred	2nd–4th (?) c.	Newton and Bouchaud, unpublished
Central Sudan	Hamadab	Sudan	Domestic (town/village)	Seeds	Charred	1st–4th (?) c.	(Fuller 2014)
Central Sudan	Meroe	Sudan	Funerary	Textile	Charred and desiccated	1st–4th c.	(Massey 1923; Griffith and Crowfoot 1934; Dunham 1963: 143, 146, 227)
Central Sudan		Sudan		Text (Royal inscription)		mid 4th c.	Ezana inscription (Littmann 1913, see comment in Zeitschrift der Deutschen Morgenländischen Gesellschaft 67:701)
Upper Nubia	Saï	Sudan	Funerary	Textile (S[b])	Desiccated	1st–4th c.	(Yvanez 2012)
Upper Egypt/ Sudan				Text (plants)		1st c.	Pliny the Elder, *Natural History*, 19.2.14, 12.21.38–22.39
Egypt?				Text (plants)		2nd c.	Julius Pollux, *Onomasticon* 6.75
Ethiopia	Adulis	Erythraea	Harbour	Text (textile imports)		1st c.	*PME*[c] 6:3.1-3 (Casson 1989)
Ethiopia	Aksum	Ethiopia	Domestic (town)	Seeds	Charred	6th–early 7th c.	(Boardman 1999; Phillipson 2000:363–69)
Ethiopia	Kidane Mehret	Ethiopia	Funerary	Textile	Desiccated	6th c.	Phillipson 2012:179
Somalia	"far-side" ports	Somalia	Harbours	Text (textile imports)		1st c.	*PME* 14:5.10-13 (Casson 1989)

(*continued*)

Table 1. (*continued*)

Region	Site	Present Country	Context	Type of cotton finds	Preservation of archaeological material	Chronology[a]	Reference
Lower Nubia	Qasr Ibrim	Egypt	Domestic/Handcraft district (fortress)	Seeds, fibres, bolls, empty capsules	Desiccated	end 1st c. BC–19th c. AD	(Rowley-Conwy 1989; Clapham and Rowley-Conwy 2007, 2009; Palmer et al. 2012)
Lower Nubia	Qasr Ibrim	Egypt	Domestic/Handcraft district (fortress)	Textiles (S)	Desiccated	end 1st c. BC–4th c. AD	(Plumley et al. 1977:45–47; Anderson and Adams 1979:39–41; Wild and Wild 2008, 2009, 2014b)
Lower Nubia	Aksha	Soudan	Funerary	Textile	Desiccated	1st c.	(Vila 1967)
Lower Nubia	Karanog	Egypt	Funerary	Textile (S)	Desiccated	2nd–3rd c.	(Griffith and Crowfoot 1934)
Lower Nubia	Ballana	Egypt	Funerary	Textile (S)	Desiccated	2nd–5th c.	(Mayer-Thurman and Williams 1979)
Lower Nubia	Qustul	Egypt	Funerary	Textile (S)	Desiccated	2nd–5th c.	(Mayer-Thurman and Williams 1979)
Lower Nubia	Wadi Qitna (Kalabsha)	Egypt	Funerary	Textile (S, Z)	Desiccated	3rd–5th c.	(Strouhal 1984)
Western Desert, Egypt	Ismant el-Kharab (Kellis)	Egypt, Dakhleh Oasis	Domestic (town/village)	Seeds, bolls, empty capsules	Desiccated	1st–4th c.	(Thanheiser 2002)
Western Desert, Egypt	Ismant el-Kharab (Kellis)	Egypt, Dakhleh Oasis	Domestic (town/village)	Textiles (S, Z)	Desiccated	2nd–4th c.	(Livingstone 2009, 2015)

(*continued*)

Table 1. (*continued*)

Region	Site	Present Country	Context	Type of cotton finds	Preservation of archaeological material	Chronology[a]	Reference
Western Desert, Egypt	Ismant el-Kharab (Kellis)	Egypt, Dakhleh Oasis	Domestic (town/village)	Text (agricultural accounts of raw cotton/fibre)		360–370	*KAB*[d] 547, 556, 558, 720, 1484 = *P. Kell*. 4.96 (Bagnall 1997)
Western Desert, Egypt	Ismant el-Kharab (Kellis)	Egypt, Dakhleh Oasis	Domestic (town/village)	Text (harvest)		237–257	*O.Kellis* 68, *O.Kellis* 69 (Worp and Hope 2004)
Western Desert, Egypt	Ismant el-Kharab (Kellis)	Egypt, Dakhleh Oasis	Domestic (town/village)	Text (list of money arrears)		4th c.	*P.Kell. 1.61* (Worp 1995)
Western Desert, Egypt	Amheida (Trimithis)	Egypt, Dakhleh Oasis	Domestic (town/village)	Seeds	Charred and desiccated	3rd c.	(Thanheiser and Walter 2015)
Western Desert, Egypt	Amheida (Trimithis)	Egypt, Dakhleh Oasis	Domestic (town/village)	Textile (S)	Desiccated	3rd c.	(Boozer 2015b)
Western Desert, Egypt	Amheida (Trimithis)	Egypt, Dakhleh Oasis	Domestic (town/village)	Text (account of raw cotton)		350–370	*O.Trim* 38, *O.Trim* 44 (Bagnall and Ruffini 2012)
Western Desert, Egypt	Douch (Kysis)	Egypt, Kharga Oasis	Temple (town/village)	Text (land registry)		164–165	*P.Iand* 7.142 (Wagner 1987; Bagnall 2008)
Western Desert, Egypt	Douch (Kysis)	Egypt, Kharga Oasis	Funerary	Seeds, leaves, twigs, capsules	Desiccated	4th c.?	(Newton unpublished data; Tallet et al. 2012:126)

(*continued*)

Table 1. (*continued*)

Region	Site	Present Country	Context	Type of cotton finds	Preservation of archaeological material	Chronology[a]	Reference
Western Desert, Egypt	Douch (Kysis)	Egypt, Kharga Oasis	Funerary	Textile	Desiccated	4th c.?	(Dunand et al. 1992:232; Tallet et al. 2012:126)
Western Desert, Egypt	Douch (Kysis)	Egypt, Kharga Oasis	Temple and fort	Text (account of raw cotton)		2nd half 4th c.	*O.Douch* 1.51 (Cuvigny and Wagner 1986), *O.Douch* 4.381 (Wagner 1999), *O.Douch* 5.537, *O.Douch* 5.634 (Wagner 2001)
Western Desert, Egypt	El-Deir	Egypt, Kharga Oasis	Temple (town/village)	Seeds, lint, capsules	Desiccated	3rd–4th c.?	(Tallet et al. 2012)
Western Desert, Egypt	El-Deir	Egypt, Kharga Oasis	Funerary, temple	Textiles (S, Z)	Desiccated	4th c.	Letellier-Willemin and Moulhérat 2006; Letellier-Willemin and Médard 2012; Tallet et al. 2012
Western Desert, Egypt	Qasr el-Sumayra	Egypt, Kharga Oasis	Domestic (town/village)	Seeds	Desiccated	1st–4th c.?	Clapham in press
Western Desert, Egypt	Ayn Umm Dabadib	Egypt, Kharga Oasis	Funerary	Textiles (S)	Desiccated	4th–5th c.	(Jones and Oldfield 2006)
Middle Egypt	Oxyrhynchus?	Egypt, Nile valley	Town/village	Text (private letter/textile)		2nd c.	*SB* 6.9025 (Bagnall 2008)
Middle Egypt	El-Hibeh	Egypt, Nile valley	Domestic?	Textiles (Z)	Desiccated	Roman?	(Ryesky 1984)

(*continued*)

Table 1. (*continued*)

Region	Site	Present Country	Context	Type of cotton finds	Preservation of archaeological material	Chronology[a]	Reference
Middle Egypt	Oxyrhynchus	Egypt, Nile valley	Town/village	Text (private letter/textile)		2nd c.	*SB* 6.9026 (Bagnall 2008)
Middle Egypt	Oxyrhynchus	Egypt, Nile valley	Town/village	Text (private letter/textile)		2nd–3rd c.	*P.Oxy* 59.3991 (Bagnall and Cribiore 2006)
Middle Egypt		Egypt		Text (account of textile)		3rd c.	*P.Lond.* 3.928
Fayum?	Karanis?	Egypt	Town/village?	Text (private letter/textile)		2nd c.	*P.Mich.* 8.500 (Winter and Youtie 1944; Bagnall 2008:24–25)
Fayum	Karanis	Egypt	Town/village	Textile	Desiccated	4–5th c.	Wilson (1933), 50 n°133
Libya	Jerma	Fezzan, Libya	Domestic (town/village)	Seeds	Charred	2nd–6th c.	(Pelling 2013)
Eastern Desert	Berenike	Egypt	Harbour	Textiles (S, Z)	Desiccated	1st–5th c.	(Wild 1997, 2013; Wild and Wild 2001, 2014a, 2014c)
Eastern Desert	Myos Hormos	Egypt	Harbour	Textile (Z)	Desiccated	late 1st-early 2nd c.	(Handley 2004; Whitewright 2007; Wild and Wild 2014c)
Eastern Desert	Al-Zarqâ (Maximianon)	Egypt	Domestic? Military fort	Seeds	Desiccated	1st–3rd c.	(Bouchaud et al. 2011:408; Newton unpublished)
Eastern Desert	Abu Sha'ar	Egypt	Harbour	Textiles (Z)	Desiccated	4th–5th c.	(Bender Jørgensen and Vogelsang-Eastwood 1991; Bender Jørgensen 2004, 2006)

(*continued*)

Table 1. (*continued*)

Region	Site	Present Country	Context	Type of cotton finds	Preservation of archaeological material	Chronology[a]	Reference
Western Arabia	Madâ'in Sâlih (Hegra)	Saudi Arabia	Domestic	Seeds	Charred	cal. 1st–4th c.	(Bouchaud et al. 2011; Bouchaud 2015)
Western Arabia	Madâ'in Sâlih (Hegra)	Saudi Arabia	Funerary	Textiles (Z)	Desiccated	1st–3rd c.	(Bouchaud et al. 2011)
Southern Arabia	Moscha Limen	Oman	Harbour	Text (textile import)		1st c.	*PME* 32:11.3 (Casson 1989)
Southern Arabia	Socotra	Yemen	Harbour	Text (textile import)		1st c.	*PME* 31:10.21-22 (Casson 1989)
Southern Near East	Jason's tomb	Israel	Funerary	Textile (netting, S?)	Desiccated	1st c.	(Rahmani 1967:93)
Southern Near East	Qumran	Israel	Cave	Textiles (Z)	Desiccated	1st c. BC–1st c. AD	(Müller et al. 2003, 2004)
Southern Near East	Khirbet Qazone	Jordan	Funerary	Textiles	Desiccated	1st–3rd c.	(Granger-Taylor 2000)
Southern Near East	Nessana	Israel	Funerary	Textiles (S, Z)	Desiccated	3rd c.	(Bellinger 1962:92)
Southern Near East	En Boqeq	Israel	Domestic? *Castellum*	Textiles (S)	Desiccated	7th c.	(Sheffer and Tidhar 1991:22–23)
Southern Near East	Tadmor (Palmyra)	Syria	Funerary	Textiles (Z)	Desiccated	1st–2nd c.	(Schmidt-Colinet 1995; Stauffer 2000)

[a]When there is no indication, the dates are AD.
[b]S or Z indicate the direction of the fibre twist
[c]*PME = Periplus Marys Erythreae*
[d]*Kellis Agricultural Account Book*

Qasr Ibrim (Lower Nubia)

Qasr Ibrim, Roman Primis, is located in Lower Nubia about 185 km south of the first cataract of the Nile. Before the building of the Aswan High Dam and the creation of Lake Nasser, the site was a promontory fort at the top of the eastern flank of the steep-sided Nile valley. It was excavated more or less continuously from 1963 to 2006 (Rose 1996, 2011, 2013). Qasr Ibrim was a major settlement and cult centre within the area controlled by the Napatan Empire during the 1st millennium BC. A Roman garrison occupied it from the end of the 1st century BC into the 1st century AD. After the Roman withdrawal in the early 1st century AD, the Meroitic state established control of Qasr Ibrim. With the collapse of the centralised Meroitic state by the mid-4th century, local polities developed in Lower Nubia, in which Qasr Ibrim continued to have a major role, and this continued after the conversion of Nubia to Christianity. Eventually Qasr Ibrim became the site of an Ottoman garrison, until it was abandoned in the early 19th century. The fortified and isolated character of the settlement and the presence of religious structures dating to almost all periods suggest a specialised urban character. The occupants may have been associated with the maintenance of state and religious institutions, and thus, in some sense, 'officials' rather than representative of a cross-section of society, the majority of whom would probably have lived in the valley (Rose and Edwards 1998; Edwards 2004; Rose 2011).

The arid environmental conditions have led to the impressive preservation of organic material. The study of the plant macroremains, mainly desiccated, was undertaken by Peter Rowley-Conwy (Rowley-Conwy 1989; Rowley-Conwy et al. 1999), then by Alan Clapham (Clapham and Rowley-Conwy 2007, 2009). The data so far highlight the changes in crops and agricultural regimes from the Napatan until the abandonment of the site. The earliest crop assemblage (Napatan) is similar to those found farther north in the Nile valley in Pharaonic Egypt, characterised by emmer wheat (*Triticum turgidum* L. subsp. *dicoccon* (Schrank) Thell), hulled barley (*Hordeum vulgare* L.), small amounts of foxtail millet (*Setaria italica* (L.) P. Beauv.) and flax (*Linum usitatissimum* L.). In the Roman period (25 BC-100 AD), a number of crops make their first appearance, albeit in a small way. These include hard wheat (*Triticum turgidum* L. subsp. *durum* (Desf.) Husn.), domesticated sorghum (*Sorghum bicolor* (L.) Moench subsp. *bicolor*), pearl millet (*Pennisetum glaucum* (L.) R. Br. subsp. *glaucum*), olive (*Olea europaea* L.), and grape (*Vitis vinifera* L.).

Cotton appears for the first time as isolated desiccated seeds and desiccated fibres with seeds (Table 2). A cotton thread attached to a heddle leash found dating to the Roman occupation produced a calibrated radiocarbon date of 100 BC–AD 70 (2 sigmas, OxA 14813) and coincides with the short term Roman occupation of the site. Two radiocarbon dates from seeds ranging from 40 BC to AD 90 (2 sigma, OxA14810), 20 BC–AD 30 (2 sigma, OxA14812) may suggest an introduction of cotton cultivation to the site during or slightly before the Roman occupation. From the Meroitic period (AD 100–350) until the 19th century, sorghum dominates the cereal remains, although barley and pearl millet remain significant. Bread wheat (*Triticum aestivum* L. subsp. *aestivum*) and lablab (*Lablab purpureus* (L.) Sweet), a tropical legume crop, are introduced during the post-Meroitic period (350–550 AD) (for more details about the

Table 2. Types and number of cotton remains from each period of occupation at Qasr Ibrim, Egyptian Nubia (Data from quantified samples only)

Period	Roman	Meroitic	Post meroitic	Christian	Islamic	**Totals**
Date range	c.25BC–AD100	AD100–350	AD350–550	AD550–1400	AD1400–1812	
Number of samples	**13**	**39**	**41**	**17**	**2**	**112**
Remain type						
Desiccated						
Complete bolls	0	1	0	0	0	1
Fibres with seeds	4	9	14	1	1	29
Seed free fibres	0	5	7	0	0	12
Pericarp (=empty capsules)	0	2	6	0	0	8
Pericarp fragments	8	75	68	41	2	194
Seeds	67	322	393	10	1	793
Seed fragments	707	569+	284	80	19	1090+
Cotyledons	4	32	9	0	0	
Total desiccated	**790**	**1016+**	**781**	**132**	**23**	
Charred						
Seeds	0	8	29	0	0	37
Seed fragments	1	131	0	1	0	133
Total charred	**1**	**139**	**29**	**1**		

archaeobotanical finds, see Clapham and Rowley-Conwy (2007). Cotton finds are also very well attested during the Meroitic and Post-Meroitic periods.

Cotton plant remains are mainly present in the form of desiccated and charred fragmented and whole seeds (Table 2). Most of them have no long fibres and correspond, at least for the desiccated ones, to a post-ginning phase. We cannot exclude that the charring process has destroyed the original lint. Some of the desiccated cotton seeds have long fibre remains still attached to the seed coat and may indicate a pre-ginning stage. Plant remains, rarely encountered in archaeological contexts, are represented among the desiccated assemblage in the form of complete bolls (including immature bolls), empty capsules (pericarp) and free long fibres (Fig. 2).

In addition to archaeobotanical study, the remarkable preservation of the plant remains has led to their examination through various biomolecular analyses in order to determine the history and origins of certain crops such as sorghum (Rowley-Conwy et al. 1999; Shaw et al. 2001) and six-row barley (Palmer et al. 2009; Smith et al. 2014). Ancient DNA analyses (transposable elements composition) conducted on cotton seeds stratigraphically dated to the 4th century AD allowed to determine the species *G. herbaceum*, of African origin (Palmer et al. 2012).

Fig. 2. Desiccated complete boll/capsule of cotton from Qasr Ibrim (Lower Nubia), including perianth remains. Scale: 5 mm

Muweis (Central Sudan)

Muweis is a small town located on the right bank of the Nile, 170 km north of Khartoum, in Central Sudan (western Butana). The city belonged to the Island of Meroe (a term coined by Strabo), a semi-desert landscape between the Nile, the Atbara and the Blue Nile rivers that was the heartland of the Meroitic kingdom (300 BC-AD 350) (Baud 2010b). Meroe, the capital, lies 50 km to the north of Muweis. Excavations directed by M. Baud (2006–2012) and M. Millet (2013-present) have revealed the existence of institutional, religious, domestic and industrial activities, each located in different areas of the town (Baud 2010a, 2014; Millet 2013). The region is subject to limited monsoon rains, so climatic conditions are more humid than in northern Nubia (present mean annual rainfall at Khartoum = 162 mm, Walsh 1991:25) and organic materials not as well-preserved as in Qasr Ibrim. The seeds and fruit elements considered in this paper come from occupation layers relating to potential domestic activities (sector 3) occurring during the Late Meroitic period (2nd-mid 4th centuries AD). The plant remains, recovered by dry-sieving (2, 1 and 0.5 mm meshes), are all charred. The study is still in progress and only preliminary results, focused on economic plants from seventeen samples (73 l of sediment) can be presented here (Fig. 3).

Sorghum and pearl millet grains dominate the archaeobotanical assemblage of economic plants. Winter cereals, hulled barley and emmer, as well as pulses (lens, *Lens culinaris* Medik.; cowpea, *Vigna unguiculata* (L.) Walp.) appear in less than half of the studied contexts. Fruit trees mainly comprise the Christ's thorn jujube tree (*Ziziphus spina-christi* (L.) Desf.) and, to a much lesser extent the desert date (*Balanites*

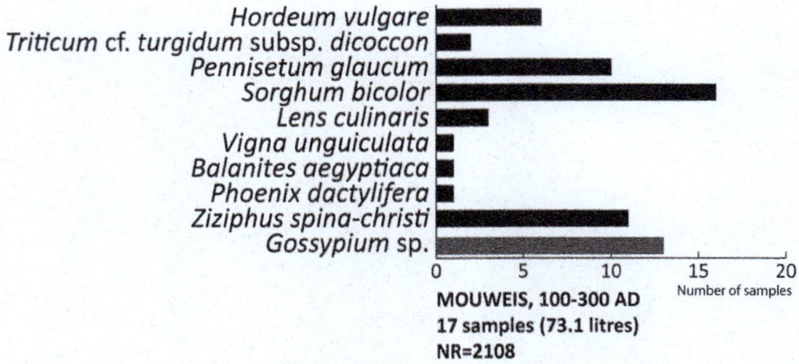

Fig. 3. Preliminary results of the analysis of charred seeds and fruits from Muweis (Central Sudan). Occurrence of the main economic taxa on 17 samples of the Late Meroitic period (2nd–4th c. AD), sector 3.

aegyptiaca Delile) and the date palm (*Phoenix dactylifera* L.). Cotton is attested in about 75% of the samples. The direct radiocarbon date of two cotton seeds show that they belong, at the earliest, to the middle of the 2nd century AD (Table 3). Two deeper layers, still dating to the Late Meroitic period and also containing cotton seeds, have been radiocarbon dated using sorghum grains. The preliminary results indicate that cotton is present at Muweis during the 1st–2nd century AD. Currently, no earlier large archaeobotanical assemblage has been investigated from this site. Charred cotton remains (Fig. 4) are mainly represented by fragmented half-seeds, seed coat and cotyledons (93% of the total number of cotton remains), as well as some whole seeds (7%) (Table 4), rarely covered with short fibre remains.

Table 3. Radiocarbon dating of plant materials from Muweis (Central Sudan), sector 3

Lab n°	US n°	Material	Stratigraphic periods	^{14}C Age (BP)	95.4%[a]	
					From	To
UBA-30480	3016	*Gossypium* sp., seed	Late Meroitic	1720 ± 50	145	423
UBA-30482	3034	*Gossypium* sp., seed	Late Meroitic	1755 ± 40	142	388
UBA-30483	3044	*Sorghum bicolor*, seed	Late Meroitic	1835± 40	77	319
UBA-30485	3071	*Sorghum bicolor*, seed	Late Meroitic	1920 ± 40	1	214

[a]Calibration IntCal.13 (Reimer et al. 2013)

Fig. 4. Cotton finds from Muweis (Central Sudan). Charred whole cotton seed (left), broken cotton seeds (right).

Table 4. Detail of charred cotton remains from Muweis (Central Sudan), sector 3, Late Meroitic

	1st–3rd centuries AD
Number of samples containing cotton	13
Gossypium sp., whole seed	16 (7%)
Gossypium sp., seed fragment	209 (93%)

Madâ'in Sâlih (North-Western Arabian Peninsula)

The city of Madâ'in Sâlih, ancient Hegra, is located in north-western Saudi Arabia, on a vast sandy plain surrounded by mountain massifs (Nehmé et al. 2006; Courbon 2008). Occupied since the 4th–3rd century BC (Durand and Gerber 2014), it was densely inhabited from at least the 2nd–1st centuries BC to the mid-4th century AD. From the 1st century BC to the 1st century AD, the town was at the southern limit of the Nabataean kingdom, on the route between Petra—the Nabataean capital—and *Arabia Felix*, before being integrated into the Roman Province Arabia (*Arabia Petraea*) after the annexation of the Nabataean kingdom in 106 AD. The work of the Franco-Saudi project, directed by L. Nehmé, D. al-Tahli and F. Villeneuve (2008-present) has revealed the existence of a residential area that was continuously inhabited from the earliest period, 4th–3rd century BC, to Late Antiquity (Nehmé et al. 2006, 2011). Monumental tombs carved into sandstone cliffs, mainly used during the 1st–3rd centuries AD, surround the residential area (Nehmé 2015). Low rainfall in the region, ranging between 50 and 100 mm per year, is compensated by the important presence of groundwater. It was largely exploited through wells, likely built during the Nabataean period, which were used to irrigate the cultivation area located to the north of the ancient city (Nehmé et al. 2006:59; Courbon 2008:49). The site is regularly subject to winter flash floods, which has prevented the preservation of organic material, with the notable exception of several monumental tombs where textile and leather pieces, as well as plant jewellery have been recovered (Bouchaud et al. 2015).

The archaeobotanical samples were taken from domestic assemblages (dumps, occupation layers, fireplaces) excavated inside the residential area. All the material is charred and was obtained through sieving and flotation using 2 and 0.5 mm meshes. The earliest layers (4th/3rd to mid. 1st centuries BC) are still under study. The preliminary observation of these early samples shows the presence of date palm, indeterminate cereals and fig remains. The samples from the following periods (from mid. 1st century BC to mid. 4th century AD) have been more thoroughly investigated (64 samples representing 775 l of sediment) and demonstrate the existence of more diversified crops. During the Classical Nabataean (mid. 1st century BC–early 2nd century AD), hard wheat and hulled barley of 2-row and 6-row varieties (*Hordeum vulgare* subsp. *distichon* (L.) Körn/*H. vulgare* L. subsp. *vulgare*) are well attested. These two cereals, like the date palm, remain dominant until the end of the site's occupation. The Nabataeo-Roman occupation (early 2nd-end. 3rd centuries AD) is characterised by the appearance of Mediterranean fruit trees, olive, and grapevine. At that time, cotton appears in one third of the studied samples. The end of the occupation (first half of the 4th century AD) presents a similar taxonomical diversity. Cotton seeds are present in more than two-thirds of the samples (Fig. 5). Most cotton remains are whole seeds. The radiocarbon dating of two of specimens (Table 5) confirms that cotton was present during the Nabataeo-Roman period, although we cannot exclude

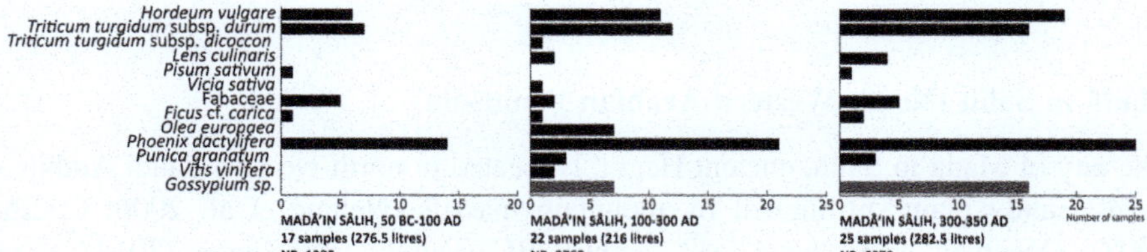

Fig. 5. Results of the analysis of charred seeds and fruits from Madâ'in Sâlih (Saudi Arabia). Occurrence of the main economic taxa from the Classical Nabataean to the Late Antique periods, residential area.

Table 5. Radiocarbon dating of plant material found at Madâ'in Sâlih (Saudi Arabia)

Lab n°	US n°	Material	Stratigraphic periods	^{14}C Age (BP)	95.4%[a]	
					From	To
SacA 31284/ Lyon 9754	80124	*Gossypium* sp., seed	Nabataeo-Roman	1815 ± 30	126	322
SacA 31285 (Lyon-9755)	80106	*Gossypium* sp., seed	Nabataeo-Roman	1850 ± 30	85	235

[a]Calibration IntCal.13 (Reimer et al. 2013)

that it was present earlier (Bouchaud 2015). Fragmented seeds constitute 6% of the cotton assemblage for the 2nd–3rd centuries AD and 39% for the last occupation period of the site (Table 6).

Table 6. Detail of charred cotton remains from Madâ'in Sâlih, Saudi Arabia

	2nd–3rd centuries AD	First half 4th century AD
Number of samples containing cotton	7	16
Gossypium sp., whole seed	30 (94%)	156 (61%)
Gossypium sp., seed fragment	2 (6%)	98 (39%)

Dakhleh Oasis: Kellis and Trimithis

The large settlement of Kellis (Islamic Ismant el-Kharab) is situated in the centre of the Dakhleh Oasis in the Western Desert of Egypt. Excavations started in 1986, under the direction of A.J. Mills and Colin Hope as part of the Dakhleh Oasis Project. They have revealed occupation mainly from the 1st–4th centuries AD, though there is evidence of earlier (Ptolemaic) activities. The climate at Kellis is hyperarid with a present-day mean annual precipitation of 0.7 mm. Underground aquifers feed to the surface through natural vents, springs, and bores. The exploitation of these aquifers developed with the Ptolemaic period introduction of advanced irrigation technology. The village expanded during the Roman period and comprised residential areas, as well as several temples, Roman baths, followed by early Christian churches, and extensive industrial areas. The excavations have yielded a rich archive of inorganic and organic remains, among them papyri, textiles, basketry and food remains, as well as the famous *Kellis Agricultural Account Book*, a collection of wooden tablets containing a set of agricultural accounts over a period of five years from the late 4th century AD (Bagnall 1997; Thanheiser 2002). Spindle whorls and loom weights were found in every domestic structure excavated at Kellis, which suggests that spinning and weaving were carried out within the households (Bowen 2001, 2010).

Archaeobotanical samples were taken from within habitation areas occupied from the 1st century AD to c. 360/380 AD, as well as in the temple complex and churches. Most samples were sieved through a 0.5 mm mesh and all significant matrices were also screened through larger meshes to recover rarer items. Approximately 700 samples, representing more than 10,000 plant remains, have been analysed to date.

The plant macroremains found, mainly desiccated, show that the agricultural life during the 1st–4th centuries was based on various winter and summer crops, including free-threshing wheat, barley and pearl millet, and pulses, especially lentil and fava bean (*Vicia faba* L.). Fruit tree production is dominated by date palm and by Mediterranean fruits such as olive, fig (*Ficus carica* L.), grape and minor taxa such as almond (*Prunus dulcis* (Mill.) D. A. Webb), pomegranate (*Punica granatum* L.), and citron (*Citrus* cf. *medica* L.). Vegetables and herbs remains are numerous and include onion (*Allium cepa* L.), garlic (*Allium sativum* L.), fennel (*Foeniculum vulgare* L.), rosemary (*Rosmarinus officinalis* L.) and coriander (*Coriandrum sativum* L.). Oil and fibre plants are mainly represented by safflower (*Carthamus tinctorius* L.), flax/linen and cotton. The majority of these crops, including cotton, are listed as locally grown in the *Kellis Agricultural Account Book* mentioned above (For more details, see Thanheiser 2002; Thanheiser et al. 2016).

Cotton remains from Kellis consist of desiccated seeds, whole and fragmented, as well as bolls and empty capsules. Seeds are much more abundant than bolls. Cotton remains always occur together with another summer crop, pearl millet, which is present in greater proportions (Thanheiser et al. 2016).

Amheida, known as Trimithis during the Roman Period, is located in the western part of the Dakhleh oasis, some 20 km north-west of Kellis. The site has a long occupational history and is dominated by a large hill surmounted by a temple that was built and rebuilt over many centuries. Amheida reached its greatest size under Roman rule (approximately 1st–4th centuries AD) (Boozer 2015a). House B2, which yielded the archaeobotanical remains described below was located in the northeast sector of the city, which was composed of domestic and industrial structures. Sediment samples and hand-picked specimens were taken from sealed deposits rich in organic material, securely dated to the 3rd century AD, and less reliable contexts such as open courtyards. Sediment samples were sieved using a 0.5 mm mesh. Plant remains are charred and desiccated. The organic material is less well-preserved than in Kellis, probably because of more humid local conditions. Nevertheless, the archaeobotanical inventory of house B2, comprising more than 17,000 items, contains almost all plants typical for a household of the 3rd century AD in the Dakhleh Oasis. The assemblage includes a variety of winter and summer staple crops, especially hard wheat and hulled barley along with emmer, pearl millet, lentil, and fava bean. Vegetables, herbs, and spices include coriander, dill (*Anethum graveolens* L.), celery (*Apium graveolens* L.) and other items mainly originating in the Near East. Grape, olive, fig, and date are the most common fruits, but rare remains of other fruits and nuts, such as pine nut (*Pinus pinea* L.), peach (*Prunus persica* (L.) Batsch), sebesten (*Cordia myxa* L.), pomegranate, watermelon (*Citrullus lanatus* (Thunb. Matsum. & Nakai) and melon (*Cucumis melo* L.) were also found.

Oil and fibre plants include flax, sesame (*Sesamum indicum* L.), safflower, and cotton. Along with the cleaned products, processing waste such as cereal chaff and flax capsules were recorded (Thanheiser and Walter 2015). The botanical remains of cotton from the house B2 consist of 58 desiccated and 47 charred whole seeds. No capsules are present (Thanheiser and Walter 2015).

Other Archaeobotanical Finds and General Considerations

Several other sites in north-eastern Africa have provided archaeobotanical cotton remains. The Meroitic site of Hamadab in Central Sudan, to the north of Muweis, has revealed quantities of cotton seeds in 1st–3rd century AD domestic layers (work in progress, see mention in Fuller 2014:173). The Roman fort of Al-Zarqā', ancient Maximianon, in the Eastern desert of Egypt, provided 24 desiccated seeds found in trash layers mostly dated to the 1st–2nd century AD, with one item from a late 2nd to early 3rd century AD layer (Newton unpublished). The archaeobotanical survey of the oasis of Jerma, ancient *Garama* in southern Libya (Fezzan), showed the regular presence of charred complete and partial remains of cotton seeds from the 2nd to the 9th century. Direct radiocarbon dating confirms the presence of cotton by the 3rd–4th century AD (Pelling 2013:482). In the Kharga oasis, western Egypt, the necropolis of the ancient town of Kysis (present Douch) (Dunand et al. 1992; Tallet et al. 2012) and the contemporaneous urban sites of el-Deir (Tallet et al. 2012:127; Bouchaud and Clapham personal communication) and Qasr el-Sumayra (Clapham personal communication) offer desiccated cotton seeds probably dating from the 4th century AD. However, radiocarbon dating should be undertaken on the material from Kysis and el-Deir to confirm the chronology of the finds as they both come from disturbed contexts. Finally, charred seeds have been found at Aksum, Ethiopia in domestic contexts of the 6th–early 7th centuries with other Near Eastern and African annual and perennial plants (Boardman 1999)[2].

To sum up the archaeobotanical finds, the charred cotton items found in the different sites only includes seeds, whole or fragmented, whereas the desiccated material also includes capsule elements. The oldest reliable archaeobotanical find of cotton come from Qasr Ibrim in Lower Nubia and is associated to a period during or immediately after the Roman occupation (end 1st century BC–1st century AD). Cotton seeds appear at the end of the 1st century/beginning of the 2nd century AD in Nabataeo-Roman contexts in western Arabia and under Meroitic rule in Central Sudan, then in Egyptian oases in the Western Desert from the 2nd–3rd century AD onwards. Almost all the cotton records come from domestic contexts in towns and villages, and potentially indicate local cultivation. At Al-Zarqā'/Maximianon, local cultivation of cotton should not be considered because of the lack of water and the military status of the site. The seeds, perhaps used in packing material for the transport of delicate materials, must have come from elsewhere.

Textile Finds

Down the *chaîne opératoire*, it is interesting to also consider the finished cotton textiles —fabrics and clothing—as well as their trade, to obtain complementary data on the geographical and chronological scope of cotton production. However, the detailed examination and interpretation of these textile finds deserve a complete study that is beyond the scope of this paper.

Aside from isolated finds of cotton fibres in Chalcolithic Jordan (Betts et al. 1994), which should be re-examined in the light of new radiocarbon dating methods to confirm the antiquity of this isolated discovery, the majority of cotton textiles appear on both sides of the Red Sea on sites dated from the 1st century AD onwards.

[2] We decided not to detail and comment later data as it is out of the scope of the present article. Discoveries of Medieval cotton seeds in Africa and Middle East are few and localized at Myos Hormos in Egypt (Van der Veen 2011), Nauri (Fuller and Edwards 2001) and Hisn al-Bab (Clapham, unpublished) in Nubia, Dia in Mali (Murray and Cappers 2007), Volubilis in Morocco (Fuller et al. forthcoming), Chwaka in Tanzania (Walshaw 2010; cited in Van der Veen 2011:114) and various sites along the Euphrates river (Samuel 2001) and in Turkey (Miller 1998, 2010; Van der Veen 2011).

Nubia and Central Sudan present the largest corpus of cotton textiles (Fig. 1, Table 1). Most of them were found in funerary contexts and, to a lesser extent, in domestic settlements, especially in Qasr Ibrim. The majority of preserved fabrics comes from Lower Nubia, thanks to the hyper-arid climate of the region and extensive archaeological activity due to the UNESCO International Rescue Nubia Campaign before the completion of the Aswan High Dam. Outside of Lower Nubia, numerous sites in Upper Nubia, Central Sudan and the southern region of the Gezira also offer a rich corpus of tools, spindle whorls, loom weights, weaving combs and needles, demonstrating the great extent of the textile craft in these areas (Yvanez 2016). In the 1st and 2nd centuries AD, Nubian and Sudanese textiles, dominated until then by linen (as in Egypt) witnessed the important introduction of cotton, which subsequently represented more than half of textile occurrences in several major funerary and residential sites (Yvanez 2016). This is the case in Lower Nubia—at Karanog (Griffith and Crowfoot 1934), Aksha (Vila 1967), Qustul and Ballana (Mayer-Thurman and Williams 1979:36), and Qasr Ibrim (Plumley et al. 1977; Anderson and Adams 1979; Wild and Wild 2008, 2009, 2014b)—as well as in Upper Nubia and Central Sudan, at Saï (Yvanez 2012) and Meroe (Massey 1923; Griffith and Crowfoot 1934; Dunham 1963). All these sites present large quantities of cotton textiles characterised by S-spun threads and a very specific style of decorated fabrics, which prompted the first textile specialists (Griffith and Crowfoot 1934) to describe a Meroitic style and assume the Nubian origin of a large part of these specimens. Except for the Roman cotton thread from Qasr Ibrim mentioned above, none of these textiles is radiocarbon dated and most of them are attributed to the end of the Classic and Late Meroitic periods, from 1st to 4th century AD. The presence of unbaked clay loom-weights in many Meroitic settlements, as well as flat-woven starting borders preserved on a number of textiles, both point to the use of the vertical warp-weighted loom (Gleba and Mannering 2012; Wild and Wild 2014b: 76). Cotton textiles undoubtedly played a very important role under Meroitic rule. The available data for the following periods, Post-Meroitic, Christian and Islamic periods, that mainly correspond to funeral contexts and small corpuses, show that production likely declined—when the wool from sheep and camel were favoured in both Lower (Adams 2007; Wild and Wild 2014b) and Upper Nubia (Adams et al. 1999; Vogelsang-Eastwood 2001).

In Egypt, the excavation of Berenike and Myos Hormos provided an important sample of cotton textiles. During Roman times, the two sites were the main active harbours on the Red Sea and were both implicated in the intensive trade between the Roman Empire and India. The discoveries comprise fragments of cotton sails dated to the late 1st-early 2nd century AD from Myos Hormos (Handley 2004; Whitewright 2007), sail remnants and piled sleeping mats, dated to the 1st century AD, from Berenike (Wild and Wild 2001; Wild 2013; Wild and Wild 2014a, c). A smaller corpus of cotton textiles dated to the 4th century AD was also found at Abu Sha’ar (Bender Jørgensen and Vogelsang-Eastwood 1991; Bender Jørgensen 2004, 2006). Based on the technical studies of fabrics woven with Z-spun yarns, and critical analysis of written sources, it is assumed that these items came from India (see below and Wild and Wild 2014a, c). Other pieces, notably those woven from S-spun yarns, are considered of possible Nubian or Egyptian manufacture (Wild 1997; Wild and Wild 2014a:211).

One undated site in the Nile valley (el-Hibeh, Ryesky 1984) and several sites in the Great Oasis (Dakhleh and Kharga oases combined) attest the presence of cotton textiles alongside wool and linen in long-lived Egyptian settlements. These include the 2nd–3rd century AD at Kellis (Livingstone 2009, 2015) and Trimithis (Boozer 2015a) in the Daklheh Oasis and during the 3rd and 4th century AD at Kysis (Dunand et al. 1992:232; Tallet et al. 2012:126), El-Deir (Letellier-Willemin and Moulhérat 2006; Letellier-Willemin and Médard 2012) and 'Ayn Umm Dabadib (Jones and Oldfield 2006). The finds were excavated from households, funerary and cultic (temple) contexts. At Kellis, textiles were found in the same habitation areas where cotton seeds were recovered (see above), showing higher frequencies of cotton threads than linen and wool specimens. The cotton material includes S-spun curtains and other locally manufactured household textiles. The presence of characteristic textile fragments with corded (multi-strand twining) starting borders indicates the possible use of a vertical two-beam loom for the manufacture of these specific cotton fabrics. Several Z-spun cotton fabrics with blue resist-dyed patterns probably represent imported items from India (Livingstone 2009).

The cotton textile evidence from western Arabia is much less widespread than in Sudan and Egypt. Madâ'in Sâlih (Bouchaud et al. 2011, 2015), where cotton seeds were found (see above) provided a few pieces of cotton in Nabataean tombs (1st–3rd century AD), along with linen and wool fabrics. All the cotton textiles are of Z-spun yarns. At the same time, several southern Near Eastern sites witness the presence of cotton threads and textiles in funerary contexts (Fig. 1, Table 1), in regions around the Dead Sea (Bellinger 1962; Rahmani 1967; Sheffer and Tidhar 1991; see a review in Wild 1997; Müller et al. 2003, 2004), in Jordan (Granger-Taylor 2000, 2007) and especially in Palmyra, where cotton evidence seems to increase through time (Schmidt-Colinet 1995; Stauffer 2000). Both Z-spun and S-spun threads are recorded and nowhere is indicated a clear dominance of one type over the other. Apart from Madâ'in Sâlih, where the textiles' local origin was tentatively suggested (Bouchaud et al. 2011), these textiles are generally considered imported, as no other evidence, such as plant remains, attests their potential local production. The importance of cotton textiles increases from the 8th century onwards around the Dead Sea, as demonstrated by the discoveries in the Jericho region (Shamir 2015; Shamir and Baginski 2013) and in the Aravah valley (Shamir and Baginski 2014). For this later period authors are more disposed to consider that the cotton was grown locally (Shamir and Baginski 2013:82).

The Texts

Different types of written sources evoke the agricultural and textile production of cotton. They mainly refer to Egypt and in a few cases to Nubia, Sudan and the Horn of Africa. The earliest and also the most vague account, was written by Pliny the Elder who describes in the 1st century AD what he called 'wool-bearing tree', *gossypinus* or *xylon*, and states that it grows on the Arabian side of Upper Egypt and Sudan (*Natural History*, 19.14). Julius Pollux repeated this information in his Greek thesaurus of the 2nd century AD (*Onomasticon* 6.75). Because of the occasional uncertain words, it is difficult to have confidence in the geographical comments, although other lines of evidence from Lower Nubia, central Sudan and Ethiopia (see below) suggest that they are accurate. At the same time, the *Periplus Maris Erythraei* (*PME*) states that Indian cotton cloth and garments of different qualities were exported to the main ports of Arabia and Africa (see below and Casson 1989).

The earliest reliable and precise textual mention of cotton cultivation in Egypt comes from the oasis of Kharga. A land registry (*P. Iand.* 7. 142) dating to 164–165 AD, referring to lands around Kysis (present-day Douch, southern Kharga), contains one mention of cotton ("wool tree") among a list of economic trees (olive tree, date palm, unspecified fruit trees). About two centuries later, the *Kellis Agricultural Account Book* (*KAB*), documenting the income and expenditures of a land estate at Kellis, in Dakhleh oasis, contains few but informative references to cotton. The text indicates that two tenant farmers paid their rent to the landowner in cotton or, when they could not supply cotton, substituted other commodities (*KAB* 547, 556, 720; 1484). The estate allocated one *lithos* (unit of weight) of cotton to a woman for weaving (*KAB* 558) (Bagnall 1997). Besides these specific references to cotton, it is evident from the Kellis texts that household textiles were manufactured in the village. Several business accounts attest to a weaving workshop in the mid to late 4th century AD (Gardner et al. 1999) and the *KAB* mentions a clothes-weaving shop (Bagnall 1997, 153). Cotton is also mentioned in two ostraca from the same site. *O. Kellis* 68 and *O. Kellis* 69, dated to circa 237 and 257 AD respectively, are two receipts for six years of harvests mentioning substantial quantities of cotton (Worp and Hope 2004; Bagnall 2008). Cotton production is also well attested during the 4th century AD in Kharga oasis. One ostracon from the temple of Kysis (*O.Douch* 1.51) lists five names of women associated with cotton who are interpreted as spinners or weavers (Cuvigny and Wagner 1986). *O.Douch* 5.537 and *O.Douch* 5.634 mention quantities of cotton (Wagner 2001) and *O.Douch* 4.381, from the Kysis military area, is an order to deliver cotton as a payment of *Annona militaris*, here a direct requisition in kind redistributed to soldiers (Wagner 1999). Finally, two ostraca found at Amheida (Dakhleh oasis), *O.Trim* 38 and *O.Trim* 44, indicate the presence of considerable quantities of cotton (Bagnall and Ruffini 2012). Indeed, based on a calculation of equivalence between the *lithos* and the Roman pound, the maximal volume recorded (in *O.Trim* 44), 26 *lithoi*, is equivalent to more than 200 kg of cotton—a very large quantity, when considered in terms of volume (Bagnall 2008).

Beside these administrative accounts that demonstrate the local production of cotton in the Great Oasis, one administrative document and several private letters of the 2nd–3rd century AD of unknown or dubious origin attest the circulation of cotton textiles between the oases of the Western desert and the Nile valley. One mentions the sending of a cotton garment from Psobthis (probably the capital of Bahariya oasis, Western desert) to Oxyrhynchus in the Nile valley (*SB* 6.9025). Another, of unknown origin, mentions the making of garments from cotton thread (*SB* 6.9326) (see comment in Bagnall 2008). An administrative document of unknown origin, *P.Lond* 3.928, presents a list of taxed goods in which cotton appears (Bagnall 2008; Winter and Youtie 1944). One papyrus found at Oxyrhynchus (*P. Oxy*. 59.3991) mentions a cotton garment (Bagnall and Cribiore 2006). Finally, one of the private letters of a 2nd century AD archive belonging to Roman citizens (*P.Mich.* 8.500.7), found at Karanis, refers to cotton goods that would have been sent from Rome, according to the translators.

However, this interpretation has been disputed (Bagnall 2008). Despite the fragmentary and limited character of these data, their repetition highlights that cotton was available in the small oasis of Bahariya and the nomes of Oxyrhynchus (Nile valley, Middle Egypt) and Arsinoe (Fayum) (Bagnall 2008).

There are no written sources available alluding to cotton in the western Arabian Peninsula and in the southern Near East, and only one inscription has been found south of Egypt. The epigraphic record of the raid of the Aksumite king Ezana against the Meroitic kingdom, dated to the mid-4th century AD, mentions the destruction of the provisions of grain and cotton (Littmann 1913, see comment in Zeitschrift der Deutschen Morgenländischen Gesellschaft 67:701). According to geographical arguments, some scholars place these agrarian stocks to the north of the island of Meroe (Central Sudan), at the junction between the Nile and the Atbara rivers (Török 1988).

To sum up, the texts clearly indicate the importation of Indian cotton to the main ports of Arabia and Africa and being transported to Egypt and perhaps the Mediterranean world. The cultivation and processing of cotton in Kharga and Dakhleh oases, as well as the circulation of cotton textiles between these oases and the Nile valley, are visible from the mid-2nd century AD onwards. The quantities indicated in the documentation suggest the existence of substantial volumes of cotton used as payment in kind, partly for the army provisioning. Besides these Egyptian centres, only one epigraphic inscription attests to the existence of cotton stocks within the Meroitic kingdom, in Central Sudan.

Cotton Dynamics in North-Eastern Africa and Western Arabia: Facts and Limits

The distribution of archaeobotanical, textile and written records in north-eastern Africa and western Arabia highlights spatial and temporal production and trade dynamics of cotton. The evidence must be understood as hypothetical and incomplete because of various degrees of preservation of the organic remains and the methodological biases specific to each scientific discipline involved. Furthermore, much of the data are not useable because of imprecise dating, except when direct radiocarbon dating or secured chronological settings can be obtained (for example, direct dates on papyri and secure archaeological contexts). Our understanding of the different agricultural production centres varies depending on the coverage of each type of data, leading to heterogeneous degrees of knowledge across the region under study. Cotton cultivation and exploitation are securely attested when contemporary archaeobotanical, textile and textual data are all available on the same site, for instance at Kellis and Amheida, or when the textual evidence clearly refers to the local cultivation of cotton, as in Kysis. Little doubt also remains when cotton represents a high proportion of the archaeological textiles and/or archaeobotanical corpus, as evident in Qasr Ibrim, Muweis, Hamadab, Jerma and Madâ'in Sâlih.

Thus, if we exclude the dubious early record from Afyeh in Nubia, dated from the 3rd millennium BC (see above), the oldest reliable use of cotton found to date in north-eastern Africa comes from Roman Qasr Ibrim, in Lower Nubia, at the end of the 1st century BC/1st century AD. In parallel, textiles appear in several tombs in Lower and Upper Nubia as well as Central Sudan, from which technical characteristics suggest that they were made locally. We therefore assume small-scale cotton production to have occurred in several Nubian and Sudanese regions at that time, as described by Pliny. However, no plant macroremains that could have supported the textile evidence have been discovered in Central Sudan for the Early (3rd–1st centuries BC) and Classic Meroitic (1st century BC–1st century AD) periods. This situation is the result of two main factors: the occurrence of seasonal rains in these southern latitudes, and the general lack of well-preserved archaeological contexts of these periods in the region. Cotton production centres emerge in the western oases of Egypt in the mid-2nd century AD (at least) in Kharga oasis, and later in Dakhleh oasis during the 2nd–3rd century, as well as in Jerma (Libya). The chronology is similar in north-western Arabia, where cotton seeds and textiles from Madâ'in Sâlih may represent local production as early as the 1st century AD. The textiles from this site, potentially locally produced, pose a new challenge as they are all woven with Z-spun yarns, which are generally attributed to Indian manufacture. Finally, the question of the local introduction of cotton plants at that time remains open for several other areas where we only know the presence of cotton fabrics: in the Fayum, in Bahariya oasis and around Oxyrhynchus in Middle Egypt, or around the Dead Sea, all places where cotton production is well attested a few centuries later (see for an example at Jericho, Shamir and Baginski 2014).

The accumulation of different types of evidence provides an increasingly complete description of cotton history in the Old World. After India in the 3rd millennium BC (Fuller 2008), and the Mesopotamia (Muthukumaran 2016) and Persian gulf (Tengberg and Lombard 2001; Bouchaud et al. 2011) during the 1st millennium BC, several groups in north-eastern Africa and western Arabia developed centres of cotton cultivation and textile production, probably at the turn of the 1st century BC and the 1st century AD. This process becomes clearly visible after the 2nd–3rd century, perhaps suggesting an increase in cotton production at that time. As the body of evidence remains limited, we cannot affirm the preponderance of cotton fibres in the economic life of the whole region, apart from Nubia and Central Sudan where cotton received the favours of the Meroitic elites during the first four centuries AD (Mayer-Thurman and Williams 1979; Adams 2007; Yvanez 2016).

Management Practices of a Perennial Tropical Crop

The Cotton Tree

We have very little reliable information on the cultivation of cotton during Antiquity, but scattered data allow us to assume that cotton mainly grew as a perennial plant (Decker 2009; for the potential emergence of annual varieties, see Brite and Marston 2013). The use of the Greek term ἐρεόξυλον (*ereoxylon*) is quite ambiguous as it refers either to the plant or to the fruit itself. However, the mention of cotton among other perennial crops (olive, other fruit trees and palms) in the 2nd century AD papyrus from Kysis in Kharga oasis (*P. Iand.* 7. 142) clearly shows that the plant should be a perennial in this case. Apart from this mention, only more recent data can be used. Ibn al-'Awām, an Andalusian agronomist, in his 12th century treaty on agriculture, describes the cultivation of cotton based on both older treaties and local experience in Al-Andalūs. While it is not actually described as a tree, the plant is said to live several years, yielding a crop of fibres every year and, in Egypt, the plant can reach the size of a fig tree (Ibn Al-'Awām, *Kitab al-Filāha*, trad. J.-J. Clément-Mullet revised by El Faïz 2000). During the 16th century, Prospero Alpini reports that the Egyptian people imported herbaceous cotton from Syria and Cyprus whereas local Egyptian cotton is described as a shrub, needing several years to mature but producing fewer fibres than the herbaceous cotton (Prospero Alpini, *De Plantis Aegypti liber* XVIII, De Fenoyl 2007). He records in another volume that both herbaceous and perennial cottons grew in Egypt *(Historiae Aegypti Naturalis*, De Fenoyl and Sauneron 2007). Finally, cotton tree cultivation was still recently practiced in Nubia (Bond 1925), southern Sudan (Crowfoot 1931) and Ethiopia (Nicholson 1960) in the 20th century, where it was grown for periods of three to five or more years (Nicholson 1960). So, if we consider that cotton in north-eastern Africa and western Arabia grew as a tree, the use of ecological data (period of blossoming and maturing, water and soil requirement, insect resistance, etc.) belonging to modern annual cotton plant becomes irrelevant. More specifically, this raises the question of water needs and irrigation requirements.

Was Ancient Cotton a Thirsty Plant?

Modern annual cotton varieties require a minimum of 500 mm from rain or irrigation between germination and boll formation (Eyhorn et al. 2005). As far as we know, there are no available assessed ecological data concerning the perennial varieties. According to Ibn al-'Awām (*op. cit.*), cotton was watered in Spain, Sicily, Egypt, Palestine, Syria, the Hijâz and Mesopotamia (Basra). It was grown under either rainfall or irrigation in modern Ethiopia (Nicholson 1960). Old world species (*G. arboreum*, *G. herbaceum*), regardless of annual or perennial forms, are known to be more resistant to drought than the American ones (*G. hirsutum* and G. *barbadense*) (Eyhorn et al. 2005). The results from an archaeogenomic study on 4th century AD cotton seeds from Qasr Ibrim suggest that cotton, like other crop species found at this site (Palmer et al. 2009), may have been locally adapted to extreme environmental stress (Palmer et al. 2009). Even considering potential lower water needs for these old cottons, annual rainfall in north-east Africa and western Arabia does not exceed 100 mm today as was likely in the past (Walsh 1991 for Sudan, Sanlaville 2000 for Egypt and Arabia) suggesting that rain fed cotton cultivation was almost impossible. Cotton requires water input during the growing season, i.e. spring and early summer depending on the variety, and should not be waterlogged as the fruits form. In central Sudanese regions, which have a limited monsoon, as in Muweis, cotton would benefit during its growing season from irregular summer rains. Elsewhere, winter rainfall is useless for the cotton growth cycle and water management must have been carried out. In the Nile valley, as around Qasr Ibrim, Hamadab and Muweis, another factor to consider is that flooding occurred between July and November when cotton fruits are mature and therefore water should be avoided. Thus, if we take into account that cotton was probably perennial, it is likely that it was cultivated in fields away from the land susceptible to Nile flooding (Fuller 2015). In Lower Nubia, it is assumed that the spread of cotton cultivation was intrinsically linked to the development of mechanical irrigation systems, and especially to the introduction of the waterwheel or *saqia* (Wild et al. 2007; Fuller 2014, 2015). Archaeological evidence, mainly from Lower Nubia, does not indicate the presence of *saqia* before the 3rd century AD (Edwards 1996, 2004:159), about two centuries after the first explicit evidence of local cotton cultivation. This means that these first attempts would have been carried out with existing irrigation equipment, including the *shaduf*, or alternative methods. For instance, in Upper Nubia, where the Nile course often changes, agronomic observations at the beginning of the 20th century revealed that perennial cottons grew sheltered in pits dug in shallow sand covering the ancient alluvial plain. Subsequently, these cotton trees, some of them reaching ten-years of age, needed no care or watering until the harvest (Bond 1925: Figs. 4 and 5, plate VII). Although the more stable Nile course in Lower Nubia or Central Sudan makes it less suitable for that practice, it illustrates first, the existence of perennial cotton able to sustain relatively dry conditions and secondly, the human ingenuity in dealing with ecological conditions to enhance a priori unfavourable lands with a minimum of technical devices.

Beside these considerations, it seems logical to consider that the development of complex irrigation systems should enhance the productivity of crops. In Lower Nubia, the emergence of the *saqia* may be correlated with an apparent increase in cotton cultivation during the 3rd century AD (Wild et al. 2007; Clapham and Rowley-Conwy 2009; Fuller 2014, 2015). In Dakhleh oasis, irrigation systems composed of wells and canals spread during the Ptolemaic period, before the visible emergence of cotton. In the Kharga oasis, a complex irrigation system was probably established in the northern region, at El-Deir, and generalised in the southern area in the 5th century BC, which was further expanded and reconfigured during the Roman period (Chauveau 2005; Gonon et al. 2005; Bravard et al. 2016). At Madâ'in Sâlih, about one hundred wells were built during the Nabataean period (1st century BC–1st century AD) (Courbon 2008), before the arrival of cotton. No technical innovation of the irrigation system is visible during the entire occupational sequence. In light of these data, it seems quite clear that the chronology of cotton introduction does not follow nor has it had direct effects on the history of irrigation technology. It may attest to the small scale of cotton cultivation at its onset and/or adequacy of extant irrigation techniques.

Farming Calendar and Agricultural Changes

Cotton, as well as sorghum, pearl millet and cowpea are tropical or sub-tropical crops whose periods of growth, maturity and harvest occur from spring to late summer. Their introduction to north-eastern Africa and western Arabia demonstrates agricultural changes as they were added to an existing agricultural regime mostly based on winter Near Eastern annual crops, such as barley, emmer/hard wheat, flax which were sown in autumn and harvested in the spring. However, several elements demonstrate that some crops with summer flowering and fruiting periods were already under cultivation. First, a range of summer crops that are little visible in the archaeobotanical, textual and iconographical records such as onion, garlic, cabbage (*Brassica oleracea* L.), lettuce (*Lactuca sativa* L.), watermelon, chickpea (*Cicer arietinum* L.), broomcorn millet (*Panicum miliaceum* L.), foxtail millet and sesame were present in Egypt and in parts of the Near East before the 1st millennium AD (Aubaile-Sallenave 1984; Van der Veen 2011:76; Zech-Matterne et al. 2015). Castor-oil-plant (*Ricinus communis* L.), a short-lived perennial tropical plant, had already been introduced to the Western Desert oases of Egypt during the 1st millennium BC. It is recorded in the 7th century BC in Bahariya oasis and was especially significant in the agricultural economy of the southern Kharga oasis from the 5th century BC, where it may have played the role of a cash crop (Agut-Labordère and Newton 2013; Newton et al. 2013). The date palm, pollinated in February-March and harvested in August-September, was also a significant staple growing at the time of introduction of these new crops (see Bouchaud 2015 for northern Arabia; Murray 2000a, b, Tengberg and Newton (2016) for Egypt and Nubia). Thus, the emergence of new tropical plants in these regions results less in an extension of the farming season, as has been suggested before (Watson 1983), than in an intensification of the summer season farming. The various archaeobotanical examples show that this process took various forms. Tropical plants played a dominant role in Nubia and Central Sudan from the Meroitic period onwards to the detriment of winter cereal crops. For sites located in the Nile valley, where perennial and summer crops should be cultivated away from the Nile flood, it resulted in an optimisation of the already cultivated plots—for date palm cultivation for instance—and/or in an expansion of agricultural land. As far as we know, no reliable written or archaeobotanical data witness the existence of these new tropical crops on sites located in Middle and Lower Egypt. At those latitudes, they seem to be restricted to the western oases of Egypt, i.e., environments not subject to the annual flood and with year-round access to water. In Dakhleh oasis (Kellis, Trimithis), winter crops remained important even after the introduction of tropical plants. A diversification of crop inventories ensued, at least for cereals and textile production, implying risk-minimization while requiring an increase in labour for agricultural and textile processing (Clapham and Rowley-Conwy 2009; Fuller 2014; Thanheiser et al. 2016). The presence of cotton in the oasis of Madâ'in Sâlih for about two centuries shows that the plant was well adapted. However, it is the only reliable tropical crop attested at that time in the region (Bouchaud 2015), except perhaps sesame (Zech-Matterne et al. 2015). Others, like sugar cane (*Saccharum officinarum* L.), or rice (*Oryza sativa* L.), seem to develop around the Dead Sea several centuries later by the 10th–12th centuries AD (Watson 1983; Samuel 2001) whereas sorghum or pearl millet only extended into southern Arabia probably during the Medieval period (Boivin and Fuller 2009; Varisco 2009). Compared to Egyptian oases, Nubia and the Sudan, it seems that tropical crop introduction during Nabataeo-Roman times in Arabia was limited.

Three archaeological markers that have been described demonstrate important agricultural changes: new crops that defined *genetic innovation* (see Kirch 1994, summarised in Thurston and Fisher 2007:11); increase in the area of land under cultivation showing an *expansion* process (Morrison et al. 1996:587); and progress in the technical irrigation devices reflecting long-term investments (Brookfield 1984). These are all present in Lower Nubia from the 3rd century AD, and they probably accompanied broader societal changes. For instance, it has recently been proposed that agricultural changes in the Meroitic kingdom, clearly visible in the north (Lower Nubia) and assumed in the south, contributed to the collapse of the Meroitic state since they allowed regional economic independence of the regional successor states (Fuller 2014, 2015). On the contrary, there is no clear evidence of expanding farmland or new irrigation techniques in north-western Arabia, as if the oasis had "absorbed" the innovation with limited social and economic impacts.

The Ginning Process: Household or Handcraft Activity?

The presence of whole and broken seed remains in substantial quantities at several archaeological sites indicates that the first step of textile production—removing the seeds from the lint, or ginning—occurred locally. However, the seeds do not have the same meaning whether they are charred or desiccated and according to the archaeological context in which they were found.

Routes of Entry into Archaeobotanical Assemblages

The assessment of the formation processes of desiccated and charred cotton seeds aims to determine the various routes of entry into the archaeobotanical record, which may illustrate day-to-day routine behavioural practices (Van der Veen 2007). The first and most basic interpretation is to assume that after ginning, the seeds were directly thrown away in dumps (desiccated seeds) or fireplaces (charred seeds) (Fig. 6). In both cases, the fragmentation of the seed coat might result from the use of a ginning roller, from trampling after dumping (desiccated seeds) and/or from breaking during charring (charred seeds). Nevertheless, it is highly conceivable that people, most of all in an environment characterized by limited natural resources, tried to recycle waste derived from the various cotton processing steps, as they do with other agricultural by-products[3] (Anderson and Ertuğ-Yaras 1998; Van der Veen 1999; Bouchaud et al. 2012). Stems, leaves, roots, capsules and oil-rich seeds of cotton naturally secrete a toxic compound known as gossypol, which make them inedible for humans or non-ruminant animals. To our knowledge, no mechanical or non-chemical process to eliminate this toxic compound existed before the 20th century (Trolinder 2009), so we must assume that cotton by-products were likely useless in the past for cooking or feeding non-ruminant animals. Nevertheless, leaves, immature capsules and dry pericarps present in the field after the harvest, as well as seeds or pressed seed-cakes resulting from the ginning process might have been consumed by ruminants as they are in present-day small farms in Africa and India (Suttie 2004). Sheep and goat are most resistant to toxicity but small and occasional quantities seem to have no effect on cattle (Suttie 2004) and camel (Omer et al. 2008). However, no clear remains of cotton seeds have been found to date in sheep/goat dung pellets, except at Afyeh (Chowdhury and Buth 1971). Apart from the food, daily domestic oil needs in the past were much higher than today because of its use for lighting (Copley et al. 2005; Bonnéric 2012). Ancient people must have noticed the rich-oil content of the cotton seed and it would be

[3] In this light, the absence of cotton wood among the anthracological assemblage is quite surprising, most of all if ancient cotton was perennial, and deserves further investigation.

worthwhile to consider that cotton oil was used in lamps. To date, only one reference, from the Babylonian Talmud compiled around 500 AD, suggests that cotton seed oil was used, without specification, by the Jewish communities of Persian Mesopotamia (see comment in Newman 1932; Decker 2009). Ethnographic data of the middle of the 20th century in Anatolia show that cotton oil was obtained from roasted, ground and pressed seeds (Wulff 1966; Ertuğ 2000). These processes, as well as pressing activities during ginning and the use of dung from animals having ingested seeds or seed-cakes might therefore also explain the presence of fragmented seeds in the archaeobotanical assemblages. Finally, leaves, flowers, seeds and roots were still recently used in Ethiopia for medicinal purposes (Nicholson 1960:15). According to Prospero Alpini, Egyptian people of the 16th century used the mucilage from the seed to ease fever, stomach pain and cough (Prospero Alpini, *De Plantis Aegypti liber* XVIII, De Fenoyl 2007). These medicinal uses likely imply special processes that might have led the cotton waste to be deposited and preserved through time.

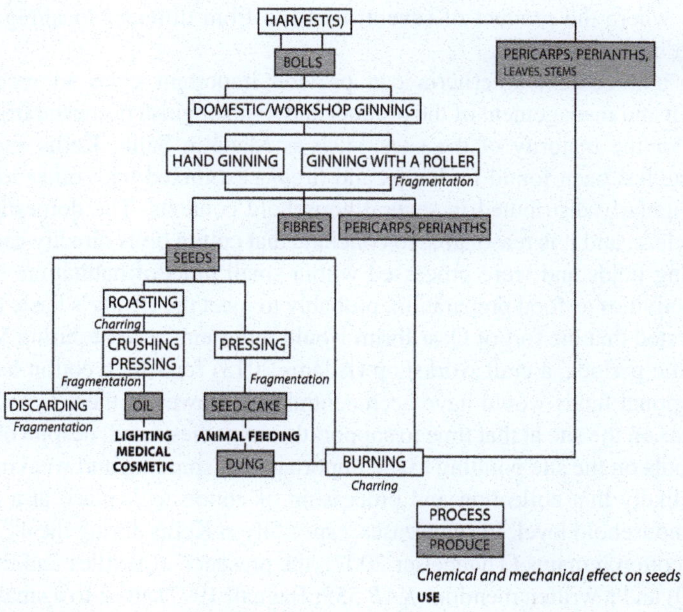

Fig. 6. Detailed diagram of the potential use of cotton seeds and various routes of entry into the archaeobotanical record.

Considering these different options[4], several paths of entry may explain the presence of desiccated and charred, fragmented and whole cotton remains in archaeological contexts (Fig. 6). All archaeobotanical assemblages presented in this paper include

[4] Accidental fires might also explain the presence of charred cotton items, but no traces of such events have been clearly recorded.

whole and fragmented cotton seeds. On Egyptian sites where climatic conditions allow the preservation of organic material (Qasr Ibrim, Kellis, Amheida), both charred and desiccated cotton items are present that suggest different routes of entry and their potential uses before being discarded. Otherwise, the proportion of fragmented charred seeds versus whole seeds is much lower at Madâ'in Sâlih than, for instance, at Muweis. If in the first case, a low rate of fragmentation cannot be used as indicator of any activities, we would suggest as a working hypothesis for future investigations that a high rate of fragmentation, as seen at Muweis or Qasr Ibrim, results from the use of the seeds before discarding.

Domestic Routine Versus Handcraft Specialisation

The presence of cotton bolls, seeds and raw fibres, despite indicating that the ginning process was taking place locally, does not necessarily mean that the site was a centre of agricultural production. Indeed, ginning occurred either within the domestic sphere, where the cotton bolls directly came from one or several local harvest, or in specialised workshops where the product of several harvests from different local/regional places might have been centralised.

Spatial and contextual criteria can provide important clues to reconstruct the organisation and management of the product and to distinguish domestic from handcraft contexts. For the majority of the sites, such as Madâ'in Sâlih, Kellis and Trimithis, cotton waste has been found in dumps and fireplaces, mixed with other food and fuel wastes, irregularly distributed in several household contexts. The domestic context is here quite clear, and it is reasonable to conclude that cotton bolls directly came from the neighbouring fields and were processed within small units of habitation, beside other domestic activities as food preparation, probably to meet the family's basic needs. It has been suggested that the fort of Qasr Ibrim would represent, at least during Meroitic and post-Meroitic periods, a craft workshop (Adams 2013) for which cotton harvests from local or regional fields would have been centralised. However, there is no evidence of storage areas on the site at that time to support this hypothesis and, despite of the density of textile tools on the site pointing to the importance of spinning and weaving activities, it is more likely that collection and processing of cotton took place at a domestic or individual household level. In some cases, especially at Kellis during the 4th century, the density of cotton remains (Thanheiser 2002), the presence of textiles and tools (Bowen 2001, 2010) and a written mention (*KAB* 558; Bagnall 1997) attest to a small textile and tailoring workshop organised by a woman, which provides evidence that such goods could be sold locally (Bowen 2001, 2010; Boozer 2015b).

The African Cotton

In 1960, G. E. Nicholson wrote: "*The precise region of Africa where the cultivated Old World cottons first arose is still a matter of much conjecture*" (Nicholson 1960:26). More than fifty years later, this issue can now be analysed in the light of new data regarding the role and diffusion of the African cotton species, *Gossypium herbaceum*.

Area of Domestication and Spread

An archaeogenomic study performed on 4th century AD cotton seeds from Qasr Ibrim has demonstrated that African cotton, *G. herbaceum*, was present in Nubia and cultivated during Antiquity (Palmer et al. 2012). This result corroborates an older hypothesis based on the analyses of Nubian textiles (Griffith and Crowfoot 1934) and refutes the theory that the rise of cotton production in north-eastern Africa was the result of imported *G. arboreum* from India (Watson 1983:34). Based on this archaeogenomic result, as well as the growing record of cotton archaeobotanical remains in Nubia and Central Sudan, and the Meroitic creation of a characteristic textile style (Yvanez 2015), it is tempting to assume that *G. herbaceum* had been consistently used since the beginning of the Nubian and Sudanese cotton production.

Even so, this assumption does not directly address the issue of where and when *G. herbaceum* was domesticated. According to the latest phylogenetic research, cultivated *G. herbaceum* was derived from a truly wild diploid ancestor, *G. herbaceum* var. *africanum*, which is the only wild taxon in the genus that bears lint. Today, this subspecies grows naturally in southern Africa, more specifically in Namibia, Botswana, South Africa, Zimbabwe and Mozambique (Hutchinson 1959; Vollesen 1987:345; Kulkarni et al. 2009:71). Hypotheses regarding its geographical origin first placed the centre of domestication in these southern regions (Hutchinson 1959). Others, following the idea of a northward migration of the wild form from southern Africa and loss of photoperiodism, postulate that domestication took place in southern Arabia (Kulkarni et al. 2009) or Ethiopia (Nicholson 1960; Kulkarni et al. 2009), where the most primitive perennial cultivar, *G. herbaceum* race '*acerifolium*," is observed (Kulkarni et al. 2009). The southern Africa and southern Arabia hypotheses are quite difficult to evaluate as no other reliable evidence has emerged to date and neither archaeobotanical data nor written sources attest to the existence of cotton before the Medieval period in both areas (Davison and Harries (1980) for South-East Africa; Varisco (2009) for Yemen). Of much interest is the eastern Africa hypothesis, because it puts the already demonstrated spread of cotton in Central Sudan and Nubia in the immediate wake of this process.

The arguments put forward for a domestication in East Africa are based on botanical observations made in the mid-20th century, showing that despite the low number of populations and varieties analysed, the phenotypic and ecological variability within the cultivar *G. herbaceum* race '*acerifolium*' is quite high, especially in southern Ethiopia (Nicholson 1960:24). Furthermore, at least four species of wild perennial *Gossypium* have been collected and recorded in the region. *G. anomalum* Wawr. & Peyer. has been observed near the Sudan-Eritrea border, *G. anomalum* Wawr. Ex Wawr. & Peyer. subsp. *senarense* (Fenzl ex Wawr & Peyer.) Vollesen in the wet bushland and grassland of western and central Sudan and *G. bricchettii* (Ulbr.) Vollesen among the *Acacia-Commiphora* woodland and bushland proper formation in eastern Ethiopia (Friis et al. 2011:54; Darbyshire et al. 2015:250). *Gossypium somalense* (Gürke) J. B. Hutch. has been recorded in all these places as well (Nicholson 1960:25; Obeid and Mahmoud 1971:190; Darbyshire et al. 2015:251). Finally, *G. longicalyx* Hutch. and B. J. S. Lee, which is thought to represent the wild form most closely related to the A-genome species (Wendel et al. 2009:13), is also present in East

Africa—in South Sudan, Ethiopia and Tanzania (Vollesen 1987:346; Wendel et al. 2009:6; Darbyshire et al. 2015:251). Although this relatively high number of wild relatives cannot be directly correlated to a centre of domestication, it highlights a rich germplasm diversity. In addition, multiplicity in cultivars is also expressed through the distinctive vernacular names for cotton in Ethiopia and most of all in the old Hamitic languages (see the linguistic analyses reported in Nicholson 1960:23). This wild and domestic diversity would explain the absence of the known wild ancestor *G. herbaceum* var. *africanum* in the region. Indeed, the ease with which intercrossing can occur in cotton suggests that in areas where cotton was domesticated or diversified, the wild ancestor of the cultivated plant may have disappeared as it evolved in association with cultivars (Watson 1983:32; Wendel et al. 2009:7).

One can postulate that the domestication process occurred somewhere in the eastern part of the sub-Saharan humid savannahs. Unfortunately, the absence of recent genetic analyses, the humid climatic conditions that led to the very poor preservation of organic material, as well as the lack of archaeological studies and written sources from those regions, restrict the possibility to verify the reliability of this option. The only historical data recovered to date consist of late records. First, the mention of cotton in the Ezana inscription (see above) shows that in the 4th century AD, the plant was known and had a specific local name in the Ge'ez language, *ṭūṭ*, and is still used in Amharic Littmann (1913). Secondly, five textile pieces found at Kidane Mehret in northern Ethiopia (Phillipson 2012:179) and a few charred seeds coming from a domestic context at Aksum (Boardman 1999; Phillipson 2000) demonstrate the presence of cotton in Ethiopia during the 6th–7th centuries AD. Likewise, the southern regions of present Sudan and South Sudan, which could link the archaeological records of Nubia and central Sudan to the potential location of *G. herbaceum* domestication, is almost devoid of data. No textile or seeds have survived in the very limited archaeological record of these humid regions. Nonetheless, the many spindle whorls found at Abu Geili, a Late Meroitic site located on the eastern shore of the Blue Nile in the Gezira area, suggest the spinning of short fibres in significant quantities, which could be wool, or, because of the regional context, most probably cotton (Yvanez 2016). While awaiting further analyses and new tools to better understand the emergence of cotton in Africa, the available data lead us to consider the Meroitic territory as an important centre of spread for the species *G. herbaceum*, whose roots may be sought in the southern humid savannahs of eastern Africa.

Trade as a stimulus for Local Production

Cotton history is a reliable example of the process described by A. Sherratt who argues that trade in consumables is one way in which crops, particularly non-staple crops, spread beyond their core areas of domestication (Sherratt 1999). Based on the spatial and chronological distribution of the available data in Africa, we postulate that the trade of Indian cotton, *G. arboreum*, passing through various localities in the Indian Ocean, stimulated local production and played an indirect role in the spread, if not the domestication, of *G. herbaceum* in sub-Saharan Africa.

This argument first relies on the *Periplus Maris Erythreae* (*PME*). This text, which describes the movement of Eastern luxuries to the ports of Egypt and the Red Sea trade, also reports a second line of trade revealed by the detailed examination of the imports and their sources that the author lists for the various ports. This concerns the trade of textiles, foods and raw materials from India to the coasts of Persia, southern Arabia and East Africa handled by Arabs or Indians, in which the merchants of Roman Egypt had no or very little part (Casson 1989:21). More specifically, the *PME* reports that Indian cotton cloth went to southern Arabia (Moscha Limên [*PME* 32:11.3]), northern Somalia ("far-side" ports [*PME* 14:5.10-13]) and the Ethiopian coast (Adulis [*PME* 6:3.1-3]). According to the description, most of the cotton cloth exported to Adulis and northern Somalia was of cheap quality and it is assumed that it was probably the same for the cotton cloth exported to southern Arabia. The finer cotton cloth and garments exported to Adulis [*PME* 6:3.3] are interpreted as higher-value items intended for the foreign colony residing there [*PME* 6:2.32] (Casson 1989:19). To date, there is very little archaeological evidence directly arising from the mentioned ports, making it impossible to demonstrate the link between these records of cheap and fine cotton imports and the emergence of a local production. Nevertheless, it is interesting to note that the earliest known production centres in western Arabia and north-eastern Africa appeared in Nabataean and Meroitic societies: both settled inland territories, engaged in overland caravan trade, and acting as intermediaries to supply certain goods to the Mediterranean world or to the ports of eastern and southern Arabia and the Horn of Africa (for the Nabataeans, Durand 2009; Terpstra 2015; for the Meroitics, Haaland 2014). The conjunction of well-established states where agriculture and trade activities are both developed and the existence of long-distance trade dynamics could have created a positive global context not only for the exchange of goods, but also for the acquisition of agricultural and economic novelties by Arab and African societies. The question remains open whether the domestication and diffusion of African cotton is correlated with the circulation of the already known cotton from India in the various trade networks. We might suggest that the circulation of cotton cloth and garments led Meroitic inhabitants to use a similar resource growing locally, i.e. G. *herbaceum*, and the Nabataeans to adopt a new plant, but this is purely speculative. However, according to this point of view and the fact that exchanges existed between India, Arabia and Africa, we must consider that the cotton cultivated in north-western Arabia could be *G. arboreum* as much as *G. herbaceum* (Bouchaud 2015).

In conclusion, it is important to remember that the spread of local production centres did not prevent Indian cotton from being imported and maintaining a prominent status through time. This is well demonstrated by the Indian items discovered in several Egyptian sites during the 1st–4th centuries, such as in Berenike or Kellis (Bender Jørgensen and Vogelsang-Eastwood 1991; Livingstone 2009; Wild and Wild 2014a, c), as well as the European evidence throughout Medieval (Mazzaoui 1981) and Modern times (Riello and Roy 2009).

Conclusion

This overview of the history of cotton in western Arabia and north-eastern Africa highlights the benefits of an interdisciplinary approach while underlining the extreme difficulty to perceive the chronological, biological, economic and sociological details of change. These limits result mainly from methodological aspects caused first by a lack of data and secondly by botanical characterisation issues. Regarding the first challenge, it is worth mentioning that the available documentation on northeast Africa and Arabia is rapidly growing, as many recent excavations conducted on Antique sites that included archaeobotanical studies have yielded new occurrences of cotton remains. Regarding the second issue, our efforts should now concentrate on the development of techniques to better identify archaeobotanical and textile remains of cotton. Increasing the number and the quality of specimens will most probably improve our understanding of the spatio-temporality of the diffusion of cotton. For now, the evidence demonstrates that cotton cultivation spread on both sides of the Red Sea around the end of the 1st millennium BC-beginning of the 1st millennium AD onwards, and that its dispersal out of its tropical region of origin demonstrates the agricultural skills of the societies that adopted it. In these regions, a key aspect of cotton production was its addition to pre-existing textile traditions dominated by flax and wool, rather than its introduction into areas devoid of textile plant resources. Another important aspect to consider is that, as far as we can see, the first attempts were not directly linked to new irrigation practices, which occurred earlier (in Egyptian oases) or later (in Nubia). It means that people introduced cotton by using existing technical knowledge of irrigation and perhaps developed alternative and original strategies in response. This leads us to consider a "bottom up" perspective (Thurston and Fisher 2007) according to which these first attempts were probably based on experimentations handled by local farmers rather than an institution. Later, the development of cotton production as observed in the Meroitic kingdom might have concurred with a progressive controlling of the product and its transformation by governing bodies, leading to an increase of production and the potential development of trade, at least during the Late Meroitic and post-Meroitic periods (Fuller 2014). In this way, Meroitic sub-Saharan cotton trade may have played a significant role in the emergence of local cotton production in the Egyptian western oases and Libyan oases during the 2nd–3rd centuries AD (Clapham and Rowley-Conwy 2009; Tallet et al. 2012).

At this stage of our knowledge, it is not possible to precisely identify where and when the two Old World cotton species were first in use: probably during the 6th–5th millennia BC in northern India, and somewhere in tropical Africa at the end of the 1st millennium BC at the latest. Nevertheless, more data coming from Arabia and north-eastern Africa marks the turn of the 1st millennium BC with the beginning of the 1st millennium AD as a key moment in cotton history, embodied by the formidable expansion of both *G. herbaceum* and *G. arboreum* utilisation.

Acknowledgements. We thank the members of the Fyssen Foundation for awarding the Grant Research (2016–2017) that supported this study.

Charlène Bouchaud and Claire Newton are grateful to Marie Millet for inviting them to participate in the French archaeological mission at Muweis, making available the unpublished radiocarbon dates and for her comments on this paper. Charlène Bouchaud also warmly thank Daifallah al-Talhi, Laïla Nehmé and François Villeneuve for having entrusted her with the analysis of the plant remains from Madâ'in Sâlih.

Alan Clapham would like to thank Dr Pamela Rose of the Österriechisches Archälogisches Institut Zweigstelle Kairo for the opportunity to work on the plant remains from Hisn al Bab, Aswan, Egypt and The Egyptian Exploration Society and Dr Pamela Rose for the chance to work on the material from Qasr Ibrim.

We also thank the two anonymous reviewers for their valuable comments on the first draft of this paper, as well as Anna Maria Mercuri and the IWAA8 team for the organisation of the conference in Modena.

References

Abasaeed AE (1992) Briquetting of carbonized cotton stalk. Energy 17(9):877–882

Adams NK (2007) Political affinities and economic fluctuations: the evidence from the textiles at Qasr Ibrim. In: Godlewski W, Lajtar A (eds) Between the cataracts, Part 2, fascicule 1. Warsaw University Press, Warsaw

Adams WY (2013) Qasr Ibrim: ballana phase (EES excavation memoir 104). Egypt Exploration Society, London

Adams WY, Adams NK, van Gerven DP et al (1999) Kulubnarti III: the cemeteries (BAR S814). Archaeopress, Oxford

Agut-Labordère D, Newton C (2013) L'économie végétale à 'Ayn-Manâwir à l'époque perse: archéobotanique et sources démotiques, Achaemenid research on texts and archaeology 2013.005. http://www.achemenet.com/pdf/arta/ARTA_2013.005-Agut-Newton.pdf. Accessed 7 May 2018

Anderson RD, Adams WY (1979) Qaṣr Ibrîm 1978. J Egypt Archaeol 65:30–41

Anderson S, Ertuğ-Yaras F (1998) Fuel fodder and faeces: an ethnographic and botanical study of dung fuel use in central Anatolia. In: Charles M, Halstead P, Glynis J (eds) Fodder: archaeological, historical and ethnographic studies (environmental archaeology). Oxbow Books, Oxford, pp 99–109

Aubaile-Sallenave F (1984) L'agriculture musulmane aux premiers temps de la conquête: apports et emprunts. J Agriculture Traditionnelle et de Botanique Appliquée 31(3–4):245–256

Bagnall RS (1997) The Kellis agricultural account book (Dakhleh Oasis Project). Oxbow Books, Oxford

Bagnall RS (2008) SB 6.9025, cotton, and the economy of the small oasis. B Am Soc Papyrologis 45:21–30

Bagnall RS, Cribiore R (2006) Women's letters from ancient Egypt, 300 BC-AD 800. University of Michigan Press, Ann Arbor

Bagnall RS, Ruffini G (2012) Amheida I: Ostraka from Trimithis, vol 1. New York University Press/Institute for the Study of the Ancient World, New York

Barber EJW (1993) Prehistoric textiles: the development of cloth in the neolithic and bronze ages with special reference to the Aegean. Princeton University Press, Princeton

Baud M (2010a) Mouweis, une ville type de la vallée? In Méroé: un empire sur le Nil. Musée du Louvre. Officina libraria, Paris, Milan, pp. 216–219

Baud M (2010b) Méroé: un empire sur le Nil. Musée du Louvre. Officina libraria, Paris, Milan

Baud M (2014) Downtown Muweis—a progress report (2007–2011). In: Welsby D, Anderson JR (eds) The fourth cataract and beyond. Proceedings of the 12th international conference for Nubian studies. Walpole, Paris

Bellinger L (1962) Textiles. In: Colt HD (ed) Excavations at nessana 1. Princeton University Press, Princeton, pp 91–105

Bender Jørgensen L (2004) A matter of material: changes in textiles from roman sites in Egypt's eastern desert. Antiquité Tardive 12(1):87–99

Bender Jørgensen L (2006) The late roman fort at 'Abu Sha'ar, Egypt: textiles in their archaeological context. Riggisberger Berichte 13:161–173

Benedict CR, Kohel RJ, Lewis HL (1999) Cotton fiber quality. In: Smith CW, Cothren JT (eds) Cotton: origin, history, technology and production. Wiley, New York, pp 269–288

Betts A, van der Borg K, de Jong A et al (1994) Early cotton in North Arabia. J Archaeol Sci 21 (4):489–499

Boardman S (1999) The agricultural foundation of the Aksumite Empire, Ethiopia: an interim report. In: Van der Veen M (ed) The exploitation of plant resources in ancient Africa. Kluwer Academic, New York, pp 137–147

Boivin N, Fuller DQ (2009) Shell middens, ships and seeds: exploring coastal subsistence, maritime trade and dispersal of domesticates in and around the ancient Arabian Peninsula. J World Prehist 22:113–180

Bond WRG (1925) Some curious methods of cultivation in Dongola Province. Sudan Notes Records 8:97–103

Bonnéric J (2012) Les huiles et leurs usages en terre d'Islam à partir du VIIe s. In: D Frère, L Hugot (eds) Les huiles parfumées en méditerranée occidentale et En Gaule (VIIIe Av.—VIIIe Ap. J.-C.) (Archéologie de l'artisanat antique 6). Centre Jean Bérard, Presses Universitaires de Rennes, Rennes, pp. 307–314

Boozer AL (2015a) Domestic archaeology and the Romano-Egyptian house. An integrated research agenda. In: AL Boozer (ed) Amheida II: a late Romano-Egyptian house in the Dakhleh Oasis. Amheida House B2. The Institute for the study of the Ancient world, New York

Boozer AL (2015b) Woven material. In: AL Boozer (ed) Amheida House B2 Amheida II: a late Romano-Egyptian house in the Dakhleh Oasis. The Institute for the study of the Ancient world, New York

Bouchaud C (2015) Agrarian legacies and innovations in the Nabataean territory. ArchéoSciences 39:103–124

Bouchaud C, Sachet I, Dal-Prà P et al (2015) New discoveries in a Nabataean tomb. Burial practices and 'plant jewellery' in ancient Hegra (Madâ'in Sâlih, Saudi Arabia). Arab Archaeol Epigr 26:28–42

Bouchaud C, Tengberg M, Dal Prà P (2011) Cotton cultivation and textile production in the Arabian Peninsula during Antiquity: the evidence from Madâ'in Sâlih (Saudi Arabia) and Qal'at al-Bahrain (Bahrain). Veg Hist Archaeobot 20:405–417

Bouchaud C, Thomas R, Tengberg M (2012) Optimal use of the date palm (*Phoenix dactylifera* L.) during Antiquity: Anatomical identification of plant remains from Madâ'in Sâlih (Saudi Arabia). In: Badal E (ed) Wood and Charcoal. Evidences for the Human and Natural History. (Saguntum-Extra 13). Universitat de Valencia, Valencia, pp. 173–185

Boulos L (2000) Flora of Egypt. Volume 2 (Geraniaceae-Boraginaceae). Al Hadara, Cairo

Bowen G (2010) Texts and textiles: a study of the textile industry at ancient Kellis. http://artsonline.monash.edu.au/ancient-cultures/texts-and-textiles-a-study-of-the-textile-industry-at-ancient-kellis/. Accessed 3 Feb 2017

Bowen GE (2001) Texts and textiles: a study of the textile industry at ancient Kellis. Artifact 24:18–28

Bravard J-P, Mostafa A, Garcier R et al (2016) Rise and fall of an Egyptian Oasis: artesian flow, irrigation soils, and historical agricultural development in El-Deir, Kharga Depression. Western Desert Egypt. Geoarchaeol 31(6):467–486

Brite EB, Marston JM (2013) Environmental change, agricultural innovation, and the spread of cotton agriculture in the Old World. J Anthropol Archaeol 32(1):39–53

Brookfield HC (1984) Intensification revisited. Pac Viewpoint 25(1):15–44

Brubaker CL, Bourland FM, Wendel JF (1999) The origin and domestication of cotton. In: Smith CW, Cothren JT (eds) Cotton: origin, history, technology and production. Wiley, New York, pp. 3–31

Bulliet RW (2009) Cotton, climate, and camels in early Islamic Iran. Columbia University Press, New York

Casson L (1989) The Periplus Maris Erythraei. University Press, Princeton

Chaudhry RM, Guitchounts A (2003) Cotton facts. Cotton Advisory Committee, Washington

Chauveau M (2005) Irrigation et exploitation de la terre dans l'oasis de Kharga à l'époque perse. Cahier de recherches de l'Institut de papyrologie et d'égyptologie de Lille 25:157–163

Chowdhury KA, Buth GM (1971) Cotton seeds from the neolithic in Egyptian Nubia and the origin of Old World cotton. Biol J Linn Soc 3(4):303–312

Clapham A, Rowley-Conwy P (2007) New discoveries at Qasr Ibrim, Lower Nubia. In: Cappers RTJ (ed) Fields of Change. Proceedings of the 4th international workshop for African archaeobotany (Groningen archaeological studies 5). Barkhuis and Groningen University Library, Groningen, pp. 157–164

Clapham A, Rowley-Conwy P (2009) The archaeobotany of cotton (*Gossypium* sp. L.) in Egypt and Nubia with special reference to Qasr Ibrim, Egyptian Nubia. In: Fairbairn A, Weiss E (eds) From Foragers to farmers. Papers in Honour of Gordon C Hillman. Oxbow Books, Oxford, pp. 244–253

Coombs CE, Woodhead AL, Church JS (2002) Report on the characterization of three fabric samples from Ismant el-Kharab. In: Hope CA, Bowen GE (eds) Dakhleh Oasis project: preliminary reports on the 1994–1995 to 1998–1999 field seasons. Oxbow Books, Oxford, pp. 115–119

Copley MS, Bland HA, Rose P et al (2005) Gas chromatographic, mass spectrometric and stable carbon isotopic investigations of organic residues of plant oils and animal fats employed as illuminants in archaeological lamps from Egypt. Analyst 130(6):860–871

Courbon P (2008) Les puits nabatéens de Madâ' in Sâlih (Arabie Saoudite). Arab Archaeol Epigr 19(1):48–70

Crowfoot GM (1924) The handspinning of cotton in the Sudan. Sudan Notes Records 7(2):83

Crowfoot GM (1931) Methods of hand spinning in Egypt and the Sudan (Bankfield Museum Notes 12). F King & Sons, Halifax

Cuvigny H, Wagner G (1986) Les Ostraca Grecs de Douch: (O. Douch). Fascicule I. 1–57 (Document de fouilles 24). Institut Français d'Archéologie Orientale, Cairo

Darbyshire I, Kordofani M, Farag I et al (eds) (2015) The plants of Sudan and South Sudan: an annotated checklist. Royal botanic gardens. Kew publishing, Kew

Davison P, Harries P (1980) Cotton weaving in South-East Africa: its history and technology. Text Hist 11(1):175–192

De Fenoyl R (ed) (2007) Plantes d'Egypte, Par Prosper Alpin, 1581–1584 (Voyageurs 22). Institut Français d'Archéologie Orientale, Cairo

De Fenoyl R, Sauneron S (eds.) (2007) Histoire Naturelle de l'Egypte, Par Prosper Alpin, 1581–1584, Vol. I et II (Voyageurs 20). Institut Français d'Archéologie Orientale, Cairo

Decker M (2009) Plants and progress: rethinking the Islamic agricultural revolution. J World Hist 20(2):187–206

Dillehay TD, Rossen J, Andres TC et al (2007) Preceramic adoption of peanut, squash, and cotton in Northern Peru. Science 316(5833):1890–1893

Dunand F, Heim J-L, Henein N et al (1992) La nécropole de douch. Douch I. Institut Français d'Archéologie Orientale, Cairo

Dunham D (1963) The royal cemeteries of Kush, Volume V: The West and South Cemeteries at Meroë. Museum of Fine Arts, Boston

Durand C (2009) The Nabataeans and oriental trade: roads and commodities (Fourth Century BC to the First Century AD). Stud Hist Archaeol Jordan 10(405):12

Durand C, Gerber Y (2014) The pottery production from Hegra/Madâ'in Sâlih (Saudi Arabia) during the Nabataean period. Preliminary results, 2008–2011. Proc Sem Arab Stud 44:153–167

Edwards DN (1996) The archaeology of the Meroitic state: new perspectives on its social and political organisation (Cambridge monographs in African archaeology 38). Tempus Reparatum, Oxford

Edwards DN (2004) The Nubian past. An archaeology of the Sudan. Londres, Routledge, New York

El Faïz M (2000) Le Livre de L'agriculture d'Ibn Al-'Auwām (Kitāb Al-Filāḥa). Actes sud, Sindbad, Paris

Ertuğ F (2000) Linseed oil and oil mills in central Turkey Flax/Linum and Eruca, important oil plants of Anatolia. Anatol Stud 50:171–185

Eyhorn F, Ratter SG, Ramakrishnan M (2005) Organic cotton crop guide. A manual for practitioners in the tropics. Research Institute of Organic Agriculture (FIBL), Frick

Friis I, Demissew S, van Breugel P (2011) Atlas of the potential vegetation of Ethiopia. Addis Ababa University Press & Shama Books, Addis Ababa

Fuller DQ (2008) The spread of textile production and textile crops in India beyond the Harappan zone: an aspect of the emergence of craft specialization and systematic trade, Linguistics, archaeology and the human past. Indus Project. Research Institute for Humanity and Nature, Kyoto, pp. 1–26

Fuller DQ (2014) Agriculture innovation and state collapse in Meroitic Nubia. In: Stevens CJ, Nixon S, Murray MA, Fuller DQ (eds) Archaeology of African plant use. Left Coast Press, Walnut Creek, pp. 165–177

Fuller DQ (2015) The economic basis of the Qustul splinter state: cash crops, subsistence shifts, and labour demands in the post-meroitic transition. In: Zach M (ed) Proceedings of the 11th international conference for meroitic studies (Beitrage zur Sudanforschung Beiheft series) Kushite World. Verein der Förderer der Sudanforschung, Vienna, pp. 33–60

Fuller DQ, Edwards DN (2001) Medieval plant economy in middle Nubia: preliminary archaeobotanical evidence from Nauri. Sudan Nubia Bull 5:97–103

Fuller DQ, Pelling P, De Varailles A (Forthcoming) Staple grains, crops and social differentiation: archaeobotanical studies at Volubilis, Marocco. In: Fentress E, Limane H (eds) Excavations at Volubilis, Marocco

Gardner I, Alcock A, Funk W-P (eds) (1999) Coptic documentary texts from Kellis, vol 1. Oxbow Books, Oxford P. Kell V

Gleba M, Mannering U (eds) (2012) Textiles and textile production in Europe from prehistory to AD 400 (Ancient textiles series 11). Oxbow Books, Oxford

Gomes RS, Wilson PN, Coates WE et al (1997) Cotton (*Gossypium*) plant residue for industrial fuel: an economic assessment. Ind Crop Prod 7(1):1–8

Gonon T, Thiers C, Wuttmann M (2005) Les qanâts d'Ayn Manawir (Oasis de Kharga, Égypte): techniques de creusement et dynamique de l'exploitation d'une ressource épuisable, de la Première Domination Perse au IIe siècle de l'ère commune. Internationales Frontinus-Symposium « Wasserversorgung Aus Qanaten-Qanate Als Vorbilder Im Tunnelbau. Frontinus Gesellschaft, Bonn, pp. 39–57

Granger-Taylor H (2000) The textiles from Khirbet Qazone (Jordan). In: Cardon D, Feugère M (eds) Archéologie des textiles, des origines au ve siècle. Actes Du Colloque de Lattes, Octobre 1999. Monique Mergoil, Montagnac, pp. 149–161

Granger-Taylor H (2007) Textiles of the Herodian, Nabataean and Roman periods from the Dead Sea area. In: Politis KD (ed) The world of the Nabataeans. Steiner-Verlag, Stuttgart, pp. 301–307

Griffith FL, Crowfoot GM (1934) On the early use of cotton in the Nile valley. J Egypt Archaeol 20(1/2):5–12

Haaland R (2014) The meroitic empire: trade and cultural influences in an Indian Ocean context. Afr Archaeol Rev 31(4):649–673

Handley FJL (2004) Quseir al-Qadim 2003: the textiles. Text Newsl 38:27–30

Hutchinson J (1959) The application of genetics to cotton improvement. Cambridge University Press, Cambridge

Jones J, Oldfield R (2006) A Kind of Wool is made by the Egyptians from a Tree. Archaeol Text Newsl 43:27–32

Kirch PV (1994) The wet and the dry: irrigation and agricultural intensification in Polynesia. University of Chicago Press, Chicago

Kulkarni VN, Khadi BM, Maralappanavar MS et al (2009) The worldwide gene pools of *Gossypium arboreum* L. and *G. herbaceum* L., and their improvement. In: Paterson AH (ed.) Genetics and genomics of cotton, (Plant Genetics and Genomics: Crops and Models 3). Springer US, New York, pp. 69–97

Leroi-Gourhan A (1971) L'homme et la matière. Albin Michel, Paris

Letellier-Willemin F, Médard F (2012) Techniques inattendues dans un fragment textile en coton, du site d'El Deir, Oasis de Kharga. Désert Ocidental Égyptien, Archaeol Text Rev, p 54

Letellier-Willemin F, Moulhérat C (2006) La découverte de coton, Oasis de Kharga. Archaeol Text Newsl 43:20–27

Littmann E (1913) Deutsche Aksum-Expedition. Bd IV. Sabaische, Griechische Und Altabessinische Inschriften. Museum of Berlin, Berlin

Livingstone R (2009) Late antique household textiles from the village of kellis in the dakhleh oasis. In: De Moor A, Fluck C (eds) Clothing the house: furnishing textiles of the 1st millennium from Egypt and neighbouring countries. Lannoo, Tielt, pp 73–85

Livingstone RJ (2015) Dress and identity in Kellis, a Roman-period village in Egypt, thesis, Monash University. Faculty of Arts. Centre for Ancient Cultures, School of Philosphical, Historical and International Studies

Lombard M (2002) Les textiles dans le monde musulman du VIIe au XIIe siècle. École des Hautes Etudes en Sciences Sociales, Paris

Massey RE (1923) A note on the early history of cotton. Sudan Notes Rec 6(2):231

Mayer-Thurman CC, Williams B (1979) Ancient textiles from nubia: meroitic, x-group, and Christian fabrics from Ballana and Qustul. Art Institute of Chicago, Chicago

Mazzaoui MF (1981) The Italian cotton industry in the later middle ages, 1100–1600. Cambridge University Press, Cambridge

Miller NF (1998) Patterns of agriculture and land use at medieval gritille. In: Redford S (ed) Archaeology of the frontier in the medieval near East (Archaeological Institute of America Monographs 3). University of Pennsylvania Museum, Philadelphia, pp. 211–252

Miller NF (2010) Botanical aspects of environment and economy at Gordion. University of Pennsylvania Museum of Archaeology and Anthropology, Philadelphia, Turkey

Millet M (2013) Mouweis: une ville de l'empire de Méroé. Bulletin de la Société Française d'Egyptologie 186–187:83–98

Morrison KD, Feinman GM, Nicholas LM et al (1996) Typological schemes and agricultural change: beyond boserup in precolonial South India [and comments and reply]. Curr Anthropol 37(4):583–608

Moulhérat C, Tengberg M, Haquet JF (2002) First evidence of cotton at Neolithic Mehrgarh, Pakistan: analysis of mineralized fibres from a copper bead. J Archaeol Sci 29(12):1393–1401

Müller M, Murphy B, Burghammer M et al (2004) Identification of ancient textile fibres from Khirbet Qumran caves using synchrotron radiation microbeam diffraction. Spectrochim Acta B 59(10):1669–1674

Müller M, Papiz MZ, Clarke DT et al (2003) Identification of the textiles from Khirbet Qumran using microscopy and synchrotron radiation x-ray fibre diffraction. In: Humbert J-B, Gunneweg J (eds) Khirbet Qumrân et'Aïn Feshkha II. Studies of Anthropology, Physics and Chemistry. Academic Press, Vandenhoeck & Ruprecht, Fribourg, pp. 277–286

Murray A, Cappers RTJ (2007) Medieval cotton and wheat finds in the Middle Niger Delta (Mali). In: Fields of Change. Proceedings of the 4th international workshop for African archaeobotany (Groningen archaeological studies 5). Barkhuis & Groningen University Library, Groningen, pp. 43–51

Murray MA (2000a) Cereal production and processing. In: Nicholson P, Shaw I (eds) Ancient Egyptian materials and technology. Cambridge University Press, Cambridge, pp. 505–536

Murray MA (2000b) Fruits, vegetables, pulses and condiments. In: Nicholson P, Shaw I (eds) Ancient Egyptian materials and technology. Cambridge University Press, Cambridge, pp. 609–655

Muthukumaran S (2016) Tree cotton (*G. arboreum*) in Babylonia. In: Foietta E, Ferrandi C, Quirico E et al (eds) Cultural and material contacts in the ancient near east. Proceedings of the international workshop 1–2 December 2014, Torino. Apice libri, Sesto Fiorentino, pp. 98–105

Nehmé L (ed) (2015) Les tombeaux nabatéens de Hégra. Académie des Inscriptions et Belles-lettres, Paris

Nehmé L, Al-Talhi D, Villeneuve F (eds.) (2011) Report on the fourth excavation season (2011) of the Madâ'in Sâlih archaeological project

Nehmé L, Arnoux T, Bessac JC et al (2006) Mission archéologique de Madâ'in Sâlih, Arabie Saoudite: Recherches menées de 2001 à 2003 dans l'ancienne Hijrâ des Nabatéens. Arab Archaeol Epigr 17:41–124

Newman RJ (1932) The agricultural life of the Jews in Babylonia between the years 200 C.E. and 500 C.E. Oxford University Press, London

Newton C, Whitbread T, Agut-Labordère D et al (2013) L'agriculture oasienne à l'époque perse dans le sud de l'oasis de Kharga (Égypte, Ve-IVe s. AEC), Revue d'ethnoécologie (4)
https://ethnoecologie.revues.org/1294. Accessed 26 May 2017

Nicholson GE (1960) The production, history, uses and relationships of cotton (*Gossypium* spp.) in Ethiopia. Econ Bot 14(1):3–36

Obeid M, Mahmoud A (1971) Ecological studies in the vegetation of the Sudan. II. The ecological relationships of the vegetation of Khartoum province. Vegetatio 23(3–4):177–198

Omer SA, Agab H, Samad HA et al (2008) Effect of feed type on some blood constituents of Sudanese growing camel (*Camelus dromedarius*) calves. Sud J Vet Sci Anim Husb 47(2):107–116

Page JT, Huynh MD, Liechty ZS et al (2013) Insights into the evolution of cotton diploids and polyploids from whole-genome re-sequencing. G3: Genes Genom Genet 3(10):1809–1818

Palmer SA, Clapham AJ, Rose P et al (2012) Archaeogenomic evidence of punctuated genome evolution in *Gossypium*. Mol Biol Evol 29(8):2031–2038

Palmer SA, Moore JD, Clapham AJ et al (2009) Archaeogenetic evidence of ancient Nubian barley evolution from six to two-row indicates local adaptation. PLoS ONE 4(7):e6301

Paterson AH, Wendel JF, Gundlach H et al (2012) Repeated polyploidization of *Gossypium* genomes and the evolution of spinnable cotton fibres. Nature 492(7429):423–427

Pelling R (2013) The archaeobotanical remains (Chap. 18) and Botanical data appendices (Chap. 28). In: Mattingly DJ, Daniels CM, Dore JN et al. (eds) The Archaeology of Fazzan, vol. 4. Survey and excavations at old Jarma (Ancient Garama) carried out by C.M. Daniels (1961–1969) and the Fazzan project (1997–2001). Society of Lybian Studies, Department of Antiquities, London, 473–94:841-852

Phillipson DW (ed) (2000) Archaeology of Axum, Ethiopia, 1993–1997, vol II. The Society of Antiquaries, London

Phillipson DW (2012) Foundations of an African civilisation. Aksum and the Northern Horn 1000 BC-AD 1300 (Eastern Africa Series). Boydell & Brewer, New York

Plumley JM, Adams WY, Crowfoot E (1977) Qasr Ibrim 1976. J Egypt Archaeol 63:29–47

Rahmani LY (1967) Jason's tomb. Israel Exploration Soc 17(2):61–100

Reimer PJ, Bard E, Bayliss A et al (2013) IntCal13 and Marine13 radiocarbon age calibration curves 0–50,000 years cal BP. Radiocarbon 55(4):1869–1887

Reis D, Vian B, Bajon C (2006) Le monde des fibres. Belin, Paris

Riello G, Roy T (2009) How India clothed the world. The world of South Asian textiles, 1500–1850 (Global Economic History Series). Brill, Leiden

Rilly C (2007) La langue du royaume de Méroé: Un panorama de la plus ancienne culture écrite d'Afrique Subsaharienne. H. Champion, Paris

Rose P (1996) Qasr Ibrim: the Hinterland survey. Egypt Exploration Society, Cairo

Rose P (2013) Qasr Ibrim: the encyclopedia of ancient history. Wiley, Chichester

Rose PJ (2011) Kirwan memorial lecture. Qasr Ibrim: The last 3000 years. Sudan Nubia 15:3–12

Rose PJ, Edwards DN (1998) Excavations at Qasr Ibrim. Sudan Nubia 2:61–73

Rowley-Conwy P (1989) Nubia AD 0–550 and the 'Islamic'agricultural revolution: preliminary botanical evidence from Qasr Ibrim, Egyptian Nubia. Archéologie du Nil Moyen 3:131–138

Rowley-Conwy PA, Deakin W, Shaw CH (1999) Ancient DNA from *Sorghum*. The evidence from Qasr Ibrim, Egyptian Nubia. In: Van der Veen M (ed) The exploitation of plant resources in ancient Africa. Kluwer Academic, New York, pp 55–61

Ryesky D (1984) The textiles. In: Wenke RJ (ed) Archaeological investigations at El-Hibeh 1980: preliminary report. Undena, Malibu, pp 78–93

Samuel D (2001) Archaeobotanical evidence and analysis. In: Berthier S (ed) Peuplement rural et aménagements hydroagricoles dans la Moyenne Vallee de l'Euphrate fin VIIe-XIXesiècle. Institut français d'études arabes de Damas, Damascus, pp 343–481

Sanlaville P (2000) Le Moyen-Orient Arabe: Le milieu et l'Homme. Nathan, Paris

Schmidt-Colinet A (1995) Palmyra: Kulturbegegnung Im Grenzbereich. Verlag Philipp von Zabern, Mainz am Rhein

Shamir O (2015) Egyptian and Nubian textiles from Qasr el-Yahud, 9th century AD. In: De Moor A, Fluck C, Linscheid P (eds) Tielt: textiles, tools and techniques of the 1st Millenium AD from Egypt and neighbouring countries. proceedings of the 8th conference of the research group 'Textiles from the Nile Valley'. Antwerp, 4–6 October 2013, Lannoo

Shamir O, Baginski A (2013) Textile hoard from Jericho Cave 38 in the Qarantal Cliff. In: Guri-Rimon O (ed) Hoards and genizot as chapters in history. Hecht Museum University of Haifa, Haifa, pp 77–88

Shamir O, Baginski A (2014) The earliest cotton Ikat textiles from Nahal'Omer Israel 650-810 CE. In: Nosch M-L, Feng Z, Varadarajan L (eds) Global textile encounters (Ancient textiles series 20). Oxbow books, Oxford, pp 65–73

Shaw CH, Deakin WJ, Rowley-Conwy PA (2001) Ancient DNA from archaeological *Sorghum* from Qasr Ibrim, Egyptian Nubia. In: Millard A (ed) Archaeological Sciences'97 Proceedings of the conference held at the University of Durham, 2nd–4th September 1997 (BAR International Series). Oxbow Books, Oxford, p. 96–99

Sheffer A, Tidhar A (1991) The textiles from the 'En-Boqeq excavation in Israel. Text Hist 22(1):3–46

Sherratt A (1999) Cash-crops before cash: organic consumables and trade. In: Gosden C, Hather J (eds) The prehistory of food: appetites for change. Routledge, London, pp 13–34

Smith CE, Stephens SG (1971) Critical identification of Mexican archaeological cotton remains. Econ Bot 25(2):160–168

Smith CW, Cothren TJ (1999) Cotton: origin, history technology and production. Wiley, New York

Smith O, Clapham AJ, Rose P et al (2014) Genomic methylation patterns in archaeological barley show de-methylation as a time-dependent diagenetic process. Sci Report 4

Stanton MA, Stewart JM, Pervical AE et al (1994) Morphological diversity and relationships in the A-Genome Cottons, *Gossypium arboreum* and *G. herbaceum*. Crop Sci 34(2):519

Stauffer A (2000) The textiles from Palmyra: technical analyses and their evidence for archaeological research. In: Cardon D, Feugère M (eds) Archéologie des textiles, des origines au Ve siècle: Actes du Colloque de Lattes, Octobre 1999. Monique Mergoil, Montagnac, pp 247–252

Stephens SG (1975) A reexamination of the cotton remains from Huaca Prieta. North Coastal Peru. Am Antiquity 40(4):406–419

Stephens SG (1976) Some observations on photoperiodism and the development of annual forms of domesticated cottons. Econ Bot 30(4):409–418

Strouhal E (1984) Wadi Qitna and Kalabsha South I: archaeology. Czech Institute of Egyptology, Prague

Suttie JM (2004) Conservation du foin et de la paille pour les petits paysans et les pasteurs (Collection FAO, production végétale et protection des plantes 29). Organisation des Nations Unies pour l'alimentation et l'agriculture, Roma

Tallet G, Gradel C, Letellier-Willemin F (2012) 'Une laine bien plus belle et douce que celle des moutons' à El-Deir (oasis de Kharga, Égypte): le coton au cœur de l'économie oasienne à l'époque romaine. In: Guédon S (ed) Entre Afrique et Égypte: Relations et échanges entre les espaces du Sud de la Méditerranée à l'époque Romaine. Ausonius, Bordeaux, pp 119–141

Tengberg M, Lombard P (2001) Environnement et économie végétale à Qal'at al-Bahreïn aux périodes Dilmoun et Tylos. Premiers éléments d'archéobotanique. Paléorient 27(1):167–181

Tengberg M, Newton C (2016) Origine et évolution de la phéniciculture au Moyen-Orient et en Égypte. In: Ruas MP (ed) Des fruits d'ici et d'ailleurs: regards sur l'histoire de quelques fruits consommés en Europe. Omnisciences, Paris, pp 83–105

Terpstra T (2015) Roman trade with the Far East: Evidence for Nabataean middlemen in Puteoli. In: De Romanis F, Maiuro M (eds) Across the Ocean: nine essays on Indo-Mediterranean trade (Columbia Studies in the Classical Tradition). Brill, Leiden, pp 73–94

Thanheiser U (2002) Roman agriculture and gardening in Egypt as seen from Kellis. In: Hope CA, Bowen GE (eds) Dakhleh Oasis project: preliminary reports on the 1994–1995 to 1998–1999 Field Seasons. Oxbow Books, Oxford, pp 299–310

Thanheiser U, Kahlheber S, Dupras T (2016) Pearl millet, *Pennisetum glaucum* (L.) R.Br. ssp. *glaucum*, in the Dakhleh Oasis, Egypt. In: Thanheiser U (ed) News from the past: progress in African archaeobotany. Proceedings of the 7th international workshop on African archaeobotany in Vienna, 2–5 July 2012 (Advances in Archaeobotany 3). Barkhuis, Groningen, pp. 115–125

Thanheiser U, Walter J (2015) Plant use in a Romano-Egyptian household in the third century CE. In: Boozer AL Amheida II: a late Romano-Egyptian house in the Dakhleh Oasis. Amheida House B2. The Institute for the study of the Ancient world, New York, pp. 375–392

Thurston TL, Fisher CT (2007) An introduction to the archaeology of subsistence intensification, innovation, and change. In: Thurston TL, Fisher CT (eds) Seeking a richer harvest (Studies in Human Ecology and Adaptation 3). Springer, New York, pp 1–23

Török L (1988) Late antique nubia, history and archaeology of the Southern Neighbor of Egypt in the 4th–6th C. AD (Antaeus 16). Archaeological Institute of the Hungarian Academy of Sciences, Budapest

Trolinder EL (2009) Genetic engineering of cotton. In: Paterson AH (ed) Genetics and genomics of cotton (Plant Genetics and Genomics: Crops and Models 3). Springer, New York, pp 187–207

Van der Veen M (1999) The economic value of chaff and straw in arid and temperate zones. Veg Hist Archaeobot 8(3):211–224

Van der Veen M (2007) Formation processes of desiccated and carbonized plant remains—the identification of routine practice. J Archaeol Sci 34(6):968–990

Van der Veen M (2011) Consumption, trade and innovation: exploring the botanical remains from the Roman and Islamic ports at Quseir Al-Qadim, Egypt. Africa Magna Verlag, Frankfurt

Varisco D (2009) Agriculture in al-Hamdānī's Yemen: a survey from early Islamic geographical texts. J Econ Social Hist Orient 52:382–412

Vila A (1967) Aksha II Le cimetière méroïtique d'Aksha. Librairie Klincksieck, Paris

Vogelsang-Eastwood G (2001) Textiles. In: Grzymski KA, Anderson J (eds) Hambukol excavations, 1986–1989. Benben, Mississauga, pp 79–83

Vollesen K (1987) The native species of *Gossypium* (Malvaceae) in Africa, Arabia and Pakistan. Kew Bull 42(2):337–349

Wagner G (1987) Les Oasis d'Égypte, à l'époque grecque, romaine et byzantine d'après les documents grecs (BiEtud 100). Institut Français d'Archéologie Orientale, Cairo

Wagner G (1999) Les Ostraca Grecs de Douch IV (Document de fouille de l'Institut français d'archéologie orientale 24). Institut Français d'Archéologie Orientale, Cairo

Wagner G (2001) Les Ostraca grecs de Douch V (Document de fouille de l'Institut français d'archéologie orientale 24). Institut Français d'Archéologie Orientale, Cairo

Walsh RPD (1991) Climate, hydrology, and water resources. In: Craig GM (ed.) Agriculture in the Sudan. Oxford University Press, Oxford

Walshaw SC (2010) Converting to rice: urbanization, Islamization and crops on Pemba Island, Tanzania, AD 700–1500. World Archaeol 42(1):137–154

Watson AM (1983) Agricultural innovation in the early Islamic world: the diffusion of crops and farming techniques, 700–1100. Cambridge University Press, Cambridge

Wendel JF, Brubaker C, Alvarez I et al (2009) Evolution and natural history of the cotton genus. In: Paterson AH (ed) Genetics and genomics of cotton (Plant Genetics and Genomics: Crops and Models 3). Springer, New York, pp 3–22

Wendel JF, Cronn RC (2003) Polyploidy and the evolutionary history of cotton. Adv Agron 78:139–186

Wendel JF, Small RL, Cronn RC et al (1999) Genes, jeans, and genomes: reconstructing the history of cotton. In: Proceedings of the International Organization of Plant Biosystematists, Amsterdam, pp. 133–161

Whitewright J (2007) Roman rigging material from the Red Sea port of Myos Hormos. Int J Naut Archaeol 36(2):282–292

Wild FC, Wild JP (2001) Sails from the Roman port at Berenike. Egypt. Int J Naut Archaeol 30(2):211–220

Wild JP (1997) Cotton in Roman Egypt: some problems of origin. Al-Rafidan 18:287–298

Wild JP (2013) The first Indian carpets—a view from Berenike. In: De Moor A, Fluck C (eds) Drawing the threads together: proceedings of the 7th conference of the research group ‘Textiles from the Nile Valley’, Antwerp, 7–9 October 2009. Lannoo, Tielt, pp. 74–85

Wild JP, Wild FC (2008) Cotton: the new wool. Qasr Ibrim study season 2008. Archaeol Text Newsl 46:23–26

Wild JP, Wild FC (2009) Qasr Ibrim study: study season 2009(49):16–18

Wild JP, Wild FC (2014a) Through Roman eyes: cotton textiles from Early Historic India. In: Bergerbrant S, Fossøy SH (eds) A stitch in time. Essays in Honour of Lise Bender Jørgensen (Gothenburg Archaeological studies 4). Gothenburg University, Gothenburg, pp. 209–235

Wild JP, Wild FC (2014b) Qasr Ibrim: new perspectives on the changing textile cultures of Lower Nubia. In: Egypt in ER O’Connell (ed) the First Millenium A.D. (British Museum Publications on Egypt and Sudan 2). Peeters Publishers, Louvain, pp. 71–80

Wild JP, Wild FC (2014c) Berenike and textile trade on the Indian ocean. In: Droß-Krüpe K (ed) Textile trade and distribution in antiquity. Harrassowitz, Wiesbaden, pp 91–110

Wild JP, Wild FC, Clapham AJ (2007) Irrigation and the spread of cotton growing in Roman times. Archaeol Text Newsl 44:16–18

Wilkinson TC (2014) Tying the threads of Eurasia. Trans-regional routes and material flows in Transcaucasia, Eastern Anatolia and Western Central Asia, c. 3000–1500 BC. Sidestone Press, Leiden

Wilson LM (1933) Ancient textiles from Egypt in the University of Michigan Collection. University of Michigan Press, Ann Arbor

Winter JG, Youtie HC (1944) Cotton in graeco Roman Egypt. Am J Philology 65:249–258

Worp KA (1995) Greek Papyri from Kellis I (Dakhleh Oasis Project 3). Oxbow Monograph, Oxford

Worp KA, Hope CA (2004) Greek Ostraka from Kellis: O. Oxbow Books, Oxford, Kellis

Wulff HE (1966) The traditional crafts of Persia: their development, technology, and influence on Eastern and Western civilizations. M.I.T Press, Cambridge

Yvanez E (2012) Les textiles des nécropoles méroïtiques de Saï. Cahiers de recherches de l’Institut de papyrologie et d’égyptologie de Lille 29:331–344

Yvanez E (2015) De fil en aiguille: aspects de l’artisanat textile méroïtique. Egypte Afrique et Orient 78:63–66

Yvanez E (2016) Spinning in Meroitic Sudan: textile production implements from Abu Geili. Dotawo: J Nubian Stud 3:153–178

Zech-Matterne V, Tengberg M, Andringa VW (2015) *Sesamum indicum* L. (sesame) in 2nd century BC Pompeii, southwest Italy, and a review of early sesame finds in Asia and Europe. Veg Hist Archaeobot 24(6):673–81

Sorghum Domestication and Diversification: A Current Archaeobotanical Perspective

Dorian Q. Fuller and Chris J. Stevens(✉)

Institute of Archaeology, University College London, London, UK
{d.fuller, c.stevens}@ucl.ac.uk

Abstract. *Sorghum bicolor*, one of the world's five most important crops, originated in Africa. While this has long been clear, accumulating data from both archaeobotany and genetics, provides the basis for a new overview on the domestication process, racial evolution, and geographical dispersal of sorghum. Archaeobotanical finds from 113 sites in Africa and Eurasia are reviewed and mapped. Of these only 16 provide identifications of probable morphological races. Domestication is evidently taking place more than 3000 years BC in the eastern Sudan near the Atbara and Gash rivers. Early domesticated race bicolor then spread to South Asia around 2000 BC and to the Niger Basin in West Africa after 1000 BC. The framework of five cultivated races remains useful, with the original domesticated race bicolor being characterized by tight-fitting hulls requiring dehusking and the other races representing subsequent parallel evolution for free-threshing and larger-grained cultivars. This took place at least three times, including race 'caudatum' focused initially on the Sahelian region race 'durra' that evolved probably in India, and race 'guinea' that evolved in forested West Africa. Early race guinea in turn produced an even more forest adapted 'mageritiferum' type that appears to be ancestral to southern African guinea and 'kafir' sorghums, implying a dispersal across the central African rainforests. In contrast other eastern African caudatums and 'bicolor' types presumably followed a savannah dispersal. In addition to the early dispersal of race bicolor from Africa to India, which was ancestral to East Asian sorghums, a later dispersal of guinea types is inferred to have taken place from southeastern Africa across the Indian Ocean.

Keywords: Archaeology · Genetics · Races · Nubia · India

Introduction

Alongside wheat, barley, maize and rice, sorghum forms one of the five dominant staple cereal crops, and with pearl millet is one of the two main cereal crops to emerge from Africa. For much of Africa, the history of sorghum is the history of early farming. Yet despite its significance it is the only one of the five major cereals whose domestication is still relatively little understood in terms of the timing, location(s) of domestication, ecological environment, and the nature and length of this process.

Some 40 years ago Harlan and Stemler (1976) wrote a seminal paper "The races of Sorghum in Africa", and together with "Variability in *Sorghum bicolor*" (de Wet et al. 1976), formed an important step in our understanding of sorghum domestication.

A. M. Mercuri et al. (Eds.): *Plants and People in the African Past*,
https://doi.org/10.1007/978-3-319-89839-1_19

Here, we revisit these papers presenting an update of the current state of knowledge, utilizing a database of archaeobotanical finds across Africa and Eurasia alongside recent insights from molecular genetics. After summarizing the variability of sorghum and genetic evidence for the relationships between different morphological and geographical populations we summarize the archaeobotanical evidence. We summarize new evidence for early cultivation and the domestication process in eastern Sudan, followed by a review of the chronology of the spread of sorghum across Africa and Asia, and the evidence for the evolution and dispersal of advanced free-threshing races of sorghum.

One of the challenges to documenting the cultural and evolutionary history of sorghum is a paucity of hard evidence in terms of archaeobotanical remains. Undoubtedly the biggest contributor to this problem has been the limited application of large scale and systematic archaeobotanical sampling programs in Africa, although these are now becoming increasingly common, with important results (e.g. Bigga and Kahlheber 2011; Giblin and Fuller 2011; Logan 2012; Crowther et al. 2016). Nevertheless, preservation on African sites is often poor, which is likely acerbated by both patterns of occupational mobility associated with the importance of pastoralism as well as high bioturbation in tropical soils. One important line of evidence has proved to be the impressions in ceramics from the use from plant harvesting or processing waste as temper (e.g. Stemler 1990; Manning and Fuller 2014; McClatchie and Fuller 2014; Beldados et al. 2018; Winchell et al. 2017). Such studies of ceramic plant fossils offers a method by which to gain some plant evidence from even the most deflated pastoral sites and can be studied from existing archives of ceramics of excavations or surveys in regions that may be less accessible to modern re-sampling.

The State of Knowledge 40 Years Ago

Research into the domestication of sorghum prior to the late 1960's had undoubtedly been hampered by complications within the taxonomy of the species. The first significant attempt to decipher the cultivated sorghums was made by Dr Otto Stapf (1917), an Austrian born botanist and pioneer of archaeobotany (Stapf 1886), during his time at the Royal Botanical Gardens in Kew. However, it was soon clear on the completion of this work that further inquiry was needed and to this effect Hugh Charles Sampson, Kew's first economic botanist, wrote to various Directors of Agriculture across the former British Empire requesting flowering and fruiting sorghum specimens that were sent to Kew (see Hill, in Snowden 1936, p. iii). The task to classify this material was handed to Joseph Davenport Snowden a recently retired economic botanist within the Department of Agriculture in Uganda. The culmination of this work was the "*Cultivated Races of Sorghum*" (Snowden 1936) a meticulously detailed work that divided cultivated sorghums into 6 sub-series, comprising 28 cultivated species, and 156 varieties.

This work was later revised and simplified by Harlan and de Wet (1972) into five basic cultivated races and 10 intermediate races, based upon their seeds, glumes and inflorescence shape (see Table 1). The taxonomy outlined by Harlan and colleagues (Harlan and de Wet 1972; de Wet et al. 1976; Harlan and Stemler 1976) also proposed that wild, weedy and cultivated sorghums were treated as a single species, *Sorghum bicolor*, with five races of the domesticated subspecies *bicolor*, and four African wild races in the subspecies *arundinaceum* (now subsp. *verticilliflorum*). A wild sorghum, currently *Sorghum propinquum* (Kunth) Hitchc. was also classified as a separate race, but was noted to be the only race found outside Africa (de Wet et al. 1976). In Table 2 we provide a simplified comparison of a binomial taxonomy, in which wild forms and domesticated forms are kept as separate species (a simpler system to follow archaeobotanically), and the single species taxonomy de Wet (1977), updated by Wiersema and Dahlberg (2007). In this paper we will follow the binomial system.[1]

The domestication of sorghum was proposed by Harlan and Stemler (1976) to have occurred in the savannah between western Ethiopia and eastern Chad, e.g. in the Republic of the Sudan from subsp. *arundinaceum* (correctly subsp. *verticilliflorum* (Steud.) de Wet ex Wiersema and Dahlberg 2007), presumably from the wild race aethiopicum, which is adapted to drier semi-desert/Sahel conditions. The oldest race they identified as bicolor on the basis of it being the least morphologically specialized, having the widest geographical distribution, its similarities to *arundinaceum* and that back-crossing the other races with wild sorghums resulted in hybrids displaying many of the bicolor characteristics (see de Wet et al. 1976).

The other four races are characterized by being free-threshing, in contrast to the tightly hulled grains of *S. arundinaceum* or race bicolor. Thus sorghum evolution parallels wheat and barley in the post-domestication selection for free-threshing forms (see Zohary et al. 2012; Fuller and Lucas 2014). Race guinea they saw as emerging in West Africa from primitive bicolors selected for tolerance of high rainfall, after which guinea spread back eastwards. Snowden (1955) speculated that wild race arundinaceum (*Sorghum arundinaceum* (Desv.) Stapf.), found in the coastal regions of Guinea and Congo, hybridized with race bicolor to give rise to the local West African Sorghums (now classed as guinea). The origin of caudatum they saw as overlapping in range with that of bicolor, broadly in the Sudan region. They associated the dispersal of caudatum with "Chari-Nile" languages (an internal grouping of Nilo-Saharan languages including East Sudanic, Central Sudanic), found today in South Sudan, parts of Ethiopia, Eritrea, Darfur and Nubia. The distribution of durra, being historically the dominant sorghum within both India and the Near East, as well as the north-eastern coastal regions of Africa and southern Sahara, suggested to Harlan and Stemler that the durra race originated outside Africa, within India or the Near East, from primitive bicolors that had been transported east in prehistory, adapting to drier regions, dispersing later back to eastern Africa. The final race kafir is largely grown by Bantu speaking peoples south of the equator. As such the race was identified as having arisen within southern Africa, potentially through hybridization with local, geographically separate wild race verticilliflorum sorghums (Shechter and de Wet 1975).

Chronologically, while bicolor was identified as the earliest race, appearing before 1000 BC they then hypothesised that guinea was most likely the second race to emerge due to its wide distribution. The third, durra, was seen as having emerged in India by

[1] It is worth noting that there remain some nomenclature differences between published taxonomy and that listed as accepted on www.theplantlist.org (accessed 28 Feb. 2016).

Table 1. The morphological races of sorghum (after Harlan and de Wet 1972)

	Race	Abbrv.	Characteristics
(1)	bicolor	(B)	Seed is elongated, long "clasping" glumes and "open" panicle/inflorescence
(2)	guinea	(G)	Grains round or broadly ovate, flattened on both dorsal and ventral sides and twisted between long "clasping" glumes. Panicles are large and open or loose. Often in wetter habitats
(3)	caudatum	(C)	Grains plano-convex flatted on the lower side "turtle-backed" and flat on the other. Glumes often shorter in length than grain.
(4)	kafir	(K)	Seed broadest above the middle – Panicles often compact. Spikelets sessile, ovate with "clasping" glumes like in guinea.
(5)	durra	(D)	Inflorescence compact. Spikelets sessile and often flattened, Seed broadest above the middle, but near spherical. Glumes wide, creased and differently colour to base. In drier habitats.

Table 2. A comparison of current binomial classification at left (followed here) and the single species taxonomy following Wiersema and Dahlberg (2007)

Selected binomial nomenclature (followed here for archaeobotanical finds)	Single species classification
Sorghum verticilliforum (Steud.) Stapf. Wild progenitor complex, mostly annual, 4 races (verticilliflorum, arundinaceum, aethiopicum, virgatum)	*Sorghum bicolor* subsp. *verticilliflorum* (syn. *S. bicolor* subsp. *arundinaceum* in de Wet (1978)) Wild progenitor complex, 4 races (as left column)
Sorghum drummondii (Steud.) Millsp. & Chase The weed complex	*Sorghum bicolor* subsp. *drummondii* Weed complex, feral sorghum or backcrosses between subsp. *bicolor* and subsp. *verticilliflorum*
Sorghum bicolor (L.) Moench. Domesticate. Diploid. Includes 5 races (cf. Table 1)	*Sorghum bicolor* subsp. *bicolor* The domesticated crop, includes 5 races.
Sorghum propinquum (Kunth) Hitchc. Perennial wild sorghum found in South and Southeast Asia. Diploid. Hybridizes with *S. bicolor*	*Sorghum propinquum* (Kunth) Hitchc
Sorghum halepense (L.) Pers. Perennial wild sorghum with creeping rhizome on damp soils in Southwest Asia through the Indus (tetraploid). Hybridizes with *S. bicolor*	*Sorghum halepense* (L.) Pers

the first millennium BC, and then spreading back to Africa. Guinea they speculated was around 2000 years ago taken directly from Ethiopia to north-west India, utilizing well established existing trade networks. That kafir was absent from India led them to postulate that this race emerged in southern Africa less than 2000 years ago, after guinea had spread from Africa to India. The most recent of the races, was suggested to be caudatum, due to its very limited distribution. At some time, probably by the early first millennium AD, they proposed that the Indian derived durra race came back west into northeast Africa via Egypt and the Sinai, a movement that led the writers of ancient Greece and Rome to speculate that sorghum came from India. The spread of race bicolor from India into China was seen to be via Burma (modern Myanmar) spreading into Indonesia and northwards into China by at least the mid first millennium AD (Hagerty 1941; Harlan and Stemler 1976). This last saw the emergence of the Chinese kaoliang, a variety of race bicolor (Snowden's *Sorghum nervosum* Bess ex. Schult.). However, there was little archaeological evidence in the 1970s with which to corroborate this framework.

The Races of Sorghum as seen from Genetics

While Snowden (1936) originally envisaged multiple domestications, Harlan and colleagues proposed a single origin for bicolor with later separation of guinea, durra, kafir and caudatum, each involving some degree of introgressions with local wild populations. A schema for understanding sorghum evolution in these terms is illustrated in Fig. 1. In recent years a large number of genetic studies have examined the division of the five main races. In the majority of studies some clear differentiation has been seen between the later races; guinea, durra, kafir and caudatum, with race bicolor showing much less clear differentiation (Deu et al. 1994, 2006; Casa et al. 2008; Mace et al. 2008; Brown et al. 2011). An earlier study demonstrated that cultivars were most closely aligned with wild populations of sorghum within central-northeast Africa (Aldrich and Doebley 1992). This study and others also note two to three strongly differentiated haplotype groups highlighting the problems of differentiating between potentially separate domestication events and later migration events with introgression from local wild populations (de Alencar Figueiredo et al. 2008; cf. Lin et al. 2012).

Genetic studies, while recognizing the separation of the races, also note divisions within them. Guinea showed separation into four clusters; two closely related in West Africa, and two related clusters in Southern Africa and Asia (Deu et al. 1994, 2006). Further the populations of sub-race margaritiferum of West Africa, with small spherical grains, clearly separated from other guineas (Deu et al. 1994, 2006; Casa et al. 2008; de Alencar Figueiredo et al. 2008; Brown et al. 2011; Ramu et al. 2013; Luo et al. 2016), showing close affinity to local wild and weedy sorghums (Billot et al. 2013; Morris et al. 2013) of wild race arundinaceum. Other studies also demonstrate a genetic affinity between guinea populations of southern and Eastern Africa with those of South Asia (Folkertsma et al. 2005), supporting a hypothesis suggesting the migration of this crop during Indian Ocean maritime trade networks from the 8th century AD onwards (see Boivin et al. 2014). Other studies show Indian guineas clustering more closely with durras (Morris et al. 2013), which might be accounted for by ongoing introgression across Indian crop populations, or both having introgression from wild *Sorghum propinquum*.

These studies demonstrate the need to reclassify some of the sub-races. One clear example is the West African kaura group (Snowden's *S. caudatum* var. *kerstingianum*) which clusters with guineas, not with durra-caudatums into which they are traditionally

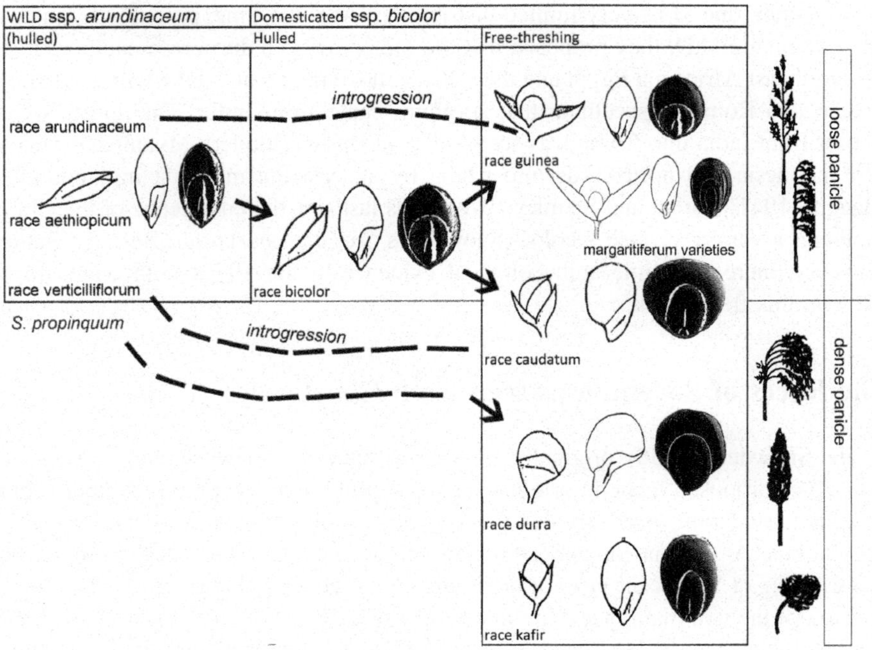

Fig. 1. The sorghum evolutionary framework relating wild, domesticated and the major domesticated races. Horizontal axis show seed dispersal from wild shattering to free- threshing; vertical axis the development of panicle form/density. Solid arrows indicate the main direction of evolution and improvement. Dashed lines indicate hypothetical introgression from wild populations.

ascribed (Brown et al. 2011). Genetically kaura sorghum is then derived from race guinea, but later evolved tighter, compacted panicles and shorter glumes, characteristic of the durra and caudatum races respectively, perhaps in response to the drier regions of northeast Nigeria, where they are most often found.

While the general implication is that genetic studies broadly support the morphological separation of the races, one of the largest and most recent studies by Billot et al. (2013) highlights the importance of geographic distribution with respect to genetic differentiation. For example, eastern Asian caudatums, durras and bicolors appear more closely related to each other than to these same races in Africa (Billot et al. 2013). What remains unclear is whether this implies that free-threshing forms evolved many times in parallel or whether the movement of free-threshing races was followed by extensive introgression with locally adapted landraces of bicolor. The latter is perhaps more parsimonious, requiring fewer selection events for free-threshing morphotypes. What is needed is an understanding of the particular alleles that create the various free-threshing morphotypes.

Durra showed strong separation in most studies (Deu et al. 2006; Perumal et al. 2007; Casa et al. 2008; Mace et al. 2008; Brown et al. 2011). Within the study conducted by Brown et al. (2011) Asian durra sorghums clustered together, but were most closely related to durras from Somalia, and Ethiopia. Other studies demonstrate similarities between northern Chinese Kaoliangs (traditionally classified as bicolor due to their persistent husks; Snowden's *S. nervosum*) with durra types, especially from Yemen (Morris et al. 2013). This might suggest that durra evolved from the same South Asian populations that dispersed to China, a process still poorly documented archaeologically, but probably dated to the early to mid-first millennium AD from northeast India via Yunnan (Hagerty 1941; Bonjean 2010).

Those assigned to the unfortunately named kafir (Snowden's *S. cafrorum*) exhibited strong separation in all studies (Deu et al. 1994, 2006; Casa et al. 2008; Mace et al. 2008; Brown et al. 2011). The low diversity seen within kafir, along with West African guineas, suggests bottlenecks relating to geographical isolation (Casa et al. 2008; see Deu et al. 2006). A study on photoperiod indicated that guinea types brought from West Africa into southern Africa, potentially gave rise to kafirs, adapted to longer-day ecosystems. Perhaps associated with Bantu-speaking groups that crossed the central African rainforest and migrated slowly southwards (cf. Klein et al. 2015). This would represent the west and southwest Bantu stream via the Sangha River Interval (sensu Grollemund et al. 2015), implying an arrival in southern Africa later than the initial sorghum associated with early Iron Age Eastern Bantu migrants.

Caudatums split into two clusters, the main one geographically centred on the Great Lakes Region of East Africa e.g. Tanzania, Malawi, Zambia, Kenya, Uganda (Deu et al. 1994, 2006), but stretching into Ethiopia and Sudan (Brown et al. 2011). However, many studies demonstrated sorghum lines assigned to caudatum outside this geographical area falling into separate groups (see Deu et al. 1994, 2006; Mace et al. 2008; Billot et al. 2013). Studies also separated out zerazera-caudatums (Snowden assigned as a form of *S. caudatum* var. *durum*), found mainly in the border region between Sudan and South Sudan but stretching southeast, from West African and Central Sudanese guinea-caudatums (probably genetically guineas) and other caudatums centred mainly on South Sudan, Uganda and northeast Congo, with some further groups in South Africa (Sukumaran et al. 2012; Casa et al. 2008).

Many studies imply at least two separate introductions of sorghum into Asia (Deu et al. 2006). The first was an early bicolor that evolved into durra and Chinese kaoliang (*S. nervosum*). The second presumably was an east African guinea variety traded from the Swahili world. The origins of these bicolors in northern Sudan would explain the genetic affinities to some central African caudatums (cf. Billot et al. 2013), which also evolved in northern Sudan. Guinea sorghums often have a hard, corneous endosperm, making the grains suited to whole-grain boiling, and these forms are popular in the tribal zones of eastern and central India (Appa Rao et al. 1996). Whether or not caudatum sorghums in India represent a third ancient introduction, a recent introduction, or rather are just morphologically similar adaptions from durra and guinea/kafir types remains to be determined. Notably, one caudatum type from India examined by Brown et al. (2011) was genetically closer to kafir, implying that the transition between guinea/kafir and caudatum morphology (i.e. a c. 45° twist in the grain primordium) represents a simple genetic shift that occurred locally in Indian guinea sorghum.

While most genetic research is congruent with the conventional view of a single major domestication episode of sorghum, followed by differentiation into races during and after geographical dispersal, a high profile study of non-shattering genetics argued for three independent domestications, two in northern Africa and one in the southeast (Olsen 2012). This is based on the identification of three alternative dominant mutations to the sorghum gene Sh1 on chromosome 1 that turn off the abscission zone at the base of the spikelet, and have differing geographical distributions. One of these mutations (Tx623) is found in kafir, bicolor and guinea sorghum races from southern Africa, another (Tx430) is largely restricted to caudatums throughout Africa, and a third (SC265) is found across all races except kafir, in both Asia and Africa (Lin et al. 2012: Table S1). Before considering the implications of this study, some caution is warranted, as early claims to have identified the key mutation for non-shattering in rice domestication (e.g. Li et al. 2006, 2007) have been overturned with evidence that the interaction of several loci is involved and that these evolved sequentially, some probably only in limited post-domestication lineages (e.g. Oryza *qsh1*) (Ishii et al. 2013; Htun et al. 2014). If multiple initial domestication episodes of race bicolor had involved other loci that also caused reduced shattering, then a higher degree of non-shattering might be expected to be seen as advanced free-threshing races evolved, much as the free-threshing *Q* gene was added to the non-shattering *Br1* mutations that was involved in hulled wheat domestication (Li and Gill 2006; Fuller and Allaby 2009: 251). Thus we predict that sorghum *Sh1* was involved in post-domestication improvements to non-shattering rather than to the initial evolution of non-shattering during race bicolor domestication. This may be further supported by the many race bicolor accessions reported as having only "moderate" non-shattering phenotypes in the USDA database (NPGS 2016), whereas only strongly shattering domesticated and non-shattering wild accessions were included in the study of Lin et al. (2012). Looking more closely at these data (Lin et al. 2012: Table S1), they can be reduced to just two pathways as the kafir type (Tx623) variant could be derived from further mutations (mainly a large deletion and 2 SNPs) to the caudatum type (Tx430) haplotype. By contrast the third variant (SC265) appears to be caused by a small number of different SNPs, shared across a range of races and regions, including widespread race bicolor accessions. This could be congruent with evolution of Sh1 twice, once within the western, the other towards the eastern parts of the zone of initial north Sudanese sorghum domestication (see archaeobotanical discussion, below). These mutations were plausibly additive to other alleles, not yet characterized, that are associated with earlier moderate non-shattering phenotypes (of which there are many varieties among Sudanese race bicolor, but none of them were included in the genetic study of Lin et al. 2012).

An Updated Archaeobotanical Framework of Sorghum Domestication

Domestication represents a number of adaptations on the part of a plant to being cultivated and harvested by people (see Harlan et al. 1973; Fuller et al. 2010, 2014). Key changes include loss of natural seed dispersal (visible archaeologically in spikelet bases), changes in germination traits (not easily visible archaeologically), and increase in grain size, most notably grain width (measurable archaeologically). To date, however, few archaeobotanical assemblages have allowed these domestication traits to be documented for sorghum.

To begin wild sorghum was likely gathered with baskets as stands of wild-type shattering plants matured. Such spikelets would still have had tight fitting husks and hence required dehusking, then winnowing, to obtain clean grain. With the evolution of non-shattering (domesticated) type plants, threshing and a further winnowing stage would be required, in order to break the rachises and secondary branches and hence separate the spikelets from the harvested panicles. Possibly winnowing/sieving practices already existed following the basket gathering of wild spikelets, in order to remove lighter plant parts and seeds from other wild species collected at the same time; the spatial patterning in wild sorghum versus other grasses in Nabta Playa contexts could be indicative of such practices (cf. Wasylokowa and Dahlberg 1999, 2001).

The development of non-shattering (domesticated) plants would have necessitated more labour input for processing, while new techniques and technologies of harvesting, such as sickles, would have been promoted. Following the trends for other cereals, cultivation would likely have selected for larger grains, leading to increased yields (Fuller et al. 2010). Subsequent evolution of the free-threshing sorghum races would have reduced and shifted the labour investment in processing. Grains would no longer have needed to be routinely dusked and winnowed, but the removal of chaff during threshing would potentially have prolonged winnowing operations following the harvest. As such practices were likely conducted seasonally and in the field, free-threshing sorghum may be less prone to preservation by charring than race bicolor. The dense panicles in some advanced races also facilitated harvesting, as well as increasing yields. Thus when considered in terms of harvesting, processing and related aspects of morphology, we can outline the morphological variation of sorghum, represented by its domestication and racial evolution (Fig. 1).

The first step in studying domestication is reconstructing the distribution of wild progenitor populations for the period in the past when cultivation is likely to have begun (Fig. 2). However, the distribution of these populations today is quite different from what they would have been at the start of the Holocene. Under the wetter conditions of the early to middle Holocene, the era of a Green Sahara (Kröpelin et al. 2008; Manning and Timpson 2014), the savannahs with sorghum were located much further north, approximately midway across the modern Sahara. This is supported by archaeobotanical finds of morphologically wild sorghum in the Western Desert of Egypt and southwest Libya (Fig. 2), dating to the early to middle Holocene, the earliest at Nabta Playa E-75-6, c. 8000 BC (Wasylikowa and Dahlberg 1999), with later finds from Farafra, c. 5800 BC, and Abu Ballas, c. 3750 BC (Barakat and Fahmy 1999). These sites provide no evidence for cultivation but it is likely that wild grains were collected (see Wasylikowa and Dahlberg 2001). Continued wild sorghum use in the 5th and 4th millennium BC, is indicated by impressions in pottery from central Sudan (Magid 1989, 2003; Stemler 1990) in the Khartoum region. These comprise Neolithic societies with evidence for domesticated sheep, goat and cattle (Gautier and van Neer 2006; Chaix and Honegger 2015). This shift in the evidence is congruous with a southerly retreat of wild sorghum stands as the Sahara dried and expanded. Possibly this involved southward migrations, and would fit with Nilo-Saharan proto-language reconstructions (Ehret 2014) and evidence for the southward spread of pastoralism. The

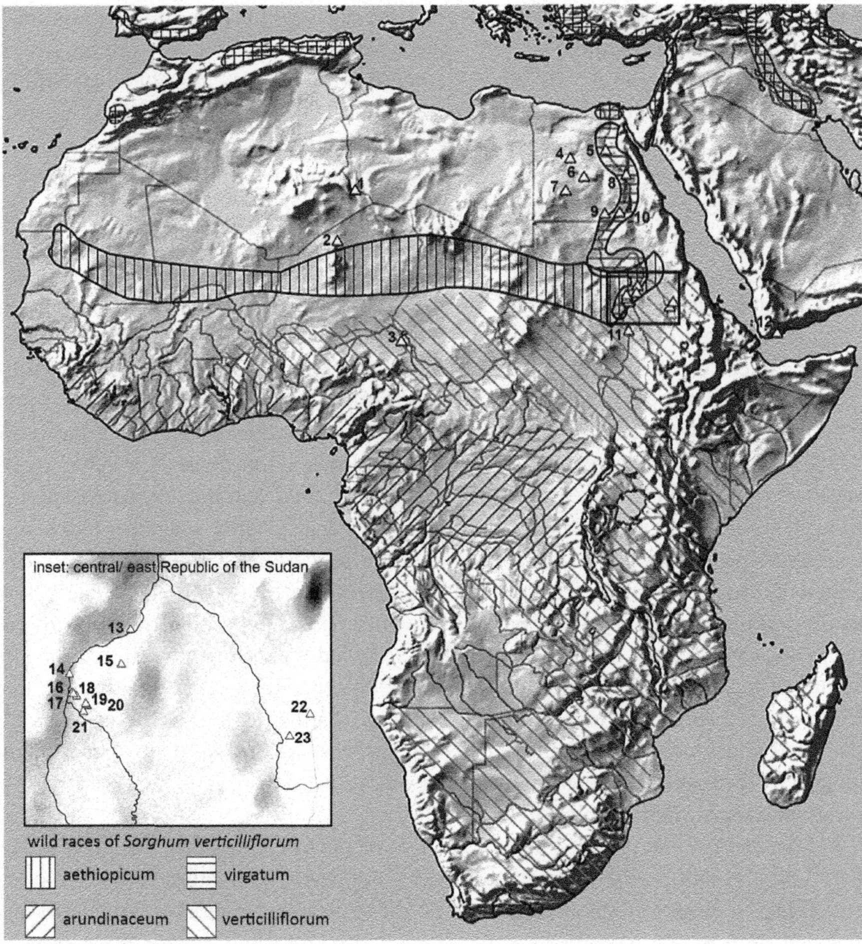

Fig. 2. Distribution of wild modern and archaeological sorghum. Hatched area represents approximate/potential modern distributions of wild sorghum races (*S. verticilliflorum*) after Harlan and Stemler (1976). Triangles indicate archaeological sites with finds of morphologically wild sorghum: 1. Takarkori (7000–6200 BC), 2. Adrar Bous (3500–1500 BC), 3. Kursakata (1000–200 BC), 4. Farafra Oasis (6060–5560 BC), 5. Amarna (1365–1345 BC), 6. Dalkhleh Oasis sites (8300–7000 BC), 7. Abu Ballas (4500–3000 BC), 8. Tutankhamun's tomb (1328 BC), 9. Nabta Playa E-75-6 (8100–7900 BC), 10. Qasr Ibrim (1000–300 BC), 11. Rabak (4500–4300 BC), 12. Sabir (1000–800 BC), 13. El Kadada (4000–3000 BC), 14. Shaheinab (4100–3300 BC), 15. Shaqadud (3500–2200 BC), 16. El Zakiab (4300–4000 BC), 17. Kadero (3800–3600 BC), 18. Um Direiwa (5000–3800 BC), 19. El Mahalab (6650–5800 BC), 20. Sheikh el-Amin (7000–6450 BC), 21. Sheikh Mustafa (4400–3300 BC), 22. Mahal Teglinos K1 (2000–1700 BC), 23. Khashm el Girba KG23 (3500–3000 BC).

evidence presumably relates to exploitation of *S. verticilliflorum* race aethiopicum, while other wild sorghums so far lack evidence for exploitation. Wild *Sorghum* sp. identified in first millennium BC Yemen (de Moulins et al. 2003) seems most likely race verticilliflorum on geographical grounds.

While Stemler (1990) identified a single impression of a possible "domesticated-type" spikelet with a torn rachilla from Um Direiwa (> 3800 BC) this might still relate to the harvesting of green or near-ripe wild sorghum. However, a larger dataset of sorghum pottery impressions has recently come to light dating to the mid to late fourth millennium BC at Khashm el Girba KG23 near Kassala (Winchell et al. 2017). These new data consist of 265 plant impressions of sorghum grains and spikelets preserved in sherds of Khordhag plain pottery of the Butana group period from site KG23 (see Winchell 2013 for archaeological details). Among these are 16 impressions from 9 sherds, that clearly preserve evidence for non-shattering sorghum spikelet bases with torn rachilla (Fig. 3a, b), while 18 other impressions spread across 13 sherds preserve wild type smooth spikelet bases (Fig. 3c, d). A further 17 impressions were identified as recognizably immature. Taken together these data suggest the harvesting of cultivated populations undergoing domestication.

These sorghum spikelets, comprising mature non-shattering types, mature shattering wild types, and immature "green" types, are consistent with the harvesting of populations still displaying asynchronous ripening. Hence, during these early stages of pre-domestication cultivation, harvesting was conducted before all plants had matured to avoid grain loss caused by the still high presence of wild-type shattering plants. This same practice has been observed during the earlier stages of pre-domestication cultivation for rice in the Lower Yangtze, where the "green" harvesting of rice resulted in charred archaeobotanical assemblages that included immature spikelet bases, wild type spikelet bases and non-shattering domesticated types (Fuller et al. 2010). These finds, together with those from the early second millennium BC at Kassala K1 (Beldados and Costantini 2011; Beldados et al. 2018) suggest a protracted period of pre-domestication sorghum cultivation from before 3500 BC to perhaps c. 1700 BC. These data are consistent with the protracted domestication episodes of 2000–4000 years documented for other cereals, including wheats, barley and rice (Fuller et al. 2014).

These new finds indicate that morphologically domesticated sorghum was present in eastern Sudanese populations, even if these types were not fully dominant before 3000 BC. They fit with the region of domestication hypothesized by Clark (1984), lying within the confluence of the Blue and White Nile, and the Atbara River, but are earlier than he anticipated. Of particular importance the Kassala evidence pre-dates finds in India, where domesticated sorghums are well established as being present by at least the early second millennium BC (Fuller 2003a; Fuller and Boivin 2009). Although grains from the Harappan site of Kunal are considered earlier (c. 2400 BC) (Saraswat and Pokharia 2003), these are not-directly dated and full details of their context are not available, so caution is warranted in accepting them as mid third millennium BC. It should be noted also that older reports of sorghum from the Arabian Peninsula are problematic. Impressions from the site of al-Raqlah in Yemen, 3000–2500 BC (Costantini 1990) are poorly illustrated and hard to accept as legitimate (Rowley-Conwy et al. 1997; Fuller 2002, 281–2; de Moulins et al. 2003; Fuller and Boivin 2009): these are too round for an early bicolor form that would be expected at

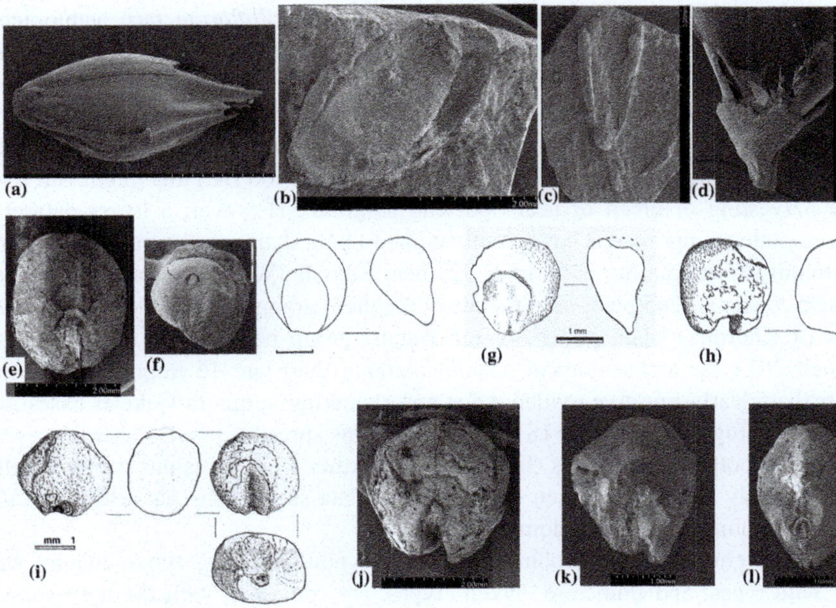

Fig. 3. Examples of archaeological sorghum relating to early domestication and racial differentiation: A. modern wild shattering spikelet of *S. verticilliflorum*; B. wild-type sorghum spikelet impression in ceramics, Khashm el Girba KG23 (studied by the authors; Winchell et al. 2017); C. domesticated, sorghum spikelet base impression with rachilla, Khashm el Girba KG23 (studied by the authors; Winchell et al. 2017); D. modern domesticated spikelet base, race bicolor, after grain removal; E. race bicolor type grain, Early Historic Paithan, Maharashtra, India (Fuller in press); F. race caudatum type grain, SEM/sketch of dorsal and lateral, Tinda B, Libya (after Pelling 2005); G. caudatum(?) type grain, Kawa, Nubia, dorsal and lateral view (after Fuller 2004); H. caudatum type grain, poorly preserved surface in dorsal and lateral views, Early Iron Age Kabusanze, Rwanda (after Giblin and Fuller 2011); I. caudatum type grain, Middle Iron Age Panga Ya Saidi, Kenya (after Shipton et al. 2013); J. durra type grain, Early Historic Paithan, Maharashtra, India (Fuller in press); K. durra type grain, Meroitic Hamadab, Sudan (authors, unpublished); L. bicolor type grain, Meroitic Hamadab, Sudan.

such an age. Clearly mis-identified are also the often cited impressions of sorghum from third millennium BC Hili in Oman (Cleuziou and Costantini (1980), Fig. 2; first questioned by Willcox (1992)), which show fruits attached on the jointed ends of flared pedicels, that are consistent with *Asphodelus* sp. (Liliaceae), a common weed of Arabian wheat and barley cultivation in macroremains assemblages of the period (Tengberg 2003). Taken together these data suggest that cultivated, at least partially domesticated, if not near fully domesticated sorghum was rapidly transferred from eastern Sudan/Eritrea to western India (Gujarat) by the start of the second millennium BC. It is not clear whether there was a geographically intermediate stage of sorghum cultivation in Yemen or southern Oman (Dhofar) (for further discussions of the mechanisms and cultural context, see Boivin and Fuller 2009). Once separated from wild gene flow and adopted into the established wheat, barley or millet cultivation regimes it is probable that the non-shattering traits became fixed in populations relatively quickly, thereby reaching full domestication more or less upon its arrival in Western India. Therefore contrary to the Haaland (1995) hypothesis there is no reason to postulate the transfer of a morphologically wild sorghum to India and a return voyage of domesticated race bicolor. Nevertheless, further systematic study of early Indian sorghum is needed to trace domestication traits and racial evolution within South Asia, although most early finds are plausibly race bicolor (e.g. Figure 3e).

Indian historical linguistics are unable to shed further light on this issue. Indic terms for sorghum (e.g. Prakrit *gajja*, Sanskrit *yavanala*) are likely derived from older roots for barley. Those in Dravidian languages possibly represent a semantic shift from an older millet term, e.g. for *Brachiaria ramosa* (Southworth 2006), as sorghum appears to reconstruct to an earlier stage of the Dravidian languages than either pearl millet or wheat (Fuller 2003b).

Several hypotheses derived from historical linguistic inferences suggest the migration of sorghum-growing farmers into various parts of Africa, although discrepancies with available archaeobotanical data remain, especially in terms of the inferred date of these dispersals. For example, while early Nilo-Saharan speakers focused on pastoralism in the eastern Sahara and Sahel, probably alongside wild sorghum harvesting, new terms for cultivation appear later in the Proto-Saharo-Sahelian stage (Ehret 2014), potentially attributed to the later middle Holocene, e.g. c. 4000 BC. As reviewed above archaeobotancial finds from the central and eastern Sudan suggest cultivation was established by 3500 BC with morphological domesticated types already present. As such sorghum cultivators a millennium or more earlier might be sought further north. However, it is hard to determine whether sorghum cultivation in this region, which generally shows local continuity through much of the earlier to middle Holocene (Winchell 2013), was brought by immigrants arriving from the north, although cattle, sheep and goat must have spread this way. Ehret (2011) also inferred that Afroasiatic speakers, from the Cushitic lineage probably had sorghum cultivation when they moved south from the Southern Atbai region into the Ethiopian highlands (by c. 3500 BC). However, it is hard at present to reconcile this with the, albeit limited, archaeobotanical data that initially show just wheat and barley in the 1st millennium BC, and only later finds of sorghum and the indigenous Ethiopian tef (D'Andrea and Wadge 2011). Nevertheless the new evidence for sorghum cultivation in the 4th millennium BC, and presumably through the early second millennium BC, in eastern Sudan might support this. Further, the arrival of domesticated sorghum in India by c. 1700 BC (Fuller 2003a; Fuller and Boivin 2009), makes it highly plausible that during the third millennium BC sorghum-growing agropastoralists were expanding through the Eritrean region and hence potentially further south into parts of Ethiopia.

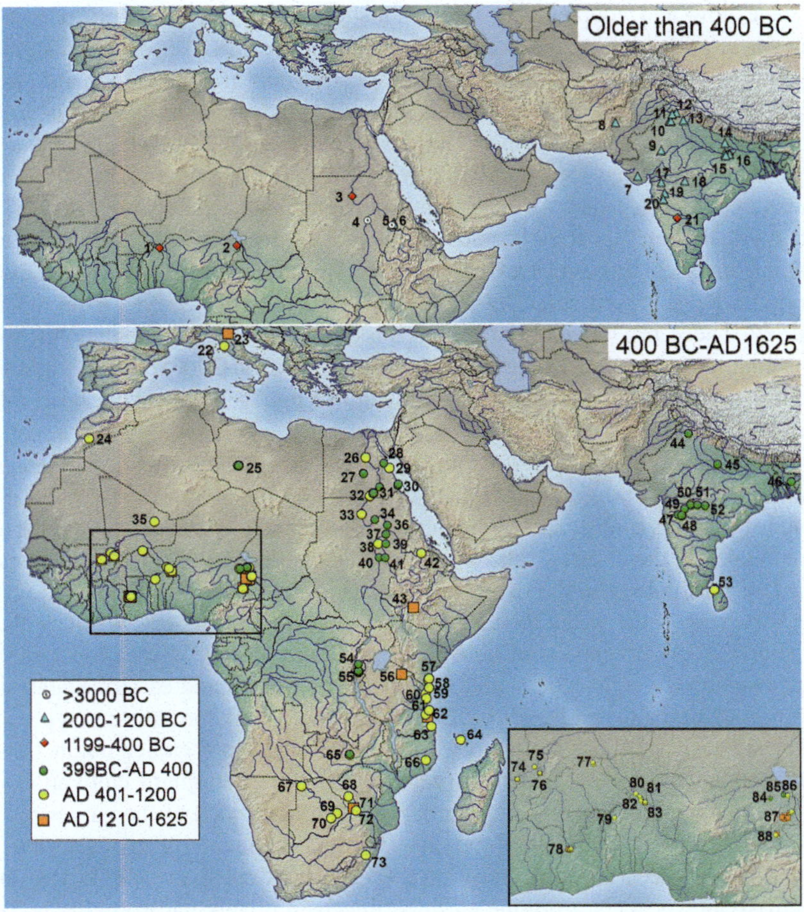

Fig. 4. Map of the distribution of archaeological domesticated sorghum in broad time slices, 1. Alibori SIII, 2. Mege, 3. Kawa, 4. Um Dereiwa, 5. Khashm el Girba, 6. Kasala K1, 7. Rojdi, 8. Pirak, 9. Ojiyana, 10. Banawali & Kunal, 11. Rohira, 12. Sanghol, 13. Hulas, 14. Imlidh-Khurd, 15. Malhar and Raja-Nala-Ka-Tila, 16. Senuwar, 17. Kaothe (intrusive?), 18. Tuljapur Garhi, 19. Daimabad, 20. Inamgaon, 21. Piklihal, 22. Miranduolo, 23. Ferrara, 24. Ingiliz, 25. Jarma & Tinda, 26. Kom el-Nana, 27. Karga Oasis, 28. Mon Claudianus, 29. Quesir al-Qadim, 30. Berenike & Shenshef, 31. Wadi Qitna, 32. Qasr Ibrim, 33. Faras East, 34. Umm Muri, 35. Essouk, 36. Dangeil, 37. Hamdab & Meroe, 38. Soda, 39. Jebel Qeili, 40. Jebel Tomat, 41. Abu Geili, 42. Axum, 43. Tuwatey Zonga Cave, 44. Sanghol, 45. Sanchankot/Ramkot, 46. Wari-Bateshwar, 47. Nevasa, 48. Paithan, 49. Bhokardan, 50. Bhon & Paturda, 51. Bhatkuli, 52. Paunar, 53. Mantai, 54. Kabusanze, 55. Kabuye (possible pollen), 56. Engaruka, 57. Mgombani, 58. Panga Ya Saidi, 59. Tumbe, 60. Fukuchani, 61. Unguka Ukuu, 62. Juani Primary School & Kilwa & Songo Mnara, 63. Mikindani sites, 64. Old Sima, 65. M'teteshi & Mondake, 66. Xakota, 67. Nqoma, 68. Leopard's Kopje, 69. Kgaswe, 70. Matlhapaneng, 71. Mapela, 72. Schroda, 73. Magogo, 74. Dia-Mara & Sorotomo, 75. Dia-Shoma, 76. Jenne-jeno, 77. Oursi hu-beero & Kolel Nord, 78. Banda 13 & 27 & 41 & Kuulo Kataa & Ngre Kataa, 79. Yohongu, 80. Pekinga, 81. Tintin, 82. Birnin Lafiya & Kantoro, 83. Madekali, 84. Elkido North & Dorota, 85. Mege & Kursakata, 86. Daima, 87. Kayam & Jidderi Saoudjo & Balda Tagamre & Tchere & Goray & Salak & Mowo & Louggero, 88. Douloumi & Be. Key sources in addition to those cited in the text, include: (Fuller 2014; Mitchell 2002; Logan 2012; Nixon et al. 2011; Ruas et al. 2011; Delneuf 1995).

Dispersal and Diversification of Sorghum: The Archaeobotany of the Races

After its establishment as a crop in parts of Sudan, e.g. Bayuda to Southern Atbai belt, morphologically domesticated sorghum dispersed and eventually diversified. Archaeobotanical finds, although still few and far between in Africa allow us to sketch the arrival time of sorghum into various regions and thereby infer how quickly it spread (Fig. 4). In some cases sorghum is likely to have diffused as part of a package of crops and animals with migrating farmers, although sorghum was also transferred across language families.

The westward dispersal of sorghum has not been tied to linguistically inferred ethnic migrations, but evidence for sorghum in northern Benin prior to the local Iron Age (c. 700 BC, see Champion and Fuller, 2018), and probable evidence for sorghum at Mege, near Lake Chad, at 800–400 BC (Bigga and Kahlheber 2011), indicate a westward diffusion in the first millennium BC. Evidence for pearl millet as early as sorghum in India, i.e. 1900–1700 BC, and possible cowpea (Fuller 2003a; Fuller and Boivin 2009) indicate that established cultural networks allowed pearl millet to diffuse eastward from West Africa probably shortly after the mid third millennium BC. Pearl millet is a crop highly suited to mobile agro-pastoralists with a short growing season and low water requirements, which may have facilitated its rapid spread eastwards over less than a millennium along the southern margins of the Sahara (Manning and Fuller 2014). These same social networks are likely to have been responsible for the less rapid spread of caudatum westwards, after free-threshing caudatum evolved in the central Sudan.

Two distinct westward trajectories can be advocated. Race bicolor likely dispersed westwards across the southern savannah, towards Nigeria, Cameroon and Benin, early in the first millennium BC, and some grains from Birnin Lafiya in Benin suggest local evolution of guinea types by the early centuries AD (Table 3), likely involving gene flow from wild race arundinceum sorghums. By contrast more northerly and Sahelian dispersal for caudatum can be postulated, presumably from origins in Sudan. This is indicated by finds at Jarma in the Fezzan in Libya by 100 BC (Fig. 3e; Pelling 2005, 2007), Jenne-Jeno, Mali by the 6th/7th century AD, and Daima near Lake Chad in the 9th or 10th century (Connah 1967, see Table 3). The still un-replicated evidence from Kawa in Nubia, 800–400 BC, might yet hint at earlier origins for race caudatum (Fig. 3f). Meanwhile the caudatum like grains from Kaothe in India probably should be regarded as intrusive, as suggested by Kajale (1990).

Moving to the south of the Ethiopian highlands evidence for sorghum is much later. Securely identified and dated finds come only from the early centuries AD in East Africa, including third or fourth century AD finds from Kabusanze (Fig. 3g) (Giblin and Fuller 2011), and from Zimbabwe, Leopard's Kopje, in the 10th or 11th Century AD (Huffman 1974; Mitchell 2002, 274). De Wet (1977) suggested the latter, based on geography rather than morphology, could represent an early southern African guinea. Along the East African coast and near shore islands, sorghum is present in

Table 3. Reports of archaeological sorghum referred to various free-threshing races

Site	Country	Age	Sorghum Race	Sources/Comments
Jarma (Old Germa)	Libya	400 BC–AD 100	Caudatum	Pelling (2005, 2007)
Daima	Nigeria	AD 800–1000	Caudatum	(Harlan and Stemler 1976; de Wet 1977)
Kabusanze	Rawanda	AD 275–400	Caudatum	Giblin and Fuller (2011)
Panga Ya Saidi	Kenya	AD 750–950	Caudatum	Id. DQ Fuller, in Helm et al. (2012), Shipton et al. (2013): see Fig. 5
Mogambani	Kenya	AD 765–980	Caudatum	Identified by DQ Fuller, in Helm et al. (2012), Shipton et al. (2013)
Old Sima	Comores Islands, Madagascar	c. 900 AD	Caudatum	the authors (see Crowther et al. 2016)
Kaothe	India	1900–1500 BC (?)/intrusive (?)	Caudatum (?)	Kajale (1990); comments in Fuller (2003b)
Jenne-Jeno	Mali	c. AD 600	Caudatum (and Bicolor)	Identified by Harlan, in McIntosh (1982)
Kawa	Sudan	800–400 BC	Durra (?)/ Caudatum (?)	Fuller (2004); high C13 raises questions about identification
Paithan	India	AD 0–400	Durra (and Bicolor)	Fuller in press
Meroe	Sudan	100 BC–AD 300	Durra (and Bicolor)	ID by Harlan, in Shinnie and Anderson (2004)
Hamadab	Sudan	50 BC–AD 350	Durra (and Bicolor)	Identified by DQ Fuller, unpublished data
Qasr Ibrim	Sudan	From AD 550	Durra (and Bicolor)	(Harlan and Stemler 1976; Rowley-Conwy 1991; Clapham and Rowley-Conwy 2007)
Soba	Sudan	From AD 600	Durra	Van der Veen and Lawrence (1991)
Quesir (Myos Hormos)	Egypt	From AD 1000	Durra (and Bicolor)	Van der Veen (2011)
Birni Lafiya	Benin	AD 200–800	Guinea (and Bicolor)	(see Champion and Fuller, 2018)

Middle Iron Age levels from the second half of the first millennium AD, e.g. at Panga Ya Saidi in southeast Kenya, Pemba, Zanzibar and Mafia Islands (Walshaw 2010; Boivin et al. 2013; Crowther et al. 2014, 2016). Many of these finds include grains resembling caudatum (Fig. 3H; Table 3), fitting with the hypotheses that caudatum, evolved in the Sudan (Harlan and Stemler 1976; Brown et al. 2011), and was transferred from Nilo-Saharan speaking farmers to the early dispersing Bantu near the southern Sudanese region as Bantu farmers moved around the rainforest (Ehret 1973; Schoenbrun 1993; Boivin et al. 2013).

The general pattern in all of these areas of East Africa is that the earliest farming is associated with an eastern Bantu crop package, including pearl millet, cowpea, baobab, and sorghum, with finger millet appearing later (Crowther et al. 2016). Small quantities of crops derived from Asia across the Indian Ocean arrived on African islands from the 8th century AD, including rice (*Oryza sativa*), mungbean (*Vigna radiata*), cotton (*Gossypium* sp.), and coconut (*Cocos nucifera*). Therefore, we lack any firm evidence for contact between tropical Asia and eastern Africa south of Ethiopia earlier than c. AD 700, despite the distribution of taro, Asian yam, and bananas in continental Africa that have been suggested to relate to much earlier trans-Indian Ocean transfers (e.g. Blench 2009; see also Fuller and Boivin 2009; Boivin et al. 2014). The presence of apparent race guinea in parts of India, has however suggested Indian Ocean transfers out of southeast Africa, later than the posited Bronze Age circum-Arabian transfer of race bicolor (see Fuller 2003a). Genetic data indicate southeast African guineas derive from margaritiferum types in the West African rainforests, which imply a dispersal from western to southern African and then across the Indian Ocean.

Archaeobotanical finds referred to as durra are restricted to regions where durra is prominent today, including peninsular India and Nubia (Sudan). Durra types are present at the early historic city of Paithan, Maharashtra (Fig. 3i) from nearly 2000 years ago, while finds from Meroe and nearby Hamdab (Fig. 3j, k) suggest a broadly similar age in Nubia (Table 3). The northerly dispersal of durra down the Nile to Lower Nubia is seen from probable intermediate durra-bicolor morphotypes at late post-Meroitic Qasr Ibrim (c. AD 500), eventually dominating in the Christian and Medieval periods (Rowley-Conwy et al. 1999; Clapham and Rowley-Conwy 2007). Nevertheless, race bicolor remained important in this region, as indicated by 12th century AD finds at Nauri near the Third cataract (Fuller and Edwards 2001), and Quesir on the Red Sea Coast (van der Veen 2011); the illustrated sorghum from Byzantine Kom el-Nana is also race bicolor (see Smith 2003). The hulled grains of race bicolor are likely more resilient under various storage conditions, and less prone to bird predation, hence still have advantages over the free-threshing races. The transfer of sorghum into Mediterranean agriculture, in central and northern Italy (e.g. Bosi et al. 2009; Buonincontri et al. 2014), notably also involved race bicolor having similar cultivation, storage and processing requirements to *Setaria italica* and *Panicum miliaceum* that were already well-established.

Sorghum Diversification and Dispersal: A Revised Framework

We are now in an improved position to outline the history of sorghum diversification and dispersal (Fig. 5) as a revision to the schema of Harlan and Stemler (1976). While genetics has revealed a more complex series of interregional relationships than was originally envisaged in the five race classification, archaeobotanical specimens that can be referred to various advanced free-threshing races, or as plausibly free-threshing, provide a framework of dates for when races differentiated and arrived in different regions (Table 3). We assume that bicolor spread along all of these paths, either accompanying or preceding the advanced free-threshing races (Fig. 5). The overall pattern suggests sorghum domestication and subsequent evolution of caudatum in the central/eastern Sudan region with dispersals east, west and southward. These dispersals likely led to introgression with local wild races, providing the genetic materials for various local ecological adaptations, including some of the variants in photoperiodicity genes that helped to alter and broaden the seasonality of sorghum. This overall pattern illustrates the importance of *introgressive capture* (sensu Larson and Fuller 2014) in structuring the genetic geography of domesticated sorghum to resemble pre-existing geographical patterning in wild diversity. It is this introgression that likely lies behind erroneous claims for multiple domestications (e.g. Shechter and de Wet 1975; Olsen 2012), as well the clustering of South and East Asian sorghum germplasm, regardless of race (Billot et al. 2013), probably through introgression of domesticated sorghum

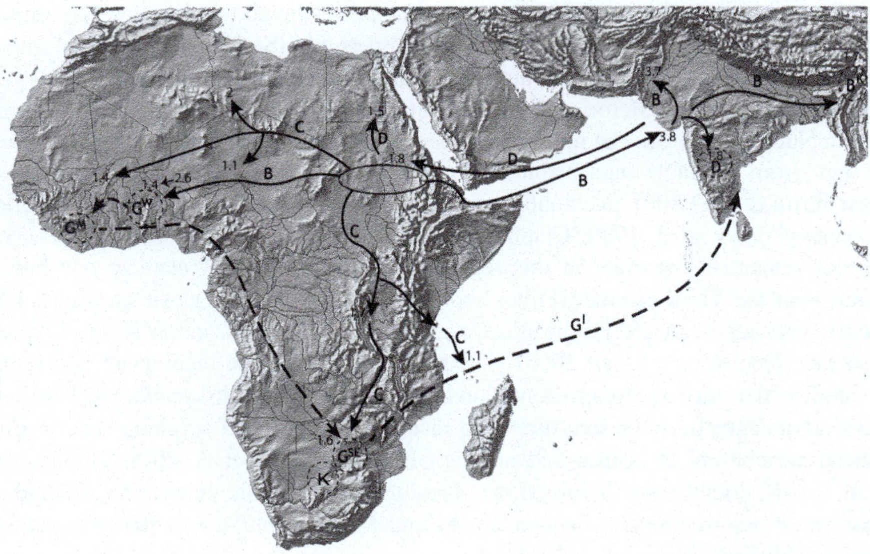

Fig. 5. A revised framework of the dispersal and diversification history of domesticated sorghum, based on the archaeobotanical and genetic evidence reviewed in this paper. Solid oval indicates zone of domestication of early race bicolor and subsequent evolution of race caudatum; dashed ovals indicate hypothetical zones for the evolution of other races or varieties. Solid arrows indicate trajectories inferred from archaeobotanical evidence, dashed arrows indicate hypothetical dispersal inferred from genetics without archaeobotanical support. Numbers indicate approximate age in years before AD 2000 for arrival time or evolutionary event. Letters on arrows indicate races or varietal groups inferred for each dispersal trajectory: B = bicolor race, B^K = Chinese Kaoliang sorghum group within race bicolor, C = caudatum race, D = durra race, G^W = race guinea, west African group, G^M = race guinea, margaritiferum varieties (= *S. margaritiferum* Stapf.), G^{SE} = race guinea, southeast African group, G^I = race guinea, Indian group, K = race kafir (= *S. caffrorum* Snowden). Note that race bicolor types should be regarded as co-dispersing along most, if not all trajectories, but these are only indicated where no advanced races are involved.

with wild *Sorghum propinquum.* The extent to which *S. propinquum* provided key adaptive attributes to cultivars requires genetic research, but it can be hypothesized to have contributed to race durra or to kaoliang sorghums (Snowden's *S. nervosum*) that extend north to c. 40° latitude in China, Korea and Japan.

The schema of sorghum dispersal and differentiation takes into account our current understanding of sorghum genetics, as well as available archaeobotanical finds, but also fits with several linguistic inferences. The later adoption of sorghum into established pearl millet farming systems across west Africa in the past 2000–2500 years is evident in linguistic evidence for a widespread cognate term for sorghum, "*kVn-*", which seems to have spread as a loan across several distinct language families and subfamilies, including West/Central Afro-Asiatic, Songhay and Saharan and several subgroups of Niger-Congo (Mande, Atlantic, Adamawa, and some West Benue-Congo languages) (Blench 2006: 215), but apparently after Bantoid languages had already diverged and moved southward with a focus on rainforest subsistence (cf. Bostoen 2014). These more woodland-focused Bantu groups (Grollemund et al. 2015), picked up pearl millet by 2500 years ago, in keeping with traditions of food preparation that included a stiff porridge of a starchy staple and production of flour, and related cooking terms (Ricquier and Bosteon 2011). The lines of margaritiferum guinea sorghums that gave rise to kafirs might also have moved via the same routes as pearl millet through the Congo, i.e. the Sangha River Interval, or, given the lack of archaeobotaical evidence, have diffused secondarily through these same societies after farming was already established, much as Asian cultivars probably did. Once established in southeast Africa, this lineage was carried via Indian Ocean trade to India to become Indian guineas. That these are popular in forested hill tracts through central and eastern Indian probably reflects the higher rainfall adaptations that ultimately derived from wild race arundinaceum, as well as their suitability to whole grain cooking as opposed to flour cuisine (Appa Rao et al. 1996).

As farmers of the eastern Bantu stream spread to the eastern/northeastern African rainforest margins, new cereals were incorporated into their subsistence, evident in loan words into the Eastern Bantu proto-language, or its early decedents, including a widespread pearl millet term (**-bèdé*), most likely from Eastern Sudanic languages around the margins of what is today South Sudan or the Central Africa Republic (Ricquier and Bosteon 2011). Similarly, terms for sorghum (**–saka*; **-pemba*; **-pila*) were adopted into Eastern Bantu or some of its subgroups in the great lakes region of eastern Africa from Central Sudanic languages (Ehret 1973; cf. Philippson and Bahuchet 1994). This is congruent with sorghum, including both races bicolor and caudatum being adopted into eastern African agriculture in the early first millennium AD.

Thus the history of early agriculture in Africa is not merely a narrative of migrating farmers but also borrowing across ethnolinguistic groups and local adaptations in part arising from capturing genetic diversity from local wild sources. Some crop varieties moved long-distances, sometimes with highly mobile economies, like those posited for the Neolithic Sahel, but also sometimes via long-distance trade, around the Bronze Age Arabian sea and later Indian Ocean.

For most of Africa sorghum is central to agriculture, overlapping with and extending beyond the range of other African cereals. It has forms nearly as tolerant of low rainfall as pearl millet and others that, like finger millet, range into higher rainfall and mountainous zones. Compared to both it is relatively more productive. In traditional East African agriculture, for example, sorghum yields from 550–1700 kg/ha, compared to 450–900 kg/ha for finger millet, and on average c. 450 kg/ha for pearl millet (Acland 1971), a range of 270–900 kg/ha is reported for West African pearl millet (Irvine 1969). Under rainfed cultivation African rice (*Oryza glaberrima*) yields 450–900 kg/ha, although 1000–3000 kg is reported for wet fields (Borlaug et al. 1996). In general then, sorghum is the most productive rainfed cereal in most African environments, at least prior to the introduction of American maize. Thus it is not surprising that sorghum cultivation is recurrently associated with sedentary farming and urbanisation in many different cultural traditions of pre-historic Africa, from the Inland Niger Delta (McIntosh 1995) to Meroitic Nubia (Fuller 2014), from Swahili islands (Boivin et al. 2013) to Engaruka (Sassoon 1967) and Great Zimbabwe (Chirikure et al. 2014); and it is unique amongst the African cereals in not just flourishing in similar environments in India but making inroads into northern temperate farming systems, in the Mediterranean or northeastern Asia.

Acknowledgements. This research has been carried out as part of the "Comparative Pathways to Agriculture" project, funded by a European Research Council advanced grant (No. 323842). We are grateful to our colleagues who have contributed to our Old World Crops Archaeobotanical Databse (OWCAD), including Leilani Lucas, Charlene Murphy, Louise Champion and Fabio Silva. We also thank Alem Beldados, Andrea Manzo, and Frank Winchell for brining to our attention new evidence from Sudan, and Alison Crowther and Nicole Boivin for bringing to our attention new samples from eastern Africa. Lastly, we thank peer-reviewers and editors for their helpful suggestions.

References

Acland JD (1971) East African crops. Longman, London

Aldrich PR, Doebley J (1992) Restriction fragment variation in the nuclear and chloroplast genomes of cultivated and wild *Sorghum bicolor*. Theor Appl Genet 85:293–302

Appa Rao S, Prasada Rao KE, Mengesha MH et al (1996) Morphological diversity in sorghum germplasm from India. Genet Resour Crop Evol 43(6):559–567

Barakat H, Fahmy el-Din AG (1999) Wild grasses as 'Neolithic' food resources in the eastern Sahara: a review of the evidence from Egypt. In: van der Veen M (ed) The exploitation of plant resources in ancient Africa. Plenum Press, New York, pp 33–46

Beldados A, Costantini L (2011) Sorghum exploitation at Kassala and its environs, Northeastern Sudan in the second and first millennium BC. Nyame Akuma 75:33–39

Beldados A, Manzo A, Murphy C et al. (2018) Evidence of sorghum cultivation and introduced West Africa crops in the second millennium BCE at Kassala, Eastern Sudan

Bigga G, Kahlheber S (2011) From gathering to agricultural intensification: archaeobotanical remains from Mege, Chad Basin, NE Nigeria. In: Fahmy AG, Kahlheber S, D'Andrea AC (eds) Windows on the African Past: current approaches to African archaeobotany. Africa Magna Verlag, Frankfurt, pp 19–45

Billot C, Ramu P, Bouchet S et al (2013) Massive Sorghum collection genotyped with SSR Markers to enhance use of global genetic resources. PLoS ONE 8(4):e59714. https://doi.org/10.1371/journal.pone.0059714

Blench RM (2006) Archaeology language and the African past. Altamira Press, Lanham

Blench RM (2009) Bananas and plantains in Africa: re-interpreting the linguistic evidence. Ethnobotany Res Appl 7:363–380

Boivin N, Crowther A, Helm R et al (2013) East Africa and Madagascar in the Indian Ocean world. J World Prehist 26(3):213–281

Boivin N, Crowther A, Prendergast M et al (2014) Indian Ocean food globalisation and Africa. Afr Archaeol Rev 31(4):547–581

Bonjean APA (2010) Origins and historical diffusion in China of major native and alien cereals. In: He Z, Bonjean APA (eds) Cereals in China. Mexico, CIMMYT, pp 1–14

Borlaug NE, Axtell J, Burton GW et al (1996) Lost crops of Africa Grains, vol I. National Academy Press, Washington DC

Bosi G, Mercuri AM, Guarnieri C et al (2009) Luxury food and ornamental plants at the 15th century AD Renaissance court of the Este family (Ferrara, northern Italy). Veg Hist Archaeobot 18(5):389–402

Bostoen K (2014) Wild trees in the subsistence economy of early Bantu speech communities: a historical-linguistic approach. In: Stevens CJ, Nixon S, Murray MA, Fuller DQ (eds) Archaeology of African Plant Use. Left Coast Press, Walnut Creek, pp 129–140

Brown PJ, Myles S, Kresovich S (2011) Genetic support for phenotype-based racial classification in Sorghum. Crop Sci 51:224–230

Buonincontri M, Moser D, Allevato E et al (2014) Farming in a rural settlement in central Italy: cultural and environmental implications of crop production through the transition from Lombard to Frankish influence (8th–11th centuries ad). Veg Hist Archaeobot 23(6):775–788

Casa AM, Pressoir G, Brown PJ et al (2008) Community resources and strategies for association mapping in sorghum. Crop Sci 48(1):30–40

Chaix L, Honegger M (2015) New data on animal exploitation from the Mesolithic to the Neolithic periods in Northern Sudan. In: Kerner S, Dann RJ, Bangsgaard P (eds) Climate and ancient societies. Museum of Tusculanum Press, Copenhagen, pp 197–209

Champion L, Fuller DQ (2018) New evidence on the development of millet and rice economies in theNiger river basin: archaeobotanical results from Benin. Plants and Peoplein the African Past 529–547

Chirikure S, Manyanga M, Pollard AM et al (2014) Zimbabwe culture before Mapungubwe: new evidence from Mapela Hill South-Western Zimbabwe. PLoS ONE 9(10):e111224. https://doi.org/10.1371/journal.pone.0111224

Clapham AJ, Rowley-Conwy PA (2007) New discoveries at Qasr Ibrim, Lower Nubia. In: Cappers RTJ (ed) Fields of change: progress in African archaeobotany. The Netherlands: Barkhuis & Groningen University Library, Groningen, pp 157–164

Clark JD (1984) Prehistoric cultural continuity and the economic change in the Central Sudan in the early Holocene. In: Clark JD, Brandt SA (eds) From hunters to farmers, the causes and consequences of food production in Africa. University of California Press, Berkley, pp 113–126

Cleuziou S, Costantini L (1980) Premiers éléments sur l'agriculture protohistorique de l'Arabie Orientale. Paléorient 6:245–251

Connah G (1967) Progress report on archaeological work in Bornu 1964–1966 with particular reference to the excavations at Daima mound. In: Northern History Research Scheme (ed), Second Interim Report. Ahmadu Bello University, Zaria, pp 20–31

Costantini L (1990) Harappan agriculture in Pakistan: the evidence of Naursharo. In: Taddei M (ed) South Asian archaeology 1987. Istituto Italiano per il Medio ed Estremo Oriente, Rome, pp 321–332

Crowther A, Horton M, Kotarba-Morley A et al (2014) Iron age agriculture, fishing and trade in the Mafia Archipelago, Tanzania: new evidence from Ukunju Cave. Azania 49(1):21–44

Crowther A, Lucas L, Helm R et al (2016) Ancient crops provide first archaeological signature of the westward Austronesian expansion. Proc Nat Acad Sci 113 (24):6635–6640. https://doi.org/10.1073/pnas.1522714113

D'Andrea AC, Wadge P (2011) T'ef (*Eragrostis tef*): A Legacy of Pastoralism? In: Fahmy AG, Kahlheber S, D'Andrea AC (eds) Windows on the African past: current approaches to African archaeobotany. Africa Magna Verlag, Frankfurt, pp 225–241

de Alencar Figueiredo LF, Calatayud C, Dupuits C et al (2008) Phylogeographic evidence of crop neodiversity in sorghum. Genetics 179(2):997–1008

de Moulins D, Phillips C, Durrani N (2003) The archaeological records of Yemen and the question of Afro-Asian contact. In: Neumann K, Buttler A, Kahlheber S (eds) Food, fuels and fields: progress in African archaeobotany. Heinrich Barth-Institut, Cologne, pp 281–299

de Wet JML (1977) Domestication of African cereals. Afr Econ Hist 3:15–32

de Wet JML, Harlan JR, Price EG (1976) Variability in *Sorghum bicolor*. In: Harlan JR, de Wet JMJ, Stemler ABL (eds) Origins of African plant domestication. Mouton Press, The Hague, pp 453–463

Delneuf MOT (1995) L'environment et les usages alimentaires en vigneur a l'époque protohistorique dans l'extreme-nord du Cameroun. In: Marliac A (ed) Milieux, Societés et Archéologues. Office de la Recherche scientifique et technique de outre-mer, Paris, pp 213–226

Deu, MD. Gonzalez-de-Leon J-C, Glaszmann I et al (1994) RFLP diversity in cultivated sorghum in relation to racial differentiation. Theor Appl Genet 88:838–844

Deu M, Rattunde H, Chantereau J (2006) A global view of genetic diversity in cultivated sorghums using a core collection. Genome 49:168–180

Ehret C (1973) Patterns of Bantu and central Sudanic settlement in central and southern Africa (1000 BC–500AD) Transafrican J Hist 3:1–71

Ehret C (2011) History and the testimony of language. University of California, Berkley

Ehret C (2014) Linguistic evidence and the origins of food production in Africa. Where are we now? In: Stevens CJ, Nixon S, Murray MA, Fuller DQ (eds) Archaeology of African plant use. Left Coast Press, Walnut Creek, pp 233–242

Folkertsma R, Rattunde HFW, Chandra S et al (2005) The pattern of genetic diversity of guinea race *Sorghum bicolor* (L.) Moench landraces as revealed with SSR markers. Theor Appl Genet 111:399–409

Fuller DQ (2002) Fifty Years of Archaeobotanical Studies in India: Laying a Solid Foundation. In: Settar S, Korisettar R (eds) Indian archaeology in retrospect, vol III. Archaeology and interactive disciplines. Publications of the Indian Council for Historical Research Manohar, New Dehli, pp 247–364

Fuller DQ (2003a) African crops in prehistoric South Asia: a critical review. In: Neumann K, Butler A, Kahlheber S (eds) Food, Fuel and Fields. Progress in Africa Archaeobotany, Africa Praehistorica 15. Heinrich-Barth-Institut, Cologne, pp 239–271

Fuller DQ (2003b) An agricultural perspective on Dravidian historical linguistics: archaeological crop packages, livestock and Dravidian crop vocabulary. In: Bellwood P, Renfrew C (eds) Assessing the languaging/farming dispersal hypothesis. McDonald Institute, Cambridge, pp 191–213

Fuller DQ (2004) Early Kushite agriculture: archaeobotanical evidence from Kawa. Sudan Nubia 8:70–74

Fuller DQ (2014) Agricultural innovation and state collapse in Meroitic Nubia: the impact of the savannah package. In: Stevens CJ, Nixon S, Murray MA, Fuller DQ (eds) Archaeology of African plant use. Left Coast Press, Walnut Creek, pp 165–178

Fuller DQ (in press) Archaeobotany. In: Kennet D, Rao JVP (eds.) *Paithan Excavations*. Memoirs of the Archaeological Survey of India, Delhi, pp 289–327

Fuller DQ, Allaby R (2009) Seed dispersal and crop domestications: shattering, germination and seasonality in evolution under cultivation. Ann Plant Rev 38:238–295

Fuller DQ, Boivin N (2009) Crops, cattle and commensals across the Indian Ocean: current and potential archaeobiological evidence. In: Lefevre G (ed) Plantes et Societes, Etudes Ocean Indien 42-43. Institut National des Langues et Civilisations Orientales, Paris, pp 13–46

Fuller DQ, Edwards DN (2001) Medieval plant economy in Middle Nubia: preliminary archaeological evidence from Nauri. Sudan Nubia 5:97–103

Fuller DQ, Lucas L (2014) Wheats: origins and development. In: Smith C (ed) Encyclopedia of global archaeology. Springer, New York, pp 7812–7817

Fuller DQ, Allaby RG, Stevens CJ (2010) Domestication as innovation: the entanglement of techniques, technology and chance in the domestication of cereal crops. World Archaeol 42(1):13–28

Fuller DQ, Denham T, Arroyo-Kalin M et al (2014) Convergent evolution and parallelism in plant domestication revealed by an expanding archaeological record. Proc Nat Acad Sci 111(17):6147–6152

Gautier A, Van Neer W (2006) Animal remains from Mahal Teglinos (Kassala, Sudan) and the arrival of pastoralism in the Southern Atbai. J Afr Archaeol 4(2):223–233

Giblin JD, Fuller DQ (2011) First and second millennium AD agriculture in Rwanda: archaeobotanical finds and radiocarbon dates from seven sites. Veg Hist Archaeobot 20(4):253–265

Grollemund R, Branford S, Bostoen K et al (2015) Bantu expansion shows that habitat alters the route and pace of human dispersals. Proc Nat Acad Sci 112(43):13296–13301

Haaland R (1995) Sedentism, cultivation and plant domestication in the Holocene middle Nile region. J Field Archaeol 22(2):157–173

Hagerty MH (1941) Comments on writings concerning Chinese sorghums. Harvard J Asia Stud 5 (3–4):234–260

Harlan JR, De Wet JML (1972) A simplified classification of cultivated sorghum. Crop Sci 12:172–176

Harlan JR, Stemler ABL (1976) The races of sorghum in Africa. In: Harlan JR, de Wet J, Stemler A (eds) Origins of African plant domestication. Mouton Press, The Hague, pp 465–478

Harlan JR, De Wet JML, Price EG (1973) Comparative evolution of cereals. Evolution 277:311–325

Helm R, Crowther A, Shipton C et al (2012) Exploring agriculture, interaction and trade on the eastern African littoral: preliminary results from Kenya. Azania 47(1):39–63

Htun TM, Inoue C, Chhourn O et al (2014) Effect of quantitative trait loci for seed shattering on abscission layer formation in Asian wild rice Oryza rufipogon. Breed Sci 64(3):199–205

Huffman TN (1974) The Leopard's Kopje Tradition. Salisbury: National Museums and Monuments (Memoir 6)

Irvine FR (1969) West African Agriculture, 3 edn, vol 2. West African Crops. Oxford University Press, Oxford

Ishii T, Numaguchi K, Miura K et al (2013) OsLG1 regulates a closed panicle trait in domesticated rice. Nat Genet 45:462–465

Kajale MD (1990) Observations on the plant remains from excavation at Chalcolithic Kaothe, District Dhule, Maharashtra with cautionary remarks on their interpretations. In: Dhavalikar MK, Shinde VS, Atre SM (eds) Excavations at Kaothe. Decan College Postgraduate Research Institute, Pune, pp 265–280

Klein RR, Miller FR, Dugas DV et al (2015) Allelic variants in the PRR37 gene and the human-mediated dispersal and diversification of sorghum. Theor Appl Genet 128(9):1669–1683

Kröpelin S, Verschuren D, Lézine A-M et al (2008) Climate-driven ecosystem succession in the Sahara: the past 6000 years. Science 320(5877):765–768

Larson G, Fuller DQ (2014) The evolution of animal domestication. Ann Rev Ecology Evolution Systematics 45:115–136

Li WL, Gill BS (2006) Multiple pathways for seed shattering in the grasses. Funct Integr Genomics 6:300–309

Li C, Zhou A, Sang T (2006) Rice domestication by reduced shattering. Science 311:1936–1939

Lin Z, Griffith ME, Li X et al (2007) Origin of seed shattering in rice (*Oryza sativa* L.). Planta 226:11–20

Lin Z, Li X, Shannon LM et al (2012) Parallel domestication of the Shattering1 genes in cereals. Nat Genet 44(6):720–724

Logan AL (2012) A history of food without history: food, trade, and environment in west-central Ghana in the second millennium AD. Unpublished PhD dissertation. University of Michigan, Ann Arbor

Luo H, Zhao W, Wang Y et al (2016) SorGSD: a sorghum genome SNP database. Biotechnol Biofuels 9:6. https://doi.org/10.1186/s13068-015-0415-8

Mace ES, Xia L, Jordan DR et al (2008) DArT markers: diversity analyses and mapping in *Sorghum bicolor*. BMC Genom 9:26

Magid AA (1989) Plant Domestication in the Middle Nile Basin: an Archaeobotanical Case Study. Cambridge Monographs in African Archaeology 35. BAR S523. British Archaeological Reports, Oxford

Magid AA (2003) Exploitation of food-plants in the early and middle Holocene Blue Nile area, Sudan and neighbouring areas, Complutum 14. Universidad Complutense de Madrid Servicio de Publicaciones, Madrid

Manning K, Fuller DQ (2014) Early millet farmers in the lower Tilemsi Valley, Northeastern Mali. In: Stevens CJ, Nixon S, Murray MA, Fuller DQ (eds) Archaeology of African plant use. Left Coast Press, Walnut Creek, CA, pp 73–81

Manning K, Timpson A (2014) The demographic response to Holocene climate change in the Sahara. Quat Sci Rev 101:28–35

McClatchie M, Fuller DQ (2014) Leaving a lasting impression: arable economies and cereal impressions in Africa and Europe. In: Stevens CJ, Nixon S, Murray MA, Fuller DQ (eds) Archaeology of African plant use. Left Coast Press, Walnut Creek, pp 259–266

McIntosh SK (1995) Paleobotanical and human osteological remains. In: McIntosh SK (ed) Excavations at Jenné-Jeno, Hambarketolo, and Kaniana (Inland Niger Delta, Mali): the 1981 season. University of California Press, Berkley, pp 348–359

Mitchell PJ (2002) The archaeology of Southern Africa. Cambridge University Press, Cambridge

Morris G, Ramu P, Deshpande SP (2013) Population genomic and genome-wide association studies of agroclimatic traits in Sorghum. Proc Nat Acad Sci 110(2):453–458

Nixon S, Murray MA, Fuller DQ (2011) Plant use at an early Islamic merchant town in the West African Sahel: the archaeobotany of Essouk-Tadmakka (Mali). Ved Hist Archaeobot 20(3):223–239

NPGS (2016) National plant germplasm system. Germplasm Resources Information Network (GRIN), Washington, USDA. http://www.ars-grin.gov/npgs/index.html. Accessed 29 Mar 2015

Olsen KM (2012) One gene's shattering effects. Nat Genet 44:616–617

Pelling R (2005) Garamantian agriculture and its significance in a wider North African context: the evidence of the plant remains from the Fazzan project. J N Afr Stud 10(3/4):397–412

Pelling R (2007) Agriculture and trade amongst the Garamantes and the Fezzanese: 3000 years of archaeobotanical data from the Sahara and its margins. PhD Thesis. UCL, London

Perumal R, Krishnaramanujam R, Menz MA et al (2007) Genetic diversity among sorghum races and working groups based on AFLPs and SSRs. Crop Sci 47:1375–1383

Philippson G, Bahuchet S (1994) Cultivated crops and Bantu migrations in central and eastern Africa: a linguistic approach. Azania 29–30(1):103–120

Ramu P, Billot C, Rami J et al (2013) Assessment of genetic diversity in the sorghum reference set using EST-SSR markers. Theor Appl Genet 126(8):2051–2064

Ricquier B, Bosteon K (2011) Stirring up the porridge: how early Bantu speakers prepared their cereals. In: Fahmy AG, Kahlheber S, D'Andrea AC (eds) Windows on the African past: current approaches to African archaeobotany. Africa Magna Verlag, Frankfurt, pp 209–224

Rowley-Conwy P (1991) Sorghum from Qasr Ibrim, Egyptian Nubia, c. 800 BC AD 1811: a preliminary study. In: Renfrew J (ed) New light on early farming. Edinburgh University Press, Edinburgh, pp 191–212

Rowley-Conwy P, Deakin W, Shaw CH (1997) Ancient DNA from archaeological sorghum (*Sorghum bicolor*) from Qasr Ibrim, Nubia: implications for domestication and evolution and a review of the archaeological evidence. Sahara 9:23–34

Rowley-Conwy P, Deakin W, Shaw CH (1999) Ancient DNA from *Sorghum*. In: van der Veen M (ed) The exploitation of plant resources in ancient Africa. Plenum, New York, pp 55–61

Ruas MP, Tengberg M, Ettahiri AS et al (2011) Archaeobotanical research at the medieval fortified site of Îgîlîz (anti-atlas, morocco) with particular reference to the exploitation of the argan tree. Vegetation History Archaeobotany 20:419–433

Saraswat KS, Pokharia AK (2003) Palaeoethnobotanical investigations at early Harappan Kunal. Pragdhara 13:105–140

Sassoon H (1967) New views on Engaruka, northern Tanzania. J African History 8(02):201–217

Schoenbrun DL (1993) We are what we eat: ancient agriculture between the Great lakes. J Afr Hist 34:1–31

Shechter Y, De Wet JMJ (1975) Comparative electrophoresis and isozyme analysis of seed proteins from cultivated races of *Sorghum*. Am J Bot 62:254–261

Shinnie PL, Anderson J (2004) The capital of kush 2: meroe excavations, 1973–1984, Meroitica 20. Harrassowitz, Wiesbaden

Shipton C, Helm R, Boivin N et al (2013) Intersections, networks and the genesis of social complexity on the Nyali coast of east Africa. Afr Archaeol Rev 30:427–453

Smith W (2003) Archaeobotanical investigations of agriculture at late antique Kom el-Nana (Tell el-Amarna). Egypt Exploration Soc, London

Snowden JD (1936) The cultivated races of *Sorghum*. Allard and Son, London

Snowden JD (1955) The wild fodder Sorghums of the secion *Eu-Sorghum*. J Linn Soc Bot (London) 55:191–260

Southworth FC (2006) Proto-Dravidian agriculture. In: Osada T (ed), Proceedings of the Pre-symposium of RIHN and 7th ESCA Harvard- Kyoto Roundtable. Research Institute for Humanity and Nature, Kyoto, pp 121–150

Stapf O (1886) Die Pflanzenreste des Hallstätter Heidengebirges. Zool -Bot Ges, Österreich

Stapf O (1917) Sorghum. In: Prain D (ed) Flora of Tropical Africa, vol 9. Gramineae (Maydeae-Paniceae). L. Reeve, Asford, pp 104–154

Stemler AB (1990) A scanning electron microscopic analysis of plant impressions in pottery from sites of Kadero, El Zakiab, Um Direiwa and El Kadada. Archaeologie du Nil Moyen 4:87–106

Sukumaran S, Xiang W, Bean S et al (2012) Associated mapping for grain quality in a diverse sorghum collection. Plant Genome 5:126–135

Tengberg M (2003) Archaeobotany in the Oman Peninsula and the role of eastern Arabia in the spread of African crops. In: Neumann K, Butler A, Kahlheber S (eds) Food, fuel and fields progress in African archaeobotany. Africa Praehistorica. Monographien zur Archäologie und Umwelt Afrikas. Heinrich-Barth-Institut, Köln, pp 229–237

van der Veen M (2011) Consumption, trade and innovation: exploring the botanical remains from the Roman and Islamic ports at Quseir al-Qadim. Egypt, Africa Magna Verglag, Frankfurt

van der Veen M, Lawrence T (1991) The plant remains. In: Welsby DA, Daniels CM (eds) Soba: archaeological research at a medieval capital on the Blue Nile British institute in Eastern Africa memoirs, London, pp 264–274

Walshaw SC (2010) Converting to rice: urbanization, Islamization and crops on Pemba Island, Tanzania, AD 700–1500. World Archaeology 42:137–154

Wasylikowa K, Dahlberg JA (1999) Sorghum in the economy of the early Neolithic nomadic tribes at Nabta Playa, southern Egypt. In: van der Veen M (ed) The exploitation of plant resources in ancient Africa. Plenum Press, New York, pp 11–32

Wasylokowa K, Dahlberg JA (2001) Sorghum remains from site E-75-6. In: Wendorf F, Schild R (eds) Holocene settlement of the Egyptian Sahara: Volume 1: the archaeology of Nabta Playa Springer/Plenum, New York, pp 578–587

Wiersema JH, Dahlberg J (2007) The nomenclature of *Sorghum bicolor* (L.) Moench (Gramineae). Taxon 56(3):941–946

Willcox G (1992) Some differences between crops of near Eastern origin and those from the tropics. In: Jarrige C (ed) South Asian archaeology 1989. Monographs in World Archaeology, vol 14, Wisconsin, pp 291–300

Winchell F (2013) The Butana group ceramics and their place in the Neolithic and post-neolithic of northeast Africa. Cambridge monographs in African archaeology 83, British archaeological report series 2459. Archaeopress, Oxford

Winchell F, Stevens CJ, Murphy C et al (2017) Evidence for sorghum domestication in fourth millennium BC eastern Sudan: spikelet morphology from ceramic impressions of the Butana Group. Curr Anthropol 58(7):673–683

Zohary D, Hopf M, Weiss E (2012) Domestication of plants in the Old World. Oxford University Press, Oxford

A Pre-Aksumite Culinary Practice at the Mezber Site, Northern Ethiopia

A. Catherine D'Andrea[1(✉)], Linda Perry[2], Laurie Nixon-Darcus[1], Ahmed G. Fahmy[3], and Elshafaey A. E. Attia[3]

[1] Department of Archaeology, Simon Fraser University, Burnaby, BC, Canada
{adandrea, ldarcus}@sfu.ca

[2] Foundation for Archaeobotanical Research in Microfossils, Fairfax, VA, USA
linda.at.the.farm@gmail.com

[3] Department of Botany and Microbiology, Helwan University, Helwan, Egypt
s.elshafaey@gmail.com

Abstract. This study integrates ethnoarchaeological, archaeological, and microbotanical evidence to investigate ancient culinary practice at the Pre-Aksumite archaeological site of Mezber (1600 BC-AD 1) in Eastern Tigrai, northern Ethiopia. Starch grains were successfully extracted from Mezber grinding stones and four taxa were identified to cf. Fabaceae and sub-groups of Poaceae. The small sample of starch extracted from the grinding stones appears to be in general agreement with the Pre-Aksumite macro-botanical record. Recovered starches were damaged by grinding and heat treatment both of which suggest they were deposited on grinding stones through food preparation activities involving grinding. Ethnoarchaeological studies of grinding stone surfaces and cooking practice were completed to provide an interpretive context for archaeological starch and grinding stone evidence. Heating damage observed on cf. Triticeae starch suggests that Mezber inhabitants were processing roasted barley flour which is a widespread culinary practice today.

Keywords: Archaeobotany · Culinary practice · Ethiopia · Ethnoarchaeology Foodways · Grinding stones · Roasted barley flour · Pre-Aksumite period Starch

Introduction

Microbotanical analysis in conjunction with studies of grinding stones offer great potential in the study of prehistoric foodways, particularly because of the possibility to associate plant species with artefacts, food processing and cooking behaviours (e.g., Adams 1988, 1993, 2014a, b; Perry 2004; Aranguren et al. 2007; Liu et al. 2010; Revedin et al. 2010). This study integrates studies of grinding stones, culinary practice, and microbotanical evidence to further our understanding of Pre-Aksumite foodways at the archaeological site of Mezber, located in Gulo Makeda administrative zone, northern Ethiopia (Fig. 1). The work was completed under the auspices of the Eastern Tigrai Archaeological Project (ETAP), which is an interdisciplinary research effort involved in ongoing archaeological survey, excavation, ethnoarchaeological and

A. M. Mercuri et al. (Eds.): *Plants and People in the African Past*,
https://doi.org/10.1007/978-3-319-89839-1_20

palaeoenvironmental research in Gulo Makeda. We aim to further our understanding of grinding stones, their role in food processing, and demonstrate how this knowledge can inform interpretations of the archaeological record. Sampling for starch and phytoliths from ancient Mezber grinding stones was completed as a pilot study to determine if microbotanical remains were preserved on grinding stone surfaces of Pre-Aksumite age and if a relationship exists between grinding surface texture and crop remains. We attempt to link these data with ethnoarchaeological evidence to suggest the possible early presence of a culinary practice involving the production of roasted barley flour.

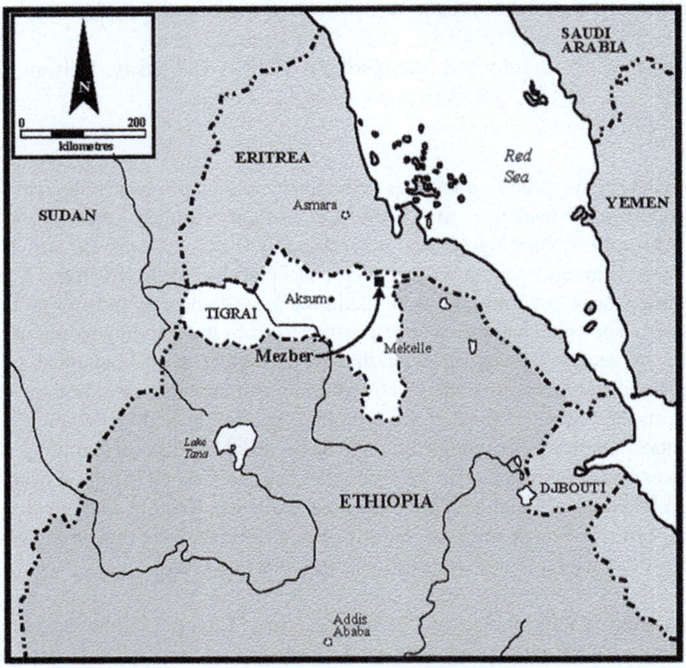

Fig. 1. Map of northern Ethiopia showing place names mentioned in the text. Map by Shannon Wood.

Background

Mezber is a Pre-Aksumite settlement in Gulo Makeda (Fig. 1). It lies on a slightly raised hillside situated in a valley bottom and covers approximately 0.83 ha. The site was discovered in 2005 during archaeological surveys completed by ETAP (D'Andrea et al. 2008a; Harrower and D'Andrea 2014). Mapping and excavations were carried out in 2007–2009 and 2011–2012 (Fig. 2). Mezber is one of only a few known Pre-Aksumite sites that have been undisturbed by later settlement. Excavations revealed an extended occupational sequence spanning the mid-second to the end of the first millennium BC. Four Pre-Aksumite phases, based on ceramics, stratigraphy and radiocarbon dating are evident with at least two building construction phases. The earliest occupation of the site is during an Initial phase (c. 1600–900 BC) followed by Early (c. 850–750 BC), Middle (c. 600–400 BC) and Late (c. 400 BC-1 AD) Pre-Aksumite phases (D'Andrea and Welton in prep.). The site has produced at least two levels of domestic field-stone architecture (Figs. 3 and 4) suggesting the presence of wealthy or elite individuals and evidence of specialized craft production in the Early and Middle phases (Peterson 2017). Pre-Aksumite peoples in the region were sedentary agro-pastoralists who practiced diverse livestock (Woldekiros and D'Andrea 2016) and crop husbandry (Beldados et al. in prep.). Excavations uncovered 108 grinding stones, which are documented for all Pre-Aksumite phases. These implements strongly resemble modern grinding equipment still used in the region today. In addition to grinding stone finds in excavated loci, a collapsed Early Phase *udo*, a table which supports grinding stones in modern households (Fig. 5a), was identified in Field E1 Locus 39 (Fig. 5b). The collapsed *udo* is positioned next to a wall, which is also commonly observed in modern residential compounds in Gulo Makeda. Other ethnoarchaeological studies of Ethiopian grinding stones have been completed (Teklu 2012; Arthur 2014), however this paper attempts to integrate ethnoarchaeological and archaeological studies of grinding stones and cuisine with archaeobotanical evidence.

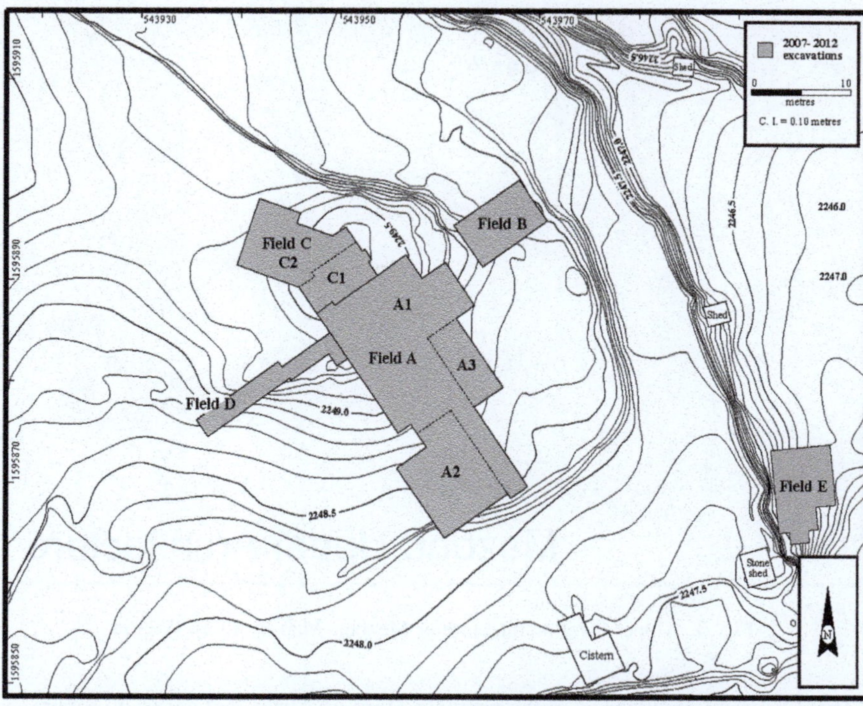

Fig. 2. Excavation units at the archaeological site of Mezber. Map by Shannon Wood.

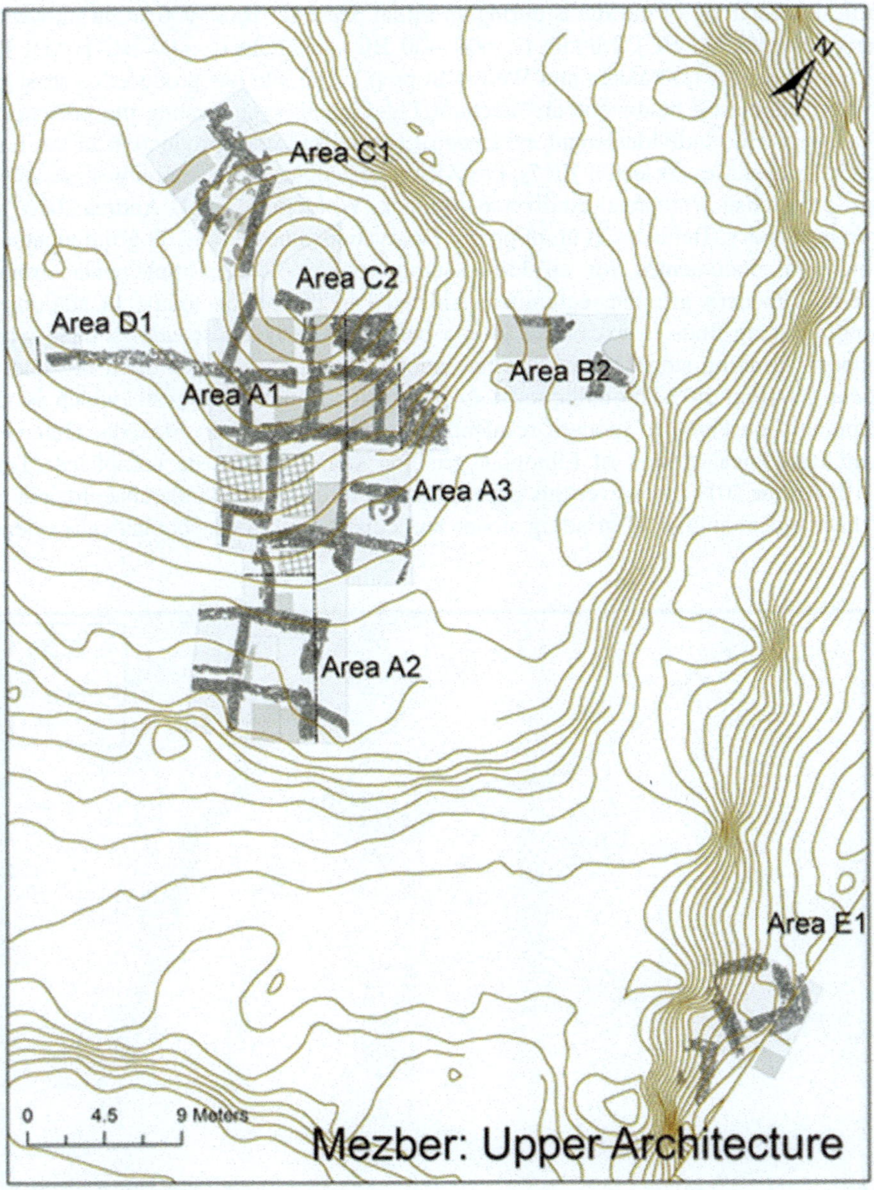

Fig. 3. Upper level architecture at Mezber. Map by Lynn Welton.

Pre-Aksumite populations of the northern Horn of Africa had access to several crop species which are likely candidates to have been processed on Mezber grinding stones. Available macrobotanical evidence for the presence of domesticated plants during the first millennium BC suggests that Near Eastern species such as barley (*Hordeum vulgare* L.), emmer (*Triticum turgidum* L. ssp. *dicoccum* (Schrank) Thell),

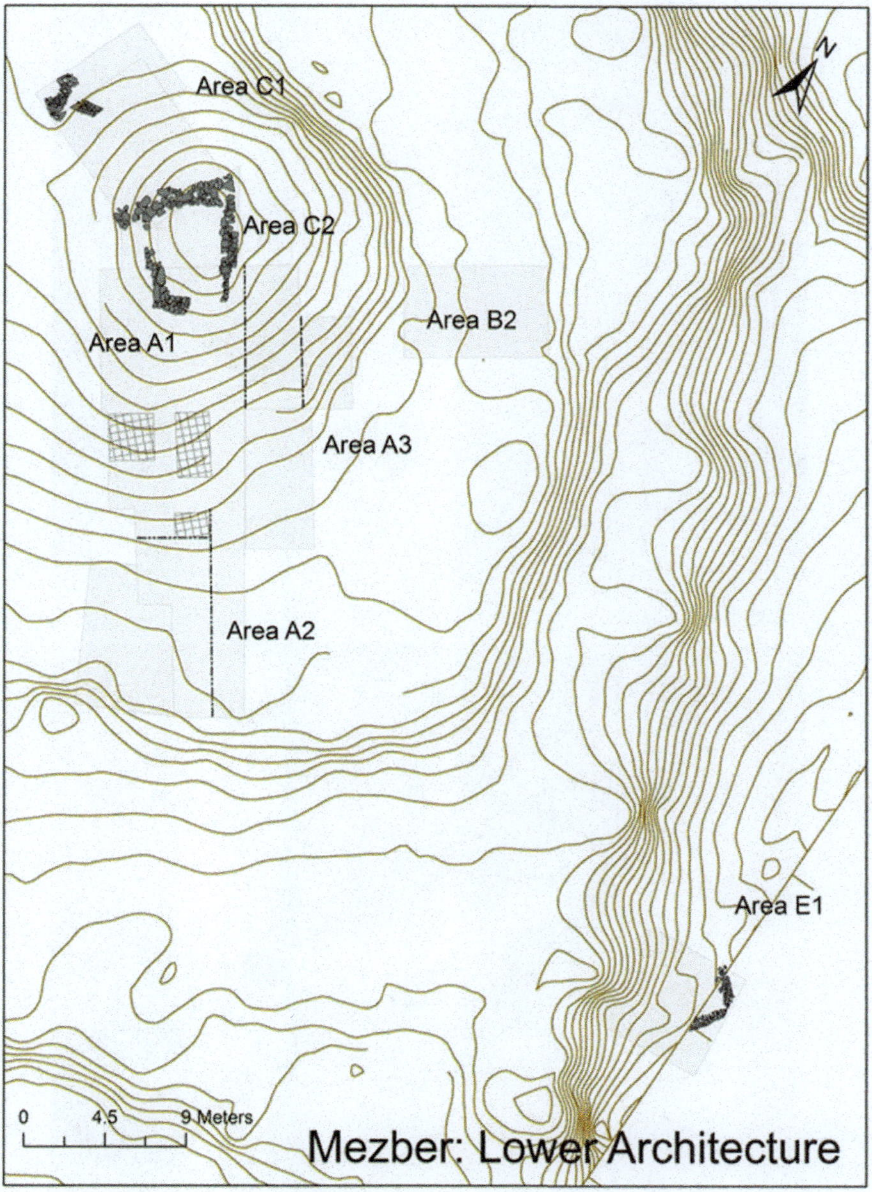

Fig. 4. Lower level architecture at Mezber. Map by Lynn Welton.

free-threshing wheat (*T. durum/aestivum*), lentil (*Lens culinaris* Medik.), and linseed (*Linum usitatissimum* L.) have an earlier presence than indigenous African crops including t'ef (*Eragrostis tef* (Zucc.) Trotter), finger millet (*Eleusine coracana* Gaertn.) and sorghum (*Sorghum bicolor* (L.) Moench) (D'Andrea 2008; D'Andrea et al. 2008b). However the Ethiopian and East African highlands were also important areas for plant

Fig. 5. Modern and ancient *maṭhan* and *udo*. a. Modern *maṭhan* quern and *madit* handstone built into a grinding table or *udo*. Image by Laurie Nixon-Darcus. b. Collapsed *udo*, Mezber Field E1 Locus 39. Image by Elizabeth Peterson.

domestication and agricultural innovation of t'ef, noog (*Guizotia abyssinica* (L.f.) Cass.), and finger millet. Archaeobotanical data further suggest the dominance of Near Eastern crops during Pre-Aksumite times with t'ef as the only indigenous domesticate identified to date (D'Andrea 2008; D'Andrea et al. 2008b; Beldados et al. in prep.). Emmer and barley have been directly dated to this period at the D site at Kidane Mehret, Aksum (Boardman 2000), where they were cultivated along with linseed. Early first millennium BC contexts at Ancient Ona sites near Asmara have produced these crops in addition to lentil (D'Andrea et al. 2008b). Mezber has produced barley and lentil directly dated to the Initial Pre-Aksumite Phase, as well as lentil and linseed, with emmer present from Early Phase contexts onward (Beldados et al. in prep.). The only indigenous African cereal documented to date for the Pre-Aksumite period is t'ef at the Mai Chiot site near Asmara, where a single grain is dated by association to the Ancient Ona Culture, during the 4th century BC (D'Andrea et al. 2008b). T'ef has also been recovered at Mezber during the Late Pre-Aksumite phase (Beldados et al. in prep.). Sorghum has not yet been identified in Pre-Aksumite contexts, however it has been recovered in Jebel Mokram Group deposits at Mahal Teglinos in the nearby Gash River region of eastern Sudan, dating to 2500–1000 BC (Beldados and Costantini 2011; Beldados et al. 2018). This archaeobotanical evidence is supported by limited stable isotopic evidence consisting of human remains recovered from the site of Etchmare East, located approximately 2.5 km to the southwest of Mezber. Analyses indicate that the single individual had a diet based on 80% C_3 (which could include wheat, barley, lentil and linseed) and 20% C_4 (t'ef, finger millet and sorghum) crops. The remains of the Etchmare East individual have been directly dated to 2290 ± 50 uncal BP (471–201 BC), which places the individual in the Late Pre-Aksumite phase at Mezber (D'Andrea et al. 2011).

Ethnoarchaeological and Archaeological Studies of Grinding Stones

Ethnoarchaeological fieldwork on grinding stone production and use was carried out in Gulo Makeda villages near Mezber (Fig. 1). A total of 52 interviews were completed during ETAP field seasons 2012 and 2013 (Nixon-Darcus 2014). Of those, 32 involved adult female advisors, and 20 were adult males. The aim was to ensure a cross section of adults and economic classes over a number of villages to allow for comparative analysis. In the Gulo Makeda region there is no running water or electricity and access to mechanical mills has been available only since the 1960s. Transportation is normally by foot and donkeys and horses are used as pack animals. The people inhabiting villages near Mezber rely almost entirely on subsistence agriculture which is supplemented by food aid through Food-For-Work and other programs. All agricultural tasks are non-mechanized including ploughing and threshing with oxen and winnowing by hand. Individuals in this area still retain first-hand knowledge of the manufacture and use of grinding stones, which remains part of the living memory (Nixon-Darcus and D'Andrea 2017).

Experts from Gulo Makeda assisted in identifying and developing a classification of archaeological grinding stones from Mezber. Grinding stones can be classified into four categories: querns are either *maṭhan* (Fig. 5a) or *madqos* (Fig. 6); and hand stones are either *madit* (companion handstone to the *maṭhan* and shown resting on the *maṭhan* in Fig. 5a) or *wedimadqos* (palm sized companion handstone to the *madqos* and shown resting on the *madqos* in Fig. 6). The generally larger *maṭhan/madit* are used to grind flour, while the thinner and more strongly concave *madqos/wedimadqos* are used to decorticate chickpeas (*Cicer arietinum* L.) and other beans, grind salt and chili peppers (*Capsicum annuum* L.). While *madqos* tend to be severely worn and thin, sometimes it is not possible to distinguish them from *maṭhan* based on dimensions alone (Nixon-Darcus and D'Andrea 2017).

Fig. 6. Modern *madqos* quern with *wedimadqos* handstone. Image by Laurie Nixon-Darcus.

Gulo Makeda consultants distinguish three surface textures of *maṭhan* and *madit* which are employed for different grains. A fine textured surface has low topographic surfaces, with rounded and flattened quartz grains and is smooth to the touch (Figs. 7, 8a). Smooth/fine surfaces are used to process small-grained cereals such as t'ef and finger millet. A coarse surface has many topographic highs and lows, raised angular quartz grains, and feels rough to the touch (Figs. 7, 8b). Coarse surfaces are preferred to grind large-grained cereals such as wheat, barley, sorghum and maize. Such surfaces, with exposed angular edges of quartz inclusions, are able to shear and tear the outer layers of larger grains when the handstone meets the quern, including removing the outer hulls for later separation during sieving (cf. Horsfall 1987:341–342). This works well for large cereal grains and in contrast a smooth/fine surface used for large grains would crush them and make it difficult to separate fine chaff from flour. Smooth/fine textured surfaces are required for grinding locally domesticated grains t'ef and finger millet because these small grains would become lodged in the deeper and wider interstices of coarse surfaces. A medium surface falls between these two extremes. When a medium texture surface was noted during ethnoarchaeological interviews, consultants advised that it had worn some from grinding larger grains, but was not quite ready for rehabilitation (re-sharpening). It was also explained that medium texture surfaces could wear a little more through grinding and the surface could become smoother/finer, ready for grinding smaller grains (Nixon-Darcus and D'Andrea 2017).

Fig. 7. Waizaro Lemlem Gebregzabier comparing coarse (left) and fine (right) textured surfaces of grinding stones. Image by A. Catherine D'Andrea.

Grinding on the *maṭhan* quern is in decline in Gulo Makeda today because of the introduction of mechanical mills, so that one *maṭhan* is sufficient for most households. In some cases women own two grinding stones with coarse and smooth/fine surface textures and several consultants stated that this was the case in the past. Many houses visited had one *maṭhan*, and to accommodate different grains, women "rehabilitate" or "re-sharpen" the surfaces of both *maṭhan* and *madit* stones to create the coarseness or smoothness/fine surface needed for different grains. In re-sharpening, a hammerstone or metal hammer is used to create a suitable surface. The technique used to prepare a smooth/fine surface consists of light quick blows close together, with the hammer head close to the grinding surface. To prepare a coarse surface, the hammer is lifted higher and brought down with more force and blows are more dispersed. In houses with one *maṭhan* quern, t'ef and finger millet are ground after larger cereal grains, because the

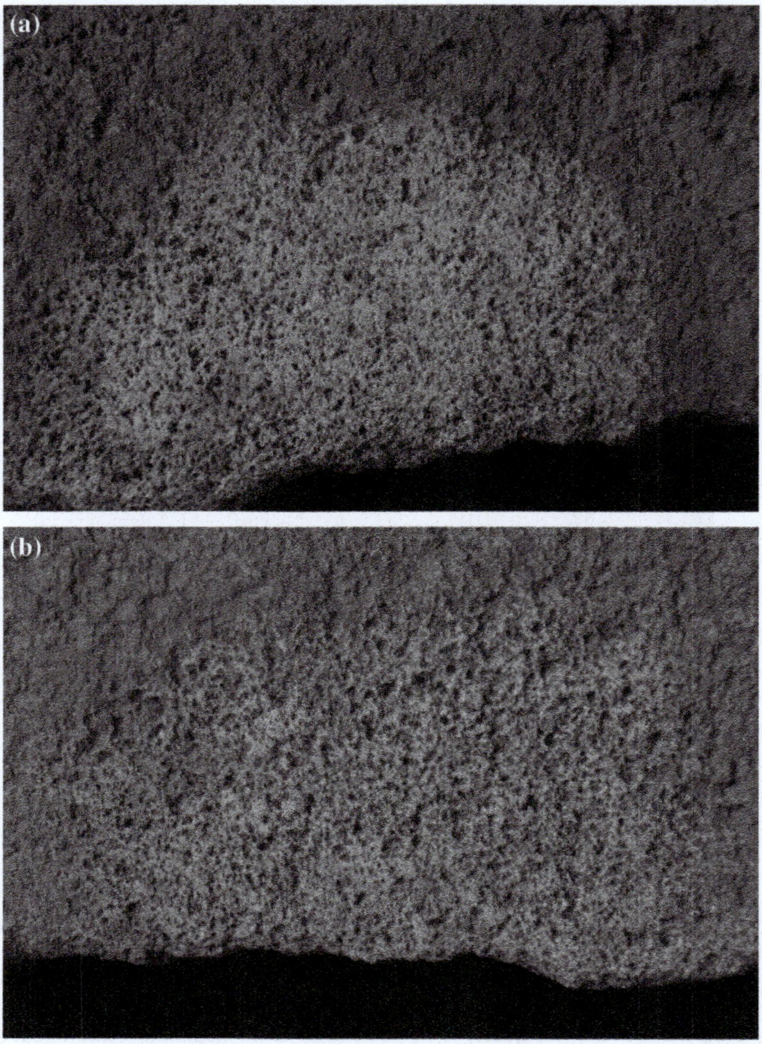

Fig. 8. Bifacial madit surfaces. Images by Laurie Nixon-Darcus. a. Bifacial *madit* with fine textured surface. Mezber, C2 Locus 30 Pail 94 #4200. b. Bifacial *madit* with coarse textured surface. Mezber, C2 Locus 30 Pail 94 #4200.

initial grinding of larger grains creates a smoother/finer grinding surface. In one case a woman had separate *madit* handstones for different grains—a heavy coarse grained *madit* for large grains, and a smaller *madit* with a smoother/finer texture surface and smaller quartz inclusions used for smaller t'ef and finger millet grains. Re-sharpening impact blows have been identified on ancient Mezber grinding stones (Fig. 9).

Fig. 9. Evidence of Re-sharpening on a *maṭhan*. Mezber C1, Locus 8, Pail 14 #410. Image by Laurie Nixon-Darcus.

An important result of this study is the identification of bifacial *madit* handstones which have one smooth/fine and a second medium to coarse surface texture. Bifacial *madit* handstones were documented in modern Gulo Makeda households and they occur in significant numbers in all Pre-Aksumite phases at Mezber. Bifacial *madit* handstones may have been preferred for use with different types of grains: coarse side for large grains, smooth or fine side for small grains. With a bifacial stone there would not be a need to constantly re-sharpen surfaces as both types are readily available. There also would not be a need for two separate *madit* stones to serve different types of grains readily, as was witnessed during interviews (Nixon-Darcus and D'Andrea 2017).

Following consultant interviews, archaeological and ethnographic grinding equipment was examined for a wide range of morphological attributes including size, shape (Wright 1992), use-surface area (Surface Coverage of SCOV, Adams 2014a) and use-wear analysis (Surface Text or STEXT, Adams 2014a:257) (Nixon-Darcus and D'Andrea 2017). For the purposes of this paper, we present results of use-wear analyses of surface texture or STEXT (Adams 2014a:257). This variable was selected in order to test the idea that different grinding surfaces may be characterized by microremains of large or small-grained cereals.

Sampling for Microbotanical Remains

Sampling for starch and phytoliths from ten Mezber grindingstones (including smooth/fine, medium, and coarse surfaces) and associated soil was completed as a pilot study to determine if microbotanical remains were preserved on grinding stone surfaces of Pre-Aksumite age and if any relationship exists between grinding surface texture and preserved crop microremains. Field sampling followed a modified version of the three-step sampling procedure recommended by Pearsall et al. (2004). Mezber grinding stones were dry brushed following excavation and too large to place into a sonicator. This meant that sampling Sediment 1 (dry brushed from artefact surface) and 3 (sonication) was not possible and only Sediment 2 was collected (Pearsall et al. 2004). Following excavations, grinding stones were not washed and stored in large plastic bags where possible. To ensure potential for future research on residues, only small sections of grinding stone surfaces were cleaned to obtain Sediment 2 (Pearsall et al. 2004). This entailed adding bottled water and scrubbing grinding surfaces with clean toothbrushes. The water was captured and stored in plastic bottles for four days after which excess water was decanted. In addition to grinding stone residues, control soil samples from the same loci surrounding the grinding stones were sampled and processed.

In the laboratory, the combined method for recovering starches and phytoliths was employed for Mezber grinding stone and soil samples following Horrocks (2005):

1. Add 2.5–3.0 cm^3 archaeological sample to 10 ml tube.
2. Add 6 ml 5% dishwashing powder solution, leave for several hours, shaking occasionally.
3. Pour sample into 50 ml beaker, swirl, pour liquid back into tube, add water to beaker and repeat to make up to 10 ml in tube, centrifuge at 3000 rpm for 3 min, decant. Dispose of settled fraction in beaker.
4. Add water, stir, centrifuge at 3000 rpm for 3 min, decant. Repeat until clear.
5. Add 3 ml 1.7 s.g. sodium polytungstate (SPT) solution, shake, centrifuge at 1500 rpm for 3 min, pipette top layer into new tube, top up with water, shake, centrifuge at 3000 rpm for 6 min, decant.
6. Add water to new tube, stir, centrifuge at 3000 rpm for 3 min, decant. Starch separation complete.
7. Wash pellet in original tube through 250 mm sieve into 50 ml beaker with <10 ml water. Pour into tube, centrifuge at 3000 rpm for 3 min, decant. Dispose of fraction on sieve.
8. Add concentrated nitric acid, place in boiling water bath. Occasionally stir in granules of potassium chlorate. When no longer red-tinted, centrifuge at 3000 rpm for 3 min, decant. Add water, stir, centrifuge at 3000 rpm for 3 min, decant.
9. Add water to top of tube, shake, let stand for 1 h, pipette off and dispose of liquid, leaving settled fraction. Repeat until clear.
10. Add 10% hydrochloric acid, stir, wait for reaction to cease, centrifuge at 3000 rpm for 3 min, decant. Add water, stir, centrifuge at 3000 rpm for 3 min, decant.
11. Add 10% potassium hydroxide solution, stir, place in boiling water bath for 6 min stirring occasionally, centrifuge at 3000 rpm for 3 min, decant. Add water, stir, centrifuge at 3000 rpm for 3 min, decant.
12. Add 3 ml 2.3 s.g. sodium polytungstate solution, shake, centrifuge at 3000 rpm for 6 min. Pipette top layer, invert pipette and rest in test tube rack, rinse tube, pipette layer back into tube, top up with water, shake, centrifuge at 3000 rpm for 3 min, decant.
13. Add water, stir, centrifuge at 3000 rpm for 3 min, decant. Phytolith separation complete.

The starch samples separated in step six were rinsed three times to eliminate excess heavy liquid, then were mounted onto clean, glass microscope slides using a mixture of 50% ultrapure water and 50% liquid white glycerin. Cover glasses were then placed over the sample, and the edges of the glass were sealed using nail polish. The phytolith extract was dried with acetone. Once completely dry, the residues were mixed with Permount and were placed onto clean, glass slides and sealed with cover glasses.

The original slides were assembled in the Archaeobotanical Laboratory at Helwan University in Cairo, and were later transported to the FARM in Virginia, USA for further study. Examination of the slides was completed at both 150x and 400x magnifications using a Zeiss Polarizing microscope, and photographs were captured at 800x magnification using a Micropublisher camera. Identifications were completed using one-to-one comparisons with modern plant materials.

Microbotanical Remains Preserved on Mezber Grinding Stones

All grinding stones and soil samples produced large quantities of phytoliths, while five of ten grinding stone samples produced identifiable starch (Fig. 10a–e). No starch was recovered from associated soil samples (Table 1). Phytolith studies are still in progress and here we report on the identification of starch. Because prepared microscope slides for starch had become desiccated prior to analysis, many of the grains could not be rotated for three-dimensional viewing. As such, identifications are limited to higher-level taxa that have morphologies consistent with those that are observable on the slides. Notably, most of the starch grains bore damage consistent with that created in laboratory experiments that replicate grinding with lithic implements (Babot 2003; Perry and Quigg 2011). This, in addition to the absence of starch in associated soil samples, supports the conclusion that recovered starch grains were derived from ancient grinding activities.

Grounding stone sample 1 (GS1) (Fig. 10a) is a Late Phase *maṭhan* quern with a medium surface texture, associated with a charcoal date of 2260 ± 30 uncal bp (397–209 BC). Samples produced four starch grains, all of which were derived from Poaceae species (Table 1). One grain has morphological features consistent with Triticeae, the tribe that includes wheat and barley (Fig. 11a). The cracking at the edges of this cf. Triticeae grain has been observed in starches subjected to grinding in laboratory experiments (Perry and Quigg 2011). This grain also appears "dirty," with small particles clinging to it, a pattern which occurs in grains that have been heated prior to grinding (Henry et al. 2009; Perry and Quigg 2011). The other three starch grains recovered from GS1 may belong to the Panicoideae group, which includes sorghum (e.g., Figure 11b, viewed under cross-polarized light). It is worth noting that the three grains are too large to be derived from finger millet. These cf. Panicoideae starches also bear damage consistent with grinding.

The other Late phase grinding stone GS6 is a *madit* handstone with a coarse surface (Fig. 10b). Recovered specimens include a compound starch, the size and morphological features of which is consistent with t'ef as viewed in available reference collections (Fig. 12a, b). However, at this point it is difficult to rule out other grasses

(a)
(b)
(c)
(d)
(e)

Fig. 10. Grinding stones sampled for microbotanical remains. Images by A. Catherine D'Andrea. a. Grinding stone GS1, Late phase *maṭhan* quern from Mezber. b. Grinding stone GS6, Late phase *madit* handstone from Mezber. c. Grinding stone GS5, Middle phase *maṭhan* quern from Mezber. d. Grinding stone GS10, Early or Middle phase *wedimadqo*s handstone from Mezber. e. Grinding stone GS4, Early phase, *madqos* quern from Mezber.

Table 1. Starch grains recovered from Mezber grinding stones.

Sample no.	Context	Phase	Associated date uncal bp	OxCal 4.2 95.4% cal BCE	Type	Texture	Identified Starch n = 13
GS1	C1 L8 P14 #1182	Late	2260 ± 30	397–209	*maṭhan*	medium	1 cf. Triticeae; 3 cf. Panicoideae
GS6	C1 L8 P14 #410	Late	2260 ± 30	397–209	*madit*	coarse	1 cf. *Eragrostis tef*; 1 Unidentified
GS5	E1 L11 P14 #1855	Middle	–	–	*maṭhan*	smooth	2 cf. Triticeae
GS10	A2 L26 P43 #1287	Early-Middle	2550 ± 40	806–542	*wedimadqos*	smooth	1 cf. Triticeae
GS4	C1 L10 P19 #649	Early	–	–	*madqos*	smooth	1 cf. Triticeae; 1 cf. Fabaceae; 2 Unidentified

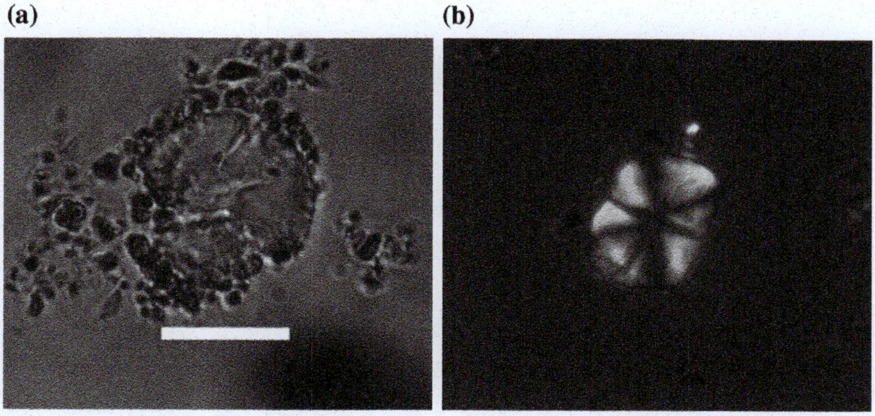

Fig. 11. Starch grains from GS1. The scale bar is 20 μ. Images by Linda Perry. a. Starch grain characteristic of the Triticeae. b. Viewed under cross-polarized light, a starch grain from cf. Panicodeae, the group that includes sorghum.

characterized by very small starch granules in compound groups. It is worth noting that the sample of finger millet starch in available comparative collections has compound groups that are much smaller in size and number of component granules than the starch found on GS6. The largest compound groups measure about 30 μ in maximum length in comparison to this much larger group that is commonplace in t'ef. There is a single, unidentifiable starch grain also present in this sample.

The Middle phase is represented by GS5, which is a *maṭhan* quern with a smooth surface (Fig. 10c). Starch from GS5 included two grains belonging to cf. Triticeae. Figure 13a was taken under white light, while Fig. 13b was taken under cross-polarized light. The damage to the extinction cross as viewed in Fig. 13b is typical of heating, and the debris adhering to the grain may be the result of parching.

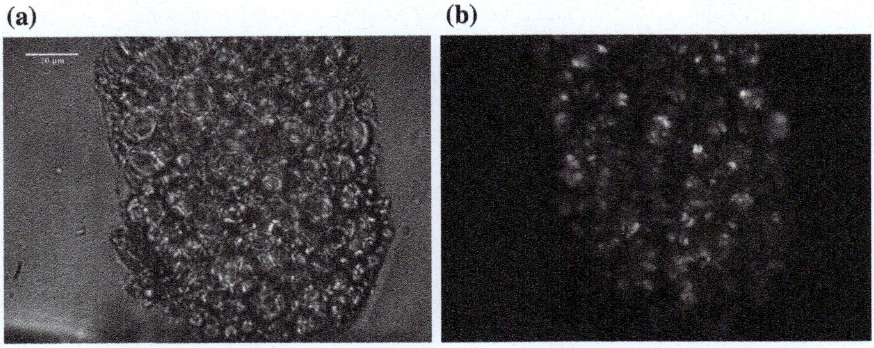

Fig. 12. Starch grains from GS6. Both micrographs represent the same compound starch grain mass. T'ef forms similar compound starch grains, and this was the only compound mass recovered from Mezber samples. Note the different scale in this image compared to other starch grain images. Images by Linda Perry. a. Viewed under white light. b. Viewed under cross-polarized light so that each, separate granule can be seen with its distinct extinction cross.

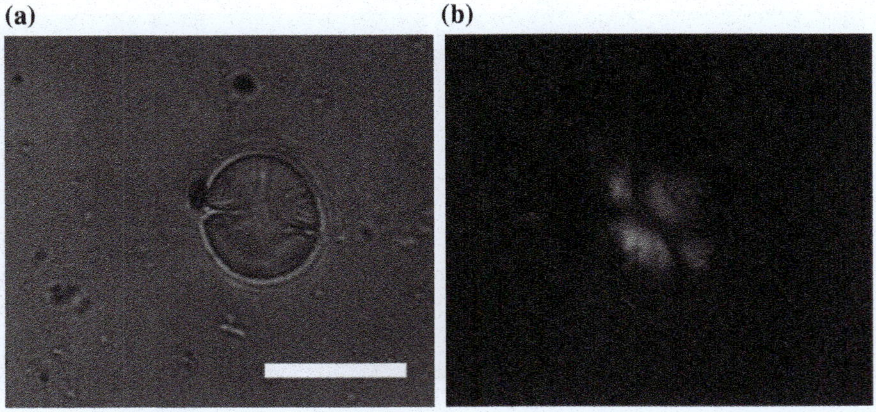

Fig. 13. Starch remains from GS5. Both A and B are micrographs of the same starch grain. The scale bar is 20 μ. Images by Linda Perry. a. Viewed under white light. b. Viewed under cross-polarized light. The damage to the extinction cross as viewed in B is typical of heating, and the debris adhering to the grain may be the result of parching.

GS10 is an Early or Middle phase *wedimadqos* handstone associated with a charcoal AMS date of 2550 ± 40 uncal bp (806–542 BC) (Fig. 10d). A single starch grain morphologically consistent with the Triticeae was recovered from GS10 (Fig. 14a, b). The smeared look and distortion of the extinction cross are probably representative of grinding damage.

The Early phase is represented by GS4, a *madqos* quern with a fine/smooth surface (Fig. 10e). It produced starch consistent with the Triticeae with cracking damage probably from grinding (Fig. 15a). In addition, starch with a gross morphology and extinction cross characteristic of cf. Fabaceae was observed (Fig. 15b). Two additional grains from GS4 are not identifiable.

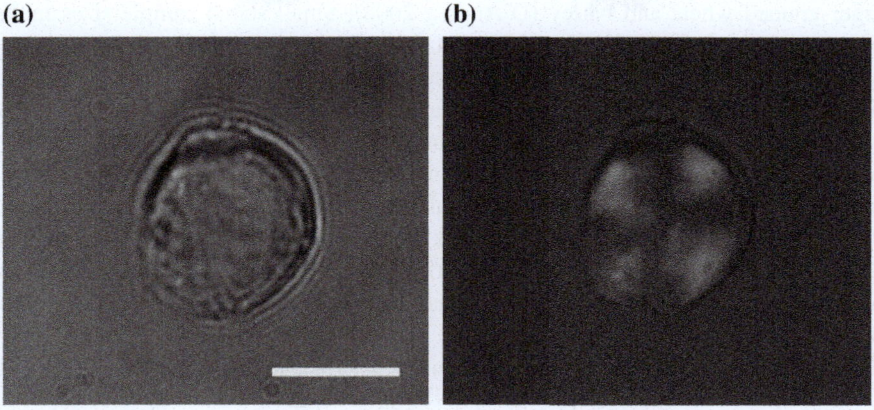

Fig. 14. Starch remains from GS10. Both micrographs are of the same starch grain. The smeared look and distortion of the extinction cross are probably representative of grinding damage. The scale bar is 20 µ. Images by Linda Perry. a. Viewed under white light. b. Viewed under cross-polarized light.

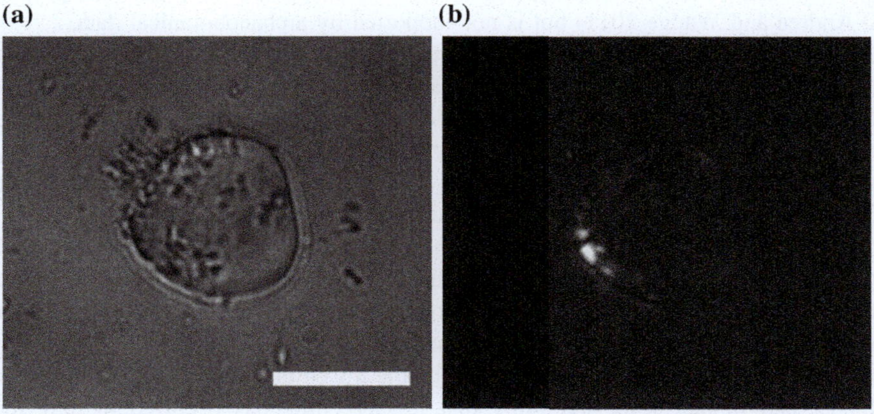

Fig. 15. Starch grains from GS4. The scale bar is 20 µ. Images by Linda Perry. a. Starch grain typical of the Triticeae, also showing cracking damage, probably from grinding. b. A starch grain under polarized light showing an extinction cross pattern that occurs in some Fabaceae.

Grinding Stone Surface Texture and Preservation of Ancient Starch

This study has been successful in the recovery of phytoliths and starch grains from grinding stones of Pre-Aksumite age from Mezber. We anticipate that additional sampling of larger numbers of grinding implements will assist in identifying plant remains to further supplement macrobotanical studies in progress (Beldados et al. in prep.). This is particularly important in the case of t'ef which may be underrepresented in the macro-botanical record because it tends to char at temperatures lower than larger-grained wheat and barley (D'Andrea 2008). Small starch sample sizes make it difficult to directly compare the Mezber starch remains with existing macro-botanical evidence, but speculations can be offered, subject to further investigation. For example, the Mezber macro-botanical record indicates that t'ef is present in the Late phase, similar to t'ef starch which has been found only on the Late phase GS6 (Table 1). Lentil and barley are both present beginning from the Initial phase (Beldados et al. in prep.) and throughout the long occupational sequence, while emmer occurs in the Early phase onward. These crops are possibilities for the cf. Triticeae and cf. Fabaceae starch identified in the Early and Middle phase grinding stones (Table 1). Potentially of great interest are the cf. Panicoideae starch grains from the Late phase GS1 which may point to the presence of sorghum in the region. To date macro-botanical evidence for sorghum has been identified only in Late Aksumite contexts at the D and K sites at Aksum (Boardman 1999:142) and finger millet in Early Aksumite (50 BC-AD 150) deposits at Bieta Giyorgis (D'Andrea 2008, 2018). The early presence of sorghum at 2500–1000 BC at Mahal Teglinos (Beldados and Costantini 2011; Beldados et al. 2018) means it is present in the wider region but at lower elevations. The question of when sorghum and possibly finger millet made the transition from Sudanese lowlands to Ethiopian highlands is unknown at present. The speculation that t'ef made this transition sometime prior to the Pre-Aksumite period in the hands of pastoralists has been proposed (D'Andrea and Wadge 2011) but is not supported by archaeobotanical data.

In this small sample of Mezber starch, there appears to be mixed results on the question of associating grinding stone surface texture with grain size. A starch grain of cf. Fabaceae was recovered on GS4 which is a *madqos* quern, and Gulo Makeda consultants stated that chick peas and other beans are normally decorticated on *madqos* querns. Also the presence of cf. Triticeae and cf. Panicoideae starch on the medium surface of GS1, a *maṭhan* quern, is consistent with ethnoarchaeological consultants who stated that large grains are ground on medium to coarse surfaces. However, fine grinding stone surfaces also yielded cf. Triticeae (GS5 and GS10), cf. Triticeae is found on a *wedimadqos* handstone (today used for spices and wet-grinding of sorghum), and the possible t'ef compound starch is found on the coarse surface of GS6 which is a *madit* handstone. Given that all but one of the 13 complete *madit* handstones from Mezber had impact scars signifying that re-sharpening has occurred, it may more reasonable to conclude that Mezber grinding stones were multipurpose implements which were employed for a variety of grains and were periodically re-sharpened and/or re-used. Different patterns of surface texture and grain may be revealed with larger numbers of samples from Mezber. We are planning such studies, as well as sampling modern ethnographic grinding stones with different surface textures for microbotanical remains.

Starch and Cooking Practice

The presence of heat damage on cf. Triticeae starch recovered from two *maṭhan* querns, one dating to the Late (GS1) and one to the Middle phase (GS5), raises the possibility that ancient inhabitants of Mezber were engaging in a culinary practice involving the roasting of barley grains prior to grinding to produce roasted barley flour. Barley, emmer, bread and durum wheat are consumed today in Ethiopia as roasted grains, called *kolo* (D'Andrea and Haile 2002; Lyons and D'Andrea 2003; Shewayrga and Sopade 2011:26), and these species are all potentially represented by the cf. Triticeae starch grains. However, Gulo Makeda consultants specify that only roasted barley grain is ground into flour (*tihini*) on a *maṭhan* quern and used in making a variety of dishes. The production of *tihini* flour and its use in the preparation of various foods has been documented in previous ETAP studies in Gulo Makeda and by food scientists working elsewhere in Ethiopia (Asfaw 1990; Kerssie and Giotom 1996; Yallew et al. 1998; Bekele et al. 2005; Nigusse 2005; Lyons 2007; Shewayga and Sopade 2011; Abraha et al. 2013). In Gulo Makeda, roasted barley flour is used in making *shamet* (a beverage), *genfo* (a firm porridge), *ga'at* (a liquid porridge), *tihlo* (a dish of dough balls made from *tihini* flour dipped in a spicy stew), and *besso* (a compressed dry paste formed into patties or balls) (cf. Shewayrga and Sopade 2011:26).

An ethnoarchaeological study of *besso* production was completed by ETAP in 2011 (Fig. 16a–e). *Besso* is a traditionally important Ethiopian food made from mixing roasted barley flour with water and oil or butter and sometimes adding salt or sugar. It is considered to be a nutritious food for travelers and is thought to relieve gastritis. *Besso* can alleviate food shortages early in the November-December harvest season because it can be made with mature grains that have not yet sufficiently dried to be made into other foods (Shewayrga and Sopade 2011). In making *besso*, previously winnowed and sieved barley grains are first pounded in a wooden mortar (*mogu*) to remove remaining hulls (Fig. 16a). After winnowing and hand-sorting to remove fine chaff (used as chicken fodder) and other impurities (discarded), the pounded grain is soaked in hot water for about 1 h, removed and stored in a burlap sack overnight, allowing excess water to drip away. The next morning, while still moist, grain is mixed with sand that has been heated on a clay (*meklogundo*) or metal griddle (*meklohachen*) (Fig. 16b, c) and stirred with a brush (*ma'ado*). The griddle may be covered with a lid (*mogu*) and sealed with cattle dung. When a slight crackling sound is heard, the barley grains are swept into a bowl-shaped depression near the oven called *mogulal* and then sieved using a coarse sieve (*mihea*). In this process, sand drops through and the roasted barley grain remains on top of the sieve (Fig. 16d). The roasted grain is re-pounded in a *mogu* mortar (Fig. 16a) and further impurities are removed by sieving and winnowing. Roasted barley before grinding is called *zighi*. The roasted grain is then ground on a *maṭhan* quern (Fig. 16e) and the resulting flour is processed using a fine sieve (*monfit*) which removes light chaff (chicken fodder) or broken grain (re-ground on *maṭhan* quern). The clean flour is known as *tihini*. To make *besso*, a water-oil mixture is heated in a small sauce pan, and *tihini* flour is added gradually while stirring quickly. This results in a soft crumbly mixture, which after cooling, can be formed into patties or balls and consumed immediately or stored (Fig. 17a, b). *Tihini* has a long shelf-life and is easily transported and mixed, making *besso* an ideal food for travelers, field workers and for use as military rations. Gulo Makeda inhabitants strongly identify with *besso*, *tihlo* and other barley-based foods. Farmers of this region consider barley to be the most ancient and culturally important of crops, even more so than t'ef. This also has been noted in other studies in northern Ethiopia where barley has been described by farmers as the "king of crops" (e.g., Shewayrga and Sopade 2011:21).

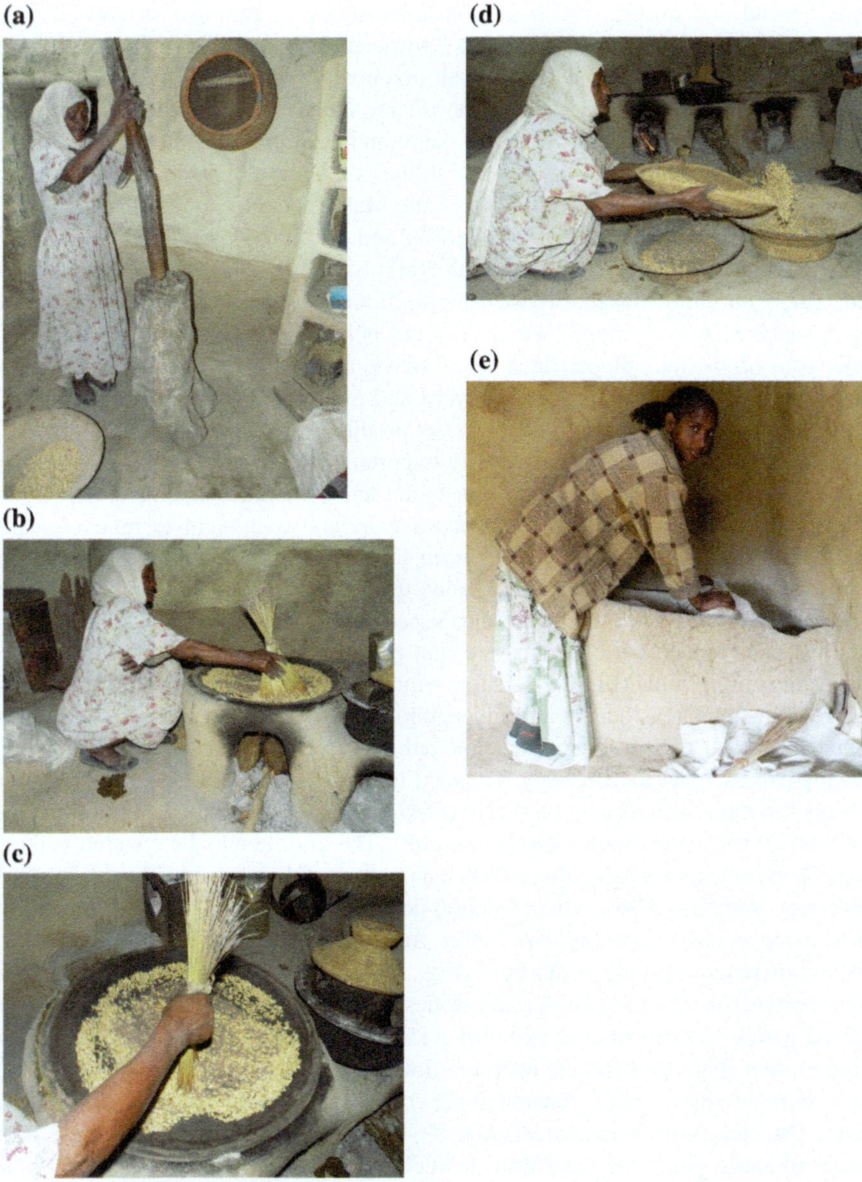

Fig. 16. Production of *besso* by Waizoro Medhin Abate. Images by A. Catherine D'Andrea. a. Pounding of cleaned barley grain. b. Heating barley grains mixed with sand on a *meklohachen* metal griddle. c. Mixing of barley grain and sand with *ma'ado* brush. d. Seiving and winnowing to remove sand. e. Grinding roasted barley grain on *maṭhan* quern.

(a)

(b)

Fig. 17. Making *Besso* patties. Images by A. Catherine D'Andrea. a. Mixing *tihini* flour with an oil and water mixture. b. *Besso* patties ready to eat.

The roasting of barley in sand, as opposed to microwave and other direct heat, creates chemical and physical transformations which facilitate food processing and improve the nutritional value of grains. Sand distributes the heat evenly to prevent grains from charring (Tashi et al. 2013:153; Rashid et al. 2015:228) and Gulo Makeda consultants stated that sand acts as a temper. Roasting in sand results in higher antioxidant activity, non-enzymatic browning and more puffing of grains (Sharma et al. 2011; Baba et al. 2016). Roasting also affects pasting property, gelatinization, colour, water absorption properties, causes damage to starch and increases ß-glucan solubility

(Sharma et al. 2011; Abraha et al. 2013). These transformations result in roasted barley flour having a high water absorption capability and foods made from it are more easily digested (Sharma et al. 2011). Gulo Makeda consultants state that roasted barley flour effectively absorbs water enabling thicker porridges to be made using relatively smaller quantities of flour. These properties also mean that *tihini* flour works well when formed into dough balls and eaten as *tihlo*. Similar observations have been made in experimental studies (Shewayrga and Sopade 2011). Consultants also add that grinding roasted barley is less strenuous than processing raw barley grains.

There is an interesting parallel of *besso* with a similar meal prepared in a different highland environment, the food known as *tsampa* of Tibet. *Tsampa* is a Tibetan and Nepalese staple food in which roasted barley flour is mixed with butter tea and consumed, like *besso,* as patties or balls (Norbu 2006:262–26; Wangdu 2016). A pre-eminent component in the diet of Tibetan peoples, *tsampa* is an ingrained symbol of "Tibetannes" in that the term "*tsampa*-eater" is a mark of Tibetan identity (Norbu 2006:263; Laurent 2015). *Tsampa* has ritual uses: it is thrown in the air during marriages and New Year celebrations and is used in making ritual cakes. The processing is strikingly similar to *besso*, where soaked barley grain is heated with sand on a metal skillet or griddle until it begins to crackle. The sand is removed by sieving and after cooling, the roasted barley is ground into *tsampa* flour. Similar to *besso*, *tsampa* is a highly appreciated traveler's food and commonly used by the military. Barley itself also has an important socio-cultural role in Tibetan society and is used in a wide variety of rituals and offerings (Norbu 2006:262–26; Wangdu 2016).

In addition to Ethiopia and Nepal, the consumption of roasted barley flour is widespread in many areas of the world, used in beverages, particularly coffee substitutes, and other foodstuffs. *Caffè d'orzo* in Italy and the traditional *máchica* drink in Ecuador are examples of beverages made from ground roasted barley grain. Similar beverages are a mixture of roasted barley powder with other plants such as chicory, for example, Barleycup®, (UK), Inka® (Poland), and pero® (Europe and USA). In Ecuador, *máchica* is prepared by roasting a mixture of barley grains and sand in a griddle, followed by grinding into a powder. *Máchica* is closely associated with an Indigenous identity to the point of being used as a form of resistance to "white foods". It is considered by Indigenous Ecuadorians as a substantive food or "food that warms you up". It also has become an Indigenous female symbol of home and hospitality (Weismantel 1988:160–61; Jamieson and Sayre 2010).

Besso (Ethiopia), *tsampa* (Nepal), and *máchica* (Ecuador) are all considered traditional foods by widely separated cultures. Despite the fact that barley is not indigenous to any of these regions, barley and barley-based foods have become not only important dietary components, but they have attained symbolic significance and become embedded in social identities. Moreover, these foods are made using a similar culinary practice involving roasting barley grain mixed with sand prior to grinding into flour. As a crop, barley has spread widely partly because of its adaptability to extreme conditions, including drought, salinity, cold, and short growing seasons (Harlan 1995:140). These traits have permitted it to flourish in widely differing environments, from the relatively arid conditions of the Near East to high elevations of the Tibetan, Ethiopian and Ecuadorian highlands, where it became integrated into Indigenous food systems and agricultural complexes. It seems that not only barley grain itself, but knowledge of the preparation and use of roasted barley flour, spread far beyond the area of origin in the Near East. Based on the identification of cf. Triticeae starch on grinding stones, we propose that the culinary practice of making roasted barley flour has a deep history in Gulo Makeda and may be traced to at least the Middle Pre-Aksumite c. 600–400 BC period in the Ethiopian highlands.

Concluding Remarks

Starch grains were successfully extracted from grinding stones dating to the Pre-Aksumite period at the archaeological site of Mezber. At least four taxa are represented in the sampled grinding stones. One set of grains bears characteristics of the Triticeae, the grass tribe that includes wheat and barley. A second group possibly represents the Panicoideae subfamily of which sorghum is a member. A third type bears characteristics of the Fabaceae family including gross morphology and the extinction cross pattern, while a fourth may represent a compound starch characteristic of t'ef. Some cf. Triticeae and cf. Panicoideae starch exhibited heat treatment and damage characteristic of grinding which suggests that they are remnants of grinding flour and cooking practice. Ethnoarchaeological and archaeological investigations of grinding stones used in processing cereals suggests that they are multi-functional implements, the surface textures of which have been repeatedly modified to accommodate the milling of different grains. It is proposed that the heat treated cf. Triticeae starch identified on grinding implements suggests that occupants of Mezber were engaged in the culinary practice of producing roasted barley flour as early as 600–400 BC. Roasted barley flour today continues to be used in preparing foods and beverages world-wide, suggesting that not only barley grain, but associated culinary knowledge has spread far from the Near Eastern area of origins.

Acknowledgements. We dedicate this paper to the memory our friend and colleague, Prof. Ahmed G. Fahmy, who initiated the laboratory analysis of microbotanical samples from Mezber grinding stones. We are greatly indebted to our Gulo Makeda consultants who shared with us their knowledge in using grinding stones and the making of *besso*. In particular we thank Waizoro Zaid Mahray, Haleka Tehwoelde Brahn Beyene, Waizoro Nigisti Hagos, Haleka Gebreselassie Gebreyesus, Waizoro Medhin Abade, Ato Hailu Hago, and Ato Mebhratu Areyhu. Our superb interpreters, Habtamu Mekonnen and Yemane Meresa, had a significant role to play in the success of ETAP fieldwork. We are grateful to the Authority for Research and Conservation of Cultural Heritage (ARCCH), and in particular to Ato Desalegn Abebaw for assistance in obtaining research permit clearances. In addition our work was supported by the Tigrai Agency for Culture and Tourism and Ato Kebede Amare. Many thanks to Shannon Wood who produced Figs. 1 and 2 and Lynn Welton who completed Figs. 3 and 4. We are grateful to Prof. Ross Jamieson and Dr. Daniela Balanzategui for discussions on the role of *máchica* in Ecuadorian indigenous cuisine. ETAP fieldwork reported herein was funded by grants awarded by the Social Sciences and Humanities Research Council, including SSHRC Standard Research Grant No. 410-2011-1646 and SSHRC Insight Grant No. 435-2014-0182.

References

Abraha A, Uhlen AK, Abay F et al (2013) Roasted barley foods: processing and varietal differences affecting *kolo* and *tihini*, traditional grain products in northern Ethiopia. Cereal Food World 58(2):71–79

Adams JL (1988) Use-wear analyses on manos and hide-processing stones. J Field Archaeol 15(3):307–315

Adams JL (1993) Toward understanding the technological development of manos and metates. Kiva 58(3):331–344

Adams JL (2014a) Ground stone analysis: a technological approach. The University of Utah Press, Salt Lake City

Adams JL (2014b) Ground stone use-wear analysis: a review of terminology and experimental methods. J Archaeol Sci 48:129–138

Aranguren B, Becattini R, Lippi RM et al (2007) Grinding flour in Upper Palaeolithic Europe (25,000 years bp). Antiquity 81(314):845–855

Arthur JW (2014) Culinary crafts and foods in southwestern Ethiopia: an ethnoarchaeological study of Gamo groundstones and pottery. Afr Archaeol Rev 31(2):131–168

Asfaw Z (1990) An ethnobotanical study of barley in the central highlands of Ethiopia. Biol Zbl 109:51–62

Baba WN, Rashid I, Shah A et al (2016) Effect of microwave roasting on antioxidant and anticancerous activities of barley flour. J Saudi Agric Sci 15(1):12–19

Babot MDP (2003) Starch grain damage as an indicator of food processing. In: Hart DM, Wallis LA (eds) Phytolith and starch research in the Australian-Pacific-Asian Regions: the state of the art. Papers from a conference held at the ANU, August 2001, Canberra, Australia, Pandanus Books, pp 69–81

Bekele B, Alemayehu F, Ladew B (2005) Food barley in Ethiopia. In: Grando S, Macpherson HG (eds) Food barley: importance, uses and local knowledge. Proceedings of the International Workshop on Food Barley Improvement. ICARDA, Aleppo, pp 53–82

Beldados A, D'Andrea AC, Manzo A (in prep) Filling the Gap: new Archaeobotanical evidence for 3rd–1st Millennium BCE Agricultural Economy in Sudan and Ethiopia

Beldados A, Costantini L (2011) Sorghum exploitation and Kassala and its environs, north eastern Sudan in the second and first millennia BC. Nyame Akuma 75:33–39

Beldados A, Manzo A, Murphy C et al. (2018) Evidence of sorghum cultivation and introduced West African crops in the second millennium BC at Kasala, Eastern Sudan

Boardman SJ (1999) The agricultural foundation of the Aksumite empire, Ethiopia. In: van der Veen M (ed) Plants and people in Africa: recent archaeobotanical evidence. Plenum Publishing Corp, New York, pp 137–147

Boardman SJ (2000) Archaeobotany. In: Phillipson DW (ed) Archaeology at Aksum, Ethiopia, 1993–7, vol 2. The British Institute in Eastern Africa and the Society of Antiquaries of London, London, pp 363–368

D'Andrea AC, Welton L (in prep) Mezber: A Pre-Aksumite Settlement in the Northern Horn of Africa

D'Andrea AC (2008) T'ef (*Eragrostis tef*) in ancient agricultural systems of highland Ethiopia. Econ Bot 62(4):547–566

D'Andrea AC (2018, in press) Palaeoethnobotanical investigations at Ona Nagast. In: Bard K, Fattovich R (eds) Archaeology of Bieta Giyorgis, Aksum. Cotsen Institute of Archaeology, UCLA

D'Andrea AC, Haile M (2002) Traditional emmer processing in highland Ethiopia. J Ethnobio 22(2):179–217

D'Andrea AC, Wadge P (2011) T'ef (*Eragrostis tef*): a legacy of pastoralism? In: Fahmy AG, Kahlheber S, D'Andrea AC (eds) Windows on the African past: current approaches to African archaeobotany. Africa Magna Verlag, Frankfurt, pp 225–241

D'Andrea AC, Manzo A, Harrower MJ et al (2008a) The Pre-Aksumite and Aksumite settlement of NE Tigrai, Ethiopia. J Field Archaeol 33(2):151–176

D'Andrea AC, Schmidt PR, Curtis MC (2008b) Palaeoethnobotanical analysis and agricultural economy in early 1st millennium BCE sites around Asmara. In: Schmidt PR, Curtis MC, Teka Z (eds) The archaeology of eritrea. Red Sea Press, Trenton New Jersey, pp 207–216

D'Andrea AC, Richards MP, Pavlish LA et al (2011) Stable isotopic analysis of human and animal diets from two pre-Aksumite/Proto-Aksumite archaeological sites in northern Ethiopia. J Archaeol Sci 38(2):367–374

Harlan JR (1995) Barley. In: Smartt J, Simmonds NW (eds) Evolution of crop plants, 2nd edn. Longman Group UK Limited, London

Harrower MJ, D'Andrea AC (2014) Landscapes of state Formation: geospatial analysis of Aksumite settlement patterns (Ethiopia). Afr Archaeol Rev 31:513–541

Henry AG, Hudson HF, Piperno DR (2009) Changes in starch grain morphologies from cooking. J Archaeol Sci 36:915–922

Horrocks M (2005) A combined procedure for recovering phytoliths and starch residues from soils, sedimentary deposits and similar materials. J Archaeol Sci 32:1169–1175

Horsfall GA (1987) Design theory and grinding stones. In: Hayden B (ed) Lithic studies among the contemporary highland Maya. University of Arizona Press, Tucson, pp 332–377

Jamieson RJ, Sayre MB (2010) Barley and identity in the Spanish colonial Audiencia of Quito: archaeobotany of the 18th century San Blas neighbourhood in Riobamba. J Anthropol Archaeol 29(2):208–218

Kerssie M, Goitom L (1996) Barley utilization. In: Gebre H, van Leur J (eds) Barley research in Ethiopia: past works and future prospects. Proceedings, First Barley Research Review Workshop. ICARDA, Addis Ababa, pp 167–170

Laurent Y (2015) The Tibetans in the making: barley cultivation and cultural representations. Rev Etud Tibétaines 33:73–108

Liu L, Field J, Fullagar R et al (2010) What did grinding stones grind? New light on Early Neolithic subsistence economy in the Middle Yellow River Valley, China. Antiquity 84(325):816–833

Lyons D (2007) Integrating African cuisines: rural cuisine and identity in Tigray, highland Ethiopia. J Soc Archaeol 7:346–371

Lyons DE, D'Andrea AC (2003) Griddles, ovens and the origins of agriculture: an ethnoarchaeological study of bread baking in highland Ethiopia. Am Anthropol 105(3):515–530

Nigusse BT (2005) Barley farming systems and types of barley end uses for human consumption as described by farmers in Eritrea. In: Grando S, Macpherson HG (eds) Food barley: importance, uses and local knowledge. ICARDA, Aleppo, pp 36–41

Nixon-Darcus L (2014) The cultural context of food grinding equipment in northern Ethiopia: an ethnoarchaeological approach. Unpublished M.A. Thesis, Simon Fraser University

Nixon-Darcus L, D'Andrea AC (2017) Necessary for life: studies of ancient and modern grinding stones in highland Ethiopia. Afr Archaeol Rev 34(2):193–223

Norbu D (2006) Economic policy and practice in contemporary Tibet. In: Sautman B, Dreyer JT (eds) Contemporary Tibet: politics, development, and society in a disputed region. M.E. Sharpe, New York

Pearsall DM, Chandler-Ezell K, Zeidler JA (2004) Maize in ancient Ecuador: results of residue analysis of stone tools from the Real Alto site. J Archaeol Sci 31:423–442

Perry L (2004) Starch analyses reveal the relationship between tool type and function: an example from the Orinoco valley of Venezuela. J Archaeol Sci 31(8):1069–1081

Perry L, Quigg JM (2011) Starch remains and stone boiling in the Texas Panhandle Part 2: identifying wild rye (*Elymus* spp.). Plains Anthropol 56:109–120

Peterson EA (2017) Development of craft specialization during the Pre-Aksumite period in Eastern Tigrai, Ethiopia: a study of hideworking traditions. Unpublished Ph.D. thesis, Simon Fraser University

Rashid U, Gani A, Shah A et al (2015) Effect of sand roasting on the antioxidant and antiproliferative activity of barley *(Hordeum vulgare*). Nutrafoods 14:227–236

Revedin A, Svoboda J, Aranguren B et al (2010) Thirty thousand-year-old evidence of plant food processing. Proc Nat Acad Sci 107(44):18815–18819

Sharma P, Gujral HS, Rosell CM (2011) Effects of roasting on barley ß-glucan, thermal, textural and pasting properties. J Cereal Sci 53(1):25–30

Shewayrga H, Sopade PA (2011) Ethnobotany, diverse food uses, claimed health benefits and implications on conservation of barley landraces in north eastern Ethiopia highlands. J Ethnobio Ethnomed 7:19–33

Tashi NT, Yawei Z, Xingquan, Z (2013) Food preparation from hulless barley in Tibet. In: Zhang G, Li C, Liu X (eds) Advances in barley sciences. Proceedings of the 11th international Barley genetics symposium. Springer, Dorkrecht, pp 151–158

Teklu G (2012) Ethnoarchaeological study of grind stones at Lakia'a in Adwa, Tigray Regional State, Ethiopia. Unpublished MA Thesis, Addis Ababa University

Woldekiros H, D'Andrea AC (2016) Early evidence for domestic chickens (*Gallus gallus domesticus*) in the Horn of Africa. Int J Osteoarchaeol. https://doi.org/10.1002/oa.2540

Wangdu Y (2016) Experience the joys of Tibetan Culture. http://www.yowangdu.com/tibetan-food/tsampa.html. Accessed 14 Mar 2016

Weismantel MJ (1988) Food, gender, and poverty in the Ecuadorean Andes. University of Pennsylvania Press, Philadelphia

Wright K (1992) A classification system for ground stone tools from the prehistoric Levant. Paléorient 18(2):53–81

Yallew A, Hailye A, Hassen H (1998) Farmers' food habit and barley utilization in northwest Ethiopia. In: Yirga C, Alemayehu F, Sinebo W (eds) Barley-based farming systems in the highlands of Ethiopia. Ethiopian Agricultural Research Organization, Addis Ababa, pp 107–112

Climate and Agrarian-Cultural Landscapes

Microbotanical Assessment of Anthropogenic Impacts in the Ngotto Forest, Central African Republic During the Last Millennium AD

Christopher A. Kiahtipes(✉)

Department of Anthropology, Southern Methodist University, Dallas, TX, USA
chris.kiahtipes@gmail.com

Abstract. Scientific understanding of past human-environment interactions in tropical Central Africa is limited by two primary problems. First, paleoenvironmental studies often focus on records that lack spatial and chronological association with documented archaeological records. Second, the range of potential human impacts on past environments is typically conflated within models of agricultural production while the other socioeconomic dimensions of human impacts linked with social complexity and specialization are poorly understood. Riparian swamps immediately adjacent to archaeological sites dating to the last five hundred years in the Central African Republic (CAR) yielded well preserved pollen and microscopic charcoal remains that document multiple phases of human settlement of the forest zone and their impacts on local vegetation. This record from the Ngotto Forest is spatially associated with a unique iron smelting center and the deposits cover major cultural transitions from the Late Iron Age through the post-Colonial era in great detail. As such, it sheds light on how intensive iron metallurgy, expanding trade networks, and, eventually, the collapse of local socioeconomic structures during French colonial rule shaped the modern composition and structure of Guineo-Congolian rain forests in CAR.

Keywords: Anthropogenic impact · Charcoal particle · Forest
Late iron age · Metallurgy · Paleoecology · Paleoenvironment
Palynology

Introduction

Archaeobotanical and paleoenvironmental methods have great potential to shed light on long-term vegetation dynamics as they relate to changing climates and human impacts in tropical Africa. This role is especially important in light of the global impacts of changing climates on forest dynamics (IPCC 2014) and the need to resolve conflicts between local stakeholders and conservation initiatives (Adams and McShane 1996). However, scholars working in Africa do not all agree on what the ecological impacts of past and present African societies are or how they shape the distribution, composition, and structure of floral diversity in the African tropics (Fairhead and Leach 1998; Sowunmi 1999; Brncic et al. 2009; Ngomanda et al. 2009; Maley et al. 2012; Bayon et al. 2013; Iles 2015; Lupo et al. 2015). Starting with a brief review of recent

A. M. Mercuri et al. (Eds.): *Plants and People in the African Past*,
https://doi.org/10.1007/978-3-319-89839-1_21

developments in this debate, I argue that the most significant constraints on our understanding of past human impacts are the lack of parallel archaeological and paleoenvironmental datasets and the coarse resolution of archaeological data in Central Africa. A more robust and realistic picture of human impacts can be constructed through examination of associated paleoenvironmental and archaeological records at a local or sub-regional level. I then present a case study of anthropogenic (human-driven) paleovegetation change associated with an intensive iron-working center located in the Ngotto Forest Reserve, Central African Republic (CAR) and its eventual collapse under the brutal forced labor regime of concessionary companies during the late 19th and early 20th centuries.

Background

Conservation of tropical rain forests is a central concern because they house an overwhelming proportion of global biodiversity and are a major terrestrial carbon sink (FAO 2010). While the rain forest biome was long assumed to be static (see discussions in Morely 2001; White 2001), a growing body of data from both Africa (Runge 2002; Elenga et al. 2004; Marchant and Taylor 2005; Brncic et al. 2009) and South America (Carson et al. 2014; Flantua et al. 2015; Piperno et al. 2015) indicate that tropical forests are responsive to both global climate change and past human activities. Scholars approach the problem of anthropogenic vegetation change in the forests of tropical Africa from a number of different methodological approaches. Synchronic studies conducted by ethnographers (Fairhead and Leach 1996, 1998) and fire ecologists (Kull and Laris 2009; Bowman et al. 2011) show that tropical Africa is one of the most burned places on the planet, yet in some cases these fires serve to stabilize forest-savanna boundaries (Fairhead and Leach 1996). What remains unknown is when this pattern of land management emerged in the past and how it is expressed in different socioeconomic organizations (Archibald et al. 2012). Archaeological studies of the settlement of African rain forests are still nascent, but it is clear that semi-sedentary populations using ceramic technologies and, albeit later, iron metallurgy settled much of the rain forest zone between 3000 and 1000 cal yr BP (Klieman 2003; Wotzka 2006; de Maret 2013). These populations are certainly related to historically observed populations of Bantu-speaking farmers that occupy much of the region today (Blench 2006; Bostoen et al. 2013). Addressing whether this pattern of forest settlement is linked with anthropogenic vegetation change through intensive farming or iron smelting is a topic of ongoing debate (Sowunmi 1999; Maley and Chepstow-Lusty 2001; Maley et al. 2012; Neumann et al. 2012a; Bayon et al. 2013; Iles 2015) because, in part, there is insufficient data to adequately model the scale or intensity of prehistoric agricultural production.

It cannot be denied that the ultimate causal mechanism driving the distribution of vegetation during the Late Holocene is global climate. Paleoenvironmental datasets from tropical Africa show a clear latitudinal distribution of the termination of the African Humid Period and onset of more variable Late Holocene precipitation regimes that follow shifting solar insolation (Shanahan et al. 2015). This pattern of climatically-induced vegetation change appears as forest replacement by wooded savannas in northern parts of the forest-woodland transition zone by 2000 BC (4000 cal yr BP; Salzmann and Hoelzmann 2005; Vincens et al. 2010) and in the southern areas of the Atlantic littoral zone as widespread forest disturbance by 500 BC (2500 cal yr BP; Maley and Brenac 1998; Vincens et al. 1998; Ngomanda et al. 2009). Records from continental environments in the interior appear to show a greater degree of forest resilience during this time, with more substantial disturbances appearing after c. AD 1000 (1000 cal yr BP; Marchant and Taylor 1998; Brncic et al. 2007, 2009; Kiahtipes et al. 2011). The diversity of vegetation response across space and time in tropical Africa argues against applying vegetation change visible in one region (e.g. Maley and Brenac 1998) across the tropical zone. While anthropogenic impacts cannot sufficiently explain vegetation change in the forest zone during the Late Holocene, this does not mean that human settlement of the forest zone does not have ecological consequences.

There are two weaknesses in the current scientific understanding of past vegetation change and anthropogenic impacts in tropical Africa. First, there is little archaeological data that allows researchers to articulate what range of impacts humans may have on forested environments. The recovery of *Pennisetum glaucum* in archaeological sites in Cameroon (Neumann et al. 2012b) and the Democratic Republic of Congo (Kahlheber et al. 2014) illustrate the potential diversity of domesticated foods and farming strategies employed by forest settlers, especially given the focus of historical linguistic research on root crops (Philippson and Bahuchet 1994). Without a clearer empirical picture of the subsistence strategies, mobility patterns, and population densities of populations settling the forest zone, it is extremely difficult to model what sorts of impacts these populations had on forest ecosystems, especially during phases of changing climate. Second, while paleoenvironmental research has made many important contributions towards documenting the dynamic response of African rain forests to climate change (see Elenga et al. 2004), these records typically sample large regional catchments that may not be sensitive to human impacts, which are expressed at smaller spatial scales. In order to develop a more robust picture of anthropogenic influences on African rain forests, it is necessary to develop case studies that combine archaeological and paleoenvironmental data to assess how particular prehistoric and historic cultural, technological, or demographic changes are linked with changing impacts on tropical vegetation cover.

Project Area

The Ngotto Forest Reserve, sometimes listed as the Mbaéré-Lobaye Reserve (Deguignet et al. 2014), encompasses a triangular shaped block of Guineo-Congolian rain forest covering 2683 km^2 (Fig. 1a). The Lobaye River bounds the reserve in the east and the Bodingué River runs along the southern border, while the Mbaéré River bisects the reserve from the northwest to the southeast. Because the Ngotto Forest sits the northern boundary between the Guineo-Congolian rain forest vegetation zone and the Sudanian transition zone (White 1983), seasonal semi-deciduous forests are one of the dominate vegetation types, but there are also extensive wet evergreen forests and swamp forests located near the floodplains of the rivers (Lejoly 1996; Yongo 2003;

Rèjou-Méchain et al. 2008). The region has been studied ethnographically (Bahuchet and Guillaume 1982; Kitanishi 1995; Hewlett 1996; Lupo and Schmitt 2005), but systematic archaeological research had not been undertaken in this area until recently (Lupo et al. 2015). Paleoenvironmental research in the region is also limited, with only one published palynological study (Kiahtipes et al. 2011) and one published stable isotope study (Neumer et al. 2008). Runge (2008) analyzed satellite imagery from the Ngotto Forest between AD 1970–1990 and discovered a surprising trend of forest growth of nearly 5%, which he attributes to humid climates and a demographic shift to urban areas in the recent past. Rèjou-Méchain et al. (2008) also suggest that the forest area is generally expanding, but attribute this to forest recovery from arid conditions at c. 2500–2000 years ago, citing Maley and Brenac (1998).

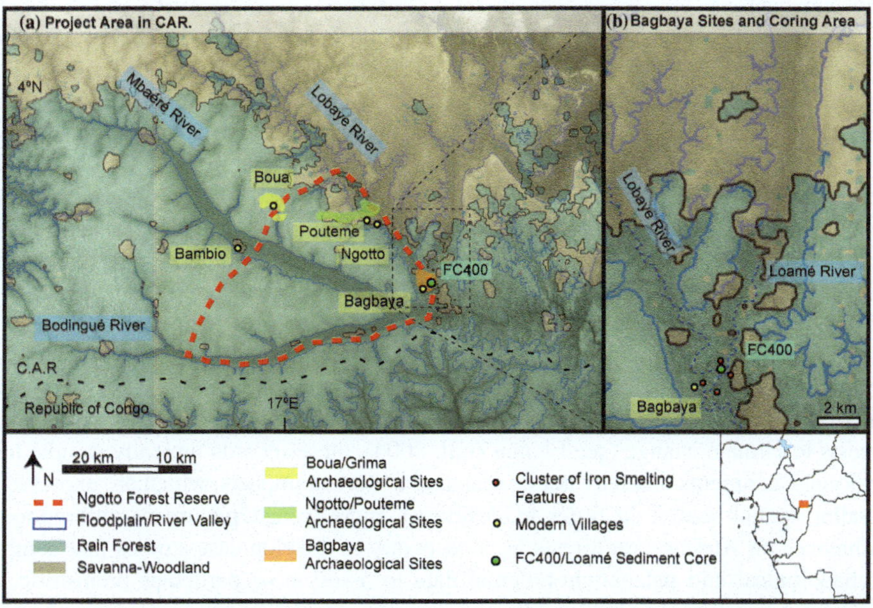

Fig. 1. (a) Overview map of the study region showing major groups of archaeological sites, locations of villages, and the FC400/Loamé River coring site. (b) View of Bagbaya village, archaeological sites, and FC400/Loamé River coring site. Background map showing major hydrological features and vegetation cover modified from DEM and MODIS data provided by NASA Land Processes Distributed Active Archive Center (LP DAAC), USGS/Earth Resources Observation and Science (EROS) Center, Sioux Falls, South Dakota (downloaded February 2016).

Aims and Methods

The primary goal of this project was to use riparian marsh deposits as a proxy paleoenvironmental record to document vegetation change in the Ngotto region that paralleled cultural changes visible in the archaeological record. These deposits are attractive because they reflect developments in the local catchment that may be sensitive to anthropogenic impacts. Riparian deposits show considerable depositional complexity where deep samples may not necessarily be old and shallow deposits may not necessarily be young. This is clearly illustrated in Runge's (2002) coring of the Mbari River floodplain located more than 500 km to the east in CAR. In the Ngotto Forest, coring activities for the research presented here focused on floodplains associated with archaeological phenomena identified through pedestrian survey and interviews with local populations. These sites were mapped, test excavated, and dated using accelerated mass spectroscopy (AMS) ^{14}C assay.

The most distinct archaeological phenomenon identified in the Ngotto Forest is the iron ore mine and smelting complex located near the village of Bagbaya (some list as Bacbaya) in the southeastern area of the Ngotto Forest (Fig. 1b). The iron ore source is a large (>150 m^2) exposure of iron-rich igneous rock located two kilometers south of the confluence of the Lobaye and Loamé rivers. This unique geologic feature has been extensively excavated in the past and based on the size of the excavated area an estimated volume of 800 m^3 of ore was removed from this location alone (Lupo et al. 2015:7). Survey of the surrounding area identified nineteen large mounds composed almost entirely of iron slag and fragments of clay pipes used in smelting (tuyeres) that measure between 1.5–2.5 m in height and 8–18 m in diameter (Lupo et al. 2015:10). AMS radiocarbon assay of seventeen charcoal samples from slag mounds, the ore mine, and other archaeological features indicates that the region was occupied by around 700 years ago and that the iron smelting features were in use between 300–50 years ago (Lupo et al. 2015:9). Estimates of slag volume and weight based on the dimensions of the mounds and an assumption that they are made up of 80% slag indicates that each of these features contain between 39 and 208 m^3 of slag weighing between 42,680–232,960 kg (Lupo et al. 2015:10). While estimating the impact of prehistoric iron production on the environment by back-calculating the volume of biomass consumed as fuel from the size of slag heaps alone (see Goucher 1981; De Barros 1986) is not sufficient evidence to demonstrate that smelting iron is responsible for deforestation (see Iles 2015), the volume of material mined and smelted at Bagbaya clearly warrants additional study.

Sediment coring took place near the modern village of Bagbaya (Fig. 1a, b; UTM 33 N, 773956 E 424104 N) in a filled channel located near the confluence of the Lobaye River and one of its smaller tributaries, the Loamé River. This area was part of a roughly linear depression and the banks of the old river and its channel were still visible on the surface. Modern vegetation at the coring location is primarily riparian swamp forest dominated by *Uapaca*. Samples were collected using a Russian peat corer that extracts deposits in 50 cm half-cylinders. Two boreholes were used and each 50 cm drive was collected with a 10 cm overlap between each drive. This ensured minimal mixing of the sampled sediments, which can occur when collecting multiple drives from a single borehole. Each drive was photographed, described, and sampled at 5 cm intervals, yielding a total of 40 samples (Fig. 2). The sediments of the core are comprised of organic-rich sandy peat, dark sandy silts mottled with organic matter, dark silty sands, and uniform gray silty sands with occasional bands of dark organic matter and charcoal.

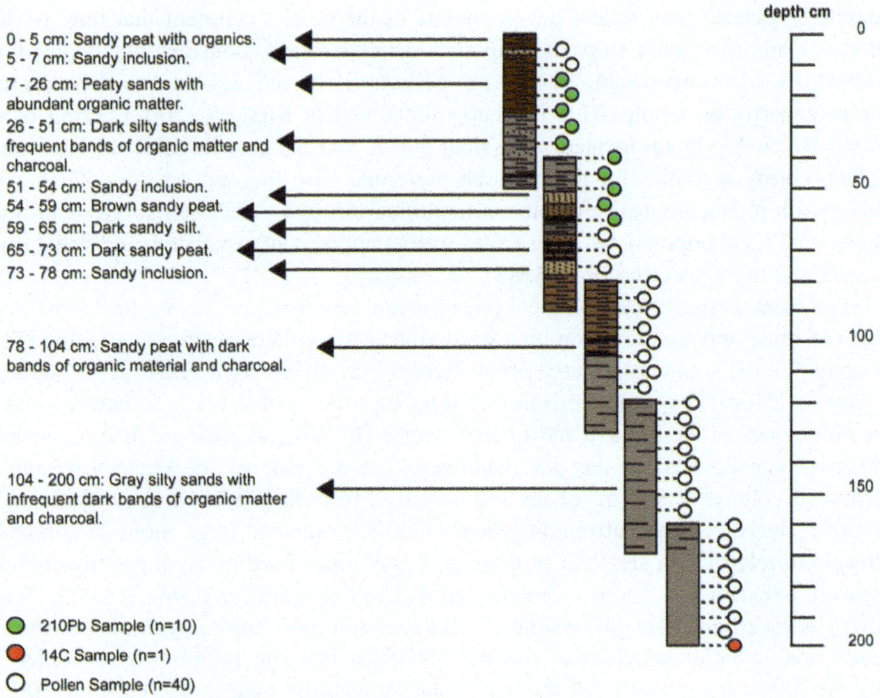

Fig. 2. Schematic of FC400/Loamé River cores showing stratigraphic intervals identified in the field and the distribution of samples collected for analysis.

The chronology of the Loamé River deposits is controlled by AMS ^{14}C assay of plant material from the base of the core and dating of $^{210}Pb/^{137}Cs$ from ten samples between 15–60 cm (Fig. 2; Table 1). Radiocarbon dating was conducted at the University of Arizona AMS radiocarbon dating laboratory and assay of $^{210}Pb/^{137}Cs$ was conducted at Microanalytica Inc (Table 1). Ages were assigned to undated samples using age-depth modeling based on the radiometric dating results. Calibrated dates and age-depth models were generated using the Clam 2.2 package (Blaauw 2010) implemented in the R statistical computing environment (R Development Core Team 2012, version 2.14.2). Calibrated dates were derived using the IntCal13 Northern Hemisphere calibration curve (Reimer et al. 2013) and age-depth models were calculated using linear interpolation of these calibrated dates over multiple iterations (N = 1000) to generate 95% confidence intervals. This method was chosen because it makes the fewest assumptions about the conditions at the coring locations and because the depositional sequences are young enough that major shifts in depositional rates or sedimentary sources are unlikely.

Table 1. Results of analysis of ^{14}C, $^{210}Pb/^{137}Cs$, and modeled ages provided by the University of Arizona AMS Radiocarbon Laboratory and Microanalytica Inc.

Results of Assay of $^{210}Pb/^{137}Cs$						
Sample	Depth cm	lab number	CRS Age	CRS error	CIC Age	CIC error
FC4.03	15	MA-4842	2011	+/- 1	2001	+/- 5
FC4.04	20	MA-4843	2005	+/- 2	1997	+/- 5
FC4.05	25	MA-4844	2001	+/- 1	1993	+/- 6
FC4.06	30	MA-4845	1996	+/- 1	1989	+/- 7
FC4.07	35	MA-5107	1990	+/- 1	1985	+/- 8
FC4.08	40	MA-5108	1983	+/- 2	1981	+/- 8
FC4.09	45	MA-5109	1974	+/- 4	1977	+/- 9
FC4.10	50	MA-5110	1958	+/- 5	1974	+/- 10
FC4.11	55	MA-5111	1936	+/- 6	1970	+/- 12
FC4.12	60	MA-5112	1915	+/- 11	1966	+/- 13

Results of AMS Radiocarbon Assay

Sample	Depth cm	Lab number	^{14}C Age	Error	Cal yr AD (2-sigma)
FC4.40	200	AA94551	345	±37	1463–1739

All 40 of the samples were sub-sampled for palynological analysis and were processed according to standard laboratory procedures (Faegri and Iversen 1979) with the addition of a density separation using a solution of zinc bromide and 10% HCl. Samples were mounted on microscope slides in glycerol and analyzed at 40x magnification using a binocular microscope. A total of 400 identifications were made per sample. Taxonomic identifications were made using reference images from the African Pollen Database (Vincens et al. 2006), published pollen atlases from West and Central Africa (Sowunmi 1973; Asseimen et al. 1974; Gosling et al. 2013), and use of the comparative collection curated at the Institut pour les Science and Evolution at Montpellier, at l'Université Montpellier II. Zonation analysis was utilized to identify groups of pollen samples (i.e. zones) from each core that stand apart from the other samples. When combined with age-depth modeling of the sample ages, this method can be used to identify important developments and a time-frame when they likely occurred. Zones were identified using the CONISS constrained hierarchical clustering method available in the rioja R package (Juggins 2015, version 0.9–6) and the vegan R package (Oksanen et al. 2012, version 2.0–3), which groups samples by Euclidean distance. The number of zones used in this analysis was decided by comparing the reductions in variance between the constrained hierarchical clustering method and a broken-stick model, which randomly splits samples into groups.

Analysis of microscopic charcoal provides a direct line of evidence for fire activity that can be evaluated in pollen slides (Carcaillet et al. 2001), but the raw abundance of microscopic charcoal is a problematic measure of local fire episodes that is susceptible to false positives (Tinner and Hu 2003). Based on the relationships between charcoal size, dispersal, and fire proximity, analysis of microscopic charcoal from pollen slides are most reliable for the detection of fire episodes when charcoal size and abundance are considered together (Asselin and Payette 2005). For the Loamé core, any microscopic charcoal was grouped into five arbitrary size classes of 10–30 μm, 30–50 μm, 50–70 μm, and >70 μm (sensu Patterson et al. 1987). A potentially informative source for discerning the location of fires in either forested or savanna contexts is the presence of charred grass cuticles and phytoliths. These are counted as part of the microscopic charcoal fraction, but are grouped separately. Distinctive cuticle features are readily visible under 40x magnification and include distinctive articulation of long cells, shape of the stomata, and the presence of barbell-shaped phytoliths; positive identification of grass cuticles were made using references to published Scanning Electron Micrograph images (Palmer and Gerbeth-Jones 1988) and SEM images curated at SMU by Dr. Bonnie Jacobs. Charcoal fragments were tallied until 30 tracer spores were encountered, which is consistent with the standard approach of counting charcoal until 20% of the tracer sum from the pollen analysis is reached (Asselin and Payette 2005).

Results

Plant matter from the bottom of the core (200 cm depth) was sent to the University of Arizona's AMS Radiocarbon laboratory for analysis and returned a date of 211–487 cal BP (AD 1463–1739). Analysis of ^{210}Pb/^{137}Cs from the upper of the core (60 cm depth) indicates that these deposits accumulated between AD 1905–2010, and is consistent with the date generated by radiocarbon dating (Table 1). Two models are available for interpreting the ^{210}Pb/^{137}Cs results. The constant initial concentration (CIC) model assumes that both the supply of ^{210}Pb and accumulation rates are constant and the constant rate of supply (CRS) model assumes only that the supply of ^{210}Pb is constant, while sedimentation rates may vary (Binford 1990). In this case the CRS method should be used because of potentially variable deposition rates of complex fluvial systems.

However, minor amounts of ^{137}Cs are present in samples that the CRS model estimates to be older than AD 1950. This is problematic because ^{137}Cs is a product of the detonation of atomic weapons and thus should not be present in samples predating AD 1950. Two possible explanations for this inconsistency are that annual flooding of the FC400 coring location produced some mixing of the deposits and allowed downward movement of ^{137}Cs into older deposits or that reworking of sediments from the surrounding massifs preferentially biased ^{210}Pb estimates towards older dates. Both of these forces are likely at play in the Lobaye floodplain, but they do not preclude assessing vegetation change from these records because the error between the ^{210}Pb and ^{137}Cs estimates is smaller (c. 50 years) than the standard errors in radiocarbon dates from this time frame (>300 years).

Pollen recovery and preservation from the Loamé River core is excellent. Indeterminate pollen grains make up between 1.5–12.3% of the pollen sum (Fig. 3) with concentrations per ml ranging between 20×10^3 and more than 200×10^3 (Fig. 4).

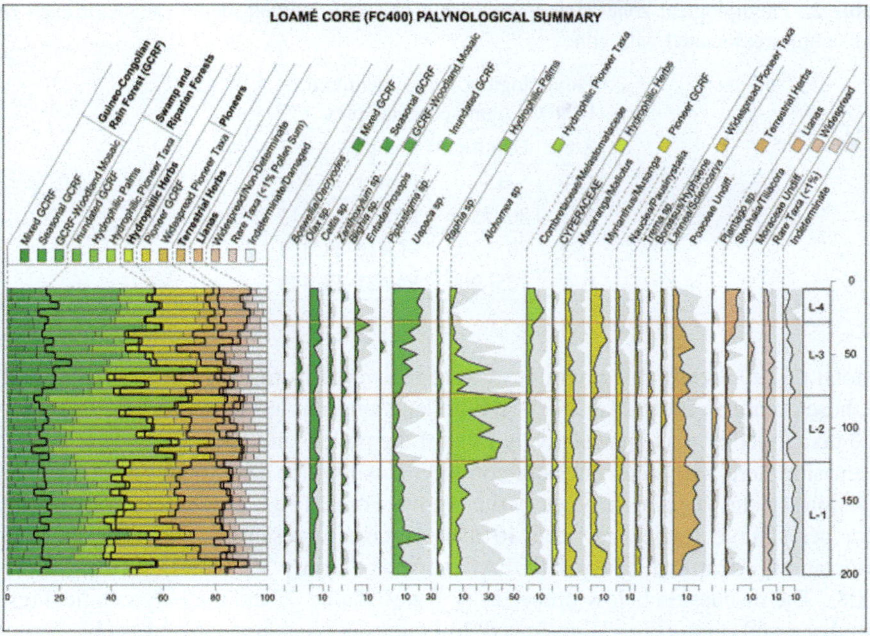

Fig. 3. Pollen diagram from the FC400/Loamé River core deposits as a percent of the total pollen sum. Shaded area is 10x exaggeration of the original values.

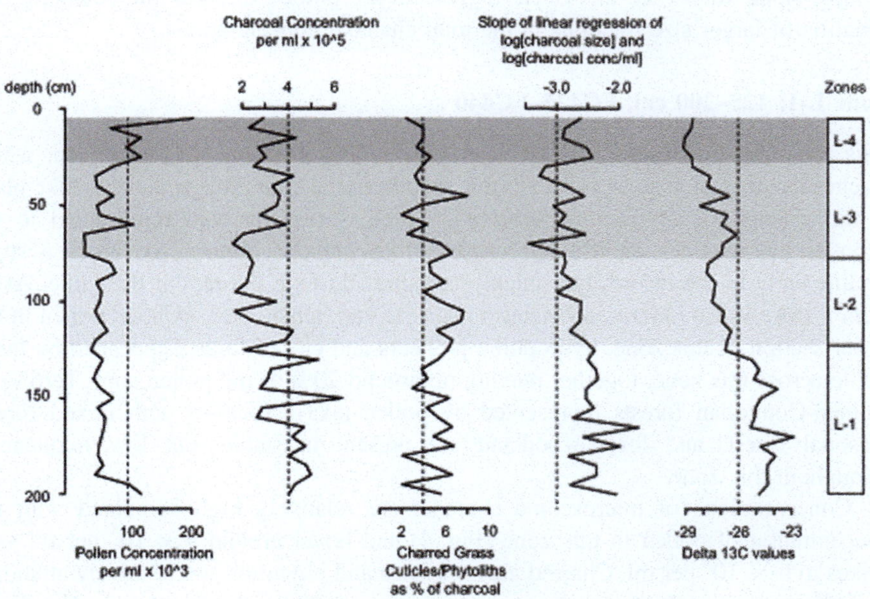

Fig. 4. Left to Right: plots showing concentration of fossil pollen, concentrations of microscopic charcoal, percent of the charcoal concentration made up of grass cuticles or phytoliths, slope values representing the distribution of charcoal sizes, and the delta 13C values for the FC400/Loamé River core samples.

Table 2. Palynological zones determined using CONISS method, depth ranges of each zone, and samples associated with zone.

Palynological zones identified in FC400/Loamé core samples		
Zone	Depth cm	Samples
L-1	5–15	FC401–FC405
L-2	30–75	FC406–FC415
L-3	80–120	FC416–FC424
L-4	125–200	FC425–FC440

A total of 190 identifiable pollen taxa were recovered from the FC400 samples and 65 of these comprise more than 1% of the pollen sum in at least one sample. Because the rare taxa together make up relatively little of the pollen sum in any sample (<10%), the discussion here will focus on more common fossil pollen taxa representative of major divisions of the vegetation types in the Ngotto Forest (Fig. 3). The diversity of pollen taxa per sample ranges between 46 and 71 taxa representing all of the ecological associations identified in the ECOFAC surveys (Lejoly 1996; Rèjou-Méchain et al. 2008). The discussion below proceeds by zones (Table 2) and the palynological results are discussed first (Fig. 3) followed by microscopic charcoal and stable isotopes (Fig. 4). Presentation of the charcoal and stable isotope results in Fig. 4 also includes an estimate of the distribution of charcoal sizes derived from a linear regression of the log value of charcoal abundances per size group on the log value of charcoal size groups; slope values for this linear regression are plotted to show the changing distribution of larger size fractions in the total charcoal concentrations.

Zone L-1: 125–200 cm, FC425–FC440

This zone captures a distinctive increase in the concentration of grass pollen, which reaches its greatest values (>15%) in this section of the core. This appears to take place at the expense of *Myrianthus/Musanga*, which is initially well represented in the deposits but declines rapidly afterwards. *Macaranga/Mallotus* also shows a small decline early in the record, but quickly rebounds through the rest of this zone. Aside from a spike in Combretaceae/Melastomataceae and, albeit later, *Uapaca* pollen in the lower section of this zone, hydrophilic pioneers and closed forests appear to be fairly stable across this zone, together making up around 20% of the pollen sum. Terrestrial Guineo-Congolian forests represented by pollen taxa associated with mixed forest, seasonal forest, and forest-woodland are present in stable, but low frequencies throughout this zone.

Concentrations of microscopic charcoal are relatively high compared with the concentration of pollen in this zone (Fig. 4) and hover around 4×10^5 per ml with spikes to 6×10^5 per ml. Charred grass cuticles and phytoliths are relatively abundant as well, reaching 6% of the charcoal concentration in the middle of L-1. The distribution of charcoal sizes through this zone shows a greater frequency of larger size classes in the charred material. Values for $\delta^{13}C$ in this zone are less negative than elsewhere in the core and suggest a greater contribution of carbon derived from C4 plants, but the overall value is still within the range of mixed forest values from other studies in the African tropics (Schwartz et al. 1996; Runge 2002).

Zone L-2: 80–120 cm, FC416–FC424

A dramatic increase in the recovery of *Alchornea* characterizes samples from the L-3 zone, where this taxon regularly makes up 30%–50% of the pollen sum. Simultaneously, *Uapaca* falls to values below 10% throughout this zone. Grasses are less abundant than in the previous zone and generally decline from the bottom to the top of L-2. Taxa that were previously uncommon become more common in this zone as well. *Plantago*, a terrestrial herb often associated with human disturbance, appears in its highest frequency among these samples. Pollen from the forest liana *Stephania/Tiliacora* also appears at frequencies of around 10%. Pioneer trees typical of the forest zone such as *Myrianthus/Musanga*, *Macaranga/Mallotus*, and *Nauclea/Pausinystalia* are present in reduced proportions (10% each) when compared with the previous zone. Pollen representing mixed Guineo-Congolian Rain Forest is uncommon in this zone, especially *Boswellia/Dacryodes*, but seasonal forest taxa *Celtis* and *Zanthoxylum* become somewhat more abundant in this zone and together make up around 10% of the pollen sum. *Blighia*, which occurs in forest-woodland mosaics and gallery forests is more consistently encountered in these samples than in samples from the previous zone.

Charcoal concentration values are lower in zone L-2 at closer to 2×10^5. This is true even in relation to the concentration of fossil pollen, which remains stable through this zone. Charred grass cuticles and phytolith frequencies remain relatively high in the lower section of this zone, but fall in the upper half to 4% of the concentration of microscopic charcoal. Charcoal size distributions from samples in zone L-2 show a gradual shift to smaller charcoal sizes between the bottom and top of the zone. More negative delta-^{13}C values from these samples argue for a predominantly C3 origin for organic carbon in these deposits.

Zone L-3: 30–75 cm, FC406–FC415

Changes in hydrophilic taxa continue to characterize the pollen spectra in this zone. The recovery of *Uapaca* and *Entada/Prosopis*, which are associated with inundated forests, steadily increases in these samples from bottom to top. *Piptostigma*, an entomophilous (insect-pollinated) member of the Annonaceae that prefers seasonally inundated swamp forests reappears in the upper half of this zone. Hydrophilic palms represented by *Raphia* appear at values similar to zone L-1 that are modestly greater than L-2. *Alchornea* frequencies are lower, but highly variable in the lower half of this zone (<10–30%) but stabilize at lower values of 10% or less in the upper half of the zone. Terrestrial forest taxa are more common and both *Boswellia/Dacryodes* and *Olax* are present in more of these samples, albeit at low values ($\sim$5%). The recovery *Celtis* and *Zanthoxylum* shows little change across this zone, but where there is change it appears to be in the direction of lower values. *Blighia* largely continues to be consistently recovered at low values (<5%), but towards the upper half of this zone is recovered less frequently. Pioneer GCRF trees show a divergent pattern with *Macaranga/Mallotus* frequencies declining and *Myrianthus/Musanga* increasing from the bottom to the top of this zone. Grass pollen values show two small peaks in the lower and upper half of this zone divided by a short set of low values (<10%), but are otherwise generally similar to values from the previous zone. *Stephania/Tiliacora* is more frequent in the uppermost samples from this zone, as well as Moraceae undiff.

Microscopic charcoal concentrations are greater in zone L-3, but this increase is paired with greater concentrations of fossil pollen. As such, this pattern likely reflects changes in the depositional environmental rather than change in fire activity on the landscape. The frequency of charred grass cuticles or phytoliths in these samples is generally lower than in the previous zones, although there is a spike to higher values a little above 50 cm. Size-class distributions of the microscopic charcoal are generally dominated by the small size fraction, but show a similar spike towards larger charcoal size a little above 50 cm. Delta-^{13}C values show a slight shift back towards less negative values in this section and indicate a marginal increase in the contribution of C4 plants to the sample's organic carbon.

Zone L-4: 5–15 cm, FC401–405

Values for swamp forest taxa continue to rebound samples from zone L-4. *Uapaca* is the dominant pollen taxon in all of these samples at values close to 30%, which follows closely after a spike in *Entada/Prosopis* to around 10% at the end of L-3. The latter returns to lower values in these samples, but Combretaceae/Melastomataceae pollen is more common here than anywhere else in the sampled deposits and reaches a maximum of 15% in the middle samples of L-4. Terrestrial forest taxa are stable, but show some small shifts in the frequency of mixed GCRF compared with seasonal GCRF taxa. *Boswellia/Dacryodes* and *Olax* are present in the L-3 samples, but are uncommon compared with values from the previous zone. *Celtis* is present in these samples at stable values of around 10%, but *Zanthoxylum* frequencies are a little lower compared with zone L-3. Pollen from pioneer trees associated with wet forests continues to show high values. This is driven by generally high frequencies of *Myrianthus/Musanga* of around 10% in these samples while *Macaranga/Mallotus* is less common than in previous zones. The frequency of *Trema* is also a little higher in this zone, reaching values comparable with the lowest section of the core. Frequencies of grass pollen plummet to consistently low values of 5% or less in these samples, while values for the liana *Stephania/Tiliacora* reach their highest values anywhere in the sampled deposits.

As with zone L-3, the L-4 samples show greater concentrations of microscopic charcoal and pollen alike, indicating further changes in the depositional environment in this zone such as ponding and reduced sedimentation rates. Of all the zones, zone L-4 has the lowest frequency of charred grass cuticles and phytoliths as a part of the total charcoal concentration. Larger charcoal sizes in these samples are indicated by the less negative values for the charcoal size distribution. Carbon derived from C3 plants continues to be the main source of organic carbon in the L-4 deposits, showing a renewed shift towards more negative values consistent with closed forest environments.

Discussion

The Loamé River (FC400) samples provide a uniquely detailed picture of vegetation change during the last 500 years in the Ngotto Forest, CAR. Because the coring site is located a short distance (<5 km) from archaeological and historical settlements it is likely that these deposits document vegetation change produced directly by the impacts of prehistoric agriculture, intensive iron smelting, and European colonization on the Bagbaya area of the Ngotto Forest. The core's chronology is also appropriate for assigning changes in the microbotanical or geochemical results to discrete historic time frames using age-depth modeling. Such data provides an opportunity to ask if there are different ecological consequences for human activities such as intensive iron production or subsistence agriculture in the past and cash-cropping and intensive resource extraction in the present. Establishing if the signal from paleoenvironmental proxies varies between these activities and the dimensions along which this variability is distributed is necessary for two reasons. First, understanding the variability in the long-term consequences of human activities in Africa's tropical forests can help bridge important gaps between conservation initiatives or NGOs and local stakeholders. Second, building case studies using spatially and chronologically associated environmental and cultural records provides a much-needed line of empirical evidence to falsify hypotheses about human-environment dynamics. Below I primarily focus on assessing the differences between swidden agriculture and intensive iron production. In order to make this assessment we first need to consider a simple model of the different ecological outcomes under different circumstances and their paleoenvironmental signals (Table 3).

Table 3. Expectations for impacts of different modes of production on paleoenvironmental proxies measured in the FC400/Loamé River core.

Production Mode	Mobility	Activities and Location	Indicators—Dry conditions	Indicators—Wet conditions
Subsistence Agriculture	Medium	Field preparation, Savanna maintenance, Terrestrial forests and savanna woodlands	+Seasonal taxa	+Forest pioneers
			+Widespread pioneers	+Grasses
			+Grasses	+Charcoal
			+Charcoal	
Specialist Iron Production	Low	Fuel production, Smelting activities, Swamp forest and riparian forests	+Swamp pioneers	
			−Closed forests	
			−Charcoal	

Variability in the ecological outcomes of settlements focused on subsistence agriculture and iron production centers is a product of their differences in mobility and the locus of vegetation modification on the landscape. Swidden agriculture and iron production also differ in the ways that farmers and metallurgists employ fire within these different economic systems. Contemporary small-scale agriculturalists regularly manipulate the timing and frequency of fire activity in forests and wooded savannas for a number of reasons, including the maintenance of productive ecotone environments and in order to move nutrients from above ground biomass back into nutrient-poor tropical soils (Kull and Laris 2009). Burning vegetation to fertilize small fields is an essential feature of swidden agriculture that exploits natural patterns in fires and its role in nutrient cycling. Villages are generally relocated after soil nutrients are locally depleted and the area goes through a fallow phase (Stone 1996). The paleoecological indicators of these activities would most likely appear as more pollen representing disturbed forests or wooded grasslands, more charred grass cuticles and a relatively greater contribution of C4 plants to sedimentary organic carbon.

Iron smelting, on the other hand, is tethered to geologic features where iron ore is available; iron production centers cannot relocate when local soil nutrients are depleted. Within the complex trading networks that linked much of the Congo Basin together during the Late Iron Age and through much of the historic era (see Klieman 2003), these populations could have used their access to iron ore to cement their position within important economic exchanges of forest products with hunter-gatherers (Bahuchet and Guillaume 1982), the production of mundane tools, and symbolic exchanges that linked political power with the regalia of iron metallurgy (de Maret 2012). Instead of exploiting the natural cycling of nutrients from biomass to soil, iron smelting re-routes forest biomass and nutrients into furnaces as charcoal fuel where they are completely volatized. In ethnoarchaeological case studies and oral histories, blacksmiths have expressed a preference for slow-growing hardwood trees common in riparian forests (Haaland 1985) including *Uapaca* species (Burkill 1985), which dominates riparian forests in the study area. If the conditions of limited residential mobility, rerouting of forest biomass into furnaces, and preferences for slow-growing swamp and riparian forest taxa are met by iron smelting centers, then we should expect *lower* rates of charcoal influx into sedimentary records and expect vegetation disturbance to shift to swamp forests instead of terrestrial forests.

The Loamé deposits are of sufficient stratigraphic integrity and the chronological resolution is fine enough to evaluate whether or not we see any such changes in paleoenvironmental proxies associated with the different archaeological and historical developments in the region (Fig. 5). The rest of the discussion focuses on comparing indices derived from indicator taxa in the pollen spectrum, measures of stable isotopes, and measures of grassland fire activity across these different spans of time (Fig. 5). The '*Alchornea* index' is calculated as the sum of *Alchornea* pollen divided by all other hydrophilic pollen taxa and highlights disturbances in the local swamp forests, while the 'grass index' tracks the abundance of grass pollen as a proportion of the terrestrial pollen sum and highlights terrestrial forest disturbance. Forest recovery and seasonality are tracked using indices using the proportion of *Celtis* and *Myrianthus*/*Musanga* pollen, respectively, within the terrestrial pollen signal. Because the former taxon is widespread in seasonal and wet forest types while the latter is limited to wet conditions, these indices track both disturbance of terrestrial forests and how changing precipitation regimes shaped forest response to disturbance during this time.

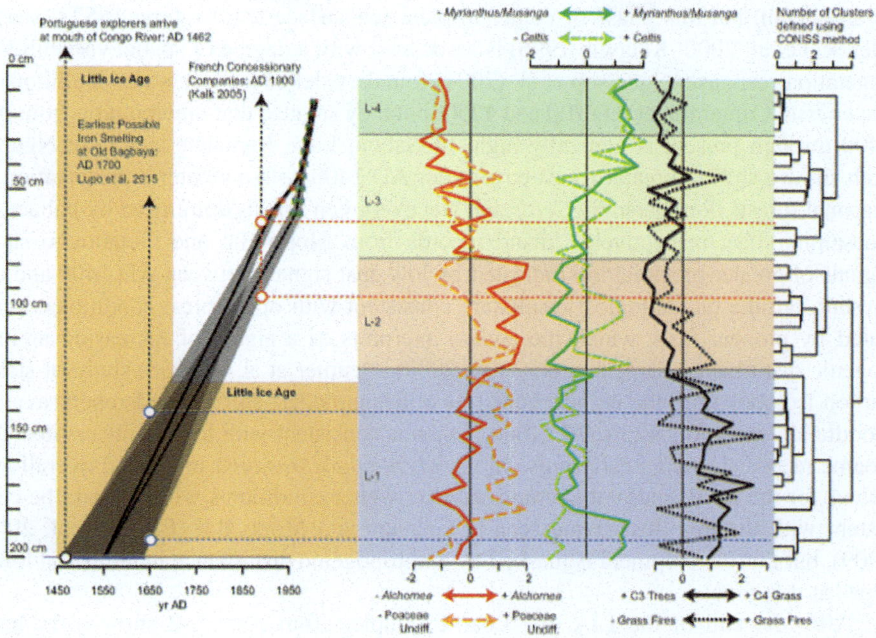

Fig. 5. Synthetic plot showing results of age-depth modeling compared with major historic events covered by the record along with standardized (z-score) indices of important taxa from the palynological analysis.

Deposits from zone L-1 capture environmental changes that span the beginning of the Atlantic slave trade during the late 15th century until the 18th century AD, when the founding population of Bagbaya arrived in the area. These events are linked, as the slave trade triggered widespread conflicts for access to trade routes and raiding interior communities for captives to be sold as slaves on the coast (Hothschild 1996). Oral histories from Bagbaya's elders indicated that they settled the region early in the 18th century when they fled slave raiding and conflict further to the south. However, archaeological evidence shows that some sites date to as early as the 15th century (Lupo et al. 2015). These sites do not contain large slag heaps and are located in wooded savannas. Regardless of their connection with the modern population of Bagbaya, it is likely that the earlier sites represent an occupation of the area that was not linked with extensive iron smelting activities. Sample indices from this time are consistent with the disturbance of terrestrial forests, a greater representation of indicators of savanna cover, and greater frequencies of fire events in the savanna woodland biome (Fig. 5). That these activities alone were sufficient to cause the encroachment of wooded savannas into previously forested areas is unlikely, as both changes in precipitation regimes and biogeographic factors such as the close proximity of savanna-woodland to this area likely contribute to its vulnerability to the impact of past human activities.

These deposits overlap with the established dates for the Little Ice Age (LIA) between AD 1400–1700 (Mann et al. 2009). Only a handful of other records show the LIA in any detail, but this phase of colder tropical sea-surface temperatures (SSTs, see de Menocal et al. 2000; Keigwin 1996) is associated with a degree of spatial variability in vegetation response. Nguetsop et al. (2004) found evidence for low water conditions at Lake Ossa, Cameroon at AD 700 and 1750, but they suggest that precipitation from AD 1400 through present is generally high. In Gabon, Lake Kamalété and Lake Nguène both show a shift to greater forest cover after AD 1400 with a greater representation of seasonal forests dominated by *Celtis* and wet evergreen forests dominated by Fabaceae-Caesalpiniodiae, respectively. Inland records from Mopo Bai and Goualougo show signals of greater precipitation indicated by low dust signals between AD 1400 and the present, but the palynological results are consistent with open forest conditions dominated by pioneer taxa, which the author interprets as a signal of increasing anthropogenic disturbance (Brncic et al. 2007, 2009). Neumer et al.'s (2008) study of stable carbon isotopes from the Ngotto Forest identifies a modest shift towards open savanna woodland vegetation after c. AD 1600, which is consistent with the Loamé results. The Loamé record shows an early pulse in the *Myrianthus/Musanga* index and overall low values for the *Celtis* index that could indicate wetter conditions, which would be consistent with the low dust signal seen at Gualago and Mopo Bai (Brncic et al. 2007, 2009), but this development is masked by anthropogenic disturbance in terrestrial forest types at this time.

After the end of the LIA, the Loamé samples from zone L-2 show a dramatic increase in the disturbance of local swamp forests and the representation of seasonal forest taxa in terrestrial contexts (Fig. 5). Impacts directly related to iron smelting are much clearer at this time, where *Uapaca*, a swamp tree that is commonly used in charcoal production (Burkill 1985) is largely replaced by the heliophilic shrub *Alchornea*. These changes take place after the settlement of Bagbaya and overlap with the radiocarbon dates from the iron ore source and iron smelting features located in the immediate vicinity of the coring site. Overall smaller charcoal size distributions, lower frequencies of savanna woodland fires, higher values on the *Celtis* index, and a lower contribution of C4 carbon at this time indicates a reduction in disturbance of the terrestrial zone and *lower* rates of fire activity in the terrestrial forest zone. These results contrast with cases of anthropogenic impacts in other paleoenvironmental studies in the tropical forest zone associated with swidden agriculturalists or agropastoralists (Marchant and Taylor 1998; Brncic et al. 2009), which are associated with a greater representation of pioneer taxa and/or a higher influx of charcoal. One last important feature of this phase of disturbance is the recovery of *Celtis* at this time, which may signal somewhat drier conditions where forest recovery is dominated by generalist taxa that are tolerant of a greater range of precipitation regimes.

An abrupt shift in the *Alchornea* index towards lower values and a rebound in the indices for both *Celtis* and *Myrianthus/Musanga* are the defining patterns of the L-3 zone. French exploration missions in Central Africa took place in earnest between AD 1891–1894 (Kalck 2005). Colonization of the Ngotto Forest by France and the arrival of concessionary companies focused on the extraction of ivory, timber, and rubber quickly followed by the early 1900s (Kalck 2005). In AD 1912 the populations of the Lobaye region staged insurrections against the concessionary companies that were suppressed through military action, which precipitated a later uprising known as the Kongo-Wara War or "Baya War" from AD 1928–1931 (Kalck 2005). These commercial economic activities resulted in catastrophic impacts on local populations through forced labor programs and the widespread disruption of systems of production and exchange (see Hothschild 1996). These events serve as a *terminus ante quem* for iron smelting activities at Bagbaya. After AD 1912 further smelting activities are unlikely, which corroborates the paleoenvironmental evidence at this time for widespread forest rebound in response to depopulation of the region.

A last shift in stable carbon isotope ratios and grass pollen during this phase perhaps signals a brief period of logging and a shift to cash cropping linked with savanna expansion after CAR coalesces as a national political entity in AD 1958 (Kalck 2005), but before the establishment of the forest reserve in the 1990s. Savanna expansion linked with cash cropping and logging in the recent past in CAR has also been noted by Aleman et al. (2013), although the vegetation changes in Ngotto differ in that they are embedded within a larger pattern of forest rebound that continues throughout the 20th century. Compared with other intervals of the Late Holocene, modern climates are generally considered to be more humid than the LIA (Ngomanda et al. 2007; Brncic et al. 2009; Neumann, et al. 2009b; Runge 2008), which is consistent with the higher values on the *Myrianthus/Musanga* index. Measurements of river flow from gauges in CAR and Kenya (Nguimalet 2014) also show an increase in precipitation and river output from AD 1935–1970 followed by generally lower precipitation and river output from AD 1970 to present. However, studies of historic documents and modern precipitation regimes reveal dry phases of anomalously low rainfall between AD 1800–1850, 1890–1950, and 1970 through present (Nicholson 2000). Studies of sedimentary charcoal and land cover data from three small lakes (Aleman et al. 2013) located to the east of the Ngotto Forest show a shift in fuel sources towards grasses and increasing anthropogenic impacts related to agricultural production (i.e. cash cropping) and demand for wood fuel after AD 1950.

The continued forest rebound from the L-4 zone through the L-3 zone provides important context for understanding recent vegetation change in the Ngotto Forest. Runge's (2008) analysis of satellite data from AD 1970–1990 indicates forests area has increased by as much as 5%, which he attributes to wetter climates and a demographic shift to urban centers such as CARs capital, Bangui. In the Loamé sequence, forest rebound during the historic era is a continuation of a nearly century-long pattern of forest rebound starting with the impacts of French colonization of the region. While the previous zone shows changes in vegetation cover potentially linked with changing land-use patterns associated with cash-cropping, logging, and fuel collecting observed elsewhere in CAR by Aleman et al. (2013), the last phase of the Loamé record may overlap with the establishment of the Ngotto Forest reserve and restrictions on these activities, which is consistent with the rather low values for all disturbance indices and greatly reduced values for fire activity and savanna cover.

Conclusions

Robust paleoenvironmental and archaeological datasets focused at the local level are needed to resolve important scholarly debates about the influence of prehistoric and contemporary African societies on the vegetation of the Guneo-Congolian rain forest.

More specifically, there is a need to build case studies evaluating how major prehistoric and historic events are reflected in changing vegetation cover and fire activity on the landscape. Analysis of pollen, microscopic charcoal, and stable carbon isotopes from the FC400/Loamé River core associated with a significant iron smelting center near the village of Bagbaya show significant changes in vegetation cover and fire activity during the last 500 years. I interpret this data as evidence of changing socioeconomic organizations and production strategies employed by the prehistoric and current peoples living in and around Bagbaya; small-scale farmers during the late 15th century AD focused their activities in terrestrial forests and wooded savannas while specialist iron smelters at Bagbaya had a greater impact on riparian forests through charcoal production. In this case, swidden agriculture and iron smelting appear to have discrete implications for vegetation cover that are discernible in paleovegetation records. Furthermore, the results discussed here show a clear signal of forest expansion starting around the start of the 20th century, which is likely the result of dramatically increased mortality and the profound cultural disruption caused by European colonization of Central Africa. Although the establishment of the Ngotto Forest Reserve after AD 1990 clearly encouraged further forest stabilization, this research illustrates how some conservation outcomes such as forest growth may be Colonial-era artifacts. Combined archaeological and paleoenvironmental approaches can build robust datasets that not only create a better understanding conservation goals and outcomes, but also illustrate the deep and important connections between plants and people in the African past.

References

Adams JS, McShane TO (1996) The Myth of wild Africa: conservation without illusion. University of California Press, Los Angeles

Aleman J, Bérangère L, Roger A (2013) Tracking land-cover changes with sedimentary charcoal in the Afrotropics. Holocene 23(12):1853–1862

Archibald S, Staver AC, Levin SA (2012) Evolution of human-driven fire regimes in Africa. Proc Nat Acad Sci 109(3):847–852

Asseimen P, Bonnefille R, Cambon-Bou G et al (1974) Pollen et Spores d'Afrique Tropicale. Centre d'Etudes de GèographieTropicale, Talence

Asselin H, Payette S (2005) Detecting local-scale fire episodes on pollen slides. Rev Palaeobot Palynol 137:31–40

Bahuchet S, Guillaume H (1982) Aka-farmer relations in the Northwest Congo Basin. In: Leacock E, Lee RB (ed) Politics and history in band societies. Cambridge University Press, Cambridge, pp 189–211

Bayon G, Dennielou B, Etoubleau E et al (2013) Intensifying weathering and land use in iron age Central Africa. Science 335:1219–1222

Binford MW (1990) Calculation and uncertainty analysis of ^{210}Pb dates for PIRLA project lake sediment cores. J Paleolimnol 3:253–267

Blaauw M (2010) Methods and code for 'classical' age-modelling of radiocarbon sequences. Quat Geochronol 5:512–518

Blench R (2006) Archaeology, Language, and the African Past. AltaMira Press, Lanham

Bostoen K, Grollemund R, Muluwa JK (2013) Climate-induced vegetation dynamics and the Bantu Expansion: evidence from Bantu names for pioneer trees (*Elaeis guineensis*, *Canarium schweinfurthii*, and *Musanga cecropioides*). CR Geosci 345:336–349

Bowman DMJS, Balch J, Artaxo P et al (2011) The human dimension of fire regimes on Earth. J Biogeogr 38:2223–2236

Brncic TM, Willis KJ, Harris DJ et al (2007) Culture or climate? The relative influences of past processes on the composition of the lowland Congo rainforest. Phi Trans R Soc B 362:229–242

Brncic TM, Willis KJ, Harris DJ et al (2009) Fire and climate change impacts on lowland forest composition in Northern Congo during the last 2580 years from palaeoecological analyses of a seasonally flooded swamp. Holocene 19(1):79–89

Burkill HM (1985) The useful plants of Tropical Africa, vol 2. Royal Botanic Gardens, Kew

Carcaillet C, Bouvier M, Fréchette B et al (2001) Comparison of pollen-slide and sieving methods in lacustrine charcoal analyses for local and regional fire history. Holocene 11:467–476

Carson JF, Whitney BS, Mayle FE et al (2014) Environmental impact of geometric earthwork construction in pre-Columbian Amazonia. Proc Nat Acad Sci 111(29):10497–10502

Deguignet M, Juffe-Bignoli D, Harrison J et al (2014) United Nations list of protected areas. http://www.unep-wcmc.org/resources-and-data/united-nations-list-of-protected-areas. Accessed 10 Jan 2016

De Barros P (1986) Bassar: a quantified, chronologically controlled, regional approach to a traditional iron production centre in West Africa. Afr J Int Afr Inst 56(2):148–174

de MenocalP, Ortiz J, Guilderson J et al. (2000) Coherent high- and low-latitude climate variability during the Holocene warm period. Science 288(5474):2198–2202

Elenga H, Schwartz D, Vincens A et al (2004) Palaeoenvironments, Palaeoclimates, and Landscape development in Atlantic Equatorial Africa: a review of Key Sites Covering the last 25 kyrs. In: Battarbee RW (ed) Past climate variability through Europe and Africa. Academic Publishers, Dordecht

Faegri K, Iversen J (1979) Textbook of pollen analysis, 4th edn. The Blackburn Press, Caldwell

Fairhead J, Leach M (1996) Enriching the Landscape: social history and the management of transition ecology in the Forest-Savanna Mosaic of the Republic of Guinea. Africa 66(1):14–36

Fairhead J, Leach M (1998) Reframing deforestation: global analyses and local realities: studies in West Africa. Routledge, London

Flantua SGA, Hooghimstra H, Vuille M et al (2015) Climate variability and human impact on the environment in South America during the last 2000 years: synthesis and perspectives. Clim Past Discuss 11:3475–3565

FAO (2010) Global forest resources assessment 2010: main report. Food and Agriculture Organization of the United Nations, Rome

Gosling WD, Miller CS, Livingstone DA (2013) Atlas of the tropical West African pollen flora. Rev Palaeobot Palynol 199:1–135

Goucher C (1981) Iron is Iron 'Til it is Rust: trade and ecology in the decline of West African Iron-Smelting. J Afr Hist 22(2):179–189

Haaland R (1985) Iron production, its socio-cultural context and ecological implications. In: Haaland R, Shinnie P (eds) African Iron Working—Ancient and Traditional. Norwegian University Press, Oslo

Hewlett BS (1996) Cultural diversity among African Pygmies. In: Kent S (ed) Cultural diversity among twentieth-century foragers: an African perspective. Cambridge University Press, Cambridge, pp 215–244

Hothschild A (1996) King Leopold's ghost. Houghton Mifflin Company, New York

IPCC (2014) Climate change 2014: synthesis report. In: Core Writing Team, Pachauri RK, Meyer LA (eds) Contribution of working groups I, II, and III to the fifth assessment report of the international panel on climate change. IPCC, Geneva, Switzerland

Iles L (2015) The role of metallurgy in transforming global forests. J Archaeol Method Theory 23(4):1219–1241

Juggins S (2015) Rioja: analysis of Quaternary Science Data, R package version (0.9.6). http://cran.r-project.org/package=rioja. Accessed 5 Feb 206

Kahlheber S, Eggert MKH, Seidensticker D et al (2014) Pearl Millet and Other Plant remains from the early iron age site of Boso-Njafo (Inner Congo Basin, Democratic Republic of the Congo). Afr Archaeol Rev 31:479–512

Kalck P (2005) Historical dictionary of the Central African Republic. Historical dictionaries of Africa no. 93 (trans: Kalck XS). The Scarecrow Press, Lanham

Keigwin LD (1996) The little ice age and medieval warm period in the Sargasso Sea. Science 274(5292):1504–1508

Kiahtipes CA, Lupo K, Schmitt DN et al. (2011) Prehistory and the present: palaeoenvironments in the northern Congo Basin. Before Farming 2011(2):1–14

Kitanishi K (1995) The exchange of forest products (*Irvingia* nuts) between the Aka hunter-gatherers and the cultivators in Northeastern Congo. Tropics 4:79–92

Klieman KA (2003) "The Pygmies Were our Compass": Bantu and batwa in the History of West Central Africa, Early Time to c. 1900 C.E. Heinemann, Partsmouth

Kull CA, Laris P (2009) Fire ecology and fire politics in Mali and Madagascar. In: Cochrane MA (ed) Tropical Fire Ecology. Praxis Publishing, Chichester, pp 171–226

Lejoly J (1996) Synthese Regionale Sur La BiodiversiteVegetale Des Ligneux Dans Les 6 Sites du Project ECOFAC en Afrique Centrale. ECOFAC Report, Bangui

Lupo KD, Schmitt DN (2005) Small prey hunting technology and zooarchaeological measures of taxonomic diversity and abundance: Ethnoarchaeological evidence from Central African forest foragers. J Anthropol Archaeol 24:335–353

Lupo KD, Schmitt DN, Kiahtipes CA et al (2015) On intensive late Holocene iron mining and production in the Northern Congo Basin and the environmental consequences associated with metallurgy in Central Africa. PLoS ONE 10(7):e0132632

Maley J, Brenac P (1998) Vegetation dynamics, palaeoenvironments, and climatic changes in the forests of Western Cameroon During the last 28,000 years B.P. Rev Palaeobot Palynol 99:157–187

Maley J, Chepstow-Lusty A (2001) *Elaeis guineensis* Jacq. (Oil Palm) fluctuations in Central Africa during the Late Holocene: climate or human driving forces for this Pioneering Species? Veg Hist Archaeobot 10:117–120

Maley J, Giresse P, Doumenge C et al (2012) Comment on "Intensifying Weathering and Land Use in Iron Age Central Africa". Science 337:1040

Mann ME, Zhang Z, Rutherford S et al (2009) Global signatures and dynamical origins of the little ice age and medieval climate anomaly. Science 326:1256–1260

de Maret P (2012) From Kinship to Kingship: an African journey into complexity. Azania Archaeol Res Afr 47(3):314–326

de Maret P (2013) Archaeologies of the Bantu expansion. In: Mitchell P, Lane P (eds) The Oxford Handbook of African Archaeology. Oxford University Press, Oxford

Marchant R, Taylor D (1998) Dynamics of montane forest in central Africa during the late Holocene: a pollen-based record from western Uganda. Holocene 8(4):375–381

Morely RJ (2001) Origin and evolution of Tropical Rain Forests. Wiley, Chichester

Neumann K, Eggert MKH, Oslisly B et al (2012a) Comment on "Intensifying Weathering and Land Use in Iron Age Central Africa". Science 337(6098):1040

Neumann K, Bostoen K, Höhn A et al (2012b) First farmers in the Central African rainforest: a view from southern Cameroon. Quatern Int 249:53–62

Neumer M, Becker E, Runge J (2008) Palaeoenvironmental studies in the Ngotto Forest: Alluvial sediments as indicates of recent and Holocene landscape evolution in the Central African Republic. In: Runge J (ed) Dynamics of forest ecosystems in Central Africa during the Holocene: past-present-future. Taylor & Francis, Leiden, pp 121–137

Ngomanda A, Jolly D, Bentaleb I et al (2007) Lowland rainforest response to hydrological changes during the last 1500 years in Gabon, Western Equatorial Africa. Quatern Res 67:411–425

Ngomanda A, Neumann K, Schweizer A et al (2009) Seasonality change and the third millennium BP rainforest crisis in southern Cameroon (Central Africa). Quatern Res 71(3):307–318

Nguetsop VF, Servant-Vildary S, Servant S (2004) Late Holocene climatic changes in west Africa, a high resolution diatom record from equatorial Cameroon. Quat Sci Rev 23:591–609

Nguimalet CR (2014) Climate change analysis across rainfall-discharge variability in selected River catchments of Kenya and the Central African Republic. In: Runge J (ed) New studies on former and recent landscape changes in Africa. Taylor & Francis Group, London

Nicholson SE (2000) The nature of rainfall variability over Africa on time scales of decades to Millennia. Global Planet Change 26:137–158

Oksanen J, Blanchet G, Kindt R et al (2012) vegan: community ecology package. R package version 2.0–3. http://CRAN.R-project.org/package=vegan

Palmer PG, Gerbeth-Jones S (1988) A scanning electron microscope survey of the epidermis of east African Grasses, V, and West African suppliment. Smithsonian Contributions to Botany no. 67. Smithsonian Institution Press, Washington DC

Patterson WA, Edwards KJ, Maguire DJ (1987) Microscopic charcoal as a fossil indicator of fire. Quat Sci Rev 6:3–23

Philippson G, Bahuchet S (1994) Cultivated crops and Bantu migrations in Central and Eastern Africa: a linguistic approach. In: Sutton JEG (ed) The growth of farming communities in Africa from the Equator Southwards, Azania special, vol XXIX–XXX, pp 103–120

Piperno DR, McMichael C, Bush MB (2015) Amazonia and the Anthropocene: what was the spatial extant and intensity of human landscape modification in the Amazon Basin at the end of prehistory? Holocene 25(10):1588–1597

Rèjou-Méchain M, Pélissier R, Gourlet-Fleury S et al (2008) Regional variation in tropical forest tree species composition in the Central African Republic: an assessment based on inventories by forest companies. J Trop Ecol 24(6):663–674

R Development Core Team (2012) R: A language and environment for statistical computing. R Foundation for Statistical Computing, Vienna, Austria. http://www.R-project.org/

Reimer PJ, Brown TA, Reimer RW et al (2013) IntCal13 and Marine13 radiocarbon age calibration curves, 0–50,000 years cal BP. Radiocarbon 55:1869–1887

Runge J (2002) Holocene landscape history and palaeohydrology evidenced by stable carbon isotope (δ^{13}C) analysis of alluvial sediments in the Mbari valley (5ºN/23ºE), Central African Republic. Catena 48:67–87

Runge J (2008) Remote sensing based forest assessment: recent dynamics (1973–2002) of forest-savanna boundaries at Ngotto Forest, Central African Republic (CAR). Paleoecol Africa 28:237–244

Salzmann U, Hoelzmann P (2005) The Dahomey Gap: an abrupt climatically induced rain forest fragmentation in West Africa during the late Holocene. Holocene 15(2):190–199

Schwartz H, de Foresta H, Mariotti A et al (1996) Present dynamics of the Savanna-Forest boundary in the Congolese Mayombe: a pedological, botanical and isotopic (^{13}C and ^{14}C) Study. Oecologia 106:516–524

Shanahan TM, McKay NP, Hughen KA et al (2015) The time-transgressive termination of the African Humid Period. Nat Geosci 8:140–144

Sowunmi MA (1999) The significance of the oil palm (*Elaeis guineensis* Jacq.) in the late Holocene environments of west and west central Africa: a further consideration. Veg Hist Archaeobot 8:199–210

Sowunmi MA (1973) Pollen grains of Nigerian plants I, Woody Species. Grana 13:145–186

Stone GD (1996) Settlement ecology: the social and spatial organization of Kofyar Agriculture. University of Arizona Press, Tucson

TinnerW, Hu FS (2003) Size parameters, size-class distribution and area-number relationship of microscopic charcoal: relevance for fire reconstruction. Holocene 13(2):291–296

Vincens A, Schwartz D, Bertaux J et al (1998) Late Holocene climatic changes in Western Equatorial Africa inferred from Pollen from Lake Sinnda, Southern Congo. Quat Res 50:34–45

Vincens A, Lézine AM, Guillaume B et al (2006) African pollen database inventory of tree and shrub pollen types. Rev Palaeobot Palynol 145:135–141

Vincens A, Buchet G, Servant M et al (2010) Vegetation response to the "African Humid Period" termination in Central Cameroon (7°N)–new pollen insight from Lake Mbalang. Clim Past 6:281–294

White LJT (2001) The African rain forest: climate and vegetation. In: Weber W, White LJT, Vedder A et al (eds) African rain forest ecology and conservation: an interdisciplinary perspective. Yale University Press, New Haven, pp 3–29

White F (1983) The vegetation of Africa: a descriptive memoir to accompany the Unesco/AETFAT/UNSO vegetation map of Africa. UNESCO, Paris

Wotzka H-P (2006) Records of activity: radiocarbon and the structure of Iron Age settlement in Central Africa. In: Wotzka H-P (ed) Beiträge zur europäischen und afrikanischen Archäologie für Manfred K. Eggert, Tübingen, Francke, pp 271–289

Yongo OD (2003) Contribution aux etudes floristique, phytogéographique et phytosociologique de la forêt de Ngotto (République Centraficaine). Acta Botanica Gallica 150(1):119–124

Evidence of Sorghum Cultivation and Possible Pearl Millet in the Second Millennium BC at Kassala, Eastern Sudan

Alemseged Beldados[1(✉)], Andrea Manzo[2], Charlene Murphy[3], Chris J. Stevens[3], and Dorian Q. Fuller[3]

[1] Department of Archaeology and Heritage Management, Addis Ababa University, Addis Ababa, Ethiopia
alemsegedb2002@gmail.com
[2] Department of Asian, African and Mediterranean Studies, University of Naples, "L'Orientale", Naples, Italy
[3] Institute of Archaeology, University College London, London, UK

Abstract. Carbonized plant remains and plant impressions in burnt clay pieces, recovered during archaeological excavation and survey of two sites in East Sudan, were subjected to archaeobotanical investigation. Analysed samples have provided evidence for plant use and cultivation of sorghum alongside the use of a range of other taxa. The results from this study illustrate that as late as the early second millennium BC, the inhabitants of Kassala were still exploiting a mixture of morphologically wild and domesticated *Sorghum bicolor*. The evidence suggests that while the domestication process of sorghum was underway, full domestication may not have been reached at this time. We can hence classify this as part of the pre-domestication cultivation stage for *Sorghum bicolor*, which can be inferred to have begun at least two thousand years earlier. Wild taxa that may also have been exploited for food include *Brachiaria* sp., *Rottboellia cochinchinensis* (itchgrass), and apparently mixed wild and domesticated *Pennisetum glaucum* (pearl millet). This study also provides the first archaeobotanical evidence for *Adansonia digitata* (baobab) in northeastern Africa. Taken together these data suggest that Kassala was part of an early core area for sorghum domestication and played an important role in the diffusion of Africa crops including pearl millet to Asia.

Keywords: *Adansonia digitata* · Agriculture · Archaeobotany
Palaeoethnobotany · Nubia · Pearl millet · *Sorghum*

Introduction

This paper presents new data on the early cultivation of sorghum (*Sorghum bicolor* (L.) Moench. *sensu stricto*, syn. *S. bicolor* subsp. *bicolor* Harlan and de Wet) and pearl millet (*Pennisetum glaucum* (L). R. Br.) in eastern Sudan, dating to the second millennium BC. These data contribute new evidence on aspects of the domestication process in sorghum. Sorghum, along with maize, rice, wheat and barley, is one of the world's five most important cereal crops. It is especially important in the dry tropics of

A. M. Mercuri et al. (Eds.): *Plants and People in the African Past*,
https://doi.org/10.1007/978-3-319-89839-1_22

Africa, India and parts of Southeast Asia, with significant production in China, parts of the Mediterranean and the Americas (Snowden 1936; Hulse et al. 1980). However, despite extensive research on the other four major cereals, little is known of the domestication process for sorghum.

Sorghum has long been recognized as originating in Africa where widespread wild populations of the *Sorghum bicolor* complex have been documented, in particular by Harlan and de Wet (de Wet and Huckabay 1967; Harland and Stemler 1976), and confirmed by more recent genetic studies (e.g. Aldrich and Doebley 1992; Billot et al. 2013; see review by Fuller and Stevens 2018). Wild populations and crops were initially divided into many different species, but later revisions of Harlan et al. (1973) and de Wet (1977) categorized them as a single species, with subsp. *verticilliflorum* (divided into various races) representing true wild populations, and subsp. *bicolor* encompassing domesticated crops. The domesticated forms were divided into 5 cultivated races, of which the hulled race bicolor is the most primitive (Wiersema and Dahlberg 2007; see also Fuller and Stevens 2018). This indicates that initially domesticated sorghum required dehusking, whereas during later selection following domestication, a range of free-threshing sorghum forms developed with the races caudatum, durra, and guinea. In this paper we use the older binomial nomenclature for simplicity, i.e. *S. verticilliflorum* for the wild progenitor and *S. bicolor* for domesticated forms. Along with pearl millet (*Pennisetum glaucum* (L.) R.Br.) and hyacinth bean (*Lablab purpureus* (L.) Sweet), sorghum is among the well-documented African domesticates that were translocated to India in prehistory between 2000 and 1500 BC (Fuller 2003; Boivin and Fuller 2009). Until recently, archaeobotanical evidence for these crops was earlier in India than in Africa (Fuller 2003; Kahlheber and Neumann 2007). But recent sampling for charred macroremains and the study of plant impressions in ceramics has revealed new evidence from Mali for domesticated pearl millet between 2500 and 1900 BC (Manning et al. 2011; Manning and Fuller 2014). This suggests a mid-Holocene domestication process in the western African Sahel prior to translocation to India, although evidence for the translocation of crops from west to east in the African Sahel or savannahs remains undocumented. The present study highlights the potential for a similar approach for sorghum, sampling for charred macroremains as well as the study of preserved plant impressions in clay. Sorghum, like pearl millet, may have first been cultivated by relatively mobile, pastoralist societies, which characterized the mid-Holocene Sahara and Sahel (e.g. Marshall and Hildebrand 2002; Garcea 2006). The sites left by such peoples often are comprised of superficial surface scatters and hence can suffer from deflation leaving very low densities or poor preservation of macroremains. In such cases, the past use of plant parts to temper clay vessels and ceramics may provide another source of evidence for domestication, especially from chaff remains.

Background on Sorghum Domestication

The domestication and dispersal history of sorghum has been debated in archaeology for many decades. Harlan (1971) inferred it was most likely wild populations of *Sorghum* subsp. *arundinaceum* race *verticilliflorum* (= *S. verticilliflorum*) in the eastern

Sahel zone of Africa (Chad, Sudan) that were first domesticated. However, a case for origins on the northern front of the Ethiopian plateau has also been argued by Doggett and Prasada Rao (1995), while some genetic evidence has suggested the possibility of three independent domestication centres (Lin et al. 2012), although this latter evidence seems more likely to relate to three separate post-domestication selection events for free-threshing sorghum varieties rather than initial domestication (see discussion in Fuller and Stevens 2018).

Harland and Stemler (1976) argued for early Holocene domestication on the basis of parity with early Holocene domesticates in other regions and a later prehistoric dispersal to India (before 1000 BC). They saw the evolution of race *durra* as occurring in India which later spread back to northeast Africa around 2000 years ago. As an alternative, Haaland (1999) suggested the possibility that wild-type sorghum cultivars had been translocated to India, domesticated and then returned to Africa 2000–2500 years ago. While this last hypothesis seemed to fit the occurrence of domesticated sorghum at Qasr Ibrim in Nubia dating from the Meroitic period onwards (100 BC–AD 100) (Rowley-Conwy et al. 1997; see Clapham and Rowley-Conwy 2007, for updated information), Fuller argued that sorghum more likely followed the domestication trajectory seen for other crops, in which directional changes took place under cultivation for an extended period of 2000–3000 years and did not require translocation as part of the process (Fuller et al. 2011; Fuller 2014a, b). In addition, reports of free-threshing durra type sorghum from Meroe (identified by Harlan, in Shinnie and Anderson 2004), as well as durra/caudatum from Kawa (Fuller 2004a) and race caudatum from southwest Libya (Pelling 2005), all at least 2000 years ago or more, would imply significantly earlier domestication; and this is further supported by recent evidence for race durra and race bicolor like sorghum remains from the Meroitic period at Hamadad (Fuller 2014a, b).

Finds of wild type sorghum from eastern Saharan sites, such as Nabta Playa, c. 7500 BC (Wasylikowa et al. 1995; Wasylikowa 2001), and wild-type impressions in Neolithic ceramics in the central Sudan near the Nile dating between 5000 and 3500 BC, seem to indicate a long persistence of gathering without domestication (Majid 1989; Stemler 1990). On the basis of such finds Clark (1984) argued for the domestication of sorghum between the early third and first millennium BC in central Sudan between the confluence of the Blue and White Nile and/or between the Nile and the Atbara River. What has been missing is systematic archaeobotanical sampling of sites of the target period for sorghum domestication, i.e. 4000–1000 BC, in these savannah to southern Sahel areas in which the proportions of wild type to domesticated types are examined in closer detail. The present paper contributes to filling this gap with new archaeobotanical data from sites in the southern Atbai region of eastern Sudan (Fig. 1), collected through dry sieving and the study of plant impressions on burnt clays.

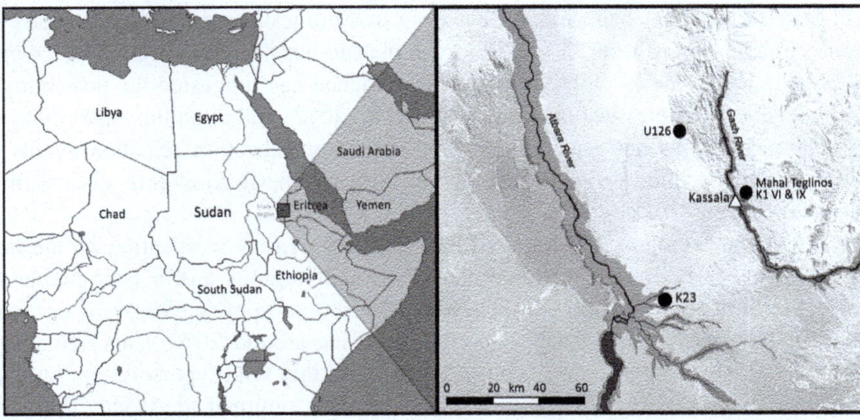

Fig. 1. Map showing the location of the study region, the modern city of Kassala and site K 1, reported here, and recently studied KG 23 [= UA 14] (Winchell 2013; Winchell et al. 2017).

The Southern Atbai: Archaeology and Archaeobotany

The chronological sequences of the southern Atbai (comprising the area between the Atbara and the Gash River and Gash Delta regions) are now well understood based on the study of plentiful archaeological remains recovered from the site of Mahal Teglinos (K 1), Kassala and its environs (Fattovich et al. 1984; Fattovich and Piperno 1986; Fattovich 1989, 1993a; Manzo et al. 2011). The cultural groups (representing phases) defined under the broad umbrella of the Atbai Ceramic Tradition can be classified into seven major entities. These are the Amm Adam Group (Pre-Saroba/Atbai *c.* 6th–5th millennia BC), the Malawiya Group (*c.* 5th–4th millennia BC), the Butana Group (*c.* 4th–3rd millennia BC), the Gash Group (*c.* 3rd–2nd millennia BC), the Mokram Group (*c.* 2nd–1st millennia BC), the Hagiz Group (*c.* mid 1st millennium BC–mid 1st millennium AD), and in the post-Atbai or Gergaf Group (mid 2nd millennium AD) (Shiner 1971; Fattovich 1989, Abbas and Jaeger 1989; Sadr 1988, 1991; Manzo et al. 2011; Manzo 2012a; Winchell 2013).

This region is of particular relevance to the history of African crops, such as sorghum and pearl millet, as it is known to be a region that was connected to long-distance trade in the Bronze Age since at least the second millennium BC (Fattovich 2012; Bard and Fattovich 2013). This is indicated by Gash group ceramics found at the Egyptian Red Sea coastal port at Mersa/Wadi Gawasis (Manzo 2010), as well as finds of Yemeni (Manzo 1997) and Egyptian ceramic types in Kassala (Manzo 1997, 2015), where in addition, personal ornaments made from Red Sea shells have been discovered in late third to early second millennium BC tombs (Manzo 2012a). Egyptian inscriptional evidence makes it clear that Mersa/Wadi Gawasis was the starting point for sea voyages to what ancient Egyptians called the land of *Punt*, since at least the Middle Kingdom (early 2nd millennium BC) (Bard and Fattovich 2013). Punt is generally identified with parts of northeastern Ethiopia and Somaliland, but may also have included Eritrea (Fattovich 2012; Bard and Fattovich 2013), and as highlighted by the finds of Mersa/Wadi Gawasis, the Egypt-*Punt* trade may have also involved South Arabia (Manzo 2010). Recent strontium and oxygen isotopes from mummified baboons of the Egyptian 20th Dynasty (ca. 1185–1070 BC) indicate that these animals were imported from the Eritrea-Ethiopia corridor or eastern Somalia (Dominy et al. 2015). In relation to southern Red Sea trade, the Gash river region represents the northern overland entry to Punt, and its sub-Saharan exotica (such as incense), from regions such as Nubia. This region provides a Sahelian (or north Savannah) route to the Red Sea and the Arabia Sea trade, which ultimately must account for the translocation of sorghum, pearl millet and hyacinth bean from the African Sahel to prehistoric India, and for Asian millets (e.g. *Panicum miliaceum* L.) back to Nubia (Boivin and Fuller 2009: 155–159). Despite the wealth of data in terms of remains of material culture from work in this region, there has been less evidence relating to subsistence patterns or the food economy of these peoples. Earlier investigations have been conducted on botanical remains in seed tempered sherds from the Shurab el Gash survey at a locality called Jebel Abu Gamal (site JAG 1), 15 km south of Kassala (Costantini et al. 1982). Costantini et al. (1983) reported the identification of imprints in ceramics as *Sorghum* cf. *bicolor*, inferred to be domesticated, and barley (*Hordeum* sp., presumably domesticated *H. vulgare sensu lato*) from the sites Shurab el Gash (SEG 9) and Jebel Abu Gamal (JAG 1), as well and *Paspalum* sp. only from Shurab el Gash, along with several desiccated fruit stones belonging to *Ziziphus* sp. All this material probably dates from between 2000 and 1000 BC. Although these finds provided evidence for farming products, since barley is not native to region, the reliably of these reports for inferring the antiquity of sorghum domestication has been debated (see Haaland 1999; Rowley-Conwy et al. 1997; Fuller 2014a, b, 2015).

While the first field seasons at Kassala were conducted between 1980 and 1995, the Italian Archaeological Expedition of the University of Naples “L’Orientale” in Eastern Sudan recently launched a second round of field seasons commencing in 2010 after a gap of 15 years (Manzo 2012b, 2013, 2014a, b, 2015, in press; Manzo et al. 2011). Reconstruction of subsistence patterns and dietary systems has been one of the major objectives of these more recent field seasons, aimed at readdressing the dearth of archaeobotanical, archaeozoological and isotopic data from previous research. As such archaeobotanical samples are now being collected routinely from both surveys and excavations. As part of this project Beldados and Costantini (2011) investigated plant imprints from excavations conducted at Mahal Teglinos K 1 excavation unit VI (15°26′ 57″ N, 36°25′57″ E). These were found in association with a Jebel Mokram Group context, probably dating to immediately after 1500 BC, as seen from a related sequence of C^{14} dates from samples collected in excavations conducted in 2014 (Manzo 2015, in press). Beldados and Costantini (2011) also conducted a pilot study on impressions in ceramics from some surface collections from regional survey, at sites UA 53, UA 14 (formerly KG 23) and UA 129 (Table 1).

This initial work found the presence of sorghum chaff and spikelet impressions (Beldados and Costantini 2011; Beldados 2015). It was thought that these represented both wild and domesticated forms of sorghum on the basis of grain size. Other impressions included *Setaria* sp., jujube (*Ziziphus* sp.) endocarps and possible cowpea (*Vigna unguiculata*). However the sorghum identifications were based on gross hulled grain morphology rather than considerations of spikelet base morphology that is

Table 1. Previously reported botanical evidence from imprints in clay or ceramics from the Southern Atbai (after Beldados and Costantini 2011; Costantini et al. 1983).

Site (period)	N = sherds/fragments	Sorghum presence	Wild/Domestic inference	Other taxa	Comments
UA 14 [KG 23 of Winchell] (3500–3000 BC)	5	3/5	3 "cultivated", 2 "wild" (based on spikelet shape)	"*Setaria* sp." in two sherds	*Setaria* identification should be revised to *Brachiaria deflexa/ramosa.* Investigations of excavated sherds (KG23) indicate a mix of wild and domesticated sorghum, and *Brachiaria* (Winchell et al. 2017)
UA 53 (2000–500 BC)	6	6/6	6 "cultivated shape" 2 "wild shape"		
UA129 (1500 BC-AD 500)	1	1/1	With "cultivated shape"		
K 1 trench VI (c. 1500 BC)	12	11/12	11 "cultivated shape" 9 "wild shape"	"Setaria sp." Ziziphus *Vigna unguiculata*	*Setaria* identification could be revised to *Brachiaria (?) Vigna* imprint needs confirmation and assessment of domestication status
Shurab el Gash (SEG 9) (2000–1000 BC)	?	Present	"cultivated"	Barley	
Abu Gamal (JAG 1) (2000–1000 BC)	?	Present	"cultivated"	Barley, *Paspalum, Ziziphus*	

diagnostic of non-shattering domesticated types. The present study considers additional material, newly studied from another trench at K 1. We also report on plant macroremains recovered from dry-sieving. Together with recently recognized evidence from the late 4th millennium BC site of KG 23 (Winchell et al. 2017; equivalent to site UA14 of Manzo), these data are a first step towards compiling a time series of archaeological sorghum populations that document the "domestication episode" in this crop (*sensu* Fuller et al. 2014).

Expectations of Pre-Domestication Cultivation and Sorghum Spikelet Morphology

Recently increased archaeobotanical interest has been focused upon documenting pre-domestication cultivation. In particular as existing empirical evidence for several crops suggests that domestication was a protracted evolutionary episode covering many hundreds of years (Fuller 2007; Fuller et al. 2010; Harris and Fuller 2014). Presently the evidence from the Near East and China demonstrates that for cereal crops these domestication periods lasted for 2500–3500 years and a millennium or more for pulses (Larson et al. 2014; Fuller et al. 2014).

In China the full process towards the domestication of rice has been documented through detailed archaeobotanical work for sites in the Lower Yangtze (Fuller et al. 2009, 2014). Pre-domestication cultivation can be seen to commence prior to 6000 BC, being established during the Kuahuqiao culture (6000–5400 BC), following an extended period of wild rice procurement starting by 7500–7000 BC (Shangshan Culture). Fully domesticated, e.g. fully non-shattering crops, began to appear between by c. 4000 BC or the early centuries of the 4th millennium BC (Fuller et al. 2014). Thus there was an extended period (at least 3000 years) characterised archaeobotanically by a mixture of non-shattering and wild-type shattering spikelet bases, with significant proportions of immature spikelets and grains. Over a period lasting two to three millennia non-shattering and immature spikelets gradually declined, while average grain breadth and thickness increased (Fuller et al. 2009, 2014). The presence of immature grains and spikelet bases reflects a stage within the exploitation of wild-type, non-shattering plants, in which the panicle did not ripen evenly, and certain harvesting strategies selected for a sizable minority of immature or green spikelets that had yet to shatter (Fuller et al. 2007, 2009). The routine collection of such intermediate forms and their threshing is expected to take place during pre-domestication cultivation, when non-shattering types increased in number, leading to changes in both harvesting and processing methods, away from simple basket harvesting of predominantly mature wild spikelets. The data reported here suggest a similar transitional assemblage characteristic of the pre-domestication cultivation period that might be expected for sorghum.

The identification of sorghum populations as wild or domesticated is best approached by examining the spikelet base evidence for the mode of seed dispersal, whether through natural abscission or human threshing (Figs. 2 and 3). This approach builds on previous studies, such as the identification of domesticated spikelets with preserved rachillae, Meroitic Jebel Tomat (Clark and Stemler 1975; see Fig. 14.4 in

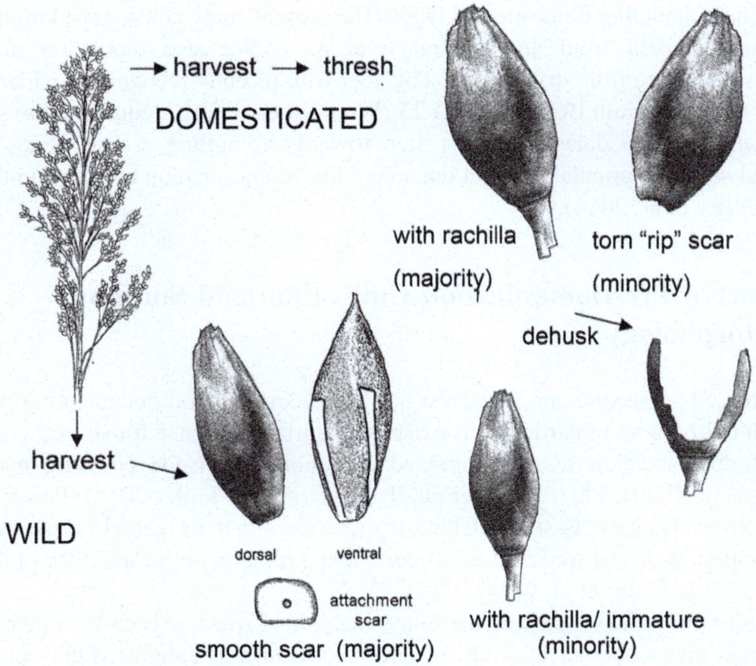

Fig. 2. Diagram illustrating the expected spikelet base types in wild and domesticated sorghum populations (drawn by DQF).

Fuller 2014a, b) or Umm Muri (Fuller 2004b), and the identification of wild, smooth-scar spikelets from Neolithic Sudanese impressions (Stemler 1990) or carbonized remains from Nabta Playa (Wasylikowa et al. 1995). We, however, have divided spikelet bases into three morphologies: smooth scars, of shattering wild types (Fig. 3a, b); those with rachilla still attached, or "torn rachilla" (Fig. 3c), which are expected to be typical of domesticated populations but may also represent green-harvested (immature) spikelets; and finally those with "rip scars" in which the base of the spikelet is torn and the rachilla is completely removed (Fig. 3d). In establishing these morphologies we have examined a range of modern reference material held in the University College London archaeobotany reference collection which included spikelets. This consisted of 9 domesticated crop accessions, 28 populations assigned to wild status, and 9 that were regarded as hybrids of feral forms of the crops (labelled here as "intermediates"), which together include 3263 spikelet bases. Three of the populations that had been collected and catalogued as "wild" may also have been free-growing feral or hybrid, as they had high proportions of domestic-like morphologies. Each reference accession was quantified in terms of spikelet bases of the three morphologies, and the frequencies of these occurrences are tabulated in Fig. 4. As is clear the persistent rachilla is most frequent in domesticated populations, while the "rip scar" is frequent or in intermediate proportions in domesticated or intermediate populations. Smooth scars clearly dominate in all wild accessions; the rare occurrence of

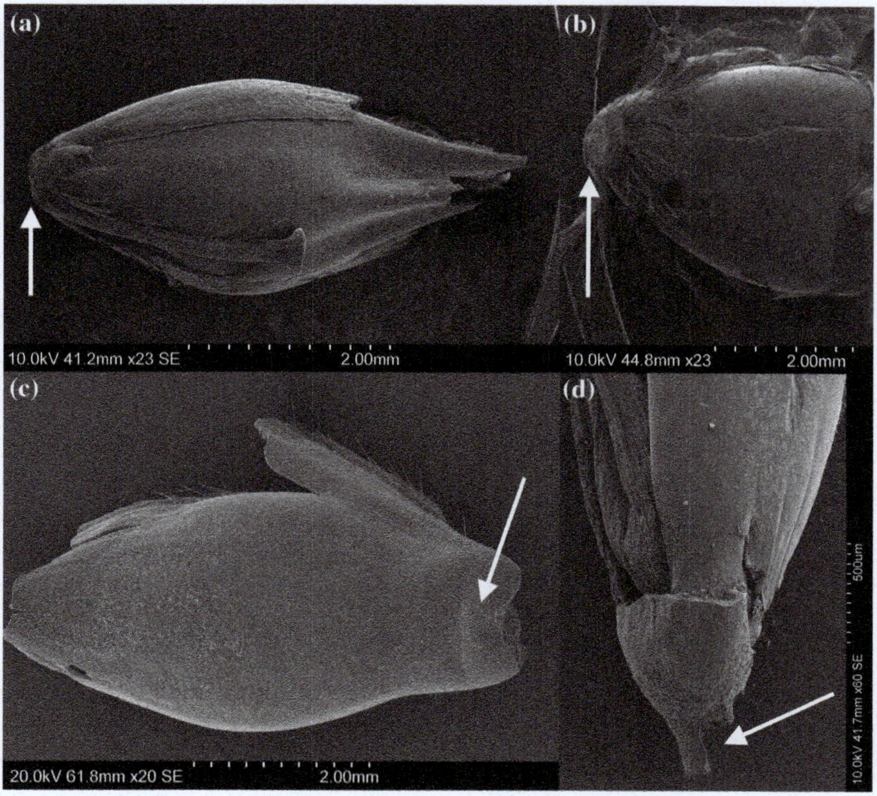

Fig. 3. Comparisons of wild and domesticated sorghum spikelet bases in modern comparative material. (a) wild *Sorghum verticilliflorum* with smooth scar, (b) close up on smooth, wild type scar, (c) rip scar type of spikelet of domesticated *Sorghum bicolor* race bicolor, (d) torn rachilla type of spikelet base of domesticated *Sorghum bicolor* race bicolor. All examples of UCL Institute of Archaeology reference collection.

domesticated type rachilla in wild populations may be the result of introgression of from domestication sorghum, since this crops has been grown for centuries to millennia in the same regions as wild populations.

Other potential characteristics of domesticated sorghum are not applied in the current study. There is a loss or reduction in awn length with sorghum domestication (see Adunga and Bekele 2013) that potentially might be detected. However no awns were identified within any of the impressions within this study. Finally, there is also some change within grain morphology, with grains becoming larger and proportionally rounder, e.g. width and breadth increase more than length.

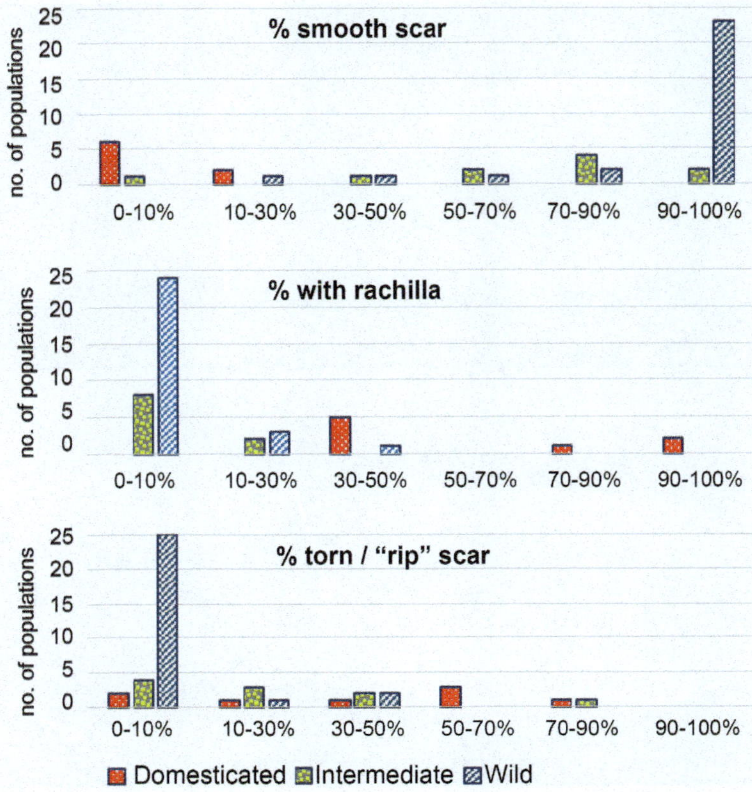

Fig. 4. The percentage of specimens within the Institute of Archaeology's sorghum collection (n = 50) with smooth scars, attached rachilla's and torn or rip scars, grouped according to their classification as wild, domesticated or intermediate.

Materials and Methods

Plant macroremains, including charred specimens and imprints on baked clays were recorded from, Mahal Teglinos K 1 Trench IX (15°26′31.55″ N 36°25′52.46″ E), however a further sampled site, UA 126 (15°38′8.61″ N, 36°10′18.03″ E) (Fig. 1), produced no evidence for sorghum or other crops. The new samples were collected at site K 1 in excavation unit IX, investigated in 2013 and 2014 (Manzo 2014a, 2015) and from site UA 126 in 2012 (Manzo 2013). Laboratory work was carried out at the Archaeobotanical Laboratory of the Institute of Archaeology, University College London by the first author (AB) together with CS, CM and DF, with UCL reference collections and recent comparative studies of cereal domestication.

K 1 IX is a 4 × 6 m excavation unit in the western sector of the site at Mahal Teglinos, on the outskirts of the Kassala, in a small valley of a granitic hill, the Jebel Taka. The aim of the excavation unit is the investigation of a previously almost unexplored Gash Group settlement in this sector of the site and the investigation of its relationship with the palaeo-lake characterizing this part of the site in ancient times. In this excavation unit, the excavation reached ca. 0.8 m in depth. Under the first stratum, characterized by some Jebel Mokram Group (early 2nd millennium BC-early 1st millennium BC) materials, a thick strata of burned soil with some stone arrangements was revealed, which yielded charred seeds relating to a living floor. These layers were placed by archaeological finds into the Gash Group Culture, and by a radiocarbon date on charcoal of 1960–1760 BC (Beta-380246: 3560 ± 30 uncal. BP), by which they can be ascribed to the latest phase of the Gash Group culture, dating to the early 2ndmillennium BC (Manzo in press). These layers yielded materials similar to those from other sectors of the site dating to the end of the Gash Group (Fattovich 1993b; Manzo 2014a, 2015).

The baked clays may come from a variety of possible sources, including linings of storage pits and ovens, or large clay "trays" similar to those from excavation unit K 1 VI (see Manzo in press). However, particularly for excavation unit K 1 IX, these baked clays would seem to come from fragments of large fixed storage vessels similar to the Nubian *gusseba* (Manzo 2014a). The baked clays from excavation unit K 1 IX were all recovered from a soil stratum covering an occupational layer, and in all cases both surfaces contained plant imprints. The designation of this late Gash Group part of the site as relating to domestic activity was further confirmed by the discovery of a plastered rounded fire-pit with some burned stones, perhaps used as baking stones, at the bottom brought to light by erosion and investigated nearby the excavation unit K 1 IX.

UA 126 II is an excavation unit at site UA126, North of Kassala, in the area endangered by the implementation of the Upper Atbara Irrigation Scheme (see Manzo et al. 2011; Manzo 2012a, 2014b). This excavation exposed features, including pits and hearths, and possible camp structures that seem to be related to the Hagiz Group (1st millennium BC–AD 1st millennium), although the reoccupation of part of the site in Gergaf Group times (AD 15th–18th century) cannot be entirely ruled out on the basis of the potsherds visible on the surface.

Dry-screening of sediments from the fills of fireplaces, pits, and other features through a 1 mm and 0.5 mm mesh were carried out as part of the excavations at all the sites. Charcoal samples recovered were small, and only up to 10 mg weight. The samples were sorted into morphotype categories by the first author and all material then examined in London with DQF, CJS and CM and compared to reference materials there.

Clay Impressions

A preliminary sorting of ceramics and clay fragments with plant impressions from Site K 1 was conducted in the laboratory of the National Museum of Khartoum by the first author. Five larger specimens of baked clays with large numbers of plant imprints were selected for further study at the archaeobotanical laboratory of the Institute of Archaeology, UCL. Clay pieces with plant impressions were examined with the aid of a binocular microscope and promising impressions were cast made using polyvinyl siloxane (as per Fuller et al. 2007; Manning et al. 2011). Cast impressions were then cut out, mounted, sputter-coated with gold examined with Hitachi S-3400 variable pressure SEM, and photomicrographs captured.

Results

The Identification of Sorghum and Other Charred Plant Remains

As stated no crop remains were recovered from UA 126II and plant macroremains consisted entirely of 58 charred globular, spherical items, around 5–8 mm, with an internal parenchymal structure and external morphology consistent with charred tubers.

Sampling at K 1-Mahal Teglinos, excavation unit IX, produced both charred macroremains and imprints in baked clay. A total number of 144 carbonized grains and fruit stones were recovered by dry sieving, all from KI IX Layer SU3 (Table 2). Edible tree fruits are represented by 66 fragments of seeds of baobab tree, *Adansonia digitata* L. (Fig. 5a, b), and 2 half-fragments of stone of jujube/Christ's thorn, *Ziziphus* sp. (Fig. 5c, d).

Table 2. Counts of identified charred plant macroremains from Site K 1 IX, sample E2 SU3.

Identification	Site K 1 IX
Adamsonia digitata. Baobab seed fragments	66
Rottboellia cochinchinensis grains	3
Ziziphus sp.	2
Immature sorghum grains with parts of spikelet preserved, or immature spikelets with preserved rachilla =*Sorghum* sp.	10
Other *Sorghum* sp. fragments of chaff without base =*Sorghum* sp.	9

From the charred remains, sorghum is represented by 19 whole or fragmented charred grains or spikelets of sorghum, none of which could be definitively placed as wild or domesticated (Fig. 6a). These grains are predominately immature, as indicated by their flattened form, and small overall size, and their papery texture and hollow centre. Two of the specimens are essentially charred florets and even included preserved the remnant stalks of the pistils (e.g. Fig. 6a, specimen at right). The identification of immature sorghum is similar to that of other immature millets (e.g. Song et al. 2013; Motuzaite-Matuzeviciute et al. 2012). These immature grains varied in size from less than 2 mm in length and 1 mm in width, to over 3 mm long and 1.75 mm wide. Many small fragments of chaff, mostly of probable *Sorghum* husk (lemma/palea) were noted, but not separately quantified and might derive from the dehusking of hulled spikelets.

In several cases, where the charred spikelet bases were still fused with the grain, they were noted to have a torn rachilla. This is to be expected with green harvesting, i.e. of immature grains, and so not necessarily indicative of genetically domesticated plants. The presence of a torn rachilla can be representative of the harvesting of domesticated mature or immature, either wild or domesticated, (green-harvested) spikelets. In the former case, the presence of well-developed spikelets containing a mature-sized, fully infilled grain, can be taken as indicating the domesticated morphotype. However, in the present study all of the preserved torn rachilla recorded are associated with charred immature, and hence probably green-harvested grains.

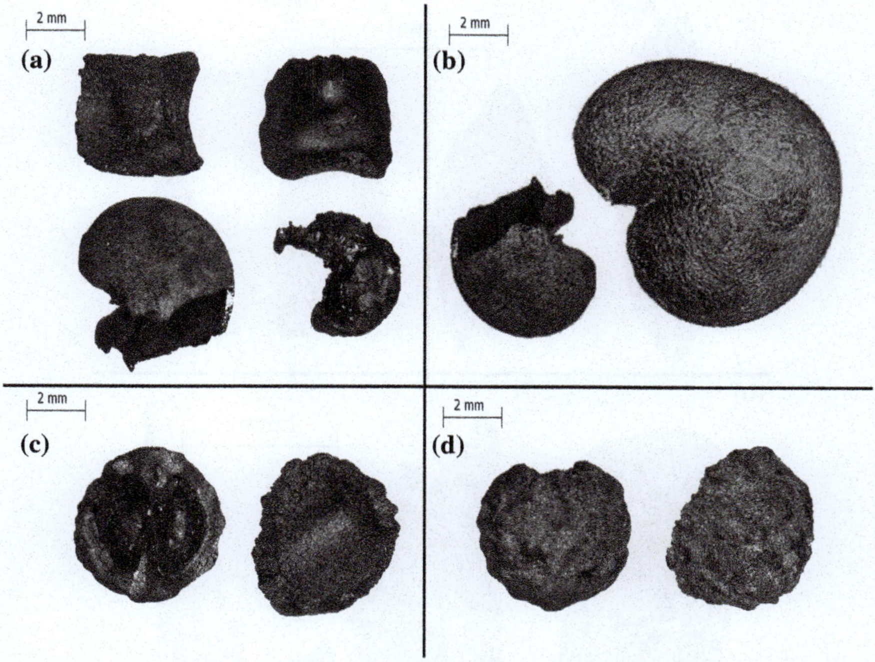

Fig. 5. Carbonized fruit seeds/stones from Kassala K 1 IX SU3; (a) Charred seed fragments of baobab (*Adansonia digitata*) (b) the largest archaeological fragment of *Adansonia digitata* (on left) compared to a modern seed of baobab (right) for comparison carbonized seed with actual grain. (c–d) Charred archaeological stones of *Ziziphus* sp. (c) the interior and (d) the exterior.

While immature remains can be related to wild or domesticated plants, their presence in quantity indicates that these sorghum plants were harvested while panicles were still partly green. This is expected to be a strategy to reduce loss of grains to wild-type seed dispersal (dehiscence), and as such we expect these to be a more prominent part of assemblages from wild-harvested, naturally shattering sorghum populations, or within the earlier stages of cultivation, where the domestication trait for non-shattering is not yet fixed. Such an expectation has been discussed and documented archaeobotanically in the context of rice domestication (Fuller 2007; Fuller et al. 2009), but the same principles should apply sorghum which has a similar panicle structure and maturation process.

Two carbonized grains and one spikelet base of *Rottboellia cochinchinensis* (Lour.) W.D. Clayton were identified from K 1 IX E2 SU3 context (Fig. 6b). This species is found commonly in the savannah regions of Africa, the southern Arabia Peninsula (e.g. Yemen, Dhofar), and parts of tropical Asia. While it is grown in the modern era as a fodder crop its presence here may be as a weed that contaminated sorghum harvests and entered the crop-processing waste, although conceivably it may have been gathered as wild grain, or entered fires through the use of animal dung fuel. Grains of *Rottboellia* have also been found associated with morphologically wild-type sorghum spikelets originating from a store at Sabir (c. 900 BC), in southern Yemen (de Moulins et al. 2003).

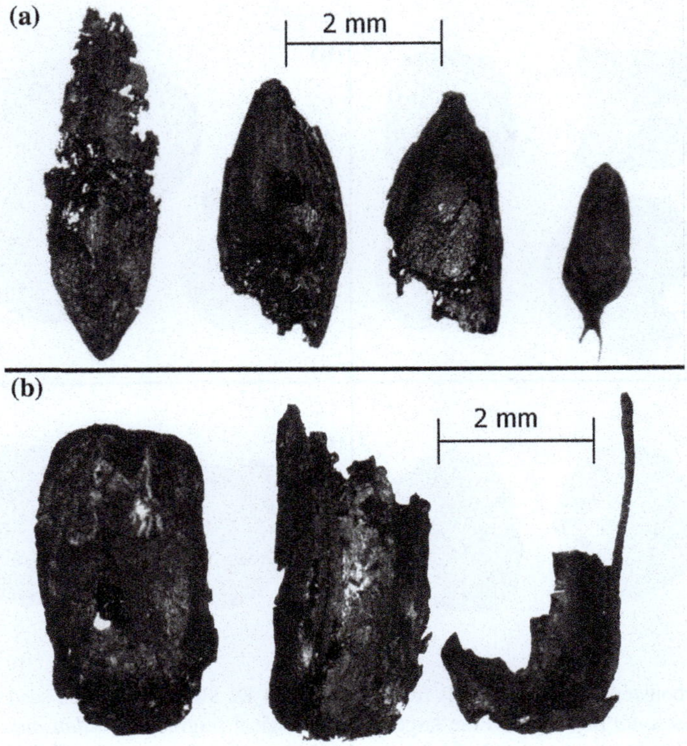

Fig. 6. Carbonized grains from K 1 IX SU3; (a) examples of immature sorghum grains and at far right a floret with remnant pistils, (b) examples of *Rottboellia cochinchinensis*.

These latter finds are important in that they indicate that morphologically wild sorghum continued to be stored and consumed into the early 1st millennium BC, and that these stores were contaminated by *Rottboellia* grains, which may have been established as a field wield, or co-harvested with sorghum earlier in the eastern Sudan Sahel zone.

Identification of Sorghum, Pearl Millet and Other Grasses from Plant Impressions

By far the most common impressions within the clays from K 1 IX were those of sorghum. A total of 78 plant remains were identified from five fragments of baked clay (Table 3). Of these, 84% of the impressions were of *Sorghum* (Fig. 7). The identification of wild and domesticated sorghum imprints was done based on observation of the spikelet bases, as outlined above. Among these 26 (39% of all sorghum remains) represent probably domesticated types, including 4 with "rip" scars (Fig. 7a, b), and 22 with remnants of torn rachillae (Fig. 7c, d). Those with a torn rachilla are recorded as probably domesticated, in particular as they are attached to what appear to be fully formed and mature spikelets. Thus among the impressions we lack evidence for immature spikelets with rachillae resembling those found in the charred macroremains.

Table 3. Result of the archaeobotanical investigation on five baked clay lumps from K 1 IX, indicating counts and percentage of identified impressions. At right is the estimate total percentage of non-shattering sorghum, estimated standard deviation and range (minimum and maximum of non-shattering forms); also shown is the total percentage of *Pennisetum*.

Taxonomic identification	Count	(%)		
Sorghum sp. (no spikelet base)	23	29		
Sorghum verticilliflorum type (smooth scar)	17	22		
Sorghum cf. *bicolor*, torn rachilla	22	28	~60 ± 7%	Domest.
Sorghum cf. *bicolor*, rip scar	4	5	(39–74%)	
Echinochloa sp.	4	5		
Brachiaria cf. *deflexa/ramosa*	4	5		
Pennisetum cf. *glaucum* (torn pedicel)	1	1		
Pennisetum cf. *monodii*, wild type (sessile)	1	1	5%	*Pennisetum*
Pennisetum sp. (indeterminate status)	2	3		

By contrast only 17 (26% of sorghum) could be identified as wild *Sorghum* cf. *verticilliflorum* type with smooth abscission scars (Fig. 7f), while 23 did not preserve spikelet bases and could only be identified as *Sorghum* sp. (Fig. 7e). Sorghum was the dominant species in the imprints, as well as in the charred remains, suggesting that this was the staple cereal. We estimate the proportion of non-shattering sorghum was likely around 60% (± 7%) of the population, and certainly no less than 39%. As such while the sorghum is likely to have been predominately of the domesticated type, the non-shattering trait would appear to have not yet been fixed in this population.

In addition to the evidence for sorghum, five impressions of *Pennisetum* with its characteristic bristles arising from the involucre below the spikelet were noted (Fig. 8a–d). *Pennisetum* can be difficult to distinguish from *Cenchrus*, in particular *Cenchrus ciliaris*, which has ciliate or hairy bristles, as opposed to spiny bristles seen in most species of *Cenchrus*. However, along with such hairs that potentially might be visible, the inner bristles on most species of *Cenchrus* are usually flatter and grooved and often connate at the base (see Kahlheber 2004; Phillips 1995: 275–278; Chemisquy et al. 2010). *Pennisetum* has rounder straight bristles. In addition, bristles of *Pennisetum glaucum* have uniseriate trichomes (see Fuller et al. 2007: Fig. 8), as opposed to multicellular trichomes found in some *Cenchrus* spp. (e.g. *C. ciliaris*, *C. biflorus*, which are among the most common in central Sudan). These uniseriate trichomes are identified by the bristle texture in the impressions. These trichomes face apically in *Pennisetum*, i.e. bristles are antorsely scabrous, whereas in some *Cenchrus* spp. bristles are retorsly scabrous (Chemisquy et al. 2010).

What is intriguing about these specimens is that they appear to indicate a mixture of morphologically wild and probably domesticated forms, although the sample size is too small to move beyond this observation (for a discussion of criteria, see Brunken et al. 1977; Manning et al. 2011; Manning and Fuller 2014). While two examples are damaged and do not preserve the involucre base and are therefore assigned to *Pennisetum* sp. (Fig. 8a), at least one example has an involucre base suggestive of a sessile non-pedicillate form (Fig. 8b), e.g. of wild type (*Pennisetum glaucum* var. *monodii*).

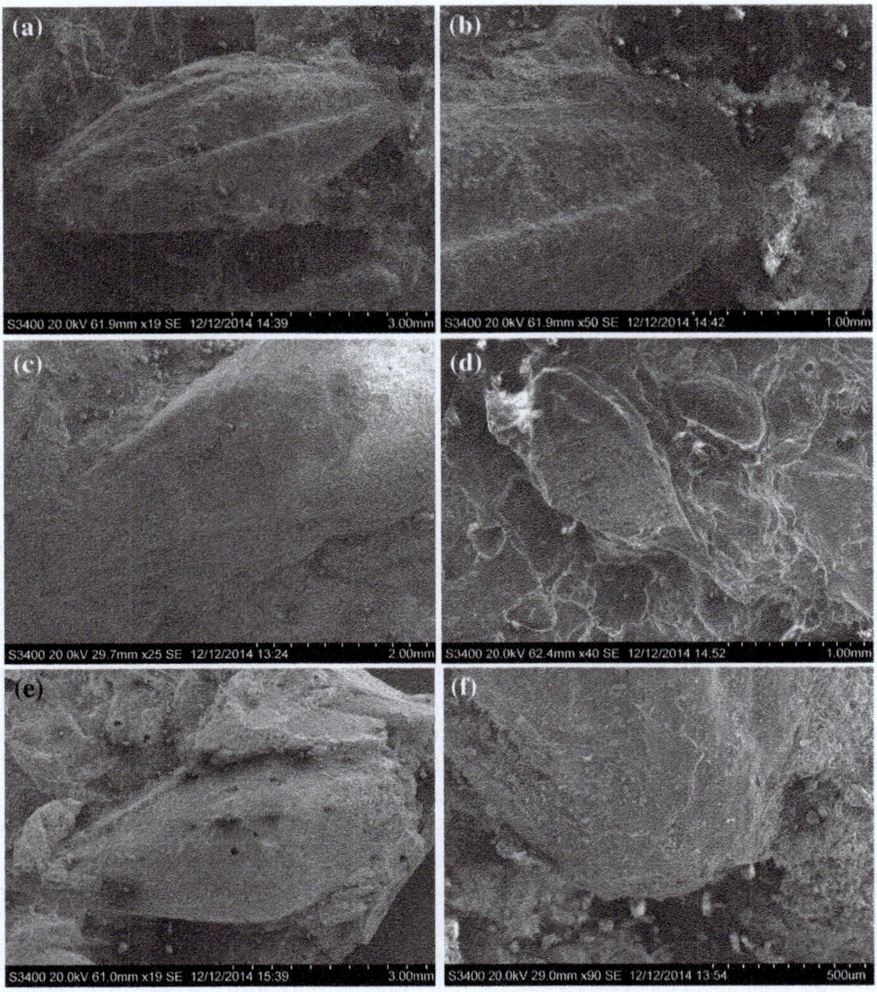

Fig. 7. Example of *Sorghum* impressions, as revealed by SEM imaging of positive casts; (a) Sorghum spikelet, ventral view with two sterile floret rachillae, basal end (at right) shows probably "rip" scar breakage pattern; (b) Close of "rip" scar in (a); (c) Sorghum spikelet, basal portion, in dorsal view, with probable remnant of torn rachilla (poorly preserved at left); (d) Sorghum spikelet, dorsal view, with torn rachilla at lower right. Note that the central part of the cast is damaged due to handling of artefact; (e) *Sorghum* sp., spikelet, damaged so that shattering or non-shattering status cannot be determined; (f) *Sorghum verticilliflorum* spikelet with smooth, wild type abscission scar.

Another appears to be a torn involucre stalk, characteristic of non-shattering, domesticated pearl millet (i.e. *Pennisetum glaucum* var. *glaucum*), although this was poorly impressed into the clay and therefore not well preserved (Fig. 8c, d). Two additional examples suggest possible wild and domesticated types. While sorghum was the dominant taxon in the imprints, these *Pennisetum* impressions might indicate that

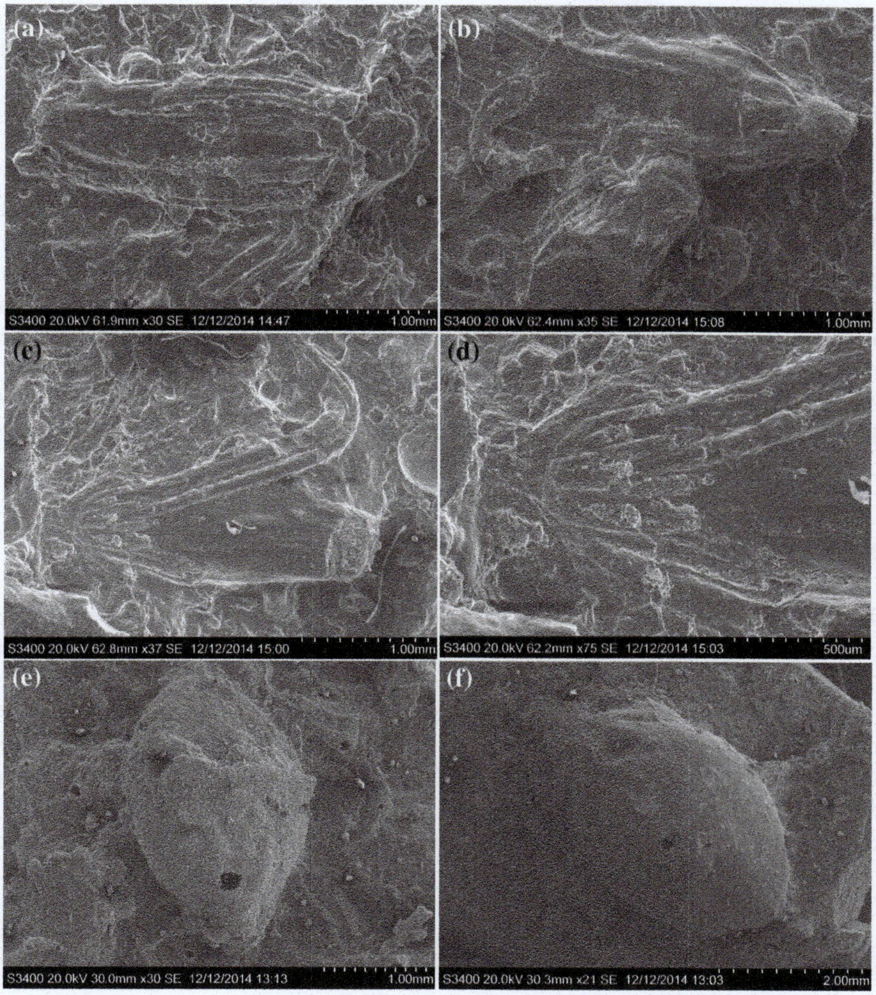

Fig. 8. Example of additional millet taxa identified from impressions; (a) *Pennisetum* sp. spikelet and involucre bristle, with involucre base obscured by damage; (b) *Pennisetum* cf. *monodii* wild type sessile involucre with bristles; (c) *Pennisetum* cf. *glaucum* domesticated type spikelet, with bristles arising from the torn involucre stalk; (d) close-up of involucre in (c); (e) *Brachiaria* cf. *deflexa* spikelet, showing characteristic rugose lemma and lemma apex beak; (f) *Echinochloa* sp. Spikelet impression (dorsal view).

pearl-millet was also in cultivation, and its threshing waste was also sometimes used in tempering clays. *Pennisetum glaucum* was likely to have been introduced from West Africa as a cultivar (see Manning and Fuller 2014) with at least partially non-shattering phenotypes in its populations.

Also recorded were 3 imprints of millets identified as probable *Brachiaria* cf. *deflexa/ramosa* (Fig. 8e), and several impressions of *Echinochloa* sp. (Fig. 8f); both of these small millets are preserved as hulled grains, i.e. in lemma and palea. They have been identified on the basis of overall shape in consultation with our reference collection of Asian and African millets and related grasses, although this collection is not comprehensive. In the case of *Brachiaria* cf. *deflexa/ramosa*, this is based on similarity of the rugose lemma sculpturing to *B. deflexa* (Schumach.) Robyns and/or *B. ramosa* (Trin.) Griseb., which distinguishes *Brachiaria* from various *Setaria* spp., previously reported from Sudan (e.g. Fuller and Edwards 2001; Fuller 2004a), and from more finely rugulose species (e.g. *B. mutica* (Schumach.) Robyns, *B. eruciformis* Griseb.). The lack of a hard, acute "beak-like" apex distinguishes this from *Brachiaria lata* (Schumach.) C.E. Hubbard, while in all characters it is very similar to *B. ramosa*, a species brought into cultivation in India (Kingwell-Banham and Fuller 2014), which might have also been wild in Africa. *B. deflexa* is perhaps a more widespread native grass in the African savannahs. Nevertheless, further comparative work is needed to accurately distinguish *Brachiaria* spp. remains in Africa to species level. These are likely to have been wild grains exploited for food, or as contaminants of harvested sorghum.

Discussion

The site at Kassala K 1 produced evidence of two important cultivars, the African cereals, sorghum and pearl millet, as well as the culturally important baobab tree (*Adansonia digitata*). Baobab is often seen as a quintessential part of the African savannah. However biogeographical and genetic studies suggest that this species was endemic to West Africa and with less diversity in eastern Africa (Pock Tsy et al. 2009). Whether baobab in eastern Sudan represents a natural range expansion, e.g. in the Pleistocene (cf. Pock Tsy et al. 2009) or a role for human agency, is unclear. The co-occurrences of baobab with pearl millet, finger millet and sorghum at East African Iron Age sites within Kenya, and in the Swahili sites of the east African islands, supports the hypothesis that it was spread, along with millet, by agro-pastoralists (e.g. Boivin et al. 2013; Shipton et al. 2013). Archaeological evidence in West Africa (Burkina Faso) associates baobab charcoal with Neolithic pearl millet farming (Kahlheber and Neumann 2007), and it seems likely the tree was managed/cultivated by Neolithic pastoralists and came to be associated with early farming villages in western Africa. This could have occurred in parallel in Sudan, or represent translocation, as likely for pearl millet. Baobab is a valuable traditional food and medicinal plant throughout Africa (Andrews 1952; Burkill 1994; Mabberley 2008), particularly in eastern and sub-Saharan Africa, as the dried fruit when still in the pod is light, storable and transportable, and associated with modern mobile pastoralists (Blench 2007). The carbonized seeds of *Adansonia digitata* recovered from Kassala (Late Gash Group context) represent the first archaeobotanical evidence of its kind for Sudan.

Pennisetum glaucum, pearl millet or bulrush millet, is an annual crop of the semi-arid regions of Africa and India. It requires lower rainfall than sorghum with the northern limit at around the 250 mm isohyet. Sorghum and pearl millet can be seen as crops that complement each other in terms of coping with the variable rainfall that characterizes the dry savannahs and Sahelian regions. The earliest archaeobotanical evidence for domesticated pearl millet, which predates previous findings from India and Africa, dates back to 4500 years ago in the Tilemsi Valley, north-eastern Mali (Manning et al. 2011; Manning and Fuller 2014). Before the identification of this domesticated morphotype in Mali, the earliest record was from India between 2000 and 1500 BC (Fuller and Boivin 2009). The recovery of *P. glaucum* from the site of Kassala (K 1) as imprints on baked clay by around 2000 BC, probably indicates that the eastern Sahel was an important area for the early dispersal of the crop. These finds suggest a possible mixture of morphologically wild and domesticated involucres, which could indicate that dispersal of this cultivar from western Africa took place before the domestication process was completed. However, wild *Pennisetum* also occurs in Sudan (Brunken et al. 1977), and so introgressions with, or recruitment of, local weedy types is also possible. In addition, the imprints are few and not ideally preserved, so further evidence is needed before we can confirm unambiguously that pearl millet was under cultivation at Kassala, or how it got there. Nevertheless, the hypothesis that Sahelian agro-pastoralists transmitted pearl millet across much of Africa already by the start of the 2nd millennium BC, would indicate quite rapid translocation after cultivation of pearl millet cultivation had been initiated.

The Upper Nile Valley in Sudan, including the Blue Nile and White Nile, and their tributaries, such as the Atbara River, have long provided evidence for both the use of wild sorghum and potentially for the earliest domesticated sorghum. The archaeobotanical work in the Kassala region presented in this paper has illustrated evidence that the domestication episode for sorghum continued into the 2nd millennium BC, and we identify this with part of the pre-domestication cultivation stage in the evolution of the sorghum crop. The admixture in the ceramic impressions of threshed, domesticated type sorghum, as well as wild-type shattering spikelets, indicates that cultivation had not yet fixed non-shattering in cultivated sorghum populations in the Kassala area. In addition is the presence of both rip scars as well as torn rachillae. While torn remnant rachillae, are more typical of domesticated sorghum, they can also occur due to green harvesting through the threshing immature spikelets. Rip scars, however, would only be expected from the threshing of mature, non-shattering (domesticated) phenotypes. Therefore the site presents clear evidence for the presence of non-shattering genotypes, and the threshing of such plants. Additionally, combined with the charred material, is evidence for a significant presence of immature and wild spikelets, which is consistent with an incompletely domesticated population, and one with asynchronous ripening. Harvesting while some proportion of the grains was immature, was probably conducted in order to reduce grain loss to dehiscence. As demonstrated with the evolution of Asian rice (*Oryza sativa*) once domestication is fully established immature spikelets largely disappear (Fuller et al. 2009, 2010, 2016; Castillo et al. 2016).

Of interest is a subtle contrast with the material from the earlier sorghum assemblage at KG 23 (3500–3000 BC) lying only 53 km to the southwest of Mahal Teglinos K 1. Imprints from KG 23 have also been identified as representing pre-domestication cultivation, but do not include rip scars (Winchell et al. 2017; Fuller and Stevens 2018). Instead they suggest some selection of non-shattering and threshed spikelets with persistent rachillae, as well as immature harvesting alongside wild types. It may be that the appearance of rip-scar spikelet bases mark a second phase in the domestication episode of sorghum, associated with more robust threshing methods and/or the rise of genes responsible for more incomplete formation of the abscission layer (e.g. stronger non-shattering traits). As genetic evidence from rice (*Oryza sativa*) has advanced, it has become clear that multiple genetic loci are involved additively in the evolution of non-shattering (Htun et al. 2014; Ishikawa et al. 2017), as opposed to early claims of a single mutation (e.g. Lin et al. 2012). It may be regarded as possible, if not likely, that the same was true of sorghum. However, we would caution that our sample size is still limited, both in total numbers and in terms of numbers of assemblages. More data, and more time points are needed to resolve the evolutionary episode of sorghum domestication and to better describe its domestication process in both time and space.

Conclusions

The new results from the Kassala region offer new data for a key geographical and chronological gap in the evidence. While the antiquity of dates for wild sorghum use in Sudan suggests a long period of collection and harvesting prior to cultivation dating back to the mid-7th millennium BC (see Fig. 1; Majid 1989; Stemler 1990), the origin of the domesticated crop has been largely undocumented. Similarly, working backwards, Butana entered history as a sorghum producing region of the Meroitic Kingdom in the second half of the 1st millennium BC (Fuller 2014a, b), but the precursors of this region and when sorghum became important have been unclear.

However, the evidence from excavation unit K 1 IX suggest that cultivation was established and domestication was well underway by 1900–1800 BC. When this is taken alongside the recent studies of imprints from site KG 23, which also indicates pre-domestication cultivation (but without rip scars) at 3500–3000 BC (Winchell et al. 2017), we can infer that at least 1500 years of the sorghum domestication process is to be sought in the archaeology of the greater Kassala region. We hypothesize that this region is a key area for sorghum domestication, calling for more intensive archaeological research and systematic archaeobotanical sampling. As recent genetic work raises the possibility of more than one domestication for sorghum (Lin et al. 2012), the current archaeobotanical evidence suggests that the area around Kassala was one of them.

The Kassala region is also important for documenting the spread of not only sorghum but also pearl millet, by connecting the Sahelian region with the Red Sea and Arabia Sea trade. Already from the 3rd millennium BC, and certainly in the early 2nd millennium BC, the Kassala region was linked to circum Red-Sea trade as indicated by imports in the region, as well as exports in southeast Egypt and Yemen (see Fattovich 1991, 2012; Manzo 1997, 2012a, b). This Red Sea network of which Kassala was a part provides the most plausible source for sorghum and pearl millet to enter the circum Arabian coastal networks that ultimately brought these crops to India early in the 2nd millennium BC.

Acknowledgements. Travel and research in London by AB, together with work by CS, CM, DF was supported by the European Research Council project "Comparative Pathways to Agriculture" (ComPAg, ERC grant agreement 323842). Fieldwork directed by AM was supported by University of Naples, "L'Orientale", the Italian Ministry of Foreign Affairs and the Italian

Ministry for University and Research grant "Futuro in Ricerca 2012" *RBFR12N6WD* "Aree di transizione linguistiche e culturali in Africa" (Cultural and linguistic transitions in Africa).

References

Abbas SAM-A, Jaeger SE (1989) The early ceramic of the Eastern Butana (Sudan). In: Krzyzaniak L, Kobusiewicz M (eds) Late prehistory of the Nile Basin and the Sahara. Poznań Archaeological Museum, Poznań, pp 473–480

Adunga A, Bekele E (2013) Geographical distribution and phenotypic diversity of wild/weedy sorghum [*Sorghum bicolor* (L.) Moench] in Ethiopia: implications for germplasm conservation and crop–wild gene flow. Plant Genet Resour-C 11(1):68–76

Aldrich PR, Doebley J (1992) Restriction fragment variation in the nuclear and chloroplast genomes of cultivated and wild *Sorghum bicolor*. Theor Appl Genet 85:293–302

Andrews FW (1952) The flowering plants of the Anglo-Egyptian Sudan. Sterculiacea-Dipsacaceae, vol II. T. Buncle & co. Ltd., Arbroath

Bard KA, Fattovich R (2013) The land of punt and recent archaeological and textual evidence from the pharaonic harbour at Mersa/Wadi Gawasis, Egypt. In: Chrisomalis S, Costopoulos A (eds) Human expeditions: inspired by Bruce Trigger. University of Toronto Press, Toronto, pp 3–11

Beldados A (2015) Paleoethnobotanical study of ancient food crops and the environmental context in North East Africa, 6000 BC-200/300 AD. Cambridge Monographs in African Archaeology 88. BAR International series 2706. Archaeopress, Oxford

Beldados A, Costantini L (2011) Sorghum exploitation at Kassala and its environs, Northeastern Sudan in the second and first millennium BC. Nyame Akuma 75:33–39

Billot C, Ramu P, Bouchet S et al (2013) Massive sorghum collection genotyped with SSR markers to enhance use of global genetic resources. PLoS ONE 8(4):e59714

Blench R (2007) The intertwined history of the silk-cotton and baobab. In: Cappers RTJ (ed) Fields of change: progress in African archaeobotany. Barkhuis & Groningen University Library, Groningen, The Netherlands, pp 1–19

Boivin N, Crowther A, Helm R et al (2013) East Africa and Madagascar in the Indian Ocean world. J World Prehist 26:213–281

Boivin N, Fuller DQ (2009) Shell middens, ships and seeds: exploring coastal subsistence, maritime trade and the dispersal of domesticates in and around the ancient Arabian Peninsula. J World Prehist 22(2):113–180

Brunken J, De Wet JMJ, Harlan JR (1977) The morphology and domestication of pearl millet. Econ Bot 31:163–174

Burkill HM (1994) The useful plants of West Tropical Africa, Vol 2, 2nd edn, Families E–I. Royal Botanic Gardens, London

Castillo C, Tanaka K, Sato YI et al (2016) Archaeogenetic study of prehistoric rice remains from Thailand and India: evidence of early japonica in South and Southeast Asia. Archaeol Anthropol Sci 8:523–543

Chemisquy MA, Giussani LM, Scataglini MA et al (2010) Phylogenetic studies favour the unification of *Pennisetum*, *Cenchrus* and *Odontelytrum* (Poaceae): a combined nuclear, plastid and morphological analysis, and nomenclatural combinations in *Cenchrus*. Ann Bot 106(1):107–130

Clapham AJ, Rowley-Conwy PA (2007) New discoveries at Qasr Ibrim, Lower Nubia. In: Cappers RTJ (ed) Fields of change: progress in African archaeobotany. Groningen Archaeological Studies 5. Barkhuis, pp 157–164

Clark JD (1984) Prehistoric cultural continuity and the economic change in the Central Sudan in the Early Holocene. In: Clark JD, Brandt SA (eds) From hunters to farmers, the causes and consequences of food production in Africa. University of California Press, Berkeley, pp 113–126

Clark JD, Stemler AB (1975) Early domesticated sorghum from Central Sudan. Nature 25:588–591

Costantini L, Fattovich R, Pardini E et al (1982) Preliminary report of archaeological investigations at the site of Teglinos (Kassala). Nyame Akuma 21:30–33

Costantini L, Fattovich R, Piperno M, Sadr K (1983) Gash Delta archaeological project: 1982 field season. Nyame Akuma 23:17–19

de Moulins D, Phillips C, Durrani N (2003) The archaeological records of Yemen and the question of Afro-Asian contact. In: Neumann K, Buttler A, Kahlheber S (eds) Food, fuels and fields: progress in African archaeobotany. Heinrich Barth-Institut, Cologne, pp 281–299

de Wet JMJ (1977) Domestication of African cereals. Afr Econ Hist 3:15–32

de Wet JMJ, Huckabay JP (1967) The origin of *Sorghum bicolor*. II. Distribution and domestication. Evolution 21:787–802

Doggett H, Prasada Rao KE (1995) *Sorghum*. In: Smartt J, Simmonds JW (eds) Evolution of crop plants, 2nd edn. Longman Group UK limited, London, pp 173–180

Dominy NJ, Ikram S, Moritz GL et al (2015) Mummified baboons clarify ancient Red Sea trade routes [Abstract]. Am J Phys Anthropol 156:122–123

Fattovich R (1989) The later prehistory of the Gash Delta (Eastern Sudan). In: Krzyzaniak L, Kobusiewicz M (eds) Late prehistory of the Nile Basin and the Sahara. Poznan Museum, Poznan, pp 481–498

Fattovich R (1991) At the periphery of the empire: the Gash Delta (Eastern Sudan). In: Davies WV (ed) Egypt and Africa. Nubia from prehistory to Islam. British Museum Press, London, pp 40–48

Fattovich R (1993a) The Gash group of the Eastern Sudan: an outline. In: Krzyzaniak L, Kobusiewicz M, Alexander J (eds) Environmental change and human culture in the Nile Basin and Northern Africa until the second millennium BC. Poznan Archaeological Museum, Poznan, pp 439–448

Fattovich R (1993b) Excavations at Mahal Teglinos (Kassala) 1984–1988. A preliminary report. Kush 16:225–287

Fattovich R (2012) The Southern Red Sea in the 3rd and 2nd millennia BC: an archaeological overview. In: Agius DA, Cooper JP, Tradakas A, Zazzaro C (eds) Navigated spaces, connected places. Proceedings of the Red Sea Project V held at the University of Exeter, 16–19 September 2010. BAR International Series 2346. Archaeopress, Oxford, pp 39–46

Fattovich R, Marks EA, Ali MA (1984) The archaeology of the Eastern Sahel, Sudan: preliminary results. Afr Archaeol Rev 2:173–188

Fattovich R, Piperno M (1986) Archaeological researches in the Gash Delta, Kassala Province (1980–81 field seasons). In: Krause M (ed) Nubische Studien. Proceedings of the symposium of the international society of Nubian studies, Heidelberg, 1982. P. von Zabern, Mainz, pp 47–53

Fuller DQ (2003) African crops in prehistoric South Asia: a critical review. In: Neumann K, Butler A, Kahlheber S (eds) Food, fuel and fields. Progress in Africa archaeobotany. Africa Praehistorica 15. Heinrich-Barth-Institut, Cologne, pp 239–271

Fuller DQ (2004a) Early Kushite agriculture: archaeobotanical evidence from Kawa. Sudan & Nubia 8:7–74

Fuller DQ (2004b) The Central Amri to Kirbekan survey. A preliminary report on excavations and survey 2003–04. Sudan & Nubia 8:4–16

Fuller DQ (2007) Contrasting patterns in crop domestication and domestication rates: recent archaeobotanical insights from the Old World. Ann Bot 100(5):903–924
Fuller DQ (2014a) Agricultural innovation and state collapse in Meroitic Nubia: the impact of the savannah package. In: Stevens CJ, Nixon S, Murray MA, Fuller DQ (eds) Archaeology of African plant use. Left Coast Press, Walnut Creek, CA, pp 165–178
Fuller DQ (2014b) Agricultural innovation and state collapse in Meroitic Nubia. In: Archaeology of African plant use, Vol 61. UCL Institute of Archaeology Publications, pp 165–177
Fuller DQ (2015) The economic basis of the Qustul splinter state: cash crops, subsistence shifts and labour demands in the Post-Meroitic transition. In: Zach MH (ed) The Kushite world. Proceedings of the 11th international conference for Meroitic studies. Beiträge zur Sudanforschung Beiheft 9. Verein der Förderer der Sudanforschung, Vienna, pp 33–60
Fuller DQ, Boivin N (2009) Crops, cattle and commensals across the Indian Ocean: current and potential archaeobiological evidence. In: Lefevre G (ed) Plantes et Societes. Etudes Ocean Indien 42–43. Institut National des Langues et Civilisations Orientales, Paris, pp 13–46
Fuller DQ, Edwards DN (2001) Medieval plant economy in Middle Nubia: preliminary archaeobotanical evidence from Nauri. Sudan & Nubia 5:97–103
Fuller DQ, Stevens C (2018) Sorghum domestication and diversification: a current archaeobotanical perspective. Plants and People in the African Past 503–528
Fuller DQ, Harvey E, Qin L (2007) Presumed domestication? Evidence for wild rice cultivation and domestication in the fifth millennium BC of the Lower Yangtze region. Antiquity 81:316–331
Fuller DQ, Qin L, Zheng Y et al (2009) The domestication process and domestication rate in rice: spikelet bases from the Lower Yangtze. Science 323:1607–1610
Fuller DQ, Allaby RG, Stevens C (2010) Domestication as innovation: the entanglement of techniques, technology and chance in the domestication of cereal crops. World Archaeol 42(1):13–28
Fuller DQ, Boivin N, Hoogervorst T et al (2011) Across the Indian Ocean: the prehistoric plants and animals. Antiquity 85:544–558
Fuller DQ, Denham T, Arroyo-Kalin M et al (2014) Convergent evolution and parallelism in plant domestication revealed by an expanding archaeological record. Proc Nat Acad Sci 111(17):6147–6152
Fuller DQ, Weisskopf AR, Castillo CC (2016) Pathways of rice diversification across Asia. Archaeol Int 19:84–96
Garcea EAA (2006) Semi-permanent foragers in semi-arid environments of North Africa. World Archaeol 38:197–219
Haaland R (1999) The puzzle of the late emergence of domesticated sorghum in the Nile Valley. In: Gosden C, Hather J (eds) The prehistory of food. pp 397–418
Harlan JR (1971) Agricultural origins: centers and noncenters. Science 174:468–474
Harland JR, Stemler ABL (1976) The races of sorghum in Africa. In: Harland JR, de Wet JMJ, Stemler ABL (eds) Origins of African plant domestication. Mouton, The Hague, pp 465–478
Harlan JR, de Wet JMJ, Price EG (1973) Comparative evolution of cereals. Evolution; Int J Org Evol 27:311–325
Harris D, Fuller DQ (2014) Agriculture: definition and overview. In: Smith C (ed) Encyclopedia of Global Archaeology. Springer-Verlag, New York, pp 104–113
Htun TM, Inoue C, Chhourn O et al (2014) Effect of quantitative trait loci for seed shattering on abscission layer formation in Asian wild rice *Oryza rufipogon*. Breeding Sci 64(3):199–205
Hulse JH, Laing EM, Pearson OE (1980) Sorghum and millets: their composition and nutritive value. Academic Press, London
Ishikawa R, Nishimura A, Htun TM et al (2017) Estimation of loci involved in non-shattering of seeds in early rice domestication. Genetica 145(2):201–207

Kahlheber S (2004) Perlhirse und Baobab – archäobotanische Untersuchungen im Norden Burkina Fasos. Unpublished Ph.D. Dissertation, Johann Wolfgang-Universitat, Frankfurt. http://publikationen.ub.uni-frankfurt.de/frontdoor/index/index/docId/4965

Kahlheber S, Neumann K (2007) The development of plant cultivation in semiarid West Africa. In: Denham TP, Iriarte J, Vrydaghs L (eds) Rethinking agriculture: archaeological and ethnoarchaeological perspectives. Left Coast Press, Walnut Creek, CA, pp 320–345

Kingwell-Banham E, Fuller DQ (2014) Brown top millet: origins and development. In: Smith C (ed) Encyclopedia of Global Archaeology. Springer, New York, pp 1021–1024

Larson G, Piperno D, Allaby R et al (2014) Current perspectives and the future of domestication studies. Proc Nat Acad Sci 111(17):6139–6146

Lin Z, Li X, Shannon LM et al (2012) Parallel domestication of the Shattering1 genes in cereals. Nat Genet 44:720–724

Mabberley DJ (2008) A portable dictionary of plants, their classification and uses, 3rd edn. Cambridge University Press, Cambridge

Majid AA (1989) Exploitation of plants in the Eastern Sahel (Sudan), 5,000–2,000 B.C. In: Krzyzaniak L, Kobusiewicz M (eds) Late prehistory of the Nile Basin and the Sahara. Poznan Museum, Pozan, pp 459–468

Manning K, Pelling R, Higham T et al (2011) 4500-year old domesticated pearl millet (*Pennisetum glaucum*) from the Tilemsi Valley, Mali: new insights into an alternative cereal domestication pathway. J Archaeol Sci 38(2):312–322

Manning K, Fuller DQ (2014) Early millet farmers in the Lower Tilemsi Valley, Northeastern Mali. In: Stevens CJ, Nixon S, Murray MA, Fuller DQ (eds) Archaeology of African plant use. Left Coast Press, Walnut Creek, CA, pp 73–81

Manzo A (1997) Les tessons "exotiques" du Groupe du Gash: un essai d'examen statistique. Cahier de Recherches de l'Institut de Papyrologie et d'Egyptologie de Lille 17(2):77–87

Manzo A (2010) Exotic ceramic materials from Mersa Gawasis, Red Sea. In: Godlewski W, Łajtar A (eds) Between the cataracts. Proceedings of the 11th conference of Nubian studies, Warsaw, 27 August–2 September 2006, vol 2.2. Polish Centre for Mediterranean Archaeology, Warsaw, pp 439–453

Manzo A (ed) (2012a) Italian archaeological expedition to the Sudan of the University of Naples L'Orientale. Report of the 2011 Field Season. Universita degli studi di Napoli "L'Orientale", Naples

Manzo A (2012b) From the sea to the deserts and back: new research in Eastern Sudan. Brit Mus Stud Anc Egypt Sudan 18:75–106. http://www.britishmuseum.org/research/publications/online_journals/bmsaes/issue_18/manzo.aspx

Manzo A (2013) The Italian archaeological expedition to the Eastern Sudan of the Università degli studi di Napoli "L'Orientale". An overview of the 2012 field season. Newsletter di Archeologia CISA 4:253–271

Manzo A (2014a) Preliminary report of the 2013 field season of the Italian archaeological expedition to the Eastern Sudan of the Università degli Studi di Napoli "L'Orientale". Newsletter di Archeologia CISA 5:375–412

Manzo A (2014b) Beyond the fourth cataract. Perspectives for research in Eastern Sudan. In: Anderson JR, Welsby DA (eds) The fourth cataract and beyond. Proceedings of the 12th international conference for Nubian studies. Peeters, Leuven-Paris-Walpole, pp 1149–1157

Manzo A (2015) Preliminary report of the 2014 field season of the Italian archaeological expedition to the Eastern Sudan of the Università degli Studi di Napoli "L'Orientale". Newsletter di Archeologia CISA 6:231–240

Manzo A (in press) The chronology of the transition between the Gash group and the Jebel Mokram group of Eastern Sudan (2nd millennium BC). In: Honneger M (ed) Proceedings of the 13th international conference of the Society for Nubian Studies

Manzo A, Coppa A, Beldados A, Zoppi V (2011) Italian archaeological expedition to the Sudan of the University of Naples 'L'Orientale': 2010 field season. Universita degli studi di Napoli "L'Orientale", Naples

Marshall F, Hildebrand EA (2002) Cattle before crops: the beginnings of food production in Africa. J World Prehist 16:99–143

Motuzaite-Matuzeviciute G, Hunt HV, Jones MK (2012) Experimental approaches to understanding variation in grain size in *Panicum miliaceum* (broomcorn millet) and its relevance for interpreting archaeobotanical assemblages. Veg Hist Archaeobot 21:69–77

Pelling R (2005) Garamantian agriculture and its significance in a wider North African context: the evidence of the plant remains from the Fazzan project. J N Afr Stud 10(3/4):397–412

Phillips S (1995) Poaceae (Gramineac). In: Hedberg I, Edwards S (eds) Flora of Ethiopia and Eretria, vol 7. The National Herbarium, Biology Department, Addis Ababa University, Uppsala University, Department of Systematic Botany, Adis Ababa and Uppsala

Pock Tsy J-ML, Lumaret R, Mayne D et al (2009) Chloroplast DNA phylogeography suggests a West African centre of origin for the baobab, *Adansonia digitata* L. (Bombacoideae, Malvaceae). Mol Ecol 18:1707–1715

Rowley-Conwy P, Deakin WJ, Shaw CH (1997) Ancient DNA from archaeological sorghum (*Sorghum bicolor*) from Qasr Ibrim, Nubia. Implications for domestication and evolution and a review of the archaeological evidence. Sahara 9:23–34

Sadr K (1988) The development of nomadism: the view from ancient North East Africa. Unpublished PhD dissertation, Southern Methodist University

Sadr K (1991) The development of nomadism in ancient Northeast Africa. University of Pennsylvania Press, Philadelphia

Shiner JL (1971) The prehistory and geology of Northern Sudan. Report to the National Science Foundation Grant GS 1192, Dallas

Shinnie PL, Anderson J (2004) The capital of Kush 2: Meroe excavations, 1973–1984, Meroitica 20. Harrassowitz, Wiesbaden

Shipton C, Helm R, Boivin N et al (2013) Intersections, networks and the genesis of social complexity on the Nyali Coast of East Africa. Afr Archaeol Rev 30:427–453

Snowden JD (1936) Cultivated races of *Sorghum*. Adlard and Sons, London

Song J, Zhao Z, Fuller DQ (2013) The archaeological significance of immature millet grains: an experimental case study of Chinese millet crop processing. Veg Hist Archaeobot 22(2): 141–152

Stemler ABL (1990) A scanning electron microscopic analysis of plant impressions in pottery from the sites of Kadero, El Zakiab, Um Direiwa and El Kadada. Archéologie du Nil Moyen 4:87–105

Wasylikowa K (2001) Site-75-6: Vegetation and subsistence of the Early Neolithic at Nabta Playa, Egypt, reconstructed from charred plant remains. In: Wendorf F, Schild R (eds) Holocene settlement of the Egyptian Sahara, The Archaeology of Nabta Playa, Vol 1. Kluwer Academic, New York, pp 544–591

Wasylikowa K, Schild R, Wendorf F, Krolik H, Kubiak-Martens L, Harlan JR (1995) Archaeobotany of the Early Neolithic site E-75-6 at Nabta Playa, Western Desert, South Egypt (preliminary results). Acta Palaeobot 35:133–155

Wiersema JH, Dahlberg J (2007) The nomenclature of *Sorghum bicolor* (L.) Moench (Gramineae). Taxon 56(3):941–946

Winchell F (2013) The Butana group ceramics and their place in the Neolithic and Post-Neolithic of Northeast Africa. Cambridge Monographs in African Archaeology 83, British Archaeological Report Series 2459. Archaeopress, Oxford

Winchell F, Stevens CJ, Murphy C, et al (2017) Evidence for sorghum domestication in fourth millennium BC Eastern Sudan: spikelet morphology from ceramic impressions of the Butana Group. Curr Anthropol 58(7):673–683

New Evidence on the Development of Millet and Rice Economies in the Niger River Basin: Archaeobotanical Results from Benin

Louis Champion(✉) and Dorian Q. Fuller

Institute of Archaeology, University College London, London, UK
championlouis@gmail.com, d.fuller@ucl.ac.uk

Abstract. The Niger River is second only to the Nile in length in Africa, and is host to dense populations of agriculturalists that supported in historical times states such as the kingdoms of Songhay and Mali. This is also the region to which the origin of the Niger-Congo language family, including its Bantu offshoot is attributed. Despite this, archaeobotanical evidence for the development of agricultural systems based on both ancient West African crops, like *Pennisetum glaucum* (L.) R. Br., *Vigna unguiculata* (L.) Walp. and *Oryza glaberrima* Steud., and crops introduced to the Niger Basin, such as *Sorghum bicolor* (L.) Moench. and *Gossypium* L. sp. has remained limited. In particular the role of multiple crop systems, that included both the wet (rice) and the dry (millets), has not been directly documented archaeobotanically. The present paper presents new archaeobotanical results from 12 sites in Benin that suggest that the rise of larger populations and population centers, like the urban site of Birnin Lafiya, developed only once agriculture diversified beyond pearl millet cultivation to include multiple cereals, as wet rice. The 12 sites are split in four time periods. Flotation results indicate that sites of the first phase (first millennium BC) were dominated by pearl millet, but included sorghum and cowpea. However by the second period (300–900 AD), rice dominated samples, correlated with increasing urbanism, a pattern congruent with existing evidence from Mali. In addition, we report evidence that probable fonio (*Digitaria* cf. *exilis* (Kippist) Staph.) also appeared first in this era of diversification, calling into question previous inferences about the antiquity of these West African millets. The third phase, 900–1400 AD, is characterized by an increase of pearl millet and a decrease of African rice. During the last time period, 1400–1950 BC, we notice a disappearance of rice and a diminution of pearl millet and sorghum. Also, the utilizations of tree fruit such as baobab (*Adansonia digitata* L.), oil palm (*Elaeis guineensis* Jacq.), and African olive (*Canarium schweinfurthii* Engl.) are in constant evolution since the second period. We conclude that agricultural diversification helped to promote urbanization and state formation in the Niger River basin, and that diversification included both use of wetter environments for rice and more marginal dry environments for millet and sorghum.

Keywords: Cotton · Domestication · Oil palm · *Oryza glaberrima* Sorghum · Urbanism

A. M. Mercuri et al. (Eds.): *Plants and People in the African Past*,
https://doi.org/10.1007/978-3-319-89839-1_23

Introduction: Previous Work on West Africa Agricultural Origins

That western Africa was host to indigenous domestications, was first argued by Murdock (1960), who essentially saw this area as a center of origin to be added to the maps of Vavilov (1950). Botanical studies through the 1960s–1970s, by Harlan and colleagues, clarified the distribution of wild progenitors of many African crops, and highlighted a distinction between three ecological zones, the Guinean forests, the west Sudanian savanna and the Sahelian Acacia savanna (see zonation in Fig. 1). The southern Guinean forest zone, or its margins, was source region of tubers (e.g. *Dioscorea cayenensis* Lam., *D. dumetorum* (Kunth) Pax), oil palm (*Elaeis guineensis*), and a more northerly open woodland savanna zone (Sudanian) was the source area of African rice (*Oryza glaberrima*), cowpea (*Vigna unguiculata*), Bambara groundnut (*Vigna subterranea* (L.) Verdc.), and the fonio millets (*Digitaria exilis* (Kippist) Stapf *Brachiaria deflexa* (Schumach.) C.E. Hubb. ex Robyns). Pearl millet origins were inferred to lie beyond this zone, further north in the Sahelian zone (Harlan 1971, 1992; Harris 1976; De Wet 1977; Fuller and Hildebrand 2013). Comparative ethnography and historical linguistics has inferred an indigenous West African vegeculture complex, focused around yams (*Dioscorea cayanensis, D. dumentorum*) but including several forest tree fruits, such as cola seeds (*Cola* spp.), African olive (*Canarium schweinfurthii*), African plum (*Dacryodes edulis* (G. Don) H.J. Lam.), and Mobola plum (*Parinari curatellifolia* Planch. Ex Benth.). These trees and tubers have a deep antiquity among Niger-Congo speaking groups, at least as far back as Proto-Benue-Congo, several proto-language grades prior to the emergence of proto-Bantu, and thus inferred to go back to perhaps 2000–3000 BC (Blench 2006; Bostoen 2014). Unfortunately there remains no archaeological hard evidence, such as from charred yam

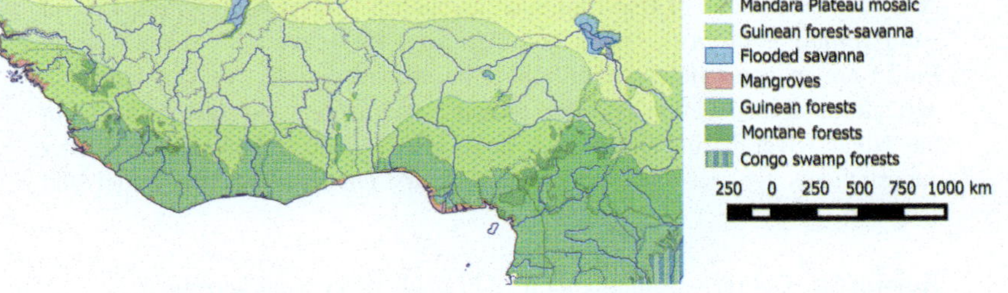

Fig. 1. Map of modern potential vegetation zones of western Africa. The zonation this map derived from GIS-shape file of Terrestrial Ecosystems of the World (Olson et al. 2001; World Wildlife Fund 2017), with zones grouped into broader categories in line with White (1983).

parenchyma or starch on ancient tools, that confirms the antiquity of yam cultivation. Among the tree fruits, use of oil palm is widely documented on archaeological sites back to the Late Stone Age, ca. 3000 BC (D'Andrea et al. 2006), with use of *Canarium* evident from 4000–5000 BC at a rock shelter in Cameroun (Lavachery 2001).

What is evident archaeologically is the penetration into the forest zone, or at least in its northern margins, by cultivators of pearl millet, which has its origins further north in the Sahel zone of Mali and/or Mauretania. Recent evidence from impressions in ceramics and a directly AMS-radiocarbon dated grain from the Tilemsi Valley suggest that domestication was well underway by 2500 BC, and completed by 2000 BC at the latest (Manning et al. 2011; Manning and Fuller 2014). Further west in Mauretania, impressions of domesticated pearl millet occur in pottery by 1700–1500 BC (Fuller et al. 2007), in what appears to be a distinct cultural tradition, therefore suggesting two regional trajectories to pearl millet cultivation in the Sahel zone (MacDonald et al. 2009; Manning and Fuller 2014). Elsewhere in Mali south of the Niger River, pearl millet dates back to c. 1800–1600 BC at Ounjougou (Ozainne et al. 2009; Eichhorn and Neumann 2014) and perhaps at the poorly-dated Windé Koroji (MacDonald 1996). While this plausibly represents the migration of millet cultivating pastoralists (plausibly Nilo-Saharan speaking, cf. Blench 2006), further south in the woodland savanna we can imagine diffusion of food production into new cultural groups. Here pearl millet is found in the Kintampo culture in Ghana, where cowpea and oil palm are added to the subsistence system, thus integrating forest and grassland subsistence systems (D'Andrea and Casey 2002; D'Andrea et al. 2006; Logan and D'Andrea 2012).

Elsewhere in the savanna zone pearl millet cultivation and villages become established over the course of the second millennium BC, from Burkina Faso to Lake Chad (e.g. Neumann et al. 2001; Höhn and Neumann 2012; Zach and Klee 2003). It is possible these go back to the early second millennium BC near Lake Chad (Klee et al. 2004). Finds in Senegal and further south in central Nigeria (the Nok Culture) indicate pearl millet cultivation by the first half of the first millennium BC (Murray and Déme 2013; Kahlheber et al. 2009), with penetration further south, in the central African rainforest zone, by the fourth century BC (Kahlheber et al. 2014). Linguistic evidence suggests that spades and axes were the most important implements for creation and maintenance of agricultural plots (see Blench 2014).

What is striking from these data is the absence of early finds of African rice (*Oryza glaberrima*), fonio millets (*Digitaria exilis*, *Brachiaria deflexa*), and sorghum, all grain crops important in the traditional agricultural diversity of western Africa. The present contribution reports archaeobotanical findings from northern Benin, along the Niger River, which provides new evidence for the diffusion of such crop diversity into West African savanna agriculture. In addition, we report evidence for adoption of cotton cultivation, and use of a number of tree fruits.

Archaeobotanical Studies from the Crossroads of Empire Project

We summarize here archaeobotanical data from twelve archaeological sites excavated as part of "Crossroads of Empires" project, directed by Anne Haour, in northern Benin (Fig. 2). The project aims to document settlement patterns and materials in the Dendi region along the Niger River, to shed new light on the development of craft production and polities in this region. Four field seasons were carried out between 2011 and 2014 (Haour et al. 2011, Haour 2013), including investigation into oral history, and studies

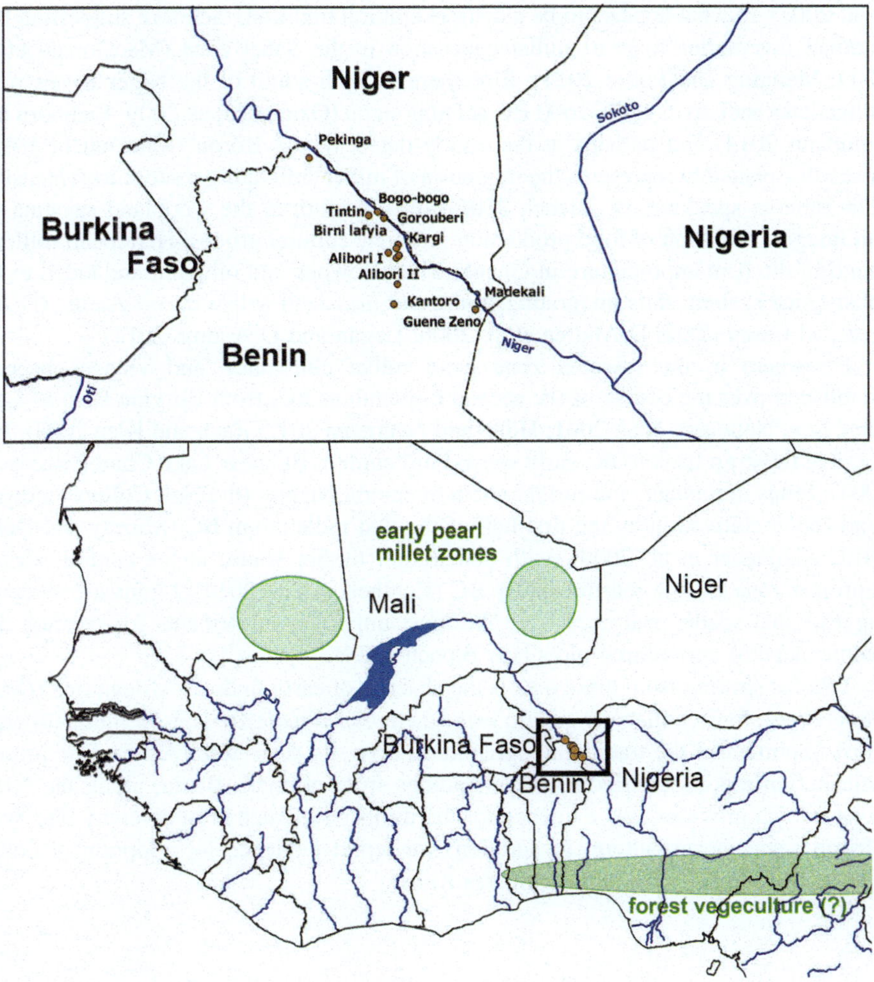

Fig. 2. Map showing the location of sites with archaeobotanical evidence discussed in this paper, and in relation to the wider western African region. Early zones of pearl millet cultivation and hypothetical West African vegeculture zone indicated (after Fuller and Hildebrand 2013; Manning and Fuller 2014).

of iron working (Robion-Brunner et al. 2015). Archaeobotanical research has been an important element in the research program to provide a better understanding of the development of the agricultural economy in this region over time, how it relates to broader regional patterns of agriculture and locally to the support of larger population centers and social complexity.

The 12 archaeological sites are mainly located in the higher reaches of the Niger River valley but still lie relatively close to the flood plain. The sites are located in the Sahelian acacia savanna area. The large alluvial floodplain offers a rich environment for settlement and agro-pastoral activities and is bordered by semi-arid woodland savanna. The climate of this area is characteristic of the West African Sahara/Sahel sub-desert area with a low level of rainfall varying from 1000 mm to less than 100 mm. There are two flood peaks during the same inundation event (Fig. 3). The first is due to the local rainfall around the end of August creating a local flood peak. The second is attributable to the Malian flood peak progression, delayed by the slow release of water from the Inland Niger Delta, which acts like a seasonal sponge, generating a second peak around February (Gallais 1959, 1967; Rouch 1950; McIntosh 1995; Van Driel 2001). During the dry period (after harvest), the flood plains are used for pastoralism and during the humid season (after the flooding) used for "décrue" agriculture (e.g. Sorghum cultivation, *Sorghum bicolor*) or rice cultivation during the flooding. The flood system can be considered as a natural hydraulic system that depends on the rising and falling phase ("crue" and "décrue") from the natural flooding of the river. Usually the principal crop plant before the flood is the rice (*Oryza glaberrima* and *Oryza sativa*) but in the Inner Niger Delta the agricultural exploitation of "décrue" is turned to millet (*Pennisetum glaucum*) and sorghum cultivation (Harlan and Pasquerleau 1969; Harris 1976).

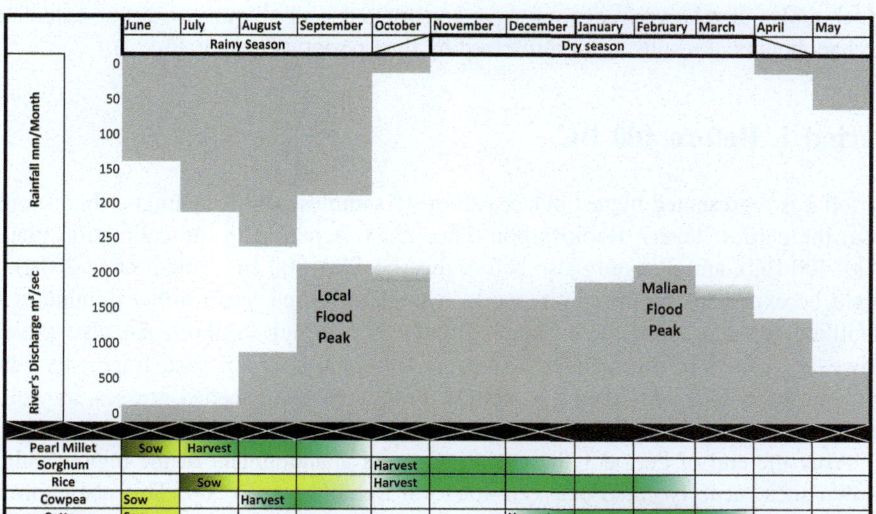

Fig. 3. Schematic representation of seasonality in the Dendi, including approximate distribution of average rainfall, river flood levels, and sowing and harvesting season. (Based on the sources noted in the texts and interviews with farmers conducted by LC.)

All 12 sampled sites were settlements. The oldest site, Alibori SIII, is the only one with Neolithic remains (i.e. with ceramics but no metal), dating from 2200–700 BC. Five other sites (Birnin Lafiya, Tintin, Alibori SII, Kantoro and Pekinga) relate to two subsequent time periods, dated from the first century to the 14th century AD. All five sites are small tells (settlement mounds). Other sites (Kozoungou, Geune Zeno, Bogo-bogo, Gourouberi, Madekali, Kargi) date to the fourth period, the last part of the second millennium AD, and were excavated as the precursors of current villages, as identified through oral history. Selected charcoal and seeds were submitted for AMS-radiocarbon dating to help establish a chronology.

Most of the excavated contexts are midden fill deposits, but two of the largest sites also produced contexts within domestic structures. The site of Birnin Lafiya, which, in light of its large area (dozens of acres) and the quantity and quality of artefacts found on its surface, is central to understanding urbanization in this region. Excavations revealed a portion of a house, a circular room paved with potsherds, dating to the 11th–13th century AD (Haour et al. 2016). The site of Tintin is also a settlement mound with evidence for houses, with seven separate potsherd pavements excavated (Champion and Haour 2013).

Over three field seasons (2012–2014) 150 archaeobotanical samples were collected (comprising 2452 litres of sediment from 29 excavation units, across the 12 sites), of which 83 have been analyzed and are summarized here. Sediments were processed manually by wash-over bucket flotation, through 0.25 mm mesh. After sorting, identification of charred seeds was made by comparing taxa with specimens in the modern reference collection housed at the Institute of Archaeology, University College London, with a focus on recognizable crops and major economic taxa. Many small seeds and seed fragments require further efforts for full identification, so results presented here should be seen as preliminary, in addition to ~70 samples still to be sorted and studied. For all the samples analyzed, whole vegetal items were recorded by count in a list of taxa, from which tables of relative frequency were constructed, and ubiquity could be calculated. Archaeobotanical results are summarized by time period (Table 1; Fig. 4).

Period 1. Before 400 BC

Period 1 is represented by test pits at Alibori (5 samples) and Kozoungou (one sample, from the earliest layer). Radiocarbon dates put this period in the calibration plateau (800–400 BC), and plausibly also before this, i.e. 900–800 BC (Haour et al. 2016). As might be expected from previous work in western Africa, pearl millet dominates the identified remains from these samples, but cowpea and sorghum are also present. However, ~60% of the recovered remains are small seeds or seed fragments which have not been fully identified but include a range of small-seeded grasses, as well as fragments of probable nutshell that have yet to be identified.

After the end of Period 1 there is an apparent abandonment of the region with no known sites securely dated between 400–300 BC and c. AD 300. This abandonment partly corresponds to the regional hiatus identified by Breunig and Neumann (2002) in sites from Burkina Faso as corresponding to an era of increased aridity (also Höhn and Neumann 2012). There is also evidence in southern Cameroun of an opening up of northern parts of the central African rainforest, starting c. 500–300 BC, which has been

Table 1. A summary of the archaeobotanical evidence from the Dendi region, Benin, in four chronological phases. Total numbers of identified specimens have been summed across sites from each phase. Sample by sample primary data in Champion and Fuller (in press).

Phases	Sites	Samples (n)	Volume (litres)	*Pennisetum glaucum*				*Oryza glaberrima*				Indet seeds	Indet fruits	*Sorghum bicolor*	*Adansonia digitata*	*Elaeis guineensis*	*Canarium schweinfurthii*	*Vigna unguiculata*	*Gossypium* sp.	*Capparidaceae*	*Digitaria* sp.	*Panicum* sp.	*Brachiaria* sp.	*Acacia* sp.	*Moraceae* sp.	
				Spikelet	involucre	Seed	Total	Lemma	Spikelet	Seed	Total															Total
Phase I	ALB-SIII KOZ-SII	6	120	0	1	20	21	0	0	0	0	183	53	15	0	0	0	7	0	0	0	0	0	0	0	269
Phase II	BLAF SIV,VIII,IX,XI,XIII	18	276	0	81	288	369	48	1144	10	1202	218	48	101	7	0	16	1	2	0	0	1	0	0	0	1965
Phase III	BLAF SIII, SIX, SX,SXIII TTK-SI PEK-SI KLI-RCI KRO-SII	47	752	0	245	2153	2398	14	879	9	902	678	310	275	101	44	10	13	2	0	3	5	1	0	0	4742
Phase IV	GOB-SI ROU-SI KOZ-SI,II GOG-SI KGI-SI KLI-SI	12	120	81	190	35	306	0	0	0	0	281	148	22	29	57	16	0	49	15	5	0	0	1	1	930
Total		83	1268	81	517	2496	3094	62	2023	19	2104	1360	559	413	137	101	42	21	53	15	8	6	1	1	1	7906

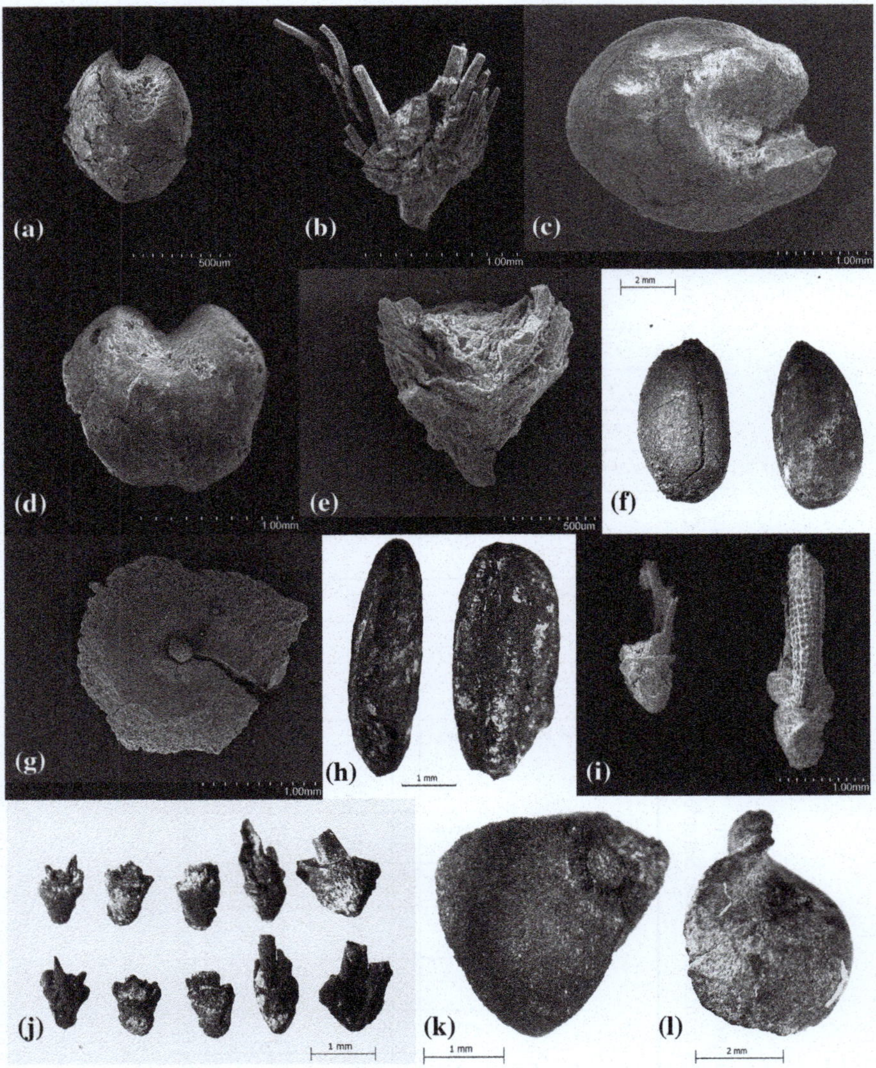

Fig. 4. Examples of identified crop and economic species remains from Dendi region archaeological sites: a. Grain of *Pennisetum glaucum* from Birnin Lafiya (BLAF SXI C17 250–260); b. Involucre base of *P. glaucum* from Birnin Lafiya (BLAF SIX C12 120 130). c. Grain of *Sorghum bicolor*, race bicolor type, from Birnin Lafiya (BLAF SIX C11). d. Grain of *Sorghum bicolor*, race guinea type, from Birnin Lafiya (BLAF SXI C17 250–260). e. Spikelet base of *Sorghum bicolor*, probable dehusking waste, domesticated race bicolor (BLAF SIX C11). f. Seeds of cotton (*Gossypium* sp.) from Madekali (KLI-14-SI Bot3); g. Seed fragment (funicular "cap") from *Gossypium* from Gourouberi (GOB SI 60–70); h. Grain of rice (*Oryza glaberrima*) in dorsal and lateral view from Birnin Lafiya (BLAF SIV Pit 3); i. Spikelet bases and husk of *O. glaberrima* from Birnin Lafiya (BLAF SXI C17 250–260); j. Spikelet bases of *O. glaberrima* from Tintin (TTK C19 355 cm); k–l. Seed fragments with hilum of baobab (*Adansonia digitata*) from Madekali and Kozungu (KLI 14-RC1 Unit 6; KOZ-14-SI-80–100 cm).

attributed to increasing seasonality with a more pronounced dry season further south which in turn promoted the expansion of savanna farming systems (Ngomanda et al. 2009; Neumann et al. 2012a, 2012b; Kahlheber et al. 2014). It is therefore plausible that abandonments of the early farming sites in Benin correspond to a period of regional environmental change and cultural readjustment. Outside of the Dendi area, this period corresponds to the emergence of urban sites at Jenne-Jeno and Dia in the Inner Niger Delta in Mali, and Gao in the Middle Niger. Presently, the earliest evidence of African rice (*Oryza glaberrima*) is from these sites (Fuller 2000, McIntosh 1995; Murray 2007a; Gallagher and McIntosh 2015).

Period 2. From AD 300 to 900

For this period, archaeobotanical results are derived from 18 samples from 5 excavation units at the urban site of Birnin Lafiya. These contexts probably represent the initial development phase of the settlement of this site. African rice (*Oryza glaberrima*) dominates the plant assemblage (more than 60% of all remains), while sorghum (<5%), pearl millet (<20%), and cowpea remain present (Fig. 4). The rice remains are dominated by charred spikelet bases, the remains of routine on-site dehusking by-products; these spikelet bases are predominantly of the non-shattering morphology of domesticated rice (e.g. Fig. 4i), which can be separated from wild rices by criteria that parallel those of Asian rice (Fuller et al. 2009). Both baobab (*Adansonia digitata*) and African olive (*Canarium schweinfurthii*) fruits are also present. It is tempting to attribute part of the success of urbanization to the highly productive cultivation of wet, riverine environments with rice, as well as the use of dry, rainfed areas for millet and sorghum. This therefore suggests that rice cultivation spread down the Niger River after its earlier establishment in the Inland Delta and around the Niger bend also associated with the rise of the first urban centres in those regions.

Period 3. From AD 900 to 1400

Several sites from the third period have been sampled, and a larger number of samples (n = 47) has been studied from this period. During this period the Dendi area was more densely occupied than in previous periods. Birnin Lafiya seems to reach its largest extent (c. 30 ha) and more intensive occupation (Haour et al. 2016). Other sedentary sites also emerge in this period, including Pekinga, Tintin (c. 1 ha), Kantoro (c. 10 ha), and Madekali. With the exception of Pekinga, all those sites gave evidence of African rice, but in a smaller proportion than during period 2 (around 20%). The pearl millet frequency is higher (around 50%), while the sorghum proportion stayed similar (around 5%). A few grains of *Digitaria* which could represent fonio (*D. exilis*) have been found, and a single occurrence of *Brachiaria* sp. Further comparative studies are needed to confirm species level identification in *Digitaria* or *Brachiaria* in this region. For the first time, a very small proportion of oil palm is noted (1%) and there are also remains of baobab (2%) (Fig. 4k–l).

This suggests expansion in the consumption of pearl millet at the transition from the first to second millennium AD. Given that the wetland area available for rice cultivation is highly constrained, we suggest that the area under wetland cultivation could not be increased from period 2 but that savanna agriculture expanded over more of the rainfed land away from the rivers.

As part of agricultural diversification in drier areas, use of marginal and poor soils is suggested by the presence of fonio (*Digitaria* cf. *exilis*). Previously, these have been suggested to be very ancient crops, with a widespread and patchy modern distribution representing relict populations (e.g. Portères 1976; Blench 2006). Alternatively, we hypothesize that the fonios are later secondary domesticates, taken into cultivation to complement pearl millet and sorghum as risk-buffering crops for poor soils. Much as oat (*Avena sativa*) and rye (*Secale cereale*) were domesticated in Europe from the Late Bronze Age/Iron Age to suit some more marginal regions (Zohary et al. 2012), we suggest that the fonio millets were taken up secondarily, probably more than once, to suit the more marginal agricultural conditions across the west African savannas. In this model, the evidence for fonio in the Mema of region of northwestern Mali (Takezawa and Cisse 2004) from the first half of the first millennium BC may be seen as a local development of cultivation in the interstices of established pearl millet agriculture and perhaps of early rice of the Inland Niger Delta.

There is also some evidence that suggests, rice cultivation may have started to be abandoned in some areas. In the site Kantoro rice is only found in the lower stratigraphic contexts. In the last occupation of Birnin Lafiya, houses of test-pits SIII and SX, no rice remains were recovered. Thus over the course of the second millennium BC prior to AD 1400 rice cultivation seems to have declined in the Dendi region.

Period 4. From AD 1400 to 1950

The final period, which includes colonial and premodern times, appears quite different from the others. Indeed, even if pearl millet is still to be the dominant grain crop of this period, evidence for trees utilization, such as baobab and oil palm, increase. Surprisingly, the presence of crops such as African rice and pearl millet decrease dramatically in this period, with rice ultimately disappearing from the assemblages. Whether or not this increase in trees represents a shift towards wetter conditions should be considered. As many older sites were abandoned in this period, it is possible there was a degree of reforestation, including more elements from the rainforest complex. However, we favor an alternative socioeconomic explanation, however, that the decrease in grain crops and the increase in tree products represents increasing importance of trade as an element in food supply. In this regard more oleaginous nuts could have been traded from the south. Certainly by this period the slave trade may have impacted labor availability and increased the potential to buy food with money, which would have necessitated less on-site crop-processing and decreased the visibility of grain crop evidence.

In this regard it is interesting that samples of this period give evidence of cotton (Fig. 4f–g), a classic cash crop which is labor intensive to process. Production of cotton would have provided a source of wealth also with which to buy cereal grain, perhaps from regions further north, and oleaginous fruits, perhaps from regions to the South.

The major difference of the archaeobotanical assemblage of Period 4 suggests that this small study region needs to be seen increasingly as part of a world system and less in terms of local agricultural subsistence production.

West African Agricultural Development: An Interpretive Summary

The new data from the Dendi region contributes to filling in the broader patterns of agricultural evolution in West Africa. We have reviewed the available data from 31 sites that are older than 100 BC (Fig. 5), and some 64 sites from the first or second millennia AD (Fig. 6). Considered in terms of broad vegetation belts (as per Fig. 1), we can summarize distinction by inter-related histories of crop use across the major biomes

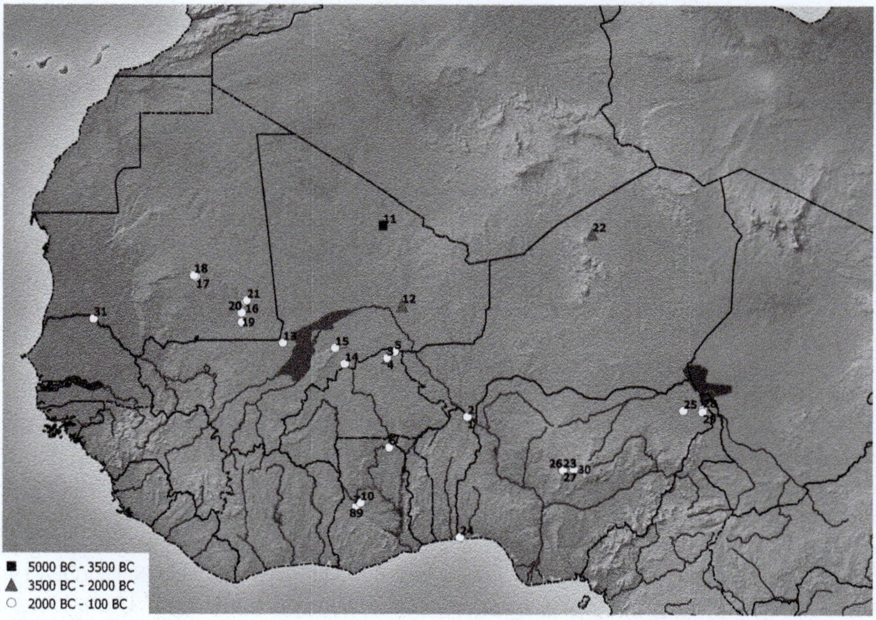

Fig. 5. Map of West African sites with archaeobotanical remains older than 2000 c y BP, sites numbered: 1, 2 Alibori SIII, Kozungu SII (the authors). 3 Corcoba (Kahlheber 2004); 4, 5 Oursi West, Ti-n-Akof (Kahlheber and Neumann 2007). 6 Birimi (D'Andrea et al. 2001); 7, 8, 9 Boase: B4C, B5C, B6B (D'Andrea et al. 2007); 10 Kintampo: K6 (Stahl 1985); 11 AZ22 (the authors, unpublished); 12 Karkarichinkat (Manning et al. 2011); 13 Kolima Sud Est (Takezawa and Cisse 2004); 14 Ounjougou: Varves West (Ozainne et al. 2009); 15 Winde Koroji (MacDonald 1996); 16 Dhar Oualata (Amblard and Pernes 1989); 17, 18 Dhar Tichitt (Munson 1976); 19, 20 Djiganyai, Oued Bou Khzama (Fuller et al. 2007); 21 Oued Chebbi (Amblard and Pernes 1989); 22 Adrar Bous (Shaw 1976); 23 Akura (Kahlheber et al. 2009); 24 Apa (Orijemie and Sowumni 2014); 25 Gajiganna (Klee et al. 2004); 26, 27 Janjala, Janruwa (Kahlheber et al. 2009); 28 Kursakata (Zach and Klee 2003); 29 Mege (Bigga and Kahlheber 2011); 30 Ungwar Kura (Kahlheber et al. 2009); 31 Walade (Murray and Déme 2013).

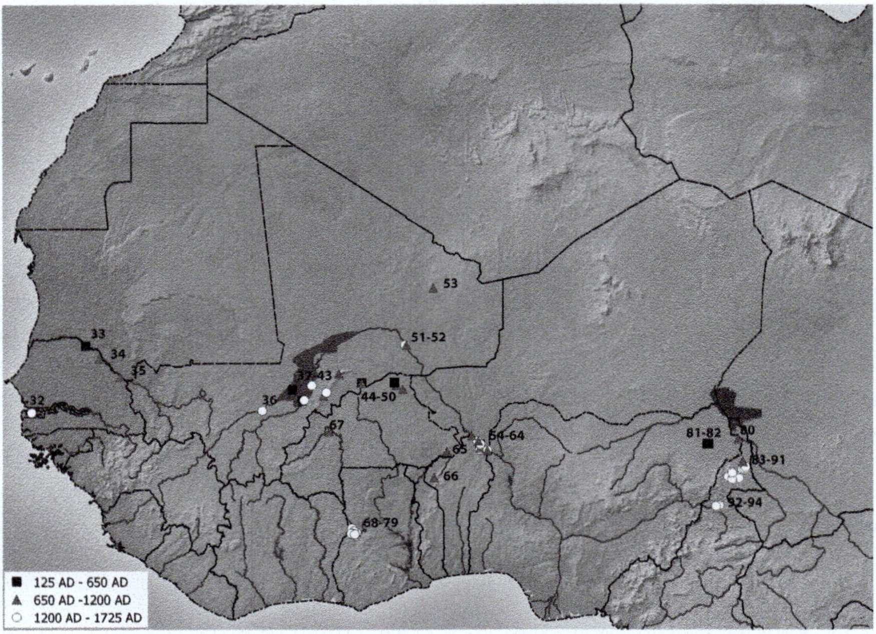

Fig. 6. Map of West African Sites with archaeobotanical crop remains younger than 2000 c y BP, sites numbered: 32 Juffure (Gijanto and Walshaw 2014); 33 Cubalel (Murray et al. 2007); 34 Sincu Bara (Murray 2008); 35 Arondo (Gallagher 1999); 36 Sorotomo (Morris 2013); 37 Dia-Shoma (Murray 2007b); 38 Jenne-jenno (McIntosh 1995); 39–40 Toguere Galia, Toguere Doupwil (Bedaux et al. 1978), 41 Tellem (Bedaux 1972), 42 Tongo Maare Diabel (MacDonald 1996, under study by LC); 43 Sadia (Huysecom et al. 2012, under study by LC); 44–50 Oursi, Oursi hu-beero, Kissi 22 & 40, Kolel Nord, Saouga, Corcoba (Kahlheber 2004); 51 Gao Saney (Gallagher App. V. in Cisse 2011 PhD); 52 Gao Gadei (Fuller 2000); 53 Essouk (Nixon et al. 2011); 54–64 Birni Lafia, Madekali Road, Madekali, Kantoro, Pekinga, Tintin, Guene-Zeno, Bogo Bogo, Kargui, Goroberi, Kozungu SI (Champion and Fuller, this paper); 65 MAS541 (Dueppen and Gallagher 2013); 66 Yohongu (Kahlheber and Neumann 2007); 67 Kirikongo (Dueppen 2008); 68–79 Banda 13, 27, 41, Kuulo Kataa, Ngre Kataa, Makala Kataa, Bui Kataa, A-212, B-112 (Logan 2012); 80 Daima (Connah 1981); 81,82 Elkido North, Dorota (Magnavita 2002); 83–91 Mongossi (Marliac 1991), Jiddere Saoudjo (Otto 1996), Kayam, Salak (Otto 1998), Tchere, Balda Tagamre (Otto 1996), Goray (Marliac 1991, Otto and Delneuf 1998), Louggero, Mowo (Delneuf and Otto 1995); 92–94 Be, Sumpa, Douloumi (David 1976, 1981);

of West Africa (Fig. 7). Early grain farming in West Africa was based on the spread of pearl millet among agropastoralists. Over time farming systems diversified and supported increasing sedentism with additional grains crops, grain legumes (especially *Vigna unguiculata*), and various tree crops, mostly with oily seeds. At present, there is no hard archaeological evidence to determine the antiquity of tropical West African vegeculture, although this plausibly preceded cereals in wetter forest regions. Sorghum, from eastern savannas of northern Sudan (see Fuller and Stevens, 2018), diffused into West Africa starting early in the first millennium BC. However, in the Dendi region, and parts of the savanna of Burkina Faso, at least some settlements were abandoned in

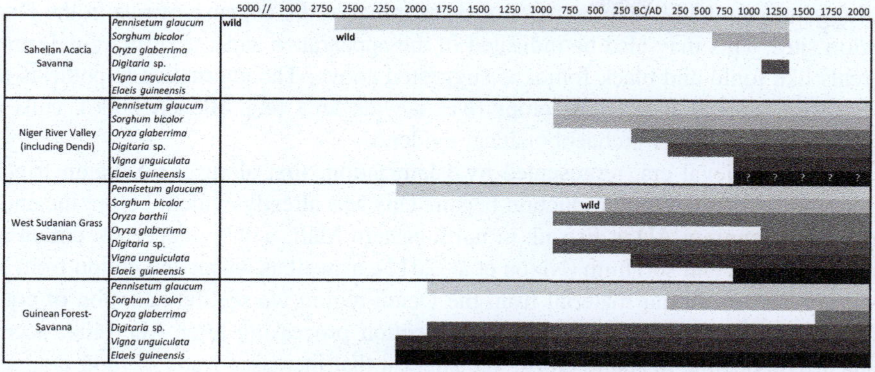

Fig. 7. Schematic timeline of the antiquity of major western African crops based on current archaeobotanical evidence across four ecological zones (see Fig. 1). Evidence from sites plotted in Figs. 5 and 6. Question marks on occurrence of *Elaeis* in the Niger River valley account for the possibility that these were imported from further south.

the middle to late part of the first millennium BC, perhaps suggesting that Neolithic cultivation regimes failed in some regions in response to climate shifts towards a more pronounced dry season. Subsequent re-establishment of agriculture tended to be more diverse, with a great range of cultivated taxa; this diversity can be suggested to have increased resilience by offering buffering against poor years in any one crop.

Another key element of agricultural diversification was the domestication and spread of African rice (*Oryza glaberrima*). While domesticated rice was present in the Inland Niger Delta before 400 BC, and probably by 800 BC (Murray 2007a), the earlier domestication of this species must have taken place in the 1000 or 1500 years prior to this, perhaps further up the Niger River or other tributary water ways to its south. Recent genomic analyses on *O. glaberrima* and *O. barthii* suggests that the maximum bottleneck associated with the domestication of *O. glaberrima* took place 1000–2000 BC (Meyer et al. 2016). Modern genetic differentiation within *O. glaberrima* also suggests that origins of the domesticated form, prior to geographical diversification should be placed somewhere between the Guinea highlands and the Inland Niger Delta, such as in western Mali (Meyer et al. 2016). Nevertheless, from the final centuries BC rice cultivation and urbanism increased in the Inland Niger Delta and began to spread downstream. The recovery of rice spikelet bases in our samples from Benin highlights the potential to track the domestication process for African rice in empirical evidence, once samples of suitable antiquity are available in the upper Niger River region or the Inland Niger Delta. Rice cultivation and urban growth took off in the Dendi area from c, AD 300. At present there is no archaeological evidence for the dispersal of rice westwards into the Guinea highlands and coast and its subsequent diversification into southern and northern Guinea groups, suggested by geographical genetic structure (Meyer et al. 2016), as well as by historical linguistic data (Fields-Black 2008).

After this period, as populations grew, agricultural production grew through increasing expansion of cultivated area and infilling of rainfed and marginal soils. This is suggested by the relative increase of pearl millet at the expense of rice in the Dendi region sites. This may also be indicated in the appearance and use of marginal, minor cereals like fonio and black fonio, as suggested above. The alternative hypothesis that these were ancient indigenous crops over a wide area (e.g. Blench 2006), calls for testing through further archaeobotanical evidence.

In the medieval era, represented by Islamization, first along the Saharan fringes, grain trade and cash crop production began. This was already evident before the end of the first millennium AD at Essouk in north-eastern Mali, with evidence for cotton and imported wheat and sorghum (Nixon et al. 2011), as well as wheat and cotton from Dia (Murray 2007b). In our material from the Dendi region we see the adoption of cotton production after AD 300 and larger scale cotton processing after AD 1400, accompanied by a decline in staple crops. We suggest that increased trade of food grains, as well as of forest fruits into the Dendi region was balanced by production and export of cotton products. Cash crops production and grain trade became increasingly enmeshed into the economic exploitation of labor, including enslavement and slave trade. Slave and food trade was intensified by the increased frequency of European trade ships that sought both food supplies and slaves from the West African coast to support European wealth creation, for example via sugar starting 15th century AD (e.g. Mintz 1985; Carney and Rosomoff 2011).

Acknowledgements. Fieldwork and archaeobotanical samples were carried out as part of the European Research Council funded "Crossroads of Empire Project" awarded to Dr. Anne Haour, University of East Anglia (ERC Grant Agreement 263747). Archaeobotanical laboratory work has been carried out by LC as part of PhD research project at University College London, while additional laboratory research, comparative databases and preparation of this paper has been supported by to the "Comparative Pathways to Agriculture" project, a European Research Council advanced grant awarded to DQF (ERC Grant Agreement 323842).

References

Amblard S, Pernes J (1989) The identification of cultivated pearl millet (*Pennisetum*) amongst plant impressions on pottery from Oued Chebbi (Dhar Oualata, Mauritania). Afr Archaeol Rev 7:117–126

Bedaux RMA (1972) Tellem, reconnaissance archéologique d'une culture de l'Ouest africain au moyen âge: recherches architectoniques. J S A 42(2):103–185

Bedaux RMA, Constandse-Westermann TS, Hacquebord L et al (1978) Recherches archéologiques dans le Delta intérieur du Niger (Mali). Extrait de Palaeohistoria 20:91–220

Bigga G, Kahlheber S (2011) From gathering to agricultural intensification: archaeobotanical remains from Mege, Chad Basin, NE Nigeria. In: Fahmy AG, Kahlheber S, D'Andrea AC (eds) Windows on the African past, Current approaches to African archaeobotany. Africa Magna, Frankfurt, pp 19–66

Blench RM (2006) Archaeology, language and the African past. Altamira Press, Lanham

Blench RM (2014) African agricultural tools. In: Stevens CJ, Nixon S, Murray MA, Fuller DQ (eds) Archaeology of African plant use. Left Coast Press, Walnut Creek, pp 243–258

Bostoen K (2014) Wild trees in the subsistence economy of early Bantu speech communities: a historical-linguistic approach. In: Stevens CJ, Nixon S, Murray MA, Fuller DQ (eds) Archaeology of African Plant Use. Left Coast Press, Walnut Creek, pp 129–140

Breunig P, Neumann K (2002) Continuity or discontinuity? The 1st millennium BC-crisis in West African prehistory. In: Lenssen-Erz T (ed) Tides of the Desert. Contributions to the archaeology and environmental history of Africa in honour of Rudolph Kuper. Africa Praehistorica, vol 14. Heinrich-Barth-Institut, Köln, pp 491–505

Carney JA, Rosomoff RN (2011) In the shadow of slavery: Africa's botanical legacy in the Atlantic world. University of California, Berkeley

Champion L, Fuller DQ (in press) Archaeobotanical Analysis. Agricultural Diversification and change in the Niger Valley. In: Haour A, Reynolds F (eds) Birnin Lafiya: 2000 years in a Dendi Village. J Afr Archaeol Monogr series, Africa Magna, Frankfurt

Champion L, Haour A (2013) Le site de TinTin Kanza, Etude préliminaire du matériel céramique du Sondage I. Nyame Akuma 79:112–120

Connah G (1981) Three thousand years in Africa. Cambridge University, Cambridge

D'Andrea AC, Casey J (2002) Pearl millet and Kintampo subsistence. Afr Archaeol Rev 19 (3):147–173

D'Andrea AC, Klee M, Casey J (2001) Archaeobotanical evidence for Pearl Millet (*Pennisetum glaucum*) in Sub-Saharan West Africa. Antiquity 75:341–348

D'Andrea AC, Logan AL, Watson DJ (2006) Oil Palm and prehistoric subsistence in Tropical Africa. J Afr Archaeol 4(2):195–222

D'Andrea AC, Kahlheber S, Logan AL et al (2007) Early domesticated cowpea (*Vigna unguiculata*) from Central Ghana. Antiquity 81:686–698

David N (1976) History of crops and peoples in North Cameroon to AD 1900. In: Harlan JR, de Wet JM and Stemler ABL (eds) Origins of African plant domestication. The Hague, Mouton, pp 223–67

David N (1981) The archaeological background of Cameroonian history. In Tardits C (ed) Contribution de la recherche ethnologique à l'Histoire des civilization du Cameroun. Colloques Internationaux du CNRS 551 Paris CNRS 79–98

Delneuf M, and Otto T (1995) L'environnement et les usages alimentaires en vigueur à l'époque protohistorique dans l'extrême-nord du Cameroun. In Marliac A (ed) Milieux, sociétés et Archéologues Paris: Office de la recherche Scientifique et Technique de Outre-Mer (Orstom), pp 213–226

De Wet JM (1977) Domestication of African cereals. Afr Econ Hist 3:15–32

Dueppen SA (2008) Reinventing equality: the archaeology of Kirikongo, Burkina Faso. Dissertation, University of Oregon

Dueppen SA, Gallagher DE (2013) Adopting agriculture in the West African savanna: Exploring socio-economic choices in first millennium CE southeastern Burkina Faso. J Anthropol Archaeol 32:433–448

Eichhorn B, Neumann K (2014) Holocene vegetation change and land use at Ounjougou, Mali. In: Stevens CJ, Nixon S, Murray MA, Fuller DQ (eds) Archaeology of African plant use. Left Coast Press, Walnut Creek, pp 83–96

Fields-Black EL (2008) Deep roots: rice farmers in West Africa and the African diaspora. Indiana University, Bloomington

Fuller DQ (2000) The Botanical Remains. In: Insoll T (ed). Urbanism, Archaeology and Trade, further observations on the Gao Region (Mali), The 1996 Field Results. BAR, Archaeopress, Oxford, pp 28–35

Fuller DQ, Hildebrand L (2013) Domesticating plants in Africa. In: Mitchell P, Lane P (eds) The Oxford handbook of African archaeology. Oxford University, Oxford, pp 507–525

Fuller DQ, Qin L, Harvey EL (2009) An evolutionary model for Chinese rice domestication: reassessing the data of the Lower Yangtze region. In: Ahn S-M, Lee J-J (eds) New approaches to prehistoric agriculture. Sahoi Pyoungnon, Seoul, pp 313–345

Fuller DQ, MacDonald KC, Vernet R (2007) Early domesticated pearl millet in Dhar Nema (Mauritania): evidence of crop processing waste as ceramic temper. In: Cappers RTJ (ed) Fields of change: progress in African archaeobotany. Barkhuis, Groningen, pp 71–76

Fuller DQ, Stevens C (2018) Sorghum domestication and diversification: a current archaeobotanical perspective. Plants and Peoplein the African Past 503–528

Gallagher D (1999) Analysis of seeds and fruits from the sites of Arondo and Ft. Senedebu, Senegal. In: Thiaw I (ed) Archaeological investigation of a long term culture change in the Lower Falemme (Upper Senegal Valley, AD 1000–1900). Dissertation, Rice University, pp 385–436

Gallagher D (2011) Palaeobotanical remains. In: Cissé M (ed), Archaeological investigations of early trade and urbanism at Gao Saney (Mali). Dissertation, Rice University, pp 350–354

Gallagher D (2012) Appendix I: preliminary results of macrobotanical analysis of seeds and fruit from Diouboye. In: Dokee CD (ed), Daily life in the Land of Bambuk: an archaeological study of political economy at Diouboye, Senegal. PhD Dissertation, University of Michigan, Ann Arbor, pp 655–661

Gallagher D, McIntosh RJ (2015) Agriculture and urbanism. In: Barker G, Goucher C (eds) The Cambridge World History, vol 2. A World with Agriculture, 12,000 BCE–500 CE. Cambridge University, Cambridge, pp 186–209

Gallais J (1959) La riziculture de plaine en Haute-Guinée, Annales de Géographie 367:207–223

Gallais J (1967) Le delta intérieur du Niger, étude de géographie régionale, IFAN-Dakar, 2 tomes, p 621

Gijanto L, Walshaw S (2014) Ceramic production and dietary changes at Juffure, Gambia. Afr Archaeol Rev 31:265–297

Haour A (2013) Mobilité et archéologie le long de l'arc oriental du Niger: pavements et percuteurs. Afriques. https://doi.org/10.4000/afriques.1134

Haour A, Banni Guene O, Gosselain O et al (2011) Survey along the Niger River Valley at the Bénin-Niger border. Nyame Akuma 76:23–32

Haour A, Nixon S, N'dah D et al (2016) The settlement mound of Birnin Lafiya: New evidence from the eastern arc of Niger River. Antiquity 90(351):695–710

Harlan JR (1971) Agricultural origins: centers and non-centers. Science 174:468–474

Harlan JR (1992) Indigenous African agriculture. In: Cowan CW, Watson PJ (eds) The origins of agriculture. Smithsonian Institution, Washington DC, pp 59–70

Harlan JR, Pasquerleau J (1969) Decrue agriculture in Mali. Econ Bot 23:70–74

Harris DR (1976) Traditional systems of plant food production and the origins of agriculture in West Africa. In: Harlan JR, De Wet JMJ, Stemler AB (eds) Origins of African plant domestication. The Hague, Mouton, pp 311–356

Höhn A, Neumann K (2012) Shifting cultivation and the development of a cultural landscape during the Iron Age (0–1500 AD) in the northern Sahel of Burkina Faso, West Africa: Insights from archaeological charcoal. Quatern Int 249:72–83

Huysecom E, Ozainne S, Mayor A et al (2012) Fouilles en Pays dogon et reconnaissance archéologique au Sénégal oriental: la 14ème année de recherches du programme "Peuplement humain et paléoenvironnement en Afrique de l'Ouest". Jahresbericht SLSA, Zürich, pp 99–172

Kahlheber S (2004) Perlhirse und Baobab: Archäobotanische Untersuchungen im Norden Burkina Fasos. Dissertation, Johann Wolfgang Goethe-Universiät

Kahlheber S, Neumann K (2007) The development of plant cultivation in semi-arid West Africa. In: Denham TP, Iriarte J, Vrydaghs L (eds) Rethinking agriculture: Archaeological and ethnoarchaeological perspectives. Left Coast Press, Walnut Creek, pp 320–345

Kahlheber S, Höhn A, Rupp N (2009) Archaeobotanical studies at Nok sites: an interim report. Nyame Akuma 71:2–17

Kahlheber S, Höhn A, Neumann K (2014) Plant and land use in southern Cameroon between 400 BCE–400 CE. In: Stevens CJ, Nixon S, Murray MA, Fuller DQ (eds) Archaeology of African plant use. Left Coast Press, Walnut Creek, pp 113–128

Klee M, Zach B, Stika HP (2004) Four thousand years of plant exploitation in the Lake Chad Basin (Nigeria), part III: plant impressions in potsherds from the Final Stone Age Gajiganna Culture. Veg Hist Archaeobot 13(2):131–142

Lavachery P (2001) The Holocene archaeological sequence of Shum Laka Rock Shelter (Grassfields, Cameroon). Afr Archaeol Rev 18:213–247

Logan AL (2012) A history of food without history: Food, trade, and environment in West-Central Ghana in the second millennium AD. Dissertation, University of Michigan

Logan AL, D'Andrea AC (2012) Oil palm, arboriculture, and changing subsistence practices during Kintampo times (3600–3200 BP, Ghana). Quatern Int 249:63–71

MacDonald KC (1996) The Winde Koroji Complex: evidence for the peopling of the Eastern Inland Niger Delta (2100–500 BC). Préhistoire, Anthropologie Méditerranéennes 5:157–165

MacDonald KC, Vernet R, Martinón-Torres M et al (2009) Dhar Néma: from early agriculture to metallurgy in southeastern Mauritania. Azania Archaeol Res Africa 44(1):3–48

Magnavita C (2002) Recent archaeological finds of domesticated *Sorghum bicolor* in the Lake Chad region. Nyame Akuma 57:14–20

Manning K, Fuller DQ (2014) Early millet farmers in the Lower Tilemsi Valley, Northeastern Mali. In: Stevens CJ, Nixon S, Murray MA, Fuller DQ (eds) Archaeology of African Plant Use. Left Coast Press, Walnut Creek, pp 73–82

Manning K, Pelling R, Higham T et al (2011) 4500-Year old domesticated pearl millet (*Pennisetum glaucum*) from the Tilemsi Valley, Mali: New insights into an alternative cereal domestication pathway. J Archaeol Sci 38(2):312–322

Marliac A (1991) De la préhistoire à l'histoire du Cameroun Septentrional. Office de la recherche Scientifique et Technique de Outre-Mer (Orstom), Paris

McIntosh SK (1995) Excavations at Jenne-jeno, Hambarketolo, and Kaniana (Inland Niger Delta, Mali), the 1981 Season. University of California, Berkeley

Meyer RS, Choi JY, Sanches M et al (2016) Domestication history and geographical adaptation inferred from a SNP map of African rice. Nat Genet 48:1083–1088

Mintz SW (1985) Sweetness and power. Viking, New York

Morris J (2013) An archaeobotanical investigation from Sorotomo, exploring the decision-making processes behind crop selection in the Middle Niger. Unpublished Master dissertation University of California, Los Angeles

Munson PJ (1976) Archaeological data on the origins of cultivation in the southwestern Sahara and their implications for West Africa. In: Harlan J, de Wet JMJ, Stamler ABL (eds) Origins of African plant domestication. The Hague, Mouton, pp 187–209

Murdock GP (1960) Staple subsistence crops in Africa. Geogr Rev 50:523–540

Murray SS (2007a) Identifying African rice domestication in the middle Niger delta (Mali). In: Cappers RTJ (ed) Fields of change: progress in African archaeobotany. Groningen Archaeological Studies 5. Barkhuis, Groningen, pp 53–61

Murray SS (2007b) Medieval cotton and wheat finds in the middle Niger Delta (Mali). In: Cappers RTJ (ed) Fields of change. Progress in African archaeobotany. Groningen Archaeological Studies 5. Barkhuis, Groningen, pp 43–51

Murray SS (2008) A report on the charred botanical remains from Sincu Bara, a Mid first millennium AD Middle Senegal Valley Site. Nyame Akuma 69:56–63

Murray SS, Déme A (2013) Early agro-pastoralism in the Middle Senegal Valley. The botanical remains from Walaldé. In: Stevens CJ, Nixon S, Murray MA, Fuller DQ (eds) Archaeology of African plant use. Left Coast Press, Walnut Creek, pp 61–97

Murray MA, Fuller DQ, Cappeza C (2007) Crop production on the Senegal River in the early first millennium AD: preliminary archaeobotanical results from Cubalel. In: Cappers RTJ (ed) Fields of change. Progress in African archaeobotany. Groningen Archaeological Studies 5. Barkhuis, Groningen, pp 63–70

Neumann K, Breunig P, Kahlheber S (2001) Early food production in the Sahel of Burkina Faso. Berichte des Sonderforschungsbereichs 268(14):327–334

Neumann K, Eggert MKH, Oslisly R et al (2012a) Comment on "Intesifying weatering and land use in Iron Age Central Afica". Science 337:1040–1041

Neumann K, Bostoen K, Höhn A et al (2012b) First farmers in the Central African rainforest: A view from southern Cameroon. Quatern Int 249:53–62

Ngomanda A, Neumann K, Schweizer A et al (2009) Seasonality change and the third millennium BP rainforest crisis in southern Cameroon (Central Africa). Quatern Res 71(3):307–318

Nixon S, Murray MA, Fuller DQ (2011) Plant use at an early Islamic merchant town in the West African Sahel: the archaeobotany of Essouk-Tadmakka (Mali). Veg Hist Arcaheobot 20(3):223–239

Olson DM, Dinerstein E, Wikramanayake ED et al (2001) Terrestrial ecoregions of the world: a new map of life on Earth. Biosci 51(11):933–938

Orijemie EA, Sowunmi MA (2014) Humans and the mangrove in Southern Nigeria. In: Stevens CJ, Nixon S, Anne-Murray M, Fuller DQ (eds) The Archaeology of African Plant Use. pp 103–112

Otto T (1996) Phyto-Archéologie de Sites Archéologiques de l'age du fer du Diamamé, Nord du Cameroun: le site de Salak.Etude de bois et de grains carbonisées. Théses et documents micro-fichées 151, Paris office de la rrecherches scientifique et technique de Outre-mer (Orstom)

Otto T (1998) Essai sur l'histoire du paysage au Diamaré pour les deux derniers millénaires. In Delneuf M, Essomba JM, Fromant A (eds) Paléo-anthropologie en Afrique Centrale: un Bilan de l'archéologie au Cameroun. Paris l'Harmattan, pp 157–64

Otto T, Delneuf M (1998) Evolution des ressources alimentaires et des paysages au nord du Cameroun: apport de l'archéologie. In Chastanet M (ed) Plantes et Paysages d'Afrique un histoire à explorer. Paris Karthala, pp 491–514

Ozainne S, Lespez L, Le Drézen Y et al (2009) Developing a chronology integrating archaeological and environmental data from different contexts: The Late Holocene sequence of Ounjougou (Mali). Radiocarbon 51:457–470

Portères R (1976) African cereals: Eleusine, Fonio, Black Fonio, Teff, Brachiaria, paspalum, Pennisetum, and African Rice. In: Harlan J, de Wet JMJ, Stamler ABL (eds) Origins of African plant domestication. The Hague, Mouton, pp 409–463

Robion-Brunner C, Haour A, Coustures M-P et al (2015) Iron production in Northern Benin: Excavations at Kompa Moussékoubou. J Afr Archaeol 13(1):39–57

Rouch J (1950) Les Sorkawa, pêcheurs itinérants du Moyen-Niger. Africa V, 20(1):5–21

Shaw T (1976) Early crops in Africa: a review of the evidence. In: Harlan J, de Wet JMJ, Stamler ABL (eds) Origins of African plant domestication. Mouton, The Hague, pp 107–153

Stahl A (1985) Reinvestigation of Kintampo 6 rock shelter, Ghana: implications for the nature of culture change. Afr Archaeol Rev 3:117–150

Takezawa S, Cisse M (2004) Domestication des cereales au Mema, Mali. In: Sanogo K, Togola T, Keita D et al (eds) XIth Congress of Panafrican Association. Prehistory and Related Fields, Acts Bamako, pp 105–121

Van Driel A (2001) Sharing a valley: the changing relations between agriculturalists and pastoralists in the Niger Valley of Benin. Research report (African Studies Centre, Leiden, Netherlands); 64/2001

Vavilov N (1950) The origin, variation, immunity, and breeding of cultivated plants. Cambridge University, Cambridge

White F (1983) The vegetation of Africa, a descriptive memoir to accompany the UNESCO/AETFAT/UNSO vegetation map of Africa (3 Plates, Northwestern Africa, Northeastern Africa, and Southern Africa, 1: 5,000,000)

World Wildlife Fund (2017) Terrestrial ecoregions of the world. http://www.worldwildlife.org/publications/terrestrial-ecoregions-of-the-world. Accessed 1 May 2017

Zach B, Klee M (2003) Four thousand years of plant exploitation in the Chad Basin of NE Nigeria II: discussion on the morphology of caryopses of domesticated Pennisetum and complete catalogue of the fruits and seeds of Kursakata. Veg Hist Archaeobot 12(3):187–204

Zohary D, Hopf M, Weiss E (2012) Domestication of plants in the Old World. Oxford University, Oxford

Pollen-Based Landscape Reconstruction and Land-Use History Since 6000 BC along the Margins of the Southern Tunisian Desert

Sahbi Jaouadi(✉) and Vincent Lebreton

Département Homme et Environnement, Muséum national d'Histoire naturelle, HNHP, UMR 7194 CNRS, Paris, France
shbjaouadi@gmail.com, vincent.lebreton@mnhn.fr

Abstract. In south-eastern Tunisia, the role of human societies in current desert margin degradation is assumed to have involved the modification of woodland vegetation over the past 2000 years, leading to the current sparse herbaceous vegetation. In this paper, we present key results of a study based on a new multiproxy record (^{14}C dates, pollen and clay mineralogy) from Sebkha Boujmel, Tunisia which reconstructs climate and human/landscape interactions over the past eight millennia. These new data emphasize the central role played by a Middle- to Late Holocene climate aridity trend that shaped landscapes by driving a gradual decline of Mediterranean woodland vegetation and the establishment of present-day semi-desert landscapes. The anthropogenic impacts of Neolithic pastoral societies remained limited, however human-mediated environmental change became more pronounced during the historical period and took place in a herbaceous landscape through agricultural activities during the Carthaginian-Roman periods. An abrupt environmental change is recorded during the 20th century. This significant change was related to socio-economic modifications in the way of life of the local populations, leading to over-exploitation of limited and fragile arid land resources through intensive agriculture and pastoral activities. Overall, the Holocene landscape history of southern Tunisia highlights the fragility of desert margins to both climate change and human impacts. The latter seems relatively limited during periods of pastoral subsistence characterized by high levels of mobility. However, sedentism associated with intensive grazing and agriculture has disrupted the delicate human/landscape equilibrium. These results highlight the importance of: (i) the socio-economic organization of the population; and (ii) historical events, in the exploitation and shaping of the landscape, while a direct link between climate and human activities is not evident.

Keywords: Agriculture · Cultural landscapes · Grazing · Holocene Anthropogenic impact · Nomadism · Palynology · Sedentism · Tunisia

Introduction

Global climate warming and its associated threats underline the significance of the impact of human activity on the Earth's eco-systems. Indeed, a new geological era, the "Anthropocene" (Crutzen 2002), and Anthropogenic biomes (Anthromes) (Ellis and

A. M. Mercuri et al. (Eds.): *Plants and People in the African Past*,
https://doi.org/10.1007/978-3-319-89839-1_24

Ramankutty 2008) have been proposed. It has been suggested that the collapse of certain ancient civilizations were "ecocides" triggered by the overexploitation and the degradation of the available environmental resources (Diamond 2005). This determinism between historical and social change on the one hand, and environmental change on the other, is doubtlessly more complex and may involve other factors such as the resilience of the socio-political structures and production systems of the populations concerned (McAnany and Yoffee 2010; Butzer 2012).

In this context, the arid steppes of central and southern Tunisia clearly illustrate issues concerning the relationships between humans and their environments and the risks which arise therefrom. In south-eastern Tunisia (Fig. 1), the Jeffara Plain is currently experiencing degradation of its vegetation cover and soils, and intensification of erosion and desertification. The latter is defined as the "land degradation in arid, semi-arid and dry sub-humid areas resulting from various factors, including climatic variations and human activities" (UNCCD 1994). Today, desertification represents an eminent threat to lasting development in south-eastern Tunisia (Floret and Pontanier 1982; Genin et al. 2006).

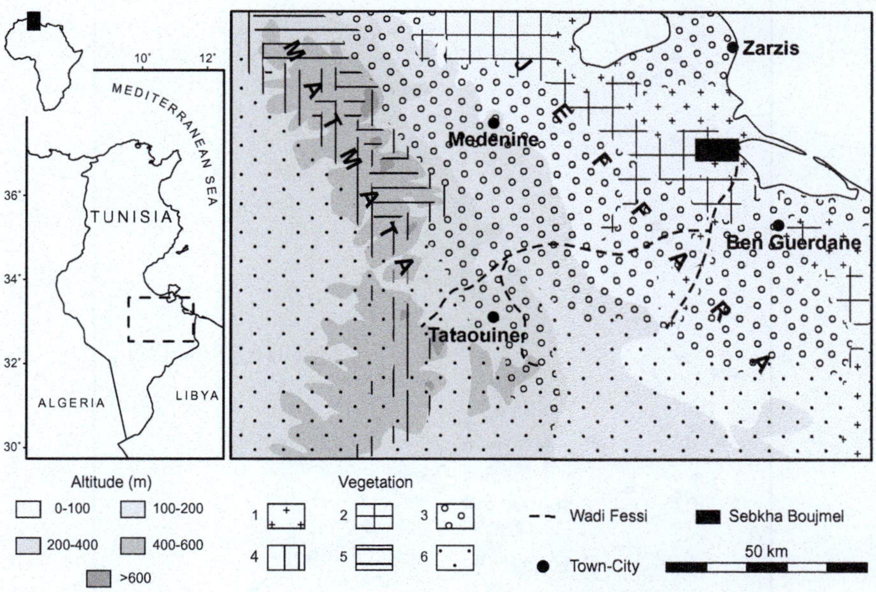

Fig. 1. Map showing the location of the study area within Africa and Tunisia and main geographical features of the Tunisian Jeffara plain including topography, the Wadi Fessi hydrographic network and main vegetation associations (after Gammar A.-M. 2008). The black rectangle marks the location of the study site: Sebkha Boujmel. Vegetation 1: Halophilous vegetation. 2: Gypsophilous steppe with *Lygeum* and *Zygophyllum*. 3: Cultivated fields (*Olea*) and steppe with *Rhanterium*, *Haloxylon scoparium* and *Artemisia*. 4: Steppe with *Haloxylon scoparium* and *Haloxylon schimittianum*. 5: Grass steppe with *Stipa tenacissima* and matorral with *Juniperus phoenicea*. 6: Desert vegetation with *Helianthemum* sp., *Calligonum*, *Gymnocarpos decander*, *Traganum nudatum*.

Archaeological remains (Fig. 2), historical texts, and the structure of the current vegetation have all been analysed and interpreted in order to infer ancient vegetation landscapes, to determine their dynamics, particularly throughout the historic period, and to explain the present degradation of the environment. Archaeological evidence for sedentary occupation associated with agricultural practices during Classical Antiquity in regions which are today arid and semi-desert has inspired a nostalgic rhetoric associating verdant, fertile landscapes with economic prosperity. The transformation of these Eden-like landscapes into steppes and semi-deserts is seen as the result of continual deforestation and over-grazing by pastoral nomadic communities. Historical texts describing Roman Africa as the "granary of Rome", or those describing the historical invasions (Vandal [429–439 AD], Arab [647–670 AD] and Hilalian [1051–1057 AD]), are evoked in order to corroborate this vision within the framework of a history characterized by significant historical and landscape upheavals between the different periods (e.g. De Planhol 1968; Idris 1968). However, when the issue of historical climatic variations is approached solely on the basis of textual evidence, it sometimes leads to contradictory conclusions being drawn from the same data (Shaw 1981).

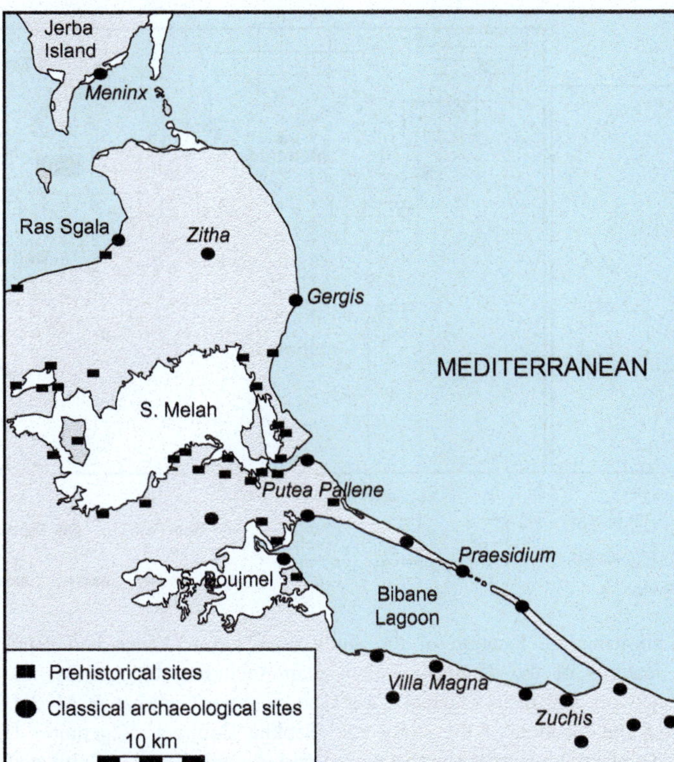

Fig. 2. Main Holocene prehistoric sites (*rammadiya*) (after Perthuisot 1975; Hannachi 2014) and Punic and Roman classical archaeological sites (after Drine 2002) in the environs of Sebkha Boujmel and nearby areas.

In the absence of archaeobotanical data and palaeoecological records, the primacy of the anthropogenic factor in environmental degradation, based solely on the interpretation of archaeological remains and literary texts, is often taken as a given and can exert a significant influence on botanical studies. The woody vegetation present in mountainous refuges is seen as a relic of dense woodland vegetation cover which would once have occupied the entire territory. The degradation of this vegetation through anthropogenic pressure during the historic period would have resulted in the present-day vegetation landscapes of southern Tunisia. Thus, Le Houérou (1959, 1969, 1995) defines a"primitive vegetation" of the steppes of central and southern Tunisia as existing since the Roman Period. The vegetation of the Jeffara would, therefore, have consisted of open woodland with Mediterranean xerophytic shrubs (*Juniperus phoenicea* L. and *Pinus halepensis* Mill.). The degradation of this woodland vegetation through anthropogenic pressure would have given rise to a grass steppe and then to the current steppic and semi-desert vegetation (Fig. 3a). A second model, proposed by Frankenberg (1986), assumes that the vegetation of southern Tunisia was more diversified since the Holocene "climate optimum". Alongside the Mediterranean semi-forest, a Saharan-Sahelian steppe developed featuring *Acacia tortilis* subsp. *raddiana* (Savi) Brenan and *Panicum turgidum* Forssk. During the Late Holocene, and especially during the historical period, the area occupied by these taxa gradually diminished under the influence of both climate aridification and human exploitation (Fig. 3b).

Recent recognition of the importance of climatic changes and their impacts on terrestrial vegetation ecosystems throughout the Holocene (Mayewski et al. 2004), the development of interpretive paradigms for the history of North Africa since Laroui's (1977) pioneering and fundamental criticisms, and data acquired through multi-disciplinary archaeological research on ancient populations of the desert margins (Barker et al. 1996; Barker 2002a; Barker and Gilbertson 2003; Barker et al. 2007) have prompted a reconsideration of these schematic interpretations. In this context, the few currently available palaeoecological data, such as the pollen sequence from Sebkha Mhabeul, indicate that semi-desert vegetation has existed in the Jeffara throughout the last two millennia (Salzmann and Schulz 1995). Moreover, critical analysis of environmental accounts suggests that laying the blame for environmental degradation on mobile pastoralists is the result of a declinist view which is closely linked to colonial ideology (Shaw 1981; Davis 2007). Studies in human geography suggest that the landscapes of the Jeffara would have been more stable under a regime of mobile pastoralism and that the marked degradation is, instead, the result of sedentism and the rise of intensive agriculture during the 20th century (Talbi 1997).

It has not been shown that woodland vegetation ever existed in the Jeffara, and if it did, its botanical composition, and the climatic and/or anthropogenic factors which led to the modification of the environment resulting in the current pre-desert steppes, have not been identified. Furthermore, the relationships between the ancient populations of North Africa and their environments and the society/climate interactions involved remain unclear (Wilson 2013). These issues can only be addressed on the basis of a record of Holocene vegetation history (Schulz et al. 2009) integrated within a multi-proxy approach (Roberts et al. 2011). Pollen analysis of natural sequences respond to these issues by allowing vegetation changes to be tracked over long periods

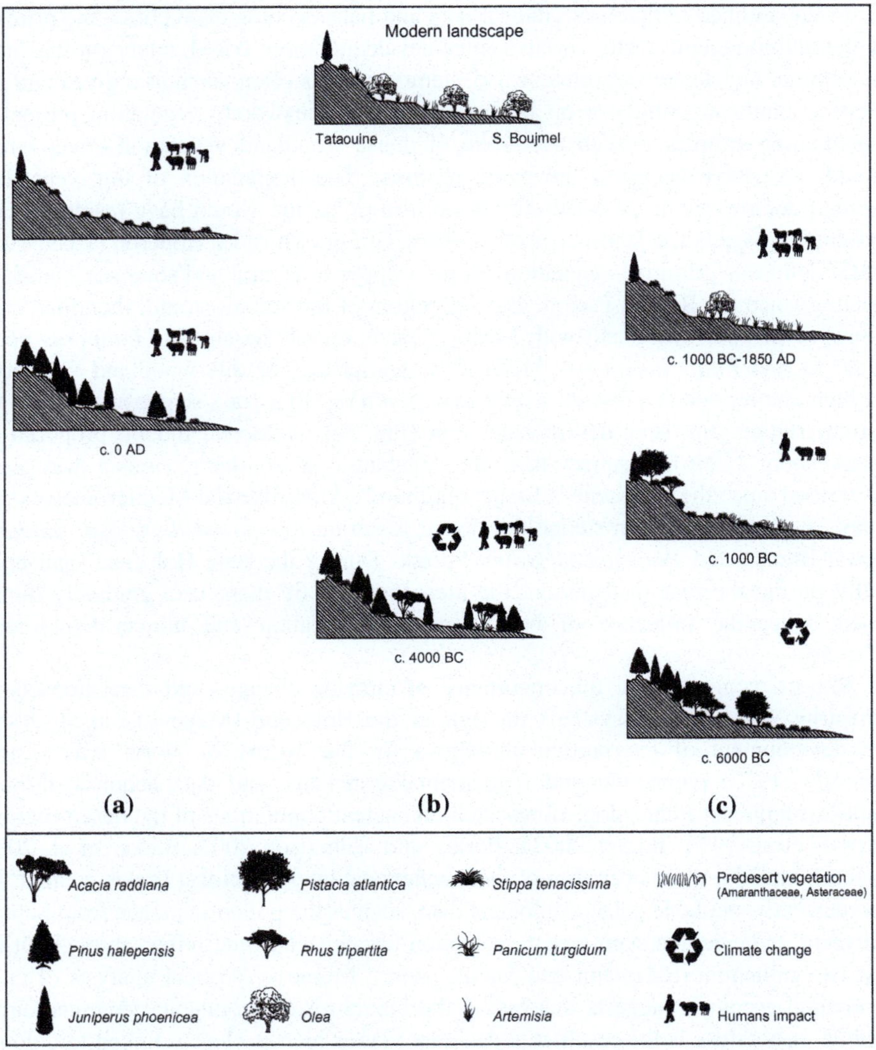

Fig. 3. Synthesis of principal models of landscape evolution, human activities and the driving factors behind landscape shaping in the Jeffara Plain during the Holocene. (a) Le Houérou (1959, 1969, 1995), (b) Frankenberg (1986), and (c) current study.

and by informing discussions regarding the respective contributions of natural forcings (climate) and anthropogenic forcings (land use) in the shaping of landscapes (Edwards et al. 2015). These pollen sequences serve to complete and moderate information originating from other archaeobotanical proxies which tend to over-emphasise anthropogenic factors on archaeological sites.

This paper aims to discuss human occupation and landscape shaping in eastern Maghreb arid lands by comparing the recently published multiproxy analyses from Sebkha Boujmel (^{14}C AMS, pollen sequence and clay mineralogy; Jaouadi et al. 2016) with the available literature. The study provides an overview of the climate and regional biogeography in order to discuss the structuring of landscapes over a long timeframe. Following an outline of the geographical context of south-eastern Tunisia and the methodologies employed, three thematics are developed: (1) the biogegraphy, vegetation ladscapes and factors determining their evolution over the course of the Holocene, compared to previously published models; (2) the impact of human societies associated with pastoral and agricultural activities, within prehistoric and historic cultural contexts of southern Tunisia and Holocene climatic variability; (3) the recent degradation of desert margins due to factors ranging from climate change to socio-economic factors.

Environmental Settings: The Modern Landscapes of South-Eastern Tunisia

South-eastern Tunisia is a semi-desert, coastal region characterized by an arid Mediterranean climate and a negative hydrological balance due to low and irregular precipitation and high levels of evaporation (Ferchichi 1996). The maritime influence of the Mediterranean Sea and the contrasting topography give rise to climatic and biogeographical differences between the mountainous areas, coastal plain and desert regions (Le Houérou 1959, 1969, 1995; Genin et al. 2006) (Fig. 1). The topography of this region is dominated by the Matmata Mountains which run north to south at an average altitude of 500 m (Fig. 1). This mountainous hinterland is characterized by average annual precipitation of between 200 and 300 mm. This climate favours the development of low shrubland vegetation (*Juniperus phoenicea* and *Rosmarinus officinalis* L.) on the summits and a grass steppe with *Stipa tenacissima* L. on the foothills (Fig. 1). These mountains become progressively lower towards the south and west where a desert climate (precipitation <100 mm/yr) and desert vegetation (*Calligonum* sp., *Ephedra* subsp. *alenda* (Stapf) Trab., *Retama raetam* (Forssk.) Webb, Amaranthaceae, Zygophyllaceae) predominate. To the east, the Jeffara coastal plain is a sub-desert zone (precipitation between 150 and 200 mm/yr) with semi-desert vegetation (*Rhanterium suaveolens* Desf., *Artemisia* sp., *Haloxylon scoparium* Pomel, *Gymnocarpos decander* Forssk.), as well as halophilic Amaranthaceae on saline soils bordering the sebkhas and coastal lagoons (Fig. 1).

Holocene Human Settlement in South-Eastern Tunisia

Surveys carried out in the environs of Sebkha Boujmel have revealed several open-air sites characteristic of the Capsian Epipalaeolithic culture (8th–5th millennia BC). These sites take the form of mounds of stone and ash mixed with terrestrial and marine mollusc shells, faunal remains and lithic debris (Fig. 2). Such sites, known as *rammadiyat*, are also reported in similar Holocene biotopes around Sebkha El Melah and further north in the coastal sebkhas of the Maharès region (Perthuisot 1975; Chenorkian et al. 2002) (Fig. 2). Thus, Capsian hunter-gatherers may have settled around lagoons and estuaries which formed along the eastern coast of Tunisia as a result of the postglacial sea-level rise. These transition zones between terrestrial and marine ecosystems offered a great diversity of natural resources and were particularly favoured for exploitation by human groups (Fa 2008). From the 6th millennium onwards, changes in techno-economic behaviour are recorded at many sites with the introduction of ceramics, and followed by the appearance of domesticated animals around the mid-6th millennium BC (Aouadi et al. 2014; Mulazzani et al. 2016). In the 5th millennium BC, nomadic pastoral societies appear to be well established in the Libyan Jeffara and at Jebel Gharbi (Barich 2014), in Cyrenaica (De Faucamberge 2016), Tunisia (Aouadi et al. 2014; Mulazzani et al. 2016), and the Atlas Mountains of northern Algeria (Roubet 2003). In the Libyan Jeffara, archaeological records suggest continuity of nomadism from the Neolithic to the historic period (Lucarini 2013).

For more recent periods, the available archaeological remains, such as the Ghomrassen rock paintings (Ben Nasr and Ghourabi 2007) and megalithic monuments (Paris and Ghaki 2010), provide little information about economic modes, habitat and the social organization of the populations. Historical texts, such as the writings of Herodotus, and archaeological data indicate the development of Phoenician settlements on the coast of the Jeffara, probably as early as the 6th century BC (Labiadh 2001; Slim et al. 2004). These sites include commercial harbors and exploitation centers for fish products (Slim et al. 2004) (Fig. 2). During the Roman period, this region experienced continuity in the occupation of Phoenician cities and a certain degree of economic development (Mattingly 1995). Historical documentation, such as Tacitus' account of the Tacfarinas War (17–24 AD) and surviving Latin inscriptions, confirm the dominance of mobile pastoralists in Tripolitania.

On the basis of inscriptions found in the mosques of the Berber mountain villages, it is assumed that the establishment of the Berber villages dates back to the period of the Hilalian invasions during the 11th century AD (Louis 1979). This economic model of occupation of southeastern Tunisia, with nomadic populations occupying the plains and sedentary farmers living in mountain villages, prevailed until the 20th century. From the period of the French protectorate (AD 1881), the control of this region resulted in a massive sedentarization policy that was systematized after independence (Abaab 1986). The Jeffara, originally a rangeland for nomadic tribes and their flocks of sheep, has recently been converted to intensive agriculture resulting in significant anthropogenic impacts. Pastoralism is an important economic activity among local populations in the Jeffara, and is principally practiced by sedentary populations who are also engaged in agriculture. Vast areas are devoted to agriculture, particularly to the cultivation of olives (*Olea europaea* subsp. *europaea* var. *europaea* L.) (Fig. 1). On the desert margins, the growing of olives outside their natural bioclimate is made possible by the ancient process of runoff water-harvesting in the mountains and foothills—*tabia* and *jessour* (Bonvallot 1986)—where olive trees are grown alongside other fruit trees (*Ficus carica* L., *Phoenix dactylifera* L., *Punica granatum* L.). In the plains, olives are grown using a system of dry-farming.

The Multiproxy Palaeoecological Data from Sebkha Boujmel

Sebkha Boujmel (33°16'N, 11°05'E, 2 m asl) is a saline depression located adjacent to Bibane Lagoon on the south-eastern coast of Tunisia (Figs. 1 and 2). Holocene lagoon sedimentation of Sebkha Boujmel was favoured by the presence of a coastal band of lithified sand and by a sea level rise associated with the Holocene transgression (Jedoui et al. 1998). This sedimentation began with the deposition of layers of lagoon sediments which progressed towards the formation of the sebkha (Lakhdar 2009). This sebkha is also the Holocene delta of the Wadi Fessi which drains the entire hinterland between the mountains and the coast (Fig. 1) (Medhioub and Perthuisot 1981). Thus, the Sebkha Boujmel sediments record pollen inputs from regional ecosystems deposited through fluvial and aeolian processes. The BJM 2 core (160 cm deep) obtained from the sebkha has been subjected to the following procedures:

- Pollen extraction from 71 sediment samples using the standard pollen chemical process (HCl 18%, HF 70% and HCl 10%) (Faegri and Iversen 1989) and a 5 µm ultrasonic filtration. An average of 300 pollen grains were counted for each sample by microscope observation, with the aid of pollen atlases (Reille 1992; Ayyad and Moore 1995) and the reference pollen collection at the Muséum National d'Histoire Naturelle of Paris for taxonomic identification. The ecology of the pollen flora was assessed based on Tunisian and Saharan floras (Le Houérou 1959, 1969, 1995; Pottier-Alapetite 1979, 1981; Floret and Pontanier 1982; Ozenda 2004). In order to track anthropogenic impacts, some pollen taxa have been grouped as nitrophilic and unpalatable taxa based on recent studies on southern Tunisia vegetation changes under pastoral activities (Tarhouni et al. 2010; Gamoun 2014).
- X-ray diffraction analysis of the clay fraction (<2 µm) of 68 samples. Clay mineralogy of sediments allows us to quantify the relative percentages of the different clay minerals (illite, kaolinite, palygorskite, smectite and chlorite). Changes in the percentages of clay minerals over time reflect variations in the provenance and dynamics of desert dust mobilization modulated by broad-scale climate trends (Bout-Roumazeilles et al. 2013). The clay geochemistry indicates that the Sebkha Boujmel clay fraction is mainly composed of kaolinite (40%), palygorskite (25%) and illite (25%). The association illite and kaolinite indicates a northern Saharan origin for the wind-blown particles (Goudie and Middleton 2001; Bout-Roumazeilles et al. 2007) while kaolinite is dominant in the central and southern Sahara (Sahelian zones). The illite to kaolinite ratio (I/K) is used to track variations in the respective contributions of Saharan and Sahelian dusts (Formenti et al. 2011).
- Eleven ^{14}C AMS dates on the sediment organic fraction. The dates have been corrected and calibrated using Calib 7.1 software under the IntCal13 calibration curve (Reimer et al. 2013). Pollen and clay mineralogy data have been inserted into a chronological framework derived from an age-depth model based on the ^{14}C AMS dates.

The multi-proxy method adopted at Sebkha Boujmel aims to combine the following elements over a period spanning the last 8000 years: (1) pollen signal for the regional vegetation; (2) clay mineralogy signal reflecting the provenance and dynamics of the desert dusts by climatic trends; and (3) pollen taxa which act as indicators of human activity (cultivated, nitrophilic, unpalatable and human introduced taxa). The combination of these proxies (Fig. 4) allows us to discuss the respective contributions of climatic and anthropogenic factors to the shaping of landscapes, particularly during the Middle Holocene when profound climatic changes are contemporary with the emergence of Neolithic societies (Mercuri et al. 2011). In parallel, economic activities deduced from pollen markers for agriculture and/or pastoralism are placed within their environmental context. This approach allows us to discuss interactions between societies and climate in desert margin areas which are extremely sensitive to climatic change and therefore, particularly constraining for human activities. The archaeobotanical, archaeological and palaeoecological data are also used to contextualise and discuss the palaeoecological information obtained.

Mid-Holocene Vegetation Landscapes: Structure and Changes

Two major phases of landscape history and associated climates can be identified for the last 8000 years (Fig. 4). Between 6000 and 3700 BC, Mediterranean trees and shrubs (*Juniperus*, *Pinus*, *Pistacia*, and *Rhus tripartita*), as well as grass steppes (Poaceae), are well represented. The presence of humidity-demanding Mediterranean plants is associated with the presence of fresh-water aquatic plants. Clay geochemistry for this period, with high percentages of palygorskite and a high I/K ratio, indicates a mix of regional inputs from Tunisian loess and the northern Saharan zone. Taken together, these elements indicate a more humid climate than exists today, as well as relative stability in the landscapes for a period of more than 2000 years. Nevertheless, xeric habitats also existed as indicated by evidence for desert vegetation (Zygophyllaceae and Amaranthaceae) (Fig. 4).

The Mediterranean woodland vegetation recorded at Sebkha Boujmel for this period is far from reflecting a verdant, wooded pre-desert, in particular on the plains which are currently pre-desert (Thinon et al. 1996; Ballouche 2002). This vegetation, which prefers a more humid climate, must have existed mainly in the mountainous hinterland where today much degraded and restricted relics survive on the wetter summits. Such woodland vegetation, especially *Juniperus* and *Pistacia*, may also have existed along the coasts near the Sebkha Boujmel. The occurrence of pollen and wood remains of *J. phoenicea* and *Pistacia lentiscus* L. on the coasts of the nearby Jerba Island, were recorded at c. 3000 BC (Damblon and Vanden Berghen 1993) and support this hypothesis.

Moreover, the significant percentages of Poaceae are indicative of grass steppes being a major component of the landscapes while *Artemisia* only occurs with low percentages. Finally, the pollen flora reflects the altitudinal zonation of the vegetation which became established under a more humid climate, with Mediterranean woodland vegetation on the mountains and foothills and grass steppes on the plains. This landscape is similar to that which can be observed today in central Tunisia along the

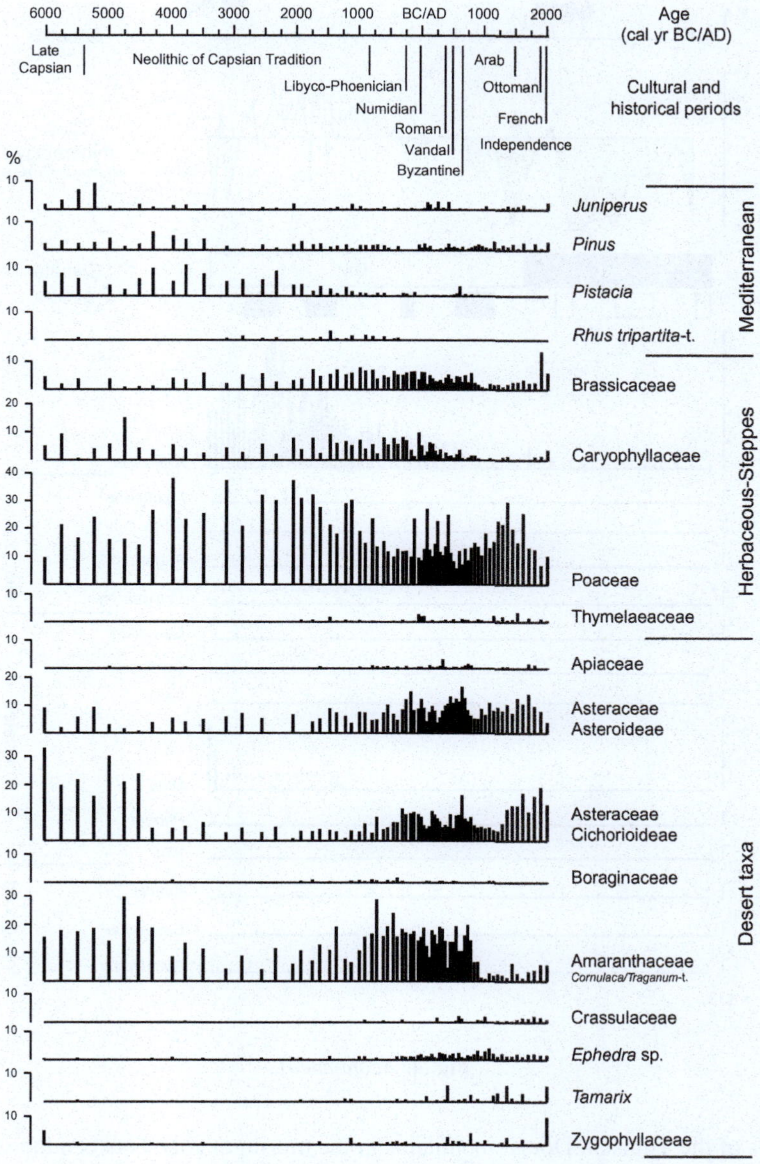

Fig. 4. Synthesis of palaeoecological data obtained from Sebkha Boujmel, showing (1) Age (cal yr BC/AD), (2) the cultural and historical periods for Tunisia, (3) changes in the percentages of Mediterranean, steppe and desert pollen taxa, (4) percentages of aquatics pollen (freshwater plants), (5) changes in the dynamics of desert sand mobilisation based on clay mineralogy analysis (The Sahara reveals local/regional provenance, while the Sahel reveals large scale sand particle mobilisation), (6) synthesis of climatic variability based on index between pollen taxa and variations in the sources and dynamics of desert sands, (7) the percentages of marker taxa for human impact with *Artemisia*, cultivated, nitrophilic and introduced taxa.

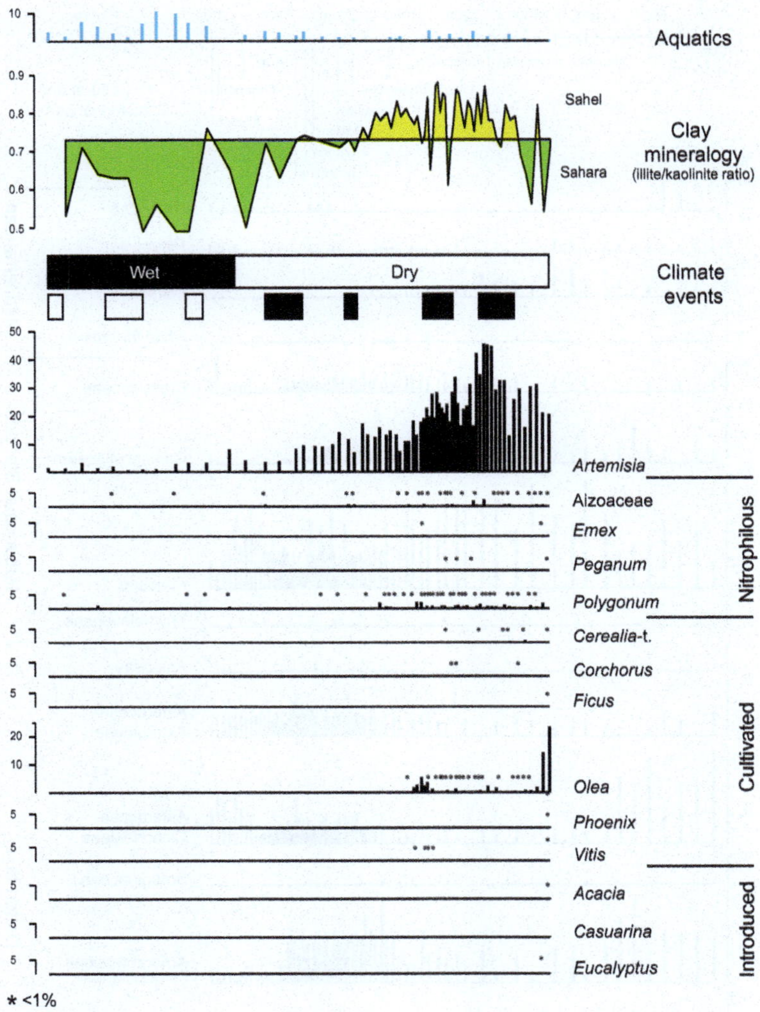

Fig. 4. *(continued)*

foothills of the Tunisian Dorsal mountains in the transition zone between the semi-arid and arid bioclimatic zones (400–300 mm/yr) (Le Houérou 1995; Schulz et al. 2009).

Around 3700 BC a progressive and profound re-shaping of regional ecosystems began, culminating in the establishment of semi-desert landscapes around 1000 BC. A rapid decline of aquatic plants is recorded between 3700 and 2600 BC, followed by a regression of Mediterranean woody species and grass steppes between 2600 and 1000 BC (Fig. 4). At the same time, there is a marked increase in steppic herbaceous species and a progression of desert vegetation (Brassicaceae, Caryophyllaceae, Asteraceae and Amaranthaceae). This phase marks an increasing and definitive aridification of the climate, with the establishment of the current pre-desert ecosystems accompanied by wider dispersion of desert dust from the central and southern Sahara indicated by the simultaneous decrease in the I/K ratio and palygorskite percentages (Fig. 4).

When compared to models previously proposed for the vegetation of southern Tunisia (Le Houérou 1959, 1969, 1995; Frankenberg 1986), the pollen data for Sebkha Boujmel confirm the presence of Mediterranean woodland vegetation, but at earlier dates and closely associated with the more humid climate of the Middle Holocene (Fig. 3). In addition, the absence of pollen indicators for a significant anthropogenic impact on regional ecosystems, coupled with the clay mineralogy data, clearly indicate that the retreat of woodland vegetation is governed by aridification of the climate. This result is corroborated by palaeoclimatic records from the Mediterranean and the Sahara (Magny et al. 2013; Cremaschi et al. 2014). The aridification of the climate recorded in the Sebkha Boujmel sequence can be explained by changes in atmospheric circulation under the influence of variations in orbital parameters and summer insolation (Jaouadi et al. 2016).

Finally, *Acacia raddiana*-type pollen does not occur in the Sebkha Boujmel sequence. This absence has also been noted in other pollen records such as the marine deposits from the Gabès Gulf (Brun 1992), the pollen diagrams for the Libyan Jeffara between 7400 and 3000 BC (Giraudi et al. 2013), and records for the island of Jerba at c. 3000 BC (Damblon and Vanden Berghen 1993). Furthermore, on Holocene archaeological sites where pollen grains of entomogamous taxa such as *Acacia* are more likely to be found, the currently available paleobotanical data (pollen, charcoal and seed remains) do not reveal the presence of *Acacia* (Renault-Miskovsky 1985), with the exception of the Capéletti Cave in Algeria where a single pollen grain is recorded (Roubet 1979). Further south in the central Sahara, available pollen data show a late extension of the *Acacia* genus concomitant with Middle Holocene climate aridification and the spread of savanna vegetation communities including several *Acacia* species (Mercuri 2008).

At Sebkha Boujmel, *Acacia* is only recorded as *Acacia cyanophylla*-type during the 20th century (Fig. 4), a period when several species of *Acacia* were introduced to Tunisia (Pottier-Alapetite 1979). Furthermore, the presence of *A. tortilis* subsp. *raddiana* today in assemblages lacking floristic homogeneity has already been highlighted (Le Houérou 1969) along with doubts regarding *Acacia* being a component of the natural vegetation of Tunisia (Damblon and Vanden Berghen 1993). However, while all this evidence points to the absence of *Acacia* in southern Tunisia during the Holocene, the low pollen production of this insect pollinated genus, and the fragile nature of its polyad, might also explain its absence from pollen records (Horowitz 1992; Davies and Fall 2001).

Human Impact on Landscapes Since the Beginning of the Neolithic

To date, very few studies have been completed on the socio-economic dynamics associated with environmental contexts of Capsian hunter-gatherers in the eastern Maghreb (8th–5th millennia BC) and the transition towards a production economy (Neolithic). In this sense, the Sebkha Boujmel sequence is important because it records the beginning of the Middle Holocene, from 6200 BC onwards. This period is characterized by greater climatic instability, with the recurrence of rapid climatic changes at a century-scale and the establishment of arid conditions of the Late Holocene (from 2200 BC onwards). These rapid climatic changes had significant impacts on the ecosystems and the availability of resources. They may have led to populations adopting a food production-based economy during the 6th millennium BC. The selective introduction of certain economic innovations from the "Neolithic package", in this case domestic animals and the emergence of caprine and ovine pastoralism in the 6th millennium BC, might indicate an adaptation to natural constraints with unpredictable precipitation limiting possibilities for agriculture in arid regions of North Africa (Barker 2002b; Garcea 2004).

Land-Use and Transformation of Landscapes by Pastoral Nomadic Societies

The development of pastoralism is evidenced in pollen data by the presence of nitrophilic taxa. These taxa are recorded discontinuously in the Sebkha Boujmel between the 6th and the 1st millennium BC, indicating a limited impact by pastoral societies on local ecosystems (Fig. 4). However, pastoral activity may have contributed to the reshaping of the steppic landscapes of the plain by engendering a progressive replacement of the grass steppe by a sagebrush steppe (*Artemisia*). In fact, pollen records allow us to conclude that *Artemisia* does not form part of desert and semi-desert ecosystems of Tunisia during the Holocene and that its record is not linked to variations in climatic conditions, be they arid or humid (Jaouadi et al. 2016). Moreover, recent studies on the dynamics of steppic vegetation (Le Houérou 1995; Genin et al. 2006) and on grazing (Tarhouni et al. 2010; Gamoun 2014) indicate that certain species of *Artemisia* (i.e. *Artemisia campestris* L.) are unpalatable pioneering plants associated with the replacement of degraded pre-desert steppes through clearance and intense pastoral activity. The concomitant occurrence of discontinuous peaks in *Artemisia* and nitrophilic taxa (Fig. 4) may ultimately reflect changes in steppic vegetation during phases of intensified pastoral activity.

From the 1st millennium onwards, the presence of nitrophilic taxa in the records is continuous, coupled with more pronounced peaks of *Artemisia* (Fig. 4). This indicates an intensification of pastoral activity reflecting adaptation to long-term climatic aridification and the development of pre-desert ecosystems. Moreover, during the last three millennia, pastoralism persists, even during periods of sedentarisation and agricultural development between the 3rd century BC and the 2nd century AD. In the Libyan Jeffara, pollen analysis of coprolites, and archaeozoological studies of sites belonging to the Romano-Libyan, Arab and Modern periods, indicate that the rearing of sheep and goats is an integral part of the exploitation of desert margins within the context of mixed agriculture in the Mediterranean region (Van der Veen et al. 1996; Hunt et al. 2001).

Another significant period in pastoral activity is revealed in the pollen diagram by high *Artemisia* values (up to 45%) between 850 and 1150 AD, i.e. the transition period between the Early and High Middle Ages (Fig. 4). This peak in *Artemisia* occurs during a globally humid episode recorded between 850 and 1450 AD, related to the Medieval Climate Anomaly (MCA). The progression of *Artemisia* reflects intensive pastoral activity and a degradation of regional vegetation within a dynamic that spans three centuries (850 to 1150 AD). During this period, significant changes occurred in the economy of the Maghreb. The combined effects of famines, wars and periods of climatic instability brought about profound economic changes and a shift towards predominately pastoral activity (Allaoua 2003). These changes culminated in the invasions by nomadic Hilalian tribes (AD 1051–1057) which constituted a final episode within a context of deep change in the economy of the Medieval Maghreb.

Use of Plant Resources and Agriculture

The use of ceramics and the emergence of mobile pastoralism over the course of the 6th millennium BC were not accompanied by the development of agriculture. At this date, hunting and gathering of wild plants and molluscs, both terrestrial and marine, are part of an economic continuum (Aouadi et al. 2014; Lucarini et al. 2016; Mulazzani et al. 2016). Available archaeobotanical data for Epipalaeolithic and Neolithic sites point to an intensification of the exploitation of plant resources, including the gathering of *Pinus halepensis* cones, *Quercus ilex* subsp. *ballota* (Desf.) Samp. acorns and the use of *Stipa tenacissima* rhizomes (Morales et al. 2015; Mulazzani et al. 2016). The gathering of wild grasses appears to have been a significant part of the exploitation of plant resources by these communities as evidenced by macro-botanical remains (Morales et al. 2015), phytolith analyses (Shipp et al. 2013), and pollen data (Jaouadi et al. 2010) for various Holocene sites in central and southern Tunisia. This exploitation of regional ecosystems by prehistoric societies does not appear to have induced sufficiently profound changes in the vegetation landscapes to render them detectable in sebkha pollen sequences.

To date, there is no well-established chronological framework documenting the emergence of agriculture and sedentary settlements, such as the Berber villages, in the mountainous hinterland or coastal and inland oases. Seed remains indicate that agriculture was already established, at the latest, by the first half of the 1st millennium BC both in the south, in the Garamantes oases of the Fazzan in Libya (van der Veen 1992; Zerboni et al. 2017), and further north in northern Tunisia at the Numidian site of *Althiburos* (Kallala et al. 2008). These data would suggest that agricultural activity was occurring in south-eastern Tunisia, but the pollen proxy is limited regarding this issue. In fact, cereals and several cultivated trees associated with oasis agriculture (*Vitis vinifera* L., *Ficus carica*, *Phoenix dactylifera*) are generally low pollen producers that tell us little about agriculture that was dispersed and practiced on a small scale within favourable oasis enclaves. On the Jeffara coast, sea shore settlements date back to the 6th century BC, a time when, according to Herodotus, the region was under the control of the Carthaginians (Labiadh 2001). These sites along the southern coasts of Tunisia and Libya were Phoenician trading settlements (emporia) which featured port facilities (*Gergis*, *Zuchis* and Ras Sgala) (Fig. 2). The sites also provide evidence of facilities for processing marine products, such as salting vats for fish and dye factories based on Murex shellfish (Drine 2002; Slim et al. 2004). These settlements do not appear to have been associated with large-scale agriculture which would otherwise be identifiable in the Sebkha Boujmel pollen diagram. It is only after 300 BC that traces of agriculture appear in the record in the form of pollen from two cultivated plants: *Olea* and *Vitis* (Fig. 4).

Olea

The wild olive tree (*Olea europaea* subsp. *europaea* var. *sylvestris* (Mill.) Lehr) forms part of the thermophilic vegetation of the semi-arid bioclimatic stage of central and northern Tunisia. Much debate centres on whether cultivated olive (*O. europaea* subsp. *europaea* var. *europaea*) was domesticated by local populations in North Africa or whether it was introduced by the Phoenicians. In Carthage, analysis of wood charcoal from Tophet (Stuijts 1991; Docter et al. 2003), along with pollen and seed remains from the Punic canal (van Zeist et al. 2001), indicate that olive growing began in the 6th century BC but remained small-scale until the 4th century BC when its expansion took place in tandem with the cultivation of other fruit trees (almond, peach, apricot and plum). From then on, the cultivated olive is believed to have played a central role in the Tunisian economy throughout the Roman Period with the development of large-scale monoculture destined for exportation. Archaeological evidence for this development includes oil presses and the discovery of the remains of African amphorae at the *Testaccio* in Rome (e.g. Camps-Fabrer 1953; Mattingly 1988). However, recent advances in the archaeology and history of Roman Africa have required a review of this speculative and schematic vision of olive cultivation and a playing down of its importance by reconsidering regional and historical specificities (Leveau 2005, 2011). Moreover, in pollen sequences which lack a firm chronological framework, peaks in *Olea* have been systematically attributed to the Roman Period (Ritchie 1984; Kolstrup 1994). These sequences are located within the natural range of the oleaster (*Olea* pollen from domesticated and wild plants is indistinguishable). Yet the development of *Olea* in central Tunisia during the Holocene, with rates of up to 35% during favourable climatic episodes (Lebreton and Jaouadi 2013; Lebreton et al. 2015), ought to be treated with a degree of caution. In southern Tunisia during the Holocene, oleasters must have been restricted to the most humid biotopes during periods of increased precipitation (Jaouadi et al. 2016). The recording of *Olea* in the Sebkha Boujmel pollen diagram during the Late Holocene, which was characterised by marked aridity, could therefore, be a reliable indicator of the wide-scale introduction of olive cultivation to the arid coastal plains. In the 4th century BC, the Periplus of Pseudo-Scylax, while describing the ports of Tripolitania also refers to the nearby island of Jerba and emphasises the richness of its soils for cereal growing. It also reports that the island's population produced olive oil solely from managed oleasters (Shipley 2011). The pollen diagram for Sebkha Boujmel also supports this observation, with a change in the forms of economic exploitation in the 3rd century BC marked by the introduction of large-scale olive cultivation (Fig. 4). Later descriptions by Polybius in the 2nd century BC, and by Strabo towards the 1st century BC, lead us to understand that there was a surplus of agricultural products to be sold in the areas surrounding the *emporia* (Slim et al. 2004). At a regional scale, the *Olea* curve in the Sebkha Boujmel diagram (Fig. 4) indicates that: (1) the introduction and development of olive cultivation occurred in the last three centuries BC, and did not consitute a monoculture (concomitant presence of *Vitis*); (2) the cultivation at this time appears to be limited in comparison to the surge in olive trees recorded from the 19th century onwards; and (3) that there is a clear regression in the percentages of *Olea* from the Roman Period onwards. The loss of political and economic importance of certain urban centres (e.g. the city of *Zitha*) during the Late-Roman Period might explain this olive tree decline (Mattingly 1995). Furthermore, regional data indicate modest olive growing to satisfy local consumption during the Roman Period as is further illustrated by the technical characteristics of Roman oil presses identified in southern Tunisia (Mrabet 2011) and by amphorae produced on Jerba (Fentress et al. 2009).

Vitis vinifera

Vitis is recorded in the Sebkha Boujmel diagram between the 2nd century BC and the 2nd century AD (Fig. 4). Several arguments converge to suggest that this viticulture might be older and might have played a vital role in the agricultural economy during Classical Antiquity. The vine (*Vitis vinifera* subsp. *sylvestris* (C.C. Gmel.) Hegi) is part of the natural vegetation of North Africa (Pottier-Alapetite 1979, 1981; Zohary 1996). A local domestication of the grapevine is, therefore, possible (Greene 1996). Furthermore, the discovery of seed remains in the Garamantes Oasis in the Libyan Sahara to the south, and in the early Numidian levels at *Althiburos* in northern Tunisia, attest to the cultivation of *V. vinifera* subsp. *vinifera* L. from the first half of the 1st millennium BC, well before the introduction of other Mediterranean cultivated trees, such as *O. europaea,* which is dated to the end of the 1st millennium BC in both regions. At Sebkha Boujmel, the recording of *Vitis* pollen grains indicates significant cultivation of this species, the pollen of which does not disperse widely. At the regional scale, the amphorae produced at *Zitha* and Jerba between the 1st century BC and the 1st century AD are identified as belonging to the Dressel 2/4 type which was destined for the exportation of wine (Bonifay 2004; Fentress et al. 2009).

During late Classical Antiquity, from the 3rd century AD onwards, the percentages of cultivated tree species remain low while nitrophilic pollen taxa, which principally reflect pastoral activity, are continuously recorded on the Jeffara Plain throughout the period of the Late Roman Empire (AD 284–429). To the east of Sebkha Boujmel, the remains of wine and oil presses, and *ostracon,* at the site of *Villa Magna* attest to the production of wine and olive oil during the 4th century AD (Drine 2002). In contrast, at the same period, the production of amphorae for the exportation of agricultural products ceases at Jerba (Fentress et al. 2009). Pollen data do not indicate any significant changes in landscape and agricultural practices over the course of the Vandal (AD 429–534) and

Byzantine Periods (AD 534–670) whereas historical and archaeological data are lacking for these periods in southern Tunisia.

Cultivation Along Desert Margins

A slight rise in humidity is recorded between 50 BC and 450 AD (Jaouadi et al. 2016) (Fig. 4). This episode is only partly contemporaneous with the development of olive and vine cultivation between the 3rd century BC and the 2nd century AD. Climatically, this episode was too insignificant to have allowed local populations to farm without using suitable irrigation techniques. Archaeological research in other desert margin regions show that the constraints of an arid, pre-desert environment could be overcome through the use of appropriate hydraulic techniques, including the control of surface run-off and the exploitation of aquifers (Barker et al. 1996; Mattingly et al. 2003). However, to date, it remains to be demonstrated if this development of agricultural activity and the necessary irrigation techniques is related to intervention by political powers or the initiative of local populations. Thus, ancient agricultural development in the Jeffara may have been related to a policy of stabilisation of the African territory by Carthage after the defeat of the first Punic war (264–241 BC) (Whittaker 1978). Alternatively, it may be linked to the weakening of Carthaginian power during the same period and the annexing of this region to the Numidian Kingdom of Massinissa in c.162 BC, as recently proposed for similar contemporary economic changes on the island of Jerba (Fentress et al. 2009). However, irrigation techniques adapted to arid and semi-arid zones are probably indigenous and independent of political powers and their intervention in these regions (Shaw 1984; Jaouadi 2013). It should also be noted that the dynamism of local nomadic populations, and the flexibility of their socio-political organisation, would have allowed intensification of agricultural production and sedentarisation by profiting from favourable political and/or economic circumstances (Grahame 1998).

For the Early Middle Ages (7th–11th centuries AD), pollen data do not indicate any major changes relative to Late Antiquity, which suggests that there was a certain continuity in land use. However, new cultivated species were introduced with the presence of a small number of *Corchorus* pollen grains attributed to the herbaceous species *Corchorus olitorius* L. (Nalta jute). This edible species was introduced by the Arabs (Jalloul 1998) and today it is grown in a number of oases in southern Tunisia, particularly in Gabès.

The *Artemisia* episode contemporary with the transition from the Early to the High Middle Ages in the 11th century AD does not appear to have been preceded by major changes in agricultural activity. It is followed by a minor increase in indicators of agricultural activity with a slight peak in *Olea* suggesting a resumption of agriculture. Pollen data indicate that the 11th century AD did not constitute a significant break, with uncultivated and unfertilised landscapes before this period and with no significant noticeable degradation afterwards. Descriptions by the Arab geographers Al-Idrissi and Al-Tijani between the 12th and the 14th centuries indicate that the Jeffara was under the control of nomadic Arab tribal factions, but they also describe numerous oases and *Ksour* on the coast as well as Berber farming villages in the mountainous hinterland (Brunschvig 1940).

Landscape Changes Since the 19th Century

The upper part of the Sebkha Boujmel pollen sequence represents the period between 1850 and the present. Significant changes in the pollen flora are evident from: (1) a rapid increase in the percentage of *Olea*; (2) an increase in certain desert taxa (*Tamarix, Nitraria, Fagonia* and *Zygophyllum*); and (3) the presence of flora introduced to Tunisia (*Acacia cyanophylla, Casuarina* and *Eucalyptus*) (Fig. 4). Here again agricultural and pastoral activities are recorded but in a different manner compared to preceding periods. The very pronounced rise in olive tree pollen reflects the importance of this emblematic tree within the vast plantations present in the landscape today. Pastoralism is evidenced by ruderal and nitrophilic taxa (*Peganum, Polygonum*), but also by the more frequent occurrence of other taxa: *Tamarix, Thymelaea* and those of the Zygophyllaceae family (*Nitraria, Fagonia* and *Zygophyllum*). The development of these unpalatable, desert-adapted plants reflects intensive pastoral activity and a degradation of the vegetation cover due to overgrazing.

All these pollen markers attest to a rapid intensification of agricultural and pastoral activities causing significant stress on the local environments and radical shifts in the relationships between human groups and their environment. These profound changes in the landscape and anthropogenic activities are linked to the drastic changes in southern Tunisian socio-economy. Indeed, up until the end of the 19th century, this region was characterised by two complementary economic systems. The first, adopted by the sedentary populations of Berber mountain villages, was based on the cultivation of fruit trees (*Ficus, Olea, Phoenix* and *Punica*) supplemented by cereals. This arboriculture was made possible by traditional systems of surface run-off management in mountainous areas and in the foothills using *jessours* and *tabias* (Bonvallot 1986). A second economic system, based on pastoralism practised within extensive, tribal nomadism or semi-nomadism, with a high degree of flock and population mobility, dominated the plains. This pastoralism was supplemented by occasional cereal cultivation, particularly of barley, which was limited to lands that are subject to flooding in favourable years. Economic complementarity and exchanges were assured through a system of clientage which guaranteed armed protection and exchanges of agricultural and animal products between sedentary farmers in the mountains and nomadic herders on the plains (Louis 1979). These two complementary forms of production and land exploitation formed an economic system which was in a state of relative balance within these territories characterised by limited and fragile resources. Since the Colonial Period, from 1881 AD onwards, the policy of sedentarisation of nomadic and semi-nomadic populations, and the introduction of new economic systems, have broken up traditional social and production structures (Abaab 1986). Widespread sedentarisation led to the transfer to private ownership of collectively held tribal lands and the establishment of intensive olive cultivation. This recent expansion of the olive has occurred at the expense of other plant groups which are barely perceptible in the Sebkha Boujmel pollen diagram. A recent study undertaken in the northern part of the Jeffara has shown that, between 1972 and 2001, the area occupied by pseudo-steppe featuring *Ziziphus lotus* L. and *Retama raetam* (Forssk.) Webb has been reduced by 91% in order to accommodate olive groves (Genin et al. 2006). Furthermore, the disappearance of collective trails, which formerly ensured a high degree of flock mobility has caused localised overgrazing with degradation of vegetation cover and the development of substitute species including taxa which are unpalatable (*Tamarix, Thymelaea, Nitraria, Fagonia* and *Zygophyllum*), thorny (*Atractylis*), or even toxic (*Peganum harmala* L.).

All these social and economic changes result in significant pressure on the environment with an accentuation of wind and water erosion. Changes in the vegetation cover, such as those recorded in the Sebkha Boujmel pollen sequence, clearly reflect these significant changes in landscapes and socio-economic systems over the course of the last century. In these arid regions, all changes underline the contrast between a nomadic or semi-nomadic way of life associated with extensive and patchy exploitation of the land on the one hand, and a sedentary way of life associated with intensive pastoral and agricultural practices on the other hand. The over-exploitation of natural resources is, without doubt, one of the principal elements in the degradation and destabilisation of these fragile, resource-poor ecosystems.

Conclusions

The pollen record from Sebkha Boujmel reveals the complexity of the long-term evolution of sub-desert regions culminating in today's landscapes in southern Tunisia. These landscapes are the product of a long history, integrating climatic and anthropogenic impacts together with various rhythms and forms of stress. The data obtained confirm the suggestions provided by botanical models regarding the presence of woodland vegetation, but with significant differences in terms of the chronology, the rhythm and the causes of changes as well as the composition and the location of this vegetation. Woodland vegetation of Mediterranean xerophytic type would only have existed on higher ground and mountain slopes with grass steppes on the plains. Recurring ecological changes in the vegetation in response to rapid climatic events in the Middle Holocene underline the fragility of this region with a continuous climatic xerophytisation, which profoundly affected landscapes. Mediterranean shrub- and woodland vegetation would thus have regressed and migrated towards the northern slopes of the mountainous hinterland. The regional micro-climate of the Matmata Mountains would have provided a refuge for this woodland vegetation, which would explain the presence today of certain relics on the more favourable summits.

Climatic changes of the Middle Holocene would have forced human societies to adapt to changes in landscapes and resources through the emergence of pastoral nomadism. In fact, pastoralism has been a perennial anthropogenic activity in this region from the Neolithic to the present day. Over the past three millennia, recorded climatic variability has been narrow in range and could not have been a determining factor in changes to landscapes and human activities. For instance, two humid phases, the Roman Humid Period and the Medieval Climate Anomaly, are in part characterised by a development of agricultural activities and a predominance of pastoralism, respectively. Instead, these facts can be explained by historical events and long-term socioeconomic dynamics on the desert margins. While pollen data do not reveal any breaks and significant changes in landscapes and associated anthropogenic activities at the time of the Vandal and Arab invasions, the period of Hilalian invasions is part of a process of landscape transformation spanning three centuries. This episode, during which *Artemisia* significantly increases, also marks a degradation of arid environments under the influence of several factors, including climate change and intense pastoral activity.

The vegetation ecosystems of southern Tunisia were, therefore, rendered more fragile during the Holocene by increasing aridity and by a long history of exploitation of natural resources by human groups. However, the last century has been characterised by changes which are more profound compared to previous phases when a relative and fragile equilibrium was maintained between human societies and their environments. The degradation recorded over the course of the last century illustrates the complexity of the interactions between human societies and their environments: these interactions are based on the social and economic organisation determining the forms and the intensity of exploitation of the natural resources. Thus, mixed agro-pastoral activities, combining limited agriculture and nomadic pastoralism, which characterised the pre-20th century phases, are replaced by intensive and localised agriculture and pastoralism. The main consequence has been the almost complete disappearance of nomadism in favour of sedentism. The shift by nomadic pastoral societies of the Maghreb towards semi-nomadism, and even their partial sedentarisation, is a well attested fact which must have occurred on several occasions in the arid and semi-arid regions of North Africa in response to various historic events or economic stresses. This oscillation between nomadism and sedentism would also have occurred on many occasions in the Jeffara. However, it would always have been temporary, followed by a return to nomadism with a continuation of the complementarity with sedentary farming populations based in the mountains. The 20th century is the only exception, with massive sedentarisation and the spatial intensification of anthropogenic activities beyond the resilience threshold of local ecosystems, resulting in their current degradation.

References

Abaab A (1986) Mutations socio-économiques de la Jeffara orientale (Sud Tunisien). Rev de l'Occident musulman et de la Méditerr 41(1):327–338

Allaoua A (2003) Retour à la problématique du déclin économique du monde musulman médiéval: le cas du Maghreb Hammadide (X–XIIe siècle). Maghreb Rev 28(1):2–26

Aouadi N, Dridi Y, Ben Dhia W (2014) Holocene environment and subsistence patterns from Capsian and Neolithic sites in Tunisia. Quatern Int 320:3–14

Ayyad SM, Moore PD (1995) Morphological studies of the pollen grains of the semi-arid region of Egypt. Flora 190:115–133

Ballouche A (2002) Entre "Sahara vert" et desert: mythes et réalités. In: Al Khatib A (ed) Le Sahara, espace de communication et d'interaction civilisationnelles dans les temps antiques. Institut des Etudes Africaines, Rabat, pp 99–126

Barich BE (2014) Northwest Libya from the early to late Holocene: New data on environment and subsistence from the Jebel Gharbi. Quatern Int 320:15–27

Barker G (2002a) A tale of two deserts: contrasting desertification histories on Rome's desert frontiers. World Archaeol 33(3):488–507

Barker G (2002b) Transitions to farming and pastoralism in North Africa. In: Bellwood P, Renfrew C (eds) Examining the farming/language dispersal hypothesis. McDonald Institute for Archaeological Research, Cambridge, pp 151–162

Barker G, Gilbertson D (2003) The archaeology of Drylands: living at the margin. In: One world archaeology, vol 39. Routledge, London, New York

Barker G, Gilbertson D, Jones B et al (eds) (1996) Farming the desert: the UNESCO Libyan Valleys archaeological survey, vol 1, Synthesis. UNESCO Publishing, Paris

Barker G, Gilbertson D, Mattingly DJ (2007) Archaeology and desertification: the Wadi Faynan landscape survey, Southern Jordan. Council for British Research in the Levant, London

Ben Nasr J, Ghourabi M-H (2007) De nouveaux documents rupestres du Sud Tunsien (Chaabit El-Maarik: Ghoumrassen). Sahara 18:154–158

Bonifay M (2004) Etudes sur la céramique romaine tardive d'Afrique. BAR International Series, vol 1301. Archaeopress, Oxford

Bonvallot J (1986) Tabias et jessour du Sud tunisien. Agriculture dans les zones marginales et parade à l'érosion. Cah ORSTOM, série Pédol 22(2):163–171

Bout-Roumazeilles V, Combourieu Nebout N, Peyron O et al (2007) Connection between South Mediterranean climate and North African atmospheric circulation during the last 50,000 yr BP North Atlantic cold events. Quat Sci Rev 26(25–28):3197–3215

Bout-Roumazeilles V, Combourieu-Nebout N, Desprat S et al (2013) Tracking atmospheric and riverine terrigenous supplies variability during the last glacial and the Holocene in central Mediterranean. Clim Past 9(3):1065–1087

Brun A (1992) Pollens dans les séries marines du Golfe de Gabès et du plateau des Kerkennah (Tunisie): signaux climatiques et anthropiques. Quaternaire 3(1):31–39

Brunschvig R (1940) La Berbérie orientale sous les Hafsides, des origines à la fin du XVe siècle. Adrien-Maisonneuve, Paris

Butzer KW (2012) Collapse, environment, and society. Proc Natl Acad Sci 109(10):3632–3639

Camps-Fabrer H (1953) L'olivier et l'huile dans l'Afrique romaine. Imprimerie Officielle, Alger

Chenorkian R, Harbi-Riahi M, Zoughlami J (2002) Atlas Préhistorique de la Tunisie, 19: Maharès. École Française de Rome, Rome

Cremaschi M, Zerboni A, Mercuri AM et al (2014) Takarkori rock shelter (SW Libya): an archive of Holocene climate and environmental changes in the central Sahara. Quat Sci Rev 101:36–60

Crutzen PJ (2002) Geology of mankind. Nature 415(6867):23–23

Damblon F, Vanden Berghen C (1993) Etude paléo-écologique (pollen et macrorestes) d'un dépôt tourbeux dans l'île de Djerba, Tunisie méridionale. Palynosci 2:157–172

Davies CP, Fall PL (2001) Modern pollen precipitation from an elevational transect in central Jordan and its relationship to vegetation. J Biogeogr 28(10):1195–1210

Davis DK (2007) Resurrecting the Granary of Rome: environmental History and French colonial expansion in North Africa. Series in Ecology and History. Ohio University Press, Ohio

De Faucamberge E (2016) Neolithic of Cyrenaica (north-east Libya): new enlightenments from recent research. Quatern Int 410:144–159

De Planhol X (1968) Les Fondements géographiques de l'histoire de l'Islam. Flammarion, Paris

Diamond JM (2005) Collapse: how societies choose to fail or succeed. Viking Press, New York

Docter RF, Smits E, Hakbijl T et al (2003) Interdisciplinary research on urns from the Carthaginian tophet and their contents. Palaeohist 43/44:417–433

Drine A (2002) Autour du lac El Bibèn: les sites d'El Mdeina et de Bou Garnin. In: Khanoussi M, Ruggeri P, Vismara C (eds) L'Africa romana. Atti del XIV convegno di studio. Sassari 7–10 dicembre 2000, vol 3. Carocci editore, pp 2001–2013

Edwards KJ, Fyfe RM, Hunt CO et al (2015) Moving forwards? Palynology and the human dimension. J Archaeol Sci 56:117–132

Ellis EC, Ramankutty N (2008) Putting people in the map: anthropogenic biomes of the world. Front Ecol Environ 6(8):439–447

Fa DA (2008) Effects of tidal amplitude on intertidal resource availability and dispersal pressure in prehistoric human coastal populations: the Mediterranean-Atlantic transition. Quat Sci Rev 27(23–24):2194–2209

Faegri K, Iversen J (1989) Textbook of Pollen analysis, 4th edn. Wiley, Chichester

Fentress E, Drine A, Holod R (2009) An Island through time: Jerba Studies, vol 1: The Punic and Roman Periods. J Rom Archaeol Suppl, vol 71. Portsmouth, Rhode Island

Ferchichi A (1996) Etude climatique en Tunisie présaharienne. Medit 7(3):46–53

Floret C, Pontanier R (1982) L'aridité en Tunisie présaharienne: climat, sol, végétation et aménagement. Trav Doc ORSTOM, vol 150. ORSTOM, Paris

Formenti P, Schütz L, Balkanski Y et al (2011) Recent progress in understanding physical and chemical properties of African and Asian mineral dust. Atmos Chem Phys 11(16):8231–8256

Frankenberg P (1986) Zeitlicher Vegetationswandel und Vegetationsrekonstruktion des "neolithischen Klimaoptimums" in der Jeffara Südosttunesiens. Abhandl der Mathemat-Naturwiss Klasse, vol 4. Akademie der Wissenschaften und der Literatur, Stuttgart

Gammar A-M (2008) Carte de la végétation de la Tunisie. In: Henia L (ed) Atlas de l'eau en Tunisie. Faculté des sciences humaines et sociales de Tunis, pp 130–133

Gamoun M (2014) Grazing intensity effects on the vegetation in desert rangelands of Southern Tunisia. J Arid Land 6(3):324–333

Garcea EAA (2004) An alternative way towards food production: the perspective from the Libyan Sahara. J World Prehist 18(2):107–154

Genin D, Guillaume H, Ouessar M et al (2006) Entre désertification et développement: la Jeffara tunisienne. IRD-Cérès, Tunis

Giraudi C, Mercuri AM, Esu D (2013) Holocene palaeoclimate in the northern Sahara margin (Jefara Plain, northwestern Libya). Holocene 23(3):339–352

Goudie AS, Middleton NJ (2001) Saharan dust storms: nature and consequences. Earth-Sci Rev 56(1–4):179–204

Grahame M (1998) Rome without Romanization: cultural change in the pre-desert of Tripolitania (First–Third Centuries AD). Oxford J Archaeol 17(1):93–111

Greene JA (1996) The beginnings of grape cultivation and wine production in Phoenician/Punic North Africa. In: McGovern P, Flemming SJ, Katz SH (eds) The origins and ancient history of wine. Food and nutrition in history and anthropology, vol 11. Gordon and Breach Publishers, Amsterdam, pp 311–322

Hannachi K (2014) Les traces des installations laissées par l'occupation humaine, durant la préhistoire, sur les îles de Kerkennah, Djerba et la presqu'île de Zarzis. Master thesis, Université de Tunis 1

Horowitz A (1992) Palynology of Arid lands. Elsevier, Amsterdam

Hunt CO, Rushworth G, Gilbertson DD et al (2001) Romano-Libyan Dryland animal husbandry and landscape: pollen and palynofacies analyses of coprolites from a farm in the Wadi el-Amud, Tripolitania. J Archaeol Sci 28(4):351–363

Idris HR (1968) L'invasion hilālienne et ses conséquences. Cah Civilis Médiév 11(43):353–369

Jalloul N (1998) Permanences antiques et mutations médiévales: agriculture et produits du sol en Ifriqiya au haut Moyen-Âge (IXe–XIIe s.). In: Khanoussi M, Ruggeri P, Vismara C (eds) L'Africa romana. Atti del XII convegno di studio. Olbia 12–15 dicembre 1996. Editrice Democratica Sarda, pp 485–511

Jaouadi S (2013) L'occupation du sol au sud de la ville de Kairouan durant l'Antiquité: Données de la feuille Pavillier (071) au 1/50000. Rev Tunis Archéol 1:93–122

Jaouadi S, Lebreton V, Bout-Roumazeilles V et al (2016) Environmental changes, climate and anthropogenic impact in south-east Tunisia during the last 8 kyr. Clim Past 12(6):1339–1359. https://doi.org/10.5194/cp-12-1339-2016

Jaouadi S, Lebreton V, Mulazzani S et al (2010) Analyses polliniques en contexte anthropisé: le cas du site holocène SHM-1 (Hergla, Tunisie centrale). Ann Univ Ferrara Sez Museol Scient Nat 6:25–32

Jedoui Y, Kallel N, Fontugne M et al (1998) A high relative sea-level stand in the middle Holocene of southeastern Tunisia. Mar Geol 147(1–4):123–130

Kallala N, Sanmartí J, Carme Belarte M et al (2008) Recherches sur l'occupation d'*Althiburos* (region du Kef, Tunisie) et de ses environs à l'époque numide. Pyrenae 39(1):67–113

Kolstrup E (1994) Late Holocene pollen records from the Segermes valley. Ne Tunisia. Hist Biol 9(1–2):131–136

Labiadh S (2001) History of the Zarzis peninsula: from ancient times to the end of the French occupation. Tunis [in Arabic]

Lakhdar R (2009) Les sédiments holocènes et les tapis microbiens du littoral du Sud-Est de la Tunisie: sédimentologie et paléoenvironnements. Université de Sfax, Sfax

Laroui A (1977) The History of the Maghrib: an interpretive essay (trans: Manheim R). Princeton Studies on the Near East. Princeton University Press, New Jersey

Le Houérou HN (1959) Recherches écologiques et floristiques sur la végétation de la Tunisie méridionale. Institut de Recherches Sahariennes, Université d'Alger, Alger

Le Houérou HN (1969) La Végétation de la Tunisie steppique (avec références aux végétations analogues d'Algérie, de Libye et du Maroc). Ann Inst Nat Recher Agronom Tunisie, vol 42

Le Houérou HN (1995) Bioclimatologie et biogéographie des steppes arides du Nord de l'Afrique: diversité biologique, développement durable et désertisation. Options méditerranéennes, Série B; Etudes & recherches, vol 10. CIHEAM, Montpellier

Lebreton V, Jaouadi S (2013) Histoire holocène de la végétation sur le littoral de la Tunisie centrale: analyse pollinique des sédiments de la sebkha-lagune Halk el Manjel. In: Mulazzani S (ed) Le Capsien de Hergla (Tunisie): culture, environnement et économie, vol 4. Reports in African Archaeology. Africa Magna Verlag, Frankfurt am Main, pp 48–56

Lebreton V, Jaouadi S, Mulazzani S et al (2015) Early oleiculture or native wild Olea in eastern Maghreb: new pollen data from the sebkha-lagoon Halk el Menjel (Hergla, Central Tunisia). Env Archaeol 20(3):265–273

Leveau P (2005) À propos de l'huile et du vin en Afrique romaine ou pourquoi "déromaniser" l'archéologie des campagnes d'Afrique. Pallas 68:77–89

Leveau P (2011) L'oléiculture en Afrique romaine: une nécessaire réévaluation. In: L'olivier en Méditerranée entre histoire et patrimoine, vol 1. Centre de Publication Universitaire, Université de la Manouba, Tunis, pp 57–86

Louis A (1979) Nomades d'hier et d'aujourd'hui dans le sud tunisien. Édisud, Aix-en-Provence

Lucarini G (2013) Was a transition to food production homogeneous along the circum-Mediterranean littoral? A perspective on the Neolithisation research from the Libyan littoral. In: Shirai N (ed) Neolithisation of Northeastern Africa. Studies in Early Near Eastern Production, Subsistence, and Environment, vol 16. *ex oriente*, Berlin, pp 149–174

Lucarini G, Radini A, Barton H et al (2016) The exploitation of wild plants in Neolithic North Africa. Use-wear and residue analysis on non-knapped stone tools from the Haua Fteah cave, Cyrenaica, Libya. Quatern Int 410:77–92

Magny M, Combourieu-Nebout N, de Beaulieu JL et al (2013) North-south palaeohydrological contrasts in the central Mediterranean during the Holocene: tentative synthesis and working hypotheses. Clim Past 9(5):2043–2071. https://doi.org/10.5194/cp-9-2043-2013

Mattingly D (1988) Oil for export? A comparison of Libyan, Spanish and Tunisian olive oil production in the Roman Empire. J Rom Archaeol 1:33–56

Mattingly DJ (1995) Tripolitania. Batsford, London

Mattingly DJ, Reynolds T, Dore J (2003) Synthesis of human activities in Fazzan. In: Mattingly DJ (ed) The archaeology of Fazzan, vol 1, Synthesis. Department of Antiquities, Tripoli and Society for Libyan Studies, London, pp 327–373

Mayewski PA, Rohling EE, Curt Stager J et al (2004) Holocene climate variability. Quat Res 62(3):243–255

McAnany PA, Yoffee N (2010) Questioning collapse: human resilience, ecological vulnerability, and the aftermath of Empire. Cambridge University Press, Cambridge, New York

Medhioub K, Perthuisot J-P (1981) The influence of peripheral sabkhas on the geochemistry and sedimentology of a Tunisian lagoon: Bahiret el Biban. Sedimentol 28(5):679–688

Mercuri AM (2008) Human influence, plant landscape evolution and climate inferences from the archaeobotanical records of the Wadi Teshuinat area (Libyan Sahara). J Arid Environ 72(10):1950–1967

Mercuri AM, Sadori L, Uzquiano Ollero P (2011) Mediterranean and North-African cultural adaptations to mid-Holocene environmental and climatic change. Holocene 21:189–206

Morales J, Mulazzani S, Belhouchet L et al (2015) First preliminary evidence for basketry and nut consumption in the Capsian culture (ca. 10,000–7500 BP): Archaeobotanical data from new excavations at El Mekta, Tunisia. J Anthropol Archaeol 37:128–139

Mrabet A (2011) Identité de la Tripolitaine occidentale: de quelques signalements archéologiques. In: Briand-Ponsart B, Modéran Y (eds) Provinces et identités provinciales dans l'Afrique romaine. Tables rondes du CRAHME, vol 6. Caen, pp 221–237

Mulazzani S, Belhouchet L, Salanova L et al (2016) The emergence of the Neolithic in North Africa: a new model for the Eastern Maghreb. Quatern Int 410:123–143

Ozenda P (2004) Flore et végétation du Sahara, 3rd edn. CNRS éditions, Paris

Paris F, Ghaki M (2010) Les monuments mégalithiques du Sud tunisien. Les Nouvelles de l'Archéologie 120–121:71–74

Perthuisot JP (1975) La Sebkha el Melah de Zarzis: genèse et évolution d'un bassin salin paralique. Trav Lab Géol, vol 9. École Normale Supérieure, Paris

Pottier-Alapetite G (1979) Flore de la Tunisie: Angiospermes-Dicotyledones-Apetales-Dialypetales. Tunis

Pottier-Alapetite G (1981) Flore de la Tunisie: Angiospermes-Dicotyledones-Gamopetales. Tunis

Reille M (1992) Pollen et spores d'Europe et d'Afrique du nord. Laboratoire de Botanique Historique et Palynologie, Marseille

Reimer PJ, Bard E, Bayliss A et al (2013) IntCal13 and Marine13 radiocarbon age calibration curves 0–50,000 years cal BP. Radiocarbon 55(4):1869–1887

Renault-Miskovsky J-R (1985) L'environnement végétal des sites préhistoriques algériens, données actuelles de la paléobotanique. L'Anthropologie 89(3):307–318

Ritchie JC (1984) Analyse pollinique de sédiments holocènes supérieurs des Hauts Plateaux du Maghreb oriental. Pollen Spores 26:489–496

Roberts N, Brayshaw D, Kuzucuoğlu C et al (2011) The mid-Holocene climatic transition in the Mediterranean: causes and consequences. Holocene 21(1):3–13

Roubet C (1979) Économie pastorale pré-agricole en Algérie orientale. Le Néolithique de tradition capsienne, l'exemple de l'Aurès. Études d'Antiquités Africaines. CNRS éditions, Paris

Roubet C (2003) "Statut de Berger" des communautés atlasiques, néolithisées du Maghreb oriental, dès 7000 BP. L'Anthropologie 107(3):393–442

Salzmann U, Schulz E (1995) Modern pollen rain and Late Holocene vegetation history of southern Tunisia. Publication occasionnelle du CIFEG 31:183–192

Schulz E, Abichou A, Adamou A et al (2009) The desert in the Sahara. Transitions and boundaries. In: Baumhauer R, Runge J (eds) Holocene Palaeoenvironmental History of the Central Sahara. Palaeoecol Afr, vol 29. CRC Press, pp 64–89

Shaw BD (1981) Climate, environment, and history: the case of the Roman Africa. In: Wrigley TML, Ingram MJ, Farmer G (eds) Climate and History. Studies in Past Climat and their impact on Man. Cambridge University Press, Cambridge, pp 379–403

Shaw BD (1984) Water and society in the ancient Maghreb: technology, property and development. Antiq Afr 20(1):121–173

Shipley G (2011) Pseudo-Skylax's Periplous: the circumnavigation of the inhabited world: text, translation and commentary. Bristol Phoenix Press, Exeter

Shipp J, Rosen A, Lubell D (2013) Phytolith evidence of mid-Holocene Capsian subsistence economies in North Africa. Holocene 23(6):833–840

Slim H, Trousset P, Paskoff R et al (2004) Le Littoral de la Tunisie: étude géoarchéologique et historique. Études d'Antiquités Africaines. CNRS éditions, Paris

Stuijts I-L (1991) Kinderoffers in de Tophet (Carthago); houtskoolonderzoek. Paleo-aktueel 2:58–61

Talbi M (1997) Action anthropique et dégradation de l'environnement aride: la désertification en Tunisie du Sud-Est. Méditerranée 86(1):25–31

Tarhouni M, Ben Salem F, Ouled Belgacem A et al (2010) Acceptability of plant species along grazing gradients around watering points in Tunisian arid zone. Flora 205(7):454–461

Thinon M, Ballouche A, Reille M (1996) Holocene vegetation of the Central Saharan Mountains: the end of a myth. Holocene 6(4):457–462

UNCCD (1994) United Nations Convention to Combat desertification in those Countries experiencing serious drought and/or desertification particularly in Africa. UNEP, Nairobi

van der Veen M (1992) Garamantian agriculture: the plant remains from Zinchecra, Fezzan. Libyan Stud 23:7–39

Van der Veen M, Grant A, Barker G (1996) Romano-Libyan agriculture: crops and animals. In: Barker G, Gilbertson D, Jones B, Mattingly D (eds) Farming the desert: The UNESCO Libyan Valleys archaeological survey, vol 1, Synthesis. UNESCO Publishing, Paris, pp 227–264

van Zeist W, Bottema S, Van der Veen M (2001) Diet and vegetation at Ancient Carthage: the archaeobotanical evidence. Groningen Institute of Archaeology, Groningen

Whittaker CR (1978) Land and labour in North Africa. Klio 60(2):331–362

Wilson AI (2013) The Mediterranean environment in Ancient History: perspectives and prospects. In: Harris WV (ed) The Ancient Mediterranean environment between science and history. Columbia Studies in the Classical Tradition, vol 39. Brill, Leiden, pp 259–325

Zerboni A, Mori L, Bosi G et al (2017) Domestic firing activities and fuel consumption in a Saharan oasis: micromorphological and archaeobotanical evidence from the Garamantian site of Fewet (Central Sahara, SW Libya). J Arid Environ 144:123–138

Zohary D (1996) The domestication of the grapevine *Vitis vinifera* L. in the Near East. In: McGovern P, Flemming SJ, Katz SH (eds) The origins and ancient history of wine. Food and nutrition in history and anthropology, vol 11. Gordon and Breach Publishers, Amsterdam, pp 23–30

Index

A

B

C

D

A. M. Mercuri et al. (Eds.): *Plants and People in the African Past*,
https://doi.org/10.1007/978-3-319-89839-1

The manufacturer's authorised representative in the EU is Springer Nature Customer Service Centre GmbH, Europaplatz 3, 69115 Heidelberg, Germany. If you have any concerns regarding our products, please contact ProductSafety@springernature.com

Printed and bound by CPI Group (UK) Ltd, Croydon, CR0 4YY

30/09/2026

02225435-0016